AEROACOUSTICS OF FLIGHT VEHICLES

Theory and Practice

Volume 1: Noise Sources

Edited by
Harvey H. Hubbard
NASA Langley Research Center
Hampton, Virginia

Published for the
Acoustical Society of America
through the
American Institute of Physics

In recognition of the importance of preserving what has been written, it is a policy of the Acoustical Society of America to have books published in the United States printed on acid-free paper.

Acoustical Society of America
500 Sunnyside Boulevard
Woodbury, NY 11797-2999

Library of Congress Cataloging-in-Publication Data

Aeronautics of flight vehicles: theory and practice / edited by Harvey H. Hubbard.
p. cm. -- (NASA reference publication; 1258)
(WRDC technical report; 90-3052)
"National Aeronautics and Space Administration, Office of Management, Scientific and Technical Information Program" -- T. p. Originally published: Washington: National Aeronautics and Space Administration, Office of Management, Scientific and Technical Information Program, 1991– <   >.
Includes bibliographical references and index.
Contents: v. 1. Noise sources -- v. 2. Noise control.
ISBN 1-56396-407-4 (set : acid-free paper). --
ISBN 1-56396-404-X (v. 1: acid-free paper). --
ISBN 1-56396-406-6 (v. 2: acid-free paper)
1. Reprinted with corrections. 2. Aerodynamic noise -- Mathematical models. 3. Airplanes -- Noise. I. Hubbard, Harvey H. II. Acoustical Society of Physics. III. Series. IV. Series: WRDC technical report; 90-3052.

TL574. N6A3 1994 94-35351
629. 132'3--dc20 CIP

10 9 8 7 6 5 4 3 2 1

Contents

Preface

The field of aeroacoustics has matured dramatically in the past two decades. Researchers have gained significant theoretical and experimental understanding of the noise generated by aircraft power plants and their components. In addition, airframe noise and interior noise have been investigated extensively. The physical understanding obtained from these efforts has resulted in the development of hardware capable of reducing community noise and of meeting strict noise certification requirements. Reductions in overall sound pressure level of 20 to 30 dB have been obtained for some types of power plants, while in the same period their installed power has increased significantly.

Current quiet flight vehicle designs are based on information reported in a multitude of journals, conference proceedings, research reports, and specialized books. Each of these scientific publications represents only incremental steps in the evolution of our present understanding of the various aeroacoustic noise generation and propagation mechanisms and procedures for noise control. There is thus a need for a reference document summarizing the current status of aeroacoustics. It is recognized that some other fine books on aeroacoustics are already available. The reader is referred to the classic handbooks by Harris on noise and vibration control; to Goldstein's "Aeroacoustics," which provides a general theoretical treatment of most aeroacoustic noise sources; to the text "Noise and Acoustic Fatigue in Aeronautics" by Richard and Mead; and to the AIAA Preprint Series volume entitled "Aerodynamic Noise." The current book represents an attempt to integrate and update the information in previous related publications, to provide a balanced viewpoint with both fundamental and applied aspects being considered, and to focus on those topics that are significant for the design and operation of quiet flight vehicles.

In July 1982, the Continuing Education Subcommittee of the Institute of Aeronautics and Astronautics (AIAA) Aeroacoustics Technical Committee identified a critical need for a reference book summarizing and interpreting the status of research in aeroacoustics. The full Aeroacoustics Technical Committee agreed with this conclusion and enthusiastically supported the concept of publishing such a book. The book would have a scope consistent with that of the Technical Committee and would include physics of noise produced by motion of fluids and bodies through the atmosphere and by chemical reaction processes; it would also include the responses of human beings, structures, and the atmosphere to aerodynamic noise. The subcommittee was then instructed to prepare an initial outline of the book for planning purposes and to procure financial support for its printing. This effort has been given generous support by the Langley, Lewis, and Ames Research Centers of the National Aeronautics and Space Administration (NASA); the U.S. Air Force Wright Research and Development Center; and the U.S. Army Aviation Systems Command.

This book is planned as a reference publication, easily readable by persons with scientific or engineering training who have completed a bachelor degree study program. It serves as an authoritative resource book for teachers, students, and researchers, but it is not designed for use directly as a textbook. It provides recommended methodology to evaluate aeroacoustics-related problems and suggests approaches to their solutions, without extensive tables, nomographs, and derivations. It is oriented toward flight vehicles and emphasizes underlying physical concepts. Theoretical, experimental, and applied aspects are covered, including the main formulations and comparisons of theory and experiment.

The preparation of the material for this book has been carried out under the general supervision of the AIAA Technical Committee on Aeroacoustics. The Committee elected the editor (Harvey H. Hubbard), two associate editors (Christopher K. W. Tam and Robert H. Schlinker), and six additional editors (Charles E. Feiler, James C. Yu, Walter K. Eversman, Marvin E. Goldstein, Robert E. Kraft, and Yung H. Yu). Donald L. Lansing and John Laufer (until his untimely death) also served for short terms. They functioned as an editorial board to establish the overall policy for the organizing, reviewing, and editing of the book. Each was selected because of his expert knowledge of at least one of the specialty areas covered in the book. They collectively comprise a team of experts who represent industry, government, and academia viewpoints.

The editorial board members chose by vote the lead authors for each chapter based on their stature and expertise in particular technical areas and on their proven ability to communicate. In all cases, contributing authors were selected and enlisted by the lead authors and on the basis of the same criteria. An outline of each chapter was first approved by the editorial board as a means of defining the overall scope of that chapter. Technical reviewers were chosen by vote of the editorial board based on their expertise of subject matter and the nature of their experience. Two to four persons were selected to provide technical reviews for each manuscript. These technical reviews were then provided to the appropriate authors as a basis for the preparation of their final manuscripts. Final editing was accomplished by Mary K. McCaskill and Thomas H. Brinkley of the NASA Langley Research Center Technical Editing Branch. This latter effort involved skilled technical editors closely associated with the publication profession. Their work included checking for accuracy, grammar, consistency of style, compliance with editorial instructions, and assembly for printing.

Authors and reviewers contributed their time for this project without receiving compensation. Draft manuscript preparation, typing, and graphics were supported partially or wholly by the participant's employer. All these contributions were vital to the success of this project and are greatly appreciated.

Supporting reference information cited in this book is limited to publications available at the time of the text preparation. No proprietary or classified information is included in order to protect the interests of authors' companies and governments. In order to enhance its utility, this book is divided into two volumes, each of which has a list of symbols, an index, and a separate glossary of terms. Reference lists for each chapter contain the key available supporting documents.

Volume 1 includes all the chapters that relate directly to the sources of flight vehicle noise: Propeller and Propfan Noise; Rotor Noise; Turbomachinery Noise; Jet Noise Classical Theory and Experiments; Noise From Turbulent Shear Flows; Jet Noise Generated by Large-Scale Coherent Motion; Airframe Noise; Propulsive Lift Noise; Combustion and Core Noise; and Sonic Boom. Volume 2 includes those chapters that relate to flight vehicle noise control and/or operations: Human Response to Aircraft Noise; Atmospheric Propagation; Theoretical Models for Duct Acoustic Propagation and Radiation; Design and Performance of Duct Acoustic Treatment; Jet Noise Suppression; Interior Noise; Flyover-Noise Measurement and Prediction; and Quiet Aircraft Design and Operational Characteristics.

This book was published initially in August 1991 as NASA Reference Publication 1258, Vols. 1 and 2, and with a companion United States Air Force Wright Research and Development Center designation as Technical Report 90-3052, Vols. 1 and 2. It was printed in soft cover and carried the appropriate NASA, U.S. Air Force, and U.S. Army logos on its covers.

In November 1993, the Acoustical Society of America agreed to reprint it in a hard-cover version but with modifications to the covers and title pages and with minor changes and corrections to the text. Elaine Moran of the Acoustical Society of America and Andrew Prince of the American Institute of Physics have coordinated the preparation of materials for this reprinting.

Technical Reviewers

Roy K. Amiet
AMI
Wooster, Ohio

Thomas F. Brooks
NASA Langley Research Center
Hampton, Virginia

Sebastien Candel
ONERA
Chatillon, France

Tony Clark
Boeing Commercial Airplane Co.
Seattle, Washington

R. D. Cuthbertson
Boeing Commercial Airplane Co.
Seattle, Washington

James H. Dittmar
NASA Lewis Research Center
Cleveland, Ohio

Michael D. Falarski
NASA Ames Research Center
Moffett Field, California

Fereidoun Farassat
NASA Langley Research Center
Hampton, Virginia

Charles Feiler
NASA Lewis Research Center
Cleveland, Ohio

Phillip R. Gliebe
General Electric Co.
Evandale, Ohio

Albert R. George
Cornell University
Ithaca, New York

John F. Groeneweg
NASA Lewis Research Center
Cleveland, Ohio

Jay C. Hardin
NASA Langley Research Center
Hampton, Virginia

David L. Hawkings
Westland Systems Assessment Ltd.
Yeovil, England

Morris P. Isom
Polytechnic University
Brooklyn, New York

Valdis Kibens
McDonnell Douglas Research Labs
St. Louis, Missouri

Eugene A. Krejsa
NASA Lewis Research Center
Cleveland, Ohio

Geoffrey M. Lilley
University of Southampton
Southampton, England

Douglas C. Mathews
United Technologies Corp.
Pratt & Whitney
East Hartford, Connecticut

Phillip J. Morris
Pennsylvania State University
University Park, Pennsylvania

S. Paul Pao
NASA Langley Research Center
Hampton, Virginia

Kumar Ramohalli
University of Arizona
Tucson, Arizona

N. N. Reddy
Lockheed Aerosystems Co.
Marietta, Georgia

A. Richard Seebass
University of Colorado
Boulder, Colorado

John M. Seiner
NASA Langley Research Center
Hampton, Virginia

Warren C. Strahle
Georgia Institute of Technology
Atlanta, Georgia

James C. Yu
NASA Langley Research Center
Hampton, Virginia

Symbols

A	cross-sectional area; acoustic admittance
b	chord
C_D	drag coefficient
C_L	lift coefficient
C_p, c_p	specific heat at constant pressure
C_v	specific heat at constant volume
c	speed of sound
D, d	diameter
d	distance, e.g., from source to receiver
F	fuel-air ratio; force
f	frequency
G_p	cross spectral density of acoustic pressure
H, h	height
h	enthalpy
I	intensity
i	imaginary number, $\sqrt{-1}$
k	wave number
L_A	A-weighted sound level
L_D	D-weighted sound level
L_{dn}	day-night average sound level
L_{EPN}	effective perceived noise level
L_{eq}	equivalent continuous sound level

L_{PN}	perceived noise level
l, ℓ	length
M	Mach number
m	mass
$\dot{m}$	mass flow ratio
N_{Pr}	Prandtl number
N_{Re}	Reynolds number
N_{St}	Strouhal number
P	power
p	sound pressure
R	reflection coefficient; acoustic resistance; gas constant; duct radius; jet radius
$\Re$	distance from arbitrary point on rotating rotor blade to observer
r	rotor radial position
S	wing area
$S(\sigma)$	Sears function
T	temperature
t	time; wing thickness
U	flight velocity
u	particle velocity; mean velocity; axial velocity
V	velocity
V_e	exit velocity of jet
X	acoustic reactance
x, r, θ	cylindrical coordinates
Z	impedance
α	sound absorption
β	$= \sqrt{M^2 - 1}$
γ	ratio of specific heats
δ_f	flap deflection

δ_{ij}	Kronecker delta
ζ	ratio of characteristic impedances
λ	wavelength
ν	viscosity
ξ	cutoff ratio
ρ	density
σ	reduced frequency of gust
ϕ	phase angle
Ω	rotor rotational rate
ω	circular frequency, $2\pi f$

Abbreviations:

BPF	blade-passage frequency
BVI	blade-vortex interaction
DNL	day-night average sound level
EPNL	effective perceived noise level
FAR	Federal Aviation Regulations
HSI	high-speed impulsive
ICAO	International Civil Aviation Organization
LEQ	equivalent continuous sound level
LL_S	Stevens loudness level
LL_Z	Zwicker loudness level
NR	noise reduction
OASPL	overall sound pressure level
PNL	perceived noise level
PWL	power level
rms	root-mean-square
SLA	A-weighted sound level
SLD	D-weighted sound level
SLE	E-weighted sound level

Symbols

SPL	sound pressure level
SWR	standing wave ratio
TL	transmission loss

1 Propeller and Propfan Noise

Lead author

B. Magliozzi
Hamilton Standard Division—
United Technologies
Windsor Locks, Connecticut

Contributing authors

D. B. Hanson
Hamilton Standard Division—
United Technologies
Windsor Locks, Connecticut

R. K. Amiet
AMI
Wooster, Ohio

Introduction

Propellers are familiar devices. Indeed, these were the first means of powering airplanes, preceding all other means of propulsion by about 40 years. Propellers were used extensively through the 1940's. Although there have been many refinements to propellers through the years, such as variable pitch and the application of composite materials to reduce weight, the general appearance of the propeller has changed little.

A propeller can be generally described as an open (unshrouded), rotating, bladed device. Although there are many differences in details among various designs and applications, such as number of blades, blade shape, and airfoil section, the noise-generating process is basically the same for all. The major propeller noise components are thickness noise (due to the volume displacement of the blades), steady-loading noise (due to the steady forces on the blades), unsteady-loading noise (due to circumferentially nonuniform loading), quadrupole (nonlinear) noise, and broadband noise. Although the relative importance of these sources depends on design and operating conditions, defining them will completely describe the acoustic signature of a propeller.

One important consideration is the effect of installation on the noise produced by a propeller. This effect is essentially the difference between the laboratory environment and the real world. It is generally assumed that in a laboratory environment conditions are ideal, that is, the propeller is operating in perfectly uniform flow. For an operational propeller, this is never the case. Propellers are always operating in a flow field that has some distortion. This can be from the wing upwash, the pylon wake, the airplane angle of attack, or the inflow turbulence. Since this distortion leads to additional noise, it is a factor which must be considered in defining the total noise of an operational propeller.

In this chapter the characteristics of propellers, their noise-generating mechanisms, propeller noise theories and calculation procedures, sound propagation effects, comparisons of predictions and measurements, and means for controlling propeller noise are described.

Description of Propellers

General Characteristics

A propeller is an open rotor having fixed or adjustable-pitch blades. The blades are designed to produce a region of low pressure on one side and high pressure on the other. The resulting forces induce air from the front and push it back, resulting in thrust. Because propellers impart a relatively small amount of velocity to a large mass of air, their efficiencies are high. However, the efficiencies of conventional propellers tend to fall off at high speed. This has led to a variation of the propeller called the propfan. The propfan is also an open rotor, but compared with conventional propellers it has a smaller diameter for a given thrust and has more blades, which are swept. To improve efficiency further, a second rotor can be added behind a propeller, resulting in a counterrotating propeller.

A typical high-performance, modern propeller in widespread use on commuter airplanes is shown in figure 1. These propellers have two to six relatively straight, narrow blades. Although this type of propeller is well-suited for moderate flight speed (below a Mach number of about 0.65), its performance decreases at higher flight speeds. The primary limitation of this propeller is associated with high drag at high speed due to blade thickness and large relative blade section speed.

The propfan, shown in figure 2, has been developed to extend the practical flight envelope of the propeller. Compared with conventional propellers, the propfan has more blades (from 6 to 12), uses thin airfoils and swept blades to provide good aerodynamic performance at high speed, and operates at much higher power loading to reduce the diameter.

One factor leading to loss of efficiency in propellers is swirl in the wake resulting from engine torque. Generation of swirl uses energy but does not contribute to thrust. The amount of swirl is related to the power loading. One concept to recover the swirl losses is to add a second blade row behind the first. This is shown in figure 3 for the propfan. The second blade row rotates in the direction opposite to that of the first, thus cancelling its swirl. This cancellation can result in performance increases of 8 to 10 percent compared with that of single rotation propellers (ref. 1).

Installation of Propellers and Propfans

In considering the noise of propellers and propfans, it is important to address the installation of these devices, as this can have a significant effect on the noise-generation process. In their simplest forms noise calculation procedures and analyses assume uniform conditions, that is, the loads on the blades are absolutely steady. In actuality, that is rarely the case. Although laboratory tests can be conducted such that the incoming flow is uniform and free of turbulence, the real environment is never as ideal. The amount of distortion is generally related to two parameters: operating conditions and installation.

Figure 1. High-performance, low-speed propeller for moderate-flight-speed commuter airplane propulsion.

Figure 2. Propfan propulsion system for high subsonic cruise speed applications.

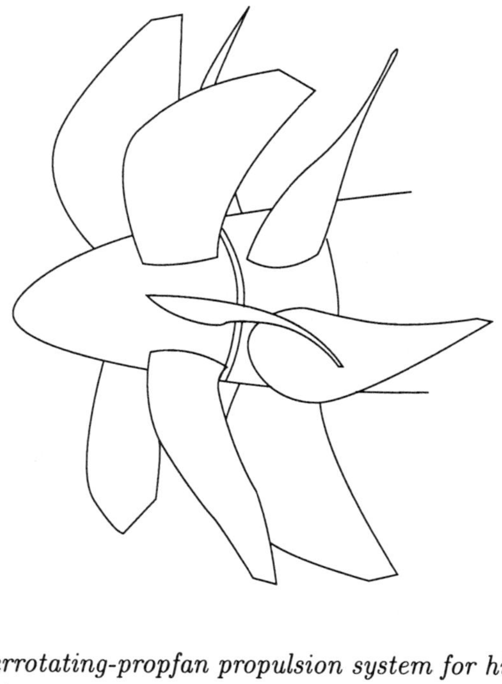

Figure 3. Counterrotating-propfan propulsion system for high subsonic cruise speed applications.

Operating conditions include static (zero flight speed) or forward flight. In flight, the propeller can be at an angle of attack. Static conditions give rise to severe inflow distortion and the resulting noise is vastly different than that from propellers in flight. This difference is discussed in more detail in a subsequent section. Angle of attack or skewed inflow causes unsteady loading, with periodicities equal to one cycle per rotation.

Regarding installations, the cleanest configuration is in front of a long, slender, axisymmetric nacelle. The worst is probably behind a wing. Intermediate configurations include wing-mounted tractors, aft-mounted tractors in front of a pylon, and aft-mounted pushers behind a pylon. These installations result in varying degrees of inflow distortion which typically results in added sources (unsteady loading noise) and increases the noise produced by the propeller.

It is thus important to evaluate the propeller as an installed system rather than as an isolated component when noise requirements are addressed. If a propeller is designed to meet the noise goals, even with a comfortable margin of error, ignoring installation effects can result in a substantial underprediction of the system noise, with the strong possibility that the airplane will not meet the noise requirements.

Propeller Noise Characteristics

Propeller noise can be classified into three categories: harmonic noise, broadband noise, and narrow-band random noise.

Harmonic noise is the periodic component, that is, its time signature can be represented by a pulse which repeats at a constant rate. If an ideal propeller with B blades is operating at constant rotational speed N, then the resulting noise appears as a signal with fundamental frequency BN. The blade-passage period is $1/BN$. Typically the generated pulse is not a pure sinusoid, so that many harmonics exist. These occur at integer multiples of the fundamental frequency. The first harmonic is the fundamental, the second harmonic occurs at twice the fundamental frequency, and so on. Figure 4 illustrates the characteristics of harmonic noise in both the time and frequency domains.

Broadband noise is random in nature and contains components at all frequencies. A typical broadband noise signal for propellers is shown in figure 5. The frequency spectrum is continuous, although there may be a "shape" to it because not all frequencies have the same amplitude.

Narrow-band random noise is almost periodic. However, examination of the harmonics reveals that the energy is not concentrated at isolated frequencies, but rather it is spread out. As illustrated in figure 6 the signal may appear periodic, but certain components do not repeat exactly with time. The frequency spectrum shows discrete components, but these spread out, particularly at the higher frequencies.

Propeller Noise-Generating Mechanisms

The mechanisms which lead to the generation of the spectral characteristics discussed above are described in this section.

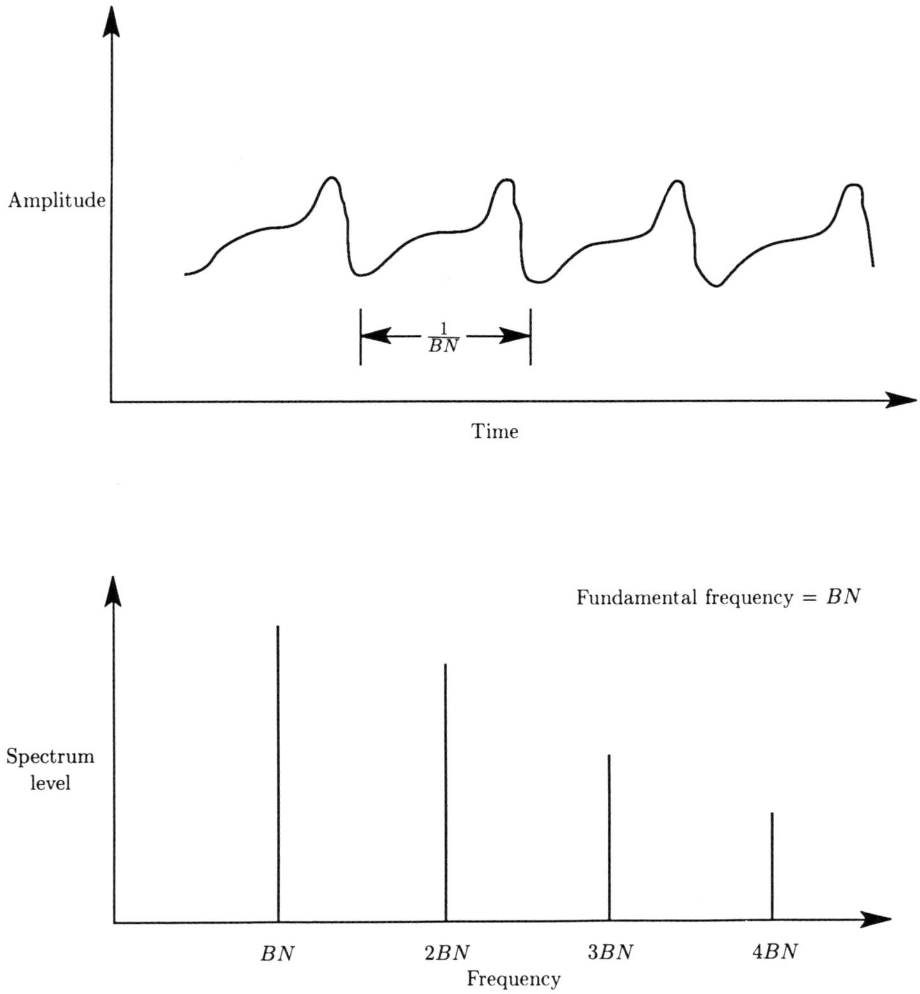

Figure 4. Characteristics of propeller rotational noise.

Steady Sources

Steady sources are those which would appear constant in time to an observer on the rotating blade. They produce periodic noise because of their rotation. Noise sources are usually divided into three categories: linear thickness, linear loading, and (nonlinear) quadrupole.

Thickness noise arises from the transverse periodic displacement of the air by the volume of a passing blade element. The amplitude of this noise component is proportional to the blade volume, with frequency characteristics dependent on the shape of the blade cross section (airfoil shape) and rotational speed. Thickness noise can be represented by a monopole source distribution and becomes important at high speeds. Thin blade sections and planform sweep are used to control this noise.

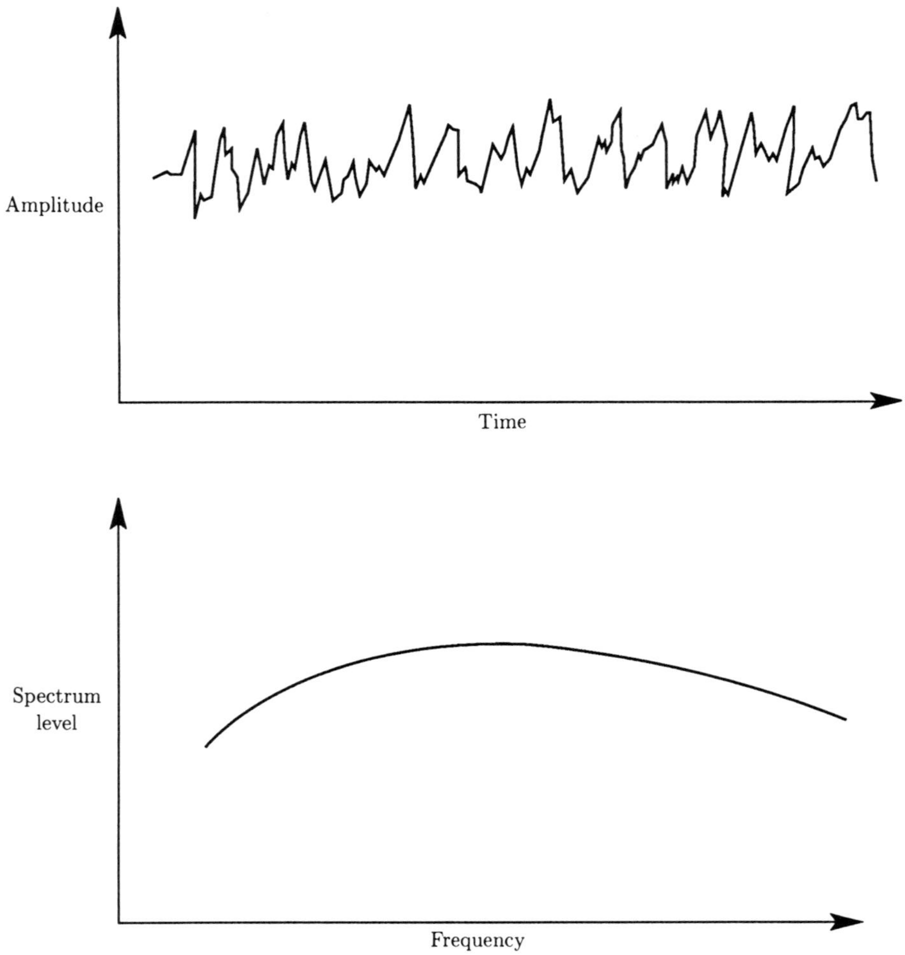

Figure 5. Characteristics of propeller broadband noise.

Loading noise is a combination of thrust and torque (or lift and drag) components which result from the pressure field that surrounds each blade as a consequence of its motion. This pressure disturbance moving in the medium propagates as noise. Loading is an important mechanism at low to moderate speeds.

For moderate blade section speed, the thickness and loading sources are linear and act on the blade surfaces. When flow over the blade sections is transonic, nonlinear effects can become significant. In aeroacoustic theory these can be modeled with quadrupole sources distributed in the volume surrounding the blades.

In principle, the quadrupole could be used to account for all the viscous and propagation effects not covered by the thickness and loading sources. However, the only practical application of this term to propeller acoustics has been its evaluation in the nonviscous flow close to the blade surface. At transonic blade section speeds the quadrupole enhances the linear thickness and loading sources and causes a noise increase for unswept, high-tip-speed propellers and helicopter rotors.

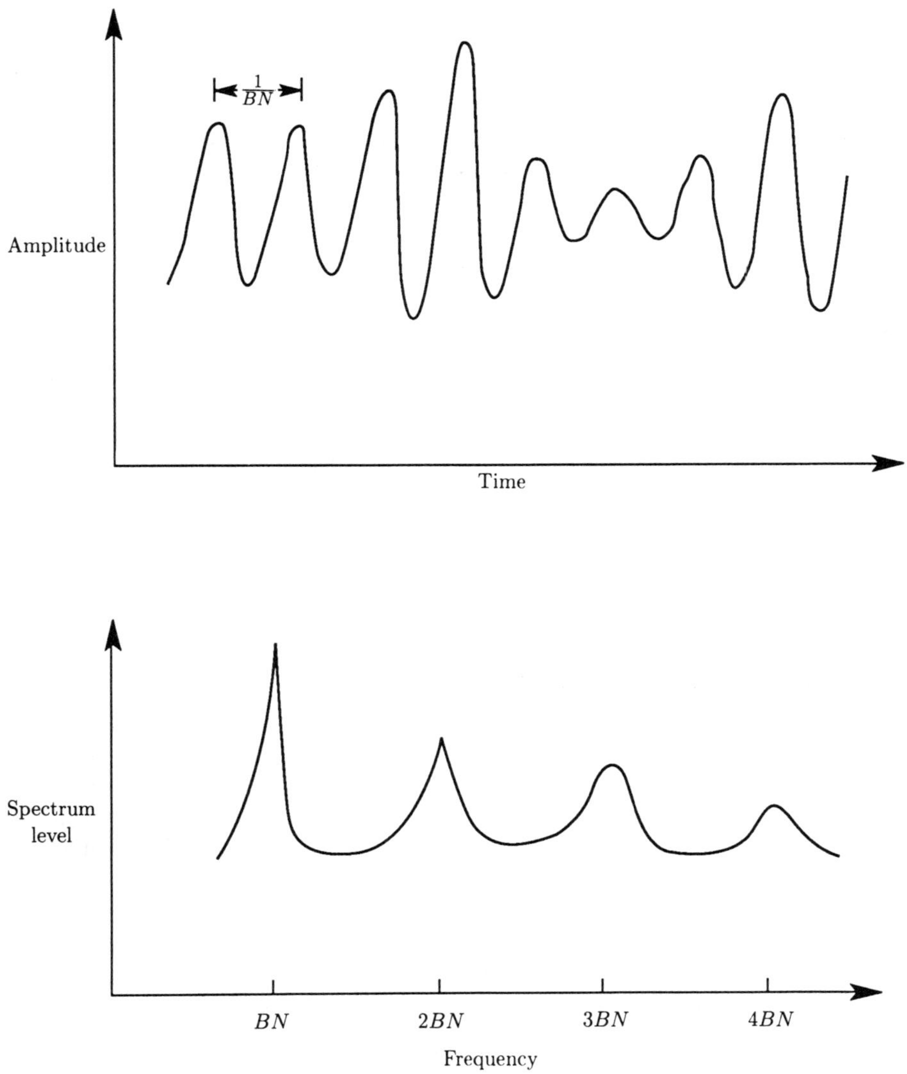

Figure 6. Characteristics of propeller narrow-band random noise.

Unsteady Sources

Unsteady sources are time dependent in the rotating-blade frame of reference. They include periodic and random variation of loading on the blades.

A typical example of periodic blade loading in propellers is the effect of shaft angle of attack. When the propeller axis is tilted relative to the inflow, each blade sees a cyclic change in local angle of attack. As a consequence, the loading on the blade varies during a revolution. The loading change may be once per revolution or several times per revolution, depending on the source of inflow distortion. All inflow distortion which is invariant with time results in blade-loading changes which repeat exactly for every propeller revolution. The resulting periodic unsteady-loading noise occurs at harmonics of blade-passage frequency.

Depending on the circumferential location of the loading disturbance relative to the observer, unsteady-loading noise can add or subtract from the steady-loading noise. The noise directivity is no longer axisymmetric and a third coordinate is needed to define it. The circumferential directivity exhibits lobes—peaks and valleys—with the number of lobes dependent on the order of the distortion and unrelated to the number of blades. For example, a propeller behind a wing might show two circumferential directivity lobes regardless of the number of blades on the propeller.

Unsteady loading is an important source in the counterrotating propeller. Although the counterrotating propeller does not contain any additional or unique sources of noise, the aerodynamic interference between the two rotors gives rise to significant levels of unsteady-loading noise which are particularly significant at low flight speeds, such as during takeoff and landing. Each front rotor blade leaves a wake which convects into the rear rotor. (This wake can be complex, consisting of downwash due to the lift on the blades, velocity deficits due to the drag of the blade sections, and tip vortices.) This convection results in a sequence of lift pulses on the rear rotor blades. Another mechanism is the potential field (due to blade loading) of the rear rotor creating a disturbance which is felt by the aft part of the front rotor blades. The magnitude of this source depends on the level of loading on the rear rotor and the spacing between the two rotors.

Because the wakes are periodic, the generated noise is also periodic. If the two rotors have the same number of blades and are operating at the same rotational speed, then the components of the steady sources and the unsteady sources are at the same frequencies and the noise spectra contain only harmonics of blade-passage frequency. However, if the number of blades of the front rotor is different from that of the rear rotor or the two rotors operate at different rotational rates, then the individual interaction components (modes) are distinct in the noise spectra.

Figure 7 shows the importance of the aerodynamic interaction in a counterrotating propeller (ref. 2). Figure 7(a) shows the spectra of single-rotating-propeller (SRP) noise at a forward directivity, near the plane of rotation, and at an aft directivity. It is readily apparent that there are no significant higher frequency harmonics. For comparison, noise spectra for a counterrotating propeller (CRP) are shown in figure 7(b). It is apparent that the counterrotating propeller has significantly higher levels of higher frequency harmonics. Figure 7(c) shows a direct comparison between the noise from the two types of propellers. In this comparison the two propellers were operating at equal tip speeds and power per rotor. Three decibels were added to the SRP noise levels to simulate the total power of the counterrotating propeller. If the two rotors of the counterrotating propeller were uncoupled, then the two spectra would be identical. In fact, the levels at the blade-passage frequency are very close. At the higher harmonics, the counterrotating propeller shows significantly higher levels. This is a direct indication of the aerodynamic interaction effects on noise in counterrotating propellers.

Aerodynamic interaction is a significant source of noise for low-speed operation. At higher flight speeds, such as during cruise, the aerodynamic interaction becomes less important because the steady sources (thickness, steady loading, and quadrupole) become dominant.

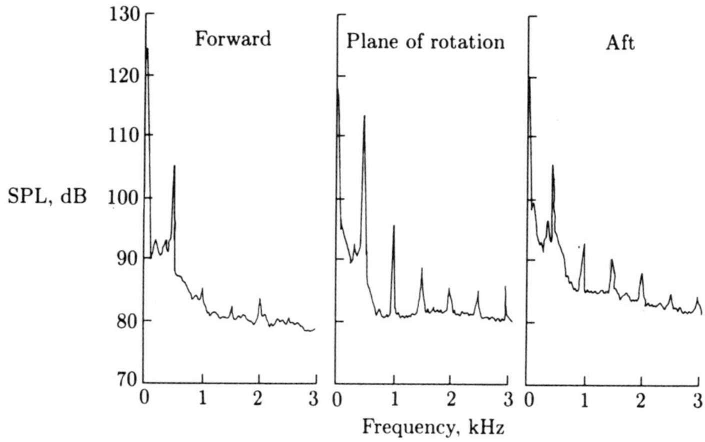

(a) Single-rotating-propeller noise spectra.

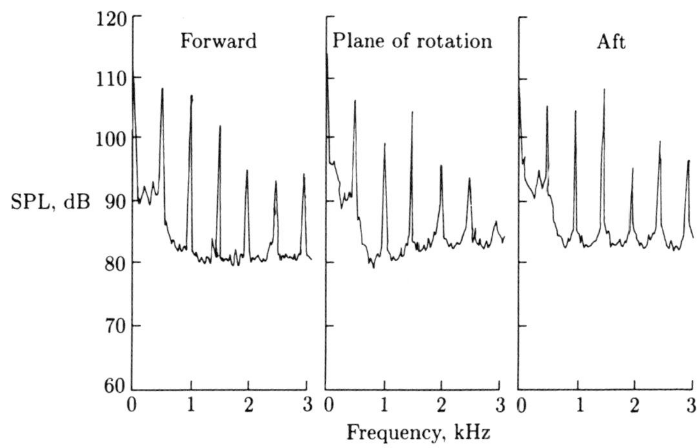

(b) Counterrotating-propeller noise spectra.

Figure 7. Aerodynamic interaction noise in counterrotating propeller. (From ref. 2.)

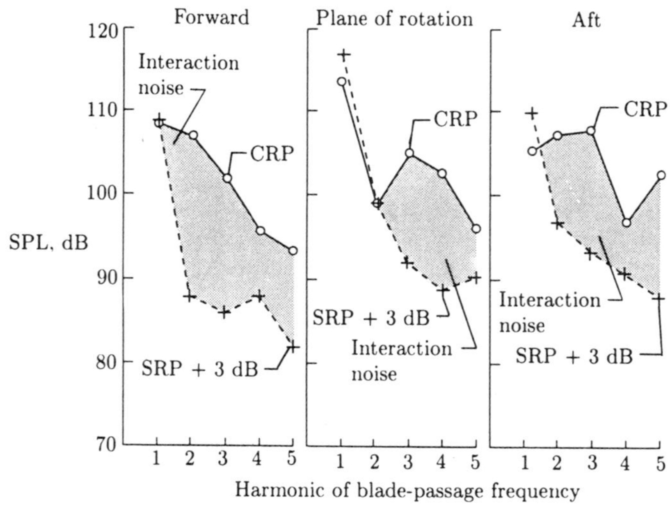

(c) Aerodynamic interaction noise.

Figure 7. Concluded.

Under certain conditions blade loading which is nearly periodic can occur. An example of this is the ingestion of a vortex, which could be induced by the propeller and attached to the fuselage or to the ground ahead of the propeller. In this example, a local distortion is induced by and drawn into the propeller. The blades chop through the distortion and a blade-loading pulse is produced. Because the distortion can persist for several propeller revolutions, the unsteady-loading noise can appear at blade-passage frequency harmonics. However, as conditions change, the location of the distortion changes and the amplitude and phase of the unsteady-loading noise change. These changes broaden the noise spectrum, as previously described for narrow-band random noise.

Random Sources

Random sources give rise to broadband noise. For propellers there are two sources which may be important, depending on the propeller design and operating conditions.

The first broadband noise source is the interaction of inflow turbulence with the blade leading edges. Because the inflow is turbulent, the resulting noise is random. The importance of this noise source depends on the magnitude of the inflow turbulence, but it can be quite significant under conditions of high turbulence at low speeds.

In the second broadband mechanism, noise is generated near the blade trailing edge. A typical propeller develops a turbulent boundary layer over the blade surfaces, which can result in fluctuating blade loading at the trailing edge. The noise is characterized by the boundary-layer properties. A related mechanism occurs at the blade tips, where turbulence in the core of the tip vortex interacts with the trailing edge.

It has been determined for full-scale propellers in flight that the broadband noise sources are relatively unimportant and do not contribute significantly to the total noise (ref. 3).

Prediction Methods for Propeller Harmonic Noise

Any of the propeller noise methods currently in use can be derived from the Ffowcs Williams and Hawkings equation (ref. 4). This fundamental equation of sound generation is attractive because it is a rigorous combination of the equations of momentum, continuity, and state into a wave equation that can be solved to varying degrees of precision by a variety of analytical methods. A scholarly survey of these methods was presented by Farassat in 1981 (ref. 5) and is recommended for readers interested in the mathematical connections between these methods and the Ffowcs Williams-Hawkings equation.

In the following sections, the noise radiation equations which are discussed were chosen to illustrate the variety of methods available, and comments on their advantages and disadvantages are included. In the case of the time-domain equations, computer programs based on the theory of Farassat are available to U.S. companies from NASA Langley Research Center. The frequency-domain equations are simple enough so that they can be coded on personal computers. Furthermore, their analytical form gives a direct indication of the influences of propeller design features on noise characteristics.

Linear Theories

As given by Farassat (ref. 5), the linear form of the Ffowcs Williams and Hawkings equation is

$$\nabla^2 p - \frac{1}{c^2}\frac{\partial^2 p}{\partial t^2} = -\frac{\partial}{\partial t}\left[\rho_o v_n |\nabla f| \delta(f)\right] + \frac{\partial}{\partial x_i}\left[l_i |\nabla f| \delta(f)\right] \tag{1}$$

where the left side is the well-known linear wave operator acting on the acoustic pressure p. The right side contains the source terms resulting from the motion of surfaces in the fluid: ρ_o is the ambient density, c is the ambient speed of sound, v_n is the local velocity of the surface normal to itself, $\delta(f)$ is the Dirac delta function, x_i is the observer position, and l_i is the ith component of the surface force. The first source term represents the effect of the blades parting the air and produces what is known as "thickness noise." The second term represents the action of the blade forces on the air and produces "loading noise."

In equation (1), the presence of the surfaces is accounted for by the factors containing f, where $f = 0$ is the equation of the blade surface. Unless very high frequencies are considered (wavelengths of the order of blade thickness), details of the airfoil section can be ignored. The source term is thus simplified, so that equation (1) becomes

$$\nabla^2 p - \frac{1}{c^2}\frac{\partial^2 p}{\partial t^2} = -\rho_o \frac{\partial q}{\partial t} + \nabla \cdot \overline{F} \tag{2}$$

where now the thickness source can be thought of as being represented by a volume distribution of sources (and sinks) of strength q. The loading source is represented by a volume distribution of doublets associated with $\overline{F}$, the force imposed by the blades on the air. Equation (2) is the linear wave equation with sources as derived, for example, in chapter 7 of reference 6 and in chapter 1 of reference 7. This source and doublet concept is equivalent to the common representation of wing aerodynamics in textbooks. For example, the sketch in figure 8, adapted from reference 8, was used to treat the thickness and loading effects of wings. The dominant sources are exactly the same for wing and propeller aerodynamics. However, in the case of propellers, the sources make noise because of unsteadiness associated with rotation and with time-dependent blade loading.

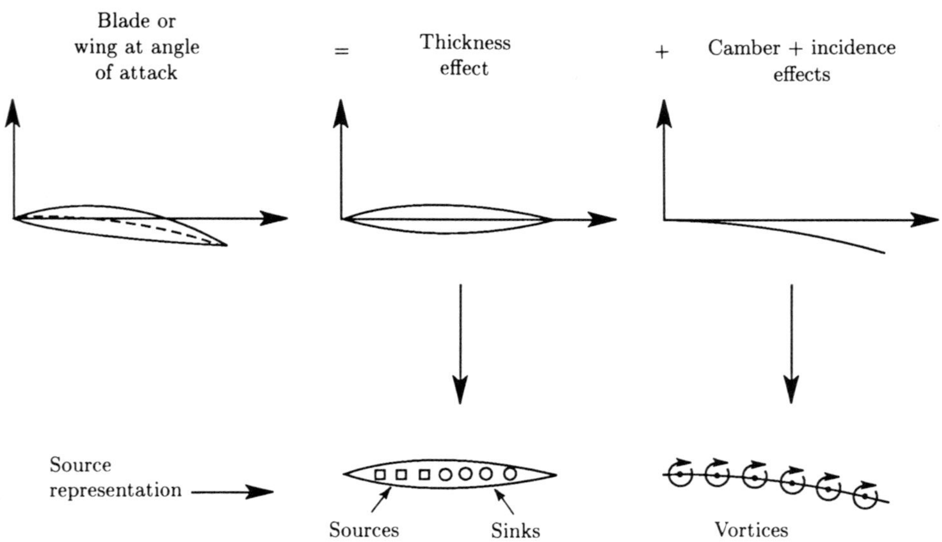

Figure 8. Decomposition of wing or blade section aerodynamics into thickness effect and camber and incidence effects. (Based on ref. 8.)

Time-domain methods are used to solve equations (1) or (2) directly in terms of the space-time variables. These methods are appealing because they can treat blade geometry with any desired level of precision. The result is the prediction of the acoustic pressure waveform $p(t)$. If noise harmonics are needed, $p(t)$ is Fourier transformed numerically.

Frequency-domain methods eliminate time from the wave equation by means of Fourier transformation. Some precision in the representation of blade geometry is usually lost through the transformation, but this loss is generally acceptable for harmonics to a fairly high order. The transformation also gives rise to Bessel functions which are indicators of radiation efficiency. Harmonics are computed one at a time and waveforms are generated by summing a Fourier series.

Time-Domain Methods

The most prolific proponent of time-domain methods for propellers and rotors has been Farassat. Papers listed in the References section can be used to trace the development of his formulations 1, 1A, 2, and 3. The preferred formulations are coded in the Aircraft Noise Prediction Program (ANOPP, ref. 9) and the Dunn-Farassat-Padula Advanced Turboprop Prediction (DFP-ATP, ref. 10) program and are described briefly below.

Formulation 1A, used in both ANOPP and DFP-ATP for subsonic source regions, gives the acoustic pressure $p(t)$ as follows:

$$\begin{aligned} 4\pi p(t) = {} & \frac{1}{c}\int_{f=0}\left[\frac{\dot{l}_i\hat{r}_i}{r(1-M_r)^2}\right]_{\text{ret}} dS + \int_{f=0}\left[\frac{l_r - l_i M_i}{r^2(1-M_r)^2}\right]_{\text{ret}} dS \\ & + \frac{1}{c}\int_{f=0}\left[\frac{l_r(r\dot{M}_i\hat{r}_i + cM_r - cM^2)}{r^2(1-M_r)^3}\right]_{\text{ret}} dS + \int_{f=0}\left[\frac{\rho_o v_n(r\dot{M}_i\hat{r}_i + cM_r - cM^2)}{r^2(1-M_r)^3}\right]_{\text{ret}} dS \end{aligned} \tag{3}$$

The dots on M_i (defined as v_i/c) and l_i denote derivatives with respect to source time, and v_i is the local velocity of the blade surface with respect to the quiescent fluid. The first three integrals give the loading noise, with the blade load vector l_i as the source. The last integral gives the thickness noise, with the surface-normal velocity v_n as the source. Also, r is the distance from a source point to the observer and M_r is the Mach number of the source toward the observer. Formulation 3 (not shown herein) is recommended by Farassat for supersonic blade sections and is coded in DFP-ATP (ref. 10). It too contains integrals in the form $\int_{f=0} [\ ]_{\text{ret}}\, dS$. The significance of this notation is that the contents of the brackets are evaluated on the surface $f = 0$ at the retarded (or emission) time and integrated over the blade surface elements dS. Thus, to compute the acoustic pressure at time t, it must be determined where every element of the surface was when it emitted the wavelet that arrives at the observer point at t. This is possible because the geometry and motion of the blades are known and it is assumed that the wavelets travel at the ambient speed of sound. When a surface is constructed by connecting all the blade edges at their retarded times, the result is distorted from the physical planform to a shape called the "acoustic planform." The process is illustrated by simple example in figure 9 (from a paper by Hanson, ref. 11), which also provides the key to the acoustic planforms shown in figure 10. The blade labeled "visual planform" has a rectangular shape and rotates in a plane with zero forward flight speed. At a tip Mach number M_t of 1.1, a sharp thickness noise pulse like the one at the right results in accordance with formulation 3 for an observer at a distance of 5 rotor diameters. The acoustic planform shown on the source disk is for the current time indicated by the dot on the waveform. Figure 10 shows the evolution of the acoustic planform with time. For this supersonic example, the planform breaks into two portions because, for some blade elements, there is more than one point on the azimuth where waves are emitted that arrive at time t. For subsonic tip speeds, the acoustic planform is in one continuous piece and is straightforward to calculate. However, the problem at supersonic speeds is surprisingly difficult and leads to significant numerical problems because the acoustic planform must be evaluated with great precision for good results.

As mentioned above, computer programs embodying Farassat's formulas are available outside NASA on a limited basis for personal and mainframe computers. For users with no desire to do their own coding, this offers a means to acquire noise prediction capability quickly. Because of the numerical difficulties associated

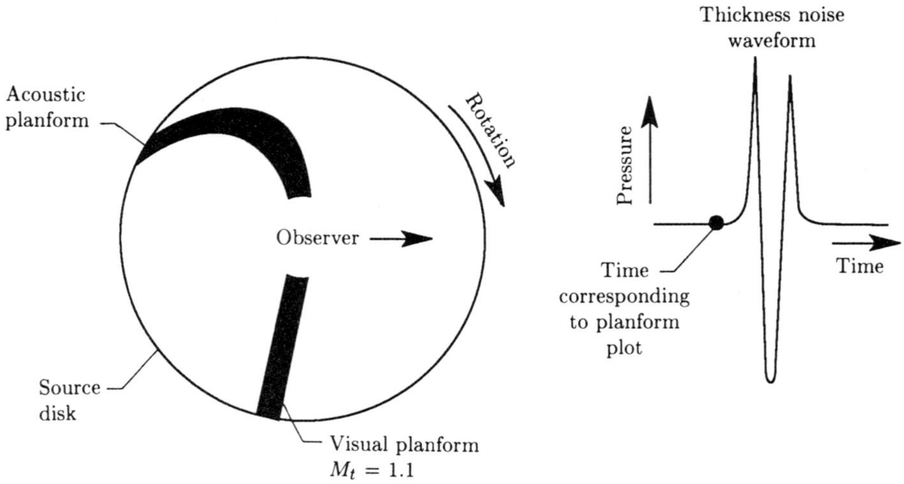

Figure 9. Key to acoustic planform plots. (From ref. 11.)

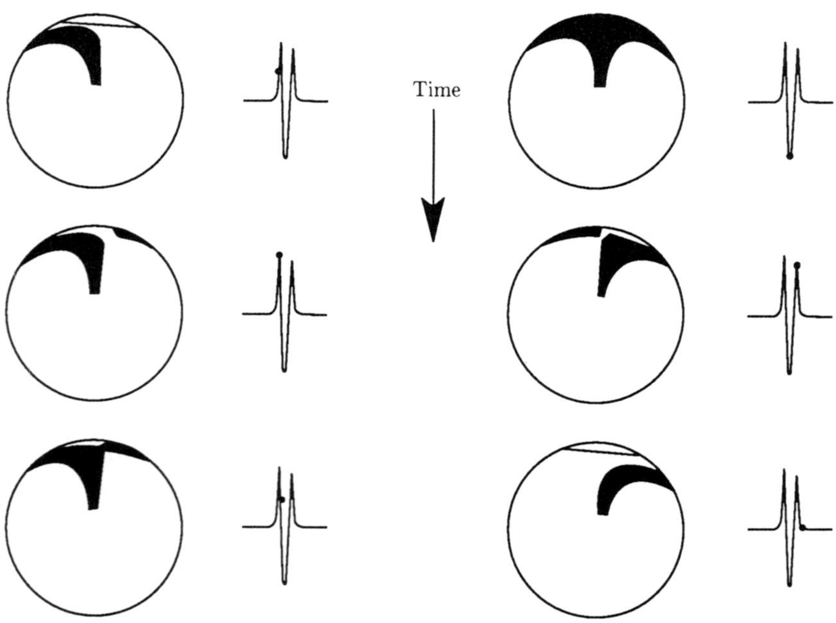

Figure 10. Evolution of acoustic planform with time at $M_t = 1.1$. (From ref. 11.)

with the acoustic planform determination and numerical differentiation, coding of equation (3) should not be attempted without careful study of references 5, 9, 12, 13, and 14.

Frequency-Domain Methods

A transformation to the frequency domain eliminates the need for computing retarded blade locations and numerical derivatives. By representing blades as helicoidal surfaces, far-field noise formulas that are easily coded on a personal computer can be derived. Effective radius versions can even be computed by hand with the help of a Bessel function table. Furthermore, these formulas give direct insight to the influence of blade geometry and operating conditions on the sound harmonics.

The first successful propeller noise theory by Gutin (ref. 15) was in harmonic form. This theory was extended by various investigators; one of these was Hanson, whose versions included effects of thickness, forward flight, and blade sweep (refs. 16 to 18). Hanson's formulas are described below in enough detail for programming. To arrive at his simple results, the approximation is made that the thickness and loading sources act on the advance helix, that is, on the surface swept out by a radial line that rotates at angular speed Ω and translates at the flight speed V. Of course, the aerodynamic loading comes from the fact that blades are at an angle of attack relative to the helical surface. However, once the loading is determined from an aerodynamic analysis, the thickness and loading sources are transferred to the advance helix for the radiation calculation. This transfer corresponds to linearization of the boundary condition to the free-stream direction in wing theory. With this simplification, the sources can be modeled with the terms on the right in equation (2), and the far-field pressure can be found from the free-space Green's function in the following form (refs. 16 and 17):

$$p(t) = \sum_{m=-\infty}^{\infty} P_{mB} \exp(-imB\Omega t) \tag{4}$$

or

$$p(t) = 2\mathrm{Re}\left[\sum_{m=1}^{\infty} P_{mB} \exp(-imB\Omega t)\right] \tag{5}$$

where $2P_{mB}$ is the Fourier transform of the pressure at the mth harmonic of blade-passage frequency for a propeller with B blades. The term P_{mB} is written as the sum of effects due to thickness (or volume) displacement P_{Vm}, drag P_{Dm}, and lift P_{Lm} so that

$$P_{mB} = P_{Vm} + P_{Dm} + P_{Lm} \tag{6}$$

Before a noise calculation can be made, the blade geometry and loading must be specified. A blade planform is specified with the parameters and nomenclature in figure 11 in terms of the chord and sweep as functions of radius ratio $z = r/r_t$, where r_t is the tip radius. Chord b is given by B_D, the ratio of chord to diameter D, and sweep by MCA/D, the ratio of mid-chord alignment to diameter. Airfoil section thickness distribution is specified by the thickness-to-chord ratio at each radius t_b, and a typical thickness distribution function $H(x)$ is shown in figure 12.

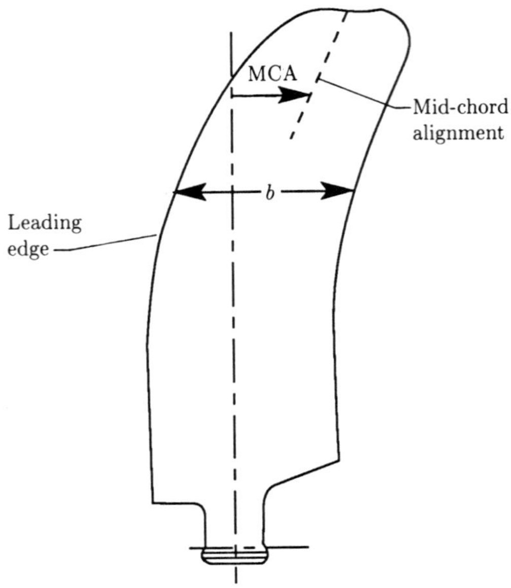

Figure 11. Blade planform defining chord b and mid-chord alignment (sweep).

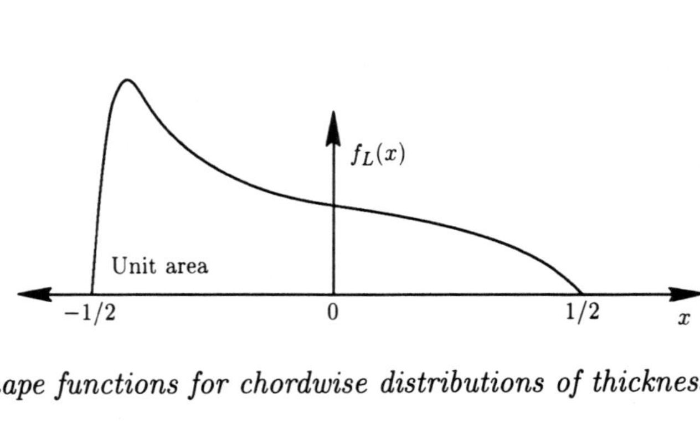

Figure 12. Shape functions for chordwise distributions of thickness and loading.

Thickness distribution function $H(x)$ is defined to be unity at the maximum thickness point. Similarly, the lift and drag distributions are given in terms of the section lift and drag coefficients C_L and C_D and the chordwise lift and drag distribution functions $f_L(x)$ and $f_D(x)$ normalized for unit area, as shown in figure 12. The terms C_L and C_D are defined such that the forces per unit spanwise distance are recovered when multiplied by $(1/2)\rho_o c_o^2 M_r^2$, where ρ_o and c_o are the ambient density and speed of sound and M_r, defined by

$$M_r^2 = M_x^2 + z^2 M_t^2 \tag{7}$$

is the section relative Mach number, with M_x and M_t defined as the flight and tip rotational Mach numbers, respectively.

With these definitions, the noise harmonics are given by (ref. 16)

$$\begin{Bmatrix} p_{Vm} \\ p_{Dm} \\ p_{Lm} \end{Bmatrix} = -\frac{\rho_o c_o^2 B \sin\theta \exp\left[imB\left(\frac{\Omega r}{c_o} - \frac{\pi}{2}\right)\right]}{8\pi(y/D)(1 - M_x\cos\theta)}$$
$$\times \int_0^1 M_r^2 \exp(i\phi_s) J_{mB}\left(\frac{mBzM_t\sin\theta}{1 - M_x\cos\theta}\right) \begin{Bmatrix} k_x^2 t_b \Psi_V(k_x) \\ ik_x(C_D/2)\Psi_D(k_x) \\ ik_y(C_L/2)\Psi_L(k_x) \end{Bmatrix} dz \tag{8}$$

where $J_{mB}(\ )$ is a Bessel function, Ψ_V, Ψ_D, and Ψ_L are source transforms, and k_x and k_y are wave numbers given by

$$k_x = \frac{2mBB_D M_t}{M_r(1 - M_x\cos\theta)} \tag{9}$$

$$k_y = \frac{2mBB_D}{zM_r}\left(\frac{M_x - M_r^2\cos\theta}{1 - M_x\cos\theta}\right) \tag{10}$$

and ϕ_s is a phase lag due to sweep:

$$\phi_s = \frac{2mBM_t}{M_r(1 - M_x\cos\theta)}\frac{\text{MCA}}{D} \tag{11}$$

Displacement normal to the planform (face alignment) also produces a phase shift (ref. 16), but that is usually weak and is not included here. The propeller (or aircraft) position is given in terms of its altitude or sideline distance y, and the retarded radiation angle θ is as sketched in the insert of figure 13 (from ref. 19). In evaluating equation (8) it may be convenient to use the fact that $\rho_o c_o^2 = \gamma p_o$, where p_o is the ambient pressure and the specific-heat ratio $\gamma = 1.4$ for air.

The retarded radiation angle θ and current (or visual) angle θ_1 are related by

$$\cos\theta = \cos\theta_1\sqrt{1 - M_x^2\sin^2\theta_1} + M_x\sin^2\theta_1 \tag{12}$$

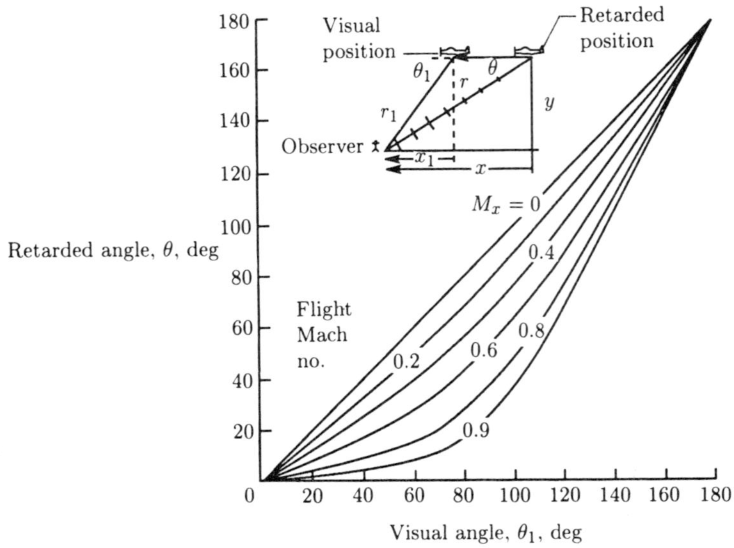

Figure 13. Relationship between retarded and visual angles. (From ref. 19.)

as plotted in figure 13. Finally, the chordwise thickness and loading distributions are given in terms of their Fourier transforms:

$$\begin{Bmatrix} \Psi_V(k_x) \\ \Psi_D(k_x) \\ \Psi_L(k_x) \end{Bmatrix} = \int_{-\frac{1}{2}}^{\frac{1}{2}} \begin{Bmatrix} H(x) \\ f_D(x) \\ f_L(x) \end{Bmatrix} \exp(ik_x x)\, dx \tag{13}$$

The above equations apply strictly only in the far field. However, they are reasonably accurate to within about a diameter of the propeller, particularly for the upper harmonics. Near-field versions of these equations are available (ref. 18). In the paragraphs below, some general properties of the radiation equations are discussed, the influence of blade geometry is explained, and some suggestions for programming are given.

In equation (8), the integrand can be considered to be the product of source terms (in the braces) times a radiation efficiency factor J_{mB}. Bessel functions of argument x and order $mB \neq 0$ have the behavior shown in figure 14. They peak for arguments about equal to order, diminish toward zero for smaller arguments, and oscillate for large arguments. In equation (8), the argument $mBzM_t \sin\theta/(1 - M_x \cos\theta)$ evaluated for radiation in the plane of rotation is $mBzM_t$. From this, it can be seen that radiation efficiency at $\theta = 90°$ is governed by zM_t, the blade section rotational Mach number. The factor $\sin\theta$ causes the noise to diminish rapidly toward the front and rear axes of the propeller and the Doppler factor, $1 - M_x \cos\theta$, shifts the directivity pattern forward.

In the source terms, it is easy to see that the thickness, drag, and lift noise components are proportional to the thickness ratio, drag coefficient, and lift coefficient, respectively. The Ψ terms represent the effect of chordwise noncompactness, that is, interference at the observer location of signals emitted from various source locations

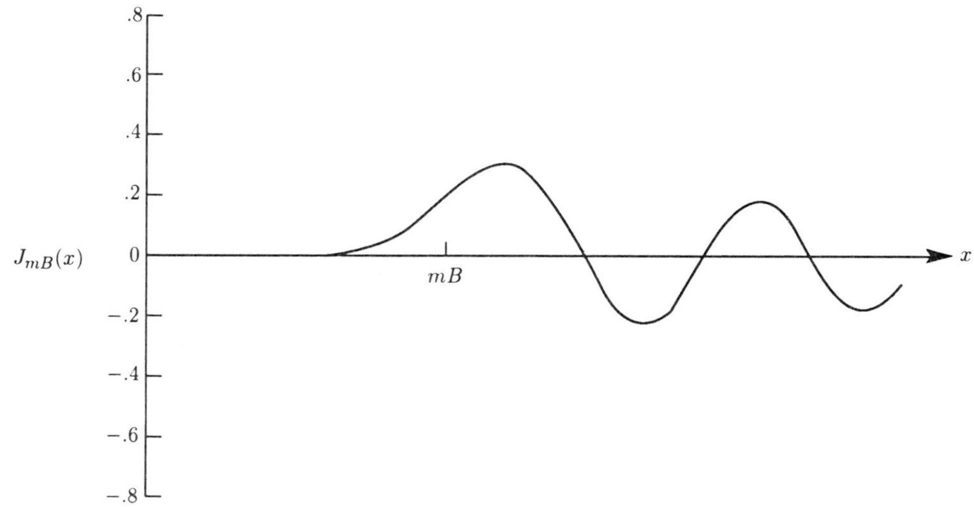

Figure 14. General behavior of Bessel function with order not equal to zero.

along the chord. The argument k_x can be considered the noncompactness parameter. Figures 15 and 16 show examples of Ψ for thickness and lift sources.

For low frequency ($k_x \to 0$), the effect of chordwise interference vanishes. As k_x increases because of increasing chord, harmonic order, or Doppler effects, significant interference occurs. For chordwise distributions, figure 15 shows that there is not much potential for reduction of thickness noise by modifying airfoil thickness distribution. However, the curves in figure 16 show that the quietest chordwise loading distributions are uniform and that peaky distributions increase the higher harmonics.

The exponential factor $\exp(i\phi_s)$ in equations (8) and (11) indicates that sweeping a blade section back along the advance helix causes the noise contribution from that radius to lag by ϕ_s. Sweep can cause significant noise reduction via the mechanism sketched in figure 17 for noise radiated at angle θ. Since only one harmonic at a time is considered, the noise from each blade element is described completely by its amplitude A_j and its phase ϕ_j in the complex notation $A_j \exp[i(\phi_j - mB\Omega t)]$. The total noise is the sum of contributions from all blade elements,

$$A_R \exp(i\phi_R) = \sum_{j=1}^{N} A_j \exp(i\phi_j) \tag{14}$$

where the common factor $\exp(-imB\Omega t)$ has been cancelled from both sides. This complex addition is easily visualized as the head-to-tail vector addition, also shown in figure 17. Because the blade sections are swept back, the phase angles of their signals lag and the vector addition plot tends to close on itself, representing the phase interference effect. The amount of noise reduction obtained with blade sweep depends on observer location and propeller operating condition. However, analytical studies (ref. 17) and test data show that peak sideline noise can be reduced by about 8 dB with blade sweep for the propfan in figure 2 at a cruise Mach number of 0.8.

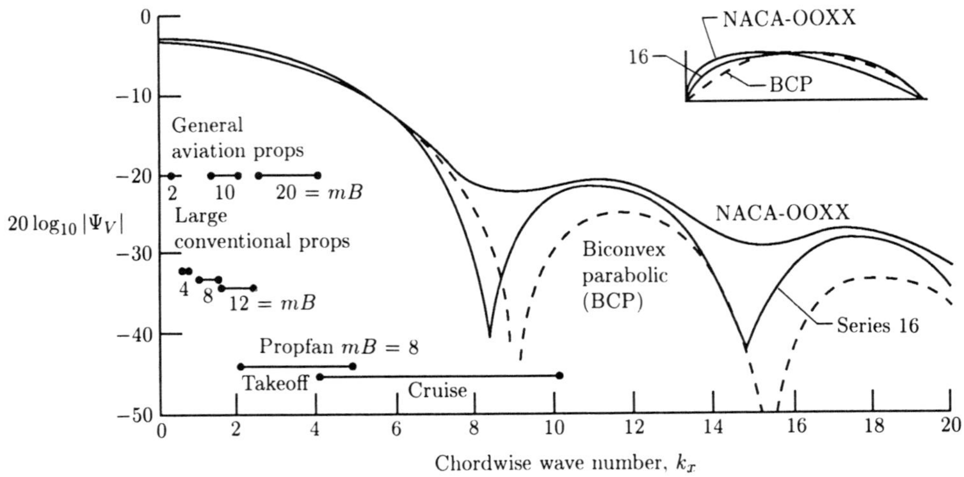

Figure 15. Reduction of thickness noise from blade element due to chordwise noncompactness. (From ref. 17.)

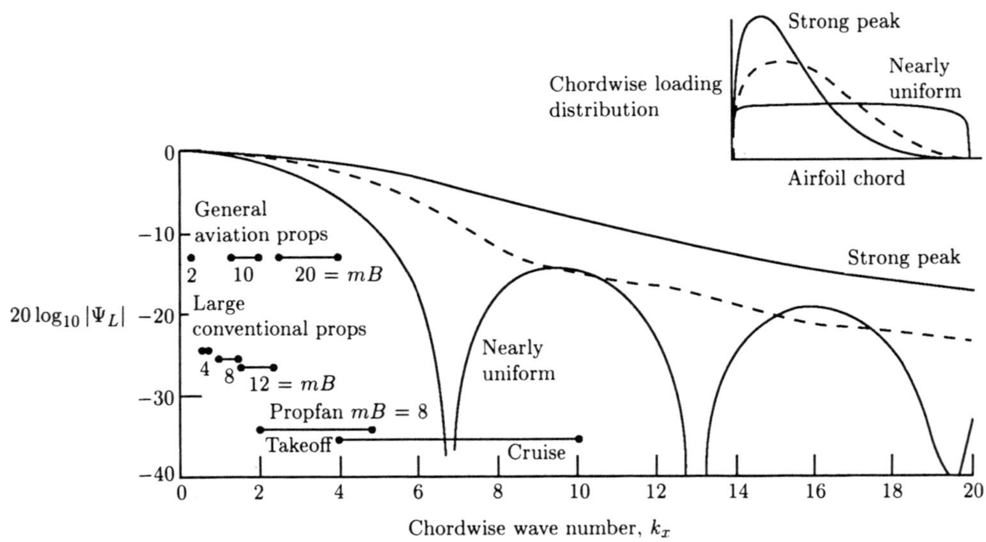

Figure 16. Reduction of loading noise from blade element due to chordwise noncompactness. (From ref. 17.)

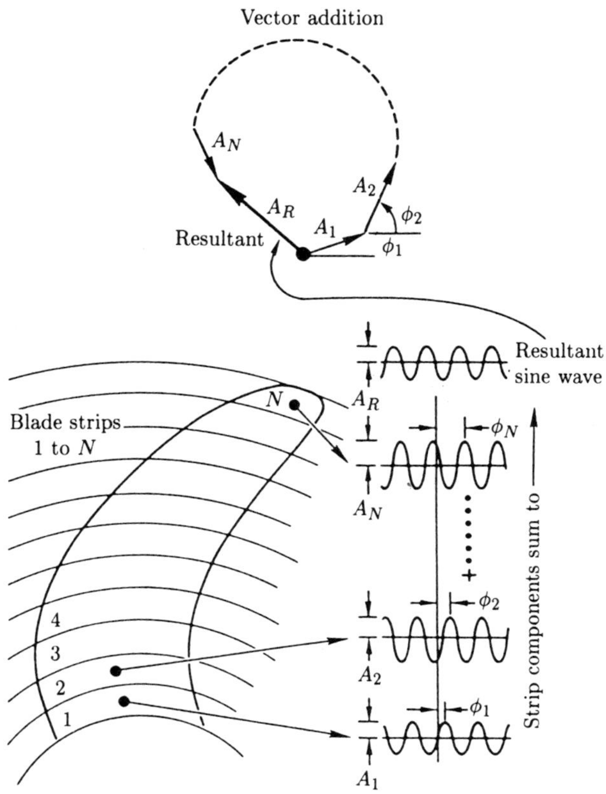

Figure 17. Conceptual benefit of blade sweep for reducing noise. (From ref. 17.)

Programming equation (8) is straightforward. However, a few comments are in order. Since the chordwise distribution needed to compute Ψ is not too critical for thickness and may not be well-known for loading, it is convenient to use simple analytical functions for $H(x)$, $f_L(x)$, and $f_D(x)$ that can be transformed analytically in equation 13. For example, a parabolic thickness distribution,

$$H(x) = 1 - (2x)^2 \qquad (|x| \leq 1/2) \tag{15}$$

transforms to

$$\Psi_V = \begin{cases} 2/3 & (k_x = 0) \quad (16a) \\ \frac{8}{k_x^2}\left[\frac{2}{k_x}\sin\left(\frac{k_x}{2}\right) - \cos\left(\frac{k_x}{2}\right)\right] & (k_x \neq 0) \quad (16b) \end{cases}$$

and a uniform lift distribution

$$f_L(x) = 1 \qquad (|x| \leq 1/2) \tag{17}$$

transforms to

$$\Psi_L = \begin{cases} 1 & (k_x = 0) \quad \text{(18a)} \\ \frac{2}{k_x} \sin\left(\frac{k_x}{2}\right) & (k_x \neq 0) \quad \text{(18b)} \end{cases}$$

These transforms should be accurate enough for most work at low harmonic order. The need for more accurate source functions can be judged from figures 15 and 16.

In the integration of equation (8), the radial step size must be chosen with some care, particularly for swept blades, since phase variations along the span increase with sweep and harmonic order. A simple expedient is to try a step size and see if the integral is adequately resolved.

Note also that the lift and drag forces defined here are the section loading components acting parallel and perpendicular to the local section advance direction as sketched to the left in figure 18. If a lifting-line theory is used to obtain the aerodynamic loading as input to the noise equations, the reference direction for C_L and C_D will probably be shifted from the advance direction by the induced angle, as shown to the right in figure 18. Correction for this shift should be made before equation (8) is used.

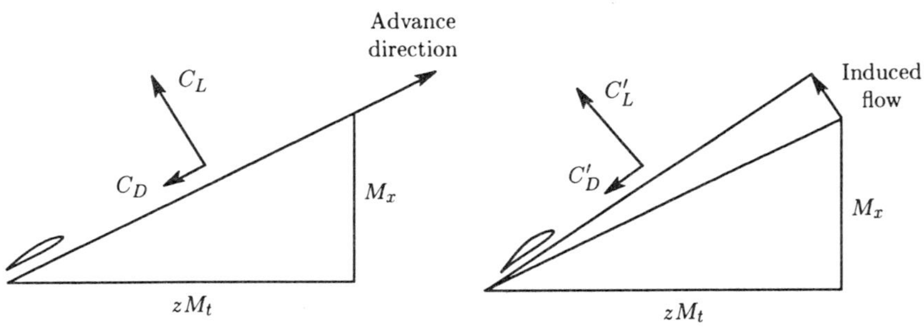

Figure 18. Reference axes for lift and drag definition.

Blade forces can also be specified in terms of thrust and torque rather than lift and drag. If it is assumed that the chordwise noncompactness factor for drag Ψ_D is the same as that for lift Ψ_L, then the two loading terms can be combined as $(P_{\text{load}})_m = P_{Lm} + P_{Dm}$ in terms of the radial gradients of thrust and torque, dT/dz and dQ/dz:

$$\begin{aligned}(P_{\text{load}})_m &= P_{Lm} + P_{Dm} \\ &= \frac{imBM_t \sin\theta}{4\pi y r_t (1 - M_x \cos\theta)} \int_{z_{\text{hub}}}^{1} \left[\frac{\cos\theta}{1 - M_x \cos\theta} \frac{dT}{dz} - \frac{1}{z^2 M_t r_t} \frac{dQ}{dz} \right] \Psi_L J_{mB} \exp(i\phi_s)\, dz \end{aligned} \quad (19)$$

The arguments of Ψ_L and J_{mB} are unchanged.

Furthermore, if there is no sweep and if Ψ_L is taken as real, as in the case of symmetrical loading functions f_L and f_D, then an effective radius version of equation (19) is obtained by setting $dT = T$ and $dQ = Q$:

$$(P'_{\text{load}})_m = \frac{mBM_t \sin\theta}{2\pi y r_t (1 - M_x \cos\theta)} \left(\frac{\cos\theta}{1 - M_x \cos\theta} T - \frac{1}{z_{\text{eff}}^2 M_t r_t} Q \right) \Psi_L J_{mB} \quad (20)$$

where we have dropped the i and multiplied by 2 so that equation (20) represents the ordinary Fourier coefficient (one-sided), as denoted by the prime on P. In terms of shaft power $W = Q\Omega$, the term in the parentheses which contains Q becomes $W/(z_{\text{eff}}^2 M_t^2 c_o)$, where z_{eff} is the effective radius, which experience shows can be taken as 0.8.

To calculate SPL, equation (20) is divided by $\sqrt{2}$ to obtain rms pressure and by the acoustic pressure reference, 4.1784×10^{-7} lb/ft^2 (20 μPa):

$$\text{SPL} = 20 \log_{10} \left[\frac{538\,673 m B M_t \sin\theta}{yD(1 - M_x \cos\theta)} \left(\frac{\cos\theta}{1 - M_x \cos\theta} T - \frac{550}{z_{\text{eff}}^2 M_t^2 c_o} W \right) \Psi_L J_{mB} \right] \tag{21}$$

where it is now understood that the thrust is T pounds, the power is W horsepower, the sideline distance is y feet, the diameter is D feet, and the speed of sound is c_o ft/sec. The Bessel function is

$$J_{mB} = J_{mB} \left(\frac{m B z_{\text{eff}} M_t \sin\theta}{1 - M_x \cos\theta} \right) \tag{22}$$

and the noncompactness factor for rectangular loading is

$$\Psi_L = \frac{\sin X}{X} \tag{23}$$

where

$$X = \frac{m B M_t B_D}{M_r (1 - M_x \cos\theta)} \tag{24}$$

and θ is the retarded radiation angle.

Unsteady sources

When the flow into a propeller is nonuniform, blade loading is unsteady and a significant increase in noise is likely to occur. Because of radiation efficiency effects described below, a small amount of unsteady loading can be the dominant noise source, particularly for low-tip-speed propellers. These effects were first recognized by Lowson (ref. 20), Wright (ref. 21), and Morse and Ingard (ref. 6), and these authors give extensive analyses, particularly for helicopter rotors and single rotating propellers at zero forward flight speed.

Noise caused by unsteady loading can be computed with either time-domain or frequency-domain methods, and both have their merits. Farassat's formula (eq. (3)) applies without modification for unsteady loading. To use it, the loading history or waveform l_r and its distribution over the blade must be determined or approximated. The proper values for l_r are then applied at the appropriate retarded blade locations. This procedure is reasonably straightforward and Farassat's program accepts unsteady-loading input. However, there are some subtleties regarding required quality of the unsteady-loading input data that must be mentioned. Because the motion of the blades shifts the source frequencies, the blade-loading waveform must be specified with adequate precision to frequencies significantly higher than the sound frequency of interest. Since the blade unsteady loading is seldom known accurately, guidance is needed for choosing the time and space resolution of the surface pressure input. The frequency-domain formulas discussed below may be used to provide such needed guidance.

Frequency-domain formulas for unsteady-loading noise appear to be more cumbersome than their steady-loading counterparts. However, they may be more convenient than time-domain formulas for users who are basically interested in noise harmonics rather than waveforms or who can approximate blade-loading harmonics without the direct time information. Furthermore, the harmonic formulas can provide valuable diagnostic information because of the frequency discrimination of spectrum analyzers. Hanson has derived radiation formulas (ref. 22) for the general case of harmonic blade loading at any frequency, whether or not it is related to the propeller rotation speed. Two special cases are presented below.

For unsteady loading, the lift coefficient can be expanded in harmonic form as follows:

$$C_L = \sum_{k=-\infty}^{\infty} C_{Lk} \exp(-i\omega_k t) \tag{25}$$

A similar expression can be derived for C_D. Equation (25) gives the lift history experienced by a blade on the radiating rotor in terms of the lift harmonics C_{Lk}, where $k = 1$ corresponds to the fundamental frequency and $k = 0$ is the steady, or mean, loading designated simply C_L previously. For the general case of radiation from a rotor with angular speed Ω_2 interfering with the flow field of another rotor with B_1 blades and rotating at Ω_1, the load frequencies on the radiating rotor are $\omega_k = kB_1(\Omega_1 + \Omega_2)$. One special case is for counterrotating propellers with equal speeds ($\Omega_1 = \Omega_2 = \Omega$) and equal numbers of blades ($B_1 = B_2 = B$). For this condition $\omega_k = 2kB\Omega$, where the factor 2 arises because of the relative speed of the rotors. The other special case is for interference with a nonrotating distortion field, where effectively $\Omega_1 = 0$ and $B_1 = 1$ so that $\omega_k = k\Omega$.

The far-field acoustic pressure (ref. 22) is

$$\begin{aligned} p(t) = \frac{-i\rho_o c_o^2 B \sin\theta}{8\pi(r_1/D)(1 - M_x\cos\theta)} \sum_{m=-\infty}^{\infty} \sum_{k=-\infty}^{\infty} \exp\left\{i\left[n\left(\phi - \frac{\pi}{2}\right) + mB\Omega\left(\frac{r}{c_o} - t\right)\right]\right\} \\ \times \int_{\text{root}}^{\text{tip}} M_r^2 \exp(i\phi_s) J_n\left(\frac{mBzM_t\sin\theta}{1 - M_x\cos\theta}\right)\left[k_x\frac{C_{Dk}}{2}\Psi_{Dk}(k_x) + k_y\frac{C_{Lk}}{2}\Psi_{Lk}(k_x)\right] dz \end{aligned} \tag{26}$$

where, for counterrotation with $B_1 = B_2 = B$ and $\Omega_1 = \Omega_2 = \Omega$, the mode order is

$$n = (m - 2k)B \tag{27}$$

the wave numbers are

$$k_x = \frac{2BM_t}{M_r}\left[\frac{m}{1 - M_x\cos\theta} - 2k\right] B_D \tag{28}$$

$$k_y = \frac{-2B}{zM_r}\left[\frac{m(M_r^2\cos\theta - M_x)}{1 - M_x\cos\theta} + 2kM_x\right] B_D \tag{29}$$

and the phase lag due to blade sweep is

$$\phi_s = \frac{2BM_t}{M_r}\left[\frac{m}{1 - M_x \cos\theta} - 2k\right]\frac{\text{MCA}}{D} \tag{30}$$

For interference with fixed distortion, the mode order is

$$n = mB - k \tag{31}$$

the wave numbers are

$$k_x = \frac{2M_t}{M_r}\left[\frac{mB}{1 - M_x \cos\theta} - k\right] B_D \tag{32}$$

$$k_y = \frac{-2}{zM_r}\left[\frac{mB(M_r^2 \cos\theta - M_x)}{1 - M_x \cos\theta} + kM_x\right] B_D \tag{33}$$

and the phase lag due to sweep is

$$\phi_s = \frac{2M_t}{M_r}\left(\frac{mB}{1 - M_x \cos\theta} - k\right)\frac{\text{MCA}}{D} \tag{34}$$

In equation (26), the first exponential gives information on the general nature of the sound field. The frequencies that appear are given by $mB\Omega$, which indicates sound at blade-passage frequency and its multiples, just as with the steady-loading formulas discussed previously. The factor $(r/c_o) - t$ indicates waves propagating radially from the source location at the ambient speed of sound c_o. The double summation shows that each loading harmonic k radiates at all the sound harmonics mB. Variation of the sound pattern in the circumferential direction is given by $n\phi$, where n is the circumferential mode order. These modes, each with n lobes, spin about the propeller axis with angular speed mB/n times the propeller shaft speed Ω. This spinning mode characteristic also occurs for compressors and turbofans, as discussed in the chapter on turbomachinery noise.

As with thickness and steady-loading noise, the radiation efficiency is governed by the ratio of argument to order in the Bessel function:

$$\xi = \frac{1}{n}\left(\frac{mBzM_t \sin\theta}{1 - M_x \cos\theta}\right) \tag{35}$$

The term ξ is similar to the cutoff ratio in turbofans. For $\xi < 1$, radiation is inefficient; for $\xi > 1$, radiation tends to be efficient because of the Bessel function properties shown in figure 14. Since $z \sin\theta/(1 - M_x \cos\ \theta)$ is of order 1, it is useful to consider the quantity

$$M_s = (mB/n)M_t \tag{36}$$

This is the spinning-mode tip Mach number. Any mode with $M_s << 1$ can be neglected in the calculation. Thus, although the summation on k in equation (26) runs formally from $-\infty$ to ∞, only a few modes for the largest $|M_s|$ need be considered. At the lower sound harmonics sometimes only one mode is required.

As an example of the above mode criterion, consider the $m = 3$ harmonic of a counterrotating propeller at $M_t = 0.7$:

$$M_s = \frac{mB}{(m - 2k)B} M_t = \frac{m}{m - 2k} M_t \tag{37}$$

The combinations of interest for $m = 3$ are

k	$m/(m-2k)$	$\lvert M_s \rvert$
−1	0.6	0.42
0	1.0	.70
1	3.0	2.10
2	−3.0	2.10
3	−1.0	.70
4	−.6	.42

Thus, only the $k = 1$ and 2 loading harmonics radiate efficiently at 3BPF. The values $k = 1$ and 2 correspond to blade loading at 2BPF and 4BPF. Through use of the above formulas, it is easy to show that for increasing sound harmonics, more and more loading harmonics radiate efficiently. Noise from low-tip-speed propellers with any source of nonuniform inflow is inevitably dominated by unsteady-loading sources at the upper harmonics because of their greater radiation efficiency.

Some insight can now be gained for the direct use of time-domain formulas for calculation of noise from unsteady loading. In equation (3), the unsteady blade loading is input numerically in discrete time steps. The size of these steps must be small enough to ensure a full and smooth representation of the loading component; otherwise, the loading signal will appear to have a higher frequency that will be strongly emphasized because of the radiation efficiency discussed above. This sensitivity is aggravated by the derivatives denoted by the dots on l_i and M_i in equation (3). The derivations must be performed numerically, a process always sensitive to smoothness of the quantity being differentiated. These problems are all manageable in principle. However, the reader is cautioned against casual application of equation (3) to the unsteady-loading problem without a thorough understanding of the numerical subtleties.

Nonlinear Effects

Blade sections of propfans and of many other high-speed propellers operate at transonic velocities. In the aerodynamics of wings and bodies, this is a regime frequently dominated by nonlinear effects. The corresponding propeller issues are discussed below under the categories of nonlinear source effects from the acoustic analogy quadrupole and full aerodynamic solutions by applying finite element methods.

At high speeds, nonlinearity may occur at the source (i.e., at the blade section) because of transonic effects. One way of dealing with this is via the quadrupole source term in the acoustic analogy. (See chapters on jet noise for definition.) The first valid analysis of the importance of the quadrupole source for high-speed rotors was made based on a two-dimensional aerodynamic calculation (ref. 23). The quadrupole contribution was compared with the linear thickness source for a propfan airfoil section; the results are shown in figure 19. These results show that quadrupole, or nonlinear, source effects are important sources of additional noise in the blade section speed range between critical Mach number (when flow over the airfoil exceeds the speed of sound) and a Mach number of 1. Quadrupole strength can be reduced to below that of linear thickness and loading sources by blade sweep so the airfoil sections operate effectively below their critical Mach numbers. To shed more light on the role of the quadrupole term in flow with solid surfaces, Blackburn examined the field of a two-dimensional wedge in fully supersonic flow (ref. 24). He was able to compute the acoustic analogy source terms exactly and found that the quadrupole was not a significant source of extra noise in this flow regime. This finding agrees with figure 19, since the Mach number in Blackburn's analysis is well to the right of the peak. However, Blackburn did find that the quadrupole term repositioned wave fronts along the shocks, rather than on the Mach surfaces as in linear theory. Quadrupole sources have been treated more extensively in the helicopter literature (see the chapter on helicopter rotor noise) because supercritical blade section speeds are common. However, for propellers and propfans, nonlinear effects are minimized by blade designs with sweep and thin airfoil sections.

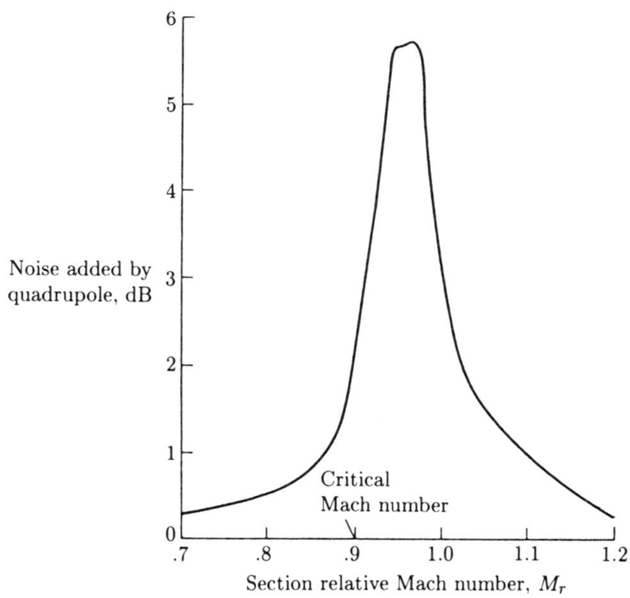

Figure 19. Increase of blade thickness sound pressure level caused by including quadrupole noise. (From ref. 23.)

Another approach to nonlinear effects is as a by-product of an aerodynamics calculation. There is considerable work on the transonic regime in progress to develop numerical methods for aerodynamic design and analysis from full-potential, Euler, and Navier-Stokes equations. For the steady-loading problem, it is tempting to believe that since these methods compute all the flow-field variables in a finite domain surrounding the propeller, the noise signal is available simply by sampling the pressure field at the appropriate locations. This, in principle, is true. However, the currently existing aerodynamic analysis methods are designed to give the best accuracy on the surfaces of the blades. For field points at a distance from the blades, predictions are degraded because the mesh sizes used in the calculation increase, numerical damping smooths the waves, and the boundary conditions at the outer edges of the computational domain are treated approximately. These problems all appear to be manageable for the steady-loading aerodynamics problem, but so far they have not been addressed for acoustic calculations. However, recent advances in computational fluid dynamics are now making this approach look more practical.

Prediction Methods for Propeller Broadband Noise

The fundamental laws of acoustics (see Curle, ref. 25, for example) state that the noise from a surface is produced by forces (dipoles) and sources (monopoles) induced on the surface to satisfy the condition of no flow through the surface. If these terms are harmonic, the noise produced is harmonic. However, if dipoles with a random time behavior are present on the surface, broadband noise is produced. (Usually there are no monopoles with a random time behavior, since this would require the surface to have a significant component of random fluctuation in position.) Random forces can be induced by several mechanisms. If significant turbulence is present in the mean stream, random forces are induced on the blades, leading to broadband noise. In the low-frequency (compact) case the entire blade is involved in the sound generation process. At higher frequencies (acoustic wavelength smaller than the chord), the noise generation becomes concentrated around the leading edge of the blade.

In the absence of inflow turbulence, any random surface forces must be self-induced. A turbulent flow moving over a plate induces unsteady surface pressures. For a uniform mean flow this turbulence can be produced in the turbulent boundary layer. If the turbulence is not in the vicinity of an edge it produces quadrupole sound, which is generally a weak generation mechanism compared with dipole sound. However, as the turbulence approaches and passes the trailing edge, the boundary conditions imposed on an eddy change: whereas the airfoil surface can support a force, the wake cannot. The result is a change in the airfoil loading as each eddy passes the edge, and sound is produced.

Two approaches to the calculation of trailing-edge noise have been developed. Since Curle has shown that the pressure field produced by the turbulence can be represented by volume quadrupole sources together with the surface monopoles and dipoles to satisfy the boundary condition on the surface, the first approach is to solve the problem of a quadrupole in the vicinity of a half-plane. Since the surface dipoles induced by the quadrupoles are the main sound-producing sources, this method can be described as a calculation of the surface forces produced by the quadrupoles, followed by the calculation of the noise. An analysis following this approach was carried out by Ffowcs Williams and Hall (ref. 26). This approach is satisfactory if the quadrupole strengths are known. However, this method presents the same kind of problems encountered in the prediction of jet noise from quadrupole distributions; in general the distribution is not known with sufficient accuracy. Furthermore, there is the possibility that other nonlinear effects, such as wave steepening (also represented by quadrupoles), are included with the sound-producing quadrupoles.

These problems led to the formulation of a second approach, which has been more extensively compared with experiment. In contrast to the above method, this new method assumes that the surface pressure produced by the convective turbulence is known. This surface pressure field is assumed to be frozen and convecting with a velocity U_c (or Mach number M_c) which may be a function of frequency. At the trailing edge, there is a change in the boundary condition; in particular, the surface pressure due to the turbulence, which is supported by the airfoil upstream of the trailing edge, is not supported by the downstream wake. This produces a fluctuating dipole force on the surface and radiates sound. Chase (ref. 27) was one of the first to employ this method for noise prediction. He assumed zero Mach number, and thus no Kutta condition is applied at the edge. A more general formulation, including a mean-flow Mach number M and the application of the Kutta condition at the trailing edge, was introduced by Amiet (ref. 28). This model is especially attractive because of its symmetrical relation with leading-edge noise. For leading-edge noise, one specifies the velocity of the incident turbulence field together with a no-flow condition through the airfoil surface; this leads to zero potential on the axis ahead of the airfoil and at the leading edge. For trailing-edge noise one specifies the incident surface pressure on upper and lower surfaces; the pressure difference is zero on the axis downstream of the trailing edge and at the trailing edge because of the Kutta condition. However, just as for the first method, this approach does not result in a compact dipole type directivity pattern. Because the fluctuating forces occur at the airfoil edge, one cannot simply replace them by the field of a compact dipole in an infinite fluid. Rather, one must include the baffling effect of the airfoil edge, the result being a modified directivity.

This model avoids the necessity of specifying the volume distribution of quadrupoles, since their effect is already included in the convecting surface pressure. Although the specification of the surface pressure is probably simpler than the specification of the volume quadrupoles, it is by no means a simple task. Amiet used surface pressure data for a flat-plate boundary layer in his calculations (ref. 28) partly because this is a simple and "classic" case. Perhaps more importantly, it was the only readily available data. (When using the following expressions for the noise, the reader should be aware that the expressions used for the surface pressure are generally obtained by curve fitting of data. Thus, they are by no means rigorous and are open to improvement.)

The far-field noise spectrum prediction S_{pp} above a flat plate for this model is (ref. 28)

$$S_{pp}(x, 0, z, \omega) = \left(\frac{\omega b z}{2\pi c_o \sigma^2}\right)^2 l_y(\omega) s |\mathcal{L}|^2 S_{qq}(\omega, 0) \tag{38}$$

where x, y, and z are, respectively, the coordinates in the axial direction, normal to the airfoil, and along the span; b is the semichord; s is the semispan; c_o is the sound speed; l_y is the spanwise correlation length, which can be a function of the radian frequency ω; $\sigma^2 \equiv x^2 + \beta^2 z^2$, where $\beta^2 \equiv 1 - M^2$; $\mathcal{L}$ is the generalized lift; and S_{qq} is the pressure spectrum on either the upper or lower surface of the airfoil near the trailing edge. (To find the total noise, the sound from both surfaces must be added.

Since the surface pressure spectra on the two surfaces are generally uncorrelated, the pressure spectra, rather than the pressure itself, are added.) In the high-frequency limit (wavelength much less than chord), the directivity factor is given by

$$\frac{z^2|\mathcal{L}|^2}{\sigma^4} = \frac{(1+M_c-M)D}{M_c r_e^2 K_x^2} \tag{39}$$

with

$$D \equiv \frac{2\cos^2(\theta_e/2)}{[1-(M-M_c)\cos\theta_e]^2(1-M\cos\theta_e)} \tag{40}$$

where $K_x = \omega b/U_c$ and r_e and θ_e are, respectively, the retarded radius from the observer to the source point and angle of the observer measured from the upstream axis. The terms r_e and θ_e are related to the actual coordinates by

$$\sigma = r_e(1-M\cos\theta_e) \qquad \text{and} \qquad x = r_e(M-\cos\theta_e) \tag{41}$$

A more general form of equations (39) and (40) for arbitrary frequency can be found (refs. 28 and 29), but since trailing-edge noise is predominantly at high frequency, the more general expression may not be needed. The spanwise correlation length l_y is defined in terms of the surface pressure spectrum as

$$l_y(\omega) = \frac{1}{S_{qq}(\omega,0)}\int_0^{\infty} S_{qq}(\omega,y)\,dy \tag{42}$$

The integration of a result from Corcos (ref. 30) leads to an expression for l_y of (ref. 28)

$$l_y \approx \frac{2.1U_c}{\omega} \tag{43}$$

For a flat plate an expression for S_{qq} can be obtained by curve fitting the data of Willmarth and Roos (ref. 31) to give

$$S'_{qq} = \frac{S_{qq}}{\frac{1}{2}(\rho_o U^2)^2(\delta^*/U)} = \frac{0.00002}{1+\tilde{\omega}+0.217\tilde{\omega}^2+0.00562\tilde{\omega}^4} \tag{44}$$

where ρ_o is the free-stream density, $\tilde{\omega} = \omega\delta^*/U$ (a form of the Strouhal number), and δ^* is the displacement thickness of the turbulent boundary layer, given approximately by (ref. 32)

$$\frac{\delta^*}{c} \approx 0.047R_c^{-1/5} \tag{45}$$

where R_c is the Reynolds number based on chord c. A reasonable value to use for the convection velocity is $U_c = 0.8U$ (ref. 28).

Equations (38) to (45) allow a calculation of the trailing-edge noise as a function of frequency and observer position, given values for U, ρ_o, c_o, chord, and span. One should be able to model trailing-edge noise for a realistic airfoil if accurate expressions for the surface pressure are known. Of the above relations, the one most subject to question is equation (44) for the spectrum of the convecting surface pressure. Although this equation may be adequate for a flat plate at zero incidence, one realizes at once that this cannot be quite correct for an actual airfoil, if only because it does not include the angle of attack. As one might expect, experimental evidence shows a noise increase as an airfoil becomes more heavily loaded. This could be due to an increase in either S_{qq} or δ^*. Also, the measurements from which equation (44) was obtained were not made near an edge. For these reasons, attempts have been made to modify equation (44) with actual airfoil surface pressure data. Thus, curve fitting the data of reference 33 gives the following general form first used in reference 34:

$$S'_{qq} = \frac{0.000666\tilde{\omega}}{1 - 5.489\tilde{\omega} + 36.74\tilde{\omega}^2 + 0.1505\tilde{\omega}^5} \tag{46}$$

In the range of interest this gives a value for S'_{qq} that is somewhat larger than that given by equation (44) for a flat plate, the maximum difference being 7.7 dB at $\tilde{\omega} = 0.18$. The curve fit is based on measurements taken on an airfoil at zero angle of attack. However, the surface pressure data are not extensive enough to assure an accurate prediction. For example, by curve fitting the data of both references 33 and 35, Chou and George (ref. 36) subsequently gave an expression for S'_{qq} for two ranges of $\tilde{\omega}$. The first expression, for $\tilde{\omega} < 0.06$, is taken from equation (46) but is increased by an overall factor of 2.6. This difference points out the uncertainty in the result. Although Chou and George used a different curve fit than equation (46) for the surface pressure, the basic model used for the noise calculation is that of references 27 and 28, in which a frozen surface pressure convecting past the trailing edge is assumed.

Obviously the problem is not solved completely until an analytical solution for the surface pressure is available. However, the above method does give an approximation to the noise produced if the surface pressure is known; it also gives an insight into the mechanism of noise production, since it relates the unsteady surface pressure to the noise produced. There is another prediction method available that dispenses with the theoretical development and concentrates on curve fitting of available trailing-edge noise data (in contrast to the above method, which curve fits surface pressure data and uses this for predicting the noise). This other prediction method uses the frequency dependence of reference 37 together with certain of the above results and gives a prediction of 1/3-octave band frequency. The result, with further details given in reference 34, is

$$\mathrm{SPL}_{1/3} = \mathrm{OASPL} + 10\log_{10}\left\{0.613\left(\frac{\tilde{\omega}}{\tilde{\omega}_{\mathrm{max}}}\right)^4\left[\left(\frac{\tilde{\omega}}{\tilde{\omega}_{\mathrm{max}}}\right)^{3/2} + 0.5\right]^{-4}\right\} \tag{47}$$

where the overall SPL is

$$\mathrm{OASPL} = 10\log_{10}\left(M^5\frac{\delta^* s}{r_e^2}D\right) + K_1 \tag{48}$$

and s is the span, $\tilde{\omega}_{\mathrm{max}}$ is the value of $\tilde{\omega}$ at the spectrum peak (usually around 0.1), and $K_1 = 141.3$. Based on comparisons of theory versus experiment (ref. 34), equations (47) and (48) give a slightly better prediction than equations (38) to (46), but equations (47) and (48) give little insight into the noise-generating process.

Equations (47) and (48) should give reasonable predictions for an airfoil with a small angle of attack. Further analysis and additional experiments are required to determine the effects of angle of attack, flow separation, etc., on the noise generation. (A more detailed analysis of the fundamental principles on which the convecting surface pressure model is based is given in ref. 28.)

For a propeller, an integral must be taken over the rotor span (i.e., the blade is treated in a stripwise manner), since each radial station moves at a different velocity. Because trailing-edge noise is random and generally has a higher frequency than the propeller rotational frequency, the overall noise spectrum for a rotating blade can be calculated by simply averaging around the azimuth the sound spectrum derived above for the case of rectilinear motion. The principle is quite straightforward, but the implementation can become somewhat involved because of the constantly changing observer position in blade-fixed coordinates (ref. 34).

Propagation Effects

The propeller noise theories previously described provide estimates of noise generated at the source without regard to any propagation effects other than spherical spreading. This section discusses the effects of Doppler frequency shift, refraction, scattering and shielding, atmospheric absorption, ground reflection, and excess ground attenuation.

Doppler Frequency Shift

This phenomenon results in a shift in the frequency perceived by an observer when the observer or the source is moving relative to the medium. The familiar train whistle is often used as an example. To an observer standing near the train tracks, the train whistle appears raised in pitch as the train approaches, and lowered as the train passes by. Similar effects can be observed in propeller noise, the most important of which is related to the measurement of airplane flyover noise.

For a moving source, stationary observer, and stationary medium the Doppler frequency shift is calculated from

$$f_o = \frac{f_s}{1 - M_x \cos\theta} \tag{49}$$

where f_o is the observed frequency, f_s is the source frequency, M_x is the flight Mach number, and θ is the angle between the line from the source to the stationary observer and the flight path at the time the sound was emitted. This equation clearly shows that for an approaching source ($\theta < 90°$) the observed frequency is raised, while for a receding source ($\theta > 90°$) the frequency is lowered.

A generalized derivation for the Doppler-shifted received frequency is given by reference 38. This derivation shows that when the medium is in motion but both the source and the receiver are stationary (as in a wind tunnel), no frequency shift occurs. For airplane flyover noise, significant differences in received frequency are expected for a case with wind compared with that for the zero-wind case.

Refraction Effects

Propagation Through the Atmosphere

Refraction generally occurs for sound propagating through the normal atmosphere because the temperature and pressure vary with altitude and present a varying characteristic impedance ρc. Although the process is continuous, it is convenient to approximate it as layers having discrete interfaces (fig. 20). For a discrete interface, rays are bent according to Snell's law, described in reference 39 as

$$\frac{\sin \theta_1}{\sin \theta_2} = \frac{c_1}{c_2} \tag{50}$$

where the incident and transmitted angles θ_1 and θ_2 are defined in figure 20 and c_1 and c_2 are respectively the speed of sound in the incident and the transmitted medium.

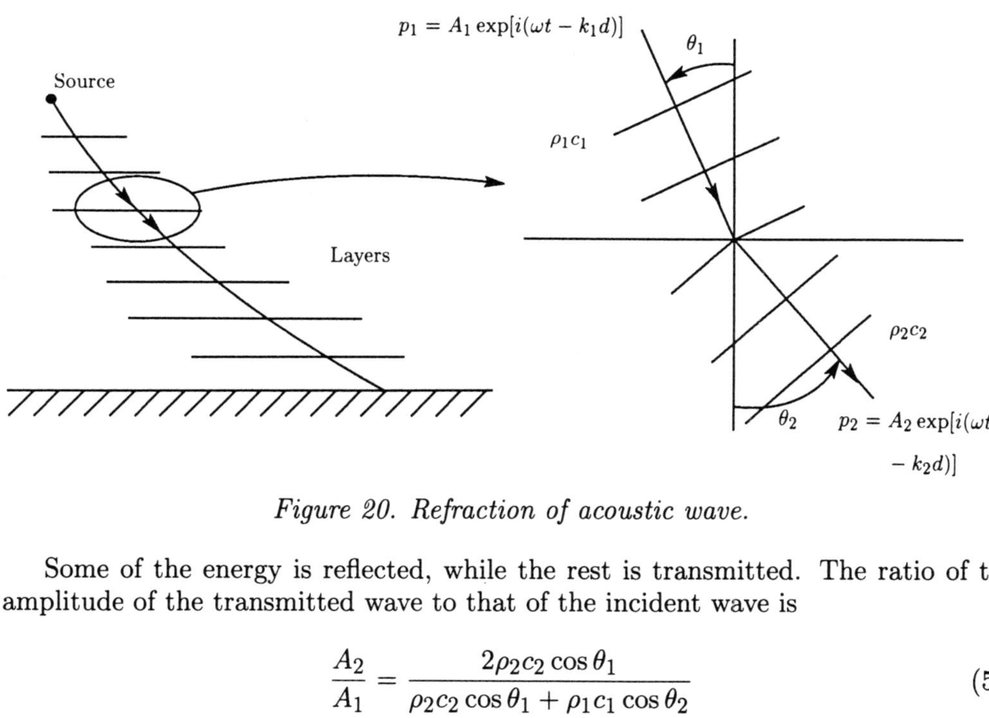

Figure 20. Refraction of acoustic wave.

Some of the energy is reflected, while the rest is transmitted. The ratio of the amplitude of the transmitted wave to that of the incident wave is

$$\frac{A_2}{A_1} = \frac{2\rho_2 c_2 \cos \theta_1}{\rho_2 c_2 \cos \theta_1 + \rho_1 c_1 \cos \theta_2} \tag{51}$$

For an acoustic plane wave, the change in sound pressure level is 10 log (Intensity ratio), where the intensity is $A^2/\rho c$. It follows that the change in the transmitted sound pressure level is

$$\begin{aligned} \Delta \text{SPL} &= 10 \log \frac{A_2^2/\rho_2 c_2}{A_1^2/\rho_1 c_1} \\ &= 10 \log \frac{4\rho_1 c_1 \rho_2 c_2 \cos^2 \theta_1}{(\rho_2 c_2 \cos \theta_1 + \rho_1 c_1 \cos \theta_2)^2} \end{aligned} \tag{52}$$

Because the sound beam is either widened or narrowed when transmitted through the interface, equation (52) does not give the ratio of total power transmitted. The change in sound power level can be calculated from

$$\Delta\text{PWL} = 10 \log \frac{4\rho_1 c_1 \rho_2 c_2 \cos\theta_1 \cos\theta_2}{(\rho_2 c_2 \cos\theta_1 + \rho_1 c_1 \cos\theta_2)^2} \tag{53}$$

Two special cases bear mentioning. For $\rho_2 c_2 \cos\theta_1 = \rho_1 c_1 \cos\theta_2$, no power is reflected and all the power is transmitted. When $c_1 < c_2$, there is a critical incidence angle θ_c, given by $\sin\theta_c = c_1/c_2$, for which the refracted ray is parallel to the interface. For incidence angles equal to or greater than θ_c no acoustic energy is transmitted into the second medium. For sound propagating from high altitudes through a normal atmosphere, the critical angle can be exceeded at large angles and no sound would be detected at the ground. This can be significant for propellers in the forward directivities because the Doppler effect shifts the source directivity forward.

It is convenient to use discrete layers, typically 100 to 300 m thick, for calculating refraction effects. Each layer is assumed to have uniform impedance represented by the mean impedance of the layer. This procedure is recommended for correcting airplane flyover noise during certification (ref. 40). It is essential to use a layered atmosphere model for propagation to the ground from high-altitude (greater than 5000 m) flight, as the cumulative effects become significant.

Propagation Through a Fuselage Boundary Layer

Refraction also occurs when sound propagates through a fuselage boundary layer because the velocity and temperature gradients in the boundary layer cause a change in the impedance encountered by the sound wave propagating through it. This effect could be important for noise impingement on a fuselage when cabin noise is being investigated or controlled.

Several analyses exist for evaluating fuselage-boundary-layer refraction effects as applied to propeller and propfan noise. Early investigations (refs. 19 and 41) addressed plane waves and two-dimensional boundary layers. Later refinements extended the analyses to propeller-type noise sources and boundary layers on cylindrical surfaces (ref. 42).

Scattering and Shielding

Fuselage Scattering

Sound incident on a cylindrical fuselage is scattered depending on the angle of incidence and the wavelength of the sound compared with the diameter of the fuselage. For normal incidence, sound at a small wavelength compared with the fuselage diameter is totally reflected. Thus, a receiver at the fuselage surface perceives a doubling of the pressure.

The analysis of reference 42 represents the fuselage as an infinitely long cylinder with infinite impedance at the surface. Scattering effects from this analysis are shown in figure 21 for the fundamental and second harmonic of a model propeller. On the side with the incident sound, pressure doubling is shown at both frequencies because the wavelengths of both tones for the model propeller are small compared with the fuselage diameter. On the opposite side, a shadow zone is shown, with a deeper shadow occurring for shorter wavelengths. Interference is shown at the edges. Note that the pattern appears rotated (i.e., the pattern is not symmetric about $\phi = 0°$) because of the rotating source. Thus the analysis is sensitive to the direction of rotation of the propeller.

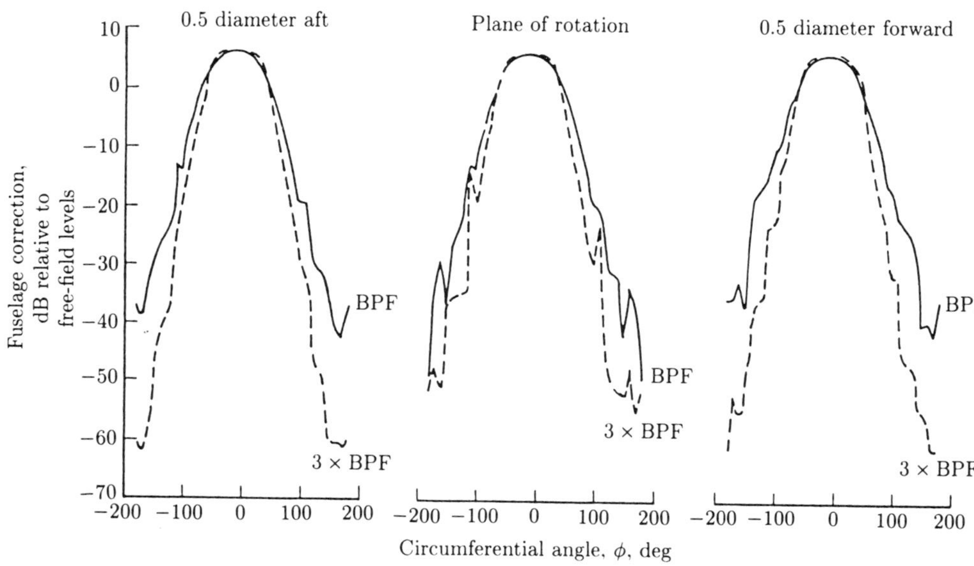

Figure 21. Calculated fuselage scattering effects at $M_x = 0.8$. (From ref. 42.)

Wing Shielding

As is the case for a fuselage, a wing can be used to provide shielding of a propfan source. Figure 22 illustrates the situation for a propfan installed on a swept wing. It is shown that for the geometry and the direction of rotation indicated, the line of sight from the advancing blade is blocked by the wing leading edge. As is the case for the fuselage, the amount of shielding depends also on the wavelength of the sound, with shorter wavelengths approaching geometric acoustic behavior.

Several analyses have been developed for shielding of sound by stationary barriers. These analyses were extended for propfans installed on wings by including the effects of flight Mach number and wing sweep in reference 43.

Atmospheric Absorption

When sound propagates over long distances through air, absorption takes place and reductions in amplitude in excess of those from simple distance effects are observed. These effects have been studied for some time and several procedures exist to calculate the effects (refs. 44 to 47). The method endorsed by the FAA and recommended for adjusting noise certification data is that of reference 47.

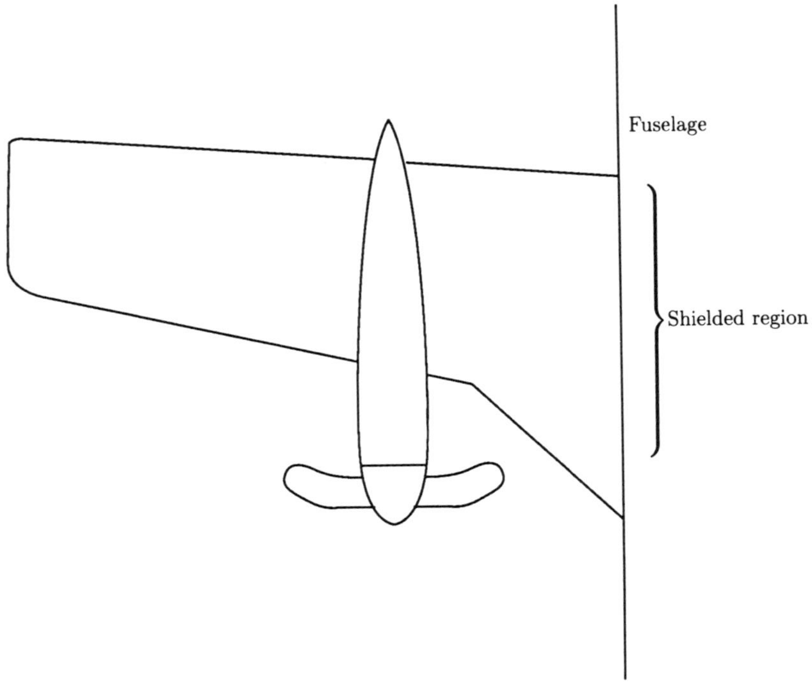

(a) Top view.

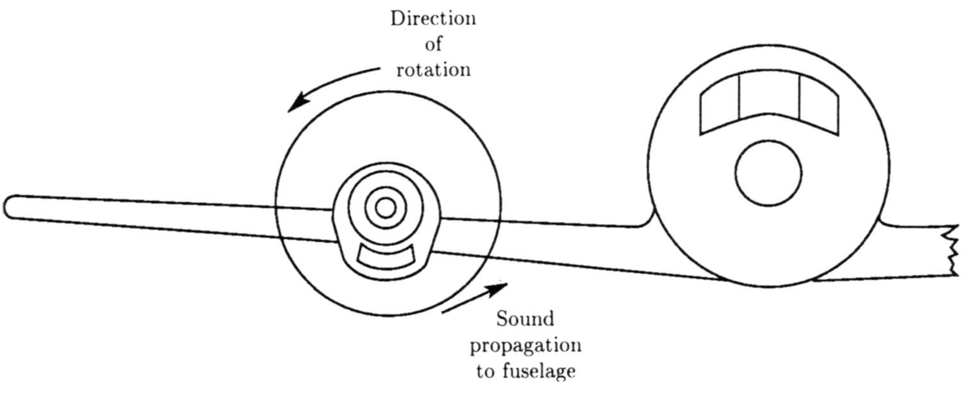

(b) Front view.

Figure 22. Shielding of propeller noise by swept wing.

This procedure presents an algorithm which allows the calculation of atmospheric attenuation, in terms of a dB attenuation per unit distance, as a function of temperature and relative humidity. These values typically vary from 0.001 dB/m at low frequencies to about 0.1 dB/m at high frequencies. Because the atmospheric attenuation has relatively little effect on low frequencies, it does not have a strong influence on propeller noise, except for the case of noise propagating to the ground from high altitudes (above 5000 m).

Ground Reflection Effects

In a typical airplane noise measurement situation, the airplane flies past a microphone which is located above a ground plane. The sound thus reaches the microphone following the direct path and a reflection from the ground plane, as illustrated in figure 23 for a simple point source S. Also shown is the equivalent image source S', which accounts for the ground reflection process. The distance traveled by the direct ray R is given by $[L^2 + (H - h)^2]^{1/2}$, whereas that of the reflected ray R' is given by $[L^2 + (H + h)^2]^{1/2}$. Note that R' is always longer than R, except in the special cases of $H = 0$ or $h = 0$, for which they are equal. Since the propagation lengths along the two paths differ, the signals arrive at the microphone with relative phase differences that cause constructive interference, when the two signals are in phase, or destructive interference, when the two signals are out of phase.

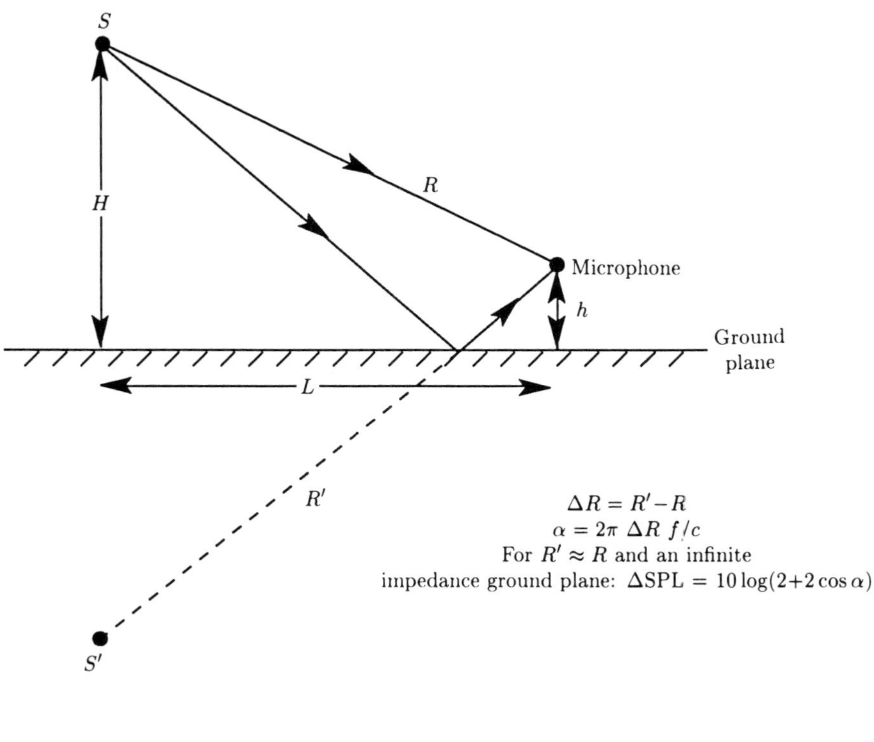

Figure 23. Ground reflection effects with image source.

For noise certification purposes, a microphone height of 1.2 m is required. At normal ambient temperatures, this height results in cancellations at approximately 80 Hz for a source directly over the microphone ($L = 0$). This value is in the range of blade-passage frequencies for many propellers. Since propeller noise is predominantly at low frequency, these effects can thus be significant.

It is difficult to interpret measurements made using a microphone located above a ground plane because of the changing ground reflection effects coupled with apparent source characteristics (e.g., directivity and Doppler shift). One method which gives good results is to use a microphone close to ground level ($h \approx 0$) over a hard surface (for high impedance). This arrangement results in the reflected signal always being in phase with the direct signal, so a nearly constant 6-dB correction (full pressure doubling) occurs over the frequency range of interest, independent of source position.

Ground reflection effects are needed for estimating the noise to be expected during noise certification, for instance. Procedures for calculating ground reflection effects can be found in references 48 to 53 and are discussed in another chapter of this book. These methods do not necessarily address tone sources. Ground reflection corrections for propeller harmonic noise should be done for small bandwidth signals at the Doppler-shifted tone frequencies. Using center frequencies of 1/3-octave bands can lead to significant errors. Experience has indicated that adjusting a ground microphone measurement of a propeller aircraft flyover to 1.2 m cannot be done with high accuracy with existing methods because of the complexity of the ground reflection process.

Excess Ground Attenuation

Excess ground attenuation, sometimes called lateral attenuation, is a term applied to discrepancies between observed levels and those expected after all other propagation effects (i.e., distance, atmospheric attenuation, and ground reflection effects) have been accounted for. This effect is usually found when measurements of flyovers are compared with those from a sideline microphone. A compendium of such measurements has been published in reference 54. A summary of these results is shown in figure 24. It is probable that a significant portion of the lateral attenuation shown in figure 24 is a result of shielding because of the apparently stronger effects for fuselage-mounted engines. It is not clear how this effect would be reflected in propeller noise. However, this can certainly be best addressed by applying wing and fuselage shielding and scattering analytical methods.

Nonlinear Propagation Effects

Open rotors generally produce intense noise levels. This is particularly true of propfans during high-speed cruise. Under these circumstances, significant nonlinearities can arise. There may be nonlinear propagation effects in addition to the nonlinear source effects discussed previously. Nonlinear propagation of propeller noise was first studied by Hawkings and Lowson (ref. 55). Since then, Barger (ref. 56), Tam and Salikuddin (ref. 57), and Lindblad (ref. 58) have pursued the subject further. All these investigators applied weak shock theory as developed for sonic booms to the propeller noise propagation problem. The analysis is applied in conjunction with a linear source theory calculation, although it is not inherently limited to this and could be matched with a nonlinear source theory or even with experimental results.

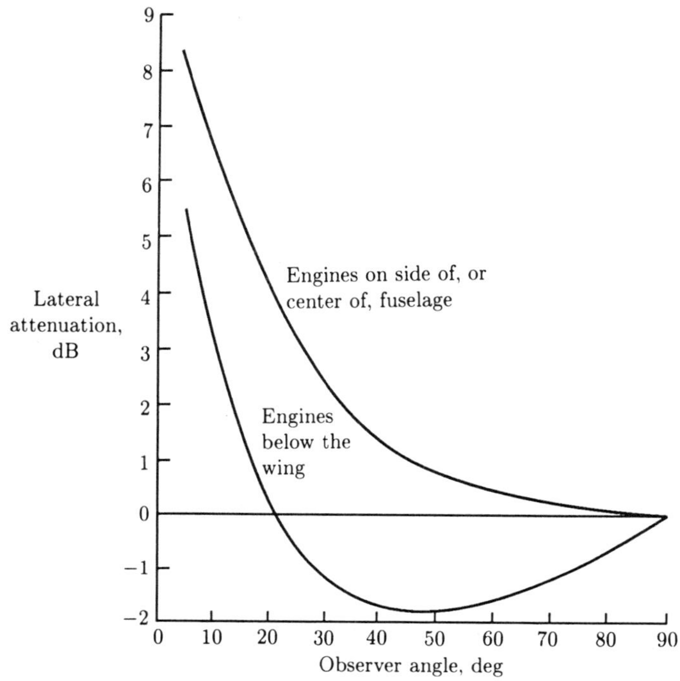

Figure 24. Typical variation of lateral attenuation with observer angle. (From ref. 54.)

The approach is to begin with an acoustic pressure waveform and apply a nonlinear propagation theory to account for the wave steepening that builds up during propagation because of finite signal amplitude. In the positive pressure peaks of the wave, the temperature and speed of sound are above ambient values so that the peaks propagate faster than the valleys.

Examples of the nonlinear propagation effects are shown in figure 25 for an unswept propfan blade. In figure 25(a) the noise pulse was computed from a linear theory equivalent to equations (3) or (28). Weak shock theory was applied to produce the wave in figure 25(b). As shown, the nonlinear propagation theory steepens the leading edges of the pulses and compares better with experimental results (fig. 25(c)) than does linear theory. A shift in energy is made from the lower frequencies to higher frequencies, although there does not appear to be a significant reduction in level at the very low frequencies. The effects described herein occur very close to a propfan. Test data show that propagation of sound from propfans follows the linear propagation laws of typical acoustic sources at distances greater than one propfan diameter.

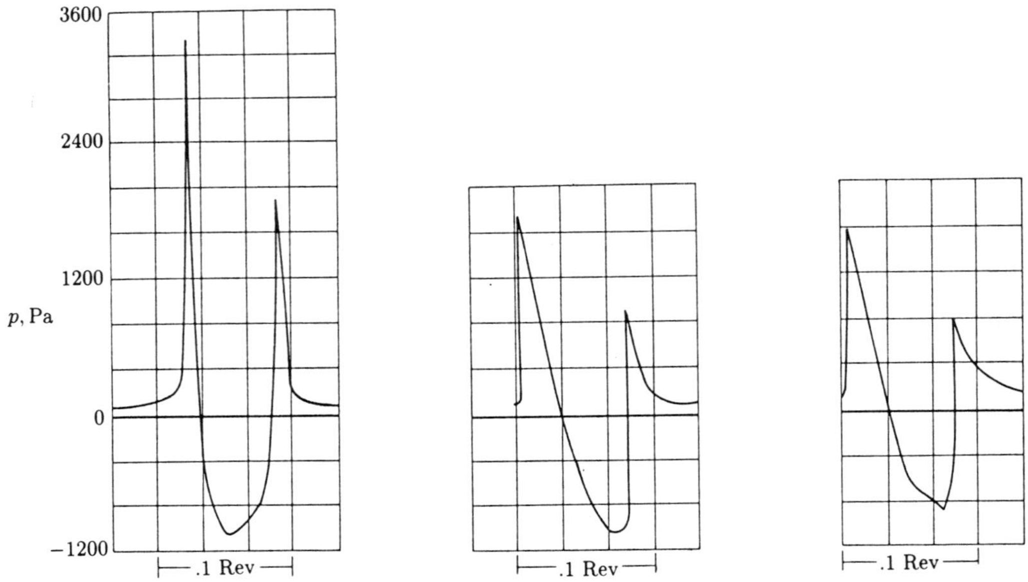

(a) Computed from linear source and linear propagation.

(b) Computed from linear source and nonlinear propagation.

(c) Test waveform; M_t = 1.18.

Figure 25. Nonlinear propagation effects computed from weak shock theory. (From ref. 56.)

Comparisons of Predictions and Measurements

Background

Before attempting any serious study of the performance of propeller noise prediction methods, one must fully understand the methods used to make the measurements. This is important because test facilities may influence the noise-generating process or propagation in a manner not modeled by the noise prediction method. Other considerations include the presence of additional sources of noise in the measurements (e.g., a drive motor) and the assurance that the propeller is operating at the conditions (i.e., blade loading, relative velocities, etc.) defined for the calculations.

Noise Measurements Under Static Conditions

As described previously, a propeller operating under static conditions encounters a great deal of nonuniform inflow, including naturally occurring turbulence in the atmosphere, ground vortices, and wakes from fuselages, wings, nacelles, or test stands. This inflow results in high levels of unsteady-loading noise that tends to dominate the higher sound harmonics. This source of noise disappears quickly when a small amount of forward speed is attained. A passenger in a propeller-driven airplane may observe high levels of noise up to the point of brake release, but the noise quickly changes as the airplane reaches a modest speed during the takeoff roll. Similar effects can be observed on static test stands in that the noise can change markedly when a modest head wind occurs.

Forward flight effects on propeller noise have been investigated (refs. 59 and 60). Figure 26 shows representative noise measured during static and flight conditions at constant propeller speed and power. As shown, the static data are dominated by high levels in the upper harmonics, but these are essentially gone in the flight data.

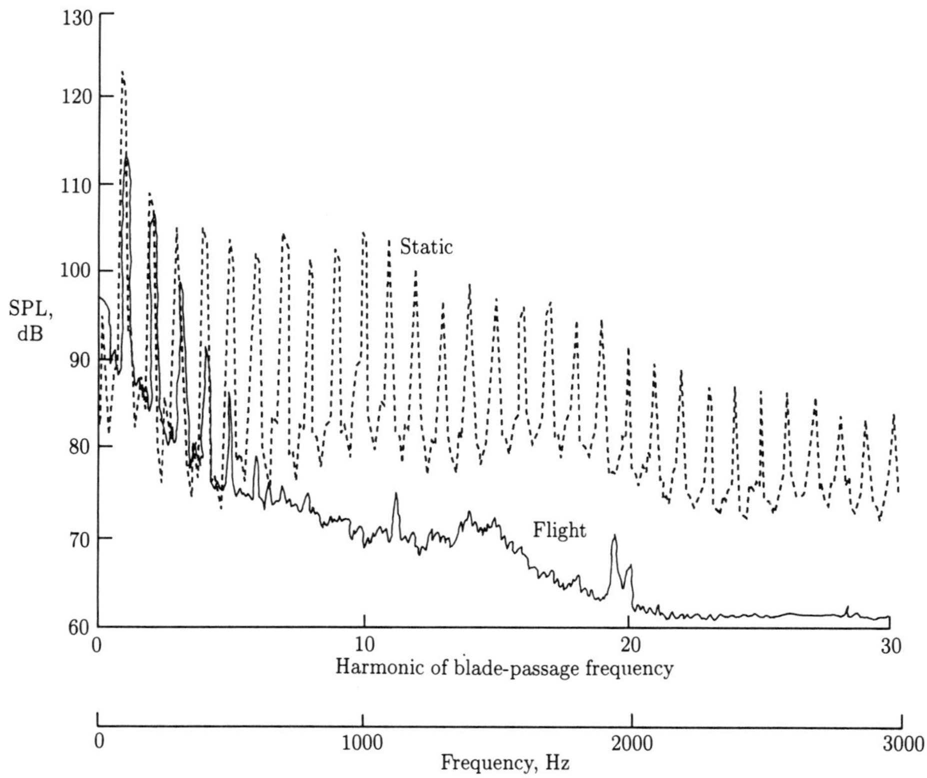

Figure 26. Effect of forward flight on propeller noise. (From ref. 59.)

Comparisons between measured and calculated levels for static propeller noise are presented in reference 60. It is concluded that a compact source calculation for fluctuating blade-loading noise and a noncompact source calculation for thickness and steady-loading noise are adequate for predicting the noise of static propellers. The prediction of static propeller noise, however, is not of great interest, as the condition is transient and is not used for noise certification or interior cabin noise control.

The noise from counterrotating propellers operating under static conditions would probably not show as much influence from turbulence ingestion effects because the higher harmonics are dominated by aerodynamic interaction between the two blade rows. Limited data exist (ref. 61), but it is not clear what components comprise the noise. It can be concluded that any static propeller data to be used for correlation purposes should be viewed with caution, as they are strongly influenced by nonuniform inflow, which is not usually well defined and often varies.

Propeller Noise Measurements in Wind Tunnels

There are two common types of wind tunnels used in acoustic research: the closed-test-section type and the open-jet type. Both offer advantages and disadvantages in regards to propeller testing.

Closed-Test-Section Wind Tunnels

Most closed-test-section wind tunnels are of the recirculating type. It is thus necessary to control the turbulence which can be introduced by the drive fan, turning vanes, recirculating wakes produced by the model, etc. If the test-section walls are metal or concrete, many reflection paths can exist, the result of which is great variability in the measured noise. This variability is particularly significant for propeller noise because of its discrete frequency components. Depending on mode characteristics of the tunnel, significant reinforcements or cancellation can occur at the harmonic frequencies (ref. 62).

One solution to this reflection problem is the use of absorptive treatment on the tunnel walls. Because of aerodynamic losses caused by flow over the treatment, the use of wedges is not practical. Flat-faced fiberglass (behind high-open-area perforated retaining plates) or polyurethane foams have been used. These appear to work reasonably well at low to moderate speeds (below Mach 0.5) for measurement locations near the peak noise (ref. 62). This type of treatment might not work well at high speed or at shallow incidence angles which occur at locations forward of the propeller plane of rotation. Acoustic qualities of wind tunnels operating at speeds above Mach 0.5 have not yet been demonstrated.

Open-Jet Wind Tunnels

Another approach is the use of open-jet wind tunnels. In this arrangement, a nozzle is typically set into the wall of an anechoic chamber. A collector is situated opposite the nozzle. When suction is applied at the collector a jet forms between the nozzle and the collector. Placing a propeller in the jet simulates flight. Because there is essentially no flow outside the jet, the chamber can be treated with acoustic wedges to provide an anechoic environment. The limitations of this scheme, however, are restricted speed (about Mach 0.5) and propagation effects through the shear layer to far-field microphones. Shear layer corrections for amplitude and directivity angles exist and have been well documented (refs. 63 to 65). The angle corrections show that the sound is refracted by the shear layer. In the forward direction, the refraction can be complete so that no sound passes through the shear layer, thus limiting the range of directivity attainable.

Propeller Noise Measurements in Flight

Measurements of noise from propellers installed on airplanes in flight should be the best data for correlation purposes because the ultimate objective of studying propeller noise is to control it in the passenger cabin and in the communities affected by noise of airplanes taking off, landing, or in flight. The measurement of propeller noise from airplanes, however, is difficult. Apart from the generally higher costs associated with flight tests, propellers installed on airplanes are subjected to inflow distortion such as angle-of-attack effects, wing upwash, nacelle blockage, engine inlet effects, and wakes from upstream disturbances which may significantly affect the propeller noise characteristics. In addition, other sources of noise are present, such as noise of the airframe and of the engines. Finally, the interpretation of airplane noise should include atmospheric attenuation, ground reflection, shielding, and Doppler shifts for propagation to the ground and fuselage reflections, refraction, and propagation through a fuselage boundary layer in the near field. (See previous discussions in this chapter.)

Comparisons in the Near Field

Predictions made with the Hanson frequency-domain method (ref. 18) and measurements made on two model propfans (ref. 66) operating over a range of tip speeds and blade loadings are shown in figure 27. The SR-2 model blades are straight whereas the SR-3 blades are swept according to recent practice. It is apparent that the level of the prediction method agrees well with test data, with the trends of noise versus tip speed being well predicted. Also, the benefits of blade sweep are shown in both the measurements and the predictions.

Figure 28 shows the measured and predicted directivity of the blade-passage frequency harmonic. This comparison indicates that all sources of noise, including the nonlinear quadrupole, are important, particularly at the forward location, when the relative Mach numbers are high. It is shown that the total noise energy is generally not equal to the sum of the component energy because of relative phase effects.

Comparable results are obtained from calculations with one of Farassat's time-domain methods. Sample correlations (ref. 10) are shown in figures 29 and 30. As shown, the general characteristics of the waveforms are predicted well by the time-domain method except for the positive peaks, which are reduced by nonlinear propagation. This is also shown in the spectrum in figure 29(c). Figure 30(c) shows better spectrum agreement than does figure 29(c), although the waveform correlation is not as good. The Hanson frequency-domain method (ref. 18) at BPF shows good agreement with measured data in both figures. In fact, it is expected that time-domain and frequency-domain methods would show nearly identical results within the realm of linear acoustics.

Comparisons in the Far Field

Measurements of the noise from a full-scale general aviation propeller were made in a large open-jet anechoic tunnel (ref. 67). These data and predicted values from a time-domain method are shown in figure 31. It is apparent that the

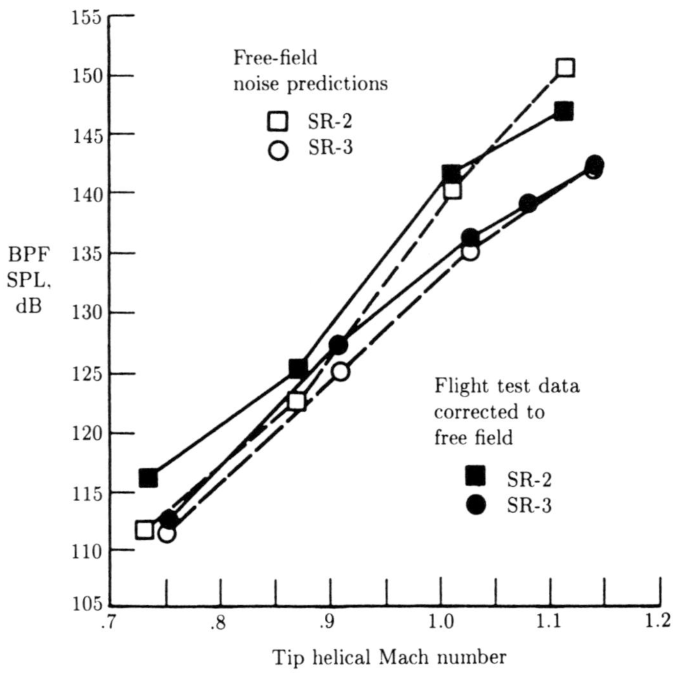

Figure 27. Measured and predicted sideline tone levels versus tip helical Mach number. Boom microphones; power coefficient, 1.9; advance ratio, 3.1. (From ref. 66.).

agreement between measurements and predictions for these cases where the propeller is operating in low turbulence, undisturbed flow is excellent.

Figure 32 shows predictions and measurements for the propeller operating on an airplane. Measurements were made with a ground-level microphone and the noise predictions included propeller angle-of-attack effects. As shown, the noise is underpredicted ahead of the overhead point and overpredicted behind the overhead point. It is conjectured that this could be caused by nonuniform inflow effects other than propeller angle of attack.

Reference 68 presents results for a model propeller operated in an open-jet facility with microphones located inside the jet to avoid shear-layer refraction effects. In addition to noise measurements, the propeller aerodynamics were measured to confirm the blade-loading distributions, which are inputs needed to calculate propeller-loading noise. Representative time-domain measurements and predictions are shown in figure 33. These plots show very good agreement between measured and predicted values. Although only waveforms are shown, it would be expected that there would be excellent agreement of harmonic data as well.

Measured and predicted values for a propfan operating at takeoff and landing conditions in a large, acoustically treated, closed-test-section wind tunnel are shown in figure 34 (ref. 69). For these predictions, the aerodynamic performance was

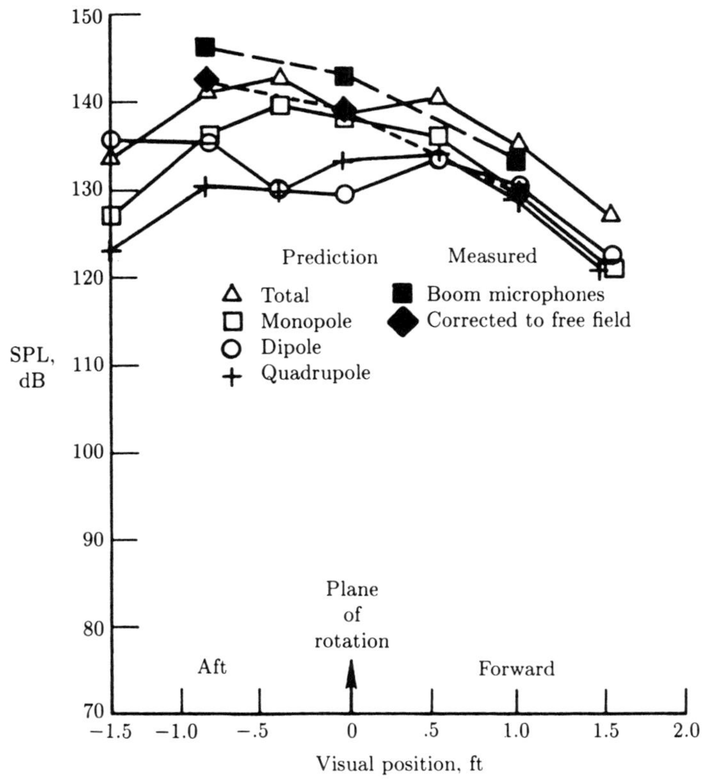

Figure 28. Measured and predicted BPF tone sideline directivity for SR-3 operating at design point condition. (From ref. 66.)

predicted with an Euler code and the noise was predicted with the time-domain method of Farassat. Figure 34 shows the predicted and measured noise for two harmonics as functions of directivity for three blade angles β. Figure 35 shows the waveforms for the three conditions in the propeller plane of rotation (at or near the peak directivity angle). There is a tendency to underpredict as the propfan blade angle is increased. This underprediction is attributed to the formation and increasing strength of a leading-edge vortex resulting from the thin, sharp-edged, swept blades at increasing blade angle. The leading-edge vortex, and also an associated tip-edge vortex, can change the blade-loading distributions significantly. If this change is not reflected in the blade-loading source distribution, then noise predictions can become inaccurate.

Noise measurements for a propeller operating on an airplane in flight were made using microphones mounted on a wingtip (ref. 59). Predictions were made using a frequency-domain method (ref. 70). Figure 36 shows these values for a propeller tip rotational Mach number of 0.77, while figure 37 shows these values for a tip

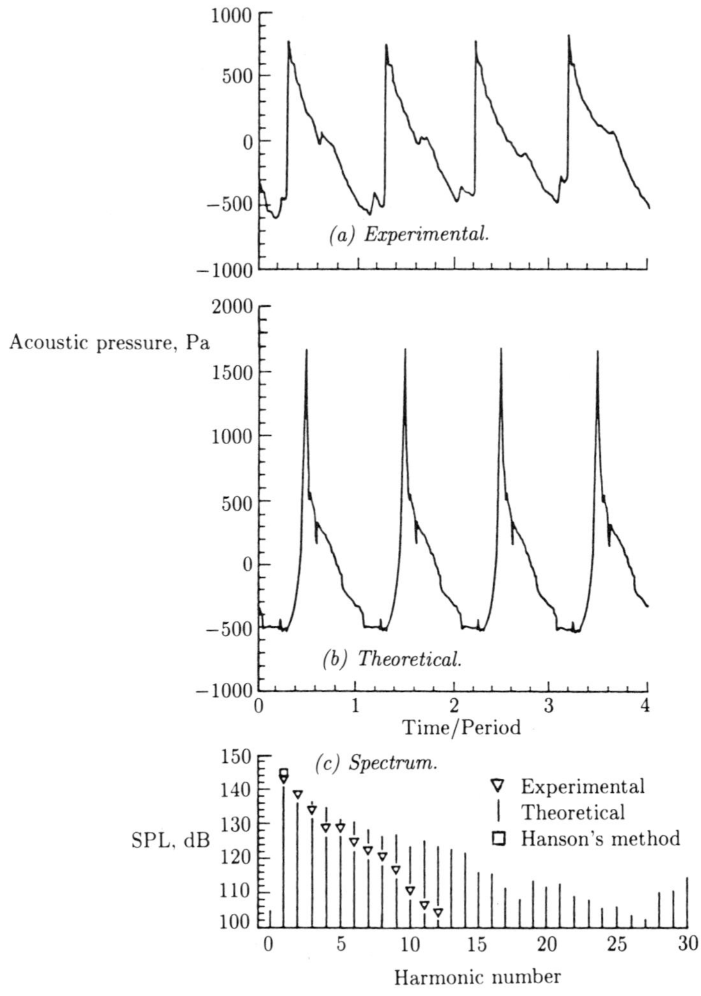

Figure 29. Measured and predicted acoustic pressure signatures and spectrum for boom microphone directly above propeller disk. Theoretical prediction corrected for boom reflection. Period, 0.991 msec; BPF = 1009.2 *Hz. (From ref. 10.)*

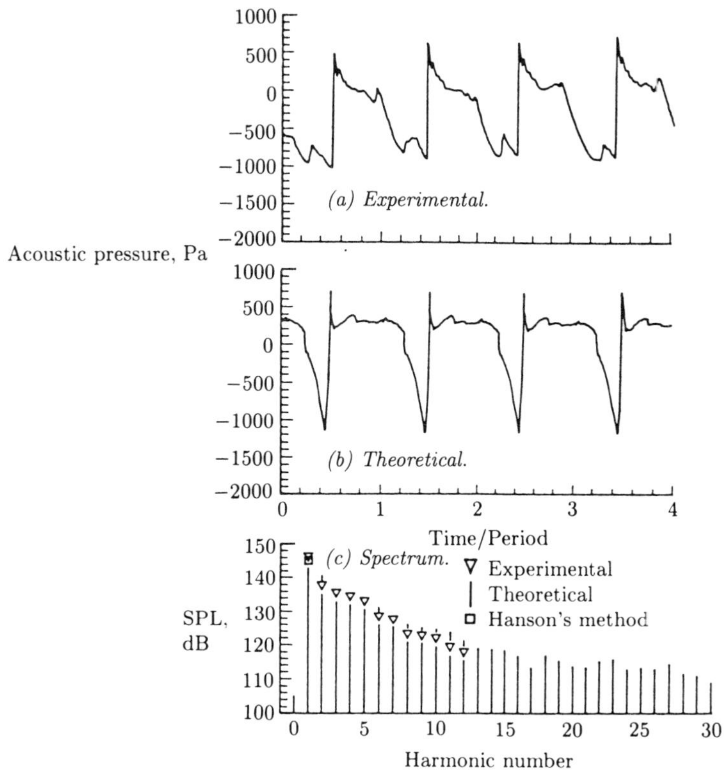

Figure 30. Measured and predicted acoustic pressure signatures and spectrum for boom microphone behind propeller disk. Boom reflection correction included. Period, 0.991 msec; BPF = 1009.2 *Hz. (From ref. 10.)*

rotational Mach number of 0.83. The agreement is quite good in the propeller plane of rotation and fair at the aft location, although the spectrum shape is well predicted. No nonuniform flow fields were included in the noise predictions. These comparisons show the importance of the thickness noise component for this propeller at these operating conditions.

General Comments

The following general observations are based on the foregoing comparisons between measurements and predictions of propeller and propfan noise.

Generally good agreement between measurements and predictions of noise can be obtained for propellers operating at low-to-moderate tip speeds at moderate loadings under ideal (undistorted) inflow conditions. For these conditions, the prediction model needs to include only linear sources, and comparable performance can be obtained with either time-domain or frequency-domain methods. It is apparent that for good loading noise calculations, the blade-loading distribution must be accurately defined.

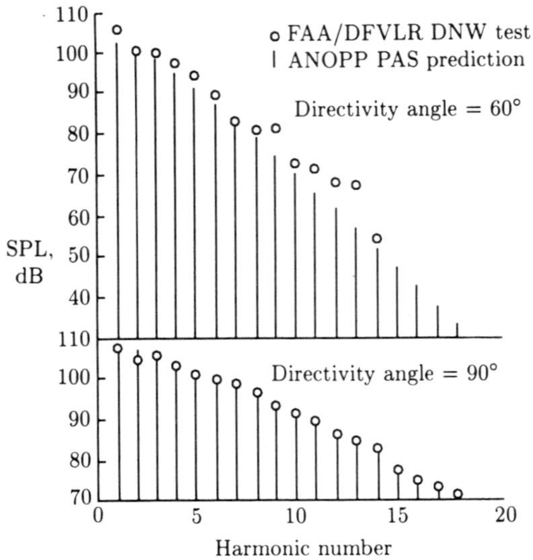

(a) First case; power coefficient, 0.028; $M_t = 0.781$.

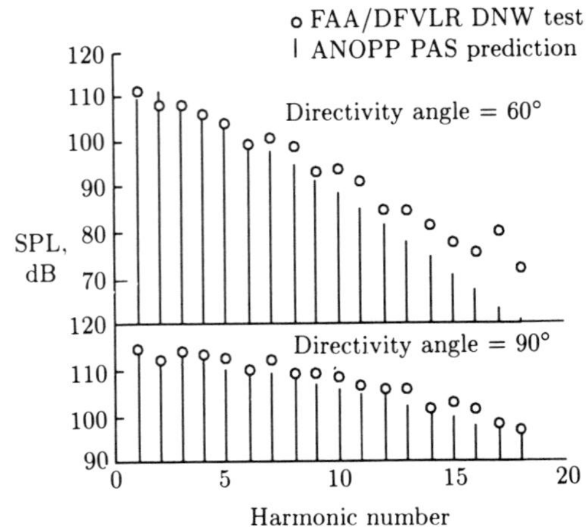

(b) Second case; power coefficient, 0.049; $M_t = 0.872$.

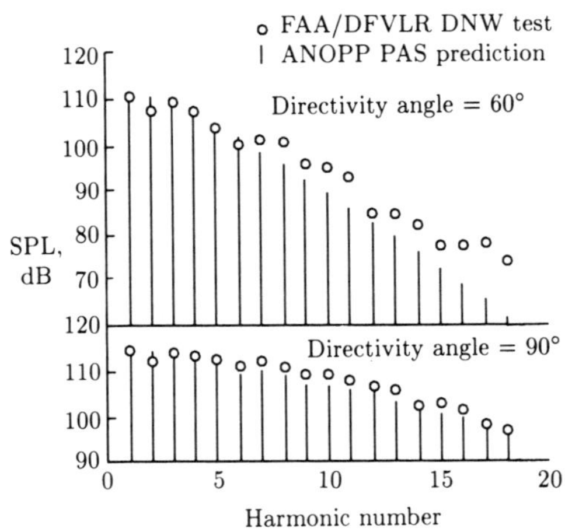

(c) Third case; power coefficient, 0.040; $M_t = 0.876$.

Figure 31. Predictions and data for full-scale general aviation propeller in large open-jet wind tunnel. (From ref. 67.)

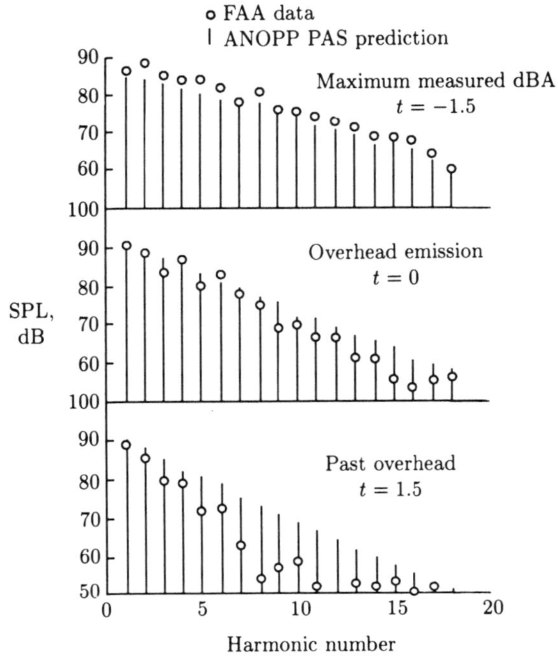

Figure 32. Measured and predicted noise spectra at three time periods for standard takeoff of Piper Lance airplane. (From ref. 67.)

The prediction of propfan noise is generally not as successful. At low speeds this is not necessarily due to failure of the same noise prediction models as above (although the existence of other sources such as a tip-edge loading is a possibility), but more likely it is due to the failure of the aerodynamic model to predict the blade-loading distribution.

During high-speed operation, additional (nonlinear) sources and/or nonlinear propagation near the propeller become apparent.

Finally, propellers and propfans installed on airplanes have other sources of noise, notably unsteady-loading noise, due to inflow distortion. Again, these effects have been included in many propeller noise prediction models, but they require a means of defining the unsteady blade loads. Accurately estimating unsteady blade loads is not easily managed by current aerodynamic methods and generally the resulting noise predictions are not as good as those for steady blade loads.

Propeller Noise Control Objectives

It is the job of the acoustician to first understand the propeller noise-generating mechanisms and then to control them using methodologies derived from theories to meet constraints demanded by airplane manufacturers. These noise constraints are based on meeting noise regulations, cabin noise comfort, airplane structural requirements, etc. In the following discussion the general noise control objectives

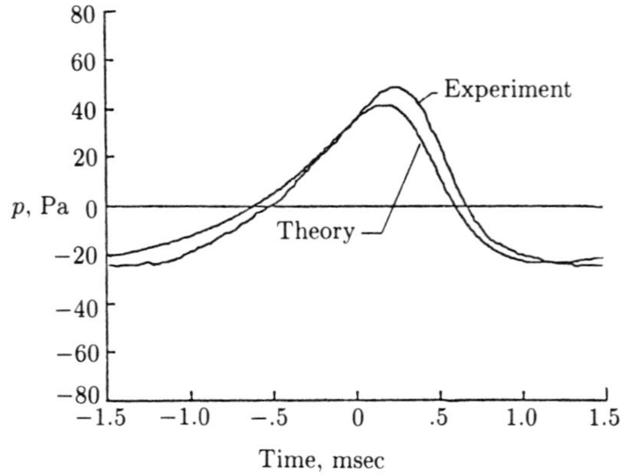

(a) −40° azimuth.

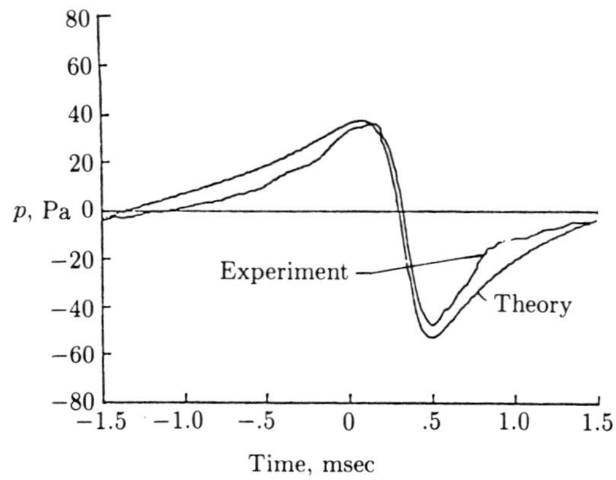

(b) 0° azimuth.

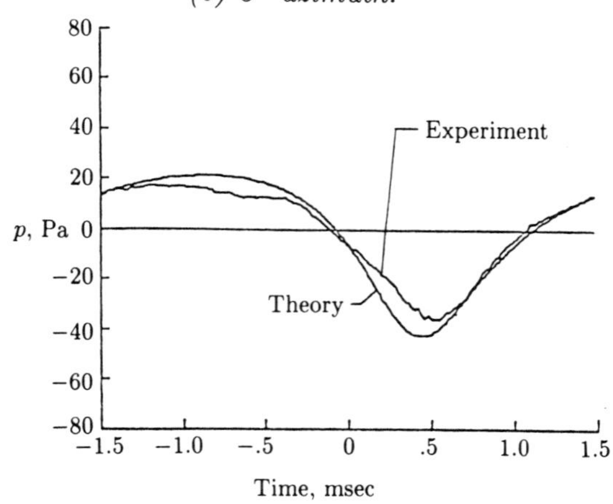

(c) 40° azimuth.

Figure 33. Measured and predicted pressure signatures at three azimuths for propeller on symmetric body. Propeller speed = 10 000 rpm. (From ref. 68.)

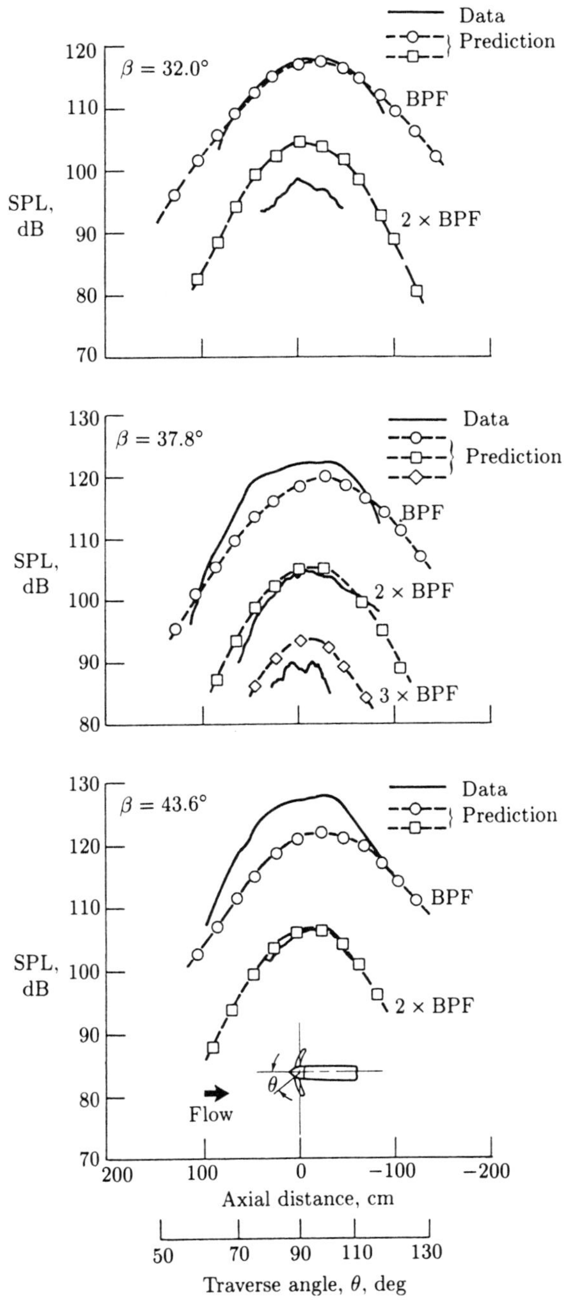

Figure 34. Measured and predicted values for SR-7A model. 1.68 m sideline; advance ratio, 0.886; $M_x = 0.2$. *(From ref. 69.)*

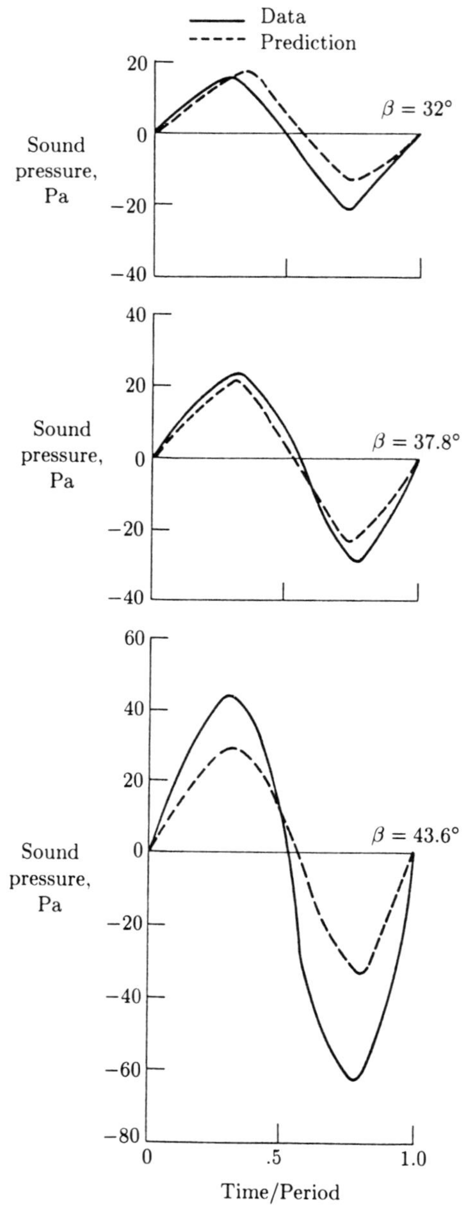

Figure 35. Waveforms in plane of propeller at three blade angles. Advance ratio, 0.886; $M_x = 0.2$. *(From ref. 69.)*

are grouped into two sections. In the first, the near-field noise requirements, which generally pertain to the cruise condition, are described. Then the far-field noise requirements, which generally apply to low-speed operation, are discussed.

Near-Field Noise Control Objectives

In a typical installation, the propeller is located adjacent to a fuselage section. Depending on propeller design, operating conditions, and proximity of the propeller to the fuselage, the acoustic excitation can produce intense fluctuating pressure loads on the surface. These loads can cause motion in the fuselage structure, with possible fatigue of the structure and noise transmission to the interior. This motion can be particularly important if the structure has response frequencies which coincide with the propeller noise frequencies. It is therefore desirable to control the propeller noise levels and to avoid excitation at structural resonances to reduce or eliminate acoustic fatigue.

For passenger comfort, the propeller noise reaching the fuselage interior space needs to be controlled. Today's airline passengers expect cabin comfort in propeller-driven airplanes to be comparable to that in turbofan-powered airplanes. This means limiting cabin noise levels to 80 dBA or less. Further, for enhanced comfort the actual propeller noise harmonics should be barely discernible. This usually implies that the propeller harmonic noise contribution is below 80 dBA and the broadband noise from fuselage boundary layer, environmental control system, etc., has comparable levels.

Far-Field Noise Control Objectives

In general, far-field noise control addresses community noise objectives. The most important of these is noise certification as set forth by the FAA (ref. 40). Additional noise requirements may be imposed by certain airports for takeoff and landing.

For an aircraft to receive certification it must satisfy noise constraints during takeoff and landing. For turbojet and transport category airplanes, these are currently defined by Stage-3 requirements described in Appendix C of Federal Aviation Regulations (FAR) Part 36 (ref. 40). Comparable requirements are imposed by the International Civil Aviation Organization (ICAO) in Annex 16 (ref. 71). In both cases the noise limits are specified at three locations, as defined in figure 38 (from ref. 72). These documents describe the procedure to be followed in certifying and specify the limits to be met, which depend on the location and the airplane gross weight.

For airplanes below 12 500 lb gross weight, the certification procedure and limits are different. These are described in Appendix F of FAR Part 36 (ref. 40). In general, this certification requires level flyover over a microphone. Adjustments are allowed for good takeoff climb airplane performance.

Although many airports have noise restrictions of various types, two airports in the United States are particularly strict. The first of these is Washington National Airport, which requires low noise for nighttime operation based on results of the FAR Part 36 certification testing. The limits, however, are based on the maximum A-weighted sound pressure level. The maximum level for takeoff is 72 dBA, while the maximum level on approach is 85 dBA. No requirements are made on sideline noise. Airplanes not meeting these requirements may not take off or land at the airport between 10 p.m. and 7 a.m.

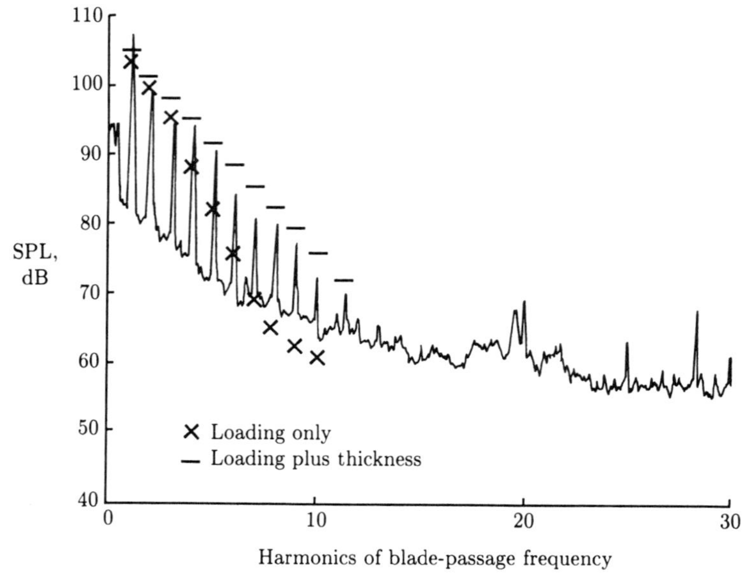

(a) In-plane location.

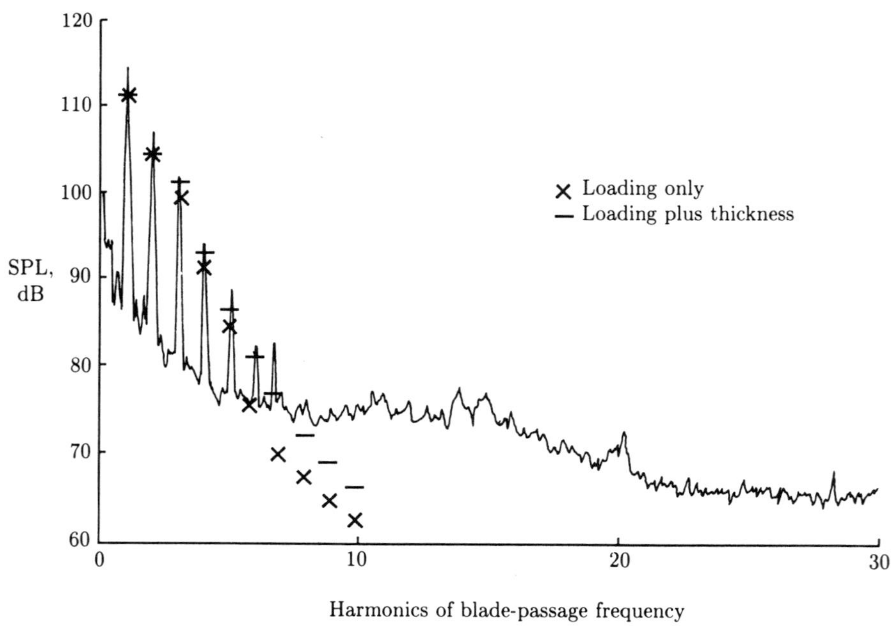

(b) Aft location.

Figure 36. Measured and predicted propeller noise levels at tip rotational Mach number of 0.77. (From ref. 70.)

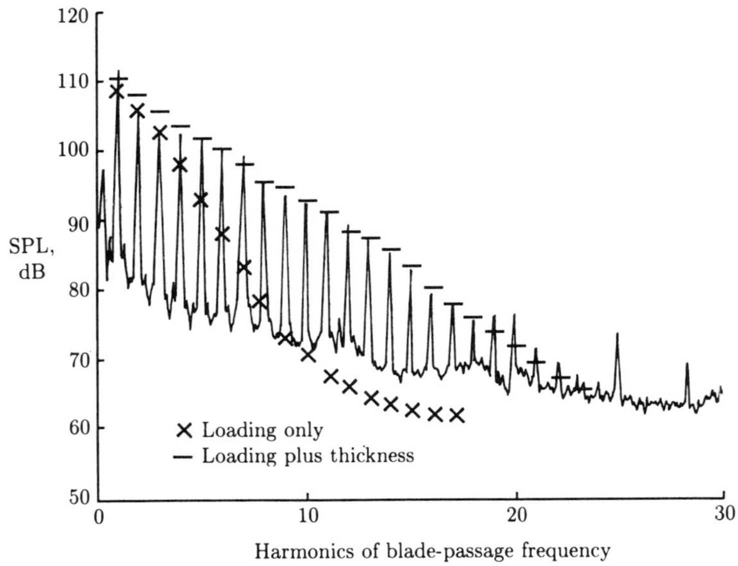

(a) In-plane location.

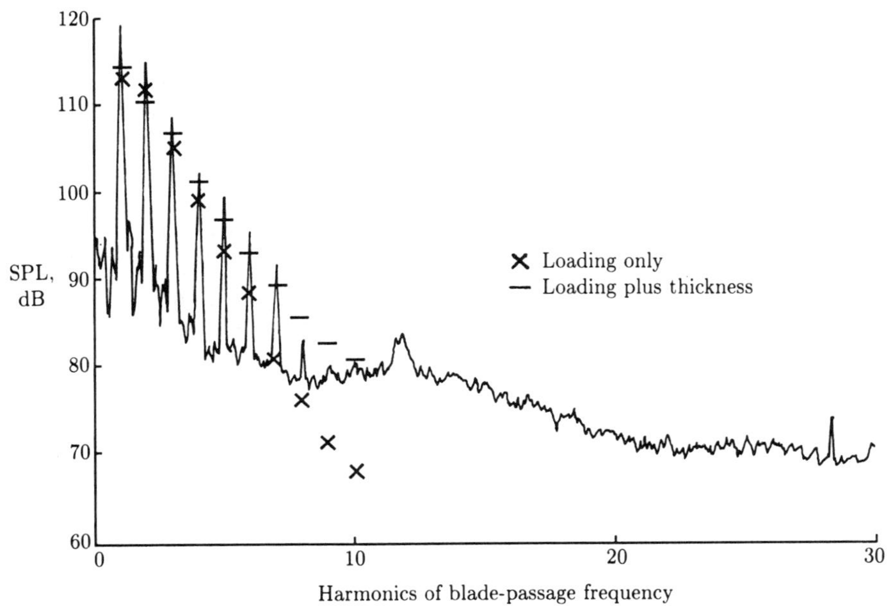

(b) Aft location.

Figure 37. Measured and predicted propeller noise levels at tip rotational Mach number of 0.83. (From ref. 70.)

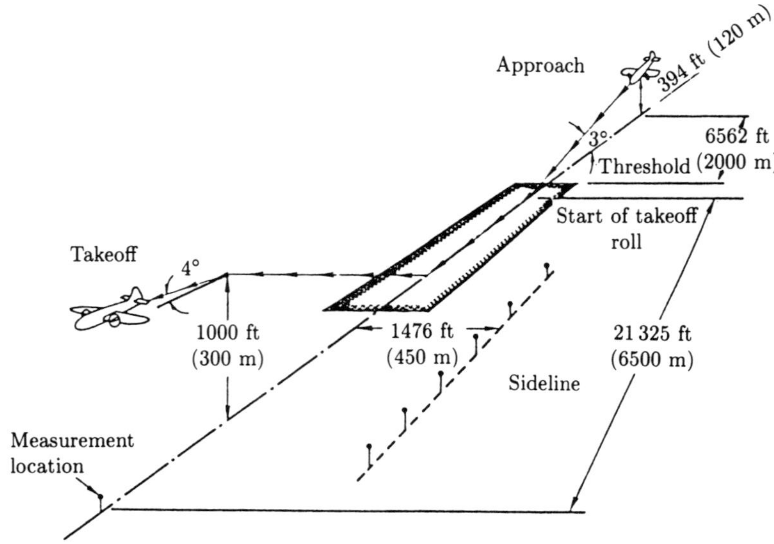

Figure 38. Noise certification measurement locations. (From ref. 72.)

The second airport having special noise rules is John Wayne Airport in Orange County, which is outside Los Angeles, California. For John Wayne Airport, the measurement units and locations are significantly different from the FAR Part 36 rules. The measurement locations can be seen in figure 39 (from ref. 72). Microphone locations are scattered within a several-mile radius of the runway.

The measurement unit at John Wayne Airport is single-event noise exposure level (SENEL). The airport requirements for varying numbers of allowed flights per day are given below.

Classification	SENEL, dBA
Unrestricted, unlimited flights	< 86
AA	86 to 89.5
A	89.5 to 100

Airplanes meeting the class A level are allowed fewer flights per day from the airport than those meeting the class AA level.

Control of Propeller Noise

It is possible to obtain guidance in controlling propeller noise by inspection of the noise prediction theories. For example, in equation (8) relative Mach number is a multiplier of the noise level. Thus, reducing blade section relative Mach number should reduce noise. In fact, for most cases that is indeed a way to reduce propeller noise—lower tip speeds almost always reduce noise. Other approaches include

Noise Monitor Station Locations

Station	Location	Distance, ft		Microphone height, ft	
		From north end (on extended C/L) runway 01L-19R	At 90° to extended C/L 01L-19R	AGL	ARL
M-1	Newport Beach, CA	10 800	200 Easterly	20	13
M-2	Newport Beach, CA	17 470	1220 Westerly	20	57
M-3	Newport Beach, CA	14 800	3570 Easterly	20	67
M-4	Santa Ana, CA	5 850	300 Westerly	43	26
M-5	Tustin, CA	28 650	175 Easterly	22	109
M-6	Santa Ana, CA	8 660	950 Westerly	25	15.5
M-7	Santa Ana, CA	8 870	900 Easterly	25	24
M-8	Newport Beach, CA	24 200	3030 Easterly	25	7
M-9	Santa Ana, CA	17 700	5350 Westerly	59	106

Remarks:

1. Length of runway 01L-19R: 5 700 ft.
2. Runway elevation: 53 ft mean sea level.
3. AGL = above ground level.
4. ARL = above runway level.
5. Runway magnetic heading: 194° 21′.

Station Location Map

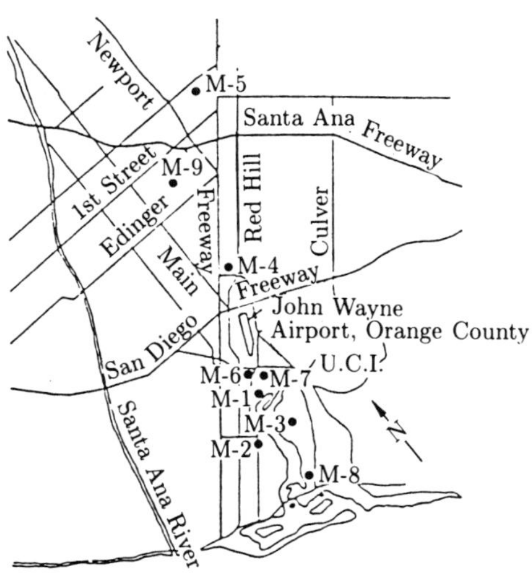

Figure 39. Noise measurement locations for John Wayne Airport. (From ref. 72.)

altering operating conditions, changing propeller design, and, for the special case of cabin noise with multiple propellers on the airplane, using a synchrophaser.

Operating Conditions

As previously mentioned, certain operating conditions can be modified to reduce noise. The most significant of these is tip speed. As can be ascertained from the governing noise equations, all sources of noise have radiation efficiencies determined by relative velocity. It is not straightforward to determine how much the noise will change with tip speed. In the case of loading noise, as an example, there are three significant parts of the noise-generating process which are affected. The first, obvious by inspection, is that relative Mach number multiplies the equation. Mach numbers also appear in the argument of the Bessel function. In this case, the effect of Mach number is not as clear, but reference to figure 14 or to Bessel function tables indicates that for subsonic tip speed and flight speed, reducing the Mach numbers reduces the noise. Finally, in order to maintain thrust (that is generally a firm requirement, as the application for propellers is generally to fly an airplane at some speed and altitude), the lift and drag coefficients change if tip speed is reduced.

Other ways to reduce propeller noise are to increase diameter or to reduce the disk loading (i.e., thrust per unit area of the propeller disk). In order to maintain thrust, operation at a lower disk loading requires a larger diameter.

Reducing tip speed has generally reduced noise for all sources. The best way to evaluate this effect is to perform calculations while observing the established ground rules (e.g., maintaining constant thrust). This is needed because the benefit depends on specific designs and baseline operating conditions. As a reference, it has been observed that for conventional propellers operating at low to moderate flight speeds, the overall noise in decibels varies as approximately 40 times the tip Mach number (ref. 73).

Reducing disk loading affects primarily loading noise. Again, evaluating the benefits requires specific calculations. As a guide, an empirical propeller noise-estimating method (ref. 73) indicates that noise varies inversely as diameter squared.

Design Parameters

It is apparent from the foregoing discussion that examination of the propeller noise equations shows areas where noise reduction benefits can be attained. For example, propeller noise could be greatly reduced by having zero-thickness blades to eliminate thickness noise, large diameters with many blades to eliminate loading noise, and large blade sweeps to eliminate quadrupole noise. Unfortunately, practical realities must also be considered. These include physical constraints, such as a certain amount of blade thickness needed for structural integrity, and practical constraints, such as a limit to the diameter for weight and installation considerations. However, general guidance can be obtained from the equations governing propeller noise. Although specific benefits must be evaluated individually and in combination for specific cases, the following is given for general guidance.

Blade Sweep

Increasing blade sweep is beneficial during high-speed cruise, when blade section relative Mach numbers are relatively high. Figure 40 shows calculated noise reduction

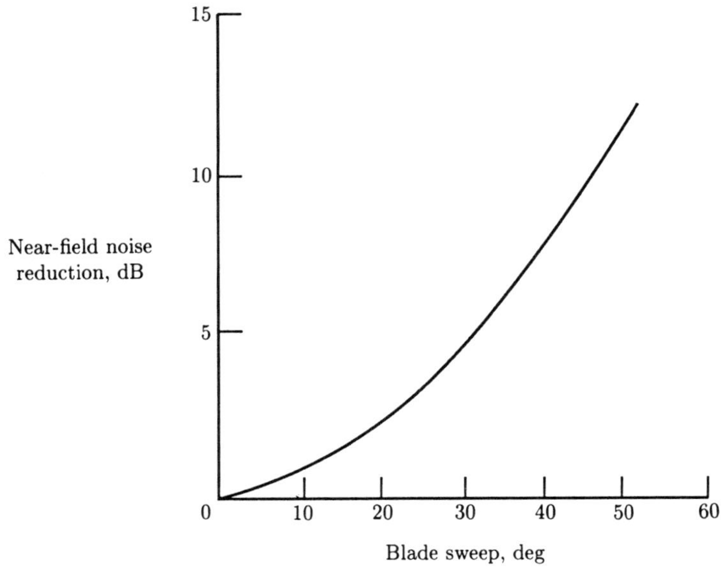

Figure 40. Noise reduction due to blade sweep calculated using frequency-domain method. $M_x = 0.8$; tip speed, 800 ft/sec; 8-bladed propfan; BPF harmonic.

for blade sweep. It is apparent that a small amount of sweep has a small effect on noise, but sweep becomes very effective at reducing noise as sweep increases. These effects were calculated for a propfan during cruise. Benefits during takeoff would be significantly less (ref. 17).

Blade Thickness

Thickness noise is significant during high-speed cruise, when blade section relative velocities are high. One means for reducing the contributions from thickness noise is to reduce the blade thickness. Actually, reductions are obtained by reducing blade thickness and chord, as it is the blade volume which factors into the source strength. The effect on spectrum depends on the shape of the airfoil. A scaled reduction in airfoil thickness at constant chord provides reduction equally at all harmonics. The noise reduction attainable varies as approximately the blade volume squared.

Reducing blade thickness also reduces quadrupole noise, but in a less predictable manner.

Blade Count

For a given thrust requirement, increasing blade count is always beneficial in reducing loading and quadrupole noise. Thus, at low-speed takeoff conditions, where loading noise dominates, reduction is obtained by increasing blade count. Although significant reduction in noise level (particularly at the higher harmonics) occurs, some of this reduction may be offset by increases in metrics which are frequency dependent, such as effective perceived noise level. This increase occurs for a given tip speed and diameter because increasing blade count raises the frequencies. In general, though, a net noise reduction can still be obtained.

Increasing blade count can raise thickness noise, depending on how it is done. Examination of equation (8) shows that simply adding blades increases the number of sources of thickness noise, with a corresponding increase in noise. If the blade volume is decreased (by reducing chord), then increasing blade count may not have as much effect on thickness noise. Again, adding blades raises the frequencies generated, so that metrics such as A-weighted overall levels commonly used in setting cabin noise limits may increase with increased blade count. This increase can be especially important during high-speed cruise, when thickness noise is an important source.

Propeller Diameter

Increasing propeller diameter reduces the blade loading. Thus, for a given thrust requirement the loading per unit area is reduced, with a corresponding reduction in loading noise. Increasing diameter is thus beneficial in reducing noise during takeoff. In addition, at low speed, propellers tend to be more efficient with increased diameter. Therefore, for a given thrust requirement less power is required, with less energy put into the system. Increased diameter can be combined with reduced tip speed for even more noise reduction.

Blade Shape

The effect on noise of blade design parameters such as twist and planform distributions is more difficult to determine by inspection because they change aerodynamic loading distribution. Although this can be done by parametric variation using a noise calculation procedure, most studies show that the noise reduction potential is small. The actual reduction to be realized depends on the starting point, but for reasonable designs the potential seems to be about 3 dB. This reduction can be realized with varying amounts of aerodynamic performance loss. The effect of blade design has a stronger impact on aerodynamic performance than on noise.

Airfoil Section

Some airfoil sections appear better for noise reduction than others. In general, however, the airfoil shape has only a small effect in the lower harmonics. Only for propfans at high speed do the airfoil shape effects appear at the lower harmonics.

The reader is cautioned that the foregoing discussion should be applied only in the context of a complete system study. Generally, the best approach requires a complete aerodynamic and acoustic methodology so that the trade-offs between noise and performance can be evaluated. Other factors such as weight, cost, and reliability must also to be considered.

Synchrophasing

Synchrophasing is not a means for reducing noise at the source, but rather it relies on phasing two or more sources to promote noise cancellation. This is done by phasing the rotation position of the blades on opposite sides of the fuselage so that the sound impinging on the fuselage has a certain phase relationship which promotes noise cancellation within the cabin. The process by which the noise cancels is too complex to define analytically to the degree sufficient to realize a reduction. All implementation of noise reduction by synchrophasing has been done experimentally and applied to cabin noise (refs. 74 and 75). Reductions of up to 15 dB may be obtained under specific conditions in limited areas of an airplane cabin, but general reductions of maximum noise throughout the cabin are less.

References

1. Mikkelson, Daniel C.; Mitchell, Glenn A.; and Bober, Lawrence J.: Summary of Recent NASA Propeller Research. *Aerodynamics and Acoustics of Propellers,* AGARD-CP-366, Feb. 1985, pp. 12-1–12-24. (Available from DTIC as AD A156 047; also available as NASA TM-83733.)
2. Magliozzi, B.: Noise Characteristics of a Model Counterrotating Prop-Fan. AIAA-87-2656, Oct. 1987.
3. Dobrzynski, Werner M.; Heller, Hanno H.; Powers, John O.; and Densmore, James E.: *DFVLR/FAA Propeller Noise Tests in the German-Dutch Wind Tunnel DNW.* DFVLR-IB-129-86/3, FAA-AEE 86-3, 1986.
4. Ffowcs Williams, J. E.; and Hawkings, D. L.: Sound Generation by Turbulence and Surfaces in Arbitrary Motion. *Philos. Trans. Royal Soc. London,* ser. A, vol. 264, no. 1151, May 8, 1969, pp. 321–342.
5. Farassat, F.: Linear Acoustic Formulas for Calculation of Rotating Blade Noise. *AIAA J.,* vol. 19, no. 9, Sept. 1981, pp. 1122–1130.
6. Morse, Philip M.; and Ingard, K. Uno: *Theoretical Acoustics.* McGraw-Hill Book Co., c.1968.
7. Goldstein, Marvin E.: *Aeroacoustics.* McGraw-Hill Book Co., c.1976.
8. Ashley, Holt; and Landahl, Marten: *Aerodynamics of Wings and Bodies.* Addison-Wesley Publ. Co., Inc., c.1965.
9. Zorumski, William E.; and Weir, Donald S., eds.: *Aircraft Noise Prediction Program Theoretical Manual—Propeller Aerodynamics and Noise.* NASA TM-83199, Pt. 3, 1986.
10. Farassat, F.; Padula, S. L.; and Dunn, M. H.: Advanced Turboprop Noise Prediction Based on Recent Theoretical Results. *J. Sound & Vibration,* vol. 119, no. 1, Nov. 22, 1987, pp. 53–79.
11. Hanson, Donald B.: Near Field Noise of High Tip Speed Propellers in Forward Flight. AIAA Paper No. 76-565, July 1976.
12. Farassat, F.: Advanced Theoretical Treatment of Propeller Noise. *Propeller Performance and Noise,* VKI-LS 1982-08, Volume 1, Von Karman Inst. of Fluid Dynamics, May 1982.
13. Farassat, F.; and Succi, G. P.: The Prediction of Helicopter Rotor Discrete Frequency Noise. *Vertica,* vol. 7, no. 4, 1983, pp. 309–320.
14. Farassat, F.: Theoretical Analysis of Linearized Acoustics and Aerodynamics of Advanced Supersonic Propellers. *Aerodynamics and Acoustics of Propellers,* AGARD-CP-366, Feb. 1985, pp. 10-1–10-15. (Available from DTIC as AD A156 047.)
15. Gutin, L.: *On the Sound Field of a Rotating Propeller.* NACA TM 1195, 1948.
16. Hanson, Donald B.: Helicoidal Surface Theory for Harmonic Noise of Propellers in the Far Field. *AIAA J.,* vol. 18, no. 10, Oct. 1980, pp. 1213–1219.
17. Hanson, Donald B.: Influence of Propeller Design Parameters on Far Field Harmonic Noise in Forward Flight. *AIAA J.,* vol. 18, no. 11, Nov. 1980, pp. 1313–1319.
18. Hanson, Donald B.: Near-Field Frequency-Domain Theory for Propeller Noise. *AIAA J.,* vol. 23, no. 4, Apr. 1985, pp. 499–504.
19. Hanson, D. B.: Shielding of Prop-Fan Cabin Noise by the Fuselage Boundary Layer. *J. Sound & Vibration,* vol. 92, no. 4, Feb. 22, 1984, pp. 591–598.
20. Lowson, M. V.: *Theoretical Studies of Compressor Noise.* NASA CR-1287, 1969.

21. Wright, S. E.: Discrete Radiation From Rotating Periodic Sources. *J. Sound & Vibration,* vol. 17, no. 4, Aug. 22, 1971, pp. 437–498.
22. Hanson, Donald B.: Noise of Counter-Rotation Propellers. *J. Aircr.,* vol. 22, no. 7, July 1985, pp. 609–617.
23. Hanson, D. B.; and Fink, M. R.: The Importance of Quadrupole Sources in Prediction of Transonic Tip Speed Propeller Noise. *J. Sound & Vibration,* vol. 62, no. 1, Jan. 8, 1979, pp. 19–38.
24. Blackburn, H. W.: Quadrupoles in Potential Flow: Two Model Problems. *J. Fluid Mech.,* vol. 116, Mar. 1982, pp. 507–530.
25. Curle, N.: The Influence of Solid Boundaries Upon Aerodynamic Sound. *Proc. Royal Soc. (London),* ser. A, vol. 231, no. 1187, Sept. 20, 1955, pp. 505–514.
26. Ffowcs Williams, J. E.; and Hall, L. H.: Aerodynamic Sound Generation by Turbulent Flow in the Vicinity of a Scattering Half Plane. *J. Fluid Mech.,* vol. 40, pt. 4, Mar. 9, 1970, pp. 657–670.
27. Chase, David M.: Sound Radiated by Turbulent Flow Off a Rigid Half-Plane as Obtained From a Wavevector Spectrum of Hydrodynamic Pressure. *J. Acoust. Soc. America,* vol. 52, no. 3, pt. 2, Sept. 1972, pp. 1011–1023.
28. Amiet, R. K.: Noise Due to Turbulent Flow Past a Trailing Edge. *J. Sound & Vibration,* vol. 47, no. 3, 1976, pp. 387–393.
29. Amiet, R. K.: Effect of the Incident Surface Pressure Field on Noise Due to Turbulent Flow Past a Trailing Edge. *J. Sound & Vibration,* vol. 57, no. 2, 1978, pp. 305–306.
30. Corcos, G. M.: The Structure of the Turbulent Pressure Field in Boundary-Layer Flows. *J. Fluid Mech.,* vol. 18, pt. 3, Mar. 1964, pp. 353–378.
31. Willmarth, W. W.; and Roos, F. W.: Resolution and Structure of the Wall Pressure Field Beneath a Turbulent Boundary Layer. *J. Fluid Mech.,* vol. 22, pt. 1, May 1965, pp. 81–94.
32. Eckert, E. R. G.; and Drake, Robert M., Jr.: *Heat and Mass Transfer, Second ed.* McGraw-Hill Book Co., Inc., 1959.
33. Yu, J. C.; and Joshi, M. C.: On Sound Radiation From the Trailing Edge of an Isolated Airfoil in a Uniform Flow. AIAA Paper No. 79-0603, Mar. 1979.
34. Schlinker, Robert H.; and Amiet, Roy K.: *Helicopter Rotor Trailing Edge Noise.* NASA CR-3470, 1981.
35. Brooks, T. F.; and Hodgson, T. H.: Prediction and Comparison of Trailing Edge Noise Using Measured Surface Pressures. AIAA-80-0977, June 1980.
36. Chou, S.-T.; and George, A. R.: Effect of Angle of Attack on Rotor Trailing-Edge Noise. *AIAA J.*, vol. 22, no. 12, Dec. 1984, pp. 1821–1823.
37. Fink, Martin R.: Noise Component Method for Airframe Noise. AIAA Paper 77-1271, Oct. 1977.
38. Roy, D.: Doppler Frequency Effects Due to Source, Medium, and Receiver Motions of Constant Velocity. AIAA-83-0702, Apr. 1983.
39. Kinsler, Lawrence E.; and Frey, Austin R.: *Fundamentals of Acoustics, Second ed.* John Wiley & Sons, Inc., c.1962.
40. *Noise Standards: Aircraft Type and Airworthiness Certification.* FAR, Pt. 36, Federal Aviation Adm., June 1974. (Consolidated reprint Aug. 12, 1985.)
41. McAninch, G. L.; and Rawls, J. W., Jr.: Effects of Boundary Layer Refraction and Fuselage Scattering on Fuselage Surface Noise From Advanced Turboprop Propellers. AIAA-84-0249, Jan. 1984.
42. Hanson, D. B.; and Magliozzi, B.: Propagation of Propeller Tone Noise Through a Fuselage Boundary Layer. *J. Aircr.*, vol. 22, no. 1, Jan. 1985, pp. 63–70.
43. Amiet, R. K.: *Diffraction of Sound by a Half-Plane in a Uniform Flow.* UTRC Rep. 86-17, United Technologies Research Center, Aug. 1986.
44. *American National Standard Method for the Calculation of the Absorption of Sound by the Atmosphere.* ANSI SI.26-1978 (ASA 23-1978), American Inst. of Physics, 1978.

45. Sutherland, Louis C.: *Development of Improved Methods for Computing Air Absorption Losses. Volume I: Evaluation of Vibrational Relaxation Frequencies From Laboratory Data.* WR 86-22 (Contract DTFA01-84-C-00003), Wyle Research, Feb. 17, 1987.
46. Marsh, Alan H.: *Evaluation of Alternative Procedures for Atmospheric Absorption Adjustments During Noise Certification.* FAA-EE-80-46, Oct. 1980.
    Volume I: Analyses and Results.
    Volume II: Computer Program.
    Volume III: Tables of Atmospheric-Absorption Losses.
47. *Standard Values of Atmospheric Absorption as a Function of Temperature and Humidity.* ARP 866A, Soc. of Automotive Engineers, Mar. 15, 1975.
48. *Acoustic Effects Produced by a Reflecting Plane.* AIR 1327, Soc. of Automotive Engineers, Inc., Jan. 15, 1976.
49. *Practical Methods To Obtain Free-Field Sound Pressure Levels From Acoustical Measurements Over Ground Surfaces.* AIR 1672B, Soc. of Automotive Engineers, Inc., June 1983.
50. Yoerkie, C. A.; and Larson, R. S.: Prediciton of Free-Field Noise Levels From Pole Microphone Measurements. AIAA-80-1058, June 1980.
51. Pao, S. Paul; Wenzel, Alan R.; and Oncley, Paul B.: *Prediction of Ground Effects on Aircraft Noise.* NASA TP-1104, 1978.
52. Zorumski, William E.: *Prediction of Aircraft Sideline Noise Attenuation.* NASA TM-78717, 1978.
53. Chessell, C. I.: Propagation of Noise Along a Finite Impedance Boundary. *J. Acoust. Soc. America,* vol. 62, no. 4, Oct. 1977, pp. 825–834.
54. *Estimation of One-Third-Octave-Band Lateral Attenuation of Sound From Jet-Propelled Airplanes.* AIR 1906, Soc. of Automotive Engineers, Inc., Oct. 1985.
55. Hawkings, D. L.; and Lowson, M. V.: Theory of Open Supersonic Rotor Noise. *J. Sound & Vibration,* vol. 36, no. 1, Sept. 8, 1974, pp. 1–20.
56. Barger, Raymond L.: *Theoretical Prediction of Nonlinear Propagation Effect on Noise Signatures Generated by Subsonic or Supersonic Propeller- or Rotor-Blade Tips.* NASA TP-1660, 1980.
57. Tam, Christopher K. W.; and Salikuddin, M.: Weakly Nonlinear Acoustic and Shock-Wave Theory of the Noise of Advanced High-Speed Turbopropellers. *J. Fluid Mech.,* vol. 164, Mar. 1986, pp. 127–154.
58. Lindblad, Ingemar A. A.: Asymptotic Analysis of the Transonic Region of a High-Speed Propeller. AIAA-89-1077, Apr. 1989.
59. Magliozzi, B.: *The Influence of Forward Flight on Propeller Noise.* NASA CR-145105, 1977.
60. Pegg, R. J.; Farassat, F.; and Magliozzi, B.: Some Measured and Calculated Effects of Forward Velocity on Propeller Noise. ASME Paper 77-GT-70, Mar. 1977.
61. Hubbard, Harvey H.: *Sound From Dual-Rotating and Multiple Single-Rotating Propellers.* NASA TN 1654, 1948.
62. Glover, B. M., Jr.; Plunkett, E. I.; and Simcox, C. D.: Noise Testing of an Advanced Design Propeller in the Boeing Transonic Wind Tunnel With and Without Test Section Acoustic Treatment. AIAA-84-2366, Oct. 1984.
63. Amiet, Roy K.: Correction of Open Jet Wind Tunnel Measurements for Shear Layer Refraction. AIAA Paper 75-532, Mar. 1975.
64. Ross, R.; Young, K. J.; Allen, R. M.; and Van Ditshuizen, J. C. A.: Acoustic Wave Propagation Through the Shear Layer of the DNW Large Open Jet Wind Tunnel. AIAA-83-0699, Apr. 1983.
65. Herkes, W. H.; Strout, F. G.; and Ross, R.: Acoustic Evaluation of DNW Free Jet Shear Layer Correction Using a Model Jet. AIAA-83-0757, Apr. 1983.
66. Brooks, Bennett M.; and Mackall, Karen G.: Measurement and Analysis of Acoustic Flight Test Data for Two Advanced Design High Speed Propeller Models. AIAA-84-0250, Jan. 1984.
67. Weir, D. S.; and Powers, J. O.: Comparisons of Predicted Propeller Noise With Windtunnel and Flyover Data. AIAA-87-0527, Jan. 1987.

68. Succi, George P.; Munro, David H.; and Zimmer, Jeffrey A.: Experimental Verification of Propeller Noise Prediction. AIAA-80-0994, June 1980.
69. Nallasamy, M.; Woodward, R. P.; and Groeneweg, J. F.: High Speed Propeller Performance and Noise Predictions at Takeoff/Landing Conditions. AIAA-88-0264, Jan. 1988.
70. Magliozzi, B.: *V/STOL Rotary Propulsor Noise Prediction Model—Ground Reflection Effects and Propeller Thickness Noise.* Rep. No. FAA-EE-82-15, Aug. 1982.
71. *Environmental Protection—Annex 16 to the Convention on International Civil Aviation. Volume I—Aircraft Noise, First ed.* International Civil Aviation Organization, Nov. 26, 1981. (As amended Nov. 21, 1985.)
72. Parzych, D. J.; Magliozzi, B.; and Metzger, F. B.: Prop-Fan/Turboprop Acoustic Terminology. SAE Tech. Paper Ser. 871839, Oct. 1987.
73. *Prediction Procedure for Near-Field and Far-Field Propeller Noise.* AIR 1407, Soc. of Automotive Engineers, Inc., May 1977.
74. Johnston, J. F.; Donham, R. E.; and Guinn, W. A.: Propeller Signatures and Their Use. AIAA-80-1035, June 1980.
75. Magliozzi, B.: Synchrophasing for Cabin Noise Reduction of Propeller-Driven Airplanes. AIAA-83-0717, Apr. 1983.

2 Rotor Noise

Lead author

F. H. Schmitz
NASA Ames Research Center
Moffett Field, California

Introduction

Types of Rotorcraft

Rotorcraft is the generic label attached to vehicles that utilize unducted rotors to create enough lift to achieve hovering flight out of ground effect. The most common aircraft in this category is the helicopter, which has proved to be the most efficient hovering heavier-than-air vehicle. Helicopters were first introduced in the 1940's and have steadily evolved into useful operation vehicles. They can be divided into two subclasses: those that use a single-rotor system for lifting and a smaller rotor system (or other control device) for yaw control, or those that use counterrotating tandem, side-by-side, or coaxial rotors for lifting and differential torque for yaw control (fig. 1). In the past four decades, the growth of these vertical-lift aircraft has been phenomenal. They have become an integral part of the military and are used in a multitude of civilian tasks where hovering flight is a necessity.

The goal of design engineers has always been to improve the usefulness and productivity of the helicopter by increasing its forward-flight performance. Unfortunately, forcing a rotor to fly through the air sideways, or in nonaxial flight, is not done easily. Aerodynamic considerations have limited the performance of pure helicopters to 150 to 200 knots in $1g$ flight. During a normal rotor-blade revolution in high-speed nonaxial flight, transonic flow on the rotor advancing blade can cause large drag, vibration, and noise effects, while dynamic stall on the retreating blade can cause similar effects. To overcome these high-speed-flight limitations, new types of rotorcraft are being developed that have nearly the hovering efficiency of the helicopter, but convert to an airplane-like configuration to achieve higher speed forward flight.

The tilt-rotor aircraft (fig. 1) is a promising vehicle in this class. In hover, its rotors are thrusting upward like a helicopter. It accelerates to forward velocity by rotating the rotors forward, creating excess thrust. The decreasing component of vertical thrust is then carried by a wing as the vehicle becomes airplane-like in its operation. After many years of successful research, tilt-rotor aircraft that can hover efficiently and still cruise at up to 300 knots have been built and are ready to go into production. The tilt-wing and stopped-rotor/X-wing aircraft are

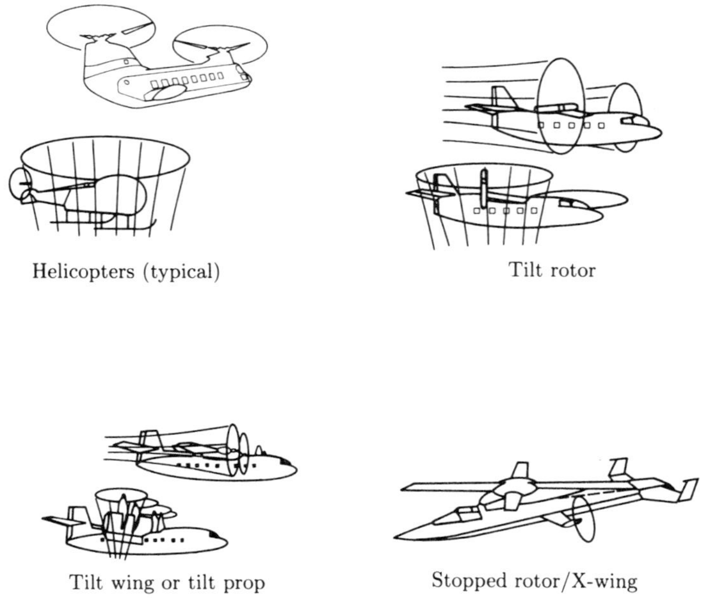

Figure 1. Current and promising rotorcraft configurations.

two other promising concepts that use rotors (or propellers) for hovering flight but convert to an airplane configuration to achieve even higher forward airspeeds (fig. 1). Unfortunately, the higher airspeeds of all these nonhelicopter configurations usually degrade the hovering performance of the vehicle, a trade-off dictated by the laws of physics and engineering. The added complexity necessary to achieve high-speed forward flight costs weight and thus reduces hovering performance.

Each of these different aircraft, which comprise a portion of the generic rotorcraft class, perform different specific missions well. If hovering efficiency is desired, then the helicopter is best. If cruise efficiency is valued and hovering time is kept to a minimum, then vehicles such as the tilt-rotor, tilt-wing, and stopped-rotor/X-wing aircraft are the better choice.

Within the generic rotorcraft class, unducted rotors are flown in a variety of operating states, including the limiting cases of axial and nonaxial flight. The axial-flight condition (i.e., normal propeller state) occurs in helicopters and other rotorcraft that are operating in hover or in a pure vertical climb or descent. It also occurs for tilt-rotor or tilt-wing aircraft when they are operating in the airplane-like configurations. The nonaxial flight states that are experienced by rotorcraft set them apart from other vehicles. The asymmetrical velocities experienced by the blades as they traverse the rotor disk and the proximity of the rotor wake under many flight conditions cause most of the aerodynamic and, hence, noise problems. Helicopters, in particular, spend much of their time operating in nonaxial flight very close to the wake shed from their rotor system. Tilt-rotor and tilt-wing aircraft do also when operating in their helicopter modes of flight, and additionally they must transit through expanded envelopes as they convert to airplane flight.

Overview

The noise that emanates from this class of rotorcraft as they operate under propeller and helicopter flight conditions has been a ubiquitous source of annoyance and has helped others detect, classify, and determine the position of rotary-wing vehicles for many years. In the 1940's and 1950's, research into the mechanisms of propeller noise was pursued with vigor. However, the phenomenal success of the jet engine decreased the importance of propeller-driven aircraft and consequently deemphasized rotorcraft aeroacoustic research. In the early 1960's, the dramatic use of the helicopter by the U.S. military rekindled interest in rotorcraft acoustics. Focusing mostly on aural detection of helicopters, researchers began a new assault on the rotorcraft noise problem which has existed in one form or another up until today. The emphasis has recently shifted to commercial certification requirements with the introduction of government-regulated noise rules. However, military detection still plays an important part of all research and development efforts.

In this chapter, the physical characteristics and sources of rotorcraft noise as they exist today are presented. Emphasis is on helicopter-like vehicles, that is, on rotorcraft in nonaxial flight. The specific noise sources of propeller-driven aircraft are covered in another chapter, and although they are similar in many cases to rotorcraft noise, they will not be treated in the context of propeller noise here. First, the mechanisms of rotor noise are reviewed in a simple physical manner for the most dominant sources of rotorcraft noise. With simple models, the characteristic time- and frequency-domain features of these noise sources are presented for idealized cases. Full-scale data on several rotorcraft are then reviewed to allow the reader to easily identify the type and extent of the radiating noise. Methods and limitations of using scaled models to test for several noise sources are subsequently presented. Theoretical prediction methods are then discussed and compared with experimental data taken under very controlled conditions. Finally, some promising noise reduction technology is reviewed.

Rotorcraft Noise Sources and Their Physical Origins

Noise Spectrum of a Helicopter With a Single Main Rotor

One of the most widely discussed rotorcraft aeroacoustic topics of the past decade has been the way rotor noise sources are classified (refs. 1 to 5). When you first hear a helicopter, you are most always impressed by the harshness and periodicity of the noise. This usually occurs when a rotorcraft is descending or maneuvering in a terminal area or when it is flying at high speed in a helicopter configuration. These loud, sharp, periodic sounds are labeled *impulsive noise* and clearly distinguish rotorcraft noise from other types of noise. In fact, there is a milder form of periodic noise, *rotational noise*, that is also distinguishable on rotorcraft. It has its origins in axial-flight (propeller) aircraft and arises because the rotor is creating thrust and torque and because its blades must displace air as they move through space. One might guess that these two different-sounding noise sources are related mathematically because they are both periodic in nature. While this is true, the aerodynamic origins of the sounds are quite different. These origins serve as the real classifiers of the resulting noise.

Broadband noise is also a source of noise on most rotorcraft. The whooshing sound usually associated with the start-up of rotorcraft typifies this type of noise. It is also noticeable when the helicopter is hovering or flying overhead at relatively low altitudes. Broadband noise has its origins in interaction of the moving blade with turbulence shed either from the blade itself, from previous rotor blades, or from the atmosphere. It is usually important at lower tip Mach numbers, where the other forms of rotor noise do not dominate the spectrum.

Early analysis equipment for acoustic signatures was not as sophisticated as today's digital technology. Noise measurements made with present-day technology are done with modern electronic computer-based equipment that can be used to measure and process extremely narrow bandwidths of data. Sophisticated signal analyses using fast Fourier transform (FFT) and signal averaging techniques yield accurate power spectra and time-history data. For rotorcraft with constant rotor speeds, it is possible to literally "pull out" periodic signals from random or broadband noise sources. This is illustrated in figure 2, wherein a 50-Hz and a 1-Hz analysis of similar data are illustrated. Because the periodic signal levels are independent of the bandwidth while the broadband noise decreases with bandwidth (10 log bandwidth ratio), the periodic noise emerges from apparently broadband noise as the bandwidth of the analysis is reduced. This technique works well as long as the periodic event is truly periodic. If changes in rotor speed (ref. 6), in frequency because of Doppler effects, in rotor-wake positions, or in distances between the microphone and the observer are allowed, then higher frequencies of periodic noise can smear across the narrow bandwidths and begin to appear to be broadband noise. In effect, the distinction between "broadband noise" and periodic noise sources can blur in the frequency domain when a basically periodic phenomenon is somewhat unsteady. Because it only takes small changes to cause this effect, it is suspected that many previously reported cases of broadband noise were really unsteady periodic noise. When some of these factors were analytically accounted for in the data analysis of flyover aircraft, noise levels that were previously attributed to broadband noise were reclassified as harmonic noise (ref. 7). More recent data taken under very controlled conditions have shown that periodic noise dominates the helicopter noise spectra under most flight conditions.

The noise spectrum of a hovering single-rotor helicopter with its various sources of noise is shown in figure 3. Main-rotor, tail-rotor, broadband, and other noise sources are identified, although there is some controversy as to whether the broadband noise shown in this figure is truly broadband noise or whether it is nonstationary periodic noise. Each noise source is truly a contributor to the radiated acoustic signature, but only a few sources actually dominate on most rotorcraft. This results because most rotorcraft manufacturers design their machines to be as totally efficient as possible while meeting the requirements of safety, low vibration, etc. A by-product of this process is the fact that the tip Mach numbers in hover range between 0.6 and 0.7 on most rotorcraft of today. This compromise uses the full aerodynamic capability of the rotors without encountering severe compressibility effects over the design flight envelope and without compromising the structural integrity of the aircraft. Because the hover tip Mach numbers are relatively high, impulsive and rotational noise sources usually dominate the spectra of rotorcraft. In this chapter, we shall focus

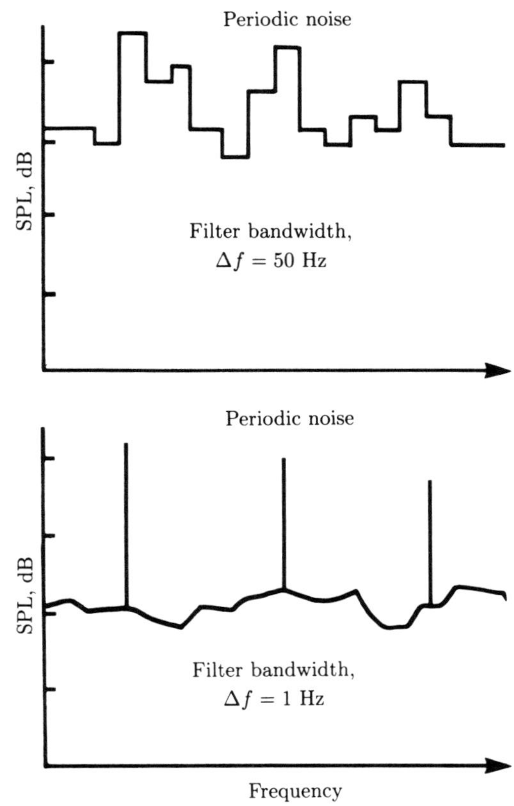

Figure 2. Effect of bandwidth reduction on spectra of rotorcraft noise.

our attention on the loudest noise sources in the belief that, if they are mitigated, much of the rotorcraft noise and annoyance problem will be as well.

Governing Acoustic Equation

Blade-Fixed Coordinates

Most of the material that is discussed in this chapter can be mathematically represented by the following general well-known integral equation which governs the noise radiated from a body in arbitrary motion:

$$4\pi c_0^2 \rho'(\mathbf{x}, t) = \frac{\partial^2}{\partial x_i\, \partial x_j} \iiint \left[\frac{T_{ij}}{\Re |1 - M_{\Re}|} \right]_\tau dV(\boldsymbol{\eta}) - \frac{\partial}{\partial x_i} \iint \left[\frac{P_{ij} n_j}{\Re |1 - M_{\Re}|} \right]_\tau dS(\boldsymbol{\eta}) + \frac{\partial}{\partial t} \iint \left[\frac{\rho_0 v_n}{\Re |1 - M_{\Re}|} \right]_\tau dS(\boldsymbol{\eta}) \tag{1}$$

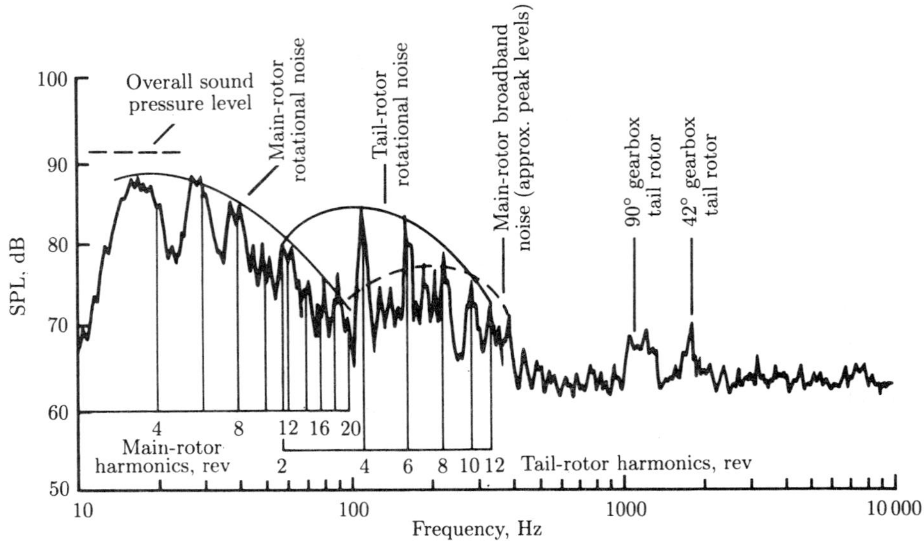

Figure 3. External noise spectrum of UH-1A. (Based on ref. 8.)

where

c_0	speed of sound
T_{ij}	$= \rho u_i u_j + P_{ij} - c_0^2 \rho' \delta_{ij}$
P_{ij}	$= p' \delta_{ij}$
$\Re$	$= \|\boldsymbol{\eta} - \mathbf{x}\|$
$\mathbf{x}$	observer position
$\boldsymbol{\eta}$	source position
t	observer time
τ	source time
$M_\Re$	Mach number of source in observer's direction
dV	elemental volume in a reference frame fixed to the body
dS	elemental area in a reference frame fixed to the body
p'	pressure on the blade surface that acts on the surrounding medium, $p - p_0$
ρ'	$= \rho - \rho_0$
u_i, u_j	components of fluid velocity in directions x_i and x_j
n_j	unit normal outward from the surface
v_n	velocity of surface in the normal direction

δ_{ij}	Kronecker delta
ρ	fluid density
ρ_0	fluid density at rest
p	fluid pressure
p_0	fluid pressure at rest

This equation was derived in reference 9 and has been expanded upon by many researchers (refs. 10 to 12). Far-field acoustic density is explicitly expressed in terms of integrals over the body surface and the surrounding volume in a *reference frame moving with the body surface.* For rotorcraft applications the blade itself is considered to be the moving body, so the reference frame for equation (1) is in blade-fixed coordinates moving with the rotating blades. Note that equation (1) is in the strictest sense a nonlinear integral equation over all space. Often, the right-hand-side integrals are assumed to be bounded and finite and basically independent of the acoustic pressure. Under these conditions, all three terms in equation (1) can be interpreted as sources of rotorcraft noise: the first term represents noise due to fluid stress and becomes important at high Mach numbers; the second term represents the noise due to blade surface pressures pushing on the fluid; the third term describes the noise that is caused by the blade displacing fluid as it traverses its circular path. For acoustics, it is normally assumed that $p' = c_0^2\rho'$, so the left-hand side of equation (1) can be interpreted as acoustic pressure.

The circular blade path of each rotor blade causes much of the apparent complexity of rotorcraft noise calculations. All sources must be tracked in this circular path, with particular attention paid to source and receiver time of emission and reception, respectively. This is largely a geometric problem, but one of considerable complexity. Fortunately, the computer thrives on such tasks and makes these laborious computations quite easily. This does not, however, eliminate the need for a solid physical understanding of the rotorcraft noise problem.

A sketch of the geometry of a simple hovering rotor is shown in figure 4. Depicted are steady force (lift and drag dipoles) and steady thickness (monopole) sources on a single blade. The distance $\Re$ between an arbitrary point on the rotating blade and the observer is also shown. These steady force and thickness effects can be thought of as rotating dipoles and monopoles, respectively, and are described mathematically in this blade-fixed coordinate system by the second and third terms of equation (1). According to this equation, the radiated noise due to steady force is simply a spatial derivative of the summation of force source terms at the correct retarded time, and the radiated noise due to steady thickness is simply a time derivative of the summation of thickness source terms taken at the correct retarded time. In essence, a simple linear three-dimensional wave equation is being solved. The retarded time operator keeps track of source emission times τ and receiver times t.

$$\tau + \frac{\Re}{c_0} = t \tag{2}$$

A source of sound emitted at an earlier time $\tau = t - \frac{\Re}{c_0}$ at a distance $\Re$ away from an observer must travel $\frac{\Re}{c_0}$ sec to reach an observer at time t. The factor $|1 - M_{\Re}|$ in the denominator of all the terms in equation (1) is the well-known Doppler factor and is a direct result of choosing to describe the acoustics of rotors in the moving-blade frame of reference. The Doppler factor strongly increases the magnitude of each term of equation (1) as $M_{\Re}$ approaches 1.0. When $M_{\Re} = 1.0$, equation (1) becomes singular and requires special numerical treatment. Fortunately, most conventional helicopters do not fly with tip Mach numbers of 1.0; $M_{\Re}$ for high-speed advancing flight is typically not greater than 0.9. Under most cruising helicopter flight conditions ($M_{\Re} = 0.85$), the second two terms of equation (1), which are the important contributors to the noise radiation, can be evaluated in a simple, straightforward manner.

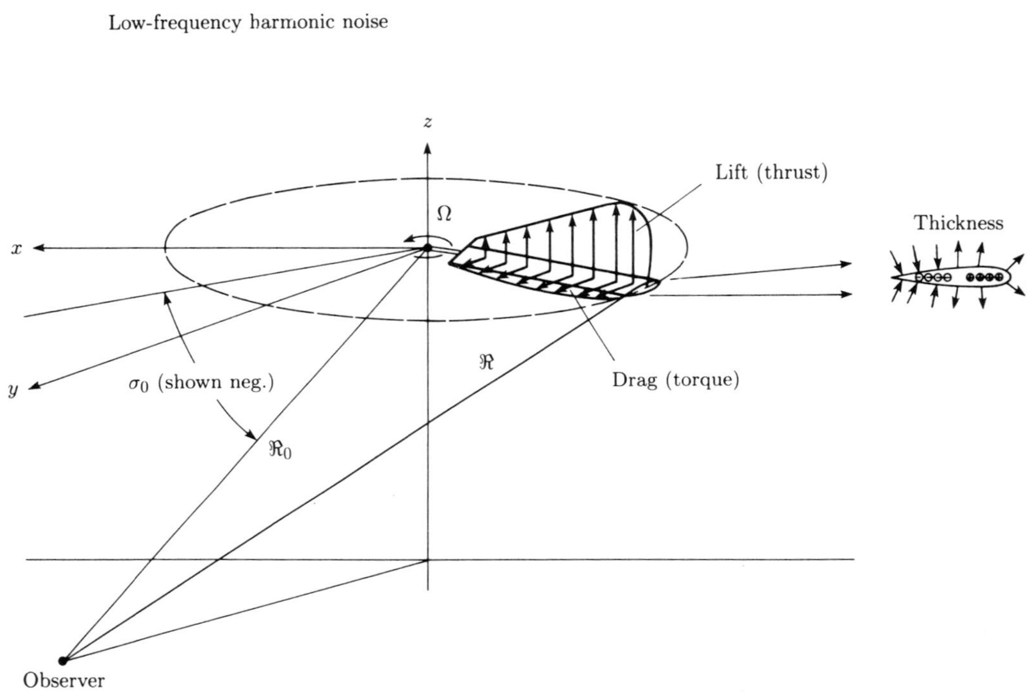

Figure 4. Geometry of simple hovering rotor.

Space-Fixed Coordinates

There is an equivalent representation of the noise generation process that illustrates the role of the circular geometry and highlights the fact that a simple, three-dimensional wave equation is being solved. First, the distributed sources are represented by equivalent point sources. For simplicity, consider only rotating point forces (force/length). Then, instead of describing these point sources as rotating sources, they are viewed as an entire disk of stationary sources that lie in a plane described by the rotating blade and bounded by the tip of the blade itself, as shown in figure 5. These stationary sources are then "switched on and off" at the appropriate times as the blade reference line passes over that particular position in space.

The resulting solution of the wave equation (only force effects considered) becomes

$$4\pi p'(\mathbf{x},t) = -\frac{\partial}{\partial x_i}\iint \left[\frac{F_i}{\Re}\right]_\tau d\Lambda \tag{3}$$

where $F_i = P_{ij}$ is the force/area exerted on the fluid and Λ is the area of the fixed-space source distribution. Although this equation looks as if the Doppler factor has been eliminated from the analysis, it has not. It reappears as the derivative of the switching functions and this derivative must be accounted for in this analysis. Most early researchers (refs. 13 to 15) developed their analyses using the fixed-space description of the wave equation given in equation (3). Either approach is still useful today, as they are equivalent. However, treating distributed sources as effective point sources is only valid when the distribution of source strengths is not important. In general, this occurs when the speed of the sources is much less than the speed of sound. The problem is said to be "compact" and distributed sources can be acoustically represented as point sources. The fixed-space representation of this problem can be extended to noncompact acoustic problems as well (ref. 16).

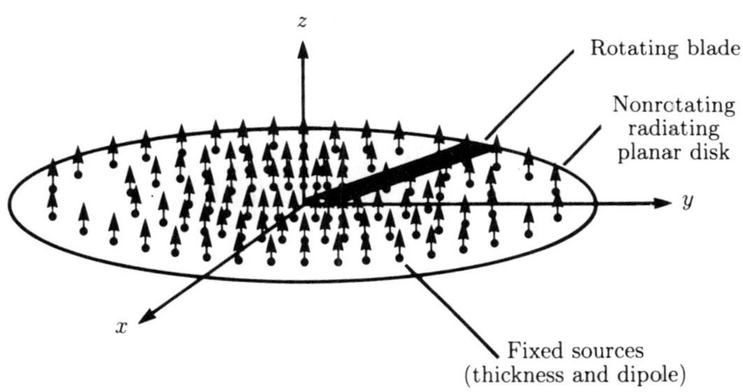

Figure 5. Fixed-space representation of classical thickness and loading acoustic sources.

Hovering Harmonic Noise

Both time- and frequency-domain results are shown in figure 6 for steady loading of a radial distribution of dipole forces and monopole thickness effects for a representative one-bladed hovering helicopter with a tip Mach number of 0.65. In general, a simple pulse is produced for each blade during one rotor revolution. The dominant pulse characteristics are controlled by those parts of the rotor disk that have the highest Mach number in the direction of the observer. For the in-plane microphone positions, thickness noise dominates the pressure time history.

Thickness noise

The actual shape of the thickness noise source can be demonstrated by considering the tip of a single rotating blade. Choosing a blade-fixed coordinate system and

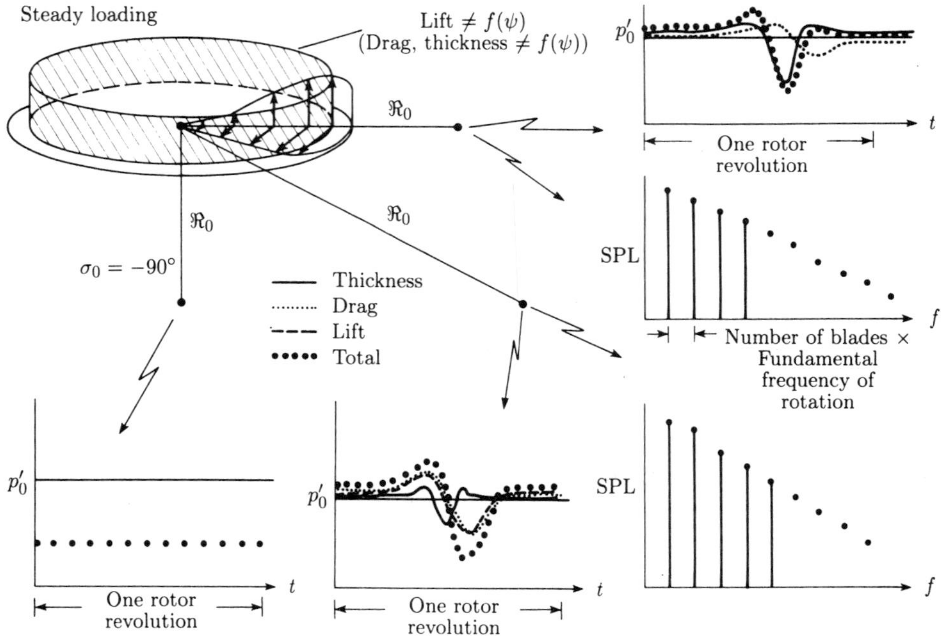

Figure 6. Relative contributions of steady loading and thickness effects to helicopter rotor noise.

rewriting the third term of equation (1) for a far-field observer, we obtain

$$p'(\mathbf{x}, t) = \frac{1}{4\pi}\frac{\partial}{\partial t}\iint \left[\frac{\rho_0 v_n}{\Re|1 - M_{\Re}|}\right]_\tau dS(\boldsymbol{\eta}) \tag{4}$$

The simplest way of describing this integration is to divide the tip section of the blade into two chordwise panels. The first panel is composed of "sources" and the second of "sinks," as shown in figure 7 for a single-bladed rotor. The strength of the source is equal to the mass flux of fluid being displaced by the blade section as it moves through space. For the single source shown, the mass flux is equal to $\rho_0 v_n$ and is positive for the forward portion of the blade section. The sink is simply the negative source and represents the mass flux of the fluid which is necessary to represent the rear portion of the rotor-blade section. In these heuristic arguments, it is important to remember that each singularity must travel a slightly different path to the observer location and therefore will arrive at different retarded times.

One of the most interesting aspects of the evaluation of the thickness integral

$$\iint \left[\frac{\rho_0 v_n}{\Re|1 - M_{\Re}|}\right]_\tau dS(\boldsymbol{\eta})$$

is that the integrand is a function of $\frac{1}{\Re|1-M_{\Re}|}$, which depends upon the observer location. The factor $\frac{1}{|1-M_{\Re}|}$ represents the Doppler amplification of acoustic signals and is a strong function of $M_{\Re}$, the Mach number of the moving source or sink in the radiation direction. As shown in figure 8, for an observer in the disk plane $M_{\Re}$

becomes a maximum when azimuthal angle $\psi \approx 90°$. Thus, we would expect the thickness noise peak to originate near $\psi \approx 90°$.

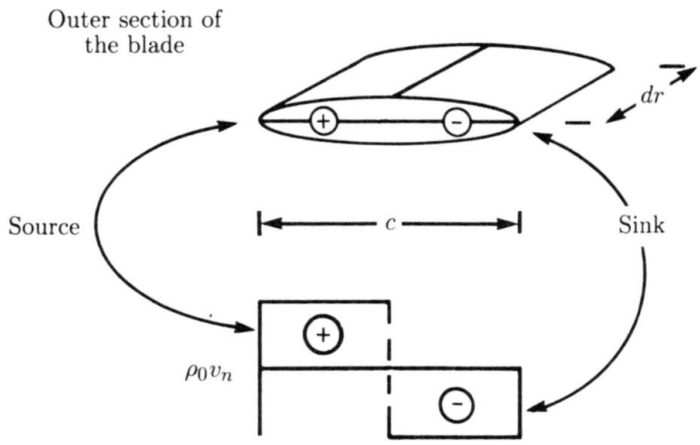

Figure 7. Simple source and sink representation of blade thickness noise. (From ref. 5.)

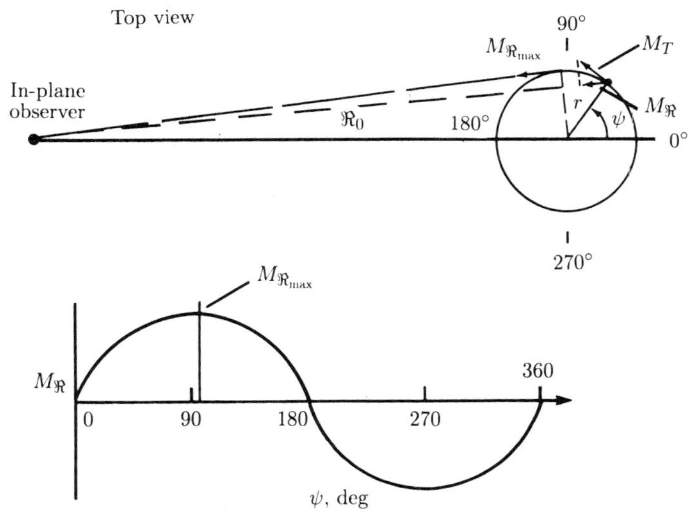

Figure 8. Doppler amplification geometry. $R_0/r = 20$; $M_T = 0.8$. *(From ref. 5.)*

Now let's sketch a graphical outline of the integration for the in-plane observer located directly ahead of the rotor (fig. 9). First, consider the simple source ($\rho_0 v_n = \oplus$). Then $\iint \left[\frac{\oplus}{\Re|1-M_\Re|}\right]_\tau dS(\boldsymbol{\eta})$ becomes as indicated in figure 9. Similarly, the integral of the simple sink ($\rho_0 v_n = \ominus$) becomes the same curve shifted (delayed) in observer time $c \sin\psi/2\Omega r$ sec, where c is the rotor chord, Ω is the rotor rotational rate, and r is the radial position. For a fixed observer at large distances from the singularities ($\Re_0/r > 6$),

$$t_{\text{source}} \approx t_{\text{sink}} + \frac{c \sin\psi}{2\Omega r}$$

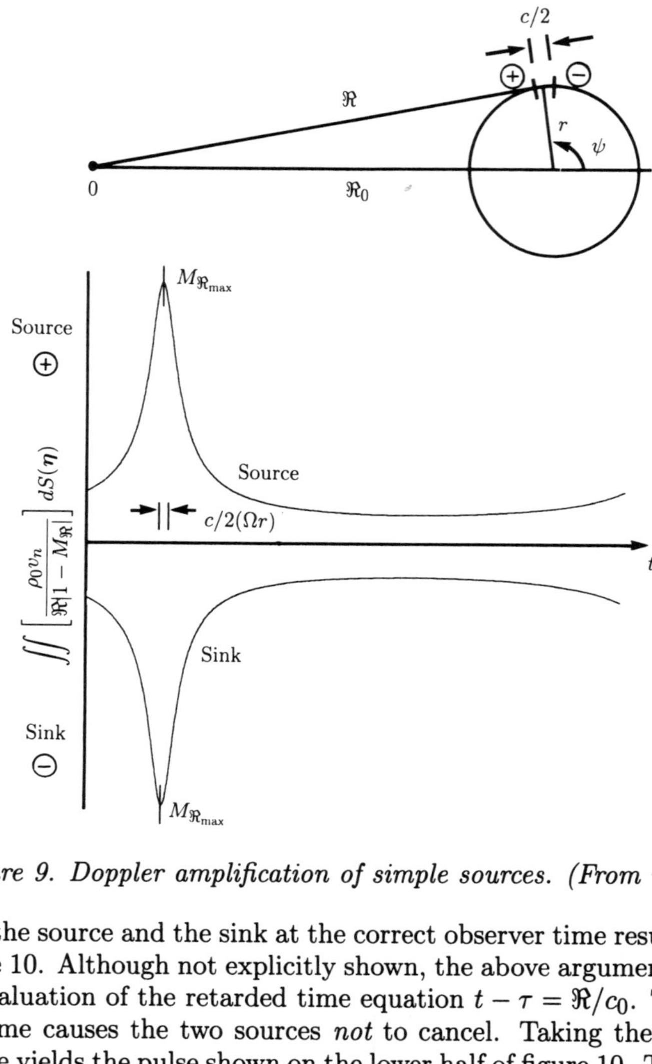

Figure 9. Doppler amplification of simple sources. (From ref. 5.)

Adding both the source and the sink at the correct observer time results in the upper curve of figure 10. Although not explicitly shown, the above arguments depend upon the correct evaluation of the retarded time equation $t - \tau = \Re/c_0$. The simple shift in observer time causes the two sources *not* to cancel. Taking the derivative with respect to time yields the pulse shown on the lower half of figure 10. This is the major mechanism of linear thickness noise and it is characterized by a large negative pulse. In much of the early literature, the sign of the thickness pulse was often mistakenly thought to be positive.

Adding many sources and sinks to accurately model the blade thickness distribution along the blade chord and radius does not change the basic shape of the radiated acoustic thickness pulse. However, as the hovering tip Mach number M_T increases, the amplitude of the negative thickness noise pulse increases quite rapidly. For most hovering rotorcraft, these simple linear arguments work well below $M_T = 0.85$. Above this value, flight data reveal that nonlinear effects begin to play a large role in the in-plane acoustic radiation.

Steady-Force Noise

A similar set of heuristic arguments can be used to illustrate the noise produced by rotating steady dipoles (forces). With the second term of equation (1) used as the

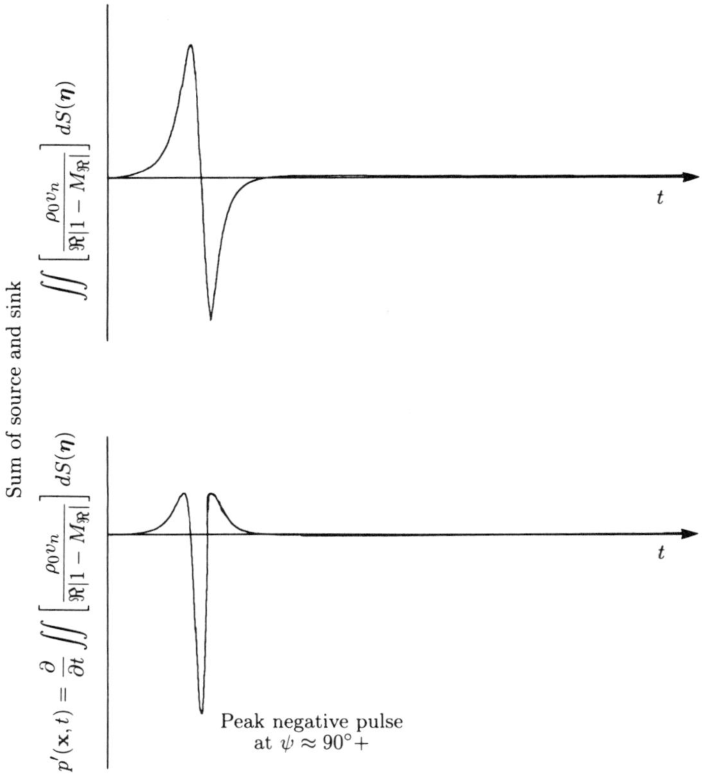

Figure 10. Simple source and sink far-field acoustic pressures. (From ref. 5.)

mathematical basis for the arguments, an entirely different characteristic time history is produced, as shown in figure 6 for steady in-plane drag and out-of-plane thrust forces. Both pulse shapes are basically asymmetrical in character, very different than the nearly symmetrical thickness noise pulse. In the plane of the rotor, the acoustic pulse amplitude is controlled by the in-plane drag of the rotor. For typical hovering rotorcraft tip Mach numbers, the in-plane peak amplitude is about the same level as the negative peak of the symmetrical thickness noise pulse. However, noise due to steady in-plane drag decreases as the observer moves above or below the tip-path plane of the rotor. Noise produced by the steady thrust of a hovering rotor also has a similar asymmetrical character but does not radiate to an observer located in the tip-path plane of the rotor (fig. 6). However, at observer positions above or below the rotor tip-path plane, steady thrust becomes the dominant contributor to the measured noise while the contribution of blade thickness is lessened. Below the rotor tip-path plane, the noise due to steady thrust and drag tends to be additive in phase. Above the rotor, noise due to steady thrust changes sign and tends to cancel the in-plane drag radiation. At the on-axis positions, the distance between any rotating source and the observer is a constant $\Re$. It follows that $M_\Re$, the Mach number of the source in the direction of the observer, is also constant. Therefore, for steady forces, all the terms under the integrals in the second term of equation (1) are constant. Consequently, no acoustic radiation is predicted for the on-axis positions depicted in figure 6.

Another important feature of all these pulse shapes is their relatively smooth time history. They represent the summation of steady-source terms amplified by their respective Doppler factors and summed at the correct retarded time. As long as the tip Mach number of the rotating source remains ≤ 0.7, no sharp impulses are expected. If a Fourier analysis is applied to the resulting time history, a sequence of Fourier series coefficients are generated that rapidly decrease in amplitude with increasing harmonic number, as illustrated in figure 6. Because the noise-generating mechanism is periodic, the amplitude of the power spectrum pulse is independent of the analysis bandwidth. If the rotorcraft has B blades instead of just the single blade considered so far, B equally spaced pulses would result. In the frequency domain, the fundamental frequency of the rotor noise would now become B times the fundamental rotation rate of the rotor.

These very simple arguments explain the physical origins of the low-frequency harmonic noise of most rotorcraft and propeller-driven vehicles. Analytical expressions describing this phenomenon were first developed over 40 years ago by Gutin (ref. 13) using equation (3) in a fixed-space reference frame. For ease in analytical calculations, the thrust and drag (torque) of the rotor were assumed to act along a radial distribution of points, as depicted in figure 4. For an observer in the far field, the expression for the acoustic pressure of the mBth harmonic becomes

$$p'_{mB} = \frac{mB\Omega}{2\pi c_0 \Re_0} \int_0^R \left(-\frac{dT}{dr}\cos\sigma + \frac{c_0}{\Omega r}\frac{dD}{dr} \right) J_{mB}\left(\frac{mB\Omega}{c_0} r \sin\sigma \right) dr \tag{5a}$$

where

$J_n(X)$	Bessel function of the first kind of order n and argument X
m	harmonic number
B	number of equally spaced rotor blades
r	rotor radial position
R	rotor radius
Ω	rotor rotational rate
c_0	undisturbed speed of sound
$\Re_0$	distance between the rotor hub and the observer
$\frac{dT}{dr}, \frac{dD}{dr}$	radial distribution of thrust and drag of the rotor
σ	elevation angle of observer with respect to the rotor plane (see fig. 4)

Gutin further simplified his analysis by assuming that the loading could be concentrated at a point of effective action along the rotor radius r_e. Integrating equation (5a), his expression for the root-mean-square (rms) acoustic pressure becomes

$$p'_{mB} = \frac{mB\Omega}{2\sqrt{2}\pi c_0 \Re_0}\left(-T_e \cos\sigma + \frac{c_0 D_e}{\Omega r_e}\right) J_{mB}\left(\frac{mB\Omega}{c_0} r_e \sin\sigma\right) \tag{5b}$$

where r_e is the effective point of action of the thrust and torque and T_e and D_e are the total thrust and drag of the rotor. Gutin found that for low orders of harmonics, choosing $r_e \approx 0.8R$ yielded good results.

The radiation effectiveness of these steady noise sources increases dramatically when the tip Mach number of the rotor is raised toward 1.0. As we shall see later, the resulting noise becomes distinct and sharp in the time domain, causing a slow or nearly nonexistent roll-off in the harmonic character of the radiated noise.

Unsteady-Force Noise

Another important source of rotorcraft noise is the unsteady rotor-blade forces. These forces can occur in both axial and nonaxial flight and can be very efficient sources of radiated noise. They can be divided into two classes: those unsteady forces which are periodic in nature and are fundamentally related to the aerodynamic events associated with the periodic flow states of the rotor, and those unsteady forces which are not periodic in nature. In the latter case, the aerodynamic events are random, causing random forces and a type of broadband noise radiation.

Unsteady periodic forces usually abound on the modern rotorcraft and are efficient generators of harmonic noise. Unsteady-harmonic-force noise can be further subdivided according to its inherent frequency content: low-frequency harmonic noise is due to low-frequency aerodynamic events, and high-frequency harmonic noise is due to near-impulsive but periodic aerodynamic events.

Low-frequency harmonic noise radiation is a result of low-frequency harmonic variations in the lift and drag of each rotor blade as it traverses the rotor disk. Figure 11 graphically illustrates noise for one blade of a hovering rotor and, for simplicity, depicts the contribution of lift at one radial location to the radiated noise. The smoothly varying loading shown can occur in hover to some degree. Fuselage interference, nonuniform downwash (or upwash), wind, and cyclic (first-harmonic) control piloting input all create low-frequency loading harmonics in near hovering flight. As with steady forces, the distributed noise sources are Doppler shifted, the result being that much of the energy of the unsteady periodic forces is strengthened in the same direction as the movement of the source. As illustrated in figure 11, for an observer 15° under the disk plane the acoustic waveform exhibits features that are spread out over the period, an indication that unsteady forces can contribute to the noise at all azimuthal positions. Also, on the axis of rotation the unsteady forces now radiate noise. Even though the radial source point is at the same distance from the observer, the time-varying nature of the resulting unsteady forces generates radiated noise on the rotor axis. When viewed in the frequency domain, these low-frequency acoustic phenomena appear as additional harmonics of noise. Instead of falling off rapidly, the harmonics now fall off more slowly and obey no real pattern, as illustrated in figure 11.

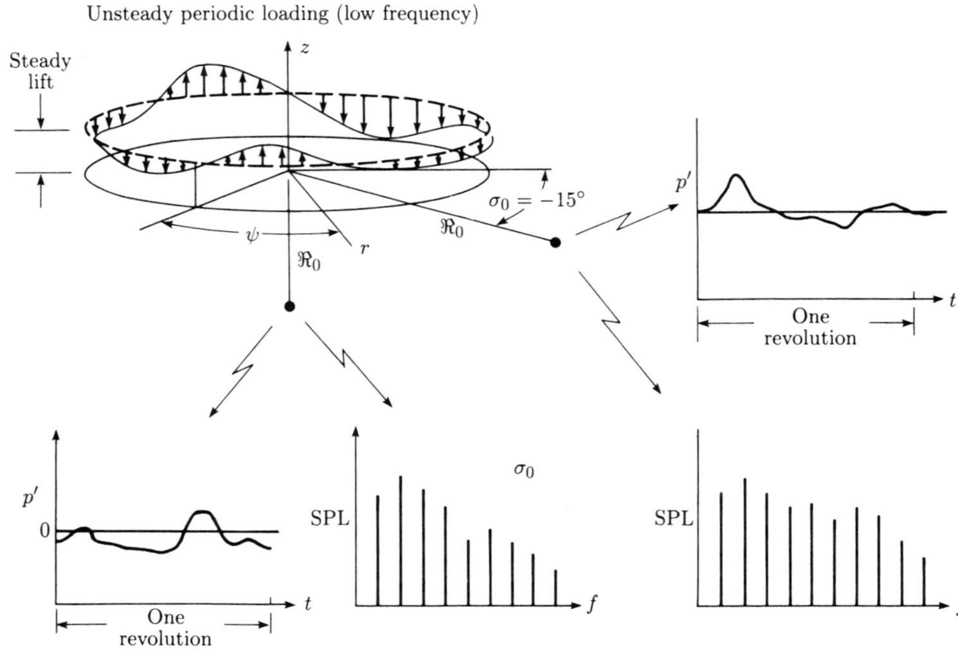

(*a*) $\sigma_0 = -90°$. (*b*) $\sigma_0 = -15°$.

Figure 11. Relative contributions of low-frequency unsteady loading to helicopter rotor noise.

Forward-Flight Harmonic Noise

The aeromechanics of a rotor in nonaxial flight are quite complicated and are the subject of much research. As shown in figure 12, there is a basic asymmetry in the velocity field of a rotor blade in forward flight. At $\psi = 90°$, the helicopter forward velocity adds to the relative velocity over the blade due to the rotor-blade rotation, while on the retreating side ($\psi = 270°$) the helicopter forward velocity reduces the relative velocity. If a perfectly rigid rotor blade were fixed to the rotor hub and if the blade pitch angle were not changed as a function of ψ, uncontrolled rolling moments would be produced by the differences in lift due to this velocity asymmetry.

The modern rotorcraft has flexible rotor blades and may or may not have flapping hinges that allow the rotor blades to flap in response to moments about the rotor hub. Blade flapping in response to the unbalanced rolling moments due to the velocity asymmetry of forward flight alters the local effective angle of attack of each blade section. In general, reductions in blade angle of attack occur on the advancing side of the disk and increases in angle of attack occur on the retreating side. These changes in effective angle of attack cause a reduction in lift on the advancing side of the disk and an increase in lift on the retreating side. When integrated in ψ and along the blade span, these changes in blade lift help reduce the unbalanced rolling moment. In addition, simple (first-harmonic) cyclic control is normally used to help balance moments about the rotor hub and to control the rotor orientation in space.

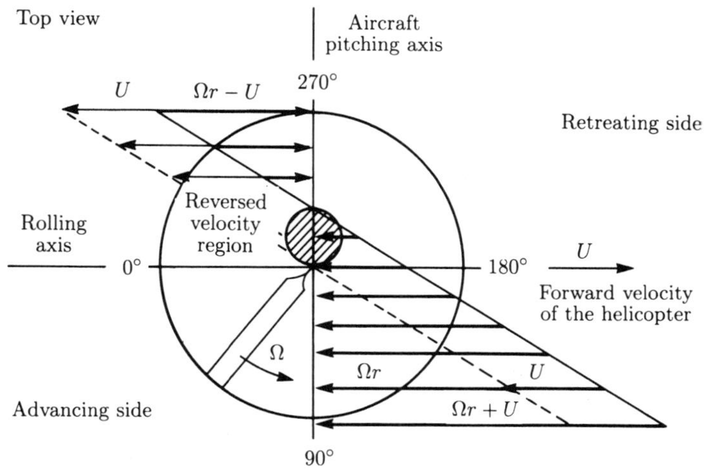

Figure 12. In-plane velocity asymmetry for rotor in nonaxial forward flight.

All these effects and others not discussed in this simple description of rotor control alter and influence the local aerodynamic force field of a rotor in nonaxial flight. The net effect is to produce a complex periodic distribution of rotor air loads, an example of which is shown in figure 13. These unsteady periodic blade forces are rich in low-frequency harmonics. Depending upon the particular flight condition, the forces can also contain high-frequency (impulsive) air loads.

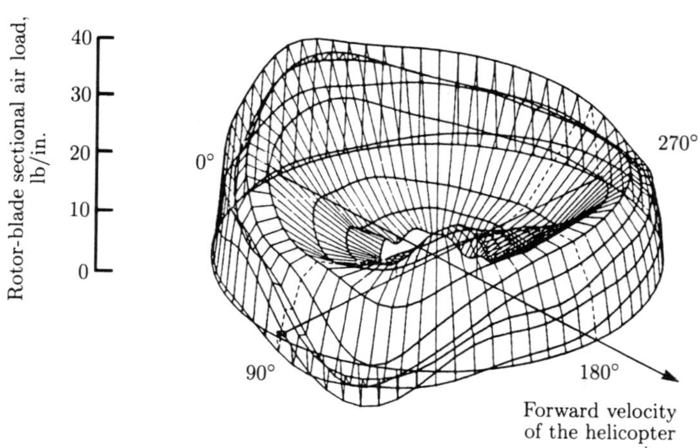

Figure 13. Air loads of rotor in forward flight. (From ref. 17.)

Low-Frequency Noise—Thickness and Force

The predominant mechanisms of low-frequency harmonic noise for a helicopter in forward flight are quite similar to those in hover. However, the geometry of the moving rotor affects the Doppler factors, the retarded-time equation, and the velocity field that the blade experiences, and must be accounted for in equation (1).

For thickness noise, the primary governing parameters are the blade thickness distribution and the advancing-tip Mach number $M_{AT} = (U + \Omega R)/c_0$, where U is the flight velocity. A symmetrical negative pulse shape is characteristic of this noise source which is quite similar to the hovering thickness-noise pulse shapes previously described.

Low-frequency harmonic force noise in forward flight is governed by the low-frequency air loads on the rotor. As with thickness noise, changes in Doppler factors and in the retarded time because of the rotorcraft nonaxial velocity must be accounted for. There is no characteristic pulse shape for this source of harmonic noise. The shape depends predominantly on the character of the harmonic air loads of the rotor.

High-Speed Impulsive (HSI) Noise

As the advancing-tip Mach number of the helicopter approaches transonic values (0.9), the negative peak of the forward-flight thickness-noise pulse shape grows dramatically in amplitude and dominates the waveform time history in the plane of the rotor. The negative pulse becomes quite narrow and impulsive in character, radiating large amounts of in-plane acoustic energy. Further increases in advancing-tip Mach number cause dramatic changes in waveform pulse shape and further increase the harmonic content of the radiation noise. This extreme of thickness noise is called high-speed impulsive (HSI) noise and is the dominant source of rotor harmonic noise when it exists. HSI is discussed in some depth subsequently in this chapter.

Blade-Vortex Interaction (BVI) Noise

Another source of high-frequency unsteady periodic loading noise is also one of the most important sources of rotor radiated noise. This noise is due to impulsive aerodynamic events that occur at deterministic locations around the rotor azimuth. These impulsive events are most likely to occur when the rotor is in nonaxial translation and the tip vortices from preceding blades interact with the following blades. A very simple sketch of this phenomenon is depicted in figure 14. A sudden impulse is produced near the leading edge of the rotor and generates an impulsive noise that radiates away from the rotor. This impulsive event contains many harmonics of radiated noise and is considered by many people to be the major source of annoyance for rotorcraft.

The qualitative characteristics of blade-vortex interaction (BVI) noise can be shown with simple two-dimensional heuristic arguments. The arguments are presented in the time domain so that acoustic events can be ordered in azimuth angle ψ and finally in observer time t for a given microphone location. Consider the top view of a two-bladed helicopter rotor at an advance ratio μ of 0.145 ($\mu = U/\Omega R =$ Forward velocity/Rotor-tip speed), which is shown in figure 15. The epicycloid-like patterns were derived from a "free-wake" computer code (ref. 18).

We know from theoretical considerations that most of the radiated noise is generated near the rotor tip. We also would expect BVI noise to occur when the rotor blade (outer 20 to 30 percent) passes close to the trailing-tip vortices. As shown in figure 15, there are seven possible BVI's (labeled 1 to 7). The strength of each interaction is governed by the local strength of the tip vortex, the core size of the tip vortex, the local interaction angle of the blade and the vortex line, and

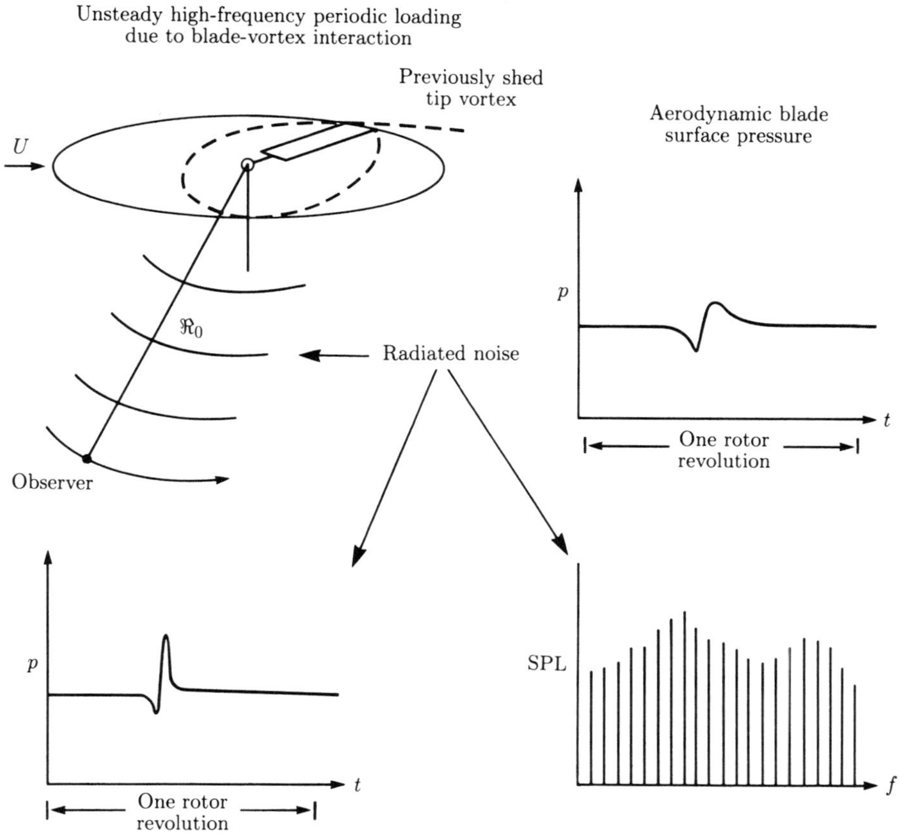

Figure 14. Contribution of high-frequency airloads (impulsive events) to helicopter rotor noise.

the vertical separation between the vortex and the blade. In general, the induced velocity of the rotor disk tends to make the tip vortices pass under the rotor disk in level, steady-state flight conditions for $\mu \approx 0.145$, as depicted in figure 16. However, if the rotor operates in steady descending flight, then the positive inflow (upflow) tends to force the epicycloid-type pattern into the rotor disk plane and causes strong blade-vortex interactions.

The net result of such considerations is shown in figure 17 for the AH-1 helicopter. A map of the regions where BVI encounters occur is shown as a function of the helicopter rate of climb. Notice that for this helicopter, the seven possible BVI encounters do not all occur at the same rate of climb and hence may not all radiate noise under the same rotor operating conditions. Of these seven potential BVI encounters, a few are known to radiate very strong impulsive noises. Consider interactions 1 to 4, which are all on the advancing side of the rotor disk and occur during descending flight. Interaction 3 in particular is an encounter in which the blade and the vortex are almost parallel during the interaction and is known to be a major source of BVI noise. In this case, simple two-dimensional arguments can be used to estimate the correct shape of the advancing-blade acoustic pulse (fig. 18).

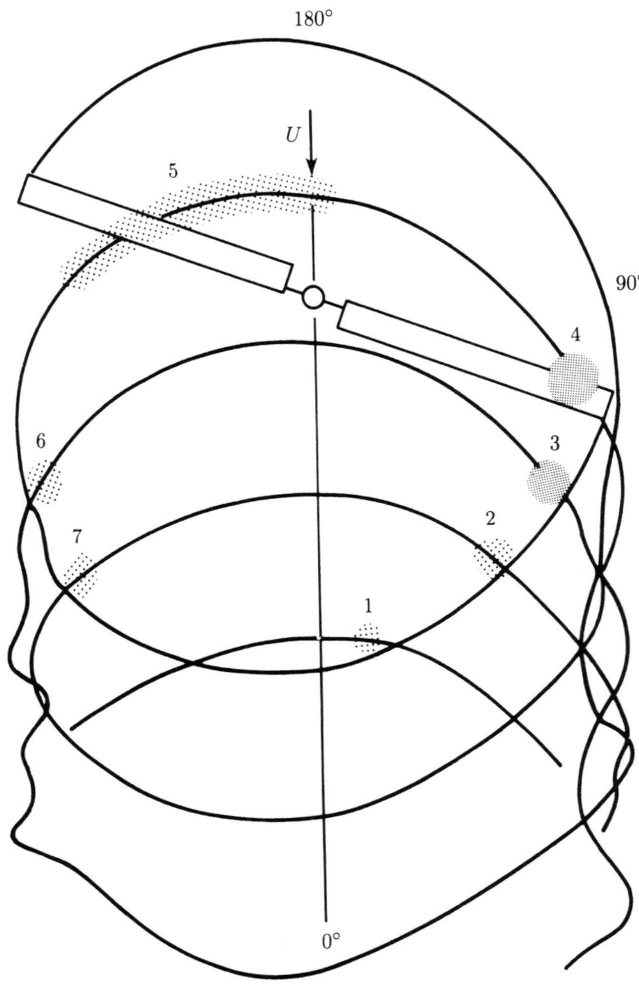

Figure 15. Blade-vortex intersections during partial-power descent. (From ref. 18.)

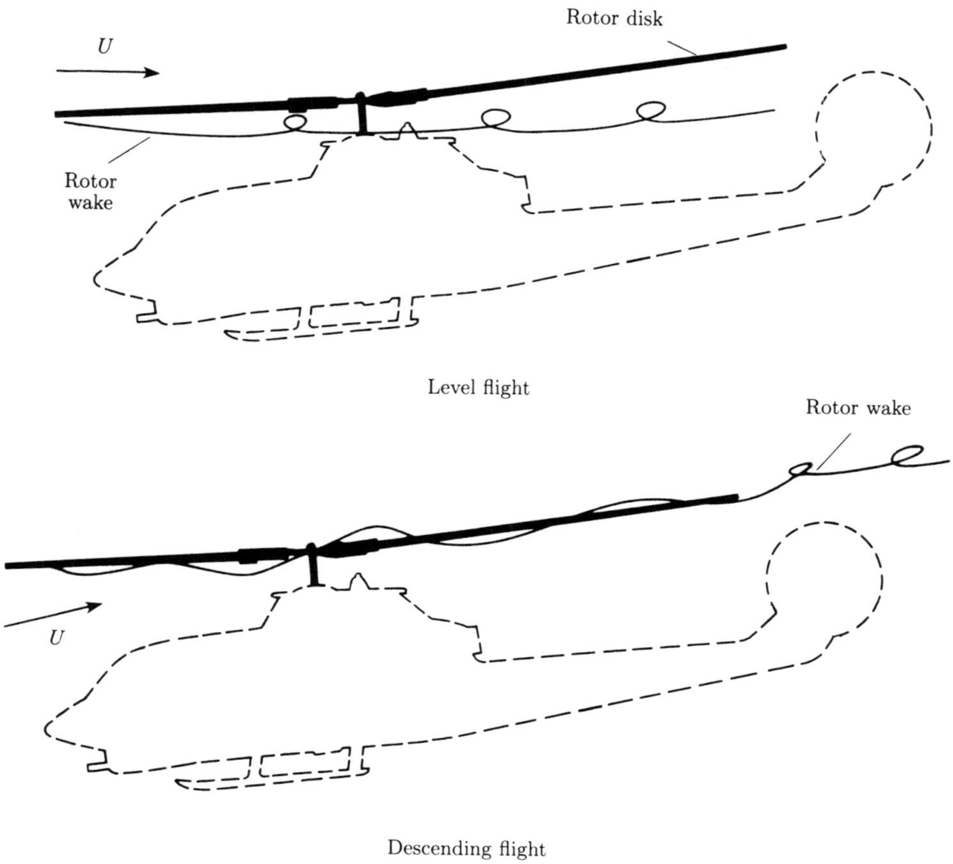

Figure 16. Flight-path effects on BVI noise. (From ref. 5.)

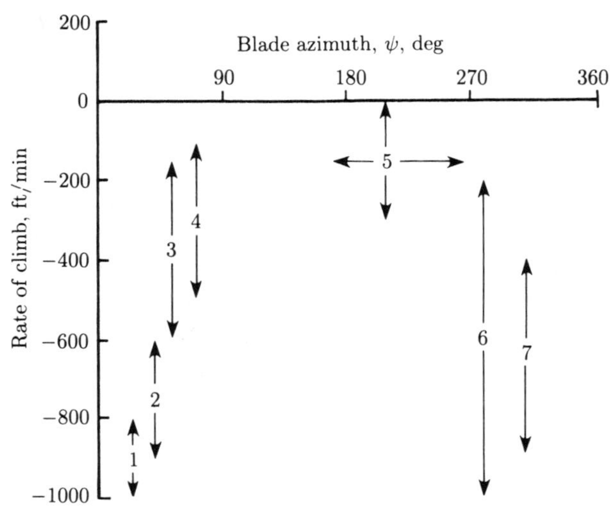

Figure 17. Regions of blade-vortex interaction encounters. (From (ref. 18.)

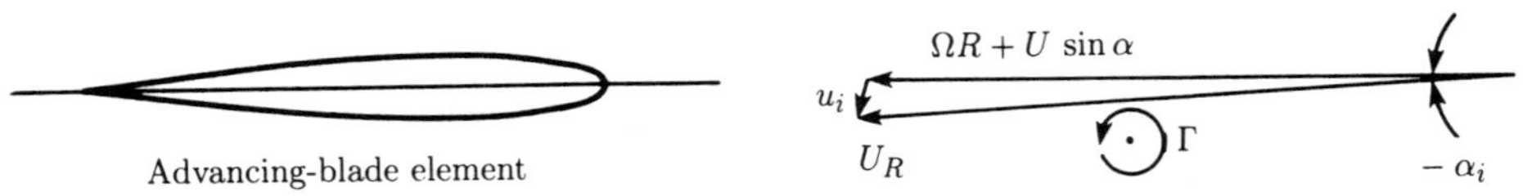

Figure 18. Advancing-blade element encountering a two-dimensional vortex. (From ref. 18.)

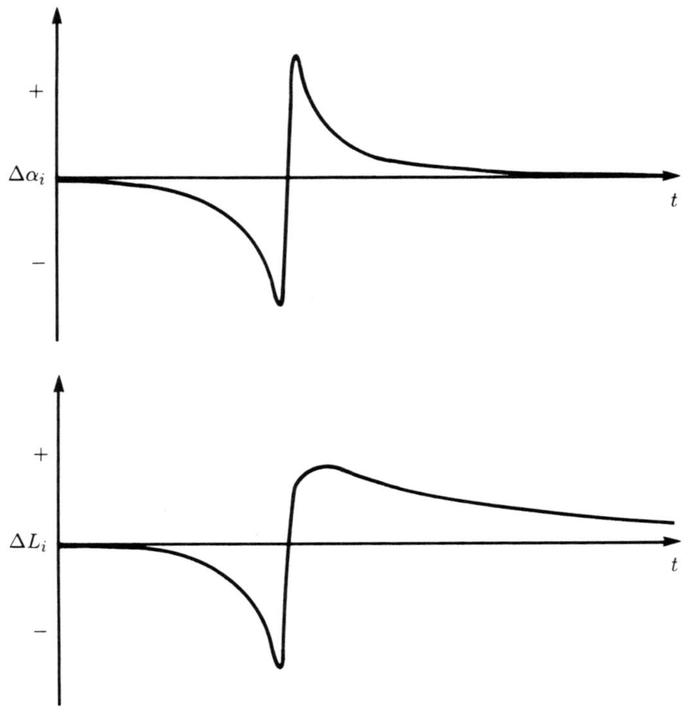

Figure 19. Changes in angle of attack α_i and section lift L_i due to two-dimensional BVI. (From ref. 18.)

A sketch of a possible angle-of-attack time history as the vortex passes near the airfoil is shown in the upper part of figure 19. The time scale shown has been stretched so that the character of the radiated noise can be illustrated. For incompressible flow, this will result in a net positive lift versus time on the rotor, which is shown on the lower part of figure 19. In these simple two-dimensional arguments, the entire blade is assumed to feel the presence of the changing angle of attack. The resulting time-varying force field is impulsive in nature. The radiated noise is given by the second term of equation (1),

$$p'(\mathbf{x}) = -\frac{1}{4\pi}\frac{\partial}{\partial x_i}\iint \left[\frac{P_{ij}n_j}{\Re|1 - M_{\Re}|}\right]_\tau dS(\boldsymbol{\eta}) \tag{6}$$

With the entire blade treated as a single radiating body (an acoustically compact body) and radiation to the far field, this expression can be rewritten as (ref. 11)

$$p'(\mathbf{x},t) \approx \frac{1}{4\pi}\frac{\partial}{\partial t}\frac{1}{c_0}\left[\frac{\Delta L \cos\zeta}{\Re|1-M_{\Re}|}\right]_{\tau} \tag{7}$$

where

$P_{ij}n_j$	$\approx \Delta L\vec{n}\cdot\vec{\Re} = \Delta L\cos\zeta$
ΔL	local sectional lift of a blade
ζ	angle between the surface normal in the direction of the force on the fluid and a line from the point of the applied force to the observer (see fig. 20)

Equation (7) plus the lift time history govern the shape of the BVI noise. Similar to the case for thickness noise, the Doppler amplification alters the magnitude of the radiation force field, but not the basic character. Thus, the shape of the radiated acoustic pressure becomes that shown in figure 21.

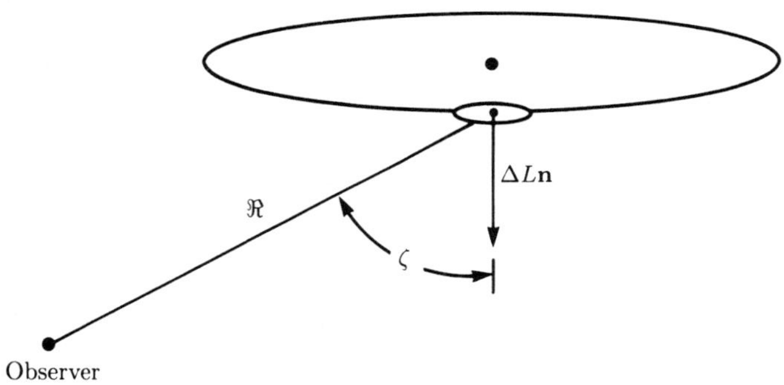

Figure 20. Geometry for far-field observer. (From ref. 5.)

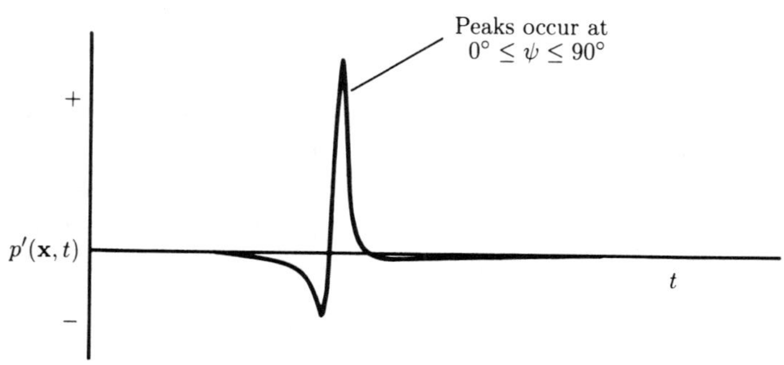

Figure 21. Acoustic pressure signature of advancing BVI.

The net effect of BVI disturbances on the advancing side of the rotor disk is acoustic radiation of a sequence of predominantly positive spikes similar to that of figure 21. These near discontinuities are of varying strengths and occur between $\psi = 0°$ and $\psi = 90°$. For the observer in the far field, these positive-pressure impulses will generally arrive before the large negative thickness noise pulse, which forms near $\psi \approx 90°$. Notice that the acoustic radiation decreases as the observer approaches the plane of the rotor disk (i.e., $\cos\zeta \to 0$).

Now consider the BVI on the retreating side of the rotor disk (interactions 5 to 7). Again, using our simplified two-dimensional qualitative model, we have the geometry in figure 22.

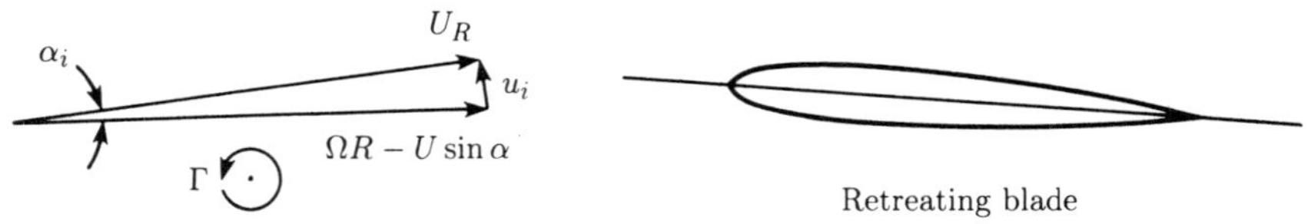

Figure 22. Geometry for BVI on retreating side of rotor disk. (From ref. 18.)

For retreating BVI, the sign of the approaching vortex is opposite to that for BVI on the advancing side. Therefore, by similar arguments, the net effect is a predominantly negative radiated acoustic signature for each BVI on the retreating side, as shown in figure 23. The time of arrival of most of the negative pressure pulses is different than that of the positive BVI pulses. For the two-bladed-rotor epicycloid pattern shown in figure 15, the far-field observer will see retreating BVI's 6 and 7 occur later in time than the thickness noise pulse. As these simple arguments demonstrate, both the sign and the timing of the acoustic pulses can often help isolate the origins of the radiated impulsive noise. Obviously, these simple qualitative arguments do not tell us many of the more interesting details. However, they do help us interpret measured impulsive noise acoustic data.

These sharp acoustic events of BVI and HSI noise are subjectively quite loud and tend to set the noise acceptance of this class of vehicles. When viewed in the frequency domain, many harmonics of periodic noise are present that can be equal to or greater than the amplitude of the fundamental.

Broadband Noise

There is another class of noise associated with rotorcraft that is more "broadband" in nature and as such is labeled "broadband noise." It can be one of the important contributors to the subjective assessment of rotor annoyance in situations where impulsive noise is notably absent. A variety of mechanisms are responsible for generating broadband noise. All the mechanisms have the common characteristic of tending to generate continuous acoustic spectra. These spectra result when the rotor blades interact with the turbulent inflow to the rotor arising because of rotor-blade wakes, blade boundary layers, or the ambient atmospheric turbulence in which the rotor operates. Figure 24 (ref. 19) lists the sources of broadband noise as blade self-noise sources and turbulence-ingestion noise sources.

Turbulence-ingestion noise is a form of broadband noise because the unsteady pressure fluctuations are randomly distributed in time and location. This noise is generated when blades interact with atmospheric turbulence and is somewhat similar to the noise measured on propellers (as discussed in another chapter). At low frequencies which are due to large turbulence eddies, stretching the eddies as they are ingested into a hovering rotor can form them into long shapes which are cut

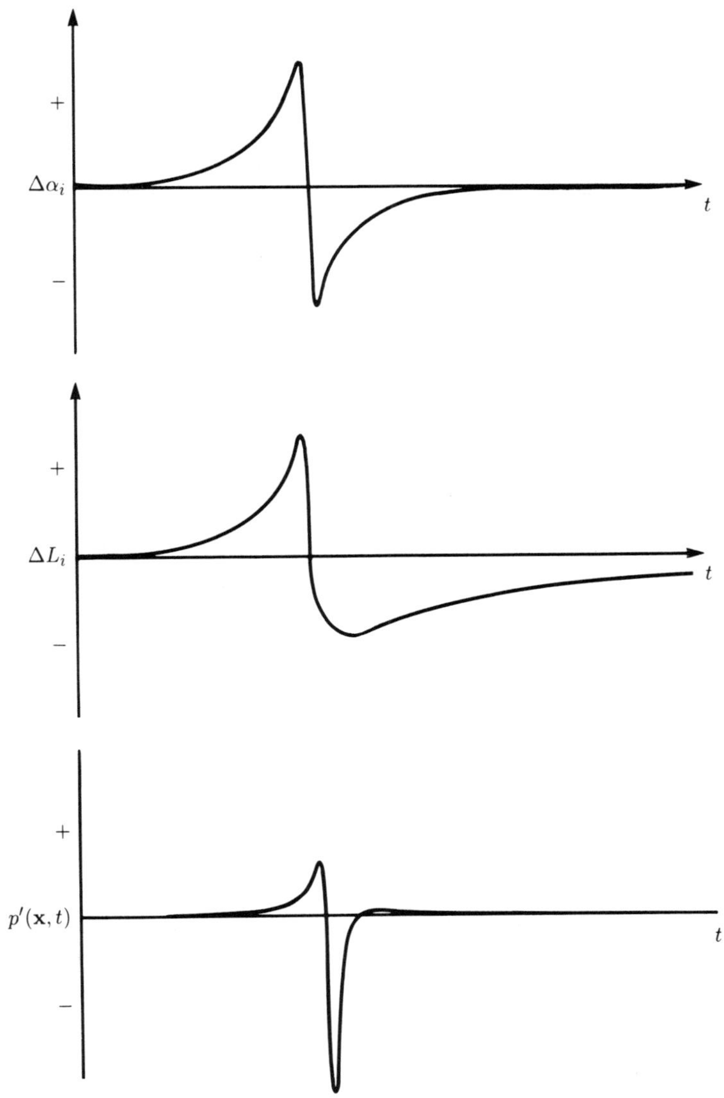

Figure 23. Angle-of-attack, lift, and acoustic pressure time histories of retreating BVI.

several times by the rotor blades, each cut creating a small-time-duration, impulsive event, as shown schematically in figure 25. These stretched eddies are cut at various random stations throughout the disk. As each stretched eddy is cut a number of different times in some nearby locations, the broadband signal displays humps at blade-passage frequencies and harmonics, and this chopping of the eddies creates a "peak-valley" shaped spectrum (ref. 20). The longer and more stretched out an eddy is, the more times it is cut at a similar location in the disk and the narrower the peak of the noise associated with it. However, except near hover, this elongation of large eddies is weak and gives a peak-valley spectrum shape only at low harmonics. At higher frequencies, the small size of the eddies does not enable each eddy to be

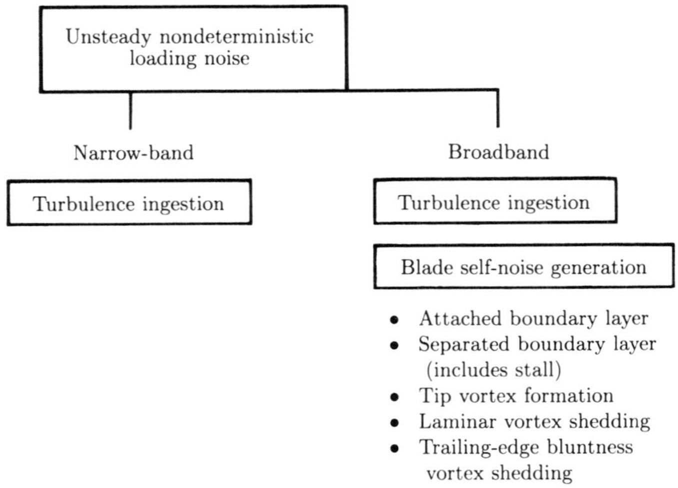

Figure 24. Categorization of helicopter rotor broadband noise. (Based on ref. 19.)

cut more than one time, and thus the broadband noise signal becomes quite smooth, as illustrated in figure 25.

The whooshing sound of rotors is governed by the higher frequency part of the turbulence-ingestion noise and blade self-noise generation. It is most noticeable on helicopters or propellers during start-up or shutdown in the near acoustic field. A shed wake system, consisting mostly of shed vorticity, induces a changing force field on the rotor that swishes through the air. This sound is thought to be important only when all other sources of noise are mitigated, or at very low tip Mach numbers atypical of normal rotorcraft flight.

Because of the number of sources present, the dominant contributors to particular portions of a rotor spectrum are a matter of controversy. Identifying noise as being discrete or broadband often can depend upon the researcher's viewpoint and how the data are processed and interpreted. This can make the identification and quantification of the noise through measurement very difficult, even if the conditions under which the noise is taken are almost ideal. As previously discussed, a typical narrow-band plot may or may not include a burst of tone-like noise which, on the average, may not be periodic but random in nature. The plot will have a tone-like character at the lower frequencies and become broadband at higher frequencies. However, small changes in rotor speed (ref. 6), rotor and wake unsteadiness, unsteadiness in microphone-to-source distances, and changing Doppler effects also cause discrete noise to appear broadband in character at higher frequencies. Thus, it is quite possible to measure what might look like a broadband noise spectrum of a basically periodic phenomenon. Many such interpretations of full-scale flight data were made in this manner in the past.

More recent research under carefully controlled conditions has clarified the problem (ref. 21). New measurement methods use narrow-band spectral analysis and supplementary diagnostics to more clearly distinguish between truly broadband noise sources and randomized periodic noise sources.

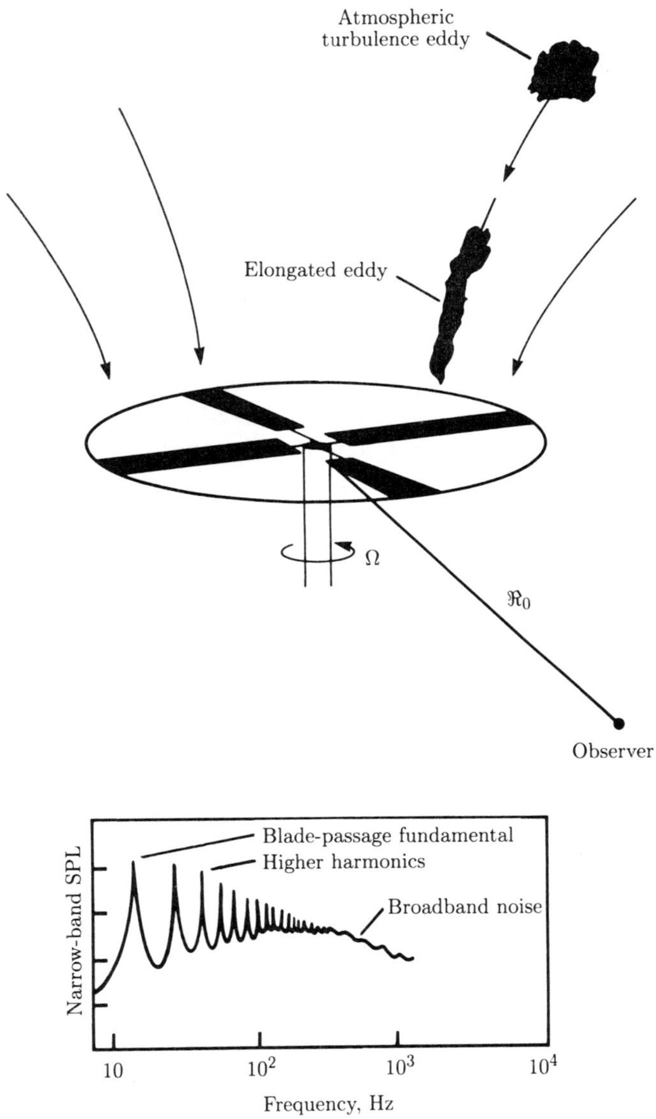

Figure 25. Origins of turbulence-ingestion noise.

Some Measured Data

Hover

One of the most difficult tasks in rotorcraft acoustics is to measure the radiated noise under carefully controlled conditions. Although it is relatively easy to measure rotorcraft noise, it is much more demanding to specify or carefully control all the parameters that can affect the radiated noise during the measurement process. For example, most sources of noise are affected by the aerodynamic state of the rotors. This in turn is controlled by the performance of the rotor in or out of ground effect, the pilot's ability to hold a steady hover, and atmospheric turbulence. In addition, the proximity of the ground, type of ground vegetation, and ambient wind and temperature effects also can distort the measured acoustic signal. Controlling all these parameters on a full-scale helicopter has proven to be almost impossible, although several very clever, near-perfect experiments have been attempted (refs. 22 and 23).

One particularly intriguing and illustrative experimental arrangement is shown in figure 26 for an OH-6A helicopter in simulated hovering flight (ref. 24). The complete helicopter was mounted on a specially developed quiet test rig that allowed the main rotor, tail rotor, and engine to be run separately or together. The recording microphone was in the acoustic far field, 7.6 main-rotor diameters d_{mr} from the rotor hub in the nearly in-plane position. The measured sound pressure level versus frequency for the main rotor alone and for the complete helicopter are shown in figures 27 and 28. As discussed previously, the low-frequency main-rotor harmonic noise decreases rapidly with increasing harmonic number. Notice, too, that there are many harmonics of the main rotor (over 50).

The hump in the curve in figure 27 above a frequency of 600 Hz is caused by a ground reflection which reinforces and destroys the harmonic decay according to the wavelength of the emission. The complete plot of OH-6A helicopter SPL versus frequency is strongly influenced by tail-rotor harmonic noise, as shown in figure 28. The higher tail-rotor rotational rate causes higher frequency tones and multiples thereof which dominate the spectrum at frequencies above 100 Hz. This particular set of data is typical of rotorcraft with a hovering tip Mach number M_T of about 0.6. At higher values of M_T, the SPL falls off less rapidly with harmonic number. It is also worthwhile to note that the data shown here were taken under ideal conditions. The rotor speed was held precisely at the desired value, the helicopter was fixed in space,

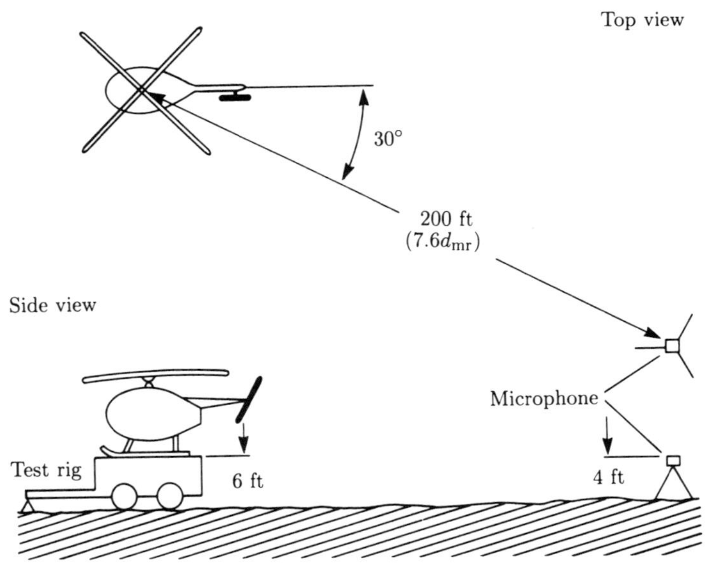

Figure 26. Microphone height and location relative to OH-6A test helicopter mounted on special test rig. (From ref. 24.)

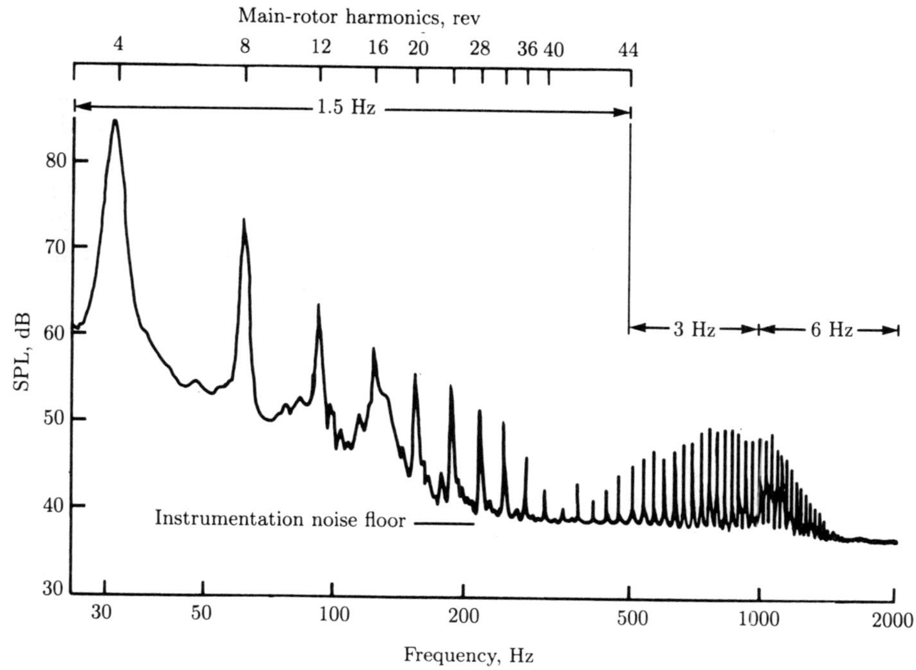

Figure 27. Narrow-band spectrum plot for OH-6A helicopter—main rotor only (4-bladed). (From ref. 24.)

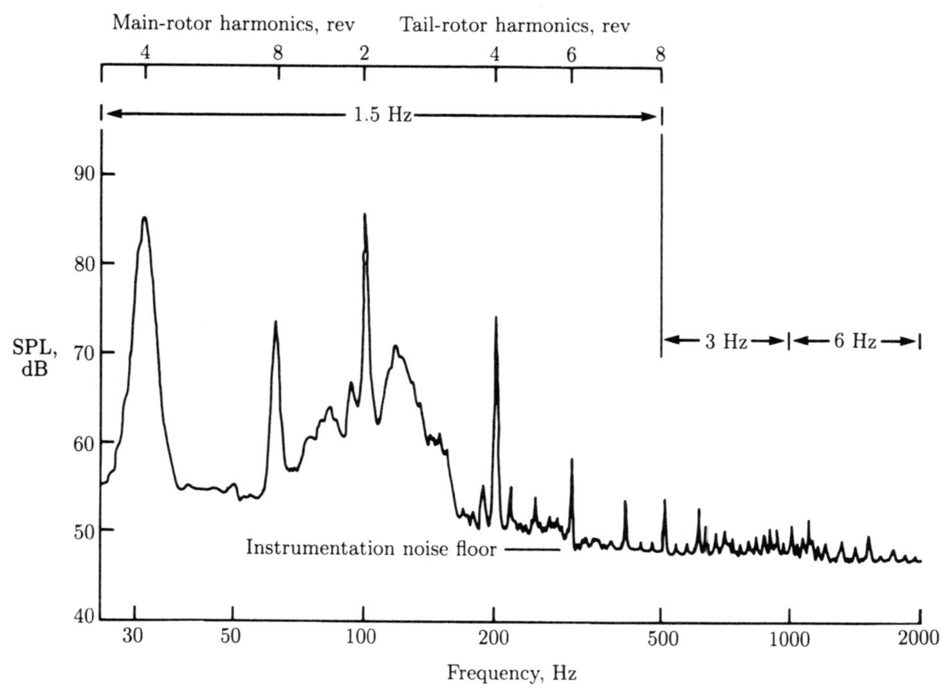

Figure 28. Narrow-band spectrum plot for complete OH-6A helicopter in simulated hover. (From ref. 24.)

and data were taken only under "no wind" conditions. If any of these parameters were allowed to vary, then the periodicity of the event might appear to change. This, in turn, would have the effect of broadening the discrete harmonic spectrum into a more broadband spectrum, especially at higher frequencies. Therefore, the same data taken on a real hovering helicopter might appear to have only very low-frequency harmonic noise with a more broadband character at higher frequencies. Nevertheless, the low-frequency harmonic character of the radiated acoustic field is typical of almost all rotorcraft. As discussed, blade thickness and steady forces cause most of this low-frequency noise to radiate to the acoustic far field.

Forward Flight

Rotorcraft impulsive noise has also been very carefully measured in some benchmark experiments (refs. 25 and 26). Data have been taken with an in-flight measurement technique whereby the measurement microphone is flown in formation with the subject helicopter, as shown in figure 29. The major advantages of gathering data in this manner are (1) no ground reflections, (2) long and steady data samples, and (3) helicopter flight conditions and directivity profiles which are easily explored. A relatively quiet aircraft was chosen as the measurement platform to keep the background noise beneath the signal level of the helicopter. Fortunately, the impulsive noise signal levels of most rotorcraft are quite large, a fact which makes this an excellent data-gathering method for this type of noise. The data shown in figures 30 to 36 were measured on the UH-1H helicopter, which is known to radiate BVI impulsive noise and HSI noise.

The helicopter flight conditions which were investigated for the UH-1H helicopter are shown in figure 30. High-speed impulsive noise was measured in high-speed forward flight, and BVI impulsive noise was measured in moderate-speed forward but descending flight. Also illustrated in figure 30 are contours of BVI noise as heard in the helicopter cabin. In early experiments, it was thought that noise which was

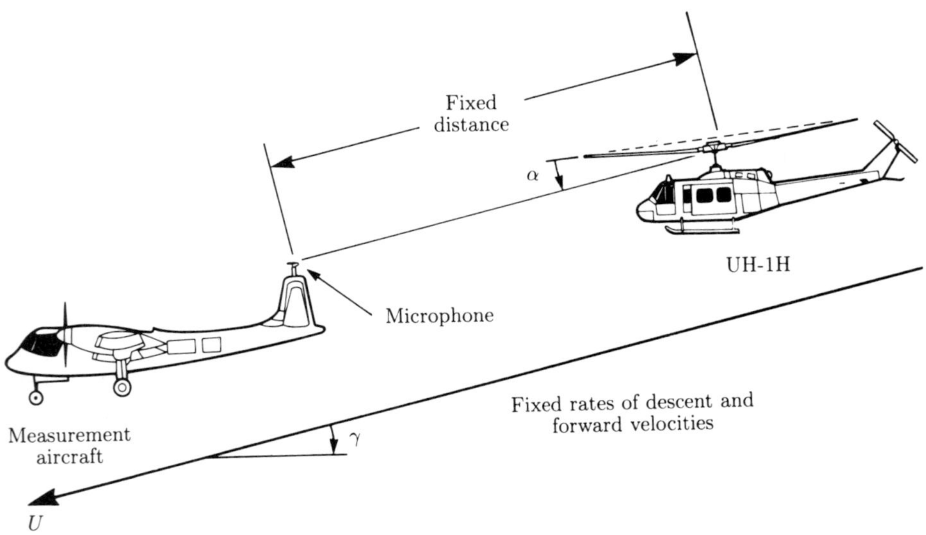

Figure 29. Technique for in-flight acoustic measurement. (From ref. 5.)

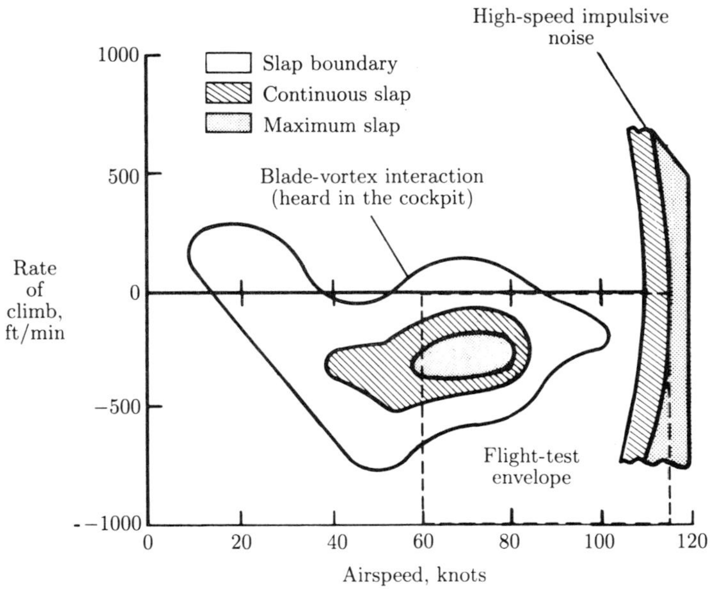

Figure 30. Impulsive-noise boundaries for UH-1H helicopter. (From ref. 5.)

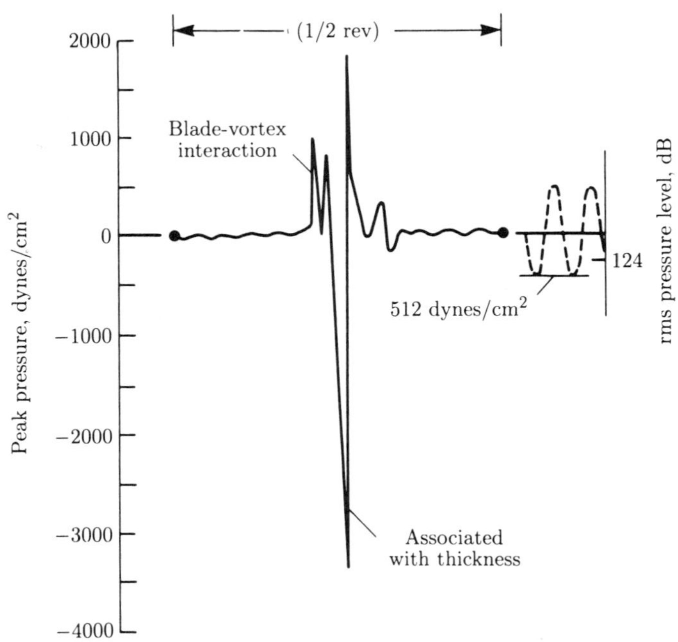

Figure 31. Composite illustration showing dominant UH-1H acoustic waveform features. (From ref. 5.)

heard in the cabin of the helicopter was a good indicator of when BVI impulsive noise was being radiated to the acoustic far field. This in-flight measurement technique confirmed that BVI noise is radiated when it is heard in the cabin. However, the

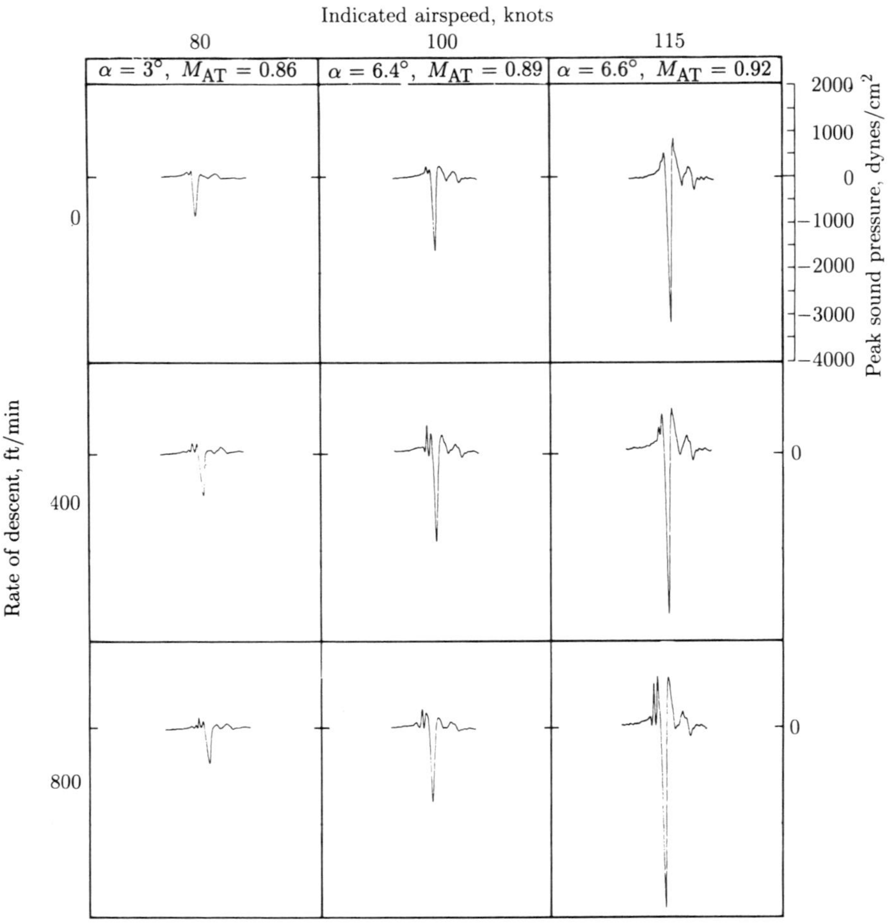

Figure 32. Averaged acoustic signature of UH-1H impulsive noise for 1/2 revolution versus forward airspeed and rate of descent. (From ref. 5.)

technique also showed that BVI noise can radiate in other directions and, because of geometry, cannot be heard in the cabin. Therefore, if a pilot were to fly so as to minimize the cabin impulsive noise, he might still be radiating BVI noise to ground observers.

It was generally observed from the measured data that the far-field acoustic waveform radiated by each blade was composed of multiple pulses. As many as three distinct pressure disturbances could be repetitively identified in the acoustic waveform. For identification of this waveform structure, an idealized composite drawing of the acoustic waveform showing this multipulse composition is presented in figure 31. This figure illustrates peak pressure amplitude of the acoustic signature versus one half revolution (one blade passage) in time, with time increasing from left to right. The peak pressure amplitude scale used here is an absolute scale measured in dynes per square centimeter. On this scale, a sinusoidal-shaped waveform with a peak pressure amplitude of 512 dynes/cm^2 would exhibit a root-mean-square (rms) SPL of 124 dB.

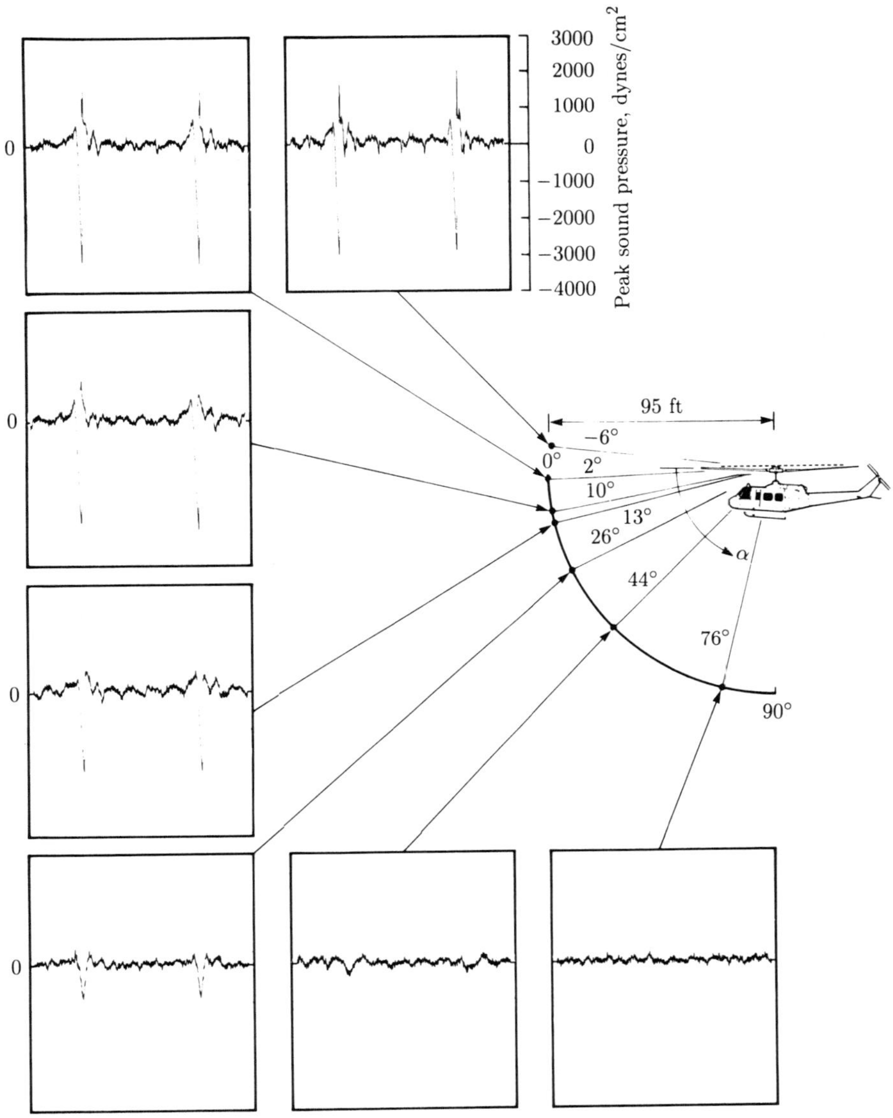

Figure 33. Longitudinal acoustic directivity for UH-1H in level flight at 115 *knots IAS.* $\beta = 0°$. *(From ref. 5.)*

The composite waveform model illustrates three predominant pressure characteristics observed in the data. They are shown in the same relative sequence and approximate pulse width that are characteristic of the measured data. Typically, the sequence begins with one or two successive positive increases in pressure ("triangular" pulse shape in fig. 31). These positive-pressure peaks are followed by a large, near-triangular negative-pressure pulse. At high advance ratios and high advancing-tip Mach numbers, the negative-pressure pulse increases in amplitude more slowly than its subsequent rapid positive pulse, and the waveform is represented more by a sawtooth, or half-triangular, pulse. Finally, an extremely narrow positive-pressure

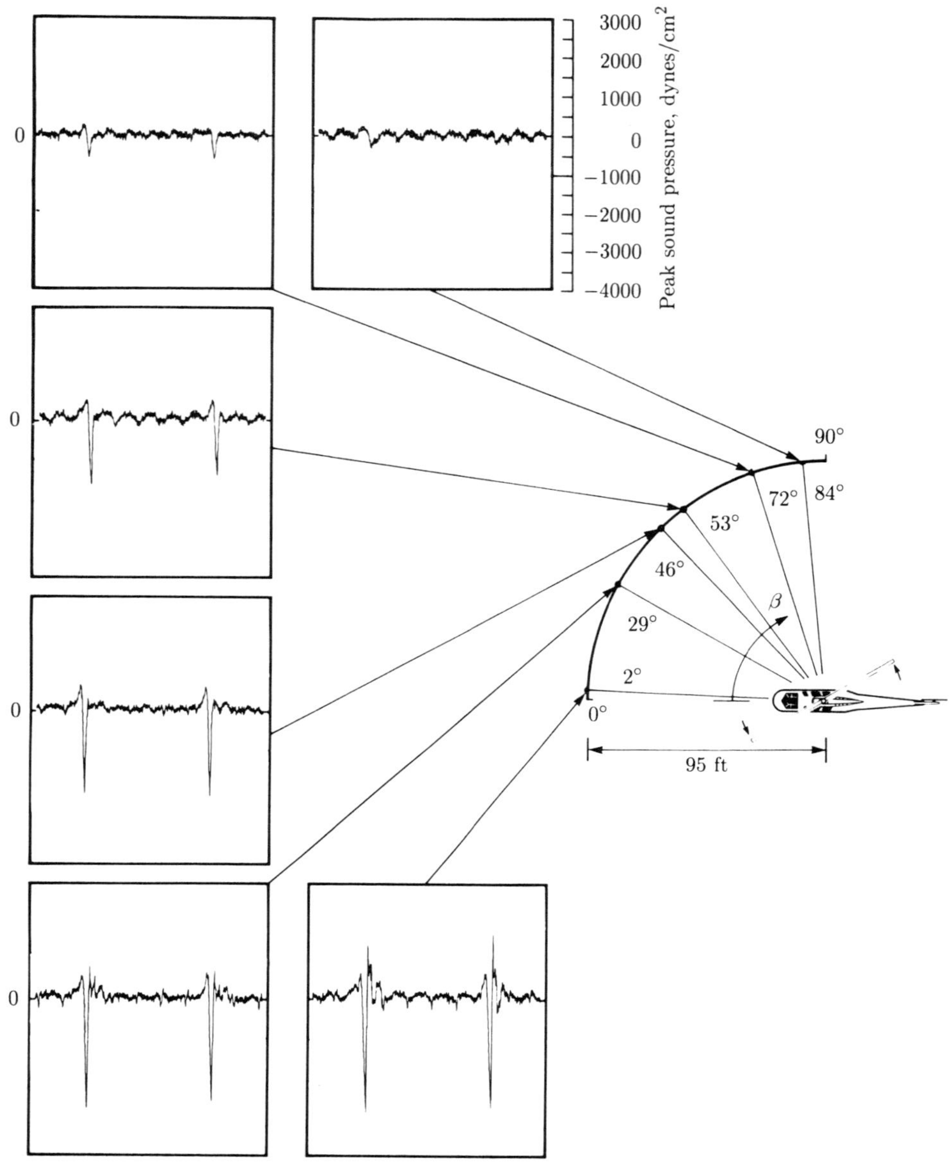

Figure 34. Lateral acoustic directivity for UH-1H in level flight at 115 *knots IAS.* $\alpha = 7°$. *(From ref. 5.)*

spike sometimes follows immediately after or as a result of the extremely rapid increase in pressure.

With the qualitative arguments presented at the beginning of this chapter, it is possible to trace the origins of the noise. As indicated in figure 31, the negative pulse is associated with thickness effects. It occurs in source coordinates at about $\psi \approx 90°$. The initial positive pulses are a direct result of blade-tip–vortex interaction on the

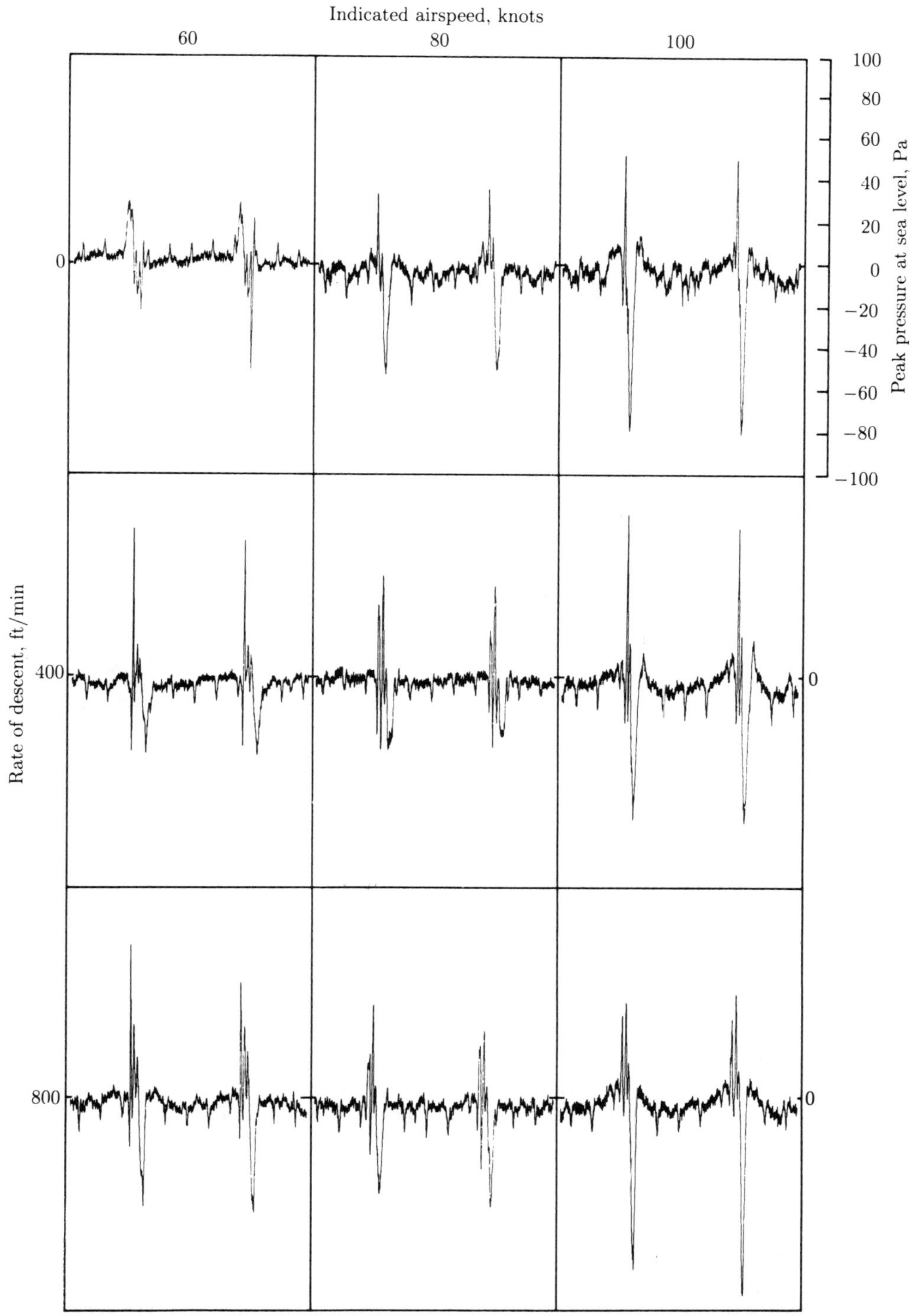

Figure 35. Unaveraged acoustic signatures of UH-1H as function of forward airspeed and rate of descent. (From ref. 5.)

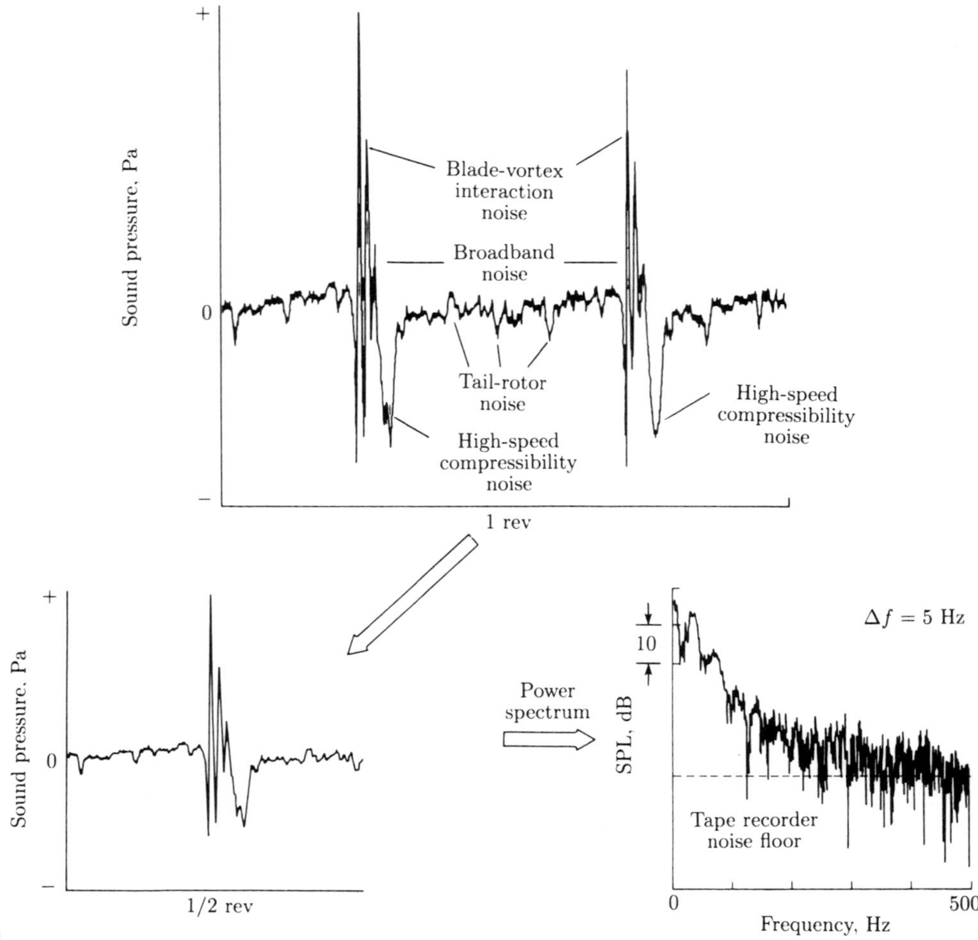

Figure 36. Time history and related power spectrum of UH-1H BVI impulsive noise. (From ref. 26.)

advancing side of the rotor disk. As we have shown, they occur before the thickness-noise impulse (at $\psi \approx 0^\circ$ to 90°). The subsequent rapid decrease in pressure is really just a manifestation of intense thickness noise. It occurs when the thickness noise (and its associated aerodynamics) is so large that local shocks on the blade radiate to the far field. In this latter case, nonlinear terms need to be added to the simple linear calculations to predict the acoustic far field.

In-Plane Noise

Figure 32 presents a performance matrix of measured in-plane acoustic data at an indicated airspeed (IAS) of 80 to 115 knots and rates of descent of 0 to 800 ft/min. To show the data trends more clearly, the acoustic waveforms for each condition were averaged 128 times. The resulting acoustic waveforms, corresponding to one blade passage, were recorded at a nominal hub-to-microphone separation distance of 95 ft, with the microphone positioned directly ahead of the helicopter and nearly within the plane of the rotor tips ($\alpha \approx 0°$). Each of the acoustic time histories has the same amplitude scale, shown with the upper right waveform in figure 32.

The peak amplitude of the large negative-pressure pulse is strongly dependent upon the forward speed of the helicopter. (The advancing-tip Mach number M_{AT} is the important governing nondimensional parameter.) Although the width of the negative pulse appears to decrease slightly with increasing speed, no consistent trends in amplitude or pulse width could be deduced with changes in descent rate. It is interesting to note that under level-flight conditions at all airspeeds, no impulsive noise was heard in the cabin, an indication that, for all flight conditions tested, the pilot was unaware that the helicopter was radiating that part of the impulsive noise waveform associated with the negative-pressure peak.

At the high forward speed of 115 knots, the large negative-pressure peak, when measured nearly in-plane, is followed by a positive-pressure pulse which varies from blade to blade. This extremely rapid rise in pressure documented herein was so intense that it was heard directly in the cockpit of the measuring aircraft over and above the aircraft's own internal noise levels. However, no apparent blade slap was heard in the cabin of the helicopter at any IAS above 100 knots, regardless of rate of descent. To the pilot of the helicopter, a moderate increase in vibration level was the only noticeable effect, even though the UH-1H was radiating tremendous amounts of acoustic energy. Blade slap was heard in the cabin under partial-power descents at forward speeds below 100 knots. Blade slap appeared to be most intense within the helicopter cabin at about 80 knots IAS at a rate of descent of 400 ft/min. The occurrence of this cabin noise correlates with the positive-pressure pulses which precede the large negative-pressure pulse on the acoustic waveforms. As discussed previously, these positive-pressure pulses are sensitive to rates of descent and resulting rotor-wake geometry, thus confirming that these pulses are a direct result of blade-tip–vortex interaction.

Directivity

Directivity profiles of the UH-1H impulsive noise at an IAS of 115 knots and a rate of descent of 0 ft/min are presented for a sweep of microphone positions in figures 33 and 34. The longitudinal angle α was measured from a line drawn between the rotor hub and the microphone to the rotor-tip-path plane, and the lateral angle β was measured from the line between the hub and microphone to the forward-velocity vector. In this high-speed level flight condition, the measured acoustic pulse consists of a large-amplitude negative pressure followed by a rapidly increasing positive-pressure pulse. The negative-pressure peak is predominantly caused by transonic thickness effects. In the longitudinal plane (fig. 33) the pulse reaches its maximum level near the in-plane positions of the rotor disk but decreases rapidly to roughly half this amplitude at $\alpha = 13°$ and continues to decrease uniformly with increasing α until it is hardly discernible above background noise levels at $\alpha = 44°$. In the lateral plane, the negative-pressure pulse decays less rapidly in plane than out of plane as β is increased. The pulse is approximately half amplitude at $\beta = 53°$ and is still discernible to the side of the helicopter ($\beta = 72°$). Although the helicopter pilot cannot hear any blade slap noise associated with the negative-pressure pulse, an observer who is generally in the path of an approaching helicopter, in regions that are effectively in the helicopter's tip-path plane, will hear impulsive noise caused by transonic thickness effects.

The extremely sharp positive-pressure pulse which follows the large negative-pressure pulse exists in a narrow angular region near and above the rotor-tip-path plane directly ahead of the helicopter. This sharp, near discontinuous pulse is attributable to weak radiating shock waves emanating from each rotor blade and is responsible for very intense radiated noise annoyance levels.

Blade-Vortex Interaction (BVI) Noise

As shown in figure 32, blade-vortex interaction (BVI) impulsive noise, sometimes called blade slap, is a strong function of rate of descent. Since this noise is predominantly due to rapid variations in lift, it will be increased relative to thickness noise at microphone locations which are not in-plane. Directivity profiles of BVI noise show this predominantly dipole (force) noise to be a maximum ahead of the helicopter at 30° to 45° under the rotor-tip-path plane. Laterally, BVI amplitudes and pulse shapes are a strong function of advance ratio, depending critically on the alignment geometry of the interaction between the blade and previously shed tip vortices (refs. 27 and 28). Positive-pressure pulses which originate on the advancing blade radiate forward, while negative-pressure pulses which originate on the retreating blade tend to radiate rearward. Figure 35 presents unaveraged signatures for a matrix of flight conditions for a microphone located ahead of the helicopter and approximately 30° beneath the rotor-tip-path plane. The wide negative-pressure pulse is indicative of high-speed impulsive (HSI) noise, and the predominantly positive-pressure pulses depict impulsive noise resulting from blade-tip–vortex interactions originating on the advancing side of the rotor disk. As shown in the sequence of large positive-pressure pulses, BVI noise plays a larger role than HSI noise in the UH-1H helicopter at this 30° down position. As discussed previously, rate of descent and forward airspeed have a large effect on the character of the BVI pulse that is generated.

This dominance of BVI noise can be seen most easily by isolating and expanding a typical BVI pulse at the $\alpha = 30°$ microphone position. The data were gathered using the in-flight measurement technique with a "quiet" YO-3A aircraft as the measurement platform. As shown in figure 36, BVI noise, high-speed compressibility noise, and tail-rotor noise are all identifiable for one characteristic period of data.

It can be shown that the distribution of energy in each pulse into harmonic levels is primarily determined by the character of each repeated pulse. The power spectral density of a typical pulse is the envelope of the power spectrum of that same pulse repeated at the characteristic periodic interval. With this reasoning, the first half-period of the pulse is plotted in the lower left of figure 36. The corresponding power spectrum (5-Hz bandwidth) is shown in the lower right of the same figure. Sound power from BVI, high-speed compressibility main- and tail-rotor noise, and broadband noise for half a rotor period are all included. The noise floor of the high-frequency data (> 2500 Hz) is set by the signal-to-noise ratio of the tape recorder.

Because BVI noise is only dominant over a narrow portion of the time history shown in figure 36, it is possible to improve the signal-to-noise ratio of the BVI phenomenon and look at the more general characteristics of a typical BVI pulse shape by "time windowing" the measured pulse (ref. 26). The data were time windowed in figure 37(a) by setting the measured pulse equal to zero everywhere except during that part of the half-period dominated by advancing-blade impulsive noise. In essence, much of the power contributed from broadband and tail-rotor noise sources has been eliminated, thus improving the signal-to-noise level of the impulsive noise. The lobed character of the resulting frequency spectrum is typical of a multi-impulsive event without discontinuous first derivatives. (See also ref. 29.) It is also noteworthy that the largest sound pressure levels of this impulsive event are in the 200- to 750-Hz range.

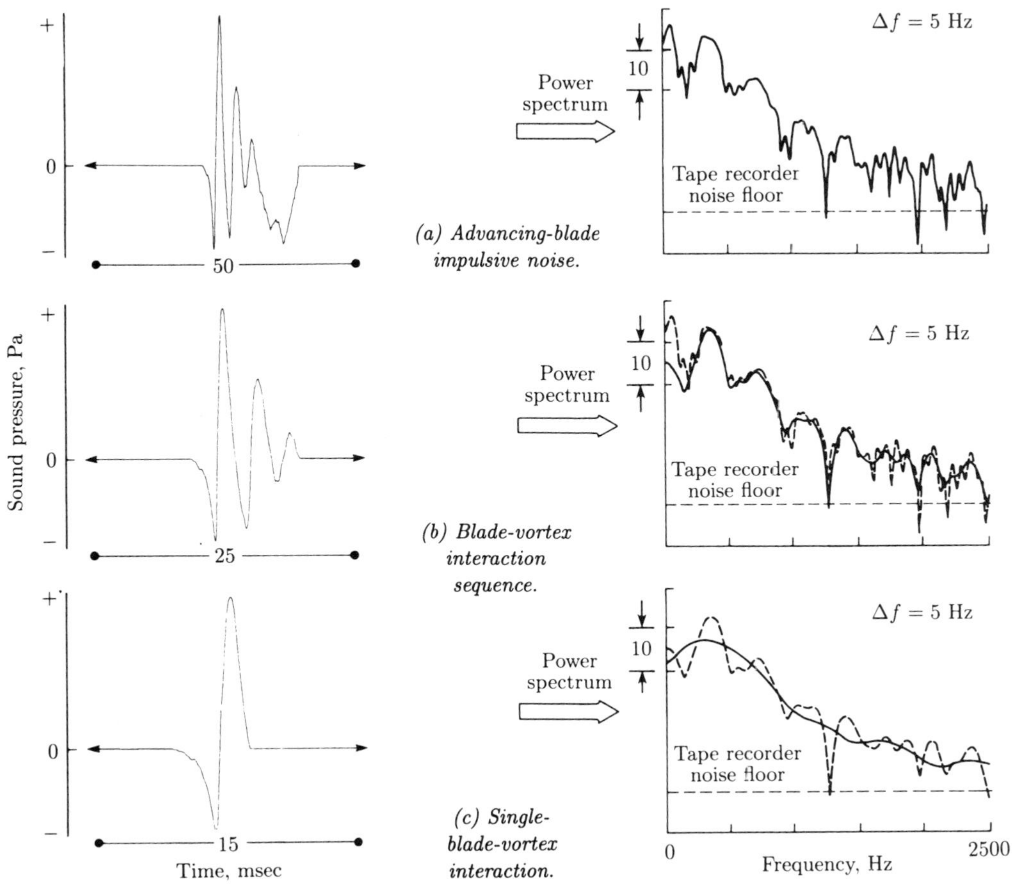

Figure 37. Time windowing of impulsive noise. (From ref. 26.)

The frequency content of BVI with the large negative thickness pulse removed is shown in figure 37(b). It is apparent that the only difference between this spectrum and the previous one (fig. 36) is in the very low-frequency range of 0 to 100 Hz. This difference represents the energy content of the high-speed compressibility noise.

Finally, when all but the largest BVI is nulled, a definite change in power spectrum results (fig. 37(c)). The many-lobed character of the spectrum has disappeared, replaced by a wide, smooth-lobed curve with noticeably less energy in the 200- to 750-Hz range. This result shows that much of the BVI energy in the 200- to 750-Hz range is a result of the multipulse character of the impulse.

All the impulsive noise data presented here were taken on the UH-1H two-bladed helicopter. Its relatively high hovering rotor-tip Mach number ($M_T = 0.73$) is responsible for the clean, high-level impulsive signals shown. The data, however, are quite typical of the more modern helicopter of today, although the level of the pulses and the regions where they occur can be quite different. The increasing importance of high-speed flight has forced the hovering tip Mach number lower to avoid high-speed compressibility problems on the advancing blade. Four or more rotor blades are common on modern configurations to reduce the operational loads. A typical time history of a 1980's four-bladed helicopter that is radiating impulsive noise is shown in figure 38 for a near in-plane microphone. High-speed compressibility (thickness) and blade-vortex interaction (BVI) noise are clearly identifiable. It is also apparent that the pulse patterns exhibit more variability from pulse to pulse, a characteristic of the more modern rotorcraft.

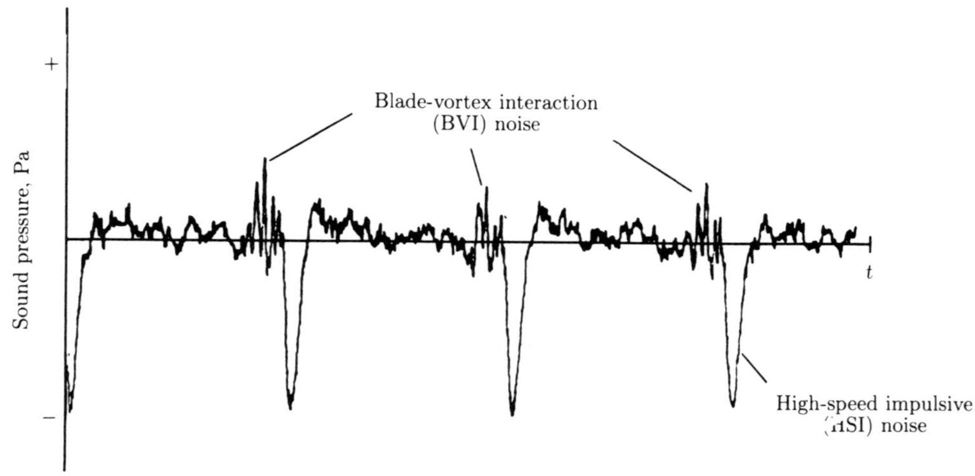

Figure 38. Acoustic signature of modern four-bladed helicopter. (Based on ref. 5.)

Broadband Noise

A typical spectrum for broadband noise is more difficult to generalize than for periodic noise. Besides the difficulty of truly separating out the periodic noise from the broadband noise, there are a large number of noise mechanisms on rotors which can be important in different parts of the acoustic frequency spectrum. These aeroacoustic source mechanisms depend upon rotor operating parameters, rotor size, and aerodynamic inflow to the rotor. They are due to various aeroacoustic effects, including boundary layers, separated flow, inflow turbulence, and nonuniform inflow. On full-scale rotors, these broadband noise sources usually become important when other impulsive periodic noise sources are absent, and then only in the mid- and high-frequency ranges.

A typical spectrum for a 2/5-scale model BO-105 rotor tested in the Duits-Nederlandse Windtunnel (DNW) aeroacoustic wind tunnel is shown in figure 39 (ref. 21) for a microphone located on the axis of the rotor in the acoustic far field. Although the data are not taken on a full-scale helicopter, they are of high quality and clearly show broadband noise. For this microphone position, noise due to steady loading is theoretically absent from the spectrum. The importance of the rotor operating state is clearly shown in the overall broadband noise levels at mid frequencies. Operating the rotor under flight conditions of mild descent increases noise levels, while pushing the rotor-wake system away from the helicopter in climbing flight does the opposite. This mid-frequency broadband noise has recently been called blade-wake interaction noise and is thought to be due to the turbulence associated with the rotor-wake system. It may also be due, in part, to the randomness of the discrete rotor-wake system itself and therefore be a type of blade-vortex interaction noise.

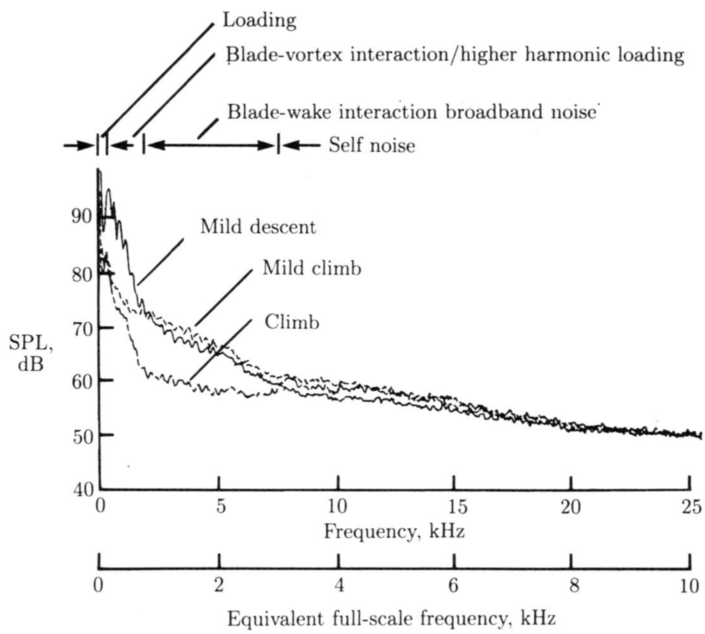

Figure 39. Typical overhead noise spectra of 2/5-scale BO-105 model rotor. (Based on ref. 21.)

At higher frequencies, above 4 kHz full scale, the broadband levels are much less dependent upon rotor inflow. Levels are also reduced as much as 40 dB from the peak low-frequency levels. However, their subjective annoyance is greater because of the sensitivity of the human ear to tones near 3 kHz. Fortunately, at larger measurement distances, these high-frequency tones are dissipated quite rapidly, leaving the predominantly low- and mid-frequency sources to control far-field annoyance levels.

The situation changes somewhat for the smaller rotors necessary for antitorque control on single-rotor helicopters. Tail rotors have small chord-based Reynolds numbers and can, under the right laminar flow conditions, induce a Karman-vortex-like high-frequency shedding into the tail-rotor wake. This phenomenon also induces unsteady periodic forces on each airfoil element, causing each element to radiate high-frequency periodic noise. Because the frequency of the shedding phenomenon is governed by the local Strouhal number of the flow, the resulting tail-rotor noise consists of a distribution of tone-like noises. This normally very high-frequency noise has been successfully mitgated by tripping the blade-surface boundary layers from laminar to turbulent flow (ref. 30.)

Scaling Rotor Noise

As mentioned previously, precise measurements of full-scale rotor noise sources are difficult to obtain from flight tests. An alternative method of gathering acoustic data uses the wind tunnel to simulate flight. In the wind tunnel, the rotor system can be flown quite precisely under carefully controlled conditions with the microphones rigidly fixed at known distances from the rotor. However, aside from the very large wind tunnel at the NASA Ames National Full-Scale Aerodynamics Complex (NFAC), there are very few wind tunnels where helicopter rotors can be tested at full scale. In addition, it is usually necessary to be in the acoustic far field of the source of interest for meaningful acoustic measurements. For low-frequency harmonic noise, this requirement leads to hemispherical microphone measurements at distances of 3 to 6 rotor radii from the hub of the rotor—typically 75 to 150 ft for a 50-ft-diameter rotor—a feat difficult to achieve at all measurement locations, even in the NFAC. It is also necessary to ensure that acoustic reflections from nearby surfaces, as well as standing acoustic waves in the tunnel test section, are minimized so that the source acoustic signal is not distorted. The perfect measurement space is said to be "anechoic" (without echoes), although in reality this anechoic condition is seldom achieved over the entire frequency spectrum. Anechoic conditions are especially difficult to achieve for full-scale rotor systems that rotate slowly and radiate much of their acoustic energy as low-frequency harmonic noise.

Wind tunnel testing of scale model rotors tends to mitigate the size and measurement quality problems of full-scale rotors. Because the rotor diameter is smaller, it is much easier to place the microphones in the acoustic far field. The smaller diameter rotor must turn at a faster rate to duplicate full-scale aerodynamic events. This raises the frequency content of the harmonic noise levels and makes a near-anechoic space easier to achieve. For these reasons, much of the experimental noise research of today uses scale model rotors which are tested in acoustically treated wind tunnels. However, *scale model testing is only valid if the acoustic phenomena of interest are in fact duplicated at model scale.* Because most of the external noise generated by rotorcraft arises from aerodynamic source mechanisms, this implies that the local aerodynamics of the model and the full-scale rotor systems must be the same. This also implies that the structural dynamics of the rotor blades may also play a role in the acoustic radiation of rotorcraft by causing changes in local blade aerodynamics.

Scaling Relationships

The conditions under which rotorcraft noise can be scaled are derived by following the standard procedures of dimensional analysis (refs. 27 and 31). The scaling objective is to rewrite the governing integral equation (eq. (1)) in nondimensional form. To this end, the nondimensional parameters are defined below.

Nondimensional time:

$$\bar{t} = \frac{t}{2\pi/\Omega}, \text{ observer time, where } \Omega \text{ is the rotational rate of the rotor}$$

$$\bar{\tau} = \frac{\tau}{2\pi/\Omega}, \text{ source time or retarded time}$$

Nondimensional geometry:

$$\bar{r} = \frac{r}{R}, \ \bar{\Re} = \frac{\Re}{R}, \ d\bar{S} = \frac{dS}{R^2}, \ d\bar{V} = \frac{dV}{R^3}$$

Mach number:

$M \equiv \dfrac{U}{c_0}$, Mach number of the flow over the blade in a blade-fixed (rotating) coordinate system

$M_T \equiv \dfrac{\Omega R}{c_0}$, rotational (hovering) tip Mach number of the blade in a ground-based inertial coordinate system

$$\frac{1}{|1 - M_{\Re}|} \equiv \text{Doppler factor}$$

Pressure coefficient:

$c_p'(\bar{\mathbf{x}}, \bar{t}) \equiv \dfrac{p'(\bar{\mathbf{x}}, \bar{t})}{\rho_0 c_0^2}$, acoustic pressure coefficient

$C_{p_{ij}} \equiv \dfrac{P_{ij}}{\rho_0 U^2}$, pressure coefficient

With these definitions, equation (1) becomes

$$\frac{p'(\bar{\mathbf{x}}, \bar{t})}{\rho c_0^2} \equiv c_p'(\bar{\mathbf{x}}, \bar{t}) = \frac{1}{4\pi} \left\{ \frac{\partial^2}{\partial \bar{x}_i \, \partial \bar{x}_j} \iiint \left[\frac{C_{Q_{ij}}}{\bar{\Re}|1 - M_{\Re}|} \right]_{\bar{\tau}} d\bar{V}(\bar{\boldsymbol{\eta}}) - \frac{\partial}{\partial \bar{x}_i} \iint \left[\frac{C_{p_{ij}} n_j M^2}{\bar{\Re}|1 - M_{\Re}|} \right]_{\bar{\tau}} d\bar{S}(\bar{\boldsymbol{\eta}}) + \frac{M_T}{2\pi} \frac{\partial}{\partial \bar{t}} \iint \left[\frac{M\lambda}{\bar{\Re}|1 - M_{\Re}|} \right]_{\bar{\tau}} d\bar{S}(\bar{\boldsymbol{\eta}}) \right\} \tag{8}$$

where

$$C_{Q_{ij}} = \frac{\rho}{\rho_0} m_i m_j + C_{p_{ij}} M^2 - \frac{\rho}{\rho_0} \delta_{ij}$$

$$m_i = \frac{u_i}{c_0}$$

$$m_j = \frac{u_j}{c_0}$$

$$\bar{t} = \bar{\tau} + \frac{\overline{\Re}}{2\pi} M_T$$

and

$$v_n = U\lambda$$

where λ is the local surface slope of the rotor blade.

Equation (8) defines a nondimensional acoustic pressure coefficient at a measurement point in terms of nondimensional parameters. Given unique values of all the nondimensional parameters on the right-hand side of equation (8), a unique value of $c_p'(\mathbf{x}, t)$ is ensured. However, it should be noted that other governing nondimensional parameters are implicitly defined in this process.

This equation may be used to develop scaling procedures and rules for rotor testing. Consider two different-sized but geometrically similar rotors of radius R, one full scale and the second $1/\gamma$ scale. Let the scale factor

$$\gamma = \frac{R}{R_m}$$

where the subscript m denotes model scale. The process of geometric scaling implies that all lengths are scaled by γ:

$$r = \gamma r_m$$

In practical terms, this implies that all model dimensions are γ times smaller than full scale and measurement microphones should be positioned γ times closer to the hub center than full-scale geometric distances.

An important nondimensional parameter for acoustic scaling is rotational tip Mach number M_T:

$$M_T = \frac{\Omega R}{c_0} = \frac{\Omega_m R_m}{c_{0_m}}$$

To hold rotational tip Mach number the same for model and full scale, the rotor-shaft rotational rate must be adjusted so that

$$\Omega_m = \frac{R}{R_m} \frac{c_{0_m}}{c_0} \Omega = \gamma \frac{c_{0_m}}{c_0} \Omega$$

A geometrical reduction to model scale by the scale factor γ for the same speed of sound must be offset by an increase of rotor-shaft rotational rate by the same factor.

Because nondimensional times must also be scaled,

$$t_m = \frac{1}{\gamma}\frac{c_0}{c_{0_m}}t \qquad \text{and} \qquad \tau_m = \frac{1}{\gamma}\frac{c_0}{c_{0_m}}\tau$$

Model-scale time decreases in relation to its full-scale counterpart.

Equation (8) also requires that the Mach number M of the aerodynamic flow field be scaled. Considering the tip of the rotor and neglecting the spanwise flow along the blade,

$$M = \frac{U}{c_0} = \frac{\Omega R + U \sin\psi}{c_0} = M_T(1 + \mu \sin\psi)$$

(A similar argument could be made at any blade radial station.) This equation implies that the advance ratio μ must be scaled, that is,

$$\mu \equiv \frac{U}{\Omega R} = \frac{U_m}{\Omega_m R_m} = \mu_m$$

Equation (8) also requires that $C_{p_{ij}}$ and $C_{Q_{ij}}$ be scaled for both model and full scale. This implies similarity in the aerodynamic flow field and scaling of rotor thrust along the blade at each azimuthal angle. This requirement is approximated by maintaining similar inflow through the rotor disk by means of similar tip-path-plane angles and rotor thrust coefficients.

The preceding formulas state the necessary conditions for rotor scaling. They do not, however, constitute sufficient conditions for all rotor acoustic scaling. This must be done on a source-by-source basis. The validity of the scaling process has been demonstrated for two specific types of rotorcraft noise: HSI noise and BVI impulsive noise (refs. 27 and 31 to 34).

High-Speed Impulsive Noise

The fact that high-speed impulsive (HSI) noise is predominantly a noncompact (sources and sinks do not completely cancel for an in-plane observer) high Mach number (compressible) event would suggest that the noise generation process is strongly controlled by Mach number. It also suggests that if the Mach numbers of the model- and full-scale rotors were matched, small models could be made to duplicate the full-scale acoustic phenomena. This fact was demonstrated in two separate wind tunnel and in-flight experiments on both the UH-1H and the AH-1G helicopter (refs. 31 and 32). The data were gathered in nearly anechoic wind tunnels using 1/7-scale models at similar nondimensional distances from the noise source. Full-scale data were obtained using the in-flight method previously described. As illustrated in figure 40, comparisons of model data with full-scale data are quite straightforward. There are no Doppler corrections and data records up to 1 minute in duration are possible at steady-state flight conditions.

From nondimensional considerations, the acoustic pressure coefficient for HSI noise is uniquely determined if the rotor, microphone geometry, time, rotor rotational rate, and local Mach number are scaled. However, most acoustic data are not

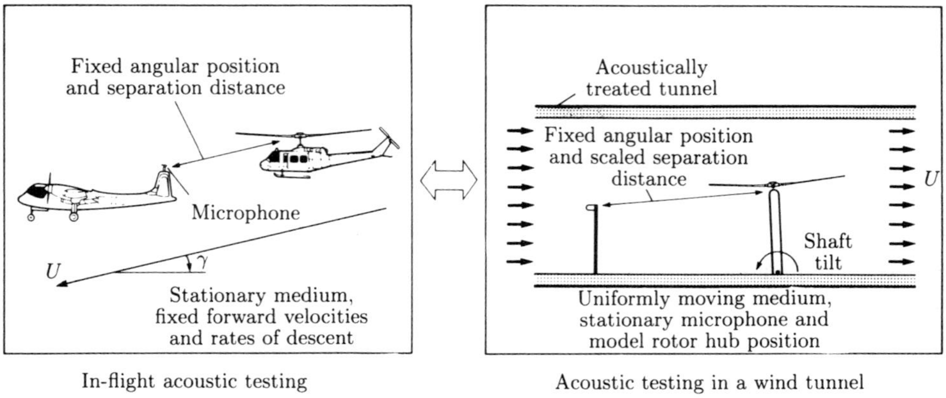

Figure 40. Equivalence between model- and full-scale acoustic testing. (From ref. 31.)

compared on a nondimensional basis. Instead, all pressures are normally referenced to sea-level standard conditions where comparisons of pressure time histories are made. For full-scale data taken at altitude, the reference pressure becomes

$$\frac{p'_{\mathrm{SL}}\left(\overline{\mathbf{x}},\bar{t}\right)}{\rho_{0_{\mathrm{SL}}}c^2_{0_{\mathrm{SL}}}}=\frac{p'\left(\overline{\mathbf{x}},\bar{t}\right)}{\rho_0 c_0^2}$$

$$p'_{\mathrm{SL}}\left(\overline{\mathbf{x}},\bar{t}\right)=\frac{\rho_{0_{\mathrm{SL}}}c^2_{0_{\mathrm{SL}}}}{\rho_0 c_0^2}p'\left(\overline{\mathbf{x}},\bar{t}\right)=\frac{p_{0_{\mathrm{SL}}}}{p_0}p'\left(\overline{\mathbf{x}},\bar{t}\right)$$

For model-scale data, the reference pressure becomes

$$p'_{m_{\mathrm{SL}}}\left(\overline{\mathbf{x}},\bar{t}\right)=\frac{(p_{0_m})_{\mathrm{SL}}}{(p_{0_m})}p'_m(\overline{\mathbf{x}},\bar{t})$$

Figure 41 presents model- and full-scale data taken under similar conditions. (See following table for conditions.) The model-scale acoustic data were taken in

Signature	U_T, knots	Rate of descent, ft/min	μ	C_T	M_{AT}
1	71	0	0.163	0.0054	0.772
2	99	0	.244		.814
3	120	0	.270		.844
4	146	400	.330		.885
5	153	1000	.345		.896
6	72	0	.169		.769
7	96	0	.222		.807
8	118	0	.276		.842
9	143	400	.330	↓	.878
10	150	1000	.348	.0053	.896

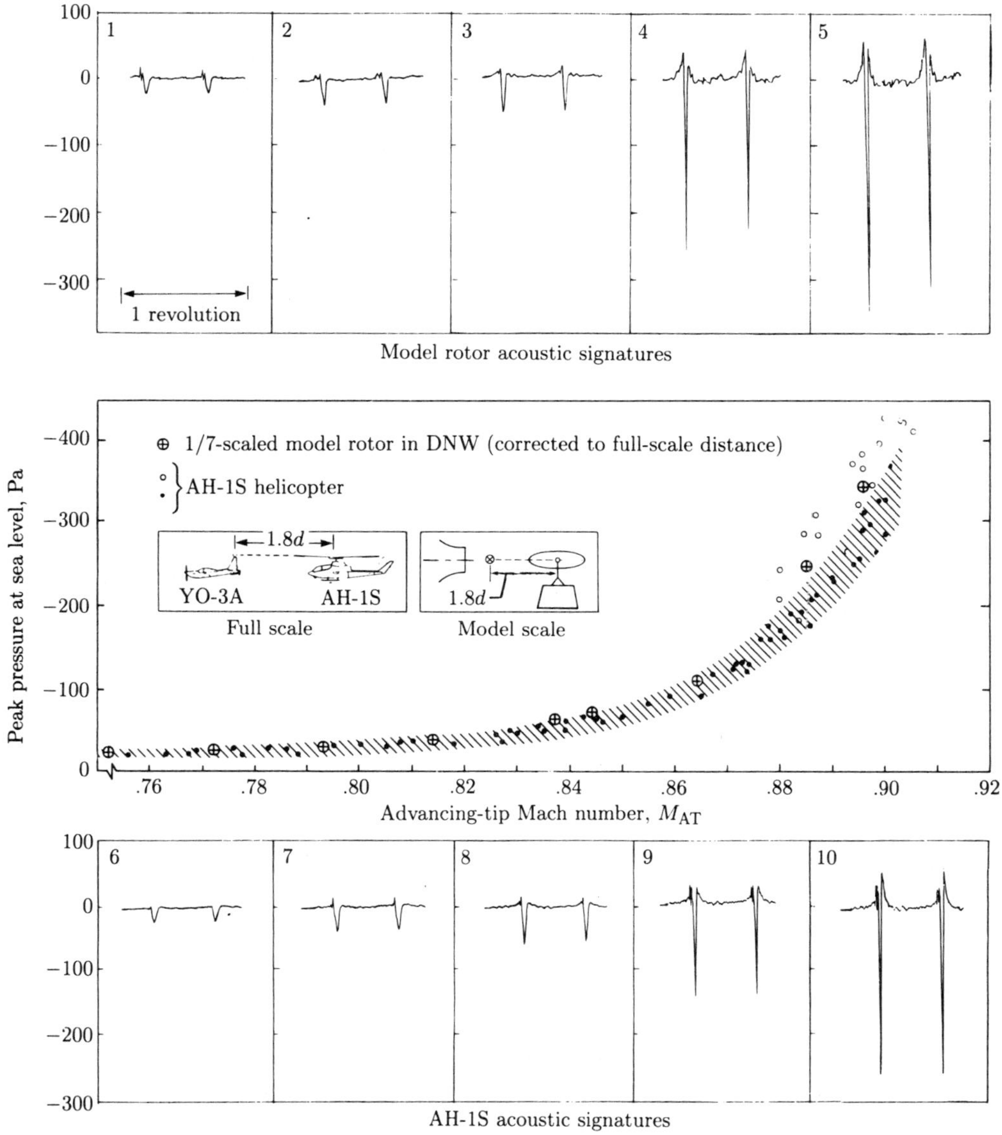

Figure 41. Model- and full-scale acoustic pressures for in-plane microphone 3.6 rotor radii ahead. (From ref. 5.)

the DNW anechoic wind tunnel and are of very high quality (refs. 32 and 35). The in-flight acoustic data which are shown were taken with a specially designed quiet aircraft. The pulse shapes for both the model- and the full-scale rotor have been averaged for comparison purposes.

For completeness, all four governing nondimensional parameters were duplicated: advancing-tip Mach number, advance ratio, thrust coefficient, and tip-path-plane angle. Excellent agreement of amplitudes and pulse shapes is demonstrated over a wide range of advancing-tip Mach numbers. Also shown is the sensitivity of peak sound pressure levels to advancing-tip Mach number. Advance ratio also plays an important role by guaranteeing that the local Mach numbers of the model and of the full-scale rotor are similar at all azimuth positions. Thrust coefficient C_T and tip-path-plane angle α_{TPP} have secondary influences at these in-plane microphone positions (ref. 31).

Blade-Vortex Interaction Noise

Scaling BVI impulsive noise is a more difficult task. In addition to blade geometry, nondimensional distances, advancing-tip Mach number, and advance ratio, it is also imperative that rotor thrust coefficient and tip-path-plane angle be duplicated (refs. 27 and 33). As shown previously, guaranteeing that advance ratio is matched uniquely determines the in-plane geometry between the rotor blade and the tip-vortex structure (fig. 15). Because advance ratio governs the large-scale BVI geometry, it plays a key role in the acoustic radiation. When viewed from above, the rotor appears to slice through the epicycloid pattern of previously shed tip vortices. The resulting locus of interactions determines the number and strength of the BVI encounters and thus strongly influences the radiated noise. Judicious matching of thrust coefficient C_T and nondimensional inflow $\mu(-\alpha_i + \alpha_{\mathrm{TPP}})$ is necessary to duplicate the pressure coefficients $C_{p_{ij}}$ of the model- and full-scale experiments. For a geometrically scaled rotor, the thrust coefficient governs the local angle of attack of the rotor blade and thus the steady-pressure field. In addition, it affects the average strength of the shed tip vortex and thus directly influences the unsteady-pressure field as well. The nondimensional inflow also affects the magnitude of the unsteady pressures by governing the vertical separation between the vortex and the rotor blade at the time of an encounter. In a rigorous sense, this parameter should scale over the portion of the rotor disk where BVI's occur. However, it is often assumed that if the geometric properties and C_T are scaled, an average value in space and time of the induced angle α_i at the rotor disk governs the interaction problem ($\alpha_i \approx C_T/\mu$). Therefore, if C_T and μ are duplicated in a model-scale test, the tip-path-plane angle (α_{TPP}) becomes the fourth nondimensional test variable.

The most rigorous test of the scalability of impulsive noise is the most direct: simply compare the character of the model- and full-scale acoustic time histories on a one-to-one basis. In addition to being a straightforward comparison, it is also helpful in identifying the occurrences of BVI's in the acoustic signatures. This phenomenological approach is illustrated in figure 42 for the AH-1S helicopter for a microphone located approximately 30° beneath the plane of the rotor tips (ref. 27). At this microphone position, BVI noise is known to be near its peak intensity, while HSI noise is reduced from its large value near the plane of the rotor disk. In the left side of figure 42, averaged and unaveraged measured acoustic time histories are shown for one rotor revolution as measured with the full-scale, in-flight technique. The helicopter and measurement aircraft were flown in formation at a 60-knot IAS partial-power descent (400 ft/min), a condition known to produce strong BVI noise. Because the measured full-scale BVI time histories were quite unsteady, two unaveraged waveforms are shown which typify maximum- and minimum-intensity BVI events. The four important nondimensional scaling parameters are listed. At this 30° microphone position, both the BVI noise and HSI noise are discernible. During advancing-blade–vortex interaction, a sequence of narrow, small negative and large positive spikes occur in the waveform just before the broader negative-pressure pulse. Scale model data (unaveraged and averaged) taken under similar nondimensional conditions are shown in the right side of the figure. Only one unaveraged waveform is shown because the measured model-scale BVI data were quite steady.

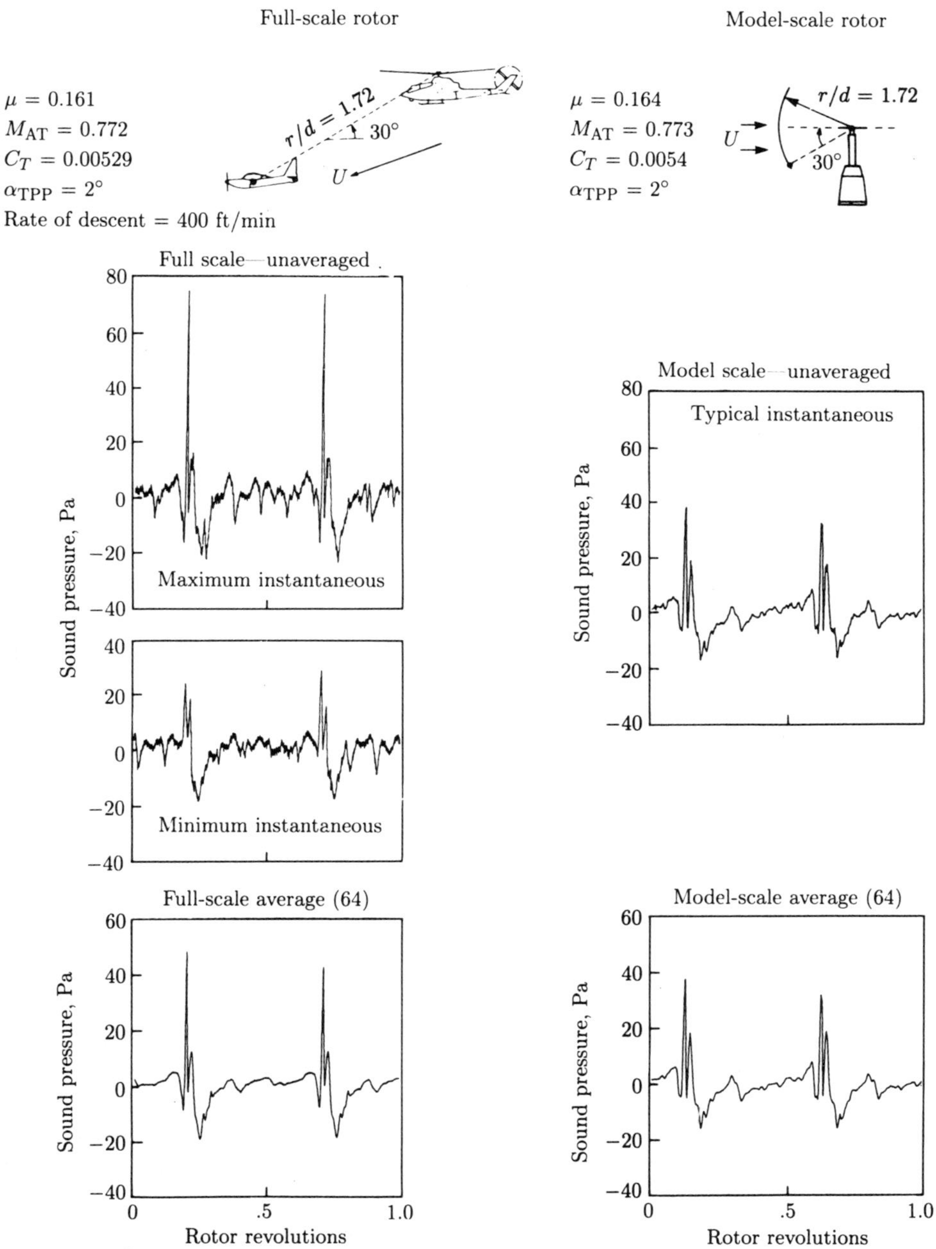

Figure 42. Unaveraged and averaged sound pressure time histories for one rotor revolution. (From ref. 27.)

The remarkable similarity in the details of the averaged pulse shapes for full-scale (left side of fig. 42) and model-scale (right side of fig. 42) experiments is evident for the advance ratio of 0.164. Scale model testing appears to faithfully reproduce the BVI noise of a full-scale helicopter. A closer look at the time histories reveals that the full-scale data have notably more narrow BVI peaks than those of the model-scale data. This is probably related to the size of the interacting tip vortex, which is related to viscous considerations only implicitly covered by these scaling relationships.

The problem becomes more apparent as advance ratio is increased in higher speed flight. As shown in figure 43, the sharp BVI pattern remains for the full-scale data, but a more broad, low-level pattern is generated for the model-scale data. It is apparent that BVI noise does not scale at these higher advance ratios. Although not conclusively proven, it is thought that local tip Reynolds number governs the size of the shed-tip-vortex filament, which ultimately determines the pulse width patterns of BVI noise.

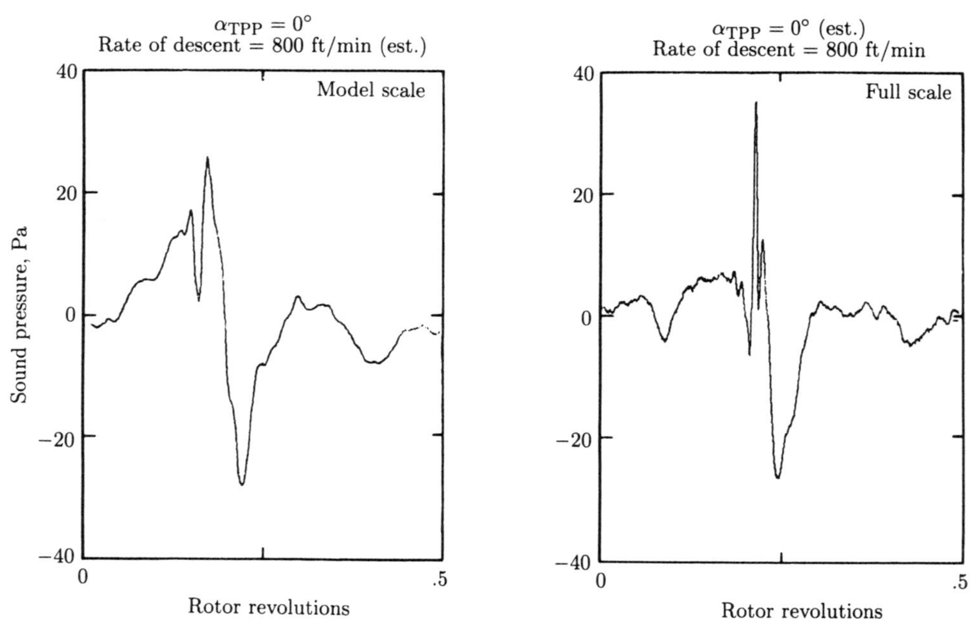

Figure 43. Model-scale and full-scale acoustic data for $\mu = 0.270$. *(From ref. 27.)*

The importance of Reynolds number and Strouhal number on aeroacoustic problems is well-known. Reynolds number is the ratio of inertial to viscous forces that classifies the aerodynamic regimes of laminar and turbulent flow. Strouhal number characterizes the frequency of unsteady-vortex shedding from blunt bodies which is itself a radiation of significant acoustic energy. Both nondimensional parameters are important for the scaling of broadband noise radiation (ref. 19).

If rotor acoustic models are made too small, a variety of problems prevent faithful acoustic scaling. Low tip Reynolds numbers cause poor representation of the rotor-tip-vortex structure. The high frequency of the model data creates new instrumentation challenges. In addition, it is quite difficult to represent the dynamic behavior of a full-scale rotor system at too small a scale. At the present time, 1/5-scale four-bladed models have become the industry standard. They are small enough to fit in most anechoic wind tunnels and yet can be dynamically scaled to the first order. However, to date no scaling results have been made to validate these model-scale results. A lingering concern is the use of tapered tips on these model-scale rotors. In such cases, the local tip Reynolds numbers become quite small and may lead to different small-scale tip-vortex filaments than those measured on the full-scale aircraft. If this observation is correct, then large-scale models may be necessary to duplicate full-scale BVI impulsive noise.

Theoretical Developments and Experimental Verification

Overview

A complete mathematical description of the sound generated by bodies in arbitrary motion was developed by Ffowcs Williams and Hawkings in 1969 (ref. 9). In essence, the basic mass and momentum equations of fluid mechanics are rewritten in wave equation form with all other quantities treated as forcing functions of the resulting integral equation. (See eq. (1).) This approach follows Lighthill's approach (ref. 36) of forcing the basic fluid mechanics equations into an "acoustic analogy." It is important to remember that this basic equation (with its many forms) is perfectly general and is applicable to all fluid mechanics as well as acoustics. If the right-side forcing functions are treated as known quantities, then equation (1) becomes much simpler; it becomes a linear wave equation with known forcing functions. This latter approach is followed in most aeroacoustic predictions of rotorcraft external noise and was discussed previously in the section entitled *Rotorcraft Noise Sources and Their Physical Origins.*

Many researchers have developed, in one form or another, valid theoretical acoustic analogies to rotorcraft noise prediction. As discussed previously, the first simple theoretical model of rotor noise was developed by Gutin (ref. 13), who recognized that steady aerodynamic forces on a propeller act as acoustic dipole sources (eq. (5)). Garrick and Watkins (ref. 14) extended this work to the case of the uniformly moving propeller. Deming (ref. 15) looked into the effect of blade thickness on the radiated noise. He replaced a symmetric airfoil with an infinite number of line pistons (sources and sinks) to match the boundary condition of no flow through the rotor airfoil surface. These simple theoretical approaches really approximated the second two terms of equation (1). However, they were developed using a coordinate system fixed in space rather than one attached to the rotor blade itself. Comparison with experiment, for the most part in the frequency domain, was very encouraging for the low harmonics of rotor noise but was lacking for higher harmonics.

Noise radiating from helicopters became important as these vehicles emerged from being research curiosities of the 1950's and began to assume new military and civilian roles of the 1960's. Quite a lot of research into the potential causes of helicopter periodic noise was initiated; the two notable efforts were made by Lowson (ref. 37) and Wright (ref. 38). Using developments based upon Lighthill's acoustic analogy, Lowson and Wright argued that in addition to steady forces (identified by propeller researchers as the cause of periodic noise), the unsteady forces that the rotor blade experiences as it traverses one revolution are very efficient radiators of periodic noise. They suggested that in order to predict the higher harmonics of radiated noise, one could use a compact source model but would need to know very high harmonics of blade loading. Although the agreement with experiment was not always consistent, their theories did show more encouraging correlation with the limited experimental frequency domain data available. Because technology had not yet made narrow-band data analysis straightforward and had not provided the large digital computer for lengthy numerical calculations and validations, important pulse shape information was not effectively used to further refine the modeling of rotorcraft noise sources.

Important differences between linear theory and time-history experimental measurements for rotorcraft whose rotors are operating at transonic tip Mach numbers were first noticed by Schmitz (refs. 39 to 41). These differences led to the realization that transonic aerodynamic effects are often important contributors to the radiating noise of rotorcraft. These effects were first predicted by using quadrupoles, in addition to monopoles and dipoles, as sources of rotorcraft noise. In essence, some of the aerodynamic details of the rotor local flow field were modeled as sources of radiating noise.

Modern electronic technology has now made quantitative time-history comparisons between theory and experiment routine. (In frequency-domain terminology, harmonic amplitude and phase are both used in the validation process.) More powerful mathematical approaches, based for the most part on equation (1), have placed much of the earlier theoretical work on a more sound mathematical basis and have extended the theory to handle noncompact sources for subsonic, transonic, and supersonic rotors. Pioneering theoretical work by Hawkings and Lowson (ref. 10), Farassat (ref. 11), Isom (ref. 12), and many others has increased the understanding of the noise generation process. This, combined with more careful measurements of the radiated noise, is leading to designs that can minimize unwanted acoustic radiation of rotors.

Hovering Harmonic Noise

The hovering rotor is a natural place to begin to compare acoustic theory with experiment. Unfortunately, the aerodynamics of a hovering rotor are far from simple, being affected by the complex wake geometry of the rotor and interference from nearby surfaces. In addition, it is necessary to test rotors in an environment which is mostly without echoes, or "anechoic," so that acoustic reflections are not measured along with the radiation noise field. The data shown in figure 44 were gathered for a model rotor in hover in a near-anechoic environment in the DNW open-jet wind tunnel (ref. 42). The rotor is a modern high-performance helicopter rotor that has been designed to be efficient in hover as well as in high-speed (200+ knots) helicopter forward flight. Unaveraged (instantaneous) and averaged time-history data are shown at several nearly in-plane microphone positions in figure 44. The waveforms are somewhat unsteady but average to a characteristic waveform for this rotor type.

Steady Thickness and Force

Predictions of noise from linear theory are also shown in figure 44 (ref. 43). Good agreement between predicted values and the averaged waveforms is demonstrated at

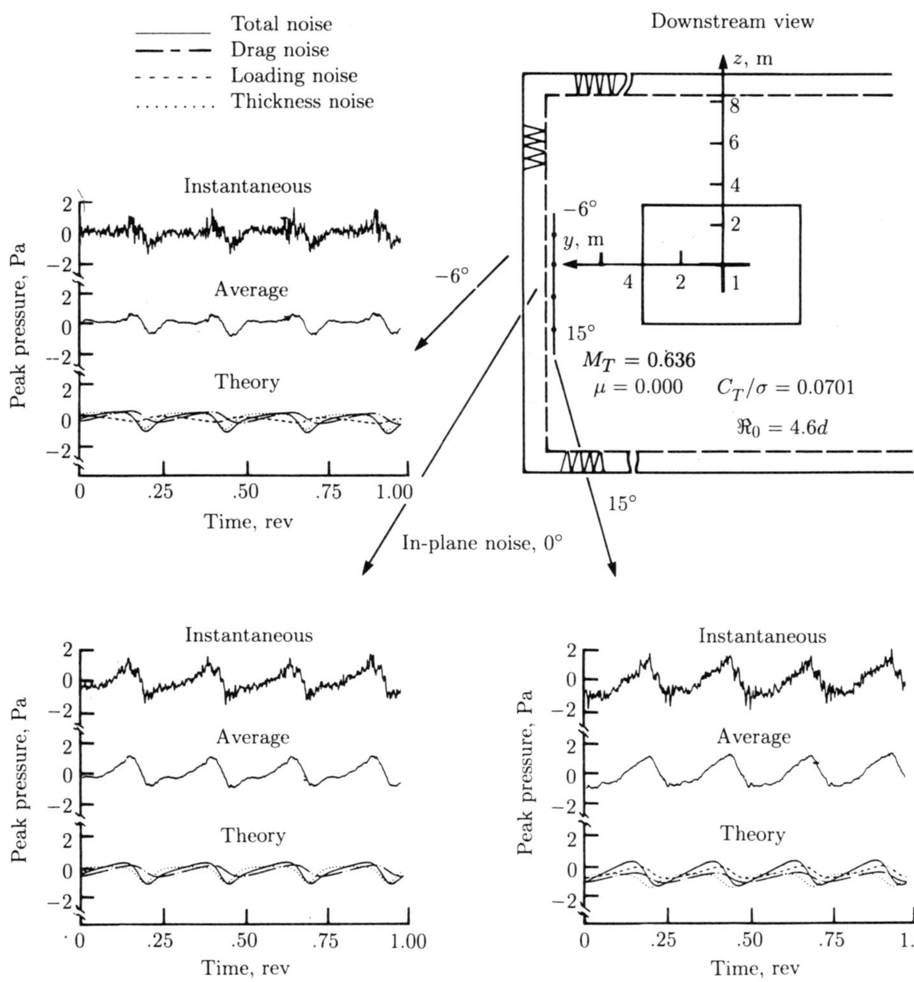

Figure 44. Measurements and predictions of isolated hover noise of model rotor.

tip Mach numbers typical of a hovering helicopter rotor. The relative importance of thickness (monopole) and force (dipole) noise sources is shown in figure 44 for this same condition. Near the plane of the rotor, both thickness and force noise are important contributors to the rotor acoustic signature in hover. However, as the observer moves farther from the in-plane position, dipole noise becomes more important as the thickness noise source decreases in level. Notice too that the character of the average waveform changes depending upon whether the microphone is above or below the rotor-tip-path plane. Below the plane of the rotor, the lift and drag contributions are in phase and add, while above the rotor plane they subtract.

The amount of unsteadiness in this hovering model rotor is typical of this type of experiment. Even in this large, open-jet acoustic tunnel, room recirculation and associated turbulence ingestion into the hovering rotor can cause the rotor to operate in a somewhat unsteady aerodynamic environment (ref. 44). This, in turn, produces unsteady blade pressures, which are needed to predict accurately the unaveraged time histories.

Linear Theory With High Tip Mach Numbers

The steady hovering problem becomes more interesting as the hovering tip Mach number M_T is increased. Thickness noise grows much more quickly than the force noise and dominates the time history of the nearly in-plane microphone positions above $M_T = 0.8$. Computation of this noise using linear theory is simply a matter of evaluating the thickness term of equations (1) and (4), because the dipole force terms do not radiate very efficiently for helicopter rotor blades operating at high (near transonic) tip Mach numbers. Because the tip Mach number never approaches 1.0, the integrable singularity in equation (4) never becomes a problem. The integration of the monopole sources is performed by dividing the rotor blade into chordwise and spanwise elements, summing each contribution, and differentiating the sum over time to yield the acoustic pressure time history at the chosen observer location. Therefore, equation (4) becomes

$$4\pi p'(\mathbf{x},t) \approx 2\frac{\partial}{\partial t}\sum_{k=1}^{l}\sum_{i=1}^{n} \rho_0 \left[\frac{v_{n_{ki}}}{\Re|1 - M_{\Re}|}\right]_{\tau} dy_{1_i}\, dy_{3_k} \tag{9}$$

where

$v_{n_{ki}}$	$= U_e \left(\frac{dy_2}{dy_1}\right)_{ki}$
U_e	velocity of each blade element
y_1	chordwise blade coordinate
y_2	coordinate normal to the mean blade chord
y_3	spanwise blade coordinate
i, k	summation indices for each blade element

Additional details describing these computational procedures can be found in reference 39. The equation may also be solved using frequency-domain procedures given in references 1, 10, and 15. Several alternative linear acoustic formulas for calculation of rotating-blade harmonic noise are reviewed in reference 45.

A key feature of the computational process is the degree to which it is dominated by Doppler amplifications at high tip Mach numbers. This can be illustrated physically by looking at the geometry of the linear acoustic process. Consider the space-fixed trajectory of a simple point source near the tip of a hovering rotor blade. Its trajectory is the circle traced by the moving blade tip. If at regular azimuth angles a pulse, depicted as a circle (a sphere in three dimensions), is emitted in space and allowed to propagate at the ambient speed of sound, these pulses form the crescent-shaped wave shown in figure 45 for a hovering rotor operating at a tip Mach number of 0.9. In effect, disturbances are propagating away from a source moving at almost the ambient speed of sound. As a result, disturbances accumulate and create Doppler amplification. As the advancing-tip Mach number increases (higher rpm), the accumulation of disturbances becomes so great as to form local shocks on the blade surface and eventually a radiating shock wave. This accumulation process is represented as a singular integral in equation (4). The process whereby shock waves on the blade surface become connected to the acoustic far field is called "delocalization" (refs. 32 and 46).

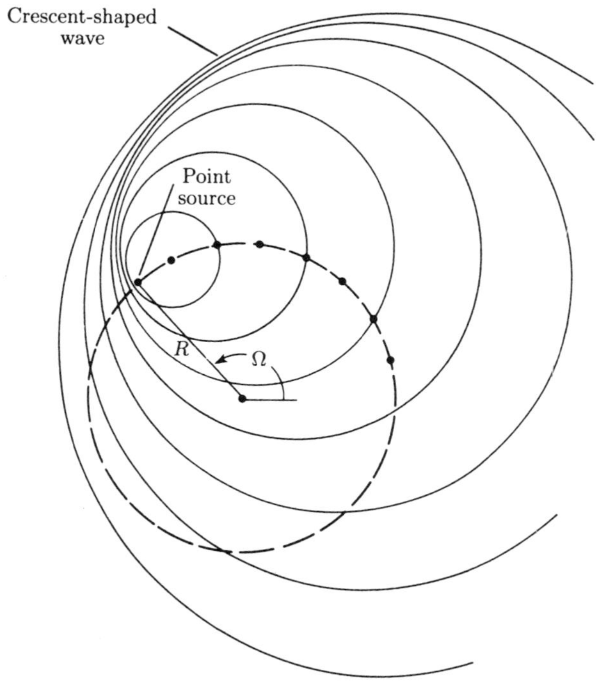

Figure 45. Linear wave amplification (Doppler effects) of rotating point source. $M_T = 0.9$. (From ref. 5.)

Linear thickness calculations for a simple hovering rotor and experimental data are shown in figure 46 for several different hover tip Mach numbers (ref. 40). The high-quality data shown here were gathered in a specially designed anechoic hover chamber. The chamber was lined with polyurethane wedges to be reflection free down to 110 Hz. The rotor was run near zero thrust and was designed with zero twist to minimize recirculation effects and to minimize thrust (dipole) sources as radiators of noise.

The striking features of the comparison between theoretical and experimental values for hover at $M_T = 0.8$ (fig. 46(a)) are the similarity in pulse shapes and the discrepancy in peak pressure levels. Only a fraction of the pressure time history is shown to facilitate the details of the comparison. Thickness-noise theory misses the measured negative-pressure peak by a factor of about 2. The comparison of theory and experiment as M_T is increased to 0.88 (fig. 46(b)) remains similar to that made at $M_T = 0.80$. The waveform shape is still generally symmetrical, and the

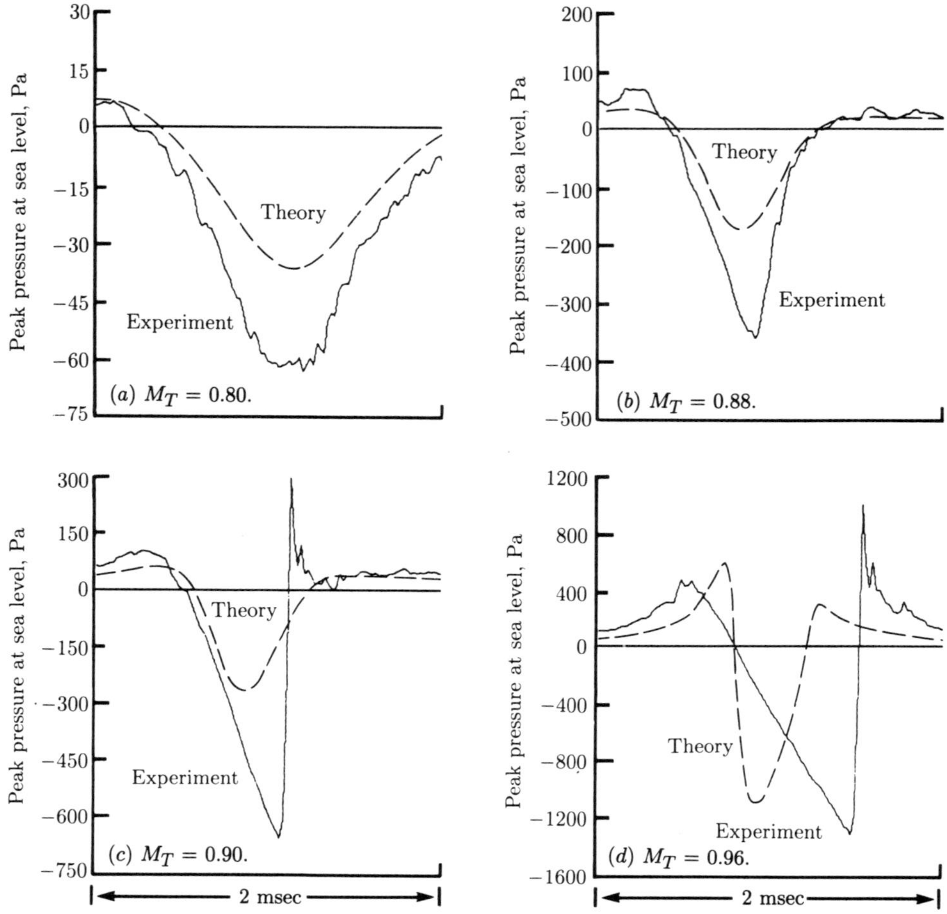

Figure 46. Theoretical and experimental pressure time histories for in-plane microphone. $\Re_0/R = 3.0$. *(From ref. 40.)*

peak negative-pressure level is underpredicted by slightly more than a factor of 2. A closer look at the waveform shows that it is just beginning to become slightly asymmetrical; the recompression part of the measured expansion wave has a slope whose magnitude is greater than the initial expansion. Linear theory values do not show this tendency.

At a hover tip Mach number of 0.90 the situation changes dramatically (fig. 46(c)). The peak negative-pressure amplitude of the measured pulse has increased substantially and the pulse shape has now lost its symmetry. The resulting sawtooth waveform is known to generate large amounts of high-intensity, high-frequency noise in the plane of the rotor. In essence, a relatively weak shock wave is radiated to the acoustic far field at a Mach number of 0.9 for this untwisted rotor with an NACA 0012 airfoil. The rotor is said to "delocalize"; the local shock waves on the surface of the transonic rotor blade are connected and, in fact, radiate to the acoustic far field. Theory again underpredicts the amplitude of the peak negative-pressure pulse by about a factor of 2. More importantly, theory does not predict any of the features of the delocalization process, totally missing the shock-like experimental waveform.

The comparison becomes even more intriguing at a hover tip Mach number of 0.96 (fig. 46(d)). The theoretical waveform is mostly symmetrical and generally smooth in shape and, thus, does not compare favorably with the measured data. In addition, theory now only slightly underpredicts the peak negative-pressure amplitude of the pulse. Also, the measured pulse width is becoming wider, whereas the linear theory predicts a narrower pulse width with increasing hover tip Mach number. In fact, the experimental pulse width (measured at zero pressure) exceeds by at least 50 percent the width expected (from linear theory) from an airfoil of chord equal to that of the model rotor tested and traveling at sonic velocity. This pulse-widening effect suggests that aerodynamic events off the rotor-blade trailing edge are contributing to the measured acoustic signature.

The difference in peak negative-pressure levels between linear monopole theory and experiment is shown more clearly in figure 47. The theoretical model does not predict the rate of increase of the peak negative-pressure level.

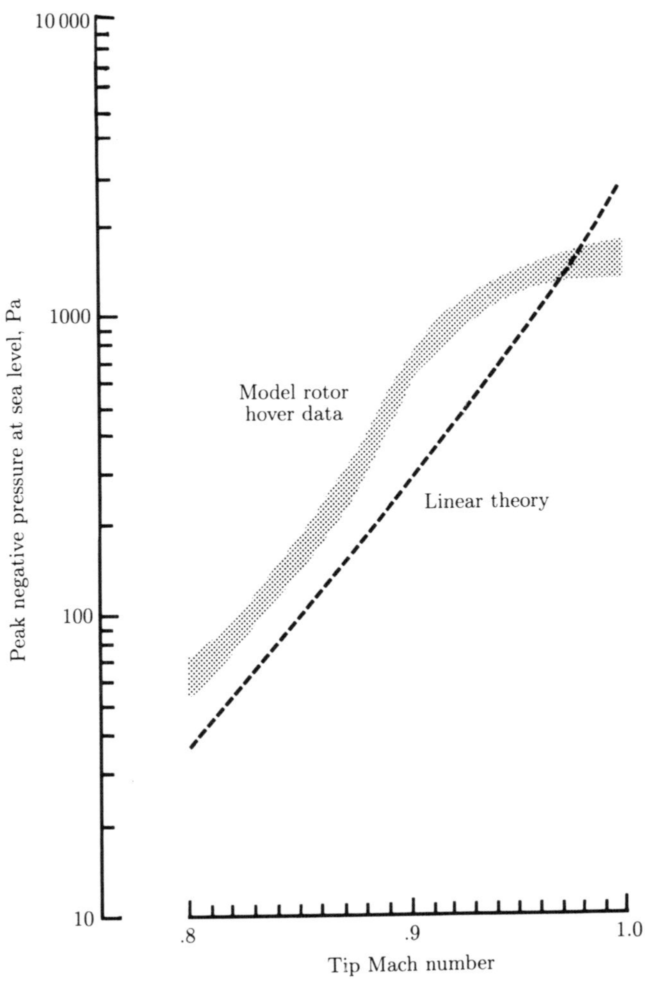

Figure 47. Theoretical and experimental peak pressures of rotor in hover. In-plane microphone; $\Re_0/R = 3.0$. *(From ref. 40.)*

It is tempting to attribute the lack of correlation with experiment to the simplicity of the linear theoretical model. Perhaps if the remaining linear dipole terms were included (the second term in eq. (1)), the agreement with measured data might be better. This was tried for a rotor with predicted values of loading and skin friction suitably chosen to act as pressure dipoles (ref. 47). The resulting theoretical time histories are almost indistinguishable from the simple linear monopole calculations previously presented.

The major conclusion from all these linear acoustic analyses is that they do not adequately describe the in-plane noise radiation processes from a high-tip-speed hovering rotor. There have been some efforts to correct this situation by improving the modeling (improving the specification of the boundary conditions) in the linear formulation. It can be argued that rotor-tip end-plate (ref. 48) and boundary-layer separation efforts increase the amplitude of the symmetrical pulse. The importance of the latter effect is shown in figure 48. This end view of an NACA 0012 rotor at near zero lift (ref. 49) is a holographic interferogram of the integrated three-dimensional flow field surrounding the tip of a hovering UH-1H model rotor at the tip Mach number of 0.9. It is quite apparent that local shock waves on the surface of the blade interact with the boundary layer to cause an enlarged separated-flow region. A rigorous treatment of this problem is not usually attempted, for it would be necessary to model the boundary-layer and separated-flow effects in equation (9). Instead, an "equivalent airfoil" comprising the original airfoil plus the outer edges of the separated-flow region is defined. This new equivalent airfoil is then used in equation (9) to define the strength of the distributed acoustic sources. If this is done, it is relatively easy to show that the peak negative-pressure calculations would increase substantially (they approximately double for each doubling of the effective airfoil chord at constant thickness). Although this effect has been known for many years, most researchers do not like to incorporate such an estimation in a "first-principles" analysis. The methods of estimating just how thick or extended the separation region is on a three-dimensional rotor in the transonic regime and how to model the equivalent airfoil for noise purposes are not well-defined or even completely understood. In addition, none of these corrections predict the development and radiation of the delocalized shock wave above a hover tip Mach number of 0.9 for a scaled UH-1H rotor. Clearly, the radiation processes at these high tip speeds are governed to a large extent by transonic effects. These must be accounted for in the theoretical modeling.

Aerodynamic Formulation With High Tip Mach Numbers

The most straightforward approach to the nonlinear acoustic problem with high tip Mach numbers might simply be to include missing terms—the quadrupoles—in the acoustic analogy formulation. However, equation (1) is in reality an integral equation which has no simple analytical solution. Some degree of approximation is necessary to proceed with this approach. These approximations rely on what we know about the problem physically. Such insight can be gained by formulating the problem as a transonic aerodynamicist would.

We begin with the classic potential equation in a space-fixed coordinate system. Assuming constant specific heats and weak shocks (i.e., negligible entropy increases),

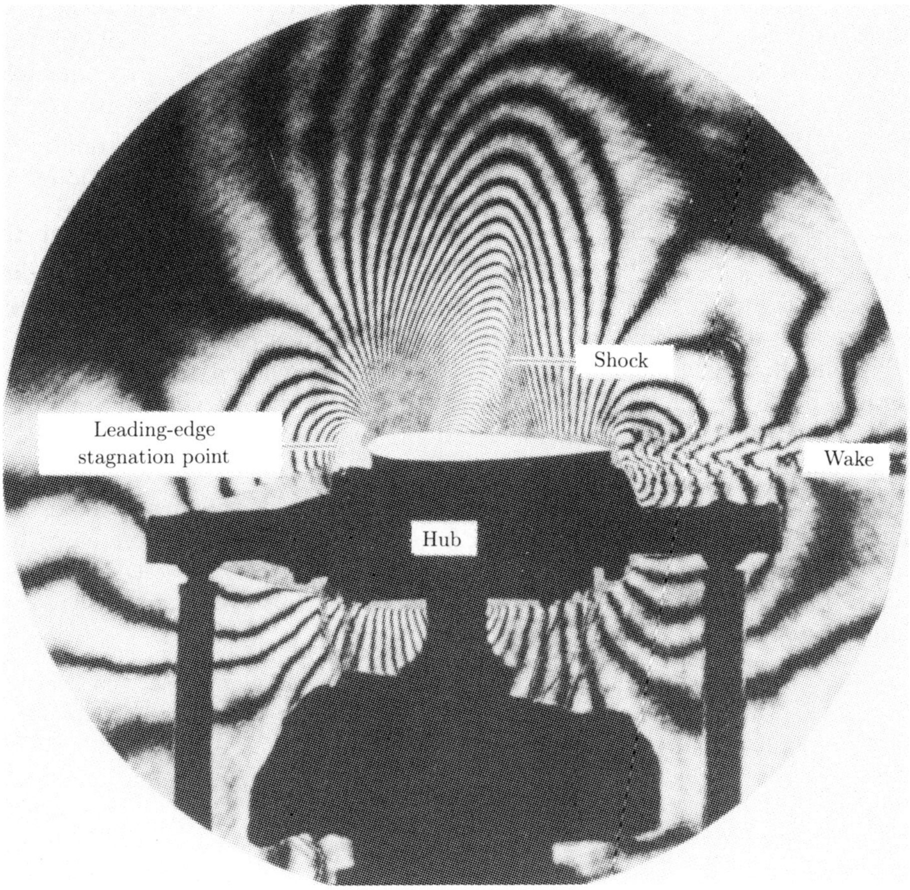

Figure 48. Interferogram recorded at ψ = 180° (chordwise view). (From ref. 5.)

we have

$$\frac{\partial^2 \phi}{\partial t^2} - c^2 \nabla^2 \phi + 2\nabla\phi \cdot \nabla \left(\frac{\partial \phi}{\partial t} \right) + \frac{1}{2} \nabla\phi \cdot \nabla \left[(\nabla\phi)^2 \right] = 0 \tag{10}$$

where ϕ represents the velocity potential and c is the local speed of sound. Fortunately, the aerodynamics of a hovering rotor are basically steady when viewed from a blade-fixed context. Therefore, by following the work in references 41, 46, 50, and 51, the governing potential equation can be transformed to blade-fixed cylindrical coordinates and expanded to second order to yield

$$\left[\Omega^2 - (c_0^2/r^2) - (\gamma+1)(\Omega/r^2)\phi_\theta\right]\phi_{\theta\theta} - 2\Omega\phi_r\phi_{r\theta} - 2\Omega\phi_z\phi_{z\theta}$$
$$= \left[c_0^2 + (\gamma-1)\Omega\phi_\theta\right]\left[\phi_{rr} + (\phi_r/r) + \phi_{zz}\right] \quad (11)$$

where Ω is the angular rotation rate, c_0 is the undisturbed speed of sound, r is the radial coordinate in the cylindrical coordinate system, and γ is the ratio of specific heats. This nonlinear but steady-state second-order partial differential equation governs the transonic aerodynamics of the hovering rotor. In addition, it governs how disturbances (acoustic waves) propagate away from this rotating coordinate system. At the present time, no closed-form solutions to this equation exist. A procedure adopted by some researchers is to solve numerically limited regions of the aerodynamic flow field (refs. 52 and 53). Others have chosen to solve the nonlinear acoustic far field using weak-shock theory (ref. 54). As we shall see, neither is a completely satisfactory solution, for the nonlinear aerodynamic and acoustic fields are interwoven.

The cylindrical coordinate system chosen is sketched in figure 49. An observer riding in this coordinate system sees a free-stream velocity that increases linearly from zero at the origin to Ωr at r. As indicated, this increasing free-stream velocity continues out past the tip of the rotor; it will be shown to be important to many of the arguments to come.

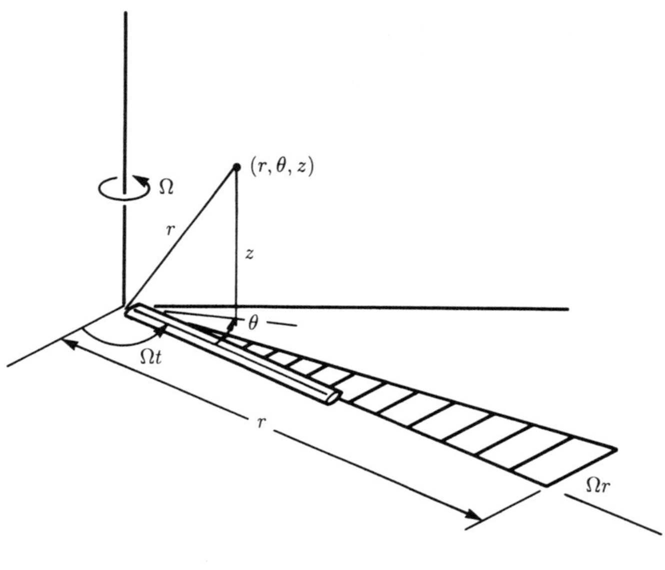

Figure 49. Cylindrical coordinate system of hovering rotor. (From ref. 5.)

Before attempting to solve equation (11), it is instructive to follow the approach of references 50 and 51 and explore the behavior of the governing equation. It is known from the theory of partial differential equations that the coefficient of $\phi_{\theta\theta}$ governs the general character of the potential equation. When $A \equiv \Omega^2 - (c_0^2/r^2) - (\gamma+1)(\Omega/r^2)\phi_\theta$ is less than zero we have elliptic behavior, and when it is greater than zero we have hyperbolic behavior. However, A takes a more recognizable form after some further manipulation. Define the local Mach number M_l as follows:

$$M_l \equiv (U_\infty + u)/c$$

The coefficient of $\phi_{\theta\theta}$ in equation (11) becomes (ref. 41)

$$A = -(c_0^2/r^2)(1 - M_l^2) \tag{12}$$

Therefore, the general behavior of the second-order transonic potential equation is governed by the local Mach number of the flow. If $M_l < 1.0$, then $A < 0$ and the governing equation is elliptic. In this case, no wave-like structure is possible. However, if $M_l > 1.0$, then $A > 0$ and the governing nonlinear partial differential equation is hyperbolic. Characteristics are then formed along which disturbances can propagate in a wave-like manner. It is also important to realize that M_l is dependent on the free-stream velocity $U_\infty = \Omega r$, the local speed of sound c, and the local perturbation velocity $u = -\phi_\theta/r$.

These ideas are quite useful when one is attempting to explain the phenomenon of delocalization for the hovering rotor experiment described previously. This connection was theoretically suggested in references 51 and 52 and has been numerically calculated (refs. 41 and 51) and experimentally verified (ref. 46). In the following paragraphs, the relationships are shown to depend on the local Mach number of flow. Three distinct cases are considered: free-stream tip Mach numbers of a hovering rotor $M_T = 0.85$, 0.88, and 0.90. Some freedom has been taken with the graphics in the interest of presenting a clear picture of the basic relationships involved. The data are the same as those reported in reference 41. In the figures that follow, the top views are sketches of events pieced together with limited experimental data, and the aft views are, for the most part, interpolations of experimental data.

Figure 50 depicts the top and aft views of shock boundaries of a rotor operating at $M_T = 0.85$. A locally supersonic region exists near the tip of the rotor. For this

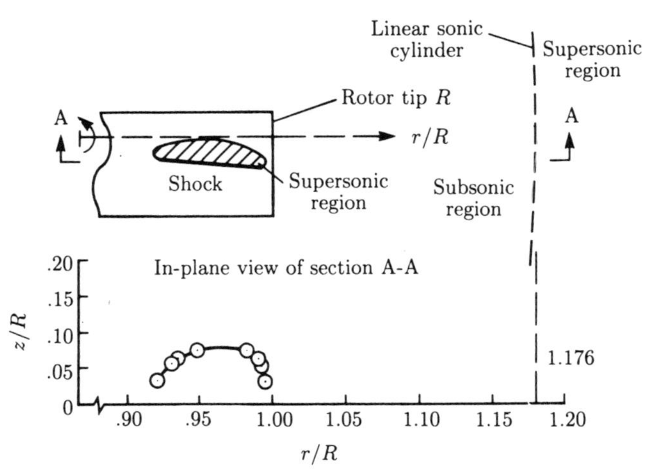

Figure 50. Top and aft views of shock boundaries of rotor at $M_T = 0.85$. (From ref. 41.)

region $M_l = [(\Omega r + u)/c] > 1.0$, even though $\Omega r/c_0$ all along the blade span is less than 0.85. The hyperbolic nature of this pocket of supersonic flow is a result of local aerodynamic nonlinearities (i.e., changes in the local speed of sound c and the local perturbation velocity u). Surrounding this locally supersonic flow is a subsonic flow region $M_l < 1.0$ in which the governing potential equation is elliptic. Waves which are embedded in the local supersonic flow region do not pass through this compressible elliptic region and hence do not radiate. However, as r increases beyond the tip of the blade, M_l again becomes greater than 1.0 because of the linearly increasing free-stream velocity field of the blade-fixed cylindrical coordinate system. For this region, $u \approx 0$ and $c \approx c_0$, so $M_l \approx \Omega r/c_0 > 1.0$.

The surface where this first happens has been called the sonic cylinder (refs. 46 and 54). At radii larger than the sonic cylinder, the equation again becomes hyperbolic and wave-like propagation is certain. The acoustic implication for this $M_T = 0.85$ case begins in the hyperbolic pocket of flow near the blade tip. Wave-like disturbances in this region terminate on the boundary of an elliptic region, where they no longer propagate in characteristic directions. The wave-like character of the inner pocket is thus broadened as information passes through the elliptic region to the sonic cylinder. These broadened disturbances are then propagated in a wave-like manner throughout the outer hyperbolic region. The result is a smoothly varying, near-symmetrical acoustic signature in the far field.

The competing phenomena become even more interesting when M_T is increased to 0.88 (fig. 51). The inner supersonic (hyperbolic) region grows and extends off the tip of the rotor, again being driven by local aerodynamic nonlinearities. At the same time, the higher free-stream tip Mach number of the rotor decreases the radius of the sonic cylinder, thus moving the outer hyperbolic region toward the rotor tip. In addition, the proximity of the linear sonic cylinder to the blade tip introduces aerodynamic nonlinearities. These nonlinearities tend to warp the sonic cylinder

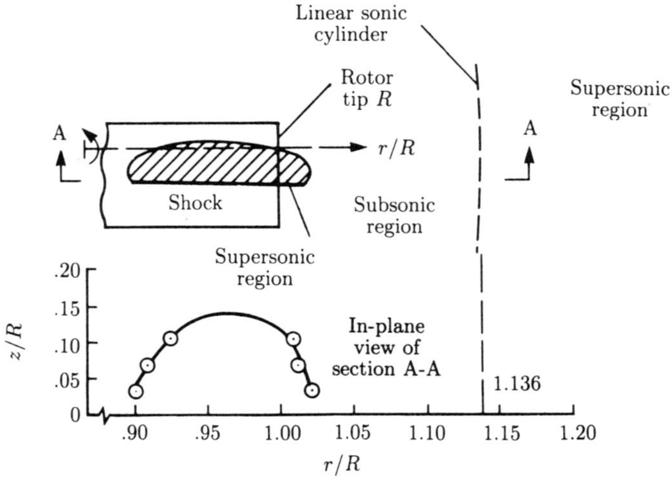

Figure 51. Top and aft views of shock boundaries of rotor at $M_T = 0.88$. *(From ref. 41.)*

inward and bring the two hyperbolic regions even closer together. However, the inner regions do not overlap, thus ensuring that locally generated waves in the inner region do not propagate along characteristics to the hyperbolic far field. The resulting acoustic signature becomes more saw-toothed in character but does not contain radiating shocks.

The last and most interesting condition, in which M_T is increased to 0.90, is sketched in figure 52. The localized inner and outer hyperbolic regions connect off the blade tip, forming one continuous supersonic region ($M_l > 1.0$). In this case, shock waves that are generated on the surface of the rotor now propagate uninterrupted radially to the acoustic far field. The resulting phenomenon (delocalization) is quite striking. The character and the intensity of the acoustic signature change dramatically. At all three of these tip Mach numbers, measured values of local Mach number support and explain the phenomenon of transonic delocalization. For a hovering UH-1H rotor, shock waves are radiated to the acoustic far field at high subsonic tip Mach numbers. The mechanism of delocalization can be further confirmed through use of computational fluid dynamics codes to predict the transonic aerodynamics of the hovering rotor (refs. 55 to 57). An example calculation for this rotor in hover is shown in figure 53 (from ref. 58). The agreement between theory and experiment is quite good, and this agreement conclusively demonstrates the interrelationship between transonic aerodynamics and high-speed rotor noise.

Quadrupole Formulation

Although the phenomenon of delocalization has been explained by simply looking at the coefficient of $\phi_{\theta\theta}$ in equation (11), predicting the radiating acoustic field

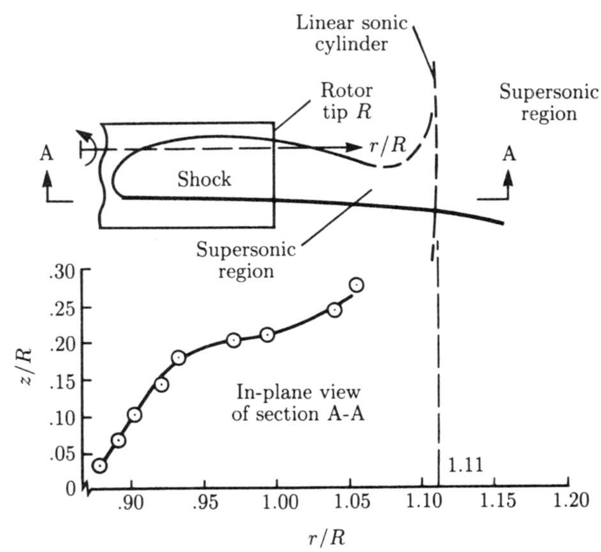

Figure 52. Top and aft views of shock boundaries of rotor and development of delocalization phenomenon at $M_T = 0.90$. (From ref. 41.)

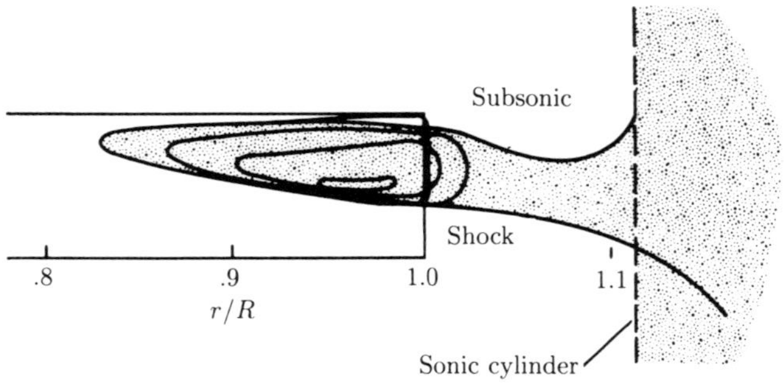

Figure 53. Tip flow field of hovering rotor. $M_T = 0.90$. *(From ref. 58.)*

is another matter. The explanations presented are themselves functions of either measured or calculated flow properties. In essence, a near-field description of the aerodynamic flow field is required before the events in the acoustic far field can be explained. Even then, acoustic nonlinearities in equation (11) may alter the waveform of the propagating wave (ref. 54). Precise calculations of the radiating sound field are dependent on the full solution of the nonlinear potential equation (ref. 50).

On the other hand, the successful explanation of the delocalization phenomenon suggests that local aerodynamic nonlinearities strongly influence the acoustic radiation problem. Therefore, a logical step in the calculation of the acoustic field is the incorporation of the near-field aerodynamic nonlinearities in the acoustic radiation equation.

Several ways of implementing these ideas have been presented in the literature. To date, the most successful procedure has been to extend the acoustic analogy procedures to evaluate the volume distributions of local aerodynamic nonlinearities, or quadrupoles (refs. 39, 41, 46, and 59). In essence, the third term in equation (1) is considered to be an important source of noise for the transonic radiation problem. As mentioned previously, evaluation of this integral equation is not directly possible; some approximations are required to make the problem manageable. Along these lines, the quadrupole term becomes simpler if we restrict our attention to the acoustic far field. Then the spatial differentiations can be easily converted to time differentiations. The first term of equation (1) becomes

$$\frac{\partial^2}{\partial x_i\, \partial x_j} \int \left[\frac{T_{ij}}{\Re|1 - M_\Re|} \right]_\tau dV \approx \frac{1}{c_0^2} \frac{\partial^2}{\partial t^2} \int \left[\frac{T_{\Re\Re}}{\Re|1 - M_\Re|} \right]_\tau dV \qquad (13)$$

where $T_{\Re\Re} = T_{ij}\Re_i \cdot \Re_j$ and $\vec{\Re}$ is the vector from the source at the retarded time to any observer in the acoustic far field. It is known from transonic computations and experimentation that the primary quadrupole regions are confined within a few chord lengths normal to the rotor plane.

For in-plane far-field radiation, the vector $\vec{\Re}$ is nearly in the blade rotational plane and is nearly parallel to the blade chordwise direction when the acoustic pressure reaches its peak level. If isentropic flow is assumed and the perturbation velocities are measured in the coordinate system given in figure 49, $T_{\Re\Re}$ becomes

$$T_{\Re\Re} = \rho_0(v_\theta^2 \cos^2\theta + 2v_r v_\theta \cos\theta \sin\theta + v_r^2 \sin^2\theta) + \frac{\gamma - 1}{2}\rho_0 \left(\frac{\Omega r}{c_0}\right)^2 v_\theta^2 \tag{14}$$

where the z-component of the perturbation velocity does not appear because of the choice of an in-plane far-field microphone position. For simplicity in the resulting calculations, it has been assumed that $\sin\theta \approx 0$ and that $u \approx v_\theta$ near the integration region of interest. This is true as long as the quadrupole field is in fact localized to a region near the rotor tip. Then,

$$T_{\Re\Re} = \rho_0 u^2 \cos^2\theta + \frac{\gamma - 1}{2}\rho_0 \left(\frac{\Omega r}{c_0}\right)^2 u^2 \tag{15}$$

where u represents the perturbation velocity along the blade chord and Ωr is the free-stream velocity of the point in the flow field being evaluated. The two terms represented in equation (15) arise from similar properties of the flow already discussed in the potential formulation. Changes in the local speed of sound and local streamwise perturbation nonlinearities are included, although the equation forms do not permit a one-to-one correspondence of terms. Equations (1), (13), and (15) describe the nonlinear far-field acoustic radiation of the transonic hovering rotor. For subsonic tip Mach numbers, numerical evaluation of the surface integrals presents no real problems. However, the volume integration of quadrupoles is not as straightforward.

The integrand in equation (13) contains the product of two terms which compete to decide the eventual magnitude of the quadrupole radiation. The first is the decaying source field represented by $T_{\Re\Re}$. This is multiplied by $1/|1 - M_\Re|$, which goes to infinity as $M_\Re$ approaches 1. Fortunately, the singularity is integrable, but it should be handled quite carefully (ref. 45). In the results summarized here, the acoustic planform technique was chosen to perform the numerical integration near $M_\Re = 1.0$. A complete discussion of the procedures and pertinent references is given in reference 41.

An evaluation of the prediction accuracy is presented herein through comparison of theoretical values with the same UH-1H hover model rotor data. Figure 54 presents the monopole and quadrupole contributions to the radiated noise at $M_T = 0.88$ (slightly before delocalization). At this tip Mach number, the shape of the quadrupole term is basically still symmetrical; however, some asymmetry is present on the pressure recovery side of the quadrupole calculation. When the monopole and quadrupole contributions are added, good correlation in amplitude and pulse shape is observed. The overall shapes of theoretical and experimental curves are still basically symmetrical in character, but the local shock structure of the transonic flow field is acting to destroy this symmetry.

At the slightly higher hover tip Mach number of 0.90, localized transonic effects cause large changes to the radiated noise field (fig. 55). Although the time history for the linear term (monopole) remains quite symmetrical in shape and substantially underpredicts the experimental data, the time history for the nonlinear term (quadrupole) changes shape dramatically and increases in amplitude. This change is a reflection of the fact that local shocks are propagating to the acoustic far

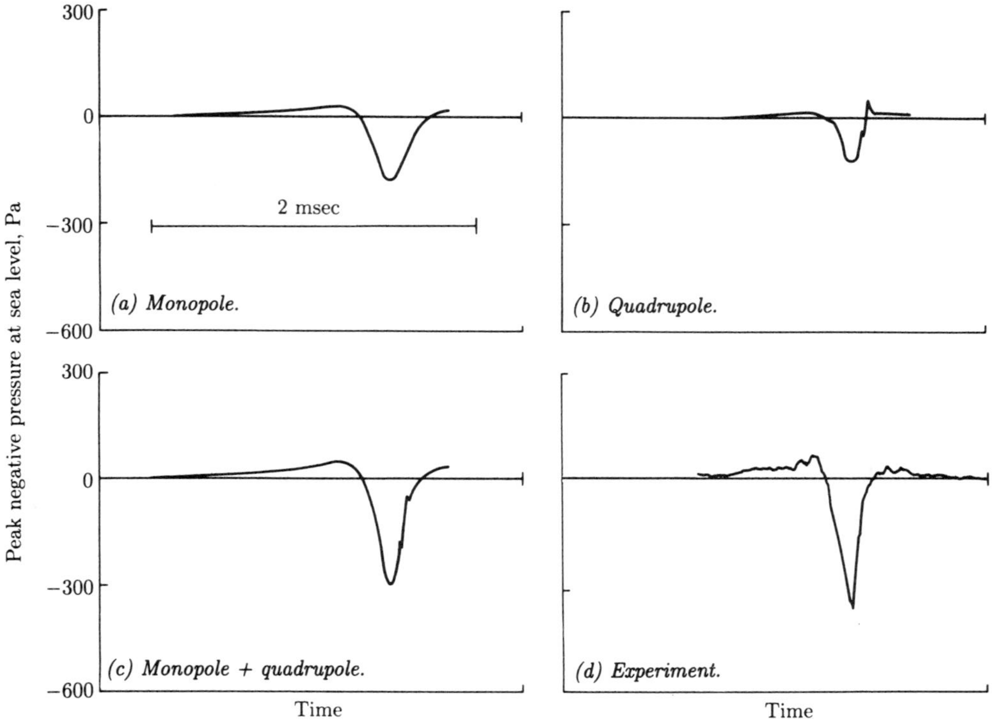

Figure 54. Theoretical and experimental time histories at $M_T = 0.88$, $\Re_0/R = 3.0$. *(From ref. 5.)*

field (delocalization). When time histories for the monopole plus quadrupole terms are compared with experimental data, good agreement in pulse shape is observed.

The accuracy with which the peak negative amplitude of the high-speed hover impulsive noise phenomenon can be predicted is illustrated in figure 56. For tip Mach numbers up to 0.9, use of quadrupole and monopole terms results in better agreement between theoretical and experimental values than use of monopole terms alone. However, at $M_T > 0.9$, amplitudes are overpredicted even though the pulse width tends to be estimated fairly accurately through delocalization. At the present time the source of this discrepancy is not understood. Additional time-history comparisons developed from a frequency-domain analysis can be found in reference 60.

Kirchhoff Formulation

Another competing method of numerically predicting the noise of the delocalization process is to map the nonlinear transonic region to a nonrotating control surface to which variations on Kirchhoff's theorem are then applied to describe the radiating acoustic field. In this first direct application of the Kirchhoff theorem, the control surface is chosen to be large enough to capture the nonlinear aerodynamic behavior of the problem, but not so large as to make numerical computation impractical. Calculations with this procedure, coupled with an existing near-field numerical code, have resulted in improved peak amplitude levels (ref. 61) but have not improved

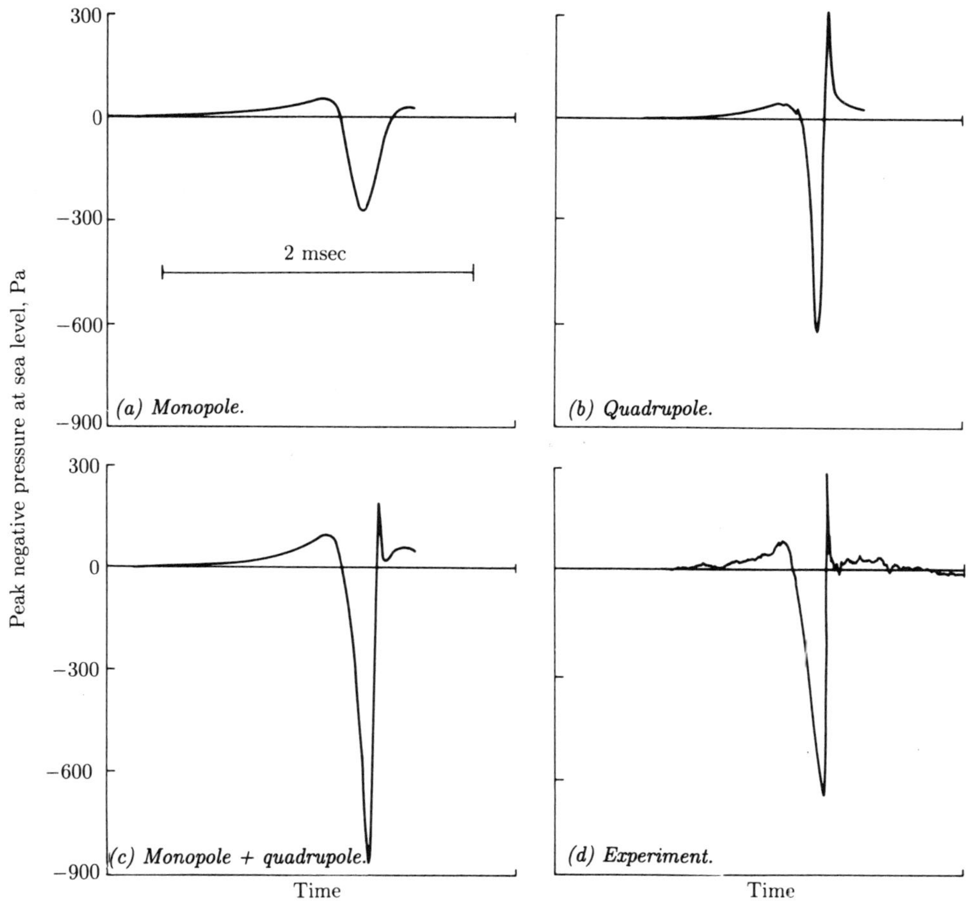

Figure 55. Theoretical and experimental time histories at $M_T = 0.90$. $\Re_0/R = 3.0$. *(From ref. 5.)*

waveform characteristics much above the delocalization Mach number. This is most likely a result of the numerical insensitivity of the transonic code at the boundary of the nonrotating control surface.

This procedure has been further extended to include the nonlinearities in the near aerodynamic field while the equivalent Kirchhoff surface is kept close enough to the blade tip to avoid computational fluid dynamics numerical computation errors. A new, expanded Kirchhoff integral has been developed which uses surface integrals of the pressure and velocity at the linear sonic cylinder to determine the acoustic pressures in the far field (ref. 61). In essence, the new method captures all the nonlinear aerodynamic effects by mapping them to the linear sonic cylinder. The sonic cylinder then becomes the sole source of all acoustic information, which is then propagated to the far field at a constant speed of sound. The resulting formula (ref. 62) contains the classic Kirchhoff's surface integrals at the sonic cylinder plus a correction factor for the local transonic effects near the blade tip.

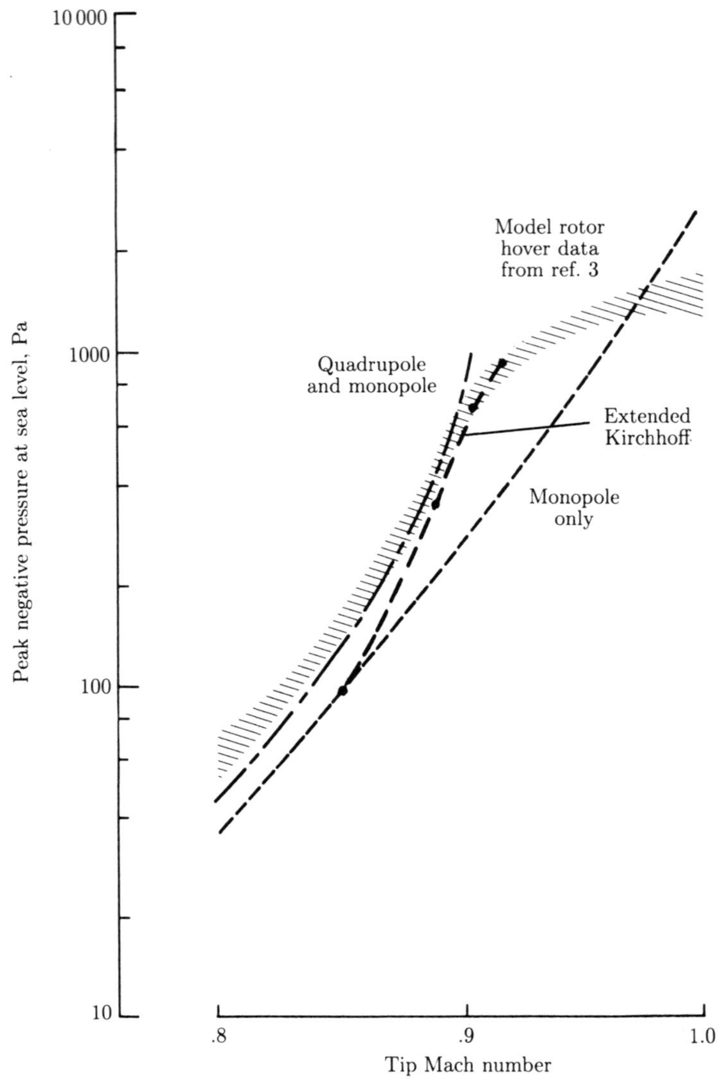

Figure 56. Theoretical and experimental pressures for rotor in hover. $R_0/R = 3.0$. (Based on ref. 5.)

Predictions from this approach are shown in figure 57 for some of the same conditions shown previously. To obtain sufficient experimental data on the sonic cylinder to validate the new method, the experiment of 1978 was run again in 1988. Near-perfect replications of the 1978 results (ref. 40) demonstrated the validity of the experimental results. Theoretical predictions of the radiated acoustic pulse correctly captured the delocalization phenomena; a relatively smooth pulse shape was predicted at $M_T = 0.88$, while a radiating shock wave was correctly predicted to form at $M_T = 0.90$ and above. However, the pulse width predicted with this extended Kirchhoff formulation appears to be in error (up to 30 percent too narrow) throughout the computed Mach number range. The source of this discrepancy is not presently understood.

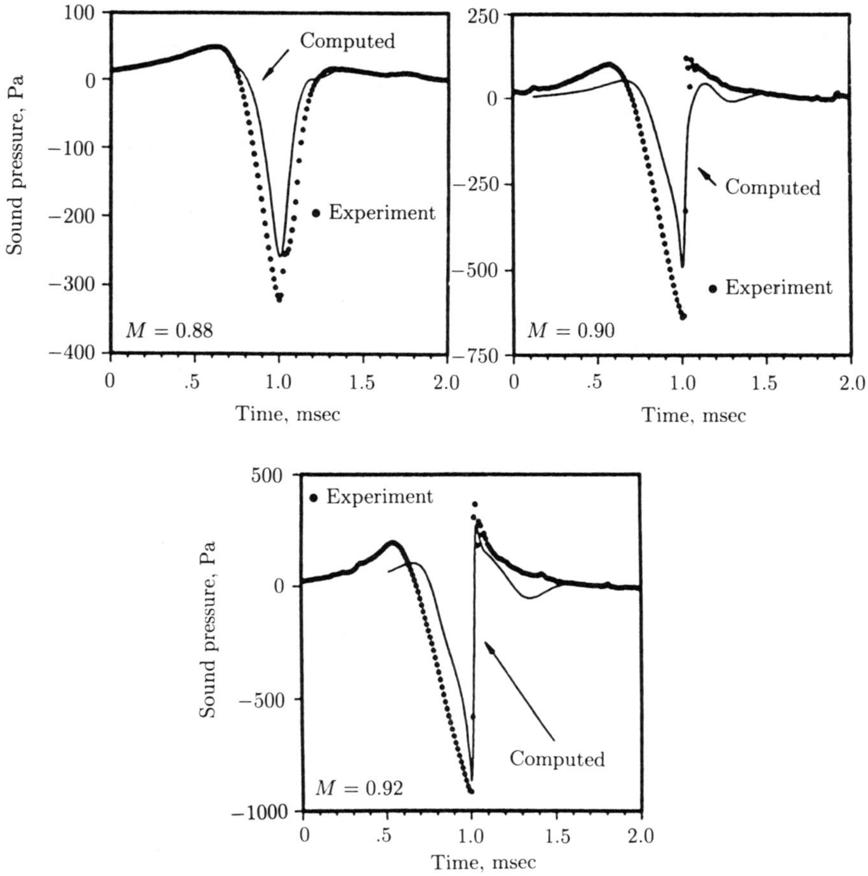

Figure 57. Predicted values from extended Kirchhoff computations and experimental values for $M_T = 0.88$, 0.90, and 0.92.

The peak negative-pressure amplitude (fig. 56) is more accurately predicted by this new method near the delocalization Mach number. In addition, the new method appears to more correctly follow the amplitude trends above $M_T = 0.90$. However, it does not adequately predict the noise amplitude at lower tip Mach numbers, a fact that is still not fully understood.

In summary, the quadrupole formulation appears to work well in hover at the delocalization Mach number and below; both amplitude and pulse shape are accurately predicted. Above the delocalization Mach number, the amplitude of the pulse and the overall pulse shape are predicted by the extended Kirchhoff formulation, but the pulse width is markedly in error.

Forward-Flight Harmonic Noise

Linear Theory for Thickness Noise

Unfortunately, neither the quadrupole nor the Kirchhoff formulation has been applied in a rigorous manner to the problem of HSI noise for helicopters in forward flight. The complexity of mapping the unsteady aerodynamic near field of a helicopter in forward flight to an observer in the far field has been too great. Instead, only computations of the linear part of the problem have been performed, as illustrated in figure 58 (from ref. 39). Measured peak pressures are plotted versus advancing-tip Mach number M_{AT}, the primary nondimensional parameter of HSI noise.

Similar to the hovering helicopter problem, linear theory substantially underpredicts the radiated noise field of a helicopter in forward flight. The predicted pressures are too low by at least a factor of 2, while the delocalization phenomenon is definitely not captured by linear theory. Clearly, nonlinear effects must be included in methods developed to predict the far-field acoustic radiation of high-speed helicopters. However, the understanding that has been developed by modeling the high-tip-speed hovering rotor has been applied, with some success, to the forward-flight rotor problem. Numerical computational fluid dynamic simulations of the advancing side of a high-speed rotor have been and are being made to help designers keep the local shock waves from delocalizing to the acoustic far field.

BVI noise

Predicting the noise that arises from blade-vortex interaction (BVI) is just as challenging as predicting HSI noise, but for different reasons. As discussed previously, the second term of equation (1) becomes an important source in the acoustic analogy formulation. If the blade pressure time histories are known and are treated as such, then the governing equation (6) becomes linear and is solved with classic techniques. Although simple in concept to solve, the linear BVI problem is still quite complex because it depends critically upon the local pressure distribution time histories of the individual rotor blades. In fact, accurate predictions of BVI noise necessitate a very high fidelity in air load predictions or measurements. Predicting air loads to the required accuracy for rotorcraft noise has not been possible to date. Instead, predictions of BVI noise have used experimental pressure distributions on the blades as input.

This type of computation has been attempted by several researchers (refs. 63 to 65). One of the most reliable sets of simultaneous pressure and acoustic data was taken in the DNW wind tunnel on an AH-1 two-bladed model rotor (ref. 27). High-frequency (up to 20,000 Hz) data were gathered on many miniature pressure transducers distributed over the blade and used as input for the resulting computations. Parameter identification methods were used to develop a continuous mathematical description of the pressures over the blade surface. This complete description of the pressure time history was then used as input in equivalent forms of equation (1) (refs. 63 and 66) and the results were compared with experimental data taken in the DNW wind tunnel. A typical comparison is shown in figure 59.

The dashed curve in figure 59(a) is the contribution of the distributions of pressure over the blade surface, while the solid curve represents the contributions of both blade pressure and thickness. For this microphone position (30° below the rotor plane), the blade pressure is the dominant source of noise.

The agreement between predicted and measured values is quite encouraging but is not quantitative enough to be able to say that linear theory can be used to completely model the BVI problem. In general, the correct number of BVI's has been predicted

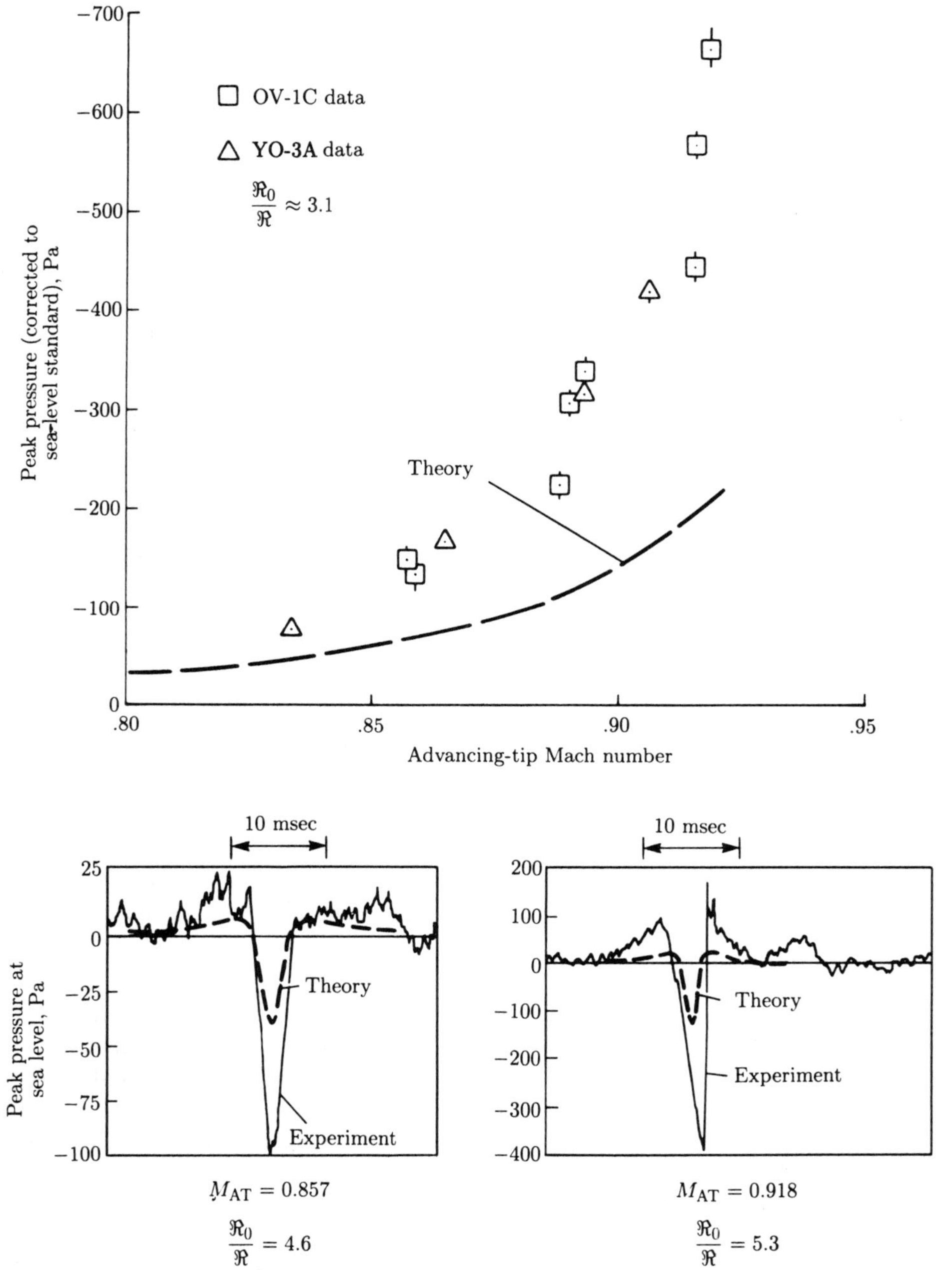

Figure 58. In-plane high-speed impulsive noise of UH-1H helicopter. (From ref. 39.)

and the width of each pulse seems representative. However, the amplitudes of the predicted pulses are different from those of the respective measured pulses by up to a factor of 2.

In addition, there is evidence that some of these pulses become shock-like in character when the BVI pulses are intense. An example of this phenomenon for an

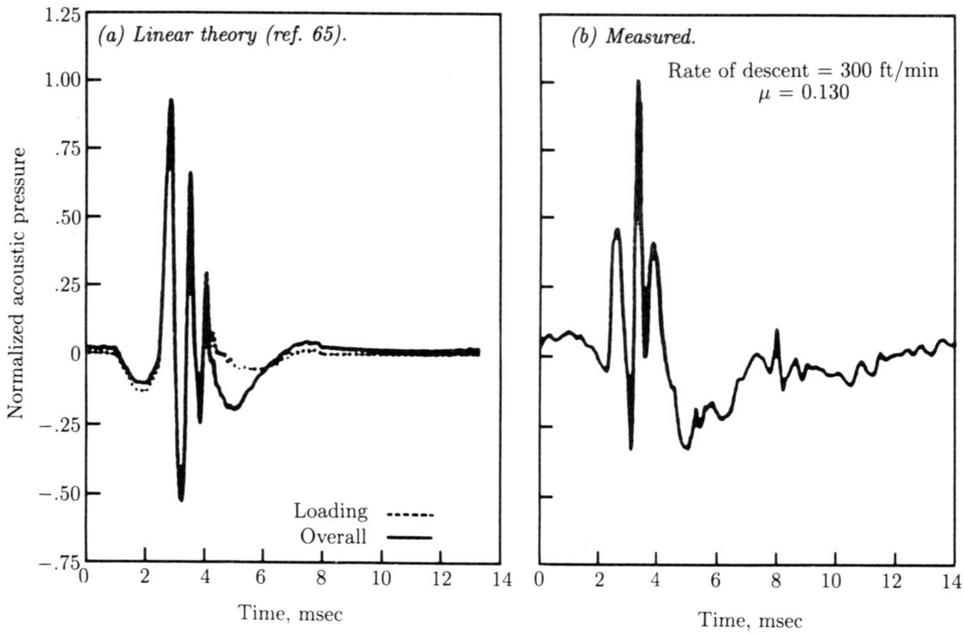

Figure 59. Blade-vortex interaction impulsive-noise time histories of AH-1 model rotor.

AH-1S helicopter that was encountering BVI impulsive noise is shown in figure 60. The two different waveforms that are shown were measured with the in-flight technique previously described for steady-state flight at an IAS of 60 knots and a rate of descent of 400 ft/min. The dashed curve represents a typical microphone pressure time history of BVI noise, while the solid curve is representative of a severe burst of BVI noise encountered during the run. The basic unsteadiness of this phenomenon in flight tests is a general characteristic of this type of noise.

The shock-like character of the solid curve is evident. The pressure rise is quite fast, but not instantaneous, even under the most severe BVI conditions measured on this helicopter. This change in pulse shape from the more symmetrical pulse shape typical of less severe BVI phenomena to a more asymmetrical pattern is responsible for an increase in subjective annoyance of about 4 PNdB. Similar to HSI noise predictions, linear theory does not predict this development of shock-like behavior for BVI noise. Unfortunately, at this time not enough high-fidelity acoustic data exist on this phenomenon to be able to assess the importance of nonlinear effects for BVI in general. It is not known whether these nonlinear effects exist only under exceptional circumstances or whether they are routinely encountered in many helicopter operations where BVI is generated.

Computational fluid dynamics calculations have recently been applied to help address a reduced class of these problems (refs. 67 to 69)—that of a rotor airfoil with a free vortex encountering two-dimensional unsteady flow (fig. 61). The vortex passes from left to right, traveling x_v chord lengths parallel to the flow at a distance of y_v chord lengths beneath the airfirfoil. That part of the complex three-dimensional BVI problem where the vortex and the rotor blade are nearly in parallel alignment has been approximated in two dimensions and the unsteady aerodynamic field near the

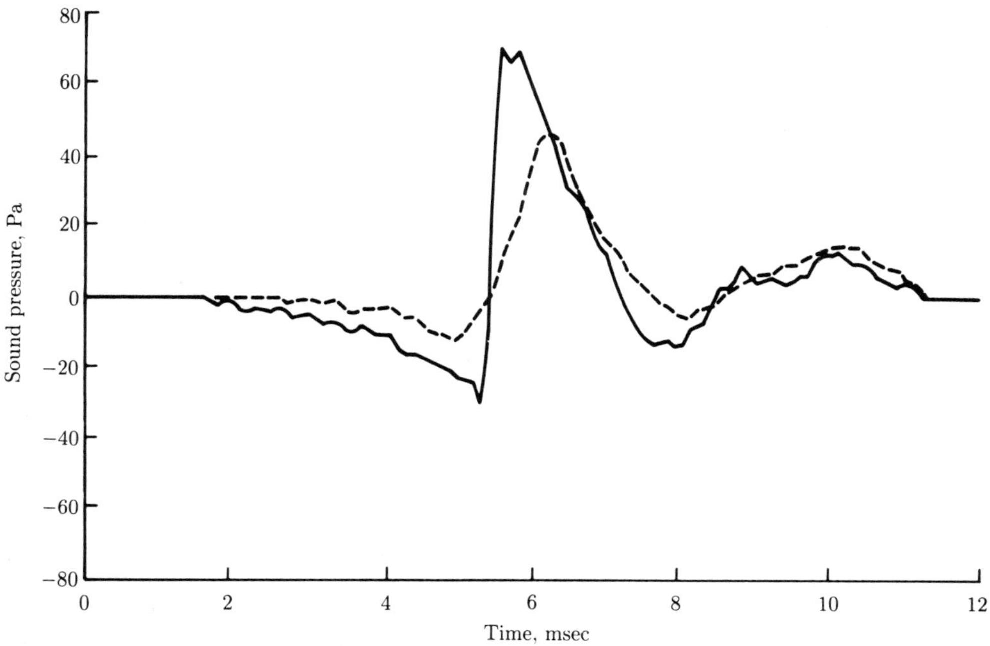

Figure 60. Typical waveform variability for measured AH-1S helicopter BVI signatures. Rate of descent, 400 ft/min; IAS = 60 knots. (Based on ref. 26.)

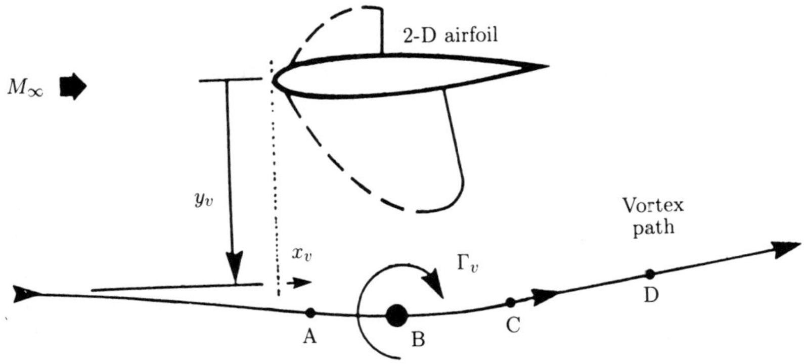

Figure 61. Two-dimensional approximation of BVI problem. (Based on ref. 70.)

airfoil has been computed. The two-dimensional far-field acoustic solution has been obtained by two separate methods: by direct numerical computation (ref. 70) and by use of a Kirchhoff solution to map the near-field aerodynamics to the acoustic far field (ref. 71). A sample computation of the vortex encounter with the airfoil is shown in figure 62 (from ref. 70). Contours of scaled pressure disturbance $(C_p - C_{p_0})\sqrt{\overline{\Re}}$ are shown, where C_p is the pressure coefficient, C_{p_0} is the undisturbed pressure coefficient, and $\overline{\Re}$ is the nondimensional distance of the source in the acoustic frame.

The development of a radiating acoustic wave is clearly evident in figure 62 as the vortex passes about a quarter of the chord beneath the airfoil. The most dense isobars exist forward and below the interaction process, an observation also made in the full-scale flight test. Although quantitative relationships between two- and three-dimensional acoustic problems are difficult, these computations can be used to develop estimates of the real BVI problem. Results to date have indicated that linearized small-disturbance simulations of the two-dimensional BVI problem do not adequately represent the aerodynamic near field. A full Navier-Stokes simulation, however, is not needed because all the important aerodynamic information which is radiated to the acoustic far field is generated near the leading edge of the airfoil. A high-order, time-accurate Euler method seems to yield the most cost-effective results.

In general, the shape of the leading edge of the airfoil seems to have the most pronounced effect on the computed far-field noise. Sharp leading edges radiate more noise than rounded ones. The importance of nonlinear aerodynamics on acoustic radiation is still being debated, with some authors claiming significant effects while others are claiming little to no measurable difference between linear and nonlinear effects in the acoustic far field. A definitive experiment to verify these methods is required.

Broadband Noise

Predicting the complete broadband noise spectra of rotorcraft is at best difficult under very controlled conditions, and under less controlled conditions it is an almost impossible task. Much of the problem occurs because it is often difficult to isolate the most important governing mechanisms for the problem at hand. The broadband acoustic radiation problem depends on details of the aerodynamic state of the rotors. These aerodynamic details include inflow turbulence characteristics, blade-wake turbulence characteristics, boundary-layer turbulence, etc. They act as input, or forcing functions, to the acoustic sources, which ultimately radiate to the acoustic far field.

Several approaches have been applied to calculate broadband acoustic radiation (ref. 72). They all use modified forms of equation (1) where the primary source term is the blade pressure. The broadband noise is due to the random forces (pressures) applied to the fluid by the pressure of the blades. (It should be noted that others have identified other broadband noise sources that are not dipole (force) in nature that may be important in high Mach number flow. These are not considered here.) One approach treats the general case of unsteady forces distributed in space, with a specialization in the rotating geometry of rotors. A second approach approximates the distributed blade forces as rotating concentrated forces (dipoles). A third approximates a rotating blade as a sequence of straight-line motions and then calculates the acoustic radiation from each blade element undergoing these linear motions.

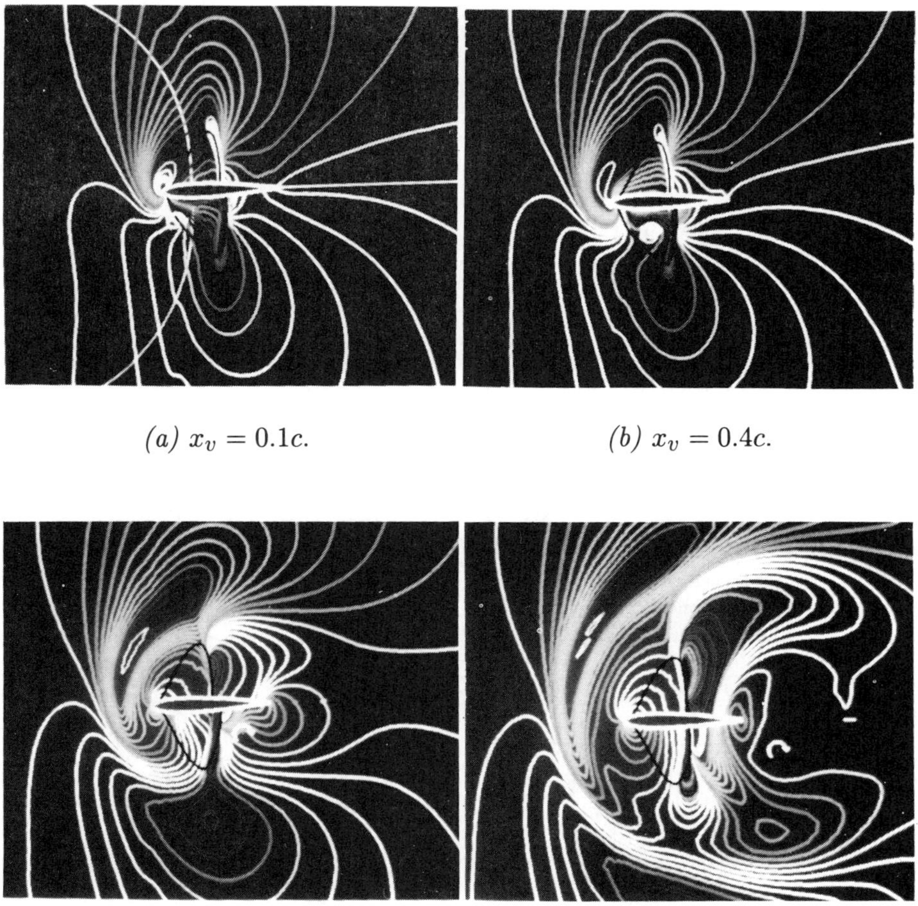

(a) $x_v = 0.1c$.

(b) $x_v = 0.4c$.

(c) $x_v = 0.8c$.

(d) $x_v = 1.2c$.

Figure 62. Pressure disturbance plots of NACA 0012 airfoil from computational Euler solution. $M_\infty = 0.8$; $y_v = -0.26c$; $\Gamma_v = 0.2$. *(From ref. 70.)*

Hover

The broadband noise due to turbulence ingestion of a hovering rotor has been studied for many years (refs. 73 and 74). Recently, a comprehensive experiment was run on a model rotor in an anechoic chamber where the radiated noise and the turbulence characteristics of the inflow to the rotor were both measured (ref. 74). These inflow characteristics were then used to generate unsteady forces (pressures) on the blade which were then used to calculate the acoustic radiation. The resulting predicted and measured values are shown in figure 63. In general, at most azimuthal locations, broadband noise theory tends to overpredict the low harmonics and underpredict the high harmonics compared with the measured noise. It is difficult to assess the contributions of steady forces and thickness at low harmonics because they were not used as input to the current theoretical model. However, the agreement between predicted and measured values at mid to high harmonics is quite reasonable. The large discrepancy at a polar angle of 90° (in the plane of the rotor) is probably

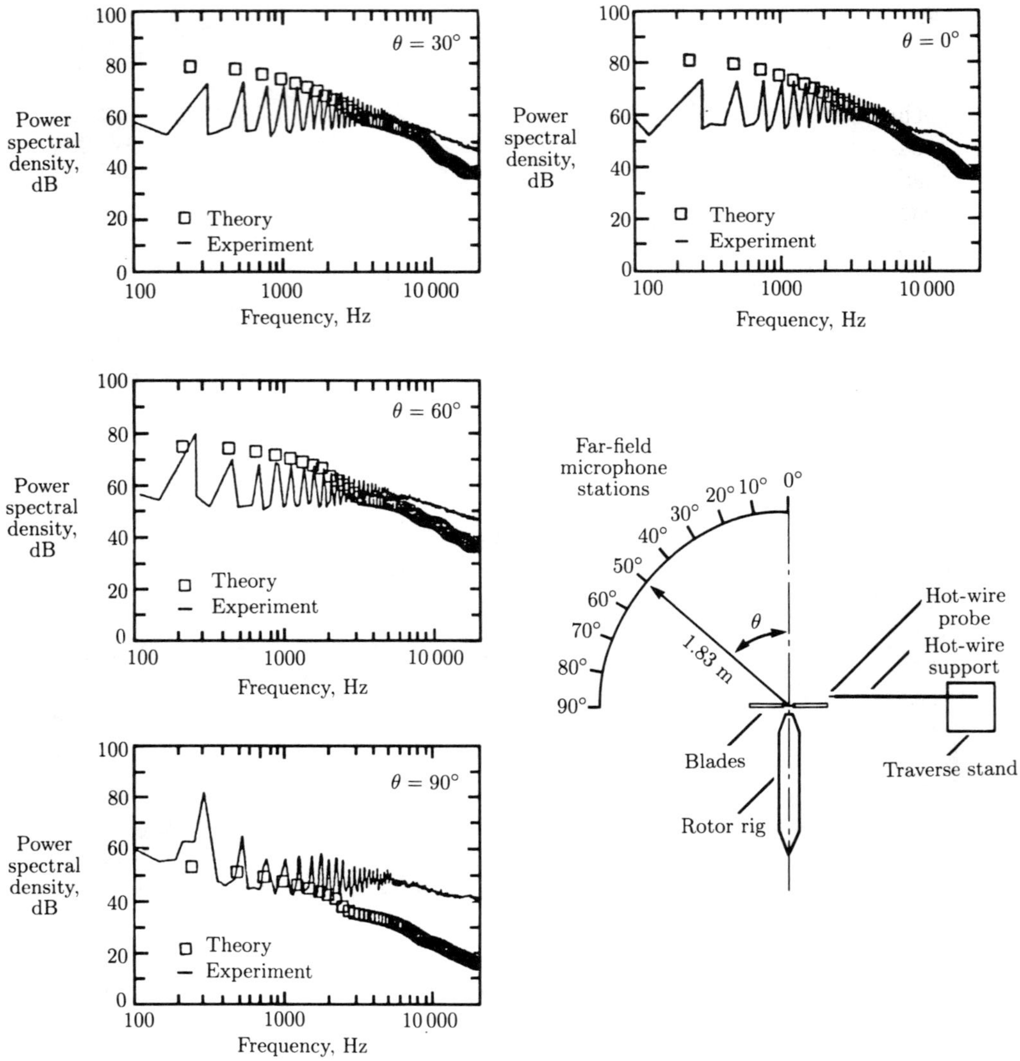

Figure 63. Predicted peak levels and measured spectra for hovering turbulence ingestion noise. (From ref. 74.)

attributable to other sources of in-plane noise which were not modeled in this approach.

Broadband noise is also produced by the self-generated turbulence in a blade boundary layer passing over the trailing edge of the blade. This effect has been theoretically modeled by solving the problem of a statistically stationary pressure field convecting past a trailing edge. The pressure distributions are integrated to obtain the fluctuating force distributions for the dipole source model used to predict the noise radiation. The resulting analysis is restricted to angles not too close to the rotor plane. A subset of this type of noise arises from blunt trailing edges. Vortex shedding is induced and creates a periodic high-frequency broadband noise. A similar phenomenon occurs when a bluff body or a laminar airfoil is placed in smooth flow.

Vortex shedding is induced and produces a peaked but continuous acoustic spectrum shape that can be predicted if the spectral content of the oscillating dipole source is known.

Another potential source of broadband noise on rotors is locally separated flow from local stall or from tip-vortex formation. Pressure fluctuations arise from large local blade angles of attack or from the boundary layer being swept around the blade tip by strong pressure gradients in that region. These fluctuations are then cast as dipole radiators and the far-field noise from these sources is computed.

Forward Flight

The prediction accuracy and importance of several of these sources of broadband noise have recently been verified in a model rotor test run in the DNW acoustically treated open-jet wind tunnel (ref. 21). A 40-percent-scale model of a BO-105 helicopter was tested in forward flight at two different operational rotor speeds: normal rotor speed (1050 rpm) and half normal rotor speed (525 rpm). At low operational tip speeds, broadband noise theory does a respectable job of predicting the shape and value of SPL's for the forward-flight condition shown in figure 64(a). The noise levels in this case are quite low, not typical of existing rotorcraft. At the higher, normal operational rotor speed, theory and experiment do not agree as well (fig. 64(b)). In addition, another source of broadband noise, called blade-wake interaction (BWI) noise, was identified and is a strong function of the rotor state. In near-level flight, large low- to mid-frequency levels of this noise were measured. In climbing flight (α_{TPP} more negative) these levels were dramatically reduced. It is postulated that this effect is caused by a wake-induced unsteadiness, which is reduced in climbing flight when the near wake of the rotor is more readily blown away from the rotor-tip-path plane. (See fig. 16.) Another source of this noise is postulated to be the interaction of the turbulent core of a tip vortex with a rotor blade. For most rotorcraft, this phenomenon is most likely to occur on the forward part of the rotor disk, where the rotor blade often intersects the path of previously shed tip vortices. The resulting unsteady blade pressures radiate broadband noise (ref. 75). It could be argued that this source of noise is always present to some degree, even during blade-vortex interaction. More careful experiments that measure the aerodynamic flow field and the acoustic radiation are needed to validate these hypotheses.

The statistical nature and multitude of potential causes of broadband noise have made theoretical prediction methods more difficult to validate than their deterministic counterparts. Discrepancies of 10 to 15 dB on full-scale aircraft in certain frequency ranges are not uncommon. However, for most rotorcraft, many of these broadband noise sources are lower in level and annoyance than the periodic main- and tail-rotor sources, and this fact makes their absolute prediction less important. In addition, the higher frequency of most broadband noise causes these sources to be rapidly attenuated by the atmosphere. However, if the tip speeds of future rotorcraft are reduced significantly, broadband noise might determine the radiation levels and annoyance of these new vehicles.

Predicting the total noise of a rotorcraft is simply a matter of summing all the important noise sources at a chosen observer position. Of course, the accuracy of this prediction is in reality no better than the accuracy of any one of the important noise sources. Under very controlled conditions, many of these sources cannot yet

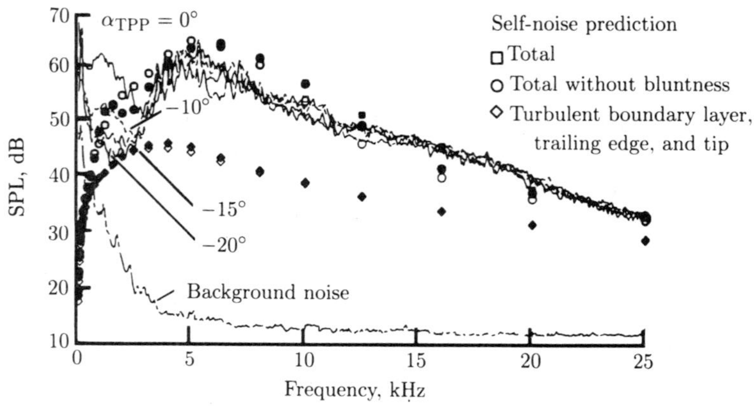

(a) Rotor speed = 525 rpm.

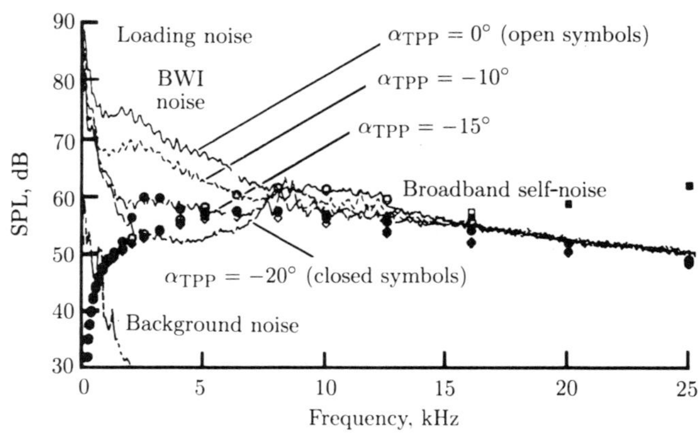

(b) Rotor speed = 1050 rpm.

Figure 64. Effect of rotor-tip-path-plane angle (α_{TPP}) for $C_T = 0.0044$ and $\mu = 0.086$ for two rotational speeds. (From ref. 21.)

be predicted to within 6 dB (factor of 2 in sound pressure), and this makes the prediction of most rotorcraft noise spectra in a typical flyby difficult at best. It is generally believed that a good part of this prediction difficulty is the inability to define adequately the aerodynamic input to the acoustic source model used to predict the noise. It is also likely that at the higher tip Mach numbers of current rotorcraft, nonlinear effects contribute to the lack of prediction accuracy. This is clearly the case for rotorcraft in high-speed nonaxial forward flight.

Rotorcraft Acoustic Design Trends

It has been known for a long time that a proven way to reduce rotorcraft noise is to simply lower the tip speed of the rotors. This technique is especially effective if the tip

Mach number of any rotating blade on the rotorcraft is originally close to 1.0. In this case, small reductions in tip Mach number can eliminate delocalization completely and greatly reduce the annoyance of the vehicle. Unfortunately, large reductions in rotorcraft tip speed are not cost-effective for most helicopters because they usually require an increase in the operational weight of the vehicle for the same payload. For example, lower tip speeds for the same input power require more blade area to carry a given payload with adequate stall margins. The added blade area increases blade weight. The rotorcraft operating at lower tip speed and the same power also requires a stronger (and usually heavier) transmission to absorb the extra torque. The lower operational tip speed usually requires more added mass at the tips to ensure a safe operational autorotation envelope, further increasing vehicle weight.

However, the most severe forms of rotorcraft noise have been mitigated to a large extent in many helicopter designs of today when compared with the UH-1 designs of the 1960's. To some extent, these noise reductions are a direct result of a design philosophy change. Instead of emphasizing hover performance at the expense of forward-flight performance to meet U.S. military requirements, newer designs have placed importance on both aspects of rotorcraft performance. In fact, in the latest designs, high-speed forward flight is a highly valued attribute of conventional rotorcraft. This increasing emphasis on high-speed flight has forced helicopter designers to lower the hovering tip Mach number, which lowers the advancing-tip Mach number as well and thus avoids compressibility effects in high-speed flight. The compressibility effects not only can cause delocalization of acoustic waves, they also can cause large increases in required power and severe vibration. Thus, it is advantageous to aerodynamicists and acousticians that tip speed be reduced to avoid delocalization in high-speed flight. Tip speeds of about 700 ft/sec, which allow forward-flight velocities of about 150 knots, are common today.

The quest to go faster but keep tip speeds high enough to yield reasonable hover and forward-flight performance has led to some blade design trade-offs on conventional rotorcraft. In particular, the tips of most new rotor blades are now thinned, tapered, and sometimes swept. All three effects tend to reduce HSI noise radiation. Thinning the tip of the rotor directly reduces thickness noise and increases the delocalization Mach number. Tapering the tip also reduces thickness noise by lessening the thickness effect. Finally, sweeping the blade tip, as in wing sweep on supersonic airplanes, tends to lower the effective tip Mach number in the tip region, thus reducing the peak noise levels and delaying and lengthening the maximum thickness pulse time event. This latter effect can alter the location of the maximum noise intensity point, focusing it more to the forward quadrant on the retreating side of the disk with increasing blade sweep.

All these design changes for high-speed blades can have significant impact for rotorcraft design and operations. Thinning the tip definitely reduces HSI noise and compressibility effects but also aggravates blade stall in both high-speed and medium-speed maneuvering flight. The thinner, sharper airfoil sections stall at smaller angles of attack and have more unfavorable pitching-moment characteristics than their thicker counterparts. On the other hand, tapering and sweeping the tips of rotors can yield aerodynamic benefits beyond reduced noise level. For example, the replacement AH-1S blade has a tapered tip that reduces HSI peak noise levels by 6 dB and also permits the AH-1S to fly 10 knots faster at the same power compared with the standard untapered blade (ref. 76).

The design of these new rotor tips (and, in fact, of the entire blade) for noise and performance is increasingly relying on computational fluid dynamics (CFD). Sophisticated models of the rotor and helicopter dynamic system are being coupled to CFD methods to model the rotor performance over a wide range of conditions. Most applications of these new methods have focused on high-speed forward flight, where there is simply no other way to predict the nonlinear aerodynamic and acoustic environment of the rotor. Tip shapes to minimize acoustic delocalization and optimize performance are being tested in wind tunnels today and will increase the cruising speed of tomorrow's conventional helicopter.

Designing rotors to minimize noise due to BVI is more challenging. Although reducing the tip speed of the rotor does significantly reduce BVI noise, this alone does not mitigate the problem enough to allow the helicopter to operate acceptably. Increasing the number of blades in the rotor system has probably been the most effective means to date of reducing this noise. For the same thrust, increasing the number of blades effectively reduces the strength of each blade-tip vortex. This, in turn, lessens the amplitude of the interaction pulses and reduces the radiated noise field. However, increasing the number of blades also raises the frequency of the BVI phenomena and therefore tends to increase the subjective annoyance levels. However, these higher frequencies can be more rapidly attenuated by the atmosphere. Of course, too many rotor blades on helicopters also cause practical engineering problems. Four and five main-rotor blades are standard practice throughout the industry today.

There have been many attempts at tailoring the tip shapes on rotor blades to reduce BVI noise (refs. 77 and 78). The concept is to spread out and diffuse the vortex so that the BVI is softened to radiate less noise. Other aerodynamic surfaces have also been added to the tips of rotors to hasten this diffusion process, with limited success to date. The best reduction method so far has simply been to taper the tip of the rotor to diffuse the concentrated bound circulation. In addition, the small chord in the tip area decreases the chord Reynolds number and makes the viscous effects more important. The BVI noise levels with the K747 tapered tip blade have shown reductions in peak annoyance levels up to 5 PNdB when compared with rectangular AH-1S blades (ref. 76).

Tail rotors have been significant sources of noise on most helicopters, especially from a community annoyance standpoint. In addition to the isolated rotor noise sources such as HSI and BVI noise, much of the tail-rotor noise is caused by unsteady velocity fluctuations (dirty inflow) passing into the tail rotor. Vortices shed from main-rotor blades and separated flows or vortex flows trailing from the fuselage and hub all create a disturbed inflow to the tail rotor that create tail-rotor noise, mostly of a harmonic nature. Designs to minimize this noise source on conventional tail rotors have focused on placing the tail rotor in as clean a flow as possible under all flight conditions. The most direct way of reducing tail-rotor noise is to remove or replace the tail rotor entirely. This has been done on some French helicopters with a fan-in-fin design to provide directional control and to counteract main-rotor torque. Of course, the fan-in-fin design has its own unique noise characteristics, a high-frequency whine that, at close ranges, can be more annoying than the tail rotor it replaces. Replacing the conventional tail rotor with a circulation-controlled boom with a small jet reaction control at its end has also been successfully demonstrated in the United States. The concept counteracts most of the main-rotor torque by generating high lift coefficients on the circulation-controlled boom, which operates in the main-rotor downwash field. The pilot controls yaw through the small jet reaction control. Noise measurements on this novel approach have been encouraging.

Broadband noise from main and tail rotors can be important in certain flight regimes. To date, no effective means have been found to reduce this source of noise from the main rotors without lowering tip speed and degrading performance. However, on tail rotors, where the chord Reynolds numbers are low, tripping the boundary layer to avoid Karman-vortex-like streets has also been used to minimize high-frequency shedding noise.

The introduction of tilt-rotor aircraft may revolutionize the rotorcraft industry and will probably change the sound of rotorcraft. High-speed impulsive noise and blade-vortex interaction noise will also govern the acoustic design of these vehicles (ref. 79). In hover, the highly twisted, heavily loaded blades will exhibit both steady and unsteady loading and thickness effects. In transition flight in the helicopter mode, HSI and BVI effects will dominate. Fortunately, in cruising flight in the aircraft mode at reduced tip speeds, very little noise will be radiated, so this will be a very quiet cruising vehicle. Near terminal areas, when the tilt rotor is in the helicopter configuration, noise levels similar to helicopter main-rotor radiation can be expected. The major design parameter to control this noise is tip speed. However, lowering the tilt-rotor hovering tip speed reduces its hovering performance, an already critical parameter for cost-effective operation of this vehicle. Operational tip speeds of 750 to 800 ft/sec are expected to yield good performance and reasonable radiated noise levels for military missions. Further reductions in tip speed may be required if commercial utilization of this new class of vehicles is to be realized.

The design of rotorcraft to minimize noise radiation has been said to be so complex a problem that both the researcher and the designer need not fear losing their jobs as technology progresses. Indeed, large noise reductions sometimes seem to be nigh impossible without sacrificing performance. Substantial progress has been made to reduce rotorcraft noise, and design tools are now available to avoid developing vehicles that are *not* good neighbors. In the next 10 years, these techniques will continue to become part of the rotorcraft design process, leading to the evolution of better performing, quieter rotorcraft.

References

1. Lowson, Martin V.: Helicopter Noise: Analysis—Prediction and Methods of Reduction. *Helicopter Aerodynamics and Dynamics,* AGARD-LS-63-5, Mar. 1973, pp. 5.1–5.37.
2. Cox, C. R.: Subcommittee Chairman's Report to Membership on Aerodynamic Sources of Rotor Noise. Preprint No. 625, 28th Annual Forum, American Helicopter Soc., Inc., May 1972.
3. George, A. R.: Helicopter Noise: State-of-the-Art. *J. Aircr.* vol. 15, no. 11, 1978, pp. 707–715.
4. White, Richard P., Jr.: The Status of Rotor Noise Technology. *J. American Helicopter Soc.,* vol. 25, no. 1, Jan. 1980, pp. 22–29.
5. Schmitz, F. H.; and Yu, Y. H.: Helicopter Impulsive Noise: Theoretical and Experimental Status. *J. Sound & Vibration,* vol. 109, no. 3, Sept. 22, 1986, pp. 361–422.
6. Brooks, Thomas F.: Effect of Signal Jitter on the Spectrum of Rotor Impulsive Noise. *Vertica,* vol. 12, no. 3, 1988, pp. 257–265.
7. Babkin, Alexander S.: *Signal Restoration of Non-Stationary Acoustic Signals in Time Domain.* NASA CR-181627, 1988.

8. Cox, C. R.; and Lynn, R. R.: *A Study of the Origin and Means of Reducing Helicopter Noise.* TCREC Tech. Rep. 62-73, U.S. Army, Nov. 1962.
9. Ffowcs Williams, J. E.; and Hawkings, D. L.: Sound Generation by Turbulence and Surfaces in Arbitrary Motion. *Philos. Trans. Royal Soc. London,* ser. A., vol. 264, no. 1151, May 8, 1969, pp. 321–342.
10. Hawkings, D. L.; and Lowson, M. V.: Theory of Open Supersonic Rotor Noise. *J. Sound & Vibration,* vol. 36, no. 1, Sept. 8, 1974, pp. 1–20.
11. Farassat, F.: *Theory of Noise Generation From Moving Bodies With an Application to Helicopter Rotors.* NASA TR R-451, 1975.
12. Isom, Morris P.: *The Theory of Sound Radiated by a Hovering Transonic Helicopter Blade.* POLY-AE/AM Rep. No. 75-4 (Contract NAS2-8399), Polytechnic Inst. of New York, May 1975.
13. Gutin, L.: *On the Sound Field of a Rotating Propeller.* NACA TM 1195, 1948.
14. Garrick, I. E.; and Watkins, Charles E.: *A Theoretical Study of the Effect of Forward Speed on the Free-Space Sound-Pressure Field Around Propellers.* NACA Rep. 1198, 1954. (Supersedes NACA TN 3018.)
15. Deming, A. F.: *Noise From Propellers With Symmetrical Sections at Zero Blade Angle, II.* NACA TN 679, 1938.
16. Wright, S. E.: Discrete Radiation From Rotating Periodic Sources. *J. Sound & Vibration,* vol. 17, no. 4, Aug. 22, 1971, pp. 437–498.
17. Hooper, W. E.: The Vibratory Airloading of Helicopter Rotors. Paper No. 46, *Ninth European Rotorcraft Forum*, Sept. 1983.
18. Tangler, James L.: Schlieren and Noise Studies of Rotors in Forward Flight. *33rd Annual National Forum,* American Helicopter Soc., Inc., May 1977, pp. 77.33-05-1–77.33-05-12.
19. Brooks, Thomas F.; and Schlinker, Robert H.: Progress in Rotor Broadband Noise Research. *Vertica,* vol. 7, no. 4, 1983, pp. 287–307.
20. Homicz, G. F.; and George, A. R.: Broadband and Discrete Frequency Radiation From Subsonic Rotors. *J. Sound & Vibration,* vol. 36, no. 2, Sept. 22, 1974, pp. 151–177.
21. Brooks, Thomas F.; Marcolini, Michael A.; and Pope, D. Stuart: Main Rotor Broadband Noise Study in the DNW. *J. American Helicopter Soc.,* vol. 34, no. 2, Apr. 1989, pp. 3–12.
22. Hubbard, Harvey H.; and Lassiter, Leslie W.: *Sound From a Two-Blade Propeller at Supersonic Tip Speeds.* NACA Rep. 1079, 1952. (Supersedes NACA RM L51C27.)
23. Leverton, J. W.: The Noise Characteristics of a Large "Clean" Rotor. *J. Sound & Vibration,* vol. 27, no. 3, Apr. 8, 1973, pp. 357–376.
24. Robinson, Frank: *Component Noise Variables of a Light Observation Helicopter.* NASA CR-114761, 1973.
25. Schmitz, F. H.; and Boxwell, D. A.: In-Flight Far-Field Measurement of Helicopter Impulsive Noise. *J. American Helicopter Soc.,* vol. 21, no. 4, Oct. 1976, pp. 2–16.
26. Boxwell, D. A.; and Schmitz, F. H.: Full-Scale Measurements of Blade-Vortex Interaction Noise. *J. American Helicopter Soc.*, vol. 27, no. 4, Oct. 1982, pp. 11–27.
27. Boxwell, D. A.; Schmitz, F. H.; Splettstoesser, W. R.; and Schultz, K. J.: Helicopter Model Rotor-Blade Vortex Interaction Impulsive Noise: Scalability and Parametric Variations. *J. American Helicopter Soc.,* vol. 32, no. 1, Jan. 1987, pp. 3–12.
28. Martin, R. M.; Splettstoesser, W. R.; Elliott, J. W.; and Schultz, K.-J.: *Advancing-Side Directivity and Retreating-Side Interactions of Model Rotor Blade-Vortex Interaction Noise.* NASA TP-2784, AVSCOM TR 87-B-3, 1988.
29. Martin, Ruth M.; and Hardin, Jay C.: The Spectral Characteristics of Rotor Blade-Vortex Interaction Noise: Experimental and Mathematical Results. AIAA-87-0251, Jan. 1987.
30. Paterson, Robert W.; Vogt, Paul G.; Fink, Martin R.; and Munch, C. Lee: *Vortex Shedding Noise of an Isolated Airfoil.* Rep. No. K910867-6 (Contract DAHC04-69-C-0089), United Aircraft Corp. Research Labs., Dec. 1971.

31. Schmitz, F. H.; Boxwell, D. A.; and Vause, C. R.: High-Speed Helicopter Impulsive Noise. *J. American Helicopter Soc.,* vol. 22, no. 4, Oct. 1977, pp. 28–36.

32. Splettstoesser, Wolf R.; Schultz, Klaus J.; Schmitz, Fredric H.; and Boxwell, Donald A.: Model Rotor High-Speed Impulsive Noise—Parametric Variations and Full-Scale Comparisons. Paper presented at 39th Annual National Forum of the American Helicopter Society (St. Louis, Missouri), May 9–11, 1983.

33. Schmitz, F. H.; Boxwell, D. A.; Dahan, C.; and Lewy, S.: *A Note on the General Scaling of Helicopter Blade-Vortex Interaction Noise.* ONERA T.P. No. 1982-32, May 1982.

34. Sternfeld, Harry, Jr.; and Schaffer, Edward: *An Investigation of Rotor Harmonic Noise by the Use of Small Scale Wind Tunnel Models.* NASA CR-166337, 1982.

35. Van Ditshuizen, J. C. A.; Courage, G. D.; Ross, R.; and Schultz, K.-J.: Acoustic Capabilities of the German-Dutch Wind Tunnel DNW. AIAA-83-0146, Jan. 1983.

36. Lighthill, M. J.: On Sound Generated Aerodynamically. I. General Theory. *Proc. Royal Soc. (London),* ser. A, vol. 211, no. 1107, Mar. 20, 1952, pp. 564–587.

37. Lowson, M. V.; and Ollerhead, J. B.: A Theoretical Study of Helicopter Rotor Noise. *J. Sound & Vibration,* vol. 9, no. 2, Mar. 1969, pp. 197–222.

38. Wright, S. E.: Sound Radiation From a Lifting Rotor Generated by Asymmetric Disk Loading. *J. Sound & Vibration,* vol. 9, no. 2, Mar. 1969, pp. 223–240.

39. Schmitz, Fredric H.; and Yu, Yung H.: Theoretical Modeling of High-Speed Helicopter Impulsive Noise. *J. American Helicopter Soc.*, vol. 24, no. 1, 1979, pp. 10–19.

40. Boxwell, D. A.; Yu, Y. H.; and Schmitz, F. H.: Hovering Impulsive Noise—Some Measured and Calculated Results. *Helicopter Acoustics,* NASA CP-2052, Part I, 1978, pp. 309–322.

41. Schmitz, F. H.; and Yu, Y. H.: Transonic Rotor Noise—Theoretical and Experimental Comparisons. *Vertica,* vol. 5, no. 1, 1981, pp. 55–74.

42. Boxwell, D. A.; Zinner, R. A.; and Kodani, H. M.: *Boeing Vertol Model 360 Rotor Acoustic Signature Measurements in the DNW—Volume 1.* AATMR TR-88-1, U.S. Army, June 1988.

43. Aggarawal, H. R.; Schmitz, F. H.; and Boxwell, D. A.: Prediction and Measurement of Low-Frequency Harmonic Noise of a Hovering Model Helicopter Rotor. *45th Annual Forum,* American Helicopter Soc., Inc., May 1989, pp. 115–126.

44. Zinner, R. A.; Boxwell, D. A.; and Spencer, R. H.: *Review and Analysis of the DNW/Model 360 Rotor Acoustic Data Base.* NASA TM-102253, USAAVSCOM TM-89-A-002, 1989.

45. Farassat, F.: Linear Acoustic Formulas for Calculation of Rotating Blade Noise. *AIAA J.,* vol. 19, no. 9, Sept. 1981, pp. 1122–1130.

46. Yu, Yung H.; Caradonna, Frank X.; and Schmitz, Fredric H.: The Influence of the Transonic Flow Field on High-Speed Helicopter Impulsive Noise. *Proceedings of the Fourth European Rotorcraft and Powered Lift Aircraft Forum, Volume 2,* Sept. 13–15, 1978, pp. 58-0–58-16.

47. Farassat, F.; Morris, C. E. K., Jr.; and Nystrom, P. A.: A Comparison of Linear Acoustic Theory With Experimental Noise Data for a Small-Scale Hovering Rotor. AIAA Paper 79-0608, Mar. 1979.

48. Farassat, F.; and Martin, R. M.: A Note on the Tip Noise of Rotating Blades. *J. Sound & Vibration,* vol. 86, no. 3, Feb. 8, 1983, pp. 449–453.

49. Kittleson, John K.; and Yu, Yung H.: *Holographic Interferometry Technique for Rotary Wing Aerodynamics and Noise.* NASA TM-84723, 1982.

50. Isom, Morris P.: *Some Nonlinear Problems in Transonic Helicopter Acoustics.* ARO 12937.1-E, U.S. Army, May 1979. (Available from DTIC as AD A069 564.)

51. Hawkings, D.: Noise Generation by Transonic Open Rotors. *Mechanics of Sound Generation in Flows,* E.-A. Muller, ed., Springer-Verlag, 1979, pp. 294–300.

52. Caradonna, Francis X.; and Isom, Morris P.: Numerical Calculations of Unsteady Transonic Potential Flow Over Helicopter Rotor Blades. *AIAA J.,* vol. 14, no. 4, Apr. 1976, pp. 482–488.

53. Caradonna, Francis Xavier: The Transonic Flow on a Helicopter Rotor. Ph.D. Diss., Stanford Univ., 1978.

54. Isom, Morris P.: Acoustic Shock Waves Generated by a Transonic Helicopter Blade. *36th Annual Forum*, American Helicopter Soc., Inc., May 1980, pp. 80-63-1–80-63-16.
55. Caradonna, F. X.; and Steger, J. L.: Implicit Potential Methods for the Solution of Transonic Rotor Flows. *Proceedings of the 1980 Army Numerical Analysis and Computers Conference,* ARO 80-3, U.S. Army, Feb. 1980, pp. 19–38.
56. Tauber, Michael E.; and Hicks, Raymond M.: Computerized Three-Dimensional Aerodynamic Design of a Lifting Rotor Blade. *36th Annual Forum*, American Helicopter Soc., Inc., May 1980, pp. 80-2-1–80-2-11.
57. Chattot, J. J.: *Calculation of Three-Dimensional Unsteady Transonic Flows Past Helicopter Blades.* NASA TP-1721, AVRADCOM TR-80-A-2, 1980.
58. Shenoy, K. Rajarama: A Semiempirical High-Speed Rotor Noise Prediction Technique. *38th Annual Forum*, American Helicopter Soc., Inc., May 1982, pp. 508–516.
59. Hanson, D. B.; and Fink, M. R.: The Importance of Quadrupole Sources in Prediction of Transonic Tip Speed Propeller Noise. *J. Sound & Vibration*, vol. 62, no. 1, Jan. 8, 1979, pp. 19–38.
60. Prieur, J.: Calculation of Transonic Rotor Noise Using a Frequency Domain Formulation. *43rd Annual Forum*, American Helicopter Soc., Inc., May 1987, pp. 469–479.
61. Isom, Morris; Purcell, Timothy W.; and Strawn, Roger C.: Geometrical Acoustics and Transonic Helicopter Sound. AIAA-87-2748, Oct. 1987.
62. Purcell, Timothy W.: CFD and Transonic Helicopter Sound. Paper No. 2, *Fourteenth European Rotorcraft Forum*, Sept. 1988.
63. Nakamura, Yoshiya: Prediction of Blade-Vortex Interaction Noise From Measured Blade Pressure. Paper No. 32, *Seventh European Rotorcraft and Powered Lift Aircraft Forum, Volume 1*, Deutsche Gesellschaft fur Luft- und Raumfahrt e. V., Sept. 1981.
64. Schultz, Klaus-J.; and Splettstoesser, Wolf R.: Prediction of Helicopter Rotor Impulsive Noise Using Measured Blade Pressures. *43rd Annual Forum*, American Helicopter Soc., Inc., May 1987, pp. 405–420.
65. Joshi, Mahendra C.; Liu, Sandy R.; and Boxwell, Donald A.: Prediction of Blade-Vortex Interaction Noise Using Measured Blade Pressures. AIAA-87-2749, Oct. 1987.
66. Brentner, Kenneth S.: *Prediction of Helicopter Rotor Discrete Frequency Noise—A Computer Program Incorporating Realistic Blade Motions and Advanced Acoustic Formulation.* NASA TM-87721, 1986.
67. George, A. R.; and Chang, S.-B.: Flow Field and Acoustics of Two-Dimensional Transonic Blade-Vortex Interactions. AIAA-84-2309, Oct. 1984.
68. Rai, Man Mohan: Navier-Stokes Simulations of Blade-Vortex Interaction Using High-Order Accurate Upwind Schemes. AIAA-87-0543, Jan. 1987.
69. Srinivasan, G. R.: and McCroskey, W. J.: Numerical Simulations of Unsteady Airfoil-Vortex Interactions. *Vertica,* vol. 11, no. 1/2, 1987, pp. 3–28.
70. Baeder, J. D.: Computatioin of Non-Linear Acoustics in Two-Dimensional Blade-Vortex Interactions. Paper No. 1-1, *Thirteenth European Rotorcraft Forum*, Sept. 1987.
71. Lyrintzis, A. S.; and George, A. R.: Far-Field Noise of Transonic Blade-Vortex Interactions. *J. American Helicopter Soc.*, vol. 34, no. 3, July 1989, pp. 30–39.
72. George, A. R.: Analyses of Broadband Noise Mechanisms of Rotors. *Recent Advances in Aeroacoustics*, Anjaneyulu Krothapalli and Charles A. Smith, eds., Springer-Verlag, c.1986, pp. 87–127.
73. Paterson, Robert W.; and Amiet, Roy K.: *Noise of a Model Helicopter Rotor Due to Ingestion of Turbulence.* NASA CR-3213, 1979.
74. Simonich, John; Schlinker, Robert; and Amiet, Roy: Experimental Assessment of a Turbulence Ingestion Noise Theory. *44th Annual Forum,* American Helicopter Soc., Inc., June 1988, pp. 731–745.
75. Glegg, Stewart A. L.: The Prediction of Blade Wake Interaction Noise Based on a Turbulent Vortex Model. AIAA-89-1134, Apr. 1989.

76. Boxwell, D. A.; and Schmitz, F. H.: In-Flight Acoustic Comparison of the 540 and K747 Main Rotors for the AH-1S Helicopter. *Production Validation Test—Government: Karman K747 Improved Main Rotor Blade*, USAAEFA Project No. 77-38, U.S. Army, Oct. 1979, pp. 65–90.
77. Stepniewski, W. Z.; and Schmitz, F. H.: Possibilities and Problems of Achieving Community Noise Acceptance of VTOL. *Aeronaut. J.*, vol. 77, no. 750, June 1973, pp. 311–326.
78. Tangler, James L.: The Design and Testing of a Tip To Reduce Blade Slap. Preprint No. 963, *31st Annual Forum*, American Helicopter Soc., Inc., May 1975.
79. Brieger, John T.; Maisel, Martin D.; and Gerdes, Ronald: External Noise Evaluation of the XV-15 Tiltrotor Aircraft. Paper presented at American Helicopter Society National 'Specialists' Meeting on Aerodynamics and Aeroacoustics (Arlington, Texas), Feb. 25–27, 1987.

3 Turbomachinery Noise

Lead author

John F. Groeneweg
NASA Lewis Research Center
Cleveland, Ohio

Contributing authors

Thomas G. Sofrin
Consultant to Pratt and Whitney Division
United Technologies
Windsor Locks, Connecticut

Edward J. Rice
NASA Lewis Research Center
Cleveland, Ohio

Phillip R. Gliebe
General Electric Company
Cincinnati, Ohio

Introduction

Typical Engine Components and Configurations

Rotating and stationary blades, often in proximity to each other, are the essence of turbomachinery as used in flight vehicle propulsion. Fans, compressors, and turbines each can generate significant tonal and broadband noise. Figure 1 is a schematic cross section of the most common propulsion system used in commercial civil aviation—the high-bypass-ratio turbofan engine. The most prominent component, the fan, whose noise-generating mechanisms are outlined in figure 1, will be the focus of much of the discussion in this chapter because it is the primary turbomachinery noise source and the physics involved illustrates the key elements of the aeroacoustics of turbomachinery. Compressor and turbine noise can be important at low power settings, particularly for the blade rows nearest the core inlet or exhaust. Other propulsion systems, such as the turbojet and turboprop, have turbomachinery configurations similar in component arrangement to the core portion of the turbofan.

Contributions to Flyover Noise

The primary concern for turbomachinery noise is the community exposure during takeoff and approach operations. Contributions of turbomachinery components to flyover noise are shown in figure 2 as taken from system noise predictions for an energy efficient engine design study (ref. 1). The fan component controls the total perceived noise at both takeoff and approach, even with suppression by substantial use of acoustic treatment. The high bypass ratio (6 in this example) accounts for the less than dominant jet noise at takeoff. Additional fan treatment would uncover the turbine at approach and could also bring compressor noise (not shown) into the picture.

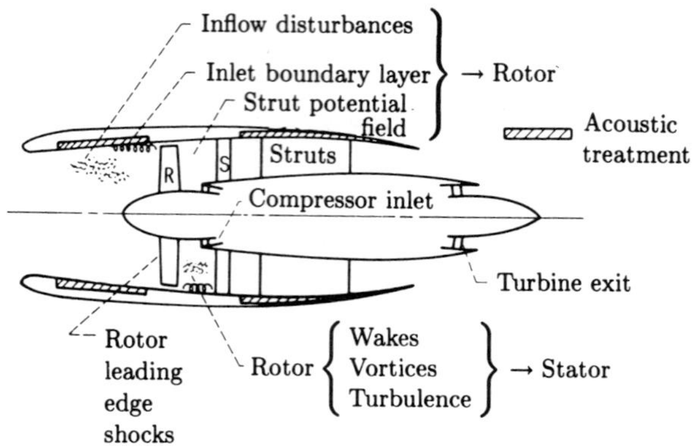

Figure 1. Schematic cross section of turbofan engine with turbomachinery noise-generating mechanisms.

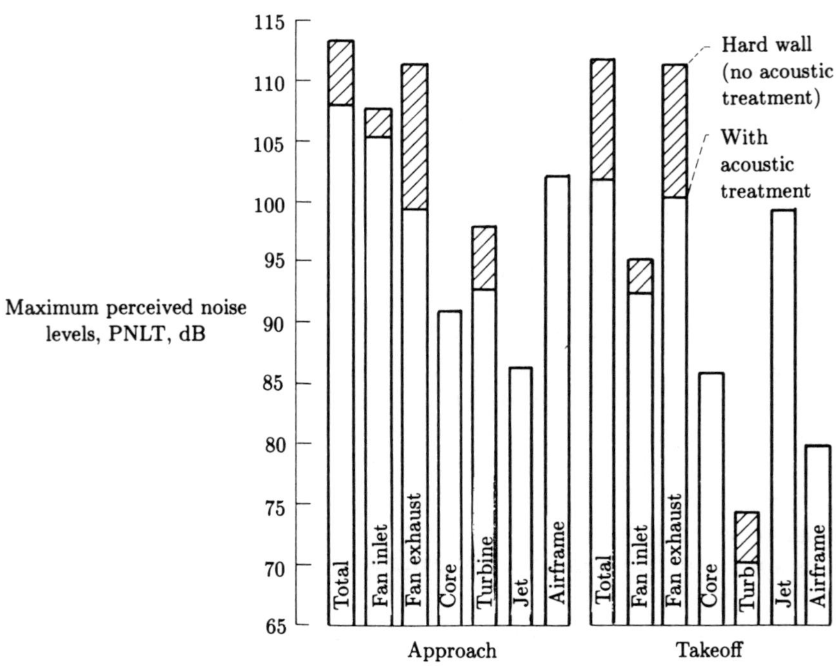

Figure 2. Component flyover noise levels for advanced turbofan. (From ref. 1.)

Operating Regimes and Spectral Content

Typical values of turbofan geometric and aerodynamic operating parameters are listed in table 1. Most recent designs favor fewer blades with wide chords. The two operating points of interest for community noise, approach and takeoff, correspond to subsonic and supersonic tip Mach numbers. Corresponding far-field spectra are very different in character, as shown by the narrow-band examples in figure 3. Subsonically, blade-passage frequency and its harmonics are superimposed on a broadband component, while supersonically, all multiples of shaft frequency appear. The latter are referred to as multiple pure tones, or "buzz saw" noise, prominently radiated from turbofan inlets during takeoff. Compressor tones radiate from the inlet and also may produce sum and difference frequencies from interaction with or scattering by the fan tones. Turbine tones radiating from the core exhaust are higher in frequency than the fan fundamental because of higher numbers of blades per stage.

Table 1. Typical Turbofan Geometric and Aerodynamic Operating Parameters

Design pressure ratio	1.5 to 1.7
Design tip Mach number	1.1 to 1.4
Tip relative Mach number	0.8 (approach) to 1.5 (takeoff)
Solidity, chord/spacing	1.0 to 1.5
Work factor	0.3 to 0.6
Blade numbers	20 to 40
Hub/tip ratio	0.4

Elements of the Generation Process

The physical processes which link unsteady aerodynamics of the turbomachinery flow field to the resultant far-field acoustic signature are shown in the flowchart in figure 4. Elements in ovals are inputs to (or outputs of) the processes in the rectangles. The four processes—blade unsteady aerodynamic response; acoustic coupling to the duct; propagation in the duct, which may contain other blade rows and have acoustically treated walls; and acoustic coupling (radiation) to the far field—have each been studied and modeled separately as convenient elements of the overall problem. A knowledge of the inputs and outputs—unsteady flow field disturbance experienced by the blades; fluctuating blade surface pressures; and duct acoustic mode content at the entrance and exit of the duct—is required to link the processes and arrive at the final output, which is far-field directivity and spectra. At supersonic relative blade velocities, a rotor-locked shock wave system appears on the blades and couples to the duct in a way which depends on nonlinearities and blade-to-blade differences. This mechanism is denoted in figure 4 by the elements enclosed by dashes in the upper right.

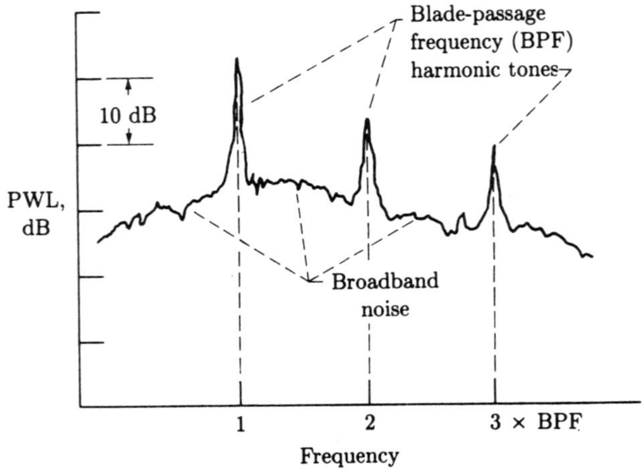

(a) Subsonic tip speed.

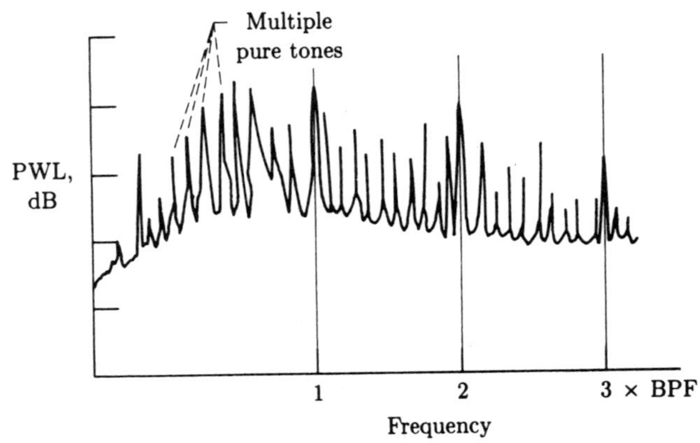

(b) Supersonic tip speed.

Figure 3. Typical turbomachinery sound power spectra.

From an experimental viewpoint, the intermediate inputs or outputs denoted in figure 4 are often missing; only acoustic measurements in the far field are available for a particular turbomachine operating point. In fact, one of the greatest hindrances to applying theories for the individual processes to practical situations has been the lack of definition of the key input-output quantities at the interfaces. Diagnostic measurements of flow disturbances, blade surface pressures, and modal content have begun to correct this deficiency.

This chapter first discusses some theories used to describe the processes in figure 4. A description of noise mechanisms which have been identified experimentally follows; this description deals in large part with the inputs and outputs. Sample applications of the theory and experiment to specific in-flight sources are followed by an overview of full-scale-engine machinery noise technology. Finally, concluding remarks indicate the significance of recent advances and point out unsolved problems requiring attention to move toward more integrated quiet designs.

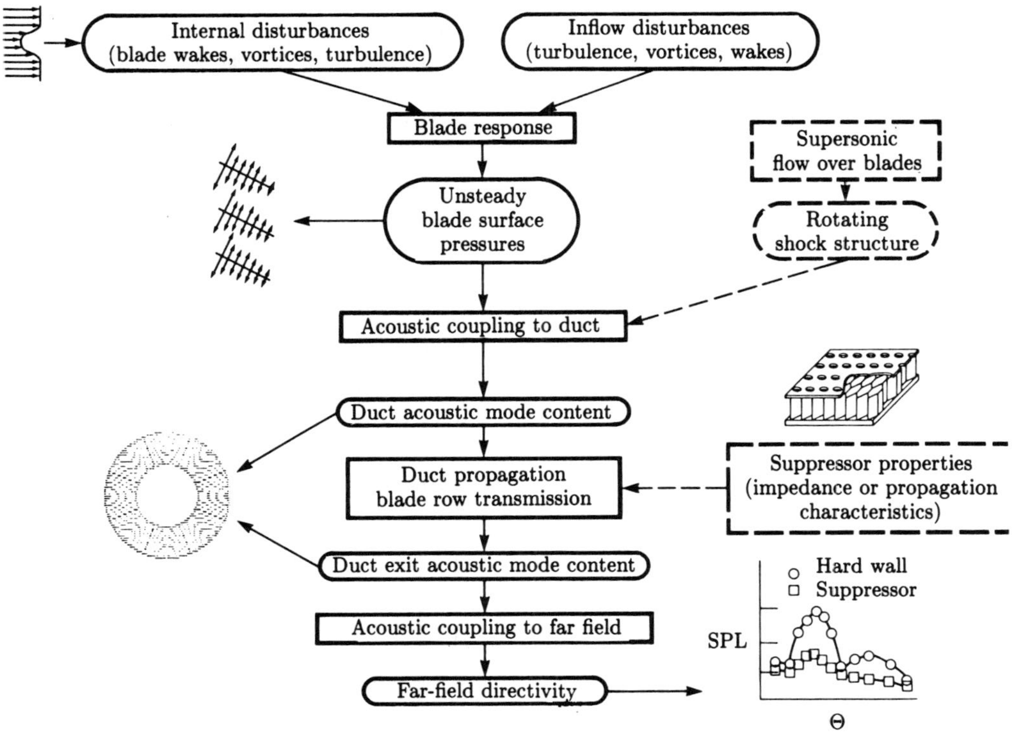

Figure 4. Turbomachinery noise-generation processes.

Process Description: Theory

Blade Pressures

Fluctuating pressures for which phase or trace speed with respect to a stationary observer is supersonic radiate sound to that observer. The origins of the fluctuating pressures on blade rows are incident vortical disturbances called gusts.

Blade Response to Periodic and Random Gusts

An incident periodic gust, "frozen" in the flow, is invariant with position as it is transported with the mean flow velocity U_∞. With respect to a particular blade row, the mathematical description takes the form

$$\mathbf{u}_\infty = \mathbf{a} \exp\left[i(\mathbf{k} \cdot \mathbf{y} - k_1 U_r t)\right] \tag{1}$$

where the vortical gust velocity vector $\mathbf{u}_\infty$ has amplitude $\mathbf{a}$ and is transported past the airfoil with relative velocity U_r. The coordinate system $\mathbf{y}$ is fixed to the blade, with the y_1 direction along the blade chord. The velocity component normal to the chord in the y_2 direction is the "upwash" and is responsible for the blade pressure fluctuations in the linearized approximation.

The fluctuating normal force per unit span on the blade F_2 is given by the expression

$$F_2 = \pi \rho_o U_r a_2 c \exp\left[i(k_3 y_3 - k_1 U_r t)\right] \mathcal{G}\left(\dot{k}_1, k_3, M_r\right) \quad (2)$$

where c is the blade chord, ρ_o is the ambient density, and $\mathcal{G}$ is the response function for a gust of wave number $\mathbf{k}$ convecting at U_r with respect to the blade. The chordwise wave number is $k_1 = \omega/U_r$ and the reduced frequency of the gust is $\sigma = \omega c/2U_r$, where ω is the gust frequency.

An array of solutions for $\mathcal{G}$ in special cases exists (refs. 2 to 7), some of which are summarized in table 2. They range from the simplest, the Sears function S for a single airfoil in incompressible flow (ref. 2), to three-dimensional gusts impinging on a cascade of thin airfoils (refs. 3 and 6). The Goldstein and Atassi case is included as the one departure from linearized analysis which examines the second-order effects of finite loading, namely, thickness, camber, and angle of attack.

Table 2. Gust-Airfoil Response Models

Investigator	Reference	2D/3D	Airfoil geometry	Flow	Description
Sears $S(\sigma)$	2	2D	Single, infinite span	Incompressible	
Goldstein $S(\sigma, M_r)$	3	2D	Single, semi-infinite chord, infinite span	Compressible	High-frequency limit
Amiet $S(\sigma, M_r)$	4	2D	Single, infinite span	Compressible	Low-frequency limit
Graham $\mathcal{G}(k_1, k_3, M_r)$	5	3D gust	Single, infinite span	Compressible	Developed relations between special cases above
Namba	6	3D	Cascade, finite span	Compressible	Annular cascade
Goldstein and Atassi	7	2D	Single	Incompressible	Effect of thickness, camber, angle of attack, and high-frequency limit
Goldstein	3	2D	Cascade, finite span	Compressible	Linear cascade

Figure 5 compares the magnitudes of the Sears function and its compressible approximations with those of the oblique gust response of Filotas (ref. 8), which includes the spanwise components of the wave number $\sigma_3 = k_3 c/2$. As shown, the spanwise gust components reduce the magnitude of the response. Phase shifts also occur.

When the incident gusts are random rather than periodic, the gust velocity expression (eq. (1)) takes the form of a continuous spectrum of vortical velocity disturbances. For a frozen gust in homogeneous turbulence, the blade-lift-power

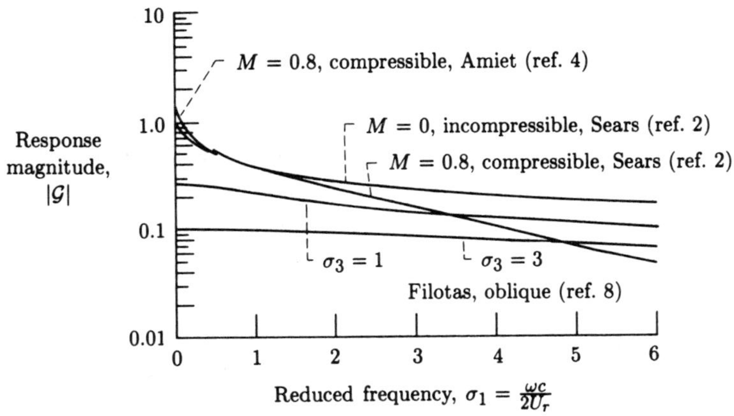

Figure 5. Airfoil gust response functions for single airfoil.

spectral density H_{22} is given by

$$H_{22}(y_3, k_3, \omega) = (\pi \rho_o c)^2 U_r \left| \mathcal{G}\left(\frac{\omega}{U_r}, k_3, M_r\right) \right|^2 \int \Phi_{22}\left(\frac{\omega}{U_r}, k_2, k_3\right) dk_2 \qquad (3)$$

where Φ_{22} is the moving-axis spectral density of the upwash velocity. (See chapter 3 of ref. 3.)

Rotor-Locked Blade Pressure Field

When the *steady* pressure field "locked" to any particular element of the rotor surface moves supersonically with respect to a stationary observer, sound is radiated to the observer. The presence of the duct around the rotor modifies the radiation condition, as is shown where coupling to the duct is discussed. If velocities relative to the rotor blades are supersonic, the rotor-locked pressure field takes on the distinctive impulsive character associated with shocks on the blades and the inlet propagation leads to formation of multiple pure tones.

Coupling to the Duct—Modes and Cutoff[1]

Knowledge of the modal structure of sound generated by blade-vane combinations and other periodic interactions is useful in several ways. Propagation in the duct can be predicted and may be controlled, sound-absorbing liners can be designed effectively, radiation directivity patterns can be estimated, and, to some extent, acoustic blade design may benefit from such information.

This section describes, in general terms, how the unsteady aerodynamic forces just discussed couple to the duct and generate acoustic modes. The structure of the analysis is best revealed if we consider a duct of any cross-section geometry having acoustic modes denoted by $\psi_q(\mathbf{r})$, where $\mathbf{r}$ is the two-dimensional position vector of

[1] Section authored by Thomas G. Sofrin.

a point in a cross section in an appropriate coordinate system. For harmonic time dependence we have

$$p(x, \mathbf{r}, t) = \text{Re}\left[P(x, \mathbf{r}) \exp(-i\omega t)\right] \tag{4}$$

The pressure p at $(x, \mathbf{r})$ due to a unit, concentrated, harmonic force at $(x_o, \mathbf{r}_o)$ is called the Green's function. This function, which may be considered an "influence coefficient," is different for unit forces in the x-direction and for forces in either of the cross-section coordinate directions. The analysis assumes that one such specific force orientation is under consideration. This Green's function is denoted by

$$G(x, \mathbf{r}|x_o, \mathbf{r}_o)$$

Instead of a single concentrated force at $(x_o, \mathbf{r}_o)$, the force is distributed over a region. For simplicity it is assumed that it is distributed over a cross-section plane at $x = x_o$. If its intensity (force per unit area) is $f(\mathbf{r}_o)$, the force on a surface element dS_o is $f(\mathbf{r}_o)\, dS_o$ and the resulting pressure P at $(x, \mathbf{r})$ is the product

$$G(x, \mathbf{r}|x_o, \mathbf{r}_o) f(\mathbf{r}_o)\, dS_o$$

Consequently, the effect of the complete force distribution is expressed by the integral over the source plane:

$$P(x, \mathbf{r}) = \int_{S_o} G(x, \mathbf{r}|x_o, \mathbf{r}_o) f(\mathbf{r}_o)\, dS_o \tag{5}$$

Expressions for the Green's functions for common duct geometries exist (e.g., ref. 9), but equation (5) does not give the desired result directly.

We are not primarily interested in the local pressure at $(x, \mathbf{r})$, but rather in the modal composition of the pressure at station x. That is, we require the *coefficients* of the modal components $\psi_q(\mathbf{r})$ at x. These can be obtained as follows.

Since $G(x, \mathbf{r}|x_o, \mathbf{r}_o)$ is the pressure resulting from unit force at $(x_o, \mathbf{r}_o)$, it can be expanded in a series of modal functions ψ_p. The source distribution can be similarly expressed. The resulting integral in equation (5) can then be easily evaluated since the ψ-functions are orthogonal. The result is automatically in the desired form for the modal composition of the acoustic pressure at x.

Accordingly, let the source distribution be represented by the series

$$f(\mathbf{r}_o) = \sum_q f_q \psi_q(\mathbf{r}_o) \tag{6}$$

The coefficients f_q depend on details of the unsteady aerodynamics and also on the modal functions ψ_q for the duct geometry. They must be obtained by numerical methods.

The expansion of the Green's function is a more complicated expression, but it has the advantage of being known for common geometries (ref. 9). It is more complicated because x and $\mathbf{r}$ must be involved as well as $\mathbf{r}_o$. The Green's function can be expanded in the following form:

$$G(x, \mathbf{r}|x_o, \mathbf{r}_o) = \sum_p G_p \psi_p(\mathbf{r}) \psi_p(\mathbf{r}_o) \exp\left\{i\left[k_{xp}(x - x_o)\right]\right\} \tag{7}$$

The G_p coefficients are known constants, generally involving frequency. Also, the axial wave number k_{xp} is frequency dependent. Above cutoff for the ψ_p mode, k_{xp} is real and a propagating wave results. At lower frequencies k_{xp} is imaginary, an indication of exponential decay.

With the expansions in equations (6) and (7), equation (5) can be arranged as follows:

$$P(x,\mathbf{r}) = \sum_q f_q \sum_p G_p \psi_p(\mathbf{r}) \int_{S_o} \psi_p(\mathbf{r}_o)\psi_q(\mathbf{r}_o)\, dS_o \exp\left\{i\left[k_{xp}(x - x_o)\right]\right\}$$

Since the modes ψ_p and ψ_q are orthogonal, the integral vanishes for all $p \neq q$. In the case of $p = q$ the integral $\int \psi_p(\mathbf{r}_o)^2\, dS_o$ is denoted by $\Lambda_q S_o$, and the resulting equation is

$$P(x,\mathbf{r}) = \sum_q \Lambda_q S_o f_q G_q \psi_q(\mathbf{r}) \exp\left\{i\left[k_{xq}(x - x_o)\right]\right\} \tag{8}$$

(The normalizing factors $\Lambda_q S_o$ are frequently included with the known coefficients G_q or the functions ψ_q, in which case eq. (8) can be written as $P(x,\mathbf{r}) = \sum_q G'_q \psi_q(\mathbf{r}) \exp\left\{i\left[k_{xp}(x - x_o)\right]\right\}$.)

Equation (8) states that the pressure amplitude of the mode ψ_q at station x is proportional to the product of f_q and G_q. This fact is intuitively clear—f_q is the strength of the qth modal distribution of the exciting force at the source and G_q is the "transfer function" relating the pressure response at station x in mode q to a unit strength force distribution in the same shape or mode at the source plane x_o.

All the factors in equation (8), with the sole exception of f_q, are fixed functions of the duct geometry and are completely independent of the aerodynamic forcing functions (although G_q is different for different force orientations such as axial, tangential, or radial). Two ways of producing but a small modal amplitude are the following:

1. Mode q is cut off at the operating frequency. Cutoff is defined such that the wave number k_{xq} is imaginary (or has an imaginary part), so that the function $\exp\left\{i\left[k_{xp}(x - x_o)\right]\right\}$ supplies an exponential attenuating factor to ψ_q. Specifically, the cutoff ratio ξ is defined as $\xi = k_o r_o / \alpha_{mn}\beta$, where α_{mn} is the duct eigenvalue for the (m,n) mode and $\beta = (1 - M_D^2)^{1/2}$, where M_D is the duct axial Mach number. For $\xi > 1$, the (m,n) mode propagates. For annular duct geometry the details of the mode functions are given in reference 10. The kinematic expression (see ref. 11) relating ξ to rotor-blade and stator-vane numbers B and V is

$$\xi_{mn} = \left|\frac{sB}{sB - kV}\right| \frac{M_t}{M^*_{mn}\beta} \tag{9}$$

where s and k are harmonic integers, M_t is the blade-tip rotational Mach number, and M^*_{mn} is α_{mn}/m (where $m = sB - kV$).

2. If mode q propagates, a way of reducing its strength is to design the aerodynamic force excitation distribution so that the coefficient f_q is small. This means that the force is distributed over the source plane in such a way that when it is resolved into a set of $\psi(\mathbf{r}_o)$ functions, the qth mode $\psi_q(\mathbf{r}_o)$ is substantially absent.

In practice (to the extent that tailoring of the force field is feasible) the objective should be to have the lower order modes (as measured by the eigenvalues) cut off. This cutoff is conventionally done by selection of blade and vane numbers and is feasible for, at most, a very few circumferential modes. Then, for the propagating modes, the *radial* distribution should be tailored. (To simplify notation, a single index has been used to identify modes, such as ψ_q. For specific ducts such as annular or cylindrical, a double subscript is used, such as $\psi_{m\mu}$. Here m specifies the "circumferential mode" number and μ indicates the associated "radial mode" number. As discussed elsewhere, the various values of $\psi_{m\mu}$ are the products of harmonic and Bessel functions.) For example, if the mode $m = 8$, $\mu = 0$ propagates, it may be found that the associated radial modes $\mu = 1$ and 2 also propagate but that $m = 8$, $\mu \geq 3$ do not. Then, if the radial force distribution is shaped such that it has only small components for $\mu = 0$, 1, and 2 (with the bulk of its energy in higher radial modes), these higher modes will decay and give the desired result. The implementation of these strategies for minimizing acoustic mode amplitudes is discussed in more detail in the section entitled *Full-Scale Engine Applications.*

Blade Row Transmission

As acoustic modes generated on a blade row propagate upstream and downstream in the engine ducts, they encounter other blade rows which both reflect and scatter the acoustic energy flux. For example, rotor viscous wakes and tip vortices interacting with a stator produce upstream-propagating modes which must traverse the rotor before reaching the inlet entrance and radiating to the far field. Some of the modal energy flux is reflected back to the stator, which must be negotiated before radiation from the exhaust duct can occur. The rotor also scatters incident modes into other circumferential orders, the result being a shift in modal energy into other harmonics of rotor blade-passage frequency. If the modal content incident on the rotor is generated from a downstream rotor turning at a different angular rate, the upstream rotor can scatter incident modes from the downstream rotor into sum and difference frequencies of the two. Methods of quantifying these phenomena will now be described.

Energy Reflection and Transmission

Table 3 summarizes blade row transmission analyses available in the literature (refs. 12 to 18). References 12 and 15 obtained results similar to the actuator disk analysis of reference 13 even though the approaches and assumptions were quite different. The overall dependence of transmission on relative Mach number along the blade chord M_r and two-dimensional cascade geometry are illustrated in figure 6 (from ref. 13). The incidence angle with respect to the blade chord is α, and the stagger angle is χ. It is worth noting that the incident wave direction is defined by the group velocity vector defining energy flux in the wave. The abscissa, $\delta = \chi - \alpha$, is the incidence angle with respect to the duct axis. Two values of δ exist where transmission is potentially high. The obvious one is where the wave is aligned with the blades ($\delta = \chi$) and the transmission is completely independent of M_r. The other potentially high transmission angle is limited to low M_r, approaches a transmission coefficient of 1 as $M_r \to 0$, and corresponds to the case where the scattering dipoles on the blade surfaces have their axes aligned with the blade chord.

Table 3. Blade Row Transmission Analyses

Investigator	Reference	Methods
Mani and Horvay	12	Wiener Hopf; 2D; $c >> \lambda$; includes scattering from one harmonic to another
Kaji and Okazaki	13	Actuator disk (includes steady loading); 2D; solidity: $s \to 0$
Kaji and Okazaki	14	Accelerator potential; 2D; finite solidity
Amiet	15	Quasi-steady Prandtl-Glauert with far-field radiation; $\lambda >> c$; $\lambda >> s$
Philpot	[a]16	Used Amiet in 2D stripwise form
Cumpsty	17	Linear kinematic arguments: scattering and production of sum and difference tones
Cumpsty	[a]18	Role of blade row transmission in rotor-stator interactions; radial scattering emphasized

[a] Heavily experimental.

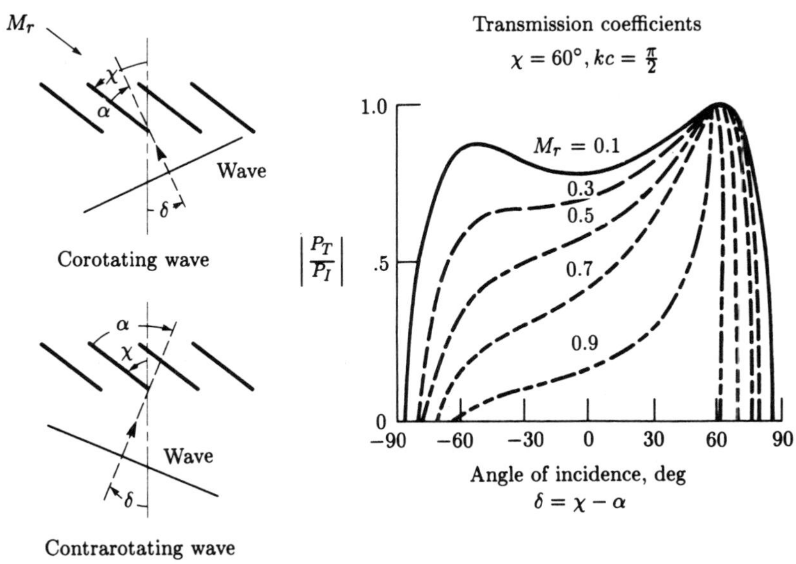

Figure 6. Blade row transmission. (Based on ref. 13.)

Since real fans have continuously varying stagger angles from hub (low χ) to tip (high χ), each blade chordwise strip transmits incident sound waves differently. Strip theory approximations (refs. 16 and 18) have been used in attempts to account for the transmission of actual annular duct modes. The equations for wave angles and transmission coefficients may be expressed in terms of cutoff ratio. For the special case where swirling flow exists, such as between rotor and stator, the cutoff ratio is more complex. In a three-dimensional rectangular approximation to a thin annular geometry,

$$\frac{1}{\xi_s^2}\ 2M_y\left(\frac{sm}{kR}\right) - M_y^2\left(\frac{sm}{kR}\right)^2 + \frac{1}{\xi^2} \tag{10}$$

where $s = -1$ for contrarotating and 1 for corotating modes, ξ_s is the cutoff ratio with swirling flow, and R is the average radius of the annulus.

Multistage Transmission

Propagation through multistage axial-flow machines having nonuniform annular ducts carrying the mean axial and swirling flow is treated in reference 19. The analysis is based on an electrical transmission line analogy (four-pole theory) where the annular duct is treated with a strip theory applicable to high hub-tip ratios. No modal distortion is considered, so λ is greater than twice the blade pitch. Large cascade coupling effects (nonlinearities) exist at low frequencies, and flow swirl affects cutoff (as noted in eq. (10)) and propagation.

Duct Propagation and Radiation[2]

Since another chapter deals with this subject in detail, including propagation in lined ducts, only a limited discussion is included here.

The emphasis in this section is on a propagation description in terms of cutoff ratio for untreated ducts. This approach is compatible with ray acoustics and has the advantage of appealing to physical intuition in terms of wave fronts and rays. The specific goal here is to offer a tool for diagnosis of turbomachinery source noise through analysis of far-field directivity or modal content at diagnostically significant locations in the engine ducts. Just as blade rows can scatter incident energy into new modes, area changes can also involve modal scattering particularly if axial gradients in the area are high.

Inlets—Upstream Propagation

If all blade row transmission effects have been accounted for, the modal content propagates to the far field through variable-area ducts carrying flow and radiates from duct openings through nonuniform flow fields. For inlet radiation two distinct flow fields are of interest: largely radial potential flow in the static test case, and flow from an inlet stream tube that is only slightly larger than the inlet diameter in the flight case. The Wiener-Hopf technique, applicable only to inlet lips of negligible thickness, has been applied to two idealized cases. One is uniform external and internal flow at the same Mach number, and the other is external flow at a constant Mach number bounding a cylinder of higher uniform Mach number extending out of the inlet. The former is an approximation to the flight case but the latter is unrepresentative of any real inlet flow. Two other approaches to analyzing inlet radiation have been followed. The first uses simplifying assumptions based on ray

[2]Section authored by Edward J. Rice.

acoustics, while the second uses a fully numerical solution incorporating the actual flow field and inlet lip geometry.

Approximate expressions for inlet radiation have been developed in terms of mode cutoff ratio ξ. The key simplification in the cutoff ratio formulation is that modes with the same ξ and with $\eta = fD/c$ propagate similarly to the far field. This has been demonstrated for radiation from a flanged duct without flow (ref. 20) and is fairly accurate for principal-lobe radiation (ref. 21). Two important duct mode propagation angles, ϕ_x and ψ_x, are defined in reference 22 as

$$\cos \phi_x = \frac{-M_D + S}{1 - M_D S} \tag{11}$$

and

$$\cos \psi_x = \frac{S\sqrt{1 - M_D^2}}{\sqrt{1 - M_D^2 S^2}} \tag{12}$$

where

$$S = \sqrt{1 - (1/\xi^2)} \tag{13}$$

and M_D is the duct axial Mach number. Here, ϕ_x and ψ_x are respectively the angles which the vector normal to the wave front and the group velocity vector make with the duct axis. The duct mode angle ψ_x, given by equation (12), closely approximates the angular location of the principal lobe in the far field (ref. 22). This conclusion was reached by inspection of the directivity coefficient appearing in the Wiener-Hopf solution for the case of uniform flow everywhere (ref. 23); an expression for the principal-lobe angle identical to equation (12) was obtained. The approximate equality of duct mode angle and far-field principal-lobe radiation angle suggests that ray acoustics arguments can be used to link the two angles for cases where the flow is not uniform.

Ray acoustics ideas have been applied to the case where far-field velocity is substantially less than inlet duct velocity, the limit being the static case, where far-field velocity is zero. Based on a ray acoustics analysis which showed that refraction in a potential flow is second order with respect to Mach number (ref. 24), the wave fronts were assumed to be unbent going from duct to far field. That is, ϕ_x was assumed to be unchanged. Since ϕ_x and ψ_x are identical if Mach number is zero, the group velocity in the far field was assumed to have been shifted. At $M_D = -0.4$ and $\xi \approx 1$ (near cutoff), the calculated radiation peak is at 66° while the group velocity in the duct propagates at $\psi_x = 90°$. A peak near 66° was observed in the far field for a nearly cut off mode generated by a controlled fan source (ref. 25). However, the agreement of this observed peak with the theory, which neglects lip shape, may be misleading. A propagation phenomenon associated with the very thick inlet lip used in the experiment may have controlled the principal-lobe location. An analysis of propagation in a variable-area duct with gentle area variation showed that mode identity is preserved (i.e., no scattering occurs, refs. 26 and 27). Thus, as a mode propagates from the inlet throat to the highlight, ξ increases causing ϕ_x and ψ_x to decrease. Recent extensions of ray theory for propagation through an irrotational flow (refs. 28 and 29) imply that it is the group velocity vector which is unchanged, not the normal to the wave fronts. The difference between the two assumptions is significant; for example, the 66° versus 90° peak near cutoff and current evidence point to preservation of group velocity as the better approach.

Numerical Model

A hybrid numerical program has been developed (ref. 30) and exercised (ref. 31) to calculate both the internal and external sound propagation for actual engine inlet geometry and flow conditions. It is a hybrid program in the sense that a finite-element method is used to calculate sound propagation within the duct and in the near field and an integral radiation method handles the sound propagation in the far field. Iteration is required to match the two solutions at the interface. A potential-flow program is used to generate the steady flow for the actual inlet geometry; boundary layers are not included. The input to the program is the pressure profile for a given mode in the annulus at the fan source. Although the combination of high Mach number and high frequency requires huge amounts of computer storage, some inlet geometry effects at modest frequencies and Mach numbers have been studied which were previously impossible to analyze.

Figure 7 contains the numerically predicted inlet tone directivity and the measured levels generated by a controlled source—a JT15D engine with inlet rods (ref. 25). A single (13, 0) mode propagates at the fan speed shown. The excellent agreement between the hybrid solution and the data is in contrast to the Wiener-Hopf solution for an infinitely thin lip. The thick lip used in the experiment (thickness-to-diameter ratio of 0.5) shifts the radiation peak toward the axis, as discussed in the preceding section, and acts as a shield to reduce the levels in the aft quadrant. The dependence of the directivity on inlet lip thickness is illustrated in figure 8, in which the shielding effect is also clearly evident. The numerical results show that the radiation peak moves aft as the lip gets thinner. At a thickness-to-diameter ratio of 0.1, the radiation pattern agrees very well with the Wiener-Hopf (zero thickness) result shown in figure 7. The hybrid program is a powerful tool for the solution of "real" inlet radiation problems.

Exhaust Radiation

In contrast to the complex inlet flow field, the exhaust flow, with mixing neglected, is much simpler. The fan exhaust may be approximated as an emerging cylindrical flow at M_D surrounded by a uniform flow at free-stream Mach number M_∞; these conditions fit the requirements for an exact Wiener-Hopf radiation solution. The approach of using ray acoustics and mode cutoff ratio to approximate a solution can also be applied with more confidence to the aft slip layer. Starting from the equation for the zero-flow flanged duct radiation, a coordinate transformation was applied to account for the duct flow, and ray acoustics arguments were applied across the slip layer (ref. 32). Single-mode aft directivity from the approximate expression is presented with results from the full Wiener-Hopf solution (ref. 33) in figure 9. The good agreement builds confidence in the simplifications used to generate the approximate solution. The Wiener-Hopf solution gives finite levels in the zone of silence, although the particular values from reference 33 are believed to be incorrect.

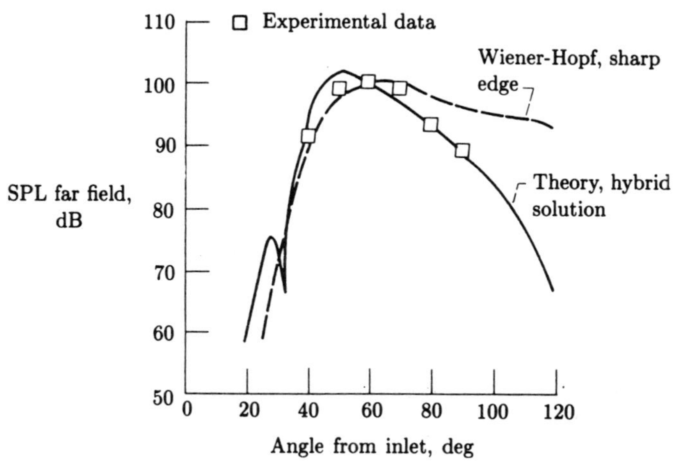

Figure 7. Theoretical and experimental single-mode inlet directivity. 3150 Hz; (13,0) mode; fan speed, 6750 rpm. (From ref. 31.)

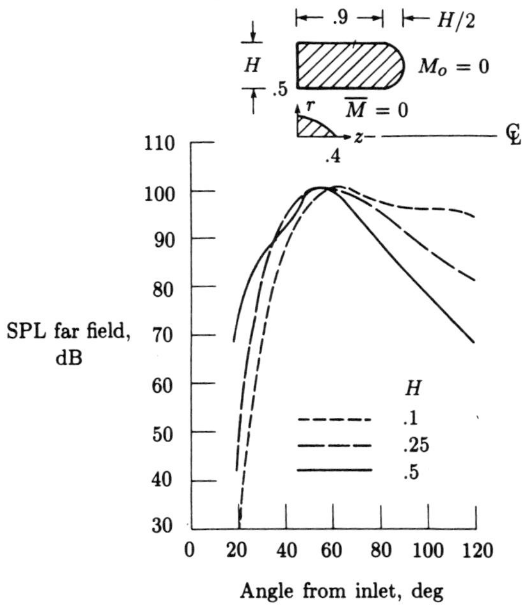

Figure 8. Effect of inlet lip thickness on single-mode directivity. $M_D = 0$; 3150 Hz; (13,0) mode; theory normalized to 100 dB at 60°. (From ref. 31.)

The location of the principal lobe in the far field ψ_{fp} is found from the approximate theory (ref. 32) to be

$$\cos\psi_{fp} = \frac{-M_D + \xi\sqrt{\xi^2 - 1}(1 - M_D^2)}{\xi(1 - M_D^2)(\xi + M_D\sqrt{\xi^2 - 1})} \tag{14}$$

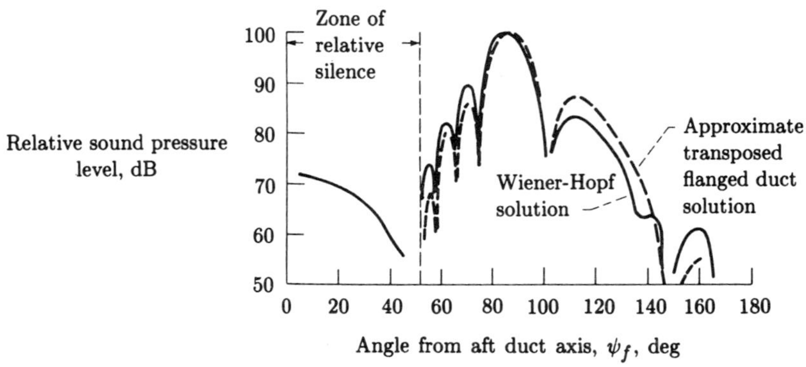

Figure 9. Approximate and exact single-mode aft directivity patterns. (8,4) mode; $M_D = 0.6$*;* $M_\infty = 0$*;* $\eta = 7.11$*. (From ref. 32.)*

for the static case ($M_\infty = 0$). For $M_D = 0.6$ and $\xi = 1$, $\psi_{fp} = 160°$ measured from the exhaust axis indicates that modes near cutoff radiate to the inlet quadrant. The analogous inlet analysis (eq. (12)) indicated that near-cutoff-mode peaks remain in the inlet quadrant. Thus, the inlet quadrant contains the near-cutoff-mode peaks no matter where the sound originates.

The locations of the peak sound pressure levels (the principal lobes) radiated from both inlet and exhaust are shown superimposed in figure 10, which relates cutoff ratio to principal-lobe angular location. Note that low cutoff ratios are associated with modal propagation nearly perpendicular to the duct axis, a situation favorable for absorption of the sound by duct linings. In contrast, high cutoff ratios are associated with nearly axial modal propagation, a situation where absorption by wall treatment is minimal. For the case illustrated in figure 10, aft duct modes radiating in the aft quadrant dominate the principal-lobe peaks in the range of angles important to flyover noise.

Broadband Noise Radiation

In the previous discussions it is implied that we are dealing with tones which are dominated by a few, or at least a reasonably limited number of, modes. This is not the case for broadband noise, which is produced by sources which are random in both time and location. All propagating modes will be energized, and the problem is to estimate the distribution of energy in the various duct modes, the number of which can be immense at the high frequencies encountered with turbofan noise. Idealized models, such as equal amplitude per mode or equal energy per mode, have been assumed for this modal distribution. Because of the random nature of the noise source, equal energy per mode is an appealing assumption. In fact, Dyer (ref. 34) has shown that a random source in a circular duct would produce equal partitioning of energy in the modes.

Because of the large number of modes carrying energy, it is also convenient to consider an integration over the modes (continuum assumed) as an approximation to the exact summation to account for the total energy. These two ideas, integration and equal energy per mode, have been combined in reference 20 to provide a very simple approximation for the far-field distribution of broadband noise, $P \approx \cos\psi$.

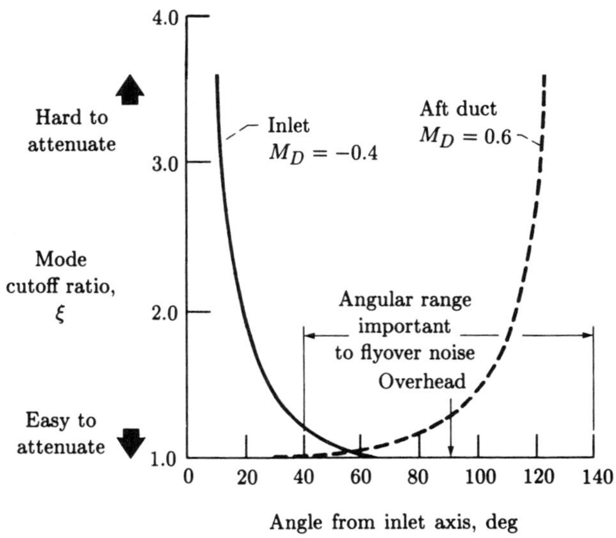

Figure 10. Location of peak sound pressure of single modes. Modes identified by cutoff ratio.

This radiation directivity has been compared with broadband inlet radiation data in reference 20 and the agreement is very good. The same approach was tried for the tonal energy which is produced by somewhat random inflow distortion. For this case the continuum idea seems to hold, but the energy is distributed more heavily toward near cutoff modes. Radiation of broadband noise from the aft duct has been treated in the same approximate manner in reference 32. Again, an approximate expression was derived which shows reasonable agreement with experimental data.

Mechanism Identification: Experiment

In any turbomachinery there are usually several contributing noise-generating mechanisms simultaneously at work. The term "mechanism identification" as used herein refers to pinpointing the blade row and origin of a particular pressure field which is unsteady when viewed in the laboratory reference frame and which is responsible for a substantial part of the radiated acoustic power. A particular flow disturbance or nonuniformity interacting with a particular blade row results in blade pressure fluctuations, portions of which couple to propagating acoustic modes in the duct. Rotor-alone steady pressure fields radiate only when the ducted rotor reaches or exceeds a rotational speed near supersonic tip speed, depending on the number of blades. The labels on the engine cross section in figure 1 indicate some of the candidate turbofan mechanisms. Flow disturbances are grouped according to the blade row with which they interact.

Flight Simulation—Inflow Control

The flow disturbances may be alternatively classified as those originating external to the engine but drawn into the inlet and those originating inside the engine. Although it has long been recognized that ingested external disturbances may control fan noise generation (ref. 35), it was the high-bypass-ratio engine flyover noise data, acquired in connection with noise certification requirements, which established that ground-test tone levels are controlled by extraneous inflow disturbances unrepresentative of flight (ref. 36). In fact, the practicality of the concept of choosing the vane-blade ratio for cutoff (refs. 10 and 11) to greatly reduce the fundamental tone was first demonstrated in rig tests and later conclusively confirmed in flight and in wind tunnels, as shown by the examples in figure 11. Both fans (figs. 11(a) and 11(b)) were designed for cutoff at subsonic tip relative Mach numbers and showed dramatic decreases in fundamental tone levels in flight or with forward velocity.

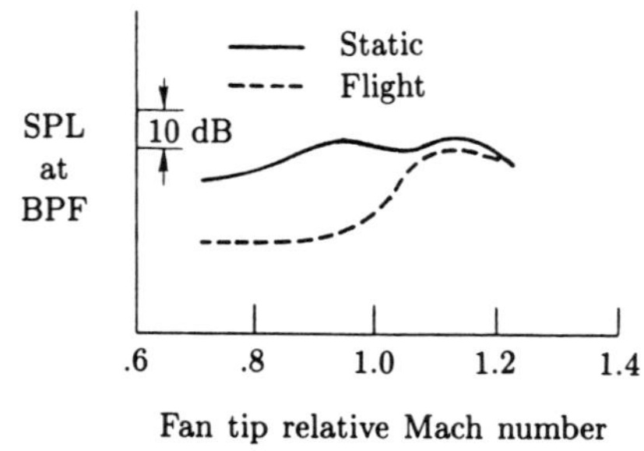

(a) High-bypass-ratio engine in flight.

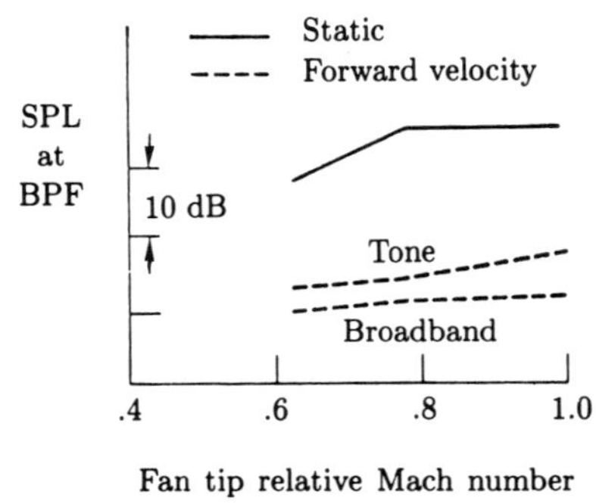

(b) Fan in anechoic wind tunnel.

Figure 11. Effect of forward velocity on fan blade-passage tone in inlet duct. (From ref. 36.)

Flight Simulation

The approach to controlling the inflow for flight simulation in static tests has evolved around the concept of inlet honeycomb—grid flow conditioners which must be acoustically transparent over the frequency range of interest. Figure 12 shows the range of inflow control devices (ICD's) investigated at the NASA Lewis Research Center (refs. 37 to 40). The sizes of the external devices, shown in figures 12(a) and 12(b), ranged from roughly 4 to 2 fan diameters D/D_{fan}. An in-duct honeycomb was aerodynamically effective but unacceptable from an acoustic transmission standpoint. The first-generation design (fig. 12(a)) drew on flow conditioning work for turbulence reduction (ref. 41) to arrive at the screen-honeycomb composite structure. The most recent version (fig. 12(b)) is reduced in size, uses honeycomb only, and employs thinner support ribs with more carefully bonded joints and cleaner attachment to the inlet lip (ref. 42). The shape conforms to an equipotential surface.

Flight data from a JT15D engine on an OV-1 test-bed aircraft (ref. 40) confirm the effectiveness of the ICD of figure 12(b), as shown in figure 13. The fundamental tone directivity with inflow control agrees well with the flight data except at the most forward angles, where the signal-to-noise ratio is low for flight.

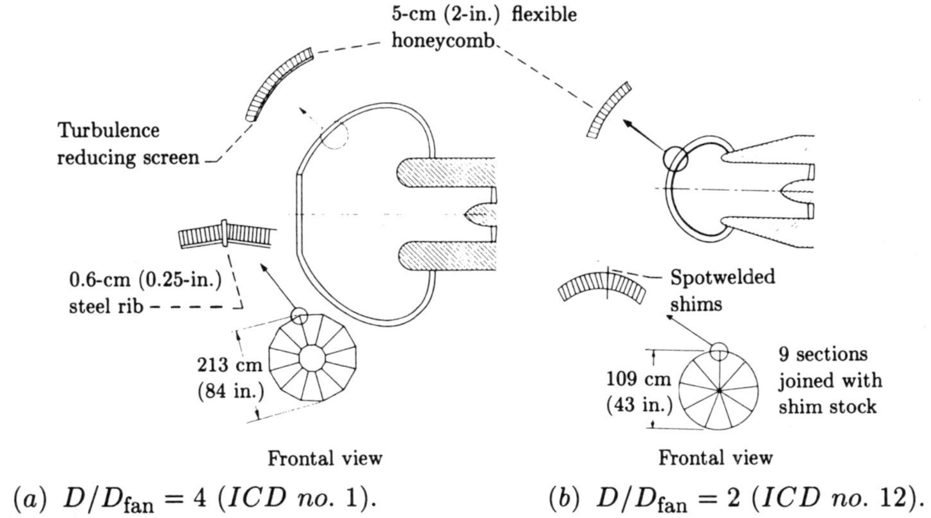

Figure 12. Inflow control devices for flight fan noise simulation. (From ref. 40.)

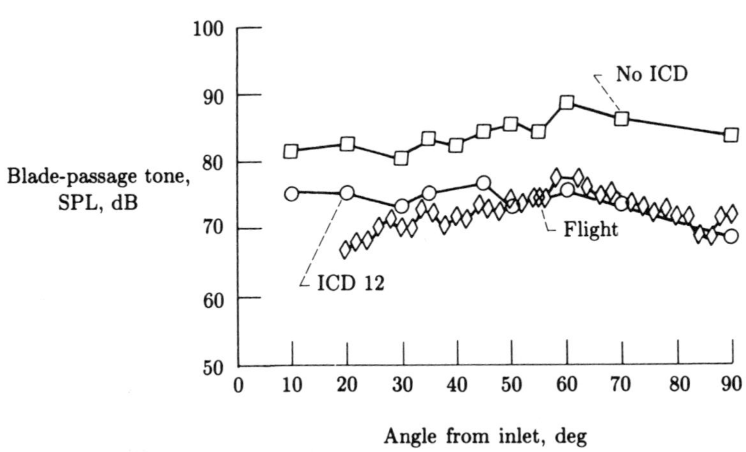

Figure 13. Effectiveness of inflow control for flight fan tone simulation. (From ref. 40.)

Substantial effort has also been applied to the inflow control problem by industry (refs. 43 to 51), and this effort includes flyover noise level comparisons with static projections (ref. 50) and development of ICD design procedures (ref. 51). The first generation of large engine ICD's, roughly 3 fan diameters in size, is currently in use. Table 4 summarizes some of these ICD configurations. Although the quantitative agreement of inflow control and flight data is still subject to some improvement, the current state of the art of static testing with inflow control does allow the study of bona fide internal sources controlling fan noise generation in flight. An alternative to ICD's is the anechoic wind tunnel (refs. 52 to 54), which has also been found to eliminate the bulk of the extraneous inlet disturbances.

Table 4. Inflow Control Devices

Facility (source)	Reference	ICD diameter, m	Ratio of ICD diameter to fan diameter	Construction	
				Honeycomb L/D	Screen or perforated plate, percent open area
Outdoor (JT9D)	43, 50	7.3	3	12	46 upstream
Outdoor	44, 45	7.3	3+	(a)	(a)
Anechoic chamber (rotor 11)	44, 45	2.0	4	8	52 spaced downstream
Outdoor (JT15D)	38, 42	.8 to 2.0	1.7 to 4.0	4 to 8	40 to 50 downstream
Anechoic chamber (QF-1, QF-13, JT15D)	39, 40	1.0 to 2.0	2.0 to 4.0	8	40 to 50 downstream
Outdoor (JT9D)	43, 50	7.3	3	8	51 upstream

[a]Similar to Boeing.

Blade Surface Pressures

Direct measurement of blade pressures has proven to be a valuable diagnostic tool for evaluating the quality of inflow to the fan and, with inflow control, for determining the residual internal sources controlling flight levels. Miniature pressure transducers mounted near the fan blade leading and trailing edges at several spanwise locations are used to continuously survey the circumferential variation of unsteady blade pressures (refs. 35, 38, and 43). This technique originally identified longitudinally persistent, circumferentially localized disturbances attributed to atmospheric turbulence elongated by the stream tube contraction in the inflow (ref. 35). Such disturbances, which may also be caused by ingested vortices, wakes, and instabilities associated with flow around the inlet lip, produce strong narrow-band random tones.

Figure 14 contains narrow-band blade pressure spectra without and with an ICD. Without an ICD (fig. 14(a)), the spectrum shows strong harmonic content at all multiples of shaft rotation frequency resulting from multiple encounters of the blade transducer with circumferentially varying flow disturbances. The additional scales on the abscissa are distortion mode number (multiple of shaft frequency) and the circumferential acoustic mode number corresponding to blade number minus distortion number. Inflow control eliminates the randomly varying and steady disturbances from the inlet flow and the corresponding bulk of the shaft harmonics disappear, as illustrated in figure 14(b). Those distortion numbers that remain are associated with periodic, internally generated flow disturbances which are fixed in space (e.g., vane potential fields) or which have fixed rotation rates with respect to the rotor (e.g., spinning acoustic modes). As a result, clues to the mechanisms governing flight noise levels are found from the prominent residual peaks.

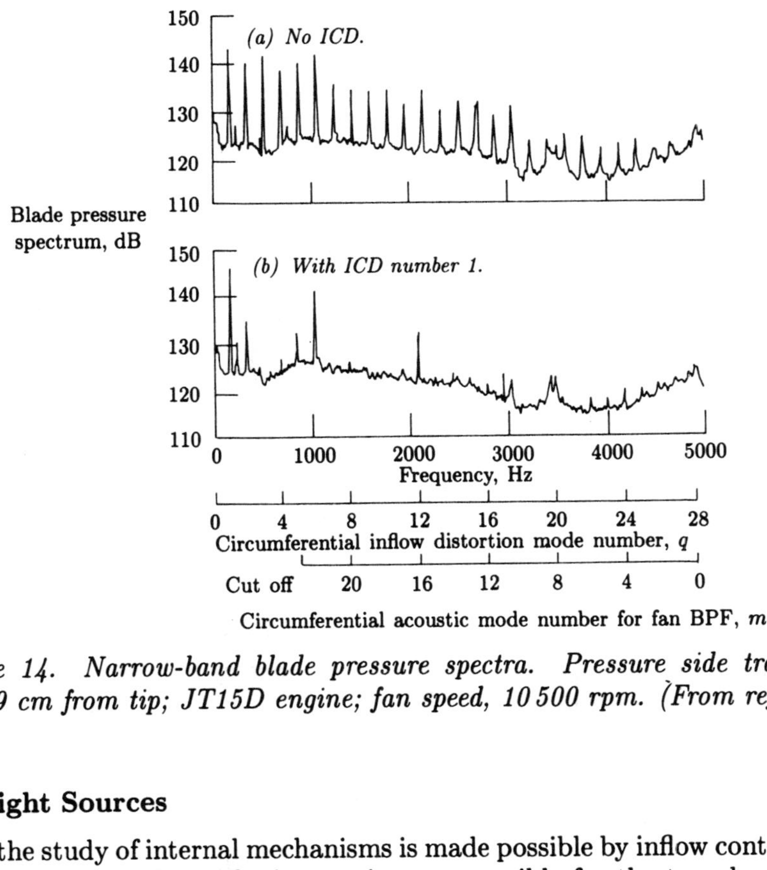

Figure 14. Narrow-band blade pressure spectra. Pressure side transducer 1.9 cm from tip; JT15D engine; fan speed, 10 500 rpm. (From ref. 38.)

In-Flight Sources

Once the study of internal mechanisms is made possible by inflow control, the task becomes one of identifying the interactions responsible for the tone levels observed over the range of engine speeds.

Rotor-Stator and Rotor-Strut Sources

Rotor wake-stator interaction remains a prime mechanism, but even with the blade-vane ratio chosen to prevent fundamental tone propagation, this interaction can still control the higher harmonics. Other interactions may also come into play.

For example, the JT15D engine exhibits a strong fan fundamental tone which appears at a speed corresponding to the start of propagation of the 22-lobed acoustic mode, as shown in figure 15 (from ref. 38). The source of the 22-lobed acoustic mode is the interaction between the 28 fan blades and the 6 structural support struts downstream of the fan stator. The blade pressure spectrum in figure 14(b) shows that a strong 6-per-revolution disturbance is sensed on the rotor. The strength of the rotor-strut interaction decreases with increased spacing between rotor and struts (ref. 55). A prime candidate for the interaction mechanism is a strut potential field extending upstream through the stators and interacting with the rotor. An alternative explanation would be the interaction of residual rotor wakes with the 6 engine struts generating the 22-lobed spinning acoustic mode, which is sensed on the rotor as a 6-per-revolution disturbance. Existing large high-bypass-ratio turbofans also contain downstream struts. Some proposed engine designs incorporate integral strut-stator vane assemblies with a potential for still more complicated interactions (ref. 56).

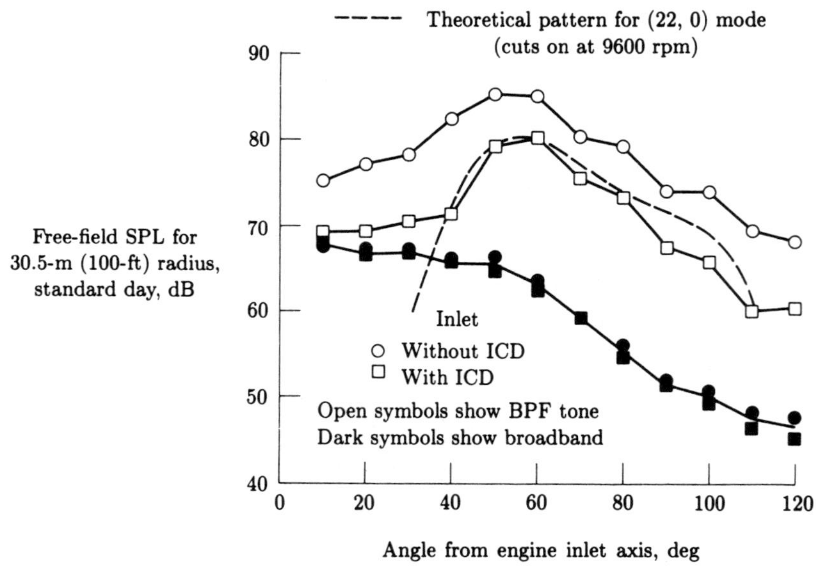

Figure 15. Directivity pattern indicating presence of 22-lobed acoustic mode due to rotor-strut interaction. JT15D fan; fan speed, 10 500 rpm. (Based on ref. 30.)

Broadband Sources

Interestingly, the broadband levels remain essentially unchanged with inflow control. This lack of change indicates that another mechanism, probably internal to the fan, controls this spectral component. Broadband levels vary strongly with fan operating point (rotor incidence angle or loading), as shown in figure 16. An empirical relationship between rotor incidence angle and forward-radiated broadband levels has been established in which broadband power level (PWL) is proportional to M_r^5 and increases 2.5 dB per degree of incidence (ref. 57). Fan-blade suction surface flow separation and interaction with the trailing edge, blade-tip interaction with the casing boundary layer, and rotor-wake-turbulence (midspan or tip) interaction with the stator are candidate mechanisms, although the last of these seems to be discounted by rotor-alone experiments (ref. 58).

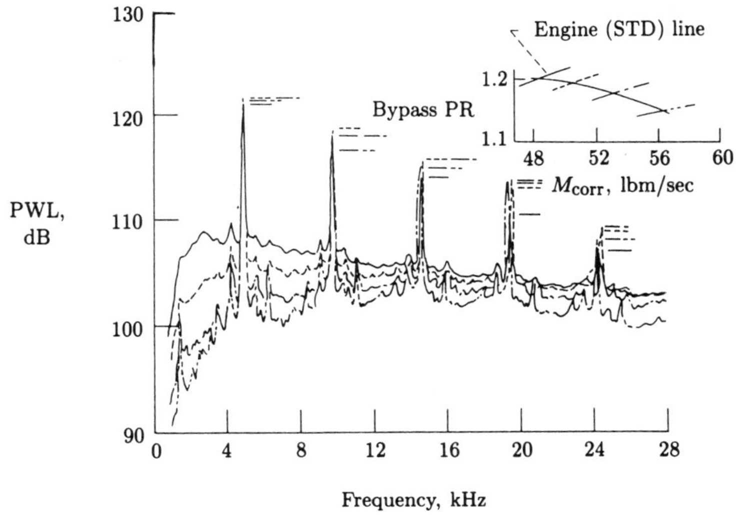

Figure 16. Variation of broadband levels with fan operating point. JT15D fan in anechoic chamber; $\Delta f = 80$ *Hz; fan speed, 10 500 rpm.*

Multiple Pure Tones (MPT's)

Although the MPT generation mechanism is clearly associated with the rotor leading-edge shocks and their blade-to-blade nonuniformity, quantitative descriptions which predict the envelope of the one-per-revolution tone multiples are lacking. This is because the detailed geometric specifics of the shock structure are unique to each particular rotor build, even for the same design, and depend on the circumferential distribution of the manufacturing tolerances in blade stagger angle or leading-edge contour. The tone spectrum depends most critically on the distribution of the intervals between shocks (ref. 59). Thus, at best, spectral predictions can be made only for an "average" fan for any particular design. As the standard deviation of the shock spacing increases, more sound power appears in the MPT's and less in the blade-passage-frequency harmonics. The higher the tip relative Mach number, the stronger the shaft lower order harmonics become (ref. 60). Some uncertainty remains as to the role of nonlinearity in the development of the spectrum at upstream locations in the inlet duct (refs. 61 and 62).

Properly designed inflow control devices are transparent to MPT's. However, there appears to be a mechanism which reduces MPT's in flight since a consistent pattern of overprediction occurs for the projection of static measurements, as illustrated in figure 17. (From ref. 40 and corroborated in ref. 50.)

Flow Disturbance Characterization—
Rotor Wakes and Vortices

To apply blade response models such as those summarized in table 2 to calculate blade pressures, a description of the incident gusts is required. Analysis of the most

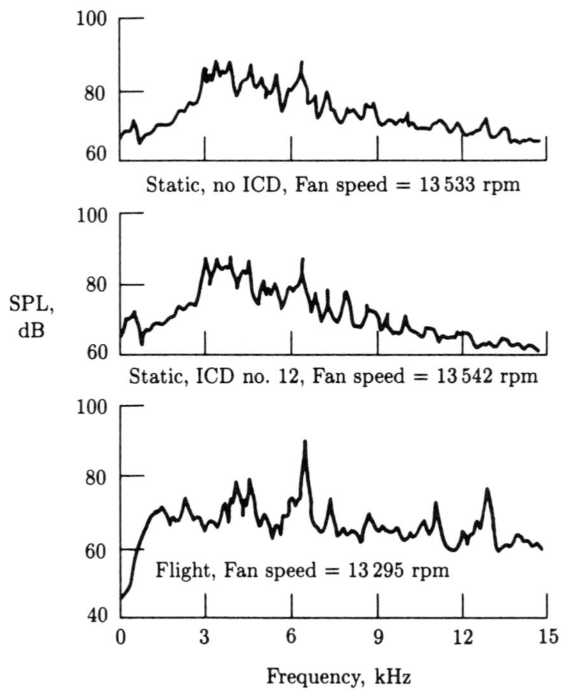

Figure 17. Supersonic tip speed noise spectra. JT15D engine; $\Theta = 70°$. (From ref. 40.)

significant mechanism, rotor-stator interaction, involves a thorough description of the rotor-produced disturbance flow field.

Rotor Wake Measurements

The need to describe rotor-blade wakes has long been recognized and a large body of wake data, including mean and turbulence properties, has been accumulated on laboratory fans. (See, e.g., refs. 63 and 64.) In addition to midspan wakes, secondary flows such as tip vortices have been recognized as potential noise contributors (ref. 65). Therefore, a linear cascade analysis including spanwise gust components has been developed to allow the relative noise contributions of tip vortices and midspan wakes to be determined (ref. 66). What is lacking is a thorough model of the total rotor downstream flow field which is linked to fan design parameters and is validated by experimental data.

Some wake data have been obtained as functions of downstream distance for a fan operated with forward velocity in an anechoic wind tunnel (ref. 67). Rotor mean wake upwash velocity profiles are shown in figure 18 as a function of spanwise position. The magnitudes vary substantially with radial location, but most significantly the profile near the tip is characterized by an extra upwash cycle between successive blades corresponding to strong secondary flows, probably a tip vortex. The variation of stator upwash harmonics, the required input to generation analyses, is shown in figure 19 as a function of downstream distance. From the complex variations observed, it must be concluded that simple Gaussian profiles which decay and spread monotonically with distance are an inadequate description of this flow field.

Semiempirical Wake Model

A semiempirical wake-vortex model has been developed specifically for use in rotor-stator noise calculations (ref. 68). Viscous wake data (refs. 63 and 64) were correlated and combined with a tip vortex model in which vortex strength depended strongly on tip clearance. Parametric investigation of the upwash spectra experienced by stator vanes showed that increasing rotor-stator spacing beyond a value at which adjacent blade wakes merged produced no appreciable reductions in upwash amplitudes. The tangential location of the tip vortex at any downstream location had an important influence on the gust harmonic content near the blade tips: odd or even harmonics could be accentuated, with midpassage locations particularly augmenting second harmonic levels. (See figs. 18 and 19.)

Mode Measurement[3]

Objectives of Mode Measurement

In turbofan engines and many other devices incorporating rotating elements, sample measurements of the acoustic frequency spectra are useful only as preliminary indications of the dominant noise sources. If, for example, sample frequency spectra for a two-stage fan display much higher levels of second-rotor harmonics than of first-rotor harmonics, the second rotor will be the obvious candidate for noise-reduction efforts. However, these spectra provide no guide to specific dominant noise-generating mechanisms involving the second rotor, such as interaction of rotor and upstream stator, interaction of rotor and downstream stator, and distortion from rotor and inflow. In order to pinpoint the source of this second-rotor noise, further information is necessary. This section deals with the type of information required (acoustic mode structure) and the means for obtaining it.

Before mode measurement techniques are described, some of the uses of modal information are summarized as follows:

1. To identify specific dominant mechanisms and thus guide noise-reduction efforts effectively
2. To isolate and measure effects of configuration modification tests, such as rotor-stator spacing, when other mechanisms are present that obscure the desired information
3. To provide detailed experimental information which can be used to evaluate theory
4. To guide the design of sound-absorbing duct liners when the source mechanisms cannot be easily modified

Requirements for Mode Measurement Tests

In turbofan engines (and for propellers and several other devices), the most direct and important type of mode structure most easily linked to machinery features, such as blade and vane numbers, is the circumferential one. The near field of a rotor at blade-passage frequency $\omega = B\Omega$ is a pressure pattern having B cycles of variation around the rotation axis. If the rotor interacts with wakes or the potential field of

[3]Section authored by Thomas G. Sofrin.

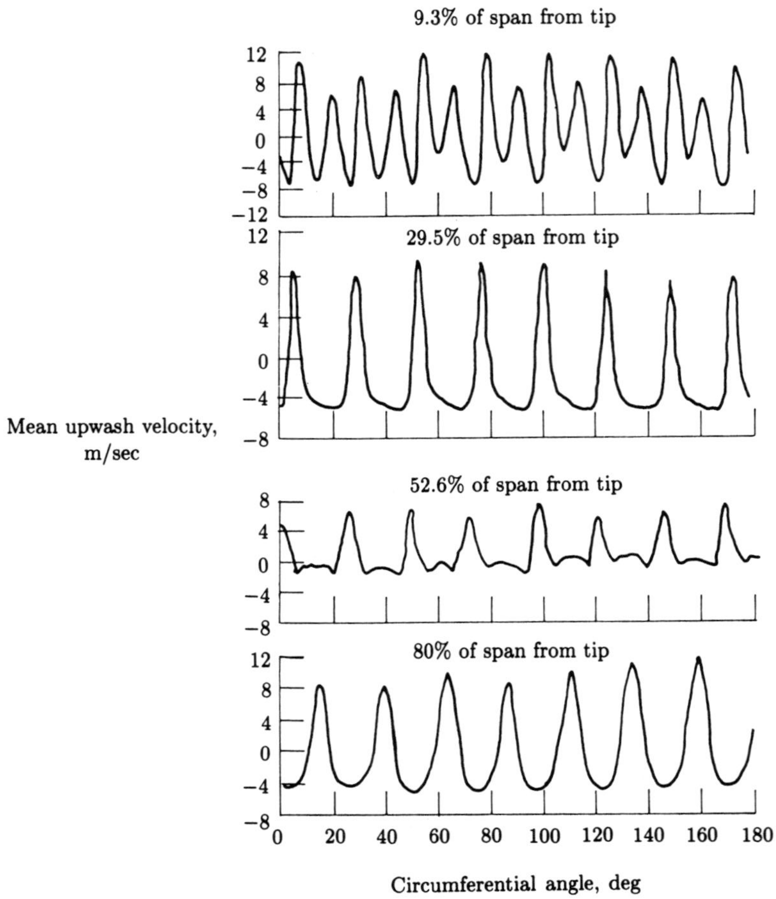

Figure 18. Rotor mean wake velocity profiles as a function of spanwise location. 1.23 rotor chords downstream; 80 percent design speed; tunnel velocity, 41 m/sec (134.5 ft/sec). (From ref. 67.)

a stator assembly or with other aerodynamic nonuniformities having, say, V cycles of nonuniformity around the duct, interaction patterns having $m = B - V$ and $m = B + V$ circumferential cycles result. These characteristic numbers, or modes, m can be used to determine immediately the source or sources of noise at any blade frequency. Harmonics of blade frequency $\omega = nB\Omega$ have similar modal patterns—the direct rotor field mode is $m = nB$, and interactions with V pure cycles of flow nonuniformity ($k = \pm 1$ in eq. (9)) have mode structures $m = nB - V$ and $m = nB + V$. Thus, if m can be measured for a given blade harmonic, the noise source $V = nB - m$ is immediately revealed.

In many applications it is sufficient to determine the m-mode structure, since this immediately identifies the significant noise sources and indicates the relevant engine components or features that require attention. Such measurements can be made conveniently with an array of flush-mounted transducers disposed circumferentially around the fan duct wall. It occasionally happens that more detailed acoustic

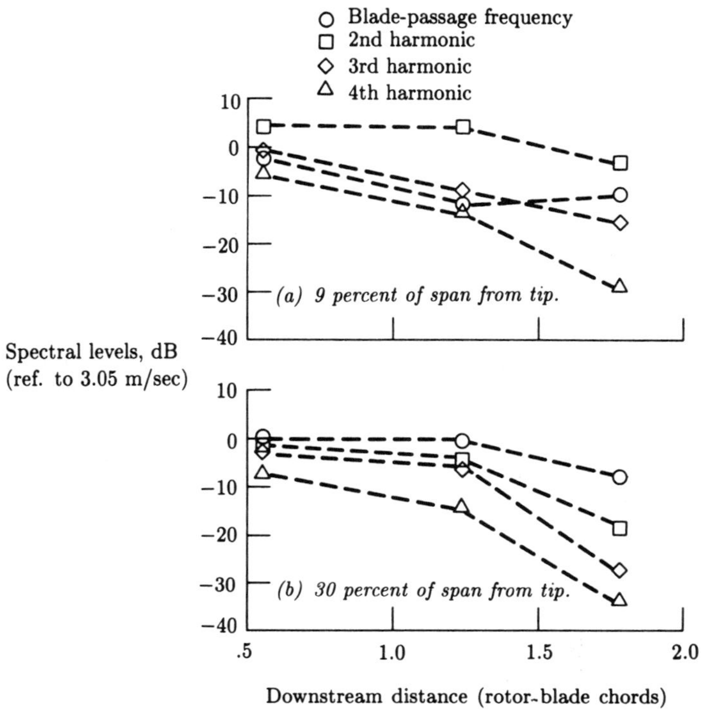

Figure 19. Mean wake harmonic levels from ensemble averaged spectra. Upwash component; 80 percent design speed; tunnel velocity, 41 m/sec (134.5 ft/sec). (From ref. 67.)

structure information is desired, such as when the spanwise source strength distribution is sought. Such radial acoustic distributions can be determined by measuring the amplitude (and phase) of the dominant m-modes at a plurality of radii. However, the mechanical equipment needed to obtain such measurements is significant, and the extraneous fan noise generated by the presence of this equipment in the airflow must be considered.

It should be mentioned that, except for unusual situations where only a few dominant modes exist, far-field measurements around the engine from front to rear are virtually useless for inferring mode structure and, thus, for identifying dominant sources. The difficulty arises because even a single m-mode pattern in the duct, associated with a single radial mode distribution, generates a complicated far-field radiation directivity pattern. If two or more radial modes, having unknown relative amplitudes and phases, are associated with the m-mode, the far-field pattern becomes yet more complex. In practice, there are usually several m-modes present, and even if one is dominant, the others further obscure the far field so that normally few source inferences are possible.

Procedures for Circumferential Mode Measurement

Determination of circumferential, or m-mode, structure obviously requires measurements in a circumferential direction, either with a fixed array of microphones or with a traversing microphone. For in-flight mode tests an array of flush-mounted transducers around the duct is usually made to suffice. When ground tests are conducted such arrays are also commonly used. Traversing systems in which circumferential surveys are made at several radii have been used in several fan tests. In all cases the basic signal processing features are similar.

The description of the simplest acoustic field (at a fixed radius), consisting of a single m-mode at harmonic n of blade frequency $B\Omega$, is

$$p(\Theta, t) = a \cos(m\Theta - nB\Omega t + \Phi)$$

Measurement of this field discloses two features: amplitude a is constant with position Θ, and the phase of the pressure $m\Theta$ varies linearly with position. It is this second feature that is significant in the identification of what mode is present. If, for example, $m = 4$, there are 4 complete 360° phase shifts in going completely around the duct. (Phase is measured with respect to some reference signal at $nB\Omega$ generated by the rotor or by a transducer fixed in the duct.) Thus, if but a single m-mode is present, very crude measurements are sufficient to identify both what it is (e.g., $m = 4$, $m = -9$, etc.) and what its amplitude is.

When two or more modes are present, the circumferential behavior of the noise can vary in a complex manner: amplitude is not constant, but can vary significantly with position. Phase may also vary circumferentially in a complex manner.

The complete pressure field at the plane of the array for some fixed radius is best described in the form

$$p(\Theta, t) = \operatorname{Re}\left\{\sum_{n=1}^{\infty} \sum_{m=-\infty}^{\infty} C_m^n \exp\left[i(m\Theta - nB\Omega t)\right]\right\} \tag{15}$$

For the nth harmonic ($p = \sum_n P^n$) we can write

$$p^n(\Theta, t) = \operatorname{Re}\left[P^n(\Theta)\exp(inB\Omega t)\right] = \operatorname{Re}\left[\sum_{m=-\infty}^{\infty} C_m^n \exp(im\Theta)\exp(-inB\Omega t)\right]$$

From this, the complex pressure $P^n(\Theta)$ is expressed simply as

$$P^n(\Theta) = \sum_{m=-\infty}^{\infty} C_m^n \exp(im\Theta) \tag{16}$$

where amplitude and phase of $P^n(\Theta)$ are measured at each transducer in the array. Before discussing this discrete case, it is helpful to examine how the mode coefficients C_m^n are obtained in the hypothetical case of a continuous measurement of the variation of P^n with Θ. The procedure involves the standard finite transform of Fourier analysis, which is applied to equation (16) to give

$$\frac{1}{2\pi}\int_0^{2\pi} P^n(\Theta)\exp(-iM\Theta)\,d\Theta = \sum_{m=-\infty}^{\infty} C_m^n \frac{1}{2\pi}\int_0^{2\pi} \exp[i(m - M)\Theta]\,d\Theta \tag{17}$$

The integral evidently vanishes for all m except the "target" value M, in which case the Θ average is unity. There follows the standard result for the mode coefficients:

$$C_m^n = \frac{1}{2\pi}\int_0^{2\pi} P^n(\Theta)\exp(-im\Theta)\,d\Theta \tag{18}$$

In principle this algorithm can be executed with an analog system involving a continuous transducer traverse in Θ, a phase shifter generating a voltage $\exp(-im\Theta)$, a multiplier, and an integrating circuit. Practically, it is much simpler and more accurate to employ digital signal processing, which implies a finite number of $P^n(\Theta)$ measurements.

In a fixed transducer array, selection of the number of transducers is limited by availability, maintenance, and recorder channel capacity. These limitations create a problem situation that is discussed subsequently.

For a fixed array of N transducers spaced $\Delta\Theta = 2\pi/N$ apart, the procedure is analogous to the continuous case and is as follows. At the jth location Θ_j, the pressure is

$$P^n(\Theta_j) = \sum_{m=-\infty}^{\infty} C_m^n \exp(im\Theta) \tag{19}$$

The discrete Fourier transform is applied, the result being

$$\frac{1}{N}\sum_{j=0}^{N-1} P^n(\Theta_j)\exp(-iM\Theta_j) = \sum_m C_m^n \frac{1}{N}\sum_{j=0}^{N-1}\exp\left[i(m-M)\Theta_j\right] \tag{20}$$

The expression

$$\frac{1}{N}\sum_{j=0}^{N-1}\exp\left[i(m-M)\Theta_j\right]$$

behaves generally as its counterpart for the continuous-measurement case in equation (17), but with one extremely important exception. Consider the following:

$$\exp\left[i(m-M)\Theta_j\right] = \exp\left[i(m-M)(2\pi/N)_j\right]$$

As before, if $m = M$, the above becomes $\exp(io) = 1$ and the average of the sum of N unit terms is unity. But if $m = M+N$, we have $\exp\left[iN(2\pi/N)j\right] = \exp(i2\pi j) = 1$ for all j. Thus, when we are trying to measure C_M^n in isolation, we also get a full contribution from any mode C_{M+N}^n that is present in the pressure field.

We can easily see that mode $m = M - N$ has the same effect. The modes $m = M \pm N$ are called the principal aliases of mode $m = M$. Other modes also are aliases of M ($m = M \pm 2N,\ M \pm 3N, \ldots$), but under ideal circumstances they have little significance. (Unfortunately the real conditions of acoustic testing are usually far from ideal.)

Aliasing is a familiar consideration in all digital signal processing. It is handled by analog prefiltering of the continuous signal prior to digitizing. This removes high-frequency components that would otherwise alias measurements of the signal properties in the lower frequency range of interest. But we are concerned here with a function $P^n(\Theta)$ which is not a time signal. Antialiasing measures here require filtering in the spatial or modal domains rather than in the time or frequency domains. It turns out that such filtering is automatically provided by the modal propagation characteristics of the duct; provided the array is reasonably distant from all sources (about one duct radius), generated modes having m in excess of some value are cut off and decay before they reach the array. Thus, the duct provides its own low-pass modal filter.

From these considerations the following procedure for selecting microphone number is established:

1. Select the highest frequency of interest—the highest harmonic at the top speed considered.
2. Determine the maximum mode number $m_{\max}$ that will propagate at this frequency.
3. Select the number of microphones as $N > 2m_{\max}$. (This assures that when measuring a mode near $m_{\max}$ there is no alias from a mode $(-m)$ near $-m_{\max}$.)

A more succinct formulation is that the separation between transducers must be less than the least circumferential half-wavelength that propagates.

Unfortunately, it is often found that this antialiasing rule leads to an unacceptably large number of transducers. In such cases it may be possible to select N judiciously, based on prior knowledge of the likely candidate noise sources, with the recognition that not *all* the modes between $-m_{\max}$ and $m_{\max}$ have significant strengths. A listing of all propagating modes that are generated by likely interactions can be made. The requirement then is to select N such that no mode m_i from interaction i is an alias of any mode m_j generated by a different candidate source j. That is, $m_i \neq m_j$ (modulo N).

With N selected according to the above conditions, the algorithm for computing modal coefficients becomes

$$C_m^n = \frac{1}{N} \sum_{j=0}^{N-1} P^n(\Theta_j) \exp(-im\Theta_j)$$

where

$$\Theta_j = j(2\pi/N) \tag{21}$$

The preceding discussion has detailed the basic principles of mode measurement in turbofan (and other) power plants. Specific techniques which have been applied range from matrix inversion of N selected wall pressure measurements to deterministic solution for N preselected modes (ref. 69) to a least-squares approach where the number of measurements is at least twice the number of modes (ref. 70). Formidable practical difficulties exist. Radial measurements upstream of the fan introduce distortions and their associated extraneous modes, and measurements on the wall alone require large numbers of microphones distributed axially and circumferentially.

A technique using an upstream rotating microphone has been formulated to overcome the problem of distortion mode generation by the probe and to reduce the number of microphones required (ref. 71). With the exception of an experiment using a wall-mounted array with an actual JT15D engine (ref. 72), published demonstrations of the techniques have been limited to relatively low-speed fans with conditions rather far removed from the turbofans of interest. An example of the wave number distributions measured on the JT15D is shown in figure 20. Direct measurement techniques require additional development and still fall in the category of research efforts and not routine tools. Thus, predictions from three-dimensional or quasi-three-dimensional analyses (refs. 73 to 76) are important sources of detailed modal information.

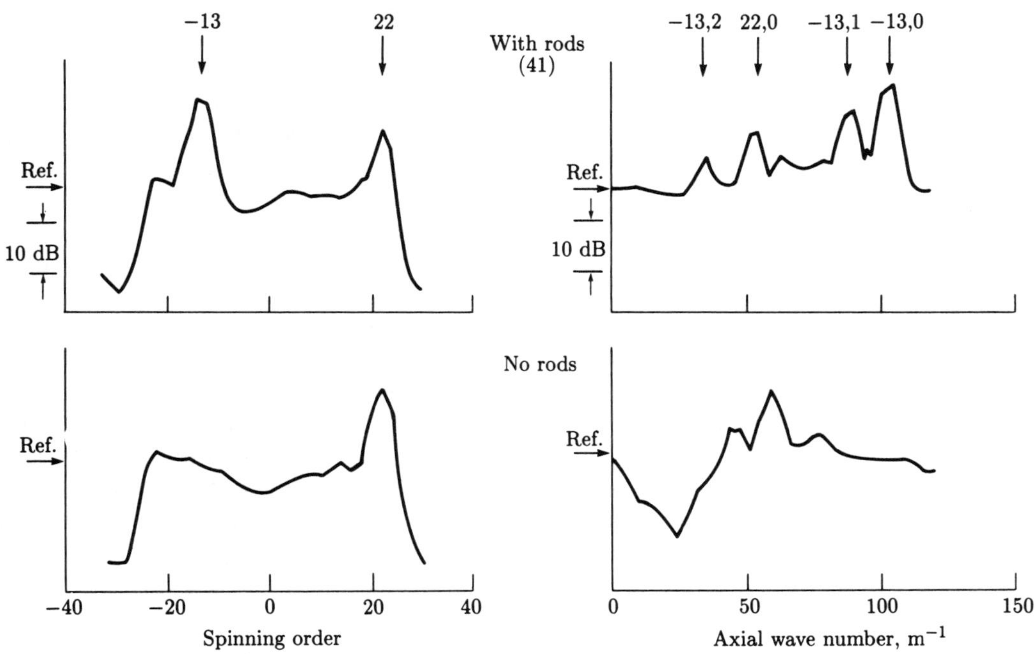

Figure 20. Inlet mode measurements on JT15D engine at 10 800 rpm. (Based on ref. 72.)

Application of Theory and Experiment to In-Flight Sources

Response of Ducted Cascade

Considerable effort has been expended to model the noncompact compressible response of a ducted cascade of blades to unsteady upwash velocities. Perhaps the most complete description available is the three-dimensional lifting-surface theory for a rotating cascade in an annular duct (ref. 6). This blade response and duct coupling analysis is the heart of specialized studies of rotor-inflow distortion (ref. 73) and rotor-stator interaction (ref. 74). These linear analyses are for the dipole-type sources at the surface of a cascade of thin (in some cases twisted) blades and represent exact solutions to the linearized continuity and momentum equations (chapter 5 of ref. 3).

Three features of these analyses are considered to be important. First, the three-dimensional approach makes possible the calculation of the circumferential and radial content of acoustic modes in annular or cylindrical ducts; the complete description of modal content is precisely the input required for successive propagation analyses. Second, cascade analysis predicts chordwise unsteady pressure distributions and integrated responses which differ substantially from single-blade results (refs. 77 and 78) that ignore blade-to-blade interactions (solidity) and the interblade phase angle of the disturbance. Third, source noncompactness, retained by calculating chordwise and spanwise pressure fluctuations, produces significant differences in calculated power compared with compact analyses. The magnitudes of the differences, which depend on incident disturbance shape and propagation direction with respect to the mean flow, are highest for single distortion modes (ref. 77). For realistic distortion profiles represented by a combination of distortion modes, the effects of noncompactness are less dramatic, with the compact analysis tending to underestimate fundamental tone power for upstream propagation and to overestimate the power propagating downstream (ref. 73).

Since full three-dimensional calculations are complex and lengthy, a quasi-three-dimensional analysis, which uses two-dimensional (strip) theory for aerodynamic response but annular duct acoustics for modal prediction, was investigated (ref. 73). The results indicate that the quasi-three-dimensional approach produces relatively small errors in power, greatly reduces computation time, and fulfills the requirement to predict annular duct acoustic modes. Consequently, the quasi-three-dimensional approach was adopted in the development of a computer program (ref. 75) which considered three types of flow disturbances: inlet turbulence, rotor mean wakes, and rotor wake turbulence. This quasi-three-dimensional approach requires further validation by data-theory comparisons. An encouraging start at validation is described in the next section.

Controlled Disturbances for Theory Validation

Predictions of three-dimensional lifting-surface tone power have been compared with fan noise data (ref. 76) for which the controlled source consisted of the fan interacting with an array of inlet distortion rod wakes. Figure 21 shows excellent agreement between the predicted total inlet fundamental tone acoustic power as a function of fan speed and the measured narrow-band tone power obtained from far-field measurements. Note the changing mix of radial mode contributions to the totals and the nonmonotonic increase with speed in both theory and data.

The predicted modal content can be used in conjunction with a Wiener-Hopf radiation analysis (refs. 33 and 79) to calculate the far-field directivity of the three propagating modes at 10 500 rpm. Individual modal and total directivities shown in figures 22(a) and 22(b), respectively, were calculated with an unpublished Wiener-Hopf code written by Y. C. Cho at NASA Lewis Research Center. The inputs to the Wiener-Hopf code were the amplitudes and phases of each mode at the inlet entrance as calculated from the analysis of reference 76. The measured directivities obtained from experiments in an anechoic chamber (ref. 40) are superimposed in figure 22(b). The shapes of the curves agree well for the shallow angles, where directivity is controlled by the principal lobe of the first mode, but agree less satisfactorily for the higher angles, where several modes contribute and the levels are sensitive to exact prediction of mode phases in addition to amplitudes.

The intermediate quantity between blade response and duct coupling is blade pressure (fig. 4). The cascade response portion of the code in reference 75 was used to calculate the chordwise magnitude of the unsteady blade pressures due to interaction with Gaussian wakes produced by upstream radial rods. As shown in figure 23, the high disturbance frequency associated with many rod wakes (in this case 41) is predicted to produce many rapid changes in pressure along the chord. Typical miniature transducer sizes are indicated near the leading and trailing edges. For high disturbance frequencies the analysis indicates that measured blade pressure amplitudes are subject to uncertainty because of finite transducer size and sensitivity to transducer location. However, experimental checks of the cascade response analysis through use of carefully controlled flow disturbances are needed.

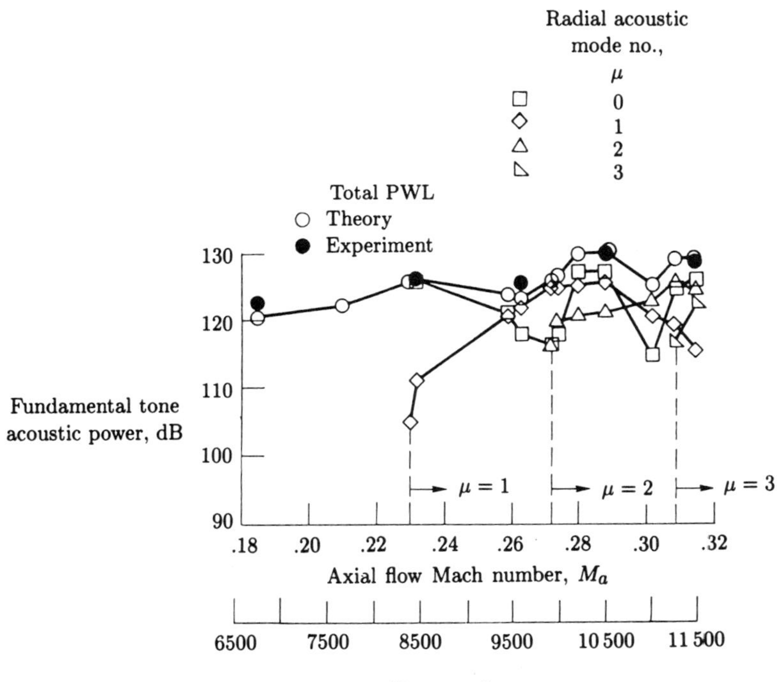

Figure 21. Relationship of speed and inlet fundamental tone acoustic power generated by 41 rod wakes interacting with JT15D fan. (From ref. 76.)

Rotor-Stator Interaction

Acoustic data are available from rotor-stator spacing experiments on the same fan as was used for the wake measurements described in the *Rotor Wake Measurements* section. Two stator-vane to rotor-blade ratios were examined, one for propagation and the other for cutoff fundamental tones. Figure 24 shows variation of the inlet narrow-band tone harmonic power level with rotor-stator spacing. (From ref. 80.) Residual levels of the fundamental for the cutoff case (25 vanes) are nearly constant, and this uniformity suggests that a weak interaction of inflow disturbance and rotor governs in this case rather than a rotor-stator interaction due to stator-vane nonuniformities sufficient to generate other propagating modes (ref. 81). Note that second and third harmonic levels for the 25-vane stator are higher than corresponding harmonics generated by the 11-vane stator, an indication of a difference in the response or coupling to acoustic modes of the two stators or both. The 11-vane stator had longer chords than the 25-vane stator in order to maintain the same solidity.

Tone powers measured in experiments on rotor-stator spacing and vane-blade ratio in an anechoic chamber with inflow control have been compared favorably with results from a two-dimensional (strip) model (refs. 82 and 83). Wake data were not acquired, so a wake model (ref. 68) was used. Although two-dimensional theory may work relatively well for power predictions, calculating far-field directivity (and, therefore, acoustic mode content) requires more sophistication in handling duct geometry and, probably, in describing the wake-vortex flow field.

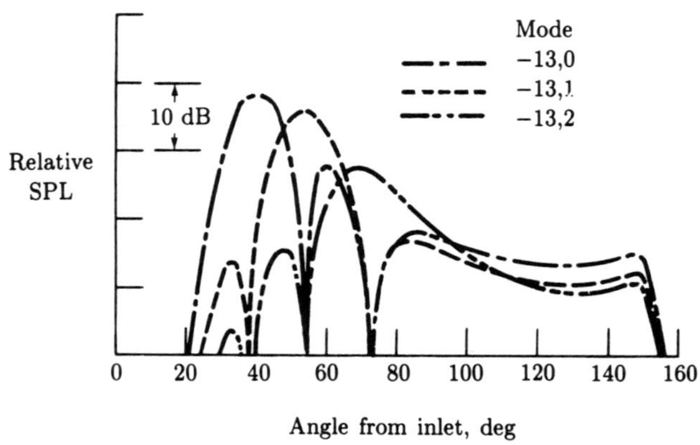

(a) Individual modal directivities.

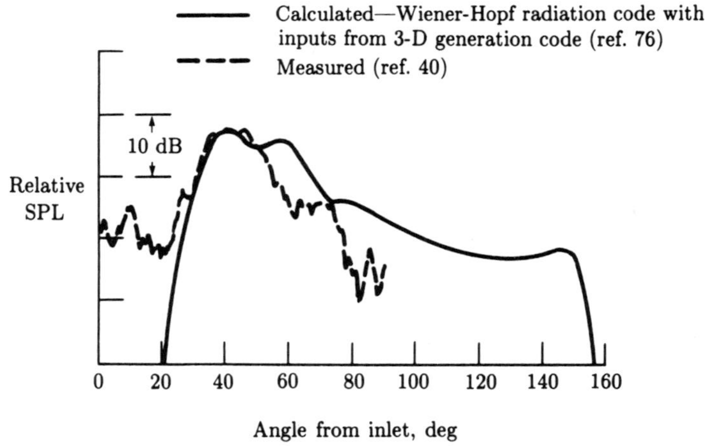

(b) Total directivities.

Figure 22. Blade-passage tone directivities with controlled source. JT15D engine; 41 rods; 10 500 rpm fan speed.

Blade Sweep To Reduce Multiple Pure Tones

Rotor-blade sweep has been investigated as a means to reduce the strong multiple pure tone (MPT) inlet source at takeoff fan speeds. A radical fan with compound leading-edge sweep was designed to keep the normal component of the blade inlet relative Mach number subsonic over the entire span (ref. 84). Except for blade end effects and the sweep reversal point, a major portion of the strong leading-edge shock system was expected to be eliminated. Figure 25 shows the measured MPT power results obtained with the swept design compared with results from a conventional, unswept fan. (Based on ref. 85.) Sweep delays the onset of MPT's and reduces the power levels over a large portion of the tip-speed range, including takeoff.

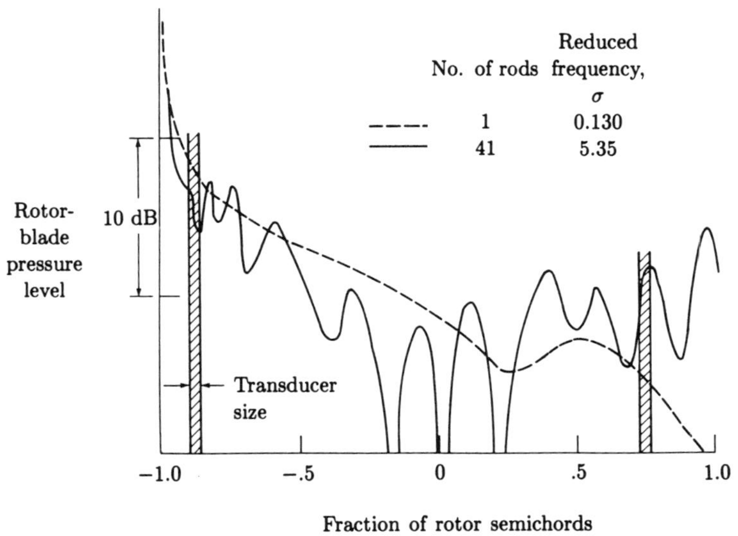

Figure 23. Calculated chordwise variation of fundamental component of rotor-blade pressure levels generated by wakes from upstream rods.

High Specific Flow

Another aspect of inlet noise generation at supersonic tip speeds concerns the observation that total tone power peaks beyond the transonic speed and then falls off. A fan designed for unusually high specific flow (220 kg/sec-m^2 (45 lb/sec-ft^2)) at a high tip speed (553 m/sec (1750 ft/sec)) exhibited a marked tone power decrease at design speed, although the results were not qualitatively different from those of other high-tip-speed designs (ref. 86). Figure 26 shows the results for this high-specific-flow fan with those for other high-tip-speed designs. The noise-reduction phenomenon appears to be only partially attributable to propagation inhibiting effects of elevated inlet Mach numbers. It may also be associated with nonlinear propagation characteristics in combination with the angle and associated strength variations of the leading-edge shocks (ref. 62).

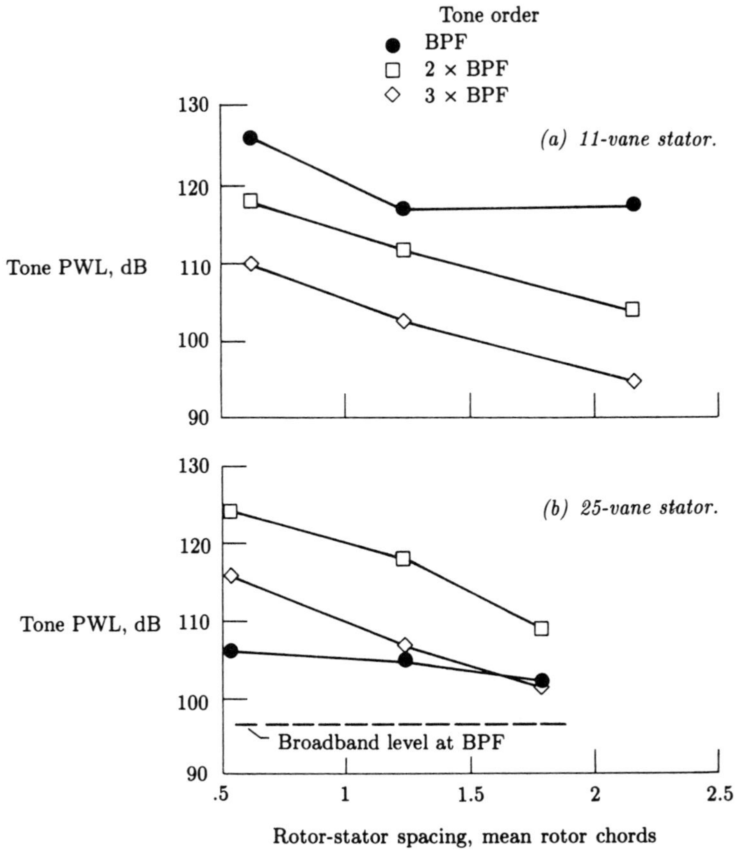

Figure 24. Variation of narrow-band tone harmonic power levels with rotor-stator spacing. 80 percent of design speed; 15 rotor blades; tunnel velocity, 41 m/sec (134.5 ft/sec). (From ref. 80.)

Full-Scale Engine Applications[4]

Introduction

There are many mechanisms which potentially contribute to the noise generation, propagation, and radiation characteristics of the turbomachinery components in an aircraft engine. These mechanisms have been discussed in some depth in the previous sections of this chapter. The purpose of this section is to describe and summarize how to apply the knowledge available to designing the turbomachinery components such that substantial noise reductions are achieved and noise regulations can be met with a minimal negative impact on engine performance, weight, manufacturing cost, complexity, and serviceability.

This section begins with a qualitative description of the spectral and directivity characteristics of a typical full-scale turbofan engine, with a breakdown of how each of

[4] Section authored by Phillip R. Gliebe.

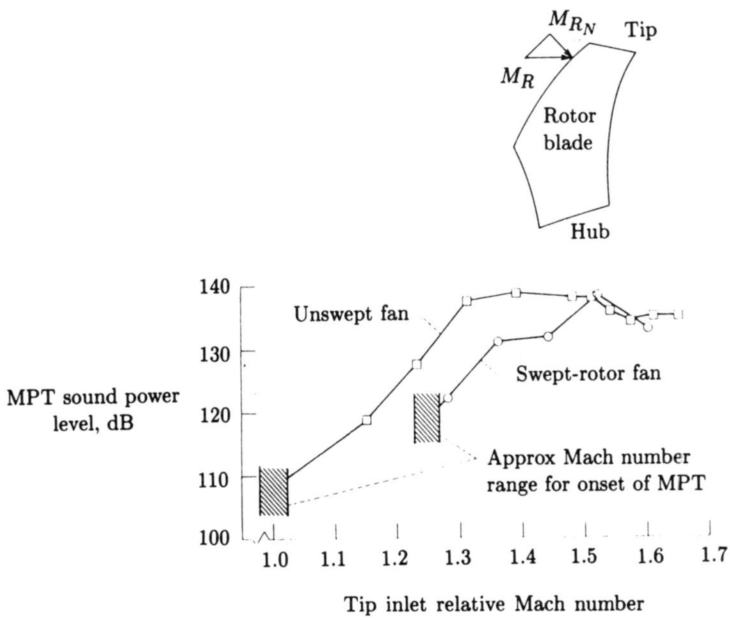

Figure 25. Effect of rotor leading-edge sweep on multiple pure tone generation. (Based on ref. 85.)

the turbomachinery components contributes to the total engine noise spectrum and directivity pattern. We then discuss methods for estimating the noise contributions of each of the turbomachinery components, including semiempirical methods and scaling from previous test results. The next logical step, if we are evaluating a new engine design for noise compliance, is to identify those engine components which require noise reduction relative to the baseline noise level estimates. Finally, we address various noise-reduction methods for each of the turbomachinery components, considering the effectiveness of the method relative to the associated penalty to the engine system.

Turbomachinery Noise Characteristics: A Qualitative Description

Typical commercial aircraft engines are of the dual-spool type having a low-pressure spool comprised of a low-pressure-ratio fan and its drive turbine and a high-pressure spool comprised of a high-pressure-ratio compressor and its drive turbine. The compressor-fan machinery is separated from the turbines by a combustion chamber. Some typical turbojet-turbofan schematic arrangements are shown in figure 27. Typically, the major turbomachinery noise contributors are the fan, the low-pressure compressor (LPC), and the low-pressure turbine (LPT). Other contributors to engine noise include the combustion process and the exhaust jet. A sample flyover balance of these components is illustrated in figure 2.

The noise characteristics of turbomachinery components in an aircraft engine are usually quantified in terms of several noise measurement parameters. These include the following:

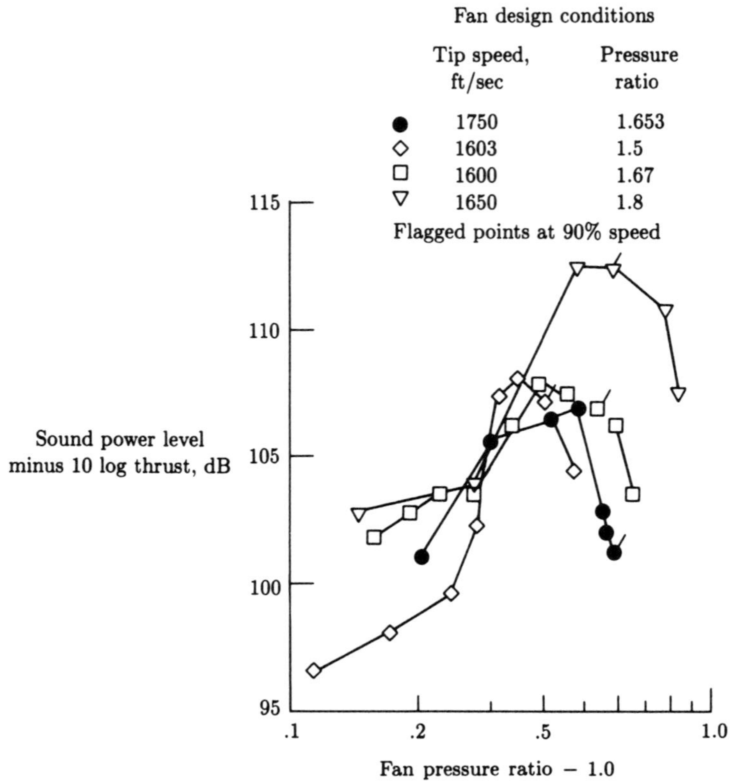

Figure 26. Inlet noise characteristics of high-tip-speed fans. (Based on ref. 86.)

1. Overall sound power level (OAPWL)
2. Sound power level spectrum (PWL(f))
3. Overall sound pressure level directivity (OASPL(Θ))
4. Sound pressure level spectrum (SPL(Θ, f))
5. An appropriate subjective noise level, e.g., perceived noise level (PNL(Θ))

These parameters are obtainable from measurements of sound pressure made with microphones placed at strategic locations around the component (or engine) during a test.

The overall sound power level (OAPWL) produced by a turbomachine is generally a function of aerodynamic and performance-related parameters such as air flow rate, tip speed, pressure ratio and/or shaft horsepower, and geometric design parameters.

The sound power level spectrum (PWL(f)) is the distribution of the generated sound over a range of audible frequencies. Typical examples of narrow-band turbomachinery noise spectra are shown in and discussed in connection with figure 3.

The directivity characteristics of the noise generated at a given frequency describe how the internally generated sound power is distributed in the radiation field at some distance away from the turbomachinery component or engine in terms of sound pressure level (SPL) measured by microphones or heard by the ear. Sound pressure level can have both azimuthal directivity and polar directivity. Azimuthal directivity describes the variation of sound pressure azimuthally, or around the machine axis. In most cases, azimuthal variations are small and can be neglected, especially for the broadband components of noise. In certain special cases, the azimuthal variation in sound pressure can be quite significant for discrete tones.

Polar directivity refers to the variation in sound pressure from inlet centerline to exhaust centerline on a constant radius arc in a fixed azimuthal plane. The polar directivity of broadband noise is usually smooth, with maximum levels occurring near the engine (or component) inlet and/or exhaust axis. The polar directivity of discrete tones can be highly irregular with several peaks and valleys referred to as lobes, the number and size of which depend on the type of tone and the source mechanism which produces it. Typical examples of polar directivity patterns are shown in figure 28.

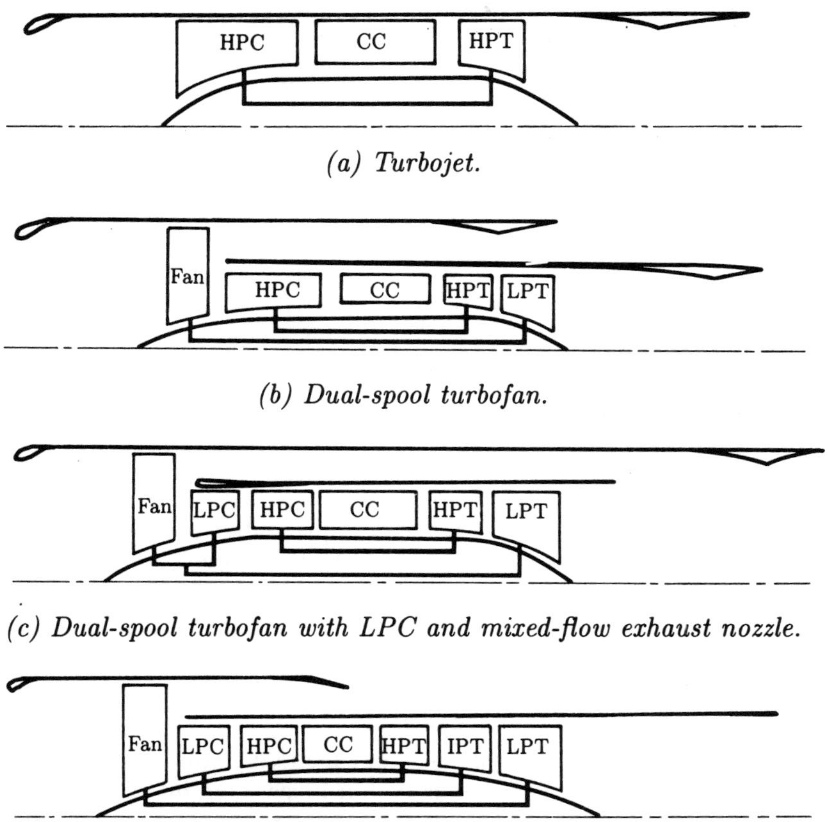

Figure 27. Typical turbojet-turbofan engine schematic arrangements.

Subjective noise levels refer to appropriately integrated or summed sound pressure levels which best represent human ear annoyance to the generated sound field. Summing is performed in the frequency domain and, often, also in the time domain. For example, perceived noise level (PNL) refers to a summation, over all 1/3-octave frequency bands, of the sound pressure levels at a given observer polar angle, with the

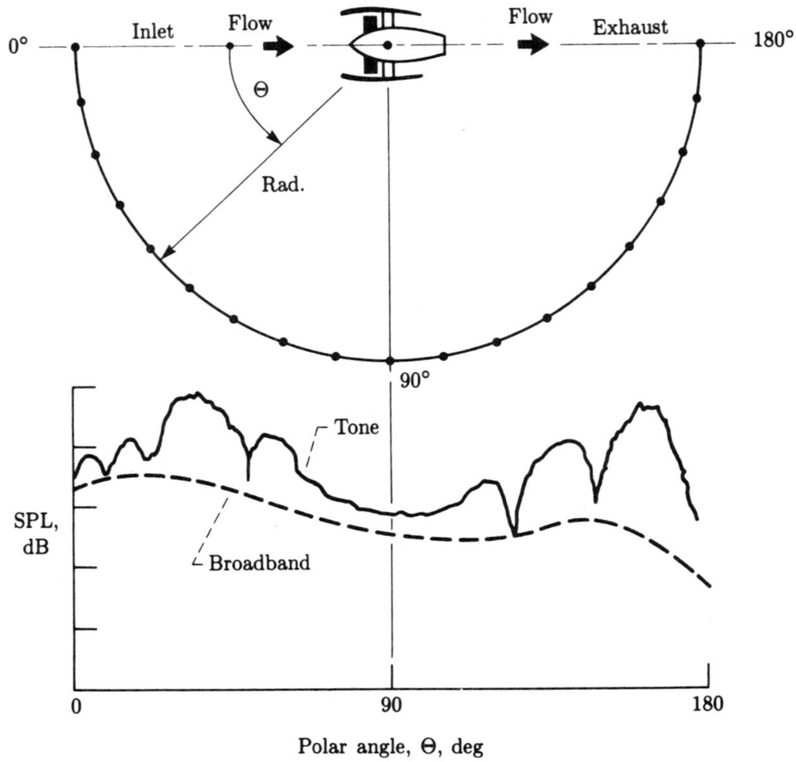

Figure 28. Typical turbomachinery sound pressure level polar directivity patterns.

level in each band weighted by a factor which represents the degree of annoyance to noise observed at that particular frequency. Effective perceived noise level (EPNL) refers to a time integration of PNL received by an observer as the noise source passes by, such as that which would occur during an aircraft flyover. It represents effects of the time duration over which a given PNL must be "endured." See references 87 and 88 for detailed descriptions of noise measurement procedures and computation methods for subjective noise level evaluation.

An example of a typical high-bypass-ratio turbofan engine noise field is shown in figure 29. The component contributions of fan noise, combustor noise, LPT noise, and jet noise are shown in figure 29 to demonstrate the dominant sources which typically control the noise in the various regions of the spectrum and directivity patterns. We see, for example, that the fan noise usually contributes the highest levels in the forward arc at midrange and high frequencies and in the aft arc at high frequencies. The turbine (LPT) only contributes in the aft arc at high frequencies. The jet dominates the low frequencies over most of the directivity arc, while the combustor contributes significantly around the sideline angles close to 120°, mostly at low to mid frequencies.

The trends shown in figure 29 are typical for bypass ratios from about 3 to 8. For low-bypass-ratio engines (mass flow ratio or bypass ratio less than about 1.0),

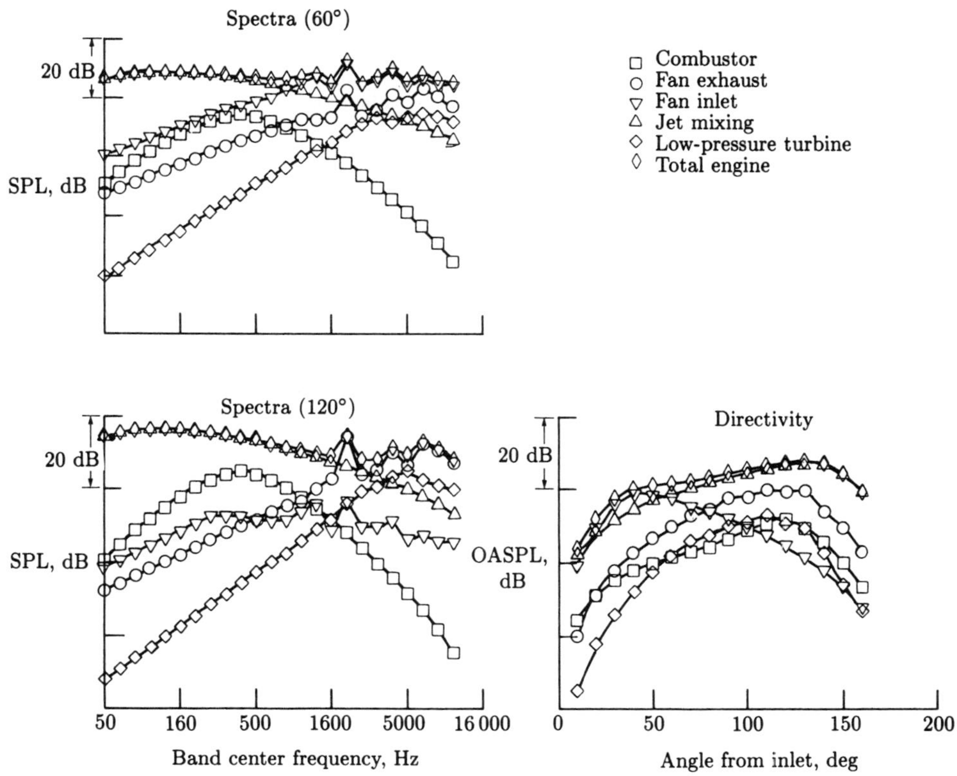

Figure 29. Typical turbofan engine component noise source contributions at takeoff power.

the jet noise is the greatest contributor to the overall noise and may actually control the total noise in the aft arc during takeoff conditions.

Preliminary Noise Estimating Procedures: Empirical Scaling of Existing Data

Turbomachinery noise characteristics can usually be estimated, at least for some of the overall level trends, through the use of empirically derived correlations and key-parameter scaling procedures. For example, Heidmann (ref. 89) developed a rather elaborate empirical prediction scheme for aircraft engine fans which takes into account a significant number of fan performance and geometric variables. A similar correlation method was developed by Kazin and Matta (ref. 90) for application to axial turbine stages.

A general formulation for the empirical correlations of turbomachinery noise characteristics which includes the correlation models of references 89 and 90 can be derived and has the following functional form:

$$\overline{P}^2 = \overline{P}_{bb}^2 + \overline{P}_{\text{tone}}^2 \tag{22}$$

where

$$\overline{P}^2_{bb} = \frac{\rho_o c_o}{4\pi R^2} W_{bb}(X_i) D_{bb} S_{bb} \tag{23}$$

and

$$\overline{P}^2_{\text{tone}} = \frac{\rho_o c_o}{4\pi R^2} W_{\text{tone}}(X_i) D_t S_t \tag{24}$$

The terms in equations (22) to (24) are defined as follows:

$W_{bb}(X_1, X_2, ..., X_n)$	broadband source acoustic power
$W_{\text{tone}}(X_1, X_2, ..., X_n)$	tone source acoustic power
$D_{bb}(\Theta, \phi)$	broadband directivity function
$D_t(\Theta, \phi)$	tone directivity function
$S_{bb}(\eta)$	broadband spectrum function
$S_t(\eta)$	tone spectrum function

Also, ρ_o is the ambient density, c_o is the ambient speed of sound, R is the distance from the source to the observer, and X_i are similarity parameters which determine the values of W_{bb} and W_{tone}.

Source Acoustic Power

The source acoustic power for the tone and broadband noise sources in turbomachinery can be expressed in terms of two basic correlating parameters for order-of-magnitude or preliminary design estimate purposes as follows:

$$\left.\begin{matrix} W_{bb} \\ W_{\text{tone}} \end{matrix}\right\} = \rho_o c_o^3 A K (\Delta T/T)^a M_t^b \tag{25}$$

where A is the inlet flow area for compressors and fans (exit flow area for turbines). The parameters K, a, and b are constants which are obtained from empirical correlations of existing data. The two basic correlating parameters are the tip-speed Mach number M_t and the loading parameter $\Delta T/T$. This loading parameter is the normalized ideal energy input (for fans or compressors) or output (for turbines) and can be expressed in terms of the turbomachine operating pressure ratio PR as follows:

$$\Delta T/T = (\text{PR})^{\frac{\gamma-1}{\gamma}} - 1 \tag{26}$$

for fans and compressors and

$$\Delta T/T = 1 - \left(\frac{1}{\text{PR}}\right)^{\frac{\gamma-1}{\gamma}} \tag{27}$$

for turbines. The Mach number is expressed in terms of rotor speed n and tip radius as follows:

$$M_t = 2\pi R_T (n/60) \big/ c_o \tag{28}$$

In equations (26) to (28), γ is the ratio of specific heats and R_T is the rotor tip radius. Examples of the correlation of the source power functions W_{bb} and W_{tone} for turbines are shown in figure 30.

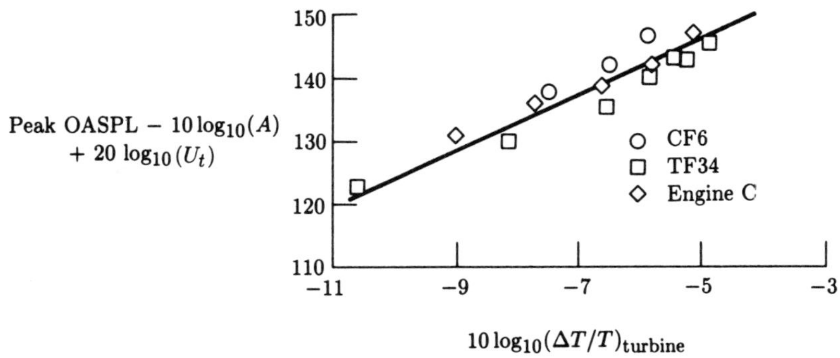

(a) Correlation for turbine noise (broadband and tone) OASPL; based on turbine overall pressure ratio; 200 ft (61-m) sideline; 120° from inlet; 1/3-octave band levels.

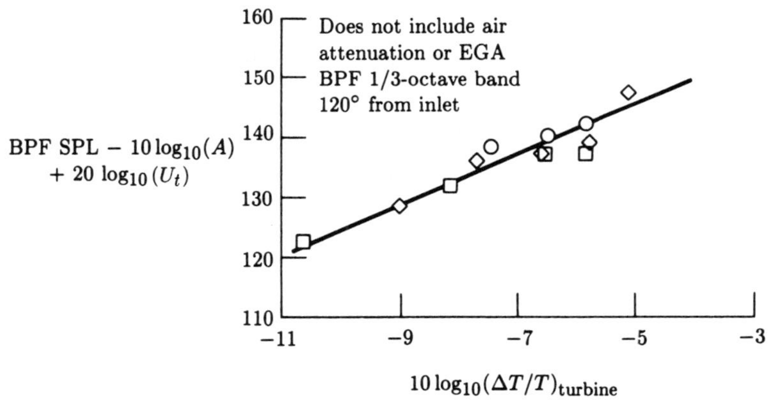

(b) Correlation for dominant tone SPL; based on turbine overall pressure ratio; 200-ft (61-m) sideline.

Figure 30. Correlation curves for turbine source power function. (From ref. 90.)

Directivity Functions

The directivity functions $D_{bb}(\Theta, \phi)$ and $D_t(\Theta, \phi)$ determine the spatial distribution of the source acoustic power. These directivity functions, as expressed in equations (22) to (24), are assumed to be independent of frequency; we shall see subsequently that this is not always the case, especially for the tone noise sources. However, for scaling purposes and order-of-magnitude estimates the assumption is not critical. These directivity functions are defined such that the original acoustic power is obtained when the sound pressure distribution is integrated over a spherical surface surrounding the source. This normalization is expressed by the following relationships:

$$\text{Power} = \int_0^{2\pi} \int_0^{\pi} \frac{\overline{P}^2}{\rho_o c_o} R^2 \sin\Theta \, d\Theta \, d\phi = W S(\eta) \tag{29}$$

or

$$\frac{1}{4\pi} \int_0^{2\pi} \int_0^{\pi} D(\Theta, \phi) \sin\Theta \, d\Theta \, d\phi = 1 \tag{30}$$

Examples of directivity correlations for turbines are shown in figure 31.

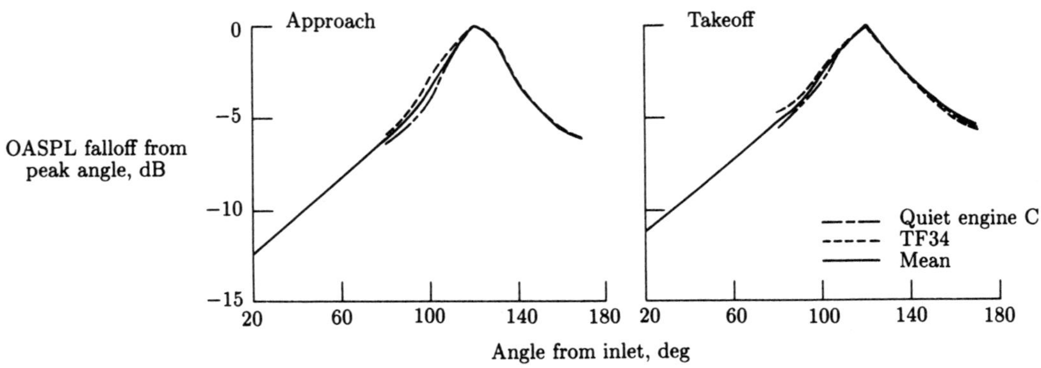

(a) Directivity for turbine noise (broadband and tone) OASPL.

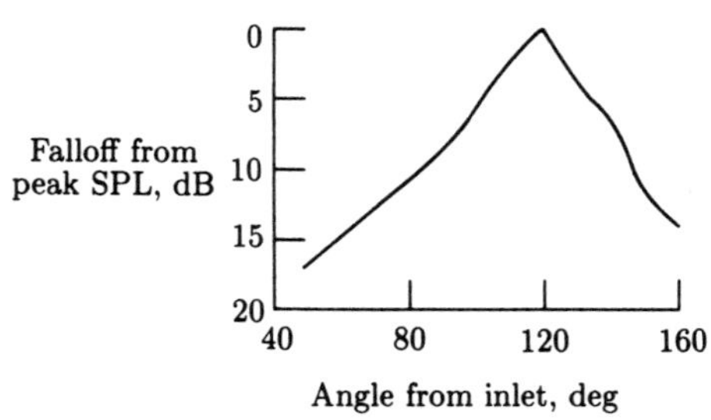

(b) Directivity for turbine tones at approach power.

Figure 31. Correlation functions for turbine directivity. 150-ft (45.7-m) arc; 1/3-octave band levels. (From ref. 90.)

Spectrum Functions

In a manner similar to that for the directivity correlations, the sound pressure spectrum shapes can be developed in a normalized fashion based on the assumption that the spectrum shape does not depend on polar directivity location. Again, as is subsequently pointed out, this is not always the case, especially for tone noise sources, but it is sufficient for scaling and preliminary design estimates. The normalization is typically done such that the summation over all frequency bands of importance in the spectrum gives a factor of unity, so the parameters W_{bb} and W_{tone} are in effect overall power levels. Hence,

$$\int S(\eta) \, d\eta = 1 \tag{31}$$

where $\eta = f/f_{\text{ref}}$. Examples of normalized spectrum functions for turbines are shown in figure 32.

Noise-Reduction Requirements

The generalized empirical correlation methods outlined in equations (22) to (31) have been developed in these or similar forms for specific applications by various engine manufacturers. The precise quantitative values of the various constants and coefficients in equations (22) to (31) are of course dependent on the data base which is utilized in the development of the correlation, and this data base is to some extent proprietary information when a specific engine manufacturer develops the correlation.

The basic approach is to take the noise characteristics data for a given family of engine designs and derive the coefficients, constants, and exponents which describe the variations in noise levels as certain key parameters are varied. Once these correlation constants have been established for each of the turbomachinery components in an engine, a preliminary assessment of the component noise levels for a new or derivative engine can be carried out by scaling the noise characteristics of a baseline engine with the correlation formulas given by equations (22) to (31).

To illustrate the process, consider the hypothetical example of designing a new low-pressure turbine (LPT) for an existing engine model to improve performance. The new LPT is to be designed to run at 5 percent higher tip speed and deliver the same shaft power to the low-pressure compressor and fan. The existing engine noise characteristics are known, and it is desired to assess the impact of the redesigned turbine on the total engine noise and to determine how much, if any, noise reduction is necessary to allow the modified engine to meet existing noise level requirements. It is assumed that the existing engine has a 2-dB margin relative to the requirements.

To begin, we need a correlation similar to that of equations (22) to (31) for the baseline engine LPT. We can use, for example, the correlation of reference 90, which is of the form

$$\text{Peak OASPL} = 40\ \log_{10}(\Delta T/T) - 20\ \log_{10} U_t + 10_{10} \log A + 164 \tag{32}$$

Comparing this expression with that of equation (25), we see that the exponents are $a = 4$ and $b = -2$. The above expression suggests that the overall noise actually decreases with increasing tip speed, contrary to intuitive expectations. This unexpected result is understood when we realize that the assumption that the turbine work does not change (i.e., that $\Delta T/T$ is constant) is what really controls the noise. To illustrate this, equation (25) can be recast in the following form, with the dimensionless work coefficient $\Delta T/U_t^2$ used as a parameter:

$$W/\rho_o c_o^3 A = K\,(\Delta T/T)^a\, M_t^b = K' M_t^{2a+b} \left(\Delta T/U_t^2\right)^a \tag{33}$$

Thus we see that the sound power varies as the $2a+b$ exponent of tip-speed Mach number when the loading parameter $\Delta T/U_t^2$ is held constant. In our example, for $a = 4$ and $b = -2$, the sound power varies as $M_t^6(2a + b = 8 - 2 = 6)$ when $\Delta T/U_t^2$ is held constant. However, in our example the loading parameter decreases between the baseline and the target engine, since the temperature drop ΔT is held constant.

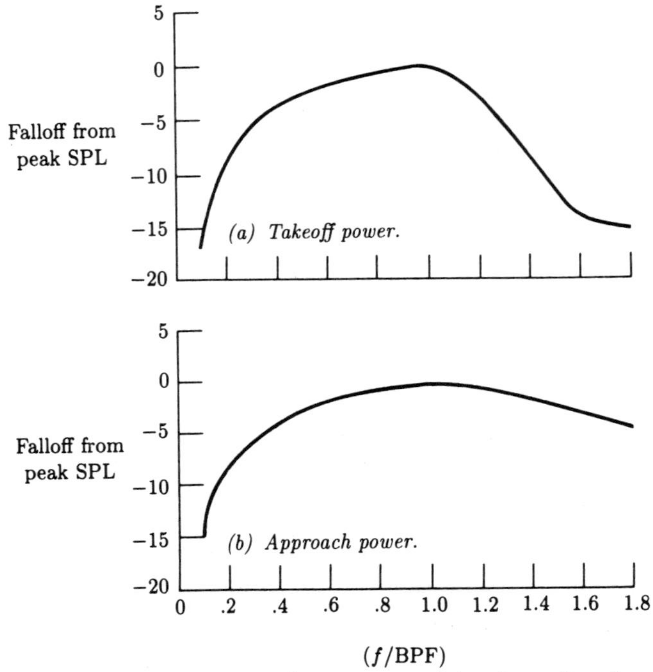

Figure 32. Correlation functions for mean turbine broadband noise spectrum. 120° from inlet; 1/3-octave band levels. (Based on ref. 90.)

The 5-percent speed increase in this case corresponds to a 10-percent decrease in loading parameter, and the net effect is a 0.4-dB decrease in overall noise level.

If, on the other hand, we wanted to examine the effect of increasing engine speed by 5 percent without redesign, the engine thrust and airflow would increase accordingly; this increase would correspond to an approximately constant loading coefficient. Hence, according to equation (32), the noise would increase by the tip-speed ratio raised to the sixth power (i.e., by ≈ 1.3 dB).

Noise-Reduction Methods

If we find that a particular component of a new or derivative engine requires a certain amount of noise reduction relative to its baseline configuration, several options are available for achieving this noise reduction. The method selected depends on several considerations, including the type of component (i.e., fan, LPC, or LPT), the cost involved, the importance of weight and complexity, and the impact on engine performance. The following sections discuss these options for each of three turbomachinery components mentioned.

Fans

For the fan of a turbofan engine, noise reduction can be achieved by either designing for noise reduction at the source or designing fan duct acoustic treatment to absorb the noise produced by the source. The topic of acoustic treatment design is treated in another chapter. However, it should be mentioned that the amount of noise suppression achieved with duct acoustic treatment is predominantly a function of the fan design characteristics. In particular, the fan tip speed and blade numbers have an influence on the achievable suppression. Thus, one can, in fact, design the fan acoustically to give the maximum possible acoustic treatment noise suppression. In general, it is desirable to have high source frequencies to provide sound wavelengths which are small compared with the treatment cavity depth and to have sound source duct mode patterns which propagate at large spiral angles relative to the duct axis (i.e., the modes are near cutoff). Blade numbers and vane numbers can be selected to provide these mode patterns. This approach tends to be a single-point design, however, as the treatment design is usually "tuned" to a particular tone or frequency band at a particular operating condition. The effectiveness of the treatment tends to deteriorate at frequencies and operating conditions away from the design condition.

As a first step in considering ways to reduce fan noise at the source, see the block diagram shown in figure 4. This diagram shows the flow of mechanisms which result in the noise radiation process, as discussed in the section entitled *Elements of the Generation Processes.* The basic idea is that any of the significant mechanisms can be characterized by a gust-type excitation which produces an unsteady, periodic force on a blade or vane, and this unsteady force generates a propagating pressure field in the fan duct, which has a certain frequency and mode pattern. If the pressure field has to pass through adjacent blade rows before radiating from the duct end, the amount of transmitted energy then depends on the mode pattern and frequency of the pressure field and the geometry of the transmitting blade row. The noise of a fan can therefore be reduced at the source by

1. Reducing the gust amplitude
2. Reducing the blade-vane response to the gust
3. Reducing the unsteady lift force amplitude
4. Reducing the efficiency of conversion of the unsteady force to acoustic energy (results in reducing coupling to duct modes)
5. Increasing the transmission loss of any "blocking" blade-vane rows

Techniques for implementing the above approaches to reducing fan noise are discussed in the following paragraphs. The discussion must necessarily be qualitative, but an attempt is made to give an order-of-magnitude estimate of the effectiveness of each technique relative to the others and to give some description of the potential penalties which might be introduced as a result of utilizing each of these techniques.

Reducing gust amplitude—rotor wakes: Fan-rotor wakes impinging on downstream stator vanes are major sources of fan noise. The fan-rotor wake velocity defect and wake turbulence act as "gusts" to the downstream stator vanes. Methods for reducing the rotor wake gust amplitudes include the following:

1. Design the rotor to operate near peak efficiency at the noise-critical operating conditions. This could be done by selecting the blade loading, camber, and incidence angles to provide minimum blade section drag coefficients and, hence, smaller wake defects. (See ref. 68.) Designing for operation at minimum incidence angle also helps reduce broadband noise, as discussed in references 57 and 58. This approach for noise reduction may not be compatible with fan performance design requirements and is dependent on a good, interactive working relationship between the fan aerodynamic designers and noise engineers. The degree of noise reduction possible is also less certain, because the precise behavior of rotor wakes is often difficult to predict, especially for highly loaded fans designed to operate at transonic tip speeds.

2. Design the rotor-stator axial spacing sufficiently large that the wake has decayed and mixed as much as possible before reaching the stator vanes. This approach can result in significant noise reductions, as demonstrated by experiments in reference 82. Examples of the variation of fan noise with spacing are illustrated in figure 33. In general, the majority of the noise reduction possible is achieved with a ratio of axial spacing to chord (upstream rotor chord projected in axial plane) of about 2.0 to 2.5. This option, although very effective in reducing noise, usually imposes a weight penalty by increasing engine length, and it may also decrease fan efficiency by as much as 1.5 percent.

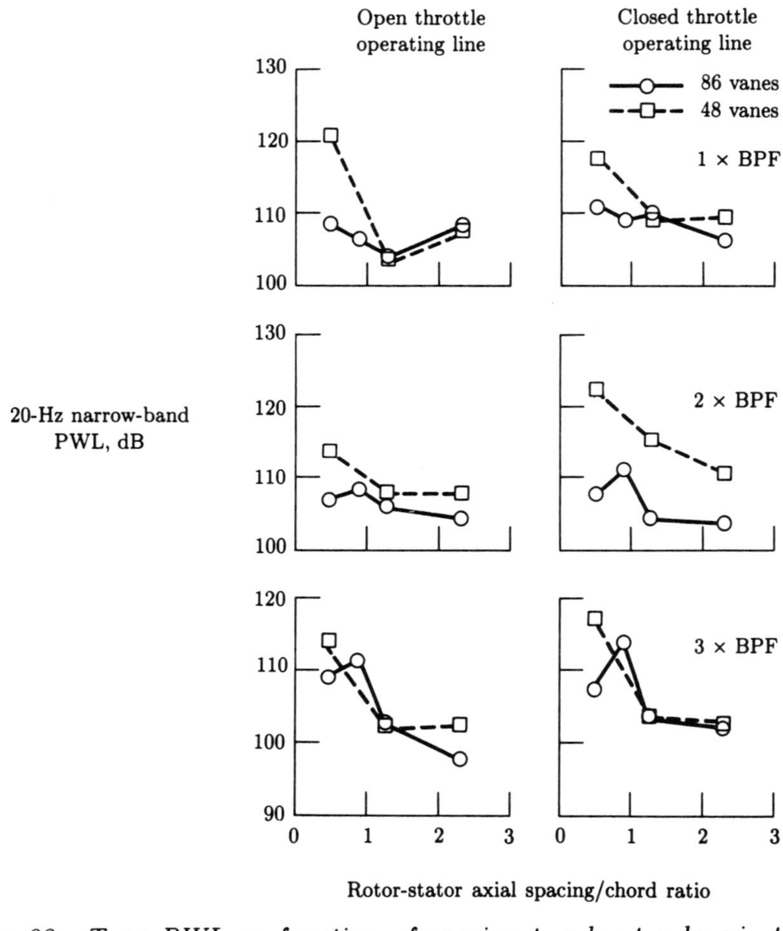

Figure 33. Tone PWL as function of spacing trends at subsonic tip speed (293 m/sec (960 ft/sec)) for single-stage fan. (From ref. 83.)

In addition to gust amplitude, axial spacing can influence other parameters important to noise generation, such as the angle of the wakes with respect to the vane leading edges, the harmonic content of the wake disturbances, and the coupling of fluctuating pressures to duct modes.

Reducing gust amplitude—strut-pylon pressure fields: A typical turbofan engine has frame struts and engine support pylons in the duct downstream of the fan stage.

These downstream struts and pylons produce static-pressure distortions which can be felt upstream in the vicinity of the fan rotor itself. The degree to which the rotor "feels" the circumferential variations in static pressure caused by these downstream obstructions is a function of the number and size of the obstructions and also depends on the spacing between the obstructions and the upstream rotor and stator. There is also an influence of the stator on this pressure field, and the stator row between the rotor and struts can act as a filter or as an amplifier of the strut pressure distortion. (See refs. 91 and 92, for example.) Methods for reducing the "gust" produced by this mechanism include the following:

1. Design the fan to have as large an axial distance between the fan rotor and the downstream struts and pylons as possible. This approach is effective but also introduces a weight penalty by increasing the length of the engine.
2. Design the stator-vane row integral to the strut and pylon assembly; tailor the vane stagger and camber angles circumferentially to produce as smooth a circumferential pressure distribution at the rotor plane as possible. This approach has been quite successful, but results in a fan stator-strut-pylon design which is quite complex and more difficult and costly to manufacture than the baseline configuration.

Of course a combination of methods 1 and 2 can be employed to attain the required noise reduction, the result being some increase in weight through an increase in axial spacing, while the stator row is designed integral with the frame struts to reduce the static-pressure distortion itself. Figure 34 shows a typical trend of measured rotor unsteady lift coefficient caused by downstream struts versus strut spacing. The figure indicates that the struts should be 4 or 5 strut widths downstream of the rotor to have a minimal impact on rotor noise. This result was taken from reference 93. An example of the effect of employing method 2 is illustrated in figure 35, which is taken from reference 56. The effect of designing the stator-vane integral with the struts is to reduce the static-pressure *distortion* seen by the rotor blades.

Interestingly, the measured noise for the integral vane-strut frame was higher for tone levels but lower for broadband levels than the noise for the separate vane-strut frame (ref. 56). Since the measured static-pressure distortion, or gust, was lower, it was concluded that the rotor wake impinging on the vane-strut combination must have produced higher fluctuating forces. Although this inference was not conclusively proven, it does suggest that care must be taken in changing the aerodynamics to reduce the noise, as it is possible to introduce an adverse effect on some other noise-generating mechanism.

Reducing blade-vane response to the gust: An additional step in the noise generation process is the rotor or stator response to the unsteady gusts generated by upstream wakes and upstream-downstream pressure field distortions. The response of a thin airfoil to a sinusoidal transverse gust is given by equation (2), a special case of which is (ref. 94)

$$C_l = 2\pi\alpha S(\sigma) \tag{34}$$

where C_l is the unsteady lift coefficient, α is the gust amplitude normalized by the mean (free-stream) relative velocity, and $S(\sigma)$ is the unsteady response function called the Sears function. The variation of $S(\sigma)$ with the reduced frequency parameter is shown in figure 5, where the reduced frequency σ is the airfoil semichord times gust frequency divided by gust velocity. It is shown in this figure that increasing reduced frequency parameter, either by increasing the airfoil chord or by reducing the gust wavelength (increasing gust frequency), tends to reduce the lift response function. Thus, for noise generated by interaction of the rotor wake and stator vane, increasing the number of rotor wakes (i.e., the number of rotor blades), increasing the rotor tip speed, and increasing the stator-vane chord are all techniques for reducing the unsteady lift response.

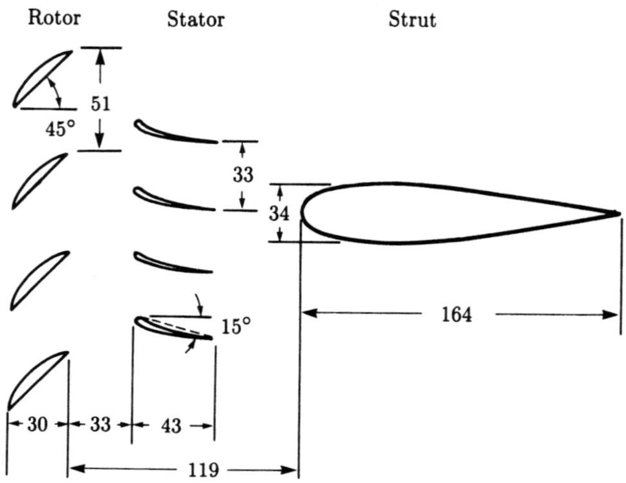

(a) Rotor-stator-strut cross section. All dimensions are in millimeters unless otherwise indicated.

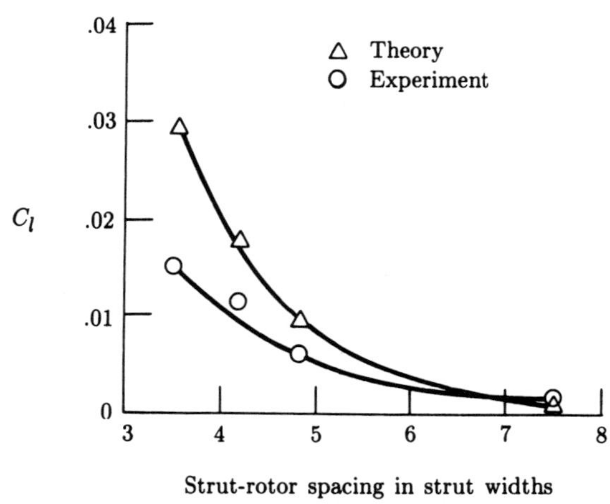

(b) Strut-induced unsteady lift coefficient C_l as function of strut-rotor spacing.

Figure 34. Measured and predicted effects of downstream strut spacing on strut-induced rotor unsteady lift. (From ref. 93.)

Usually, for practical designs, one can only affect the reduced frequency by about 10 to 25 percent by using the above techniques, and the corresponding decrease in

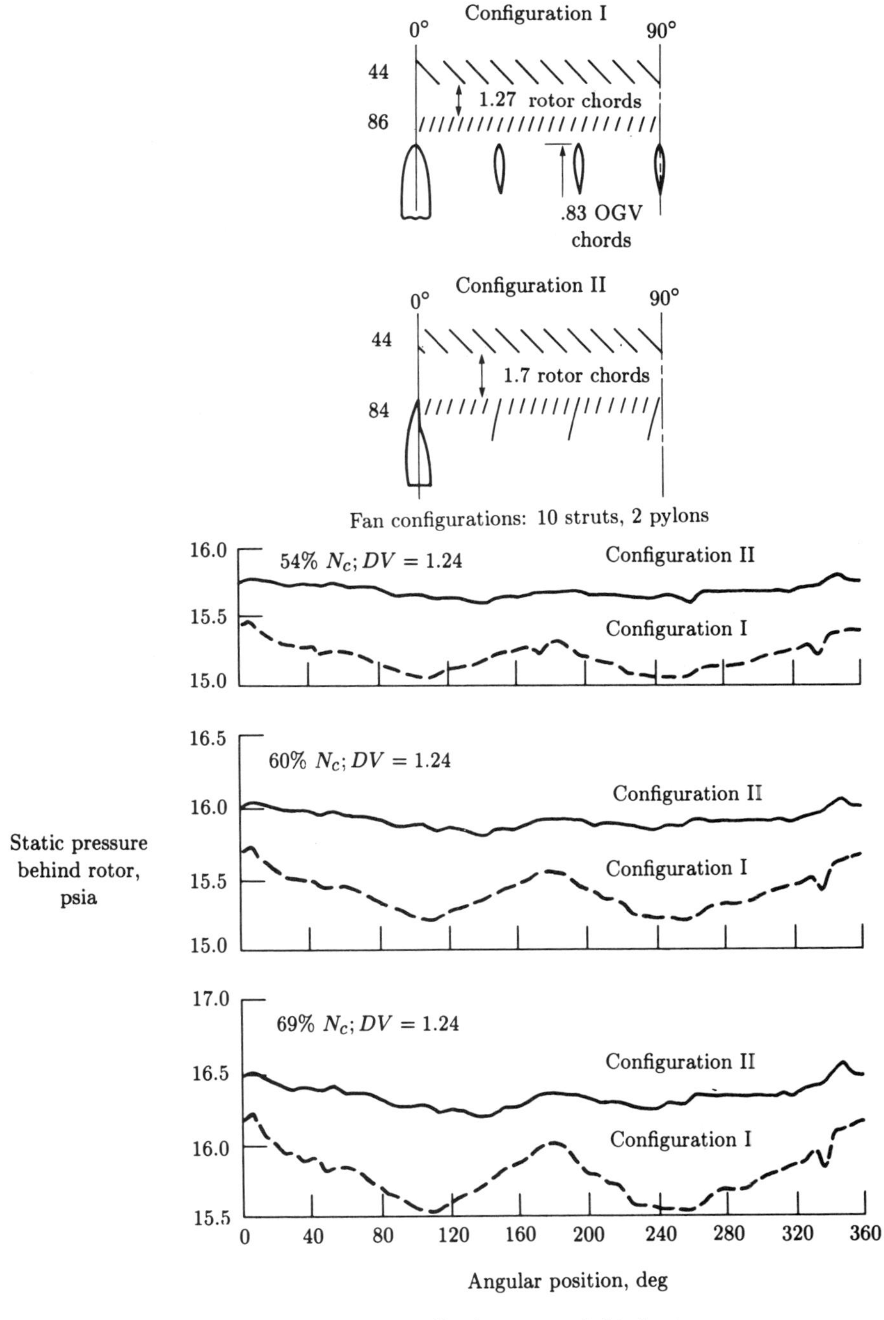

Figure 35. Rotor exit static-pressure distortions for integral vane-strut frame and separate vane-strut frame. (From ref. 56.)

lift response function is therefore going to be small, usually less than 25 percent. Hence the potential noise reductions are modest, say less than 2 dB. This gain has to be balanced against the increases in weight and decreases in performance when the merits of such a design change are assessed. For example, increasing the number of rotor blades may cause the rotor to produce regions of choked flow in the hub and decrease its efficiency and mass flow pumping capability. Also, more blades usually mean a heavier rotor.

Other parameters can have an effect on blade and vane lift response. Inlet relative Mach number has an effect, the typical trend being that the response function decreases with increasing Mach number. The Mach number dependency is less clearly understood at transonic and supersonic Mach numbers, so it is difficult to utilize Mach number as a controllable design parameter. Steady-state loading level can also affect the unsteady lift response, and analytical results aimed at understanding this effect are just beginning to emerge. Vane lean and sweep can also affect the unsteady lift response of a stator vane to rotor wake gusts. References 74 and 95 present analytical results showing the potential effects of vane sweep and lean, and the implications are that the effects are (or can be) beneficial. However, these results require experimental substantiation before one can rely on them for design guidance.

Reducing unsteady lift force amplitude: Since the absolute magnitude of the unsteady lift force produced by the gust response is essentially the unsteady lift coefficient C_l multiplied by the upstream dynamic pressure, the lift amplitude can be reduced by reduction of the upstream velocity. This may not always be an option for noise reduction, since aerodynamic design considerations may preclude any changes of this nature.

Reducing efficiency of conversion to acoustic energy: Most of the theories for noise radiation from turbomachinery stages (e.g., that of Mani, ref. 96) conclude that the sound power emitted by a blade row due to periodic excitation from adjacent blade rows or flow nonuniformities is a function of duct flow Mach number, tip Mach number, fluctuating force frequency, and the ratio of blade number to vane number. The sound power is made up of propagating pressure patterns, or modes, which propagate in a spiral path along the duct, away from the generating blade row. For a fan stage, the vane-blade number ratio V/B is usually a key parameter in selecting a low-noise fan design.

The vane-blade ratio can be selected to "cut off" certain interaction mode tone frequencies, as discussed in the section entitled *Coupling of the Duct—Modes and Cutoff*. It is usual practice to select a vane-blade ratio such that the blade-passage frequency is cut off, i.e., it produces no propagating pressure patterns in the duct. A rule of thumb derived from equation (9) for selecting vane-blade ratio based on this cutoff criterion is given by

$$\frac{V}{B} \geq 1 + \frac{M_t}{\sqrt{1 - M_D^2}} \tag{35}$$

If it is not possible to select V/B to cut off a problem tone (usually second and higher harmonics require $V/B > 4$ to achieve cutoff), an alternative is to select V/B such that the wave propagation spiral angle in the duct is as close to 90° as possible (cutoff corresponds to 90°), so that the residence time of the wave pattern in the duct is sufficiently long for the duct acoustic treatment to attenuate it as much as possible.

As another alternative, one could select blade and vane numbers to reduce duct mode coupling by aligning the directions of the prominent modes of the interaction with the vane chords, thereby putting vane dipoles at 90° to the direction preferred by the mode. Such an approach may be limited by practical blade angle and number constraints.

Increasing transmission loss of blocking blade rows: For a fan stage consisting of a rotor followed by a stator, the noise radiated forward by the stator has to pass through the rotor before radiating outside of the duct to the observer. One way to reduce the net radiation to the outside is to select a vane-blade ratio such that the mode having the highest acoustic energy suffers the highest transmission loss in passing through the rotor. This is effectively achieved by ensuring that the wave spiral angle from the stator is at as nearly a right angle as possible to the stagger angle of the rotor, as illustrated in figure 6. This concept is discussed, for example, in references 13 to 15. The same principle can apply to rotor-generated noise propagating downstream and passing through the stator.

It is possible that the fan design can be tailored to minimize the net upstream and downstream noise radiation to the outside. Both transmitted and reflected wave energy should be considered, with a reasonably accurate analytical model used for predicting these effects. Although the models proposed in references 13 to 15 are a good starting point for understanding the phenomena involved, many applications require a more general blade row transmission-reflection analytical model which includes the multiple (at least two) blade row environment effects (ref. 19).

Low-Pressure Compressors

All the noise-reduction concepts discussed above for fans apply in principle to low-pressure compressors (LPC's). In general, however, there is usually less flexibility available to the acoustic designer in terms of variations in axial spacing between blade rows, loading control, chord and vane-blade number ratio selection, etc. Usually, the most economically viable design control the acoustic designer has for an LPC is in selecting the vane-blade number ratios for the first two or three stages of the LPC to maximize the forward-radiation transmission loss. Because substantially more blades and vanes are involved with an LPC than with a fan, use of mode cutoff is usually not a practical option.

For a high-bypass engine, where the fan rotor hub flow is closely coupled to a core engine LPC or high-pressure compressor (HPC), it is sometimes the case that the fan hub rotor wakes impinging on the stator-vane row in the core duct cause higher noise levels than the rotor-stator interaction levels produced in the fan bypass duct. This increase can occur because the core stator-vane row is usually much closer to the rotor than the bypass duct stator-vane row (outlet guide vane (OGV)) and because the bypass duct has the benefit of acoustic treatment, whereas the core duct usually does not contain any treatment. In addition, the core duct stator also acts as an inlet guide vane (IGV) to the first-stage rotor of the LPC or HPC, so that it is a source of rotor-stator interaction with two rotors. Careful selection of the vane number for this core duct stator, plus the inclusion of additional axial spacing on both sides of the vane row, can be very beneficial to reducing the total compressor system noise.

Low-Pressure Turbines

Once again, the noise-reduction techniques discussed in the *Fans* section apply in principle to low-pressure turbines (LPT's). There are two features of an LPT which are unique in terms of noise-reduction options. First, the number of rotor blades is usually fairly high, say 50 to 150, so that the blade-passage tone frequencies are fairly high. This results in the higher harmonics of the blade-passage tone frequencies usually being higher than the audible range (greater than 10 000 to 20 000 Hz). Hence, only the fundamental blade-passage tones need to be considered. Second, the gas stream temperatures are usually fairly high (greater than 1000°R (283°C)) in an LPT compared with those of a fan or LPC first stage. Hence, the flow and tip-speed Mach numbers are usually fairly low. Thus, from equation (35), a cutoff condition for the fundamental blade-passage tones can be achieved with vane-blade ratios substantially less than 2.0, and this low ratio makes it easier to design for cutoff without substantial performance penalty.

A successful demonstration of the concept of designing LPT stages for cutoff was reported in reference 97. The authors of reference 97 also found that the blade row transmission losses suffered by the first two stages of the four-stage turbine were substantial, so that noise-reduction considerations were only required for the last two stages. An additional observation was that, because the loading of the last stage (i.e., $\Delta T/T$) was relatively low at approach power (where LPT noise is usually a concern), its fundamental blade-passage tone level was also low. This confirmed the loading dependency given by equation (32).

Concluding Remarks

Major Advances

This chapter summarizes key advances in experimental techniques and theoretical applications which point the way to a broad understanding and control of turbomachinery noise. On the experimental side, the development of effective inflow control techniques makes it possible to conduct, in ground-based facilities, definitive experiments on internally controlled blade row interactions. Results can now be valid indicators of flight behavior and can provide a firm base for comparison with analysis. Inflow control coupled with detailed diagnostic tools such as blade pressure measurements can be used to uncover the more subtle mechanisms such as rotor-strut interaction, which can set tone levels for some engine configurations. Initial mappings of rotor wake-vortex flow fields have provided a data base for a first-generation semiempirical flow disturbance model. Laser velocimetry offers a nonintrusive method for validating and improving the model. Digital data systems and signal processing algorithms are bringing mode measurement closer to a working tool to be frequently applied to a real machine such as a turbofan engine.

On the analytical side, models of most of the links in the chain from turbomachine blade source to far-field observation point have been formulated. Three-dimensional lifting-surface theory for blade rows including source noncompactness and cascade effects, blade row transmission models incorporating mode and frequency scattering, and modal radiation calculations including hybrid numerical-analytical approaches are tools which await further application. The more computationally demanding of these can at least serve as checks and guides for simpler design methods, and the generation physics described by the models suggests noise-reduction tactics.

Unsolved Problems

One of the phenomena most difficult to understand quantitatively, as indicated by our inability to identify and describe the dominant generation mechanism, is turbomachinery broadband noise. Experimental evidence points to a dominant internal source which has a spectral shape that is nearly independent of fan inflow conditions. Blade loading is influential, but the details have remained elusive preventing spectral prediction. Another question awaiting resolution is the relative importance of hub and tip vortex flow disturbances compared with blade wakes in generating rotor-stator interaction noise. The unknown element here seems to be the disturbance flow field description rather than the modeling of noise generation by gust-airfoil interaction. A final phenomenon offered as an example of a problem requiring further study is the observed characteristic of decreasing tone power radiated from the inlet as fan speed is increased at supersonic tip relative Mach numbers. The influences of source strength and inlet propagation need to be quantified.

Toward Integrated Quiet Designs

Two types of integration are essential to the formulation of low-noise, high-efficiency turbomachine designs. Early study of interplay between aerodynamic and acoustic analyses can help us avoid the unfortunate circumstance of attempts at noise reduction when hardware constraints are fixed and severe. But beyond this critical interdisciplinary integration, a second, deeper level of interplay remains to be fully exploited: blade row *and* duct treatment designs tailored to minimize radiated noise. Coupling of the source to the duct, scattering, absorption, and radiation can now be analyzed on a detailed modal basis. It remains for us to fully exploit and refine these tools to realize the benefits of considering the total generation-propagation process.

References

1. Owens, R. E.: *Energy Efficient Engine Propulsion System—Aircraft Integration Evaluation.* NASA CR-159488, 1979.
2. Sears, William R.: Some Aspects of Non-Stationary Airfoil Theory and Its Practical Application. *J. Aeronaut. Sci.*, vol. 8, no. 3, Jan. 1941, pp. 104–108.
3. Goldstein, Marvin E.: *Aeroacoustics.* McGraw-Hill Book Co., c.1976.
4. Amiet, Roy K.: Compressibility Effects in Unsteady Thin-Airfoil Theory. *AIAA J.*, vol. 12, no. 2, Feb. 1974, pp. 252–255.
5. Graham, J. M. R.: Similarity Rules for Thin Airfoils in Non-Stationary Flows. *J. Fluid Mech.*, vol. 43, pt. 4, Oct. 2, 1970, pp. 753–766.
6. Namba, M.: Three-Dimensional Analysis of Blade Force and Sound Generation for an Annular Cascade in Distorted Flows. *J. Sound & Vibration,* vol. 50, no. 4, Feb. 22, 1977, pp. 479–508.
7. Goldstein, M. E.; and Atassi, H.: A Complete Second-Order Theory for the Unsteady Flow About an Airfoil Due to a Periodic Gust. *J. Fluid Mech.*, vol. 74, pt. 4, Apr. 22, 1976, pp. 741–765.

8. Filotas, L. T.: *Theory of Airfoil Response in a Gusty Atmosphere, Part I—Aerodynamic Transfer Function.* UTIAS Rep. No. 139, Univ. of Toronto, Oct. 1969.
9. Morse, Philip M.; and Ingard, K. Uno: *Theoretical Acoustics.* McGraw-Hill Book Co., c.1968.
10. Tyler, J. M.; and Sofrin, T. G.: Axial Flow Compressor Noise Studies. *SAE Trans.,* vol. 70, 1962, pp. 309–332.
11. Sofrin, Thomas G.; and McCann, John C.: Pratt and Whitney Aircraft Experience in Compressor-Noise Reduction. *J. Acoust. Soc. America,* vol. 40, no. 5, Nov. 1966, pp. 1248–1249.
12. Mani, R.; and Horvay, G.: Sound Transmission Through Blade Rows. *J. Sound & Vibration,* vol. 12, no. 1, May 1970, pp. 59–83.
13. Kaji, S.; and Okazaki, T.: Propagation of Sound Waves Through a Blade Row. I. Analysis Based on the Semi-Actuator Disc Theory. *J. Sound & Vibration,* vol. 11, no. 3, Mar. 1970, pp. 339–353.
14. Kaji, S.; and Okazaki, T.: Propagation of Sound Waves Through a Blade Row. II. Analysis Based on the Acceleration Potential Method. *J. Sound & Vibration,* vol. 11, no. 3, Mar. 1970, pp. 355–375.
15. Amiet, R. K.: Transmission and Reflection of Sound by a Blade Row. AIAA Paper No. 71-181, Jan. 1971.
16. Philpot, M. G.: The Role of Rotor Blade Blockage in the Propagation of Fan Noise Interaction Tones. AIAA Paper 75-447, Mar. 1975.
17. Cumpsty, N. A.: Sum and Difference Tones From Turbomachines. *J. Sound & Vibration,* vol. 32, no. 3, Feb. 8, 1974, pp. 383–386.
18. Cumpsty, N. A.: Tone Noise From Rotor/Stator Interactions in High Speed Fans. *J. Sound & Vibration,* vol. 24, no. 3, Oct. 8, 1972, pp. 393–409.
19. Heinig, K. E.: Sound Propagation in Multistage Axial Flow Turbomachines. *AIAA J.*, vol. 21, no. 1, Jan. 1983, pp. 98–105.
20. Rice, E. J.: Multimodal Far-Field Acoustic Radiation Pattern Using Mode Cutoff Ratios. *AIAA J.*, vol. 16, no. 9, Sept. 1978, pp. 906–911.
21. Saule, A. V.; and Rice, E. J.: *Far-Field Multimodal Acoustic Radiation Directivity.* NASA TM-73839, 1977.
22. Rice, Edward J.; Heidmann, Marcus F.; and Sofrin, Thomas G.: Modal Propagation Angles in a Cylindrical Duct With Flow and Their Relation to Sound Radiation. AIAA Paper 79-0183, Jan. 1979. (Also available as NASA TM-70930.)
23. Homicz, G. F.; and Lordi, J. A.: A Note on the Radiative Directivity Patterns of Duct Acoustic Modes. *J. Sound & Vibration,* vol. 41, no. 3, Aug. 8, 1975, pp. 283–290.
24. Landau, L. D.; and Lifshitz, E. M. (Morton Hamermesh, transl.): *The Classical Theory of Fields, Revised Second ed.* Addison-Wesley Publ. Co., Inc., c.1962.
25. Heidmann, M. F.; Saule, A. V.; and McArdle, J. G.: Predicted and Observed Modal Radiation Patterns From JT15D Engine With Inlet Rods. *J. Aircr.,* vol. 17, no. 7, July 1980, pp. 493–499.
26. Cho, Y. C.; and Ingard, K. U.: Closed-Form Solution of Mode Propagation in a Nonuniform Circular Duct. *AIAA J.*, vol. 20, no. 1, Jan. 1982, pp. 39–44.
27. Cho, Y. C.; and Ingard, K. U.: Mode Propagation in Nonuniform Circular Ducts With Potential Flow. AIAA-82-0122, Jan. 1982. (Also available as NASA TM-82776.)
28. Cho, Y. C.; and Rice, E. J.: High-Frequency Sound Propagation in a Spatially Varying Mean Flow. *J. Acoust. Soc. America,* vol. 70, no. 3, Sept. 1981, pp. 860–865.
29. Amiet, R. K.: Correction of Fan Noise for Effects of Forward Flight. *J. Sound & Vibration,* vol. 89, no. 2, July 22, 1983, pp. 243–259.
30. Horowitz, S. J.; Sigman, R. K.; and Zinn, B. T.: An Iterative Finite Element-Integral Technique for Predicting Sound Radiation From Turbofan Inlets in Steady Flight. AIAA-82-0124, Jan. 1982.

31. Baumeister, K. J.; and Horowitz, S. J.: Finite Element-Integral Acoustic Simulation of JT15D Turbofan Engine. *ASME J. Vibration, Acoust., Stress, & Reliab. Design,* vol. 106, no. 3, July 1984, pp. 405–413.
32. Rice, Edward J.; and Saule, Arthur V.: *Far-Field Radiation of Aft Turbofan Noise.* NASA TM-81506, 1980.
33. Savkar, S. D.; and Edelfelt, I. H.: *Radiation of Cylindrical Duct Acoustic Modes With Flow Mismatch.* NASA CR-132613, 1975.
34. Dyer, Ira: Measurement of Noise Sources in Ducts. *J. Acoust. Soc. America,* vol. 30, no. 9, Sept. 1958, pp. 833–841.
35. Hanson, Donald B.: Spectrum of Rotor Noise Caused by Atmospheric Turbulence. *J. Acoust. Soc. America,* vol. 56, no. 1, July 1974, pp. 110–126.
36. Feiler, Charles E.; and Groeneweg, John F.: Summary of Forward Velocity Effects on Fan Noise. AIAA Paper 77-1319, Oct. 1977. (Also available as NASA TM-73722.)
37. Jones, W. L.; McArdle, J. G.; and Homyak, L.: Evaluation of Two Inflow Control Devices for Flight Simulation of Fan Noise Using a JT15D Engine. AIAA Paper 79-0654, Mar. 1979. (Also available as NASA TM-79072.)
38. McArdle, J. G.; Jones, W. L.; Heidelberg, L. J.; and Homyak, L.: Comparison of Several Inflow Control Devices for Flight Simulation of Fan Tone Noise Using a JT15D-1 Engine. AIAA-80-1025, June 1980. (Also available as NASA TM-81505.)
39. Woodward, R. P.; Wazyniak, J. A.; Shaw, L. M.; and MacKinnon, M. J.: *Effectiveness of an Inlet Flow Turbulence Control Device To Simulate Flight Fan Noise in an Anechoic Chamber.* NASA TM-73855, 1977.
40. Chestnutt, David, ed.: *Flight Effects of Fan Noise.* NASA CP-2242, 1982.
41. Loehrke, R. I.; and Nagib, H. M.: Control of Free-Stream Turbulence by Means of Honeycombs: A Balance Between Suppression and Generation. *J. Fluids Eng.,* vol. 98, no. 3, ser. 1, Sept. 1976, pp. 342–353.
42. Homyak, L.; McArdle, J. G.; and Heidelberg, L. J.: A Compact Inflow Control Device for Simulating Flight Fan Noise. AIAA-83-0680, Apr. 1983. (Also available as NASA TM-83349.)
43. Rogers, D. F.; and Ganz, U. W.: Aerodynamic Assessment of Methods To Simulate Flight Inflow Characteristics During Static Engine Testing. AIAA-80-1023, June 1980.
44. Kantola, R. A.; and Warren, R. E.: Reduction of Rotor-Turbulence Interaction Noise in Static Fan Noise Testing. AIAA Paper 79-0656, Mar. 1979.
45. Ho, P. Y.; Smith, E. B.; and Kantola, R. A.: An Inflow Turbulence Reduction Structure for Scale Model Fan Testing. AIAA Paper 79-0655, Mar. 1979.
46. Ginder, R. B.; Kenison, R. C.; and Smith, A. D.: Considerations for the Design of Inlet Flow Conditioners for Static Fan Noise Testing. AIAA Paper 79-0657, Mar. 1979.
47. Peracchio, A. A.; Ganz, U. W.; Gedge, M.; and Robbins, K.: *Studies on Proper Simulation During Static Testing of Forward Speed Effects on Fan Noise.* NASA CR-165626, 1980.
48. Ganz, Ulrich W.: *Analytical Investigation of Fan Tone Noise Due to Ingested Atmospheric Turbulence.* NASA CR-3302, 1980.
49. Gedge, M. R.: *A Design Procedure for Fan Inflow Control Structures.* NASA CR-165625, 1980.
50. Atvars, Y.; and Rogers, D. F.: The Development of Inflow Control Devices for Improved Simulation of Flight Noise Levels During Static Testing of a HBPR Turbofan Engine. AIAA-80-1024, June 1980.
51. Peracchio, A. A.: Assessment of Inflow Control Structure Effectiveness and Design System Development. AIAA-81-2048, Oct. 1981.
52. Shaw, Loretta M.; Woodward, Richard P.; Glaser, Frederick W.; and Dastoli, Benjamin J.: Inlet Turbulence and Fan Noise Measured in an Anechoic Wind Tunnel and Statically With an Inlet Flow Control Device. AIAA Paper 77-1345, Oct. 1977.
53. Heidmann, Marcus F.; and Dietrich, Donald A.: Effects of Simulated Flight on Fan Noise Suppression. AIAA Paper 77-1334, Oct. 1977.

54. Holm, R. G.; Langenbrunner, L. E.; and McCan, E. O.: *Forward Velocity Effects on Fan Noise and the Influence of Inlet Aeroacoustic Design as Measured in the NASA-Ames 40 × 80 Foot Wind Tunnel.* NASA CR-166461, 1981.
55. Woodward, Richard P.; and Balombin, Joseph R.: Tone Generation by Rotor-Downstream Strut Interaction. *J. Aircr.*, vol. 21, no. 2, Feb. 1984, pp. 135–142.
56. Ho, P. Y.: The Effect of Vane-Frame Design on Rotor-Stator Interaction Noise. AIAA-81-2034, Oct. 1981.
57. Gliebe, P. R.: The Effect of Throttling on Forward Radiated Fan Noise. AIAA Paper 79-0640, Mar. 1979.
58. Ginder, R. B.; and Newby, D. R.: An Improved Correlation for the Broadband Noise of High-Speed Fans. *J. Aircr.,* vol. 14, no. 9, Sept. 1977, pp. 844–849.
59. Pickett, Gordon F.: Prediction of the Spectral Content of Combination Tone Noise. *J. Aircr.,* vol. 9, no. 9, Sept. 1972, pp. 658–663.
60. Philpot, M. G.: The Buzz-Saw Noise Generated by a High Duty Transonic Compressor. ASME Paper 70-GT-54, May 1970.
61. Stratford, B. S.; and Newby, D. R.: A New Look at the Generation of Buzz-Saw Noise. AIAA Paper 77-1343, Oct. 1977.
62. Mathews, D. C.; and Nagel, R. T.: Inlet Geometry and Axial Mach Number Effects on Fan Noise Propagation. *Aeroacoustics: Fan, STOL, and Boundary Layer Noise; Sonic Boom; Aeroacoustic Instrumentation,* Henry T. Nagamatsu, ed., American Inst. of Aeronautics and Astronautics, c.1975, pp. 73–96.
63. Reynolds, B.; and Lakshminarayana, B.: *Characteristics of Lightly Loaded Fan Rotor Blade Wakes.* NASA CR-3188, 1979.
64. Ravindranath, A.; and Lakshminarayana, B.: *Three Dimensional Mean Flow and Turbulence Characteristics of the Near Wake of a Compressor Rotor Blade.* NASA CR-159518, 1980.
65. Dittmar, James H.: Interaction of Rotor Tip Flow Irregularities With Stator Vanes as a Noise Source. AIAA Paper 77-1342, Oct. 1977.
66. Atassi, H.; and Hamad, G.: Sound Generated in a Cascade by Three-Dimensional Disturbances Convected in a Subsonic Flow. AIAA-81-2046, Oct. 1981.
67. Shaw, Loretta M.; and Balombin, Joseph R.: Rotor Wake Characteristics Relevant to Rotor-Stator Interaction Noise Generation. *J. Aircr.,* vol. 19, no. 11, Nov. 1982, pp. 954–962.
68. Majjigi, R. K.; and Gliebe, P. R.: *Development of a Rotor Wake/Vortex Model.*
Volume I—Final Technical Report. NASA CR-174849, 1984.
Volume II—User's Manual for Computer Program. NASA CR-174850, [1984].
69. Pickett, G. F.; Sofrin, T. G.; and Wells, R. W.: *Method of Fan Sound Mode Structure Determination—Final Report.* NASA CR-135293, 1977.
70. Moore, C. J.: Measurement of Radial and Circumferential Modes in Annular and Circular Fan Ducts. *J. Sound & Vibration,* vol. 62, no. 2, Jan. 22, 1979, pp. 235–256.
71. Cicon, D. E.; Sofrin, T. G.; and Mathews, D. C.: *Investigation of Continuously Traversing Microphone System for Mode Measurement.* NASA CR-168040, 1982.
72. Joppa, P. D.: Acoustic Mode Measurements in the Inlet of a Turbofan Engine. *J. Aircr.,* vol. 24, no. 9, Sept. 1987, pp. 587–593.
73. Kobayashi, Hiroshi: Three-Dimensional Effects on Pure Tone Fan Noise Due to Inflow Distortion. AIAA Paper 78-1120, July 1978.
74. Schulten, J. B. H. M.: Sound Generated by Rotor Wakes Interacting With a Leaned Vane Stator. *AIAA J.,* vol. 20, no. 10, Oct. 1982, pp. 1352–1358.
75. Ventres, C. S.; Theobald, M. A.; and Mark, W. D.: *Turbofan Noise Generation.*
Volume 1: Analysis. NASA CR-167951, 1982.
Volume 2: Computer Programs. NASA CR-167952, 1982.
76. Kobayashi, Hiroshi; and Groeneweg, John F.: Effects of Inflow Distortion Profiles on Fan Tone Noise. *AIAA J.,* vol. 18, no. 8, Aug. 1980, pp. 899–906.

77. Kaji, S.: Noncompact Source Effect on the Prediction of Tone Noise From a Fan Rotor. AIAA Paper 75-446, Mar. 1975.
78. Fleeter, Sanford: Discrete Frequency Noise Reduction Modeling for Application to Fanjet Engines. *J. Acoust. Soc. America,* vol. 68, no. 3, Sept. 1980, pp. 957–965.
79. Lansing, Donald L.: Exact Solution for Radiation of Sound From a Semi-Infinite Circular Duct With Application to Fan and Compressor Noise. *Analytic Methods in Aircraft Aerodynamics,* NASA SP-228, 1970, pp. 323–334.
80. Woodward, Richard P.; and Glaser, Frederick W.: Effects of Blade-Vane Ratio and Rotor-Stator Spacing on Fan Noise With Forward Velocity. AIAA-81-2032, Oct. 1981. (Also available as NASA TM-82690.)
81. Sofrin, T. G.; and Mathews, D. C.: Asymmetric Stator Interaction Noise. AIAA Paper 79-0638, Mar. 1979.
82. Kantola, R. A.; and Gliebe, P. R.: Effects of Vane/Blade Ratio and Spacing on Fan Noise. AIAA-81-2033, Oct. 1981.
83. Gliebe, P. R.; and Kantola, R. A.: *Effects of Vane/Blade Ratio and Spacing on Fan Noise. Volume I—Final Technical Report.* NASA CR-174664, 1983.
Volume II—Data Supplement. NASA CR-174665, [1983].
84. Hayden, R. E.; Bliss, D. B.; Murray, B. S.; Chandiramani, K. L.; Smullin, J. I.; and Schwaar, P. G.: *Analysis and Design of a High Speed, Low Noise Aircraft Fan Incorporating Swept Leading Edge Rotor and Stator Blades.* NASA CR-135092, 1977.
85. Lucas, James G.; Woodward, Richard P.; and MacKinnon, Michael J.: Acoustic Evaluation of a Novel Swept-Rotor Fan. AIAA Paper 78-1121, July 1978.
86. Lucas, James G.; Woodward, Richard P.; and Michels, Charles J.: *Forward Acoustic Performance of a Model Turbofan Designed for a High Specific Flow (QF-14).* NASA TP-1968, 1982.
87. *Noise Certification Test and Analysis Procedures.* AC No. 36-4A, Federal Aviation Adm., Jan. 21, 1986.
88. *Noise Standards: Aircraft Type and Airworthiness Certification.* FAR Pt. 36, Federal Aviation Adm., June 1974.
89. Heidmann, M. F.: *Interim Prediction Method for Fan and Compressor Source Noise.* NASA TM X-71763, 1975.
90. Kazin, S. B.; and Matta, R. K.: Turbine Noise Generation, Reduction, and Prediction. *Aeroacoustics: Fan Noise and Control; Duct Acoustics; Rotor Noise, Volume 44 of Progress in Astronautics and Aeronautics,* Ira R. Schwartz, ed., American Inst. of Aeronautics and Astronautics, c.1976, pp. 109–138.
91. Ng, W. F.; O'Brien, W. F.; and Olsen, T. L.: Experimental Investigation of Unsteady Fan Flow Interaction With Downstream Struts. AIAA-86-1870, July 1986.
92. Nakamura, Y.; Isomura, K.; and Kodama, H.: Rotor-Strut Interaction Noise of a Model Fan. AIAA-86-1871, July 1986.
93. Taylor, A. C., III; and Ng, W. F.: Analytical Prediction of the Unsteady Lift on a Rotor Caused by Downstream Struts. ASME Paper 87-GT-145, May–June 1987.
94. Kemp, Nelson H.; and Sears, W. R.: Aerodynamic Interference Between Moving Blade Rows. *J. Aeronaut. Sci.,* vol. 20, no. 9, Sept. 1953, pp. 585–597, 612.
95. Envia, Edmane; and Kerschen, E. J.: Noise Generated by Convected Gusts Interacting With Swept Airfoil Cascades. AIAA-86-1872, July 1986.
96. Mani, Ramani: Descrete Frequency Noise Generation From an Axial Flow Fan Blade Row. *ASME J. Basic Eng.,* vol. 92, ser. D, no. 1, Mar. 1970, pp. 37–43.
97. Lavin, S. P.; Ho, P. Y.; and Chamberlin, Roger: Measurement and Prediction of Energy Efficient Engine Noise. AIAA-84-2284, Oct. 1984.

4 Jet Noise Classical Theory and Experiments

Lead author

G. M. Lilley
University of Southampton
Southampton, England

Introduction

The noise from gaseous jets has concerned man wherever they have been used. However, the advent of the jet engine as a power plant for military aircraft during the Second World War gave prominence to this problem of jet noise as a potential hazard. It became clear that unless methods could be designed to limit such noise for a given aircraft engine thrust, considerable opposition to the future use of the jet engine as a power plant for civil aircraft was likely. Hence, in the late 1940's, when the advantages of the jet engine led to its being considered as the appropriate technical and economic power plant for the future generation of civil aircraft for short-, medium-, and long-range aircraft, much research activity was initiated as to the source and causes of jet noise as well as to methods for its reduction.

It was perhaps surprising that the field of acoustics had excited little attention since the work of Lord Rayleigh in the last century. It was left to aerodynamicists to join forces with acousticians to investigate jet noise theoretically and experimentally. The subject was called *aerodynamic noise,* a marriage of acoustics with unsteady aerodynamic flow. By 1949 there had been little published work on investigations of jet noise and its generation, with the exception of some early measurements on the intensity of the far-field noise from turbulent air jets by Morley (ref. 1). These early measurements showed that the sound power is proportional to about the eighth power of the jet velocity.

Our understanding of jet noise as a study in aerodynamic noise had its foundations, however, in the work of Lighthill (refs. 2 to 5) on "sound generated aerodynamically." That work was complemented by several experimental studies (refs. 6 to 17). These experimental studies not only verified Lighthill's eighth power law, but also confirmed the other broad features of the theory relating to convective amplification with Mach number and consequent changes in directivity and spectra. Another feature of the experimental work was the early establishment by Westley and Lilley (ref. 6) of methods for jet noise reduction and the extension of these methods by Greatrex (refs. 16 and 17) to full-scale devices known as corrugated nozzles, which have been fitted to numerous jet engines powering many types of civil aircraft. The corrugated nozzle continues to be used on advanced jet engine power plants for civil aircraft for which maximum noise reduction is needed to enable compliance with aircraft noise certification legislation standards.

Lighthill's theory of aerodynamic noise is based on the exact equations of fluid flow. Lighthill showed that the energy radiated outward as sound from an unsteady fluid flow is such a small fraction of the flow kinetic energy that any approximation made in solving these equations for the fluctuating density could lead to an incorrect solution, and indeed in extreme circumstances to a solution that is physically wrong. Lighthill overcame these difficulties by the introduction of an analogy, which we refer to as *Lighthill's acoustic analogy*, in which the unsteady fluid flow is replaced by a volume distribution of equivalent acoustic sources throughout the entire flow field. In this analogy the sources are embedded in a uniform medium at rest, in which the sources may move but not the fluid. All the actual fluid flow dynamics, including the generation of noise within the flow and its interaction with the flow, are included in the strength and distribution of the equivalent acoustic source field. It is in this sense that Lighthill's theory of aerodynamic noise is exact. The theory is only predictive when the equivalent acoustic source field is known to some good approximation. Unless the properties of the unsteady flow are known, the details of the source field cannot be determined. However, good estimates can be made of the order of magnitude for the radiated noise based on the characteristic properties of the flow and empirically derived constants.

This chapter is devoted to the derivation and exploitation of Lighthill's theory of aerodynamic noise as the central pillar of all work concerned with the understanding and generation of jet noise. The subject of aerodynamic noise has undergone major changes in recent years and has attracted worldwide attention. The chapter concludes that although Lighthill's theory provides the essential framework for a full understanding of the noise generation in turbulent jets and the overall characteristics of its propagation to the far field, it is difficult to apply when acoustic interaction occurs with the flow field. This interaction involves consideration of the actual flow field and results in changes in the directivity and amplitude of the radiated sound field and its dependence on the flow speed relative to that of the external medium. The necessary modifications to the theory of aerodynamic noise to deal with flow-acoustic interaction are considered in detail by Goldstein (ref. 18) and in other chapters herein.

In studies of static jet noise, boundary-layer noise is normally absent. However, in flight the external boundary layers upstream of the jet exit and around the engine cowling radiate noise which is additional to that radiated by the jet. In many practical cases this noise can be neglected, since it is a function of flight speed and this is small compared with the jet speed.

Lighthill's theory of aerodynamic noise and its applications, as discussed in this chapter, assume all solid boundaries are absent from the flow field. The modification to the theory to include solid surfaces, and thereby to develop a theory for boundary-layer noise, was first investigated by Curle (ref. 19) and subsequently by many researchers (refs. 20 to 26). Reference should be made to these papers for the modifications to Lighthill's acoustic analogy when applied to flows containing solid boundaries.

Lighthill's Theory of Aerodynamic Noise

The Equations of Fluid Flow

The exact flow equations for a perfect gas relate to the conservation of mass, momentum, and energy and can be written, respectively, as

$$\frac{\partial \rho}{\partial t} + \nabla \cdot \rho \mathbf{v} = \rho m \tag{1}$$

$$\frac{\partial}{\partial t}\rho \mathbf{v} + \nabla \cdot (\rho \mathbf{v}\mathbf{v} - \boldsymbol{\tau}) + \nabla p = \rho g \mathbf{k} + \rho \mathbf{F} \tag{2}$$

$$\frac{\partial}{\partial t}\rho h_s + \nabla \cdot (\rho \mathbf{v} h_s - \mathbf{q} - \boldsymbol{\tau} \cdot \mathbf{v}) - \frac{\partial p}{\partial t} = \rho E + \rho \mathbf{F} \cdot \mathbf{v} \tag{3}$$

where ρ, $p, h, h_s, \boldsymbol{\tau}, \mathbf{q}$, and $\mathbf{v}$ are, respectively, the fluid density, the fluid pressure, the fluid specific enthalpy, the fluid specific stagnation enthalpy, the viscous stress tensor, the heat flux vector, and the fluid velocity vector. For a Newtonian fluid the viscous stress tensor and the heat flux vector are

$$\boldsymbol{\tau} = \mu \left(-\frac{2}{3}\mathbf{I}\nabla \cdot \mathbf{v} + \nabla \mathbf{v} + \mathbf{v}\nabla \right) \tag{4}$$

$$\mathbf{q} = -\frac{\mu}{N_{\mathrm{Pr}}}\nabla h \tag{5}$$

where μ is the fluid viscosity, N_{Pr} is the Prandtl number $\mu C_p/k$, C_p is the specific heat of the fluid at constant pressure, k is the thermal conductivity, and $\mathbf{I}$ is the unit tensor.

Equations (1) to (3) also include ρm, $\rho \mathbf{F}$, and ρE, which are, respectively, the density distributions of mass, force, and energy sources per unit volume; $\mathbf{k}$ is a unit vector in the z-direction (measured downward in the atmosphere). In problems of jet noise and turbomachinery noise the gravitational term $\rho g \mathbf{k}$ can be neglected, but it is important when one is dealing with the problem of the propagation of shock waves through the atmosphere. In studies of the noise from aircraft traveling at supersonic speeds the source terms ρm and ρE relate to the geometry of the aircraft, and in particular to its volume, while $\rho \mathbf{F}$ relates to its lift distribution. Similarly, in studies on turbomachinery noise ρm denotes the effect due to volume displacement of the rotating blades, while $\rho \mathbf{F}$ represents the equivalent aerodynamic force distribution on the blades per unit volume and includes both steady and unsteady aerodynamic loads. In studies on jet noise these source terms are absent.

The Equation for the Pressure Fluctuations

The flow equations can be reduced to a suitable form for the study of the generation and the propagation of sound. It can be shown that the convection equation for the pressure is

$$\frac{\partial^2 p}{\partial t^2} + 2\mathbf{v}\cdot\nabla\frac{\partial p}{\partial t} + (\mathbf{vv} - c^2\mathbf{I}) : \nabla\nabla p - \left(\frac{p}{\rho}\nabla\frac{s}{C_v} - g\mathbf{k}\right)\cdot\nabla p =$$

$$\rho c^2 \nabla\mathbf{v} : \mathbf{v}\nabla + \frac{1}{p}\left(\frac{\partial p}{\partial t} + \mathbf{v}\cdot\nabla p\right)^2 + \rho c^2\left[\frac{D}{Dt}\left(\frac{M}{\gamma} + \frac{E}{h}\right) - \nabla\cdot\mathbf{F}\right] \tag{6}$$

where : is the double dot product.

The entropy is defined by the usual thermodynamic relation:

$$\nabla\frac{s}{C_v} = \frac{\nabla p}{p} - \gamma\frac{\nabla\rho}{\rho} \tag{7}$$

where C_v is the specific heat at constant volume and γ is the ratio of the specific heats C_p and C_v. The equation of state for a perfect gas is

$$p = \rho R T \tag{8}$$

where R is the gas constant and T is the temperature. The speed of sound $c = \sqrt{\gamma R T}$. The linearized equation for the perturbation pressure p is found if we neglect the squares and products of all perturbation terms in equation (6) and note that $E = RT_A M$:

$$\frac{\partial^2 p}{\partial t^2} - c_0^2\nabla^2 p + \left(\gamma g - \frac{dc_0^2}{dz}\right)\frac{\partial p_0}{\partial z} = \rho_0 c_0^2\left(\frac{dM}{dt} - \nabla\cdot\mathbf{F}\right) \tag{9}$$

where c_0 is the ambient speed of sound.

In equation (9) the coordinate system is stationary in the atmosphere and can be used for the study of the propagation of sound through the atmosphere as well as for the study of noise generated in and propagated through the atmosphere from an aircraft traveling at both subsonic and supersonic flight speeds. We describe the aircraft as it is in motion with a prescribed velocity at a given altitude, where $c_0 = c_A$. When the flight speed is supersonic, shock waves generated near the aircraft (see ref. 27) propagate toward the ground and generate the sonic boom.

In problems of jet noise the atmospheric terms in equation (6) are neglected and all source and diffusive terms are omitted, but all nonlinear terms are retained:

$$L(p) = \rho c^2 \frac{\partial v_i}{\partial x_j}\frac{\partial v_j}{\partial x_i} + \frac{1}{p}\left(\frac{Dp}{Dt}\right)^2 \tag{10}$$

where the wave operator, in Cartesian tensor notation (described below), is

$$L \equiv \frac{\partial^2}{\partial t^2} + 2v_i\frac{\partial^2}{\partial t\,\partial x_i} + \left(v_i v_j - c^2\delta_{ij}\right)\frac{\partial^2}{\partial x_i\,\partial x_j}$$

We deduce, by inspection of the terms on the right-hand side of equation (10), that fluctuating vorticity and pressure fields are the major sources of aerodynamic noise. Equations (6) and (10) differ from Lighthill's equation of aerodynamic noise in that Lighthill (refs. 2 and 3) argued that the density, rather than the pressure, was the proper independent variable for the study of aerodynamic noise. Of course external to the flow, in the radiation field, the density fluctuations are directly a function of the pressure fluctuations. When the flow field is only weakly nonisentropic, we may assume p varies as ρ^γ.

Lighthill's Equation of Aerodynamic Noise

Lighthill's equation of aerodynamic noise is obtained by subtracting the divergence of equation (2) from the time derivative of equation (1) and neglecting the atmospheric and source terms. This results in the inhomogeneous wave equation:

$$\frac{\partial^2 \rho}{\partial t^2} - c_\infty^2 \, \nabla^2 \rho = A(\mathbf{x}, t) \tag{11}$$

where the source term, in rectangular Cartesian tensor notation, is

$$A(\mathbf{x}, t) = \frac{\partial^2 T_{ij}}{\partial x_i \, \partial x_j} \tag{12}$$

with $i = 1, 2, 3$. The exact expression for T_{ij} in viscous compressible flow is

$$T_{ij} = \rho v_i v_j + (p - \rho \, c_\infty^2) \, \delta_{ij} - \tau_{ij} \tag{13}$$

where $\delta_{ij} = 1, 0$ when $i = j$ and $i \neq j$, respectively, and T_{ij} is the Lighthill acoustic analogy instantaneous applied stress tensor. In the inhomogeneous wave equation (eq. (11)), source terms that involve $\partial/\partial t$, $\partial/\partial x_i$, $\partial^2/\partial x_i \, \partial x_j$, and $\partial^3/\partial x_i \, \partial x_j \, \partial x_k$ are labeled, respectively, monopole, dipole, quadrupole, and octopole. For the source distribution function given in equation (12) the source is quadrupole.

Lighthill's equation is *exact* and has the following solution for an unbounded flow:

$$\rho(\mathbf{x}, t) = \frac{1}{4\pi c_\infty^2} \iiint \frac{A(y, \tau)}{|\mathbf{x} - \mathbf{y}|} \, d\mathbf{y} \tag{14}$$

where ρ is the density fluctuation, relative to the ambient density ρ_∞, received by an observer $Q(\mathbf{x}, t)$ in the far field due to disturbances of source strength $A(\mathbf{y}, \tau)$ per unit volume generated in the flow field at $P(\mathbf{y}, \tau)$, $\tau = t - (|\mathbf{x} - \mathbf{y}|/c_\infty)$ is the retarded time, and $|\mathbf{x} - \mathbf{y}|/c_\infty$ is the time for sound to travel from the flow disturbance at $P(\mathbf{y})$ to the field point $Q(\mathbf{x})$ at the ambient speed of sound c_∞. We see that in the Lighthill acoustic analogy the acoustic source distribution $A(\mathbf{y}, \tau)$ replaces the actual fluid flow and, moreover, the sources may move, but the fluid in which they are embedded may not. As discussed above, the sources are embedded in a medium at rest having the constant properties ρ_∞ and c_∞, the same as in the fluid external to the flow.

The solution of the inhomogeneous wave equation (eq. (11)) can be obtained more generally, as shown in reference 28, but it leads to the same solution (eq. (14)) when the sources are at rest.

As stated above, the sources may move but the flow may not. Let us now consider the sources moving at a uniform velocity $\mathbf{U}$, and we define $\mathbf{M} = \mathbf{U}/c_\infty$, the so-called acoustic Mach number. We introduce a system of moving coordinates

$$\boldsymbol{\eta} = \mathbf{y} - c_\infty \mathbf{M}\tau \tag{15}$$

such that the source emits when crossing the fixed point $\mathbf{y}$ at the time τ. The solution to equation (11) in moving coordinates is then

$$\rho(\mathbf{x}, t) = \frac{1}{4\pi c_\infty^2} \iiint \frac{A(\boldsymbol{\eta}, \tau)}{|\,|\mathbf{x} - \mathbf{y}| - \mathbf{M} \cdot (\mathbf{x} - \mathbf{y})|}\, d\boldsymbol{\eta} \tag{16}$$

where τ is the retarded time. This is Lighthill's well-known result.

If the instantaneous flow properties $\rho, p, \boldsymbol{\tau}$, and $\mathbf{v}$ are known everywhere within the flow, T_{ij} and $A(\mathbf{y}, \tau)$, or $A(\boldsymbol{\eta}, \tau)$, are known everywhere, and the far-field density perturbations can be obtained by quadrature throughout the flow volume. It is assumed that $A(\mathbf{y}, \tau)$ vanishes beyond the flow boundaries and the far-field observer is at a distance that is large compared with the finite dimensions of the flow field.

This seemingly simple yet exact solution to the fluid flow equations represents one of the major advances in the solution of unsteady fluid flow problems and is one of the most significant advances in the study of acoustics following the pioneering work of Lord Rayleigh. An immediate deduction from Lighthill's theory is that at low Mach numbers the total acoustic power P_a radiated from a jet is given by

$$P_a = \frac{K \rho_j^2 A_i U_j^8}{\rho_\infty c_\infty^5} \tag{17}$$

and since the kinetic energy flux is proportional to $P_j = \rho_j A_j U_j^3$, we see that

$$\frac{P_a}{P_j} = K(\rho_j/\rho_\infty)(U_j/c_\infty)^5 \tag{18}$$

where K is a constant of the order of 10^{-5} and ρ_j, A_j, and U_j are the values of the density, cross-sectional area, and velocity at the jet exit. Thus, the total acoustic power is a small fraction of the flow kinetic energy flux.

In order to arrive at this result the integrand in equation (16) needs special treatment, and arbitrary approximations to it are not permitted. Now T_{ij} has an order of magnitude equal to that of the kinetic energy of the turbulence per unit volume, and only a small fraction of that energy escapes from the flow as noise. This noise energy is then radiated to the far field, apart from the energy which is lost by absorption in the atmosphere. The source strength is equal to the double divergence of T_{ij}, and if the retarded time were ignored, then at a large distance from the flow, where $|\mathbf{x}| \gg |\mathbf{y}|$, no matter how large the source strength, the integral taken over the flow field would be the same as over all space and would be exactly zero; to that approximation the intensity of the radiated noise would be zero. In order to avoid this physically unacceptable result, Lighthill (ref. 2) showed that

$$\left[\frac{\partial^2 T_{ij}}{\partial y_i\, \partial y_j}\right] = \frac{\partial^2 T_{ij}}{\partial y_i\, \partial y_j} + 2\,\frac{\partial^2 T_{ij}}{\partial y_i\, \partial \tau}\frac{\partial \tau}{\partial y_j} + \frac{\partial^2 T_{ij}}{\partial \tau^2}\frac{\partial \tau}{\partial y_i}\frac{\partial \tau}{\partial y_j}$$

where square brackets denote the quantity is to be evaluated at $t = \tau$.

In the far field we find the first two terms, which represent true divergences of T_{ij}, generate zero contribution to the radiated noise. Thus, it is only the third term that is responsible for the radiated sound, and it follows that

$$\rho(\mathbf{x}, t) \approx \frac{1}{4\pi c_\infty^4}\frac{x_i x_j}{x^3}\iiint \frac{\partial^2 T_{ij}(\mathbf{y}, \tau)}{\partial \tau^2}\, d\mathbf{y} \tag{19}$$

Since x_i/x represents the direction cosines of the vector joining the source point $P(\mathbf{y})$ to the far-field observer point $Q(\mathbf{x})$, we may write $(x_i x_j/x^2)T_{ij} = T_{xx}$. We deduce that the contribution to the radiated noise at $Q(\mathbf{x})$ from each source point $P(\mathbf{y})$ in the flow field involves only those components of the Lighthill stress tensor that are aligned in the direction from $\mathbf{y}$ to $\mathbf{x}$, and its amplitude is proportional to the second time derivative of T_{ij} at emission. According to Lighthill's acoustic analogy, all acoustic sources within a flow volume radiate to the far field regardless of their position with respect to the flow boundaries. In the acoustic analogy, internal acoustic sources radiate with the same efficiency as sources closer to the bounding surface.

These important results may be derived directly if the solution of Lighthill's equation is written in the form

$$\rho(\mathbf{x}, t) = \frac{1}{4\pi c_\infty^2}\frac{\partial^2}{\partial x_i\, \partial x_j}\iiint \frac{[T_{ij}(\mathbf{y}, \tau)]}{|\mathbf{x} - \mathbf{y}|}\, d\mathbf{y} \tag{20}$$

where, in the far field, this reduces to

$$\rho(\mathbf{x}, t) \approx \frac{1}{4\pi c_\infty^4 x}\iiint \frac{\partial^2 T_{xx}(\mathbf{y}, \tau)}{\partial \tau^2}\, d\mathbf{y} \tag{21}$$

as derived above.

Order of Magnitude Approximations

If typical velocity and length scales in the turbulent region of a jet are represented by u_0 and l_0, and $\omega_0 = u_0/l_0$ is a typical frequency in the turbulence, then $\omega_0 l_0/u_0$ is $O(1)$, in agreement with the experimental results of reference 29. We therefore find the following orders of magnitude:

$$\frac{\partial^2 T_{ij}}{\partial y_i\, \partial y_j} = O(\rho_0 u_0^2/l_0^2) \qquad \text{and} \qquad \frac{\partial^2 T_{ij}}{\partial \tau^2}\frac{\partial \tau}{\partial y_i}\frac{\partial \tau}{\partial y_j} = O(\rho_0 u_0^2 \omega_0^2/c_\infty^2)$$

The ratio of these two quantities, $O(u_0^2/c_\infty^2)$, represents the fraction of the flow kinetic energy escaping as sound. Since the sound intensity in the far field at $\mathbf{x}$ is proportional to ρ^2, it follows that the sound power radiated from a unit volume of turbulence is

$$O\left(\frac{\rho_0^2}{l_0}\frac{c_\infty^3}{\rho_\infty}\frac{u_0^8}{c_\infty^8}\right)$$

This is one of the more important results derived directly from the Lighthill acoustic analogy. It shows that the sound power per unit volume of the flow is proportional to the eighth power of the flow velocity.

The viscous contribution to T_{ij} is $O(\rho_0 u_0^2/R_0)$, where $R_0 = \rho_0 u_0 l_0/\mu_0$ is the Reynolds number of the turbulence. At high Reynolds numbers, $R_0 \gg 1$ and then the viscous contribution can be neglected.

Thus, a good approximation to T_{ij} in high Reynolds number flows is

$$T_{ij} = \rho v_i v_j + (p - \rho c_\infty^2)\delta_{ij} \tag{22}$$

where the pressure p, the density ρ, and the velocity components v_i and v_j are evaluated in the flow at emission points $\mathbf{y}$. As previously stated the first term has an order of magnitude of $\rho_0 u_0^2$, but we need to study the second term carefully since, at least in an isothermal flow, it appears to have an order of magnitude similar to that in the external flow, which is zero.

The Effects of Temperature (Enthalpy) Fluctuations

We need to turn to the equation of conservation of energy, which has the form of equation (3). At high flow Reynolds numbers we can neglect the diffusive terms, since we are studying the larger scale motions in the flow field responsible for turbulent mixing and not the very-small-scale turbulent eddies responsible for the viscous dissipation in the flow. Using the equation of state for a perfect gas, we find that

$$\frac{\partial p}{\partial t} = -\frac{\gamma - 1}{2}\frac{\partial \rho v^2}{\partial t} - (\gamma - 1)\frac{\partial}{\partial x_j}\rho v_j h_s \tag{23}$$

The energy equation with the diffusive terms neglected is the same equation we would have derived if the flow were assumed to be isentropic, with $Ds/Dt = 0$. We may assume equation (23) holds even when the flow is weakly nonisentropic. Thus, we find that

$$\frac{\partial}{\partial t}\frac{p - \rho c_\infty^2}{c_\infty^2} = \frac{\partial}{\partial y_j}\left(\rho v_j \frac{h_\infty - h_s}{h_\infty}\right) - \frac{\gamma - 1}{2}\frac{\partial}{\partial t}\frac{\rho v^2}{c_\infty^2} \tag{24}$$

Hence, for an inviscid flow,

$$\left[\frac{\partial^2 T_{xx}}{\partial t^2}\right] = \left[\frac{\partial^2}{\partial t^2}\rho v_x v_x\right] - \frac{\gamma - 1}{2}\left[\frac{\partial^2}{\partial t^2}\rho v^2\right] + \left[\frac{\partial^2}{\partial t^2}\left(c_\infty \rho v_x \frac{h_\infty - h_s}{h_\infty}\right)\right] \tag{25}$$

and it follows that

$$\rho(\mathbf{x},t) \approx \frac{1}{4\pi c_\infty^4 x} \iiint \left[\frac{\partial^2}{\partial \tau^2} \left(\rho v_x v_x - \frac{\gamma-1}{2}\rho v^2 + \rho v_x c_\infty \frac{h_\infty - h_s}{h_\infty} \right) \right] d\mathbf{y} \qquad (26)$$

The source terms have orders of magnitude of $\rho_0 u_0^2$, except the final term has $O\left\{\rho_0 u_0 c_\infty \left[(h_{sj}/h_\infty) - 1\right]\right\}$. This latter term possesses dipole, quadrupole, and octopole contributions that generate noise proportional to u_0^6, u_0^8, and u_0^{10}, respectively. Hence, for example, in a heated jet at low Mach numbers, where temperature fluctuations exist, the far-field noise intensity is proportional to M^6, whereas for the isothermal jet the far-field noise intensity is proportional to M^8 under similar conditions.

In reference 2,

$$\frac{\partial^2 T_{ij}}{\partial y_i\, \partial y_j} = \frac{\partial^2}{\partial y_i\, \partial y_j} \left[(\rho v_i v_j - \tau_{xx}) + (p - \rho c_\infty^2)\, \delta_{ij} \right] \qquad (27)$$

and in the case of the heated jet, Lighthill (ref. 3) argues the final term could be replaced by $\left[1 - (c_\infty^2/c^2)\right] \nabla^2 p$, where the local mean jet temperature is found from $c = \sqrt{\gamma R T}$. It is also assumed that $\nabla^2 p$ as a source of noise is quadrupole, and therefore is of similar order of magnitude to the other quadrupole sources. However, that argument is shown above not to be complete.

If we consider the equation for the fluctuating pressure instead of that for the density, then

$$\left(\frac{\partial^2}{c_\infty^2\, \partial t^2} - \nabla^2 \right) p = \frac{\partial^2}{\partial x_i\, \partial x_j} (\rho v_i v_j - \tau_{ij}) - \frac{\gamma-1}{2} \frac{\partial^2}{\partial t^2} \frac{\rho v^2}{c_\infty^2} - \frac{\partial^2}{\partial x_j\, \partial t} \frac{\rho v_j (h_s - h_\infty)}{h_\infty} \qquad (28)$$

and we are reminded that the term $h_s - h_\infty$ contains not only octopole and quadrupole terms but also a dipole term of strength proportional to $\rho v_x h'$, where h' is the temporal fluctuation in specific enthalpy. Of course, in the case of the heated jet the speed of sound inside the flow differs markedly from that outside. Thus we might expect important flow-acoustic interaction effects to result in this case, since in the real flow the convecting eddies are shielded from the ambient medium external to the flow by heated, moving fluid. Such effects are, as already stated, included in the Lighthill acoustic analogy in the form of the Lighthill stress tensor. Nevertheless, the detailed fluid mechanics of such flow-acoustic interaction are hidden in the Lighthill acoustic analogy and are better dealt with by considering the actual flow, as discussed by Goldstein in another chapter.

We note here that in the Lighthill acoustic analogy, $\rho v_i v_j$ is augmented by the quantity $-(\gamma - 1)\rho v^2/2$ even when the flow is isothermal, but the major difference between the isothermal and the heated jet comes from the dipole term involving temperature fluctuations in the jet. Provided the intensity of the temperature fluctuations does not vary with jet Mach number, the far-field noise intensity for the heated jet will be proportional to M^6 at low Mach numbers. This is confirmed in experiments.

For the heated jet, even at low Mach numbers, it can be shown using the results of references 30 and 31 that the temperature, or enthalpy, fluctuations for a flow having a turbulent Prandtl number of unity are

$$\frac{h'}{h_j} = \frac{\sqrt{h'^2}}{h_j} \approx 1.6 \frac{\sqrt{v'^2}}{U_j} \frac{(h_j - h_\infty)}{(h_j + h_\infty)} \tag{29}$$

We see that the intensity of the temperature fluctuations is proportional to the intensity of the turbulent velocity fluctuations. The temperature fluctuations become negligible for the isothermal jet. Using this result for the intensity of the temperature, or enthalpy, fluctuations leads to the following ratio of the intensities of dipole to quadrupole noise for a heated jet at low Mach numbers:

$$\frac{\text{Dipole}}{\text{Quadrupole}} = 1.6 \frac{T_j}{T_\infty} \frac{c_\infty}{U_j}$$

The switch from dipole to quadrupole dependence for the heated jet is a function of the enthalpy ratio h_j/h_∞. From the experimental data of reference 32 the switchover occurs roughly when $M_j \approx 1.6(h_j - h_\infty)/h_\infty$. The experimental data of references 33 and 34 on heated jets (both static and in flight) at low Mach numbers confirm that the far-field noise intensity varies with M^6. This result is also in agreement with the analyses of references 35 and 36. The results of reference 37 show how a prediction model for the far-field noise from a jet can be established to provide a combination of the M_j^6 and M_j^8 dependences and to provide a good fit with experimental data.

Lighthill's theory of aerodynamic noise has shown that for a jet at ambient temperature and low Mach number, the far-field noise intensity varies with M_j^8. Many experimental studies on jet noise, including those reported in reference 38, have shown a dependence of noise intensity on M_j^6 at low Mach numbers. To explain these findings, Krasil'nikova considered Lighthill's solution for a turbulent shear flow at ambient temperature. Only the first term of the Lighthill stress tensor was considered, and the source term that was taken to be dipole was itself a space derivative and therefore was quadrupole, in agreement with Lighthill's derivation. We can only assume that the experimental results at ambient temperature available to Krasil'nikova, as well as other experimental results showing an M_j^6 dependence at low Mach numbers, were all subject to "excess noise." This is discussed in the section *Experimental Considerations.*

Thus, it has been shown that the complete Lighthill stress tensor is required for modeling both cold and heated jets, and this model leads to a dependence on jet exit Mach number in agreement with experiment at low Mach numbers. However, in some flows a good approximation is $T_{ij} \approx \rho v_i v_j$, where ρ is equal to the ambient density outside the flow. In general, though, the full stress tensor is required.

The Effects of Convection

Lighthill's Acoustic Analogy in a Moving Frame

It has been stated previously that in Lighthill's acoustic analogy the equivalent sources may move but the fluid may not. In the application of Lighthill's theory to the study of the noise from turbulent jets it has been found that the dominant sources are confined to a more or less central region of the mixing layers between the jet and the surrounding ambient fluid. Thus it is a satisfactory approximation to assume that the dominant sources all convect downstream parallel to the jet axis at a more or less constant speed or, as we will deduce subsequently, at a speed that, in general, is a function of the distance from the jet exit.

It is convenient to evaluate Lighthill's integral in a frame of reference moving with the convection speed of the turbulence. If we do not do this, then the space-time correlation function, corresponding to T_{xx}, must itself contain the effects of convection, and in such a frame of reference the effects of the retarded time are large. Thus as noted by Lighthill (ref. 2), an additional advantage in effecting the quadrature in a frame of reference moving with the convection speed, is that the effects of the retarded time between the emissions from any two sources whose far-field radiation arrives at the observer simultaneously at a time t are minimized. This can be shown to be true generally, provided $M_c \cos\theta \neq 1$.

Let U_c be the convection velocity and $\mathbf{M}_c$ be the convection Mach number with reference to the external speed of sound. In studies of aerodynamic noise, it is more convenient to use this "pseudo Mach number" rather than the true Mach number in the flow, which is equal to the local speed divided by the corresponding local speed of sound. We define a system of moving coordinates

$$\boldsymbol{\eta} = \mathbf{y} - c_\infty \mathbf{M}_c \tau$$

such that the source emits as it crosses the fixed point $\mathbf{y}$ at time $t = \tau$.

When Lighthill's integral is transformed to $\boldsymbol{\eta}$-space and assuming that $A(\mathbf{y}, \tau) \equiv A(\boldsymbol{\eta}, \tau)$, we find that in the far field, as given by equation (16),

$$\rho(x,t) \approx \frac{1}{4\pi c_\infty^2 x |1 - M_c \cos\theta|} \iiint A(\boldsymbol{\eta}, \tau)\, d\boldsymbol{\eta} \tag{30}$$

where

$$\tau = \frac{t - (x/c_\infty)}{|1 - M_c \cos\theta|} + \frac{\mathbf{x} \cdot \boldsymbol{\eta}}{c_\infty x |1 - M_c \cos\theta|} \tag{31}$$

is the retarded time and the effective volume of the sources is augmented by the Doppler factor $|1 - M_c \cos\theta|$. We see that when the source is convected relative to the fixed observer, the radiation is preferentially directed in the downstream direction. The radiation in the direction $\theta = 90°$ is unchanged. The far-field density appears to be singular when $M_c \cos\theta = 1$. However, this is not the case because the entire source function is not responsible for the noise radiated to the far field. Only a very small part of the characteristics of the overall source function are responsible for sound radiation, and the detailed analyses of Lighthill and Ffowcs Williams show the sound radiation is finite in this limit when $M_c \cos\theta = 1$. Moreover, the sound intensity increases smoothly in the passage from $M_c \cos\theta < 1$ to $M_c \cos\theta > 1$, as is shown subsequently when we consider in detail the case of noise radiation from a jet at all Mach numbers.

The Fourier Transform of the Density Fluctuations

Now the Fourier transform of the far-field density fluctuations is given by

$$\overline{\rho}(\mathbf{x}, \omega) = \frac{1}{2\pi} \int_{-\infty}^{\infty} \exp(-i\omega t) \rho(\mathbf{x}, t)\, dt \tag{32}$$

where ω is the circular frequency, and hence,

$$\overline{\rho}(\mathbf{x}, \omega) \approx \frac{1}{4\pi c_\infty^2 x} \exp(-i\omega x / c_\infty) \iiint \exp(-i\mathbf{k} \cdot \boldsymbol{\eta}) \times d\boldsymbol{\eta} \int \frac{1}{2\pi} A(\boldsymbol{\eta}, \tau)\, \exp(-i\omega_D \tau)\, d\tau \tag{33}$$

where the wave-number vector of the far-field noise is

$$\mathbf{k} = -\frac{\omega \mathbf{x}}{x c_\infty}$$

and the Doppler-shifted frequency is

$$\omega_D = \omega(1 - M_c \cos\theta) \tag{34}$$

Hence, if the four-dimensional Fourier transform of $A(\boldsymbol{\eta}, \tau)$ is $\overline{A}(\mathbf{k}, \omega_D)$, then

$$\overline{\rho}(\mathbf{x}, \omega) \approx \frac{\overline{A}(\mathbf{k}, \omega_D)}{4\pi c_\infty^2 x} \exp(-i\omega x / c_\infty) \tag{35}$$

We can gain an insight into the characteristics of the radiated sound, and in particular the effects of source convection, by first considering an elementary distribution for $A(\boldsymbol{\eta}, \tau)$ that represents a line source distribution along the x_1-axis. Let us put

$$A(\boldsymbol{\eta}, \tau) = \delta(\eta_2)\delta(\eta_3) l_2 l_3\, \exp(i\omega_{D_0}\tau) \sum_{-\infty}^{\infty} A_m \exp(i2\pi m \eta_1 / l_1) \tag{36}$$

where l_1, l_2, and l_3 are length scales of the disturbance, m is the mode number, and ω_{D_0} is the source frequency. On substitution into Lighthill's integral we find

$$\overline{\rho}(x, \omega_0) \approx \frac{l_1 l_2 l_3}{2 c_\infty^2 x} \exp(-i\omega_0 x / c_\infty) A_m \tag{37}$$

where $m = -\omega_0 l_1 \cos\theta/2\pi c_\infty$. Since the source is moving at the speed U_c we see that

$$k_1 = k\cos\theta = -2\pi m/l_1 \qquad \text{and} \qquad \omega_0 = k/c_\infty = \omega_{D_0}/(1 - M_c\cos\theta)$$

Thus, in the far field, at a polar angle θ, we have a discrete Doppler-shifted frequency sound field with $\omega_0 = \omega_{D_0}/(1 - M_c\cos\theta)$, which varies with the angle θ, and only the mode $m = -\omega_0 l_1 \cos\theta/2\pi c_\infty$ can radiate, where m is an integer. We interpret this result as providing a condition that sound radiation to the far field involves only that part of the wave-number–frequency spectrum of the source function $A(\mathbf{y}, t)$ for which the phase speed ω/k of its wave components exactly equals the external speed of sound c_∞, where $k = \sqrt{k_1^2 + k_2^2 + k_3^2}$ is the wave number of the sound. We see that for this source function, the radiation changes with Mach number of the source, but its amplitude is always finite. Alternatively, for given values of ω_0, l_0, and m, sound radiation will be beamed at one angle $\theta_m = \theta$ only, where $\cos\theta_m = -2\pi m c_\infty/\omega_0 l_0$.

The Lighthill-Ffowcs Williams Theory of Convection

The special properties of the Lighthill source function, which include the second time derivative of T_{ij} at emission, generate similar preferential radiation characteristics at all Mach numbers. It can be shown that, provided $M_c\cos\theta \neq 1$,

$$\rho(\mathbf{x}, t) \approx \frac{1}{4\pi c_\infty^4 x} \iiint \frac{\partial^2 T_{xx}(\boldsymbol{\eta}, \tau)/\partial\tau^2}{|1 - M_c\cos\theta|^3}\, d\boldsymbol{\eta} \tag{38}$$

which is another of Lighthill's important results. In figure 1 the effects of convective amplification are clearly shown. Equation (38) was the starting point for the work of Ffowcs Williams (ref. 39) on the radiated noise from high-speed jets. This solution applies to a volume distribution of quadrupoles traveling at subsonic and supersonic speeds, including the case where $|1 - M_c\cos\theta| = 0$. An uncritical deduction from equation (30) would lead to the assumption that the emission of infinite sound occurs in a direction perpendicular to Mach waves. When $|\mathbf{M}| > 1$, care is needed to find the emission of finite sound in directions along and close to the normals to the Mach waves.

At supersonic convection speeds the disturbance created by the moving eddies in the jet mixing region is responsible for the creation of Mach waves and weak shock waves in the external medium. Figure 2 (from ref. 40) shows typical pictures of Mach wave radiation. However, eddies are not solid objects and they do not move at a steady speed. Thus, we must regard the eddies as possessing both a mean convection speed and some fluctuation. At subsonic and supersonic convection speeds in directions other than normal to the Mach waves, the effect of the fluctuation in convection speed is negligible. However, in the direction normal to the Mach waves, where $M_c\cos\theta = 1$, the Doppler factor in equation (38) must be replaced by

$$|(1 - M_c\cos\theta)^2 + (\alpha v_r^2/c_\infty^2)|^{1/2} \tag{39}$$

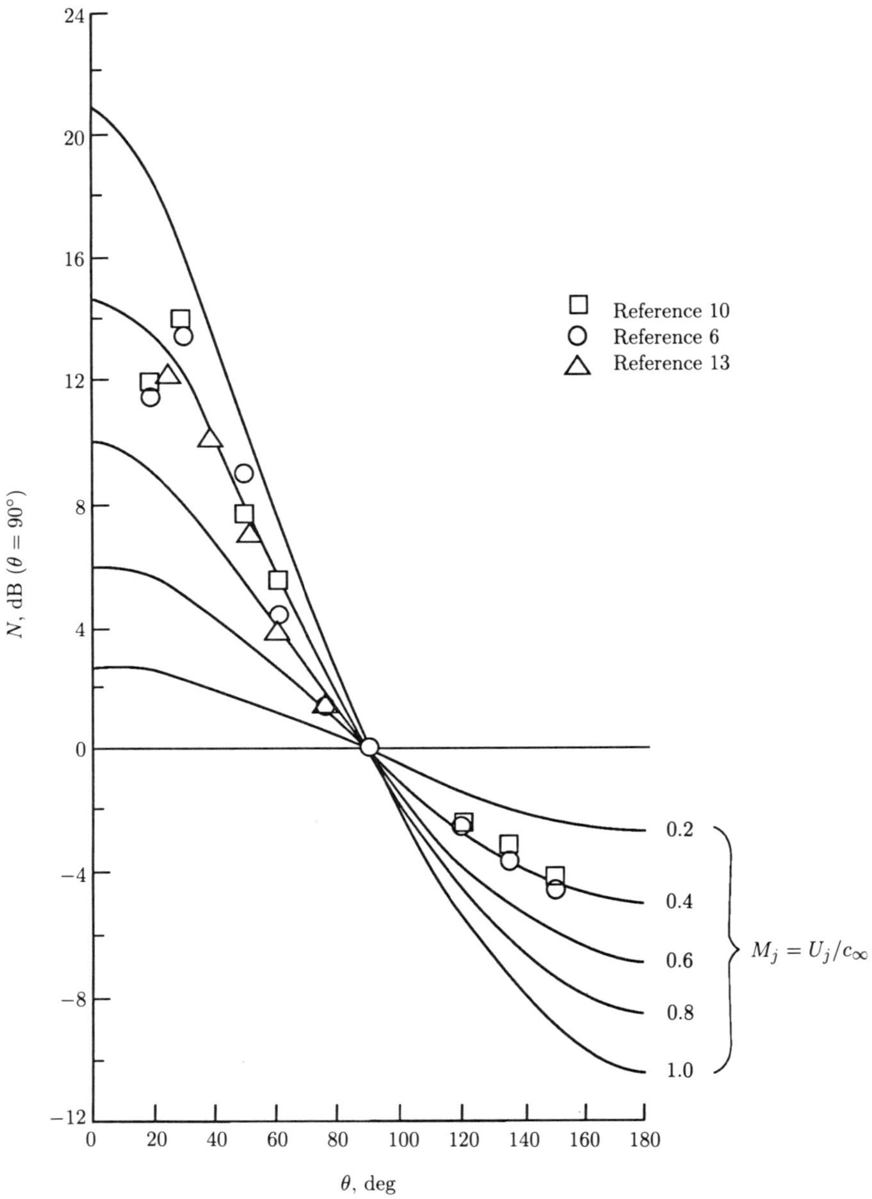

Figure 1. Directional distribution of jet noise as function of convection Mach number. $U_j = 300$ *m/sec*; $M_c = 0.62M_j$.

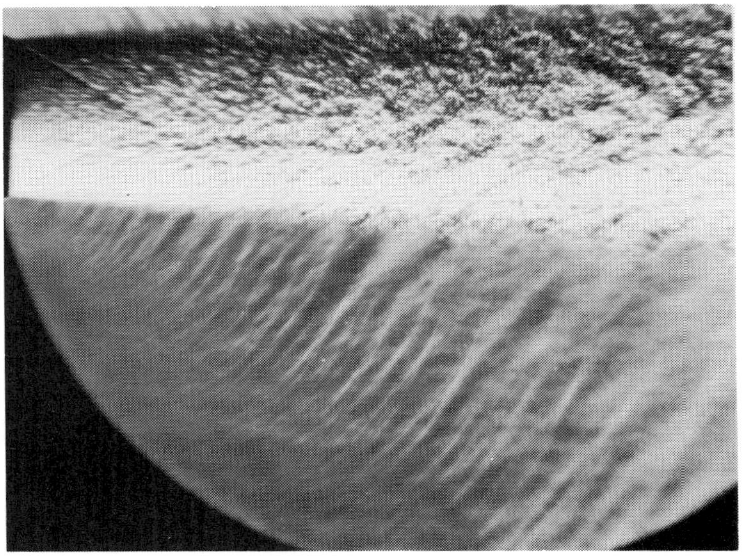

(a) Fully expanded jet.

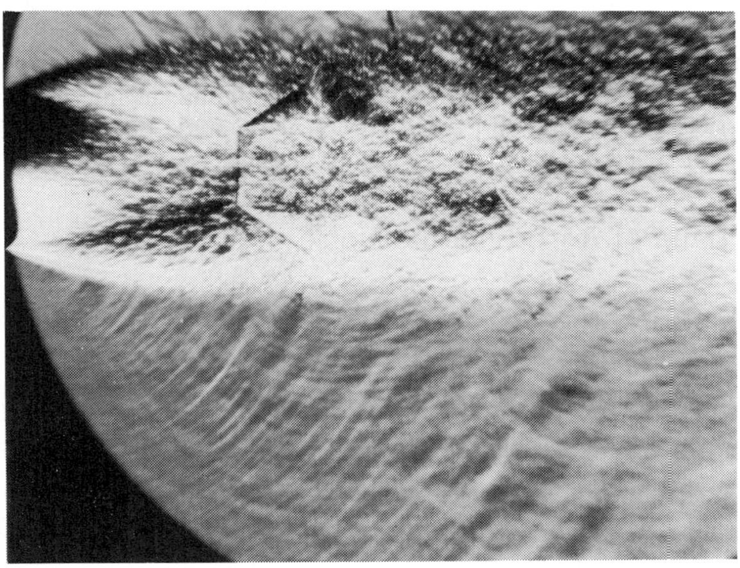

(b) Choked jet.

Figure 2. Jet at supersonic speeds showing Mach waves outside jet boundary. (From ref. 40.)

where v_r is the average fluctuating velocity component (root-mean-square value) normal to the Mach waves and α has an order of magnitude equal to the characteristic Strouhal number of the turbulence $\omega_0 l_r / v_r$, where ω_0 is the characteristic turbulent circular frequency and l_r is the characteristic turbulence length scale in a direction normal to the Mach waves. The term $\alpha v_r^2 / c_\infty^2$ is negligible except near $M_c \cos\theta = 1$. In the direction normal to the Mach waves the sound power radiated per unit volume of turbulence is proportional to

$$\frac{\overline{T^2} l_0^3}{\rho_\infty \omega_0 l_r^5}$$

where $\overline{T^2}$ is the mean-square fluctuation of the stress tensor T_{ij}, l_0 is a length scale of the correlation volume, and $\alpha v_r^2 / c_\infty^2$ is replaced by $(\omega_0 l_r / c_\infty)^2$. This result was first given in reference 39.

The Neglect of Density Fluctuations in the Flow

The Lighthill acoustic analogy provides a satisfactory foundation for the study of the sound radiation from unsteady aerodynamic flows, including turbulent jet flows, in motion at subsonic and supersonic speeds. In all the discussions relating to estimates of the magnitude of the effective source strength the fluctuations in density in the source field have been ignored. This approximation may be justified on the basis of the Morkovin (ref. 41) and Bradshaw (ref. 42) hypotheses for mean jet convection Mach numbers less than about 1.5. For jets at higher speeds the turbulent mixing region contains eddies moving supersonically relative to the ambient flow. The accompanying wavelets, or shocklets, produce significant fluctuations in density in the acoustic source region and these cannot be ignored. Further discussion of the noise from high-speed jets is given in another chapter.

The Spectrum of Aerodynamic Noise

Space-Time Correlations of the Source Function

The general expressions for the autocorrelation of the noise intensity and its spectral density at the position of the far-field observer are, respectively,

$$I(\mathbf{x}, t^*) = \frac{c_\infty^3}{\rho_\infty} \lim_{T\to\infty} \frac{1}{2T} \int_{-T}^{T} \rho(\mathbf{x}, t') \rho(\mathbf{x}, t' + t^*) \, dt' \tag{40}$$

and

$$\overline{I}(\mathbf{x}, \omega) = \frac{1}{2\pi} \int_{-\infty}^{\infty} \exp(-i\omega\tau^*) I(\mathbf{x}, \tau^*) \, d\tau^* \tag{41}$$

where ω is the frequency in the fixed frame of the observer. If we assume that the turbulent field is stationary and thus its mean properties do not vary with the time of measurement, then the space-time correlation of the second time derivative of the stress tensor may be written as

$$\frac{\partial^4}{\partial \tau^4} P_\theta(\mathbf{y}, \boldsymbol{\delta}, \tau) = \overline{\frac{\partial^2 T_{xx}}{\partial \tau_1^2}} \, \overline{\frac{\partial^2 T'_{xx}}{\partial \tau_2^2}} \tag{42}$$

where T_{xx} and T'_{xx} are, respectively, the aligned components of the Lighthill stress tensor at the source positions $\boldsymbol{\eta}$ and $\boldsymbol{\zeta} = \boldsymbol{\eta} + \boldsymbol{\delta}$, with corresponding retarded times τ_1 and τ_2, for a given observer position $\mathbf{x}$. The term P_θ is the space-time covariance of T_{ij} at a fixed point in the source region aligned with the observer situated at (x, θ), where θ is the angle relative to the direction of motion, which in the case of a jet would be along the jet axis.

Autocorrelation of the Far-Field Sound Intensity

Now $\boldsymbol{\eta} = \mathbf{y} - c_\infty \mathbf{M}_c \tau_1$, $\boldsymbol{\zeta} = \mathbf{z} - c_\infty \mathbf{M}_c \tau_2$, and $\boldsymbol{\delta} = \boldsymbol{\zeta} - \boldsymbol{\eta}$ is the spatial separation of the sources at $\boldsymbol{\eta}$ and $\boldsymbol{\zeta}$ in the moving frame, and $\boldsymbol{\eta}$ and $\boldsymbol{\zeta}$ correspond respectively to the two fixed points $\mathbf{y}$ and $\mathbf{z}$ at which emission takes place. The corresponding retarded times are τ_1 and τ_2. With τ as the difference in retarded times between the emissions at $\mathbf{z}$ and $\mathbf{y}$ and t^* as the difference in their reception times at the observer we find that

$$\tau = \frac{c_\infty t^* |\, |\mathbf{x} - \mathbf{y}| + \boldsymbol{\delta} \cdot (\mathbf{x} - \mathbf{y})|}{c_\infty |\, |\mathbf{x} - \mathbf{y}| - \mathbf{M}_c \cdot (\mathbf{x} - \mathbf{y})|} \tag{43}$$

and in the far field, where $|\mathbf{x}| \gg |\mathbf{y}|$, this reduces to

$$\tau = \frac{c_\infty t^* x + \boldsymbol{\delta} \cdot \mathbf{x}}{c_\infty x |1 - M_c \cos\theta|} \tag{44}$$

If we write the wave-number vector as $\mathbf{k} = -\omega \mathbf{x}/x c_\infty$ and note that $dt^* = (1 - M_c \cos\theta)\, d\tau$, then the far-field autocorrelation of the sound intensity is

$$I(\mathbf{x}, t^*) \approx \frac{1}{16\pi^2 \rho_\infty c_\infty^5 x^2 |1 - M_c \cos\theta|^5} \iiint d\mathbf{y} \iiint \frac{\partial^4}{\partial \tau^4} P_\theta(\mathbf{y}, \boldsymbol{\delta}, \tau)\, d\boldsymbol{\delta} \tag{45}$$

and the cross-power spectral density is

$$\overline{I}(\mathbf{x}, \omega) \approx \frac{\pi}{2\rho_\infty c_\infty^5 x^2} \iiint \omega^4 \overline{P}_\theta(\mathbf{y}, \mathbf{k}, \omega_D)\, d\mathbf{y} \tag{46}$$

where $\overline{P}_\theta$ is the four-dimensional Fourier transform of the source space-time covariance and

$$\overline{P}_\theta(\mathbf{x}, \mathbf{k}, \omega_D) = \frac{1}{(2\pi)^4} \iiint \exp(-i\mathbf{k} \cdot \boldsymbol{\delta})\, d\boldsymbol{\delta} \int_{-\infty}^{\infty} \exp(-i\omega_D \tau) P_\theta(\mathbf{y}, \boldsymbol{\delta}, \tau)\, d\tau \tag{47}$$

The Doppler-shifted noise frequency in the far field is $\omega_D = \omega(1 - M_c \cos\theta)$, where ω is the source frequency. The far-field noise intensity and its spectral density are finite at all Mach numbers. This result was first given in reference 39. An earlier approximation to this result was given by Lilley (ref. 43). The correct result for $\overline{I}(\mathbf{x},\omega)$ was given by Lighthill (refs. 4 and 5) and by Ribner (ref. 44).

If the source function P_θ has the physically possible form

$$P_\theta = \rho_0^2 u_0^4 \exp\left[(-\delta_i^2/l_i^2) - \omega_0^2\tau^2\right] \tag{48}$$

in a moving frame having the convection velocity U_c, we can easily find the part of P_θ that is responsible for the far-field radiation. We find the four-dimensional Fourier transform of P_θ and then integrate the result over all angles in wave-number space to find the average wave-number spectrum function B as a function of wave number k and frequency ω only. Here $k = |k|$ and ω is the frequency of sound in the far field and is related to the frequency in the source field ω_D by $\omega_D = \omega(1 - M_c \cos\theta)$. Thus we find

$$\overline{A}_\theta(\mathbf{y},\mathbf{k},\omega_D) = \frac{1}{(2\pi)^4|1 - M_c\cos\theta|^4}\iiint \exp(-i\mathbf{k}\cdot\boldsymbol{\delta})\; d\boldsymbol{\delta}\int \exp\left(-i\omega_D\tau\right)\frac{\partial^4}{\partial\tau^4}P_\theta(\mathbf{y},\boldsymbol{\delta},\tau)\; d\tau \tag{49}$$

and

$$B\ (\mathbf{y},\mathbf{k},\omega_D) = 2\pi\int_0^\pi \sin\theta\,\overline{A}_\theta(\mathbf{y},\mathbf{k},\omega_D)\; d\theta \tag{50}$$

Contours of constant B are plotted in figure 3 as functions of k and ω_D for several values of M_c. Also plotted is the line $\omega l_0 = k l_0 c_\infty$. It is only values of B lying on this line that contribute to the far-field radiation. At low convection Mach numbers the wavelength of the noise is four to five times the characteristic length of the energy-containing turbulent eddies, and those eddies responsible for most of the radiation are a slightly smaller scale than the energy-containing eddies. This conclusion was found independently in reference 45, an investigation of the noise radiated from isotropic turbulence, and in reference 43, which contains studies on the noise from jets. At higher convection Mach numbers approaching unity the wavelength of the noise is roughly twice the characteristic length of the energy-containing eddies, and those eddies responsible for most of the radiation are about one-third the scale of the energy-containing eddies. These results were obtained in reference 39.

Useful Definitions Used in Aerodynamic Noise Theory

In the results discussed subsequently we use a number of quantities that we define here for convenience. These are the sound intensity

$$I(\mathbf{x}) = \int_{-\infty}^{\infty} \overline{I}(\mathbf{x},\omega)\; d\omega \tag{51}$$

and the sound power

$$P = 2\pi x^2\int_0^\pi \sin\theta I(\mathbf{x},\theta)\; d\theta \tag{52}$$

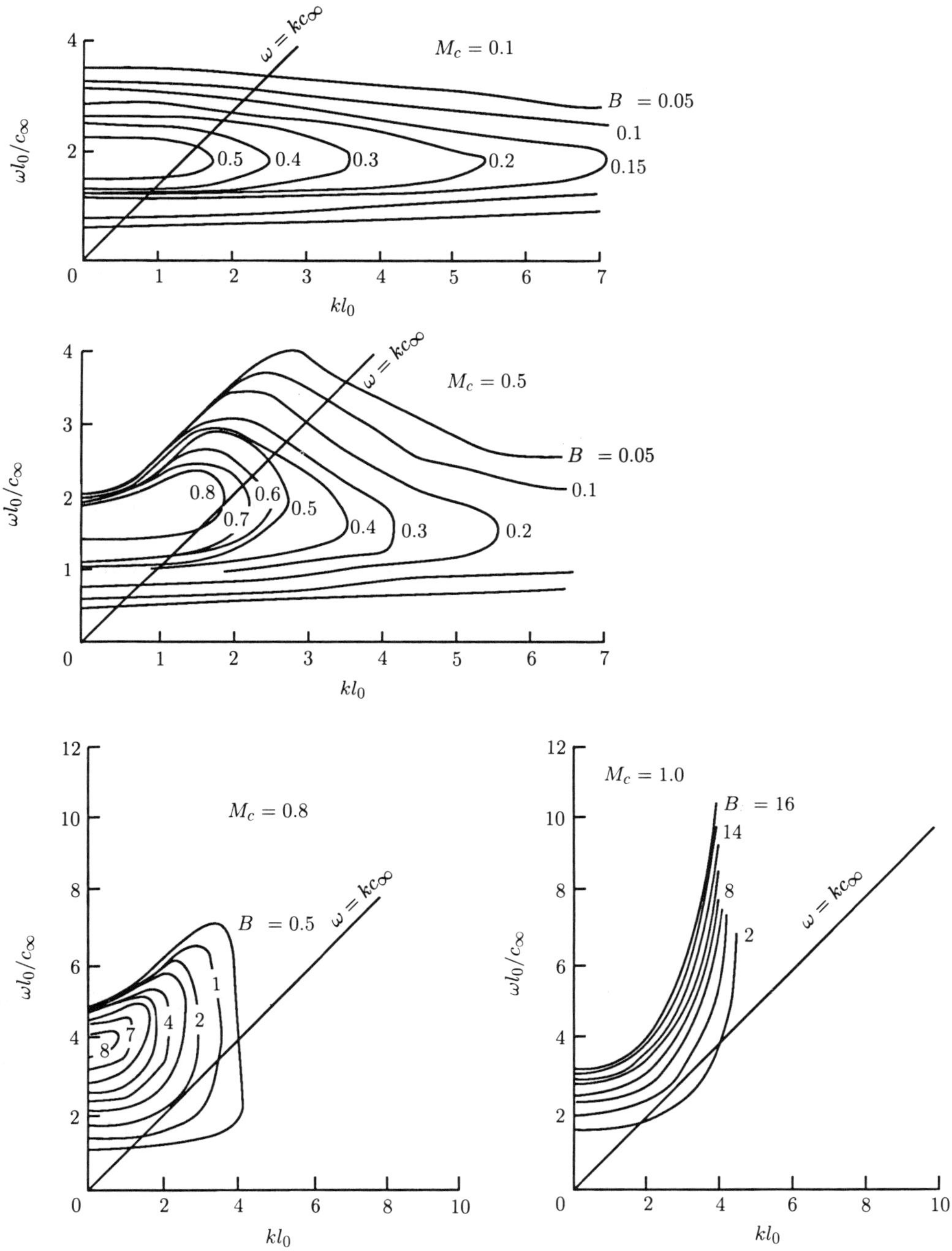

Figure 3. Isocorrelation curves of the Gaussian source function B for various values of convection Mach number.

on the assumption that the acoustic far field is cylindrically symmetric with respect to the jet axis. We define the sound pressure level in decibels as

$$\mathrm{SPL} = 10 \log_{10} \frac{\overline{p^2}}{p_{\mathrm{ref}}^2} \tag{53}$$

and the total sound power level in decibels is

$$N = 10 \log_{10} \frac{P}{P_{\mathrm{ref}}} \tag{54}$$

The Structure of a Turbulent Jet

The Initial Mixing Layer

The structure of the turbulent mixing region of a circular jet has been studied extensively by many experimentalists. The initial mixing region, from about one to four diameters from the jet exit, is similar to the two-dimensional plane mixing layer since its overall thickness b is small compared with the jet diameter D_j. For the plane mixing layer and for values of $U_1 y_1/\nu > 10^5$, where U_1 is the velocity outside the mixing region, y_1 is measured parallel to U_1, and ν is the kinematic viscosity, it is known from the measurements of reference 46 that the flow structure is self-preserving in the sense that the average properties of the turbulence and of the mean flow at any section of the mixing region are similar except for a change in scale.

In the initial mixing region, if all upstream disturbances are absent, the mixing layer exhibits characteristics of laminar flow followed by transition to turbulence at Reynolds numbers of about $U_1 y_1/\nu > 4 \times 10^5$. At low jet Reynolds numbers the disturbances associated with the most amplified instability waves can be readily visualized and their breakdown results in the formation of vortex rings and, subsequently, secondary azimuthal waves and the formation of longitudinal, or streamwise, vortices. The experiments of Crow and Champagne (ref. 47) and the theory of Michalke (refs. 48 and 49) show that the preferred wavelength for maximum spatial growth is 7δ to 8δ, where δ is the thickness of the initial shear layer. The initial region is shown clearly in figure 4, which is for a 25-mm-diameter air jet at a jet Mach number close to unity. Similar results were obtained in reference 50.

Vortex breakdown occurs with and without coalescence, or pairing, of successive vortex rings. The detailed description of convective instability and, in certain cases, absolute instability of mixing layers and their progress toward transition are interpreted and expertly summarized in reference 51.

The Turbulent Structure in a Mixing Region

Experimental evidence suggests that the vortex structures existing in the final stages of transition persist in the region where flow is fully turbulent. In addition, large vortex structures arise naturally in the turbulent flow, and further discussion on this takes place below. The question of the importance of the collapse of regular

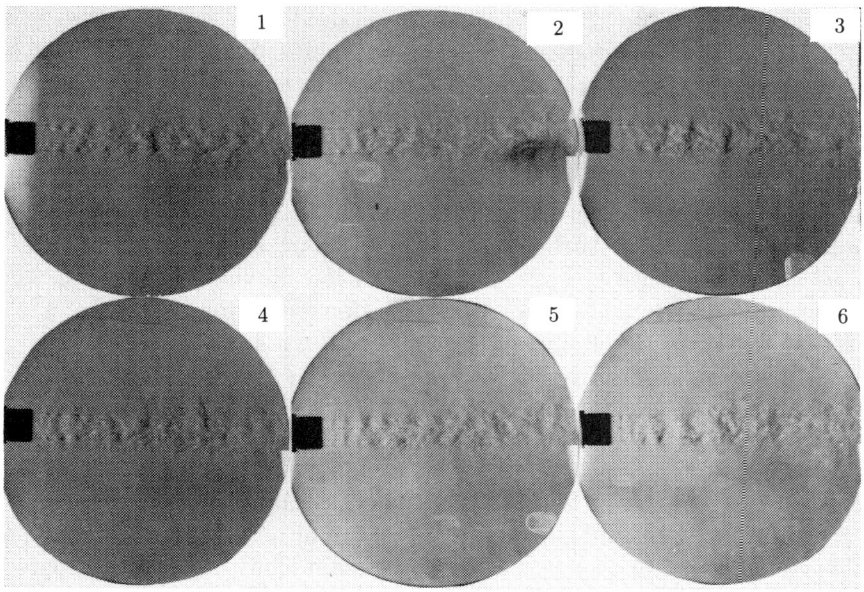

(a) Knife-edge vertical.

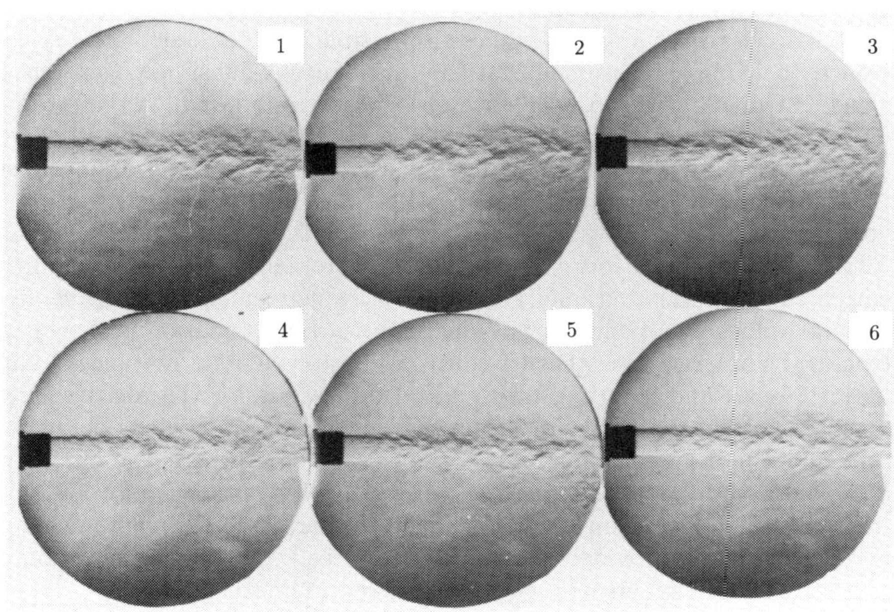

(b) Knife-edge horizontal.

Figure 4. Structure of initial mixing region of 25-mm-diameter air jet at high subsonic Mach numbers. $U_j = 250$ *m/sec; field diameter, 0.3 m; picture sequence, 0.5 msec.*

vortex rings and vortex pairing and the resulting large localized pressure fluctuations as a source of intense noise generation has been raised by many researchers (refs. 47, 52, 53, 54, and 55). This question, however, remains unanswered for jets at high Reynolds numbers, where the turbulent diffusion processes act to smear out such peaks in pressure fluctuations, although the concensus is that at subsonic jet speeds in fully turbulent flow such noise is small in amplitude compared with the noise generated by turbulent mixing. Further discussion on this topic, including the corresponding effects in supersonic flow, is found in another chapter. Certainly transition from laminar to turbulent flow is a strong source of noise in a mixing region, and measurements of noise from jets at low Reynolds numbers, where the extent of laminar flow from the jet exit to transition covers many diameters in length, do not follow the corresponding results at higher Reynolds numbers.

The turbulent structure in a jet at high Reynolds numbers is strongly inhomogeneous as a result of the spreading of the flow into the surrounding nonturbulent ambient fluid. The bounding surface of the mixing zone is highly contorted by eddies that, according to references 31 and 56, resemble the Helmholtz instability of a vortex sheet, with a growth and decay cycle. The alternation between instability and stability suggests that overall the flow is in a state of near neutral stability, and the contortions of the bounding surface allow the entrainment rate of irrotational fluid to be self-adjusting and dependent on a flow constant R_s only. The flow in a jet may be assumed to be composed of a mean velocity field $\mathbf{U}(x)$, a large eddy motion $\mathbf{u}'(x,t)$, and the main turbulent motion $\mathbf{u}''(x,t)$. The main turbulent motion includes all the small eddies down to the smallest eddies responsible for the dissipation. According to reference 31, it may be assumed that the turbulence is quasi-homogeneous at the higher end of this range down to a state of local isotropy in which the structure is near universal, which by observation is in accord with Kolmogoroff's theory. Eddies in this lower range of sizes make little contribution to the total kinetic energy of the turbulent motion. Townsend shows that the main turbulent motion is exposed to the mean shear or strain rates imposed by the mean flow gradients. As a result of rapid-distortion theory the essential anisotropic features of the main turbulent motion can be estimated, and good qualitative agreement of these values with experimental values is obtained. Thus, the main turbulent motion is shown to possess structural similarity such that its contribution to the main motion is limited to changes in velocity and length. Townsend quotes results for the relative strengths of the components of the Reynolds stress tensor for different regions of the jet and shows the differing degrees of anisotropy that exist between the initial mixing region and the flow downstream of the potential core. All these results are for low subsonic Mach numbers. However, many of these structural parameters remain unchanged at higher Mach numbers, provided the jet is shock free. A detailed discussion of the turbulent structure in supersonic jets is considered in another chapter.

Turbulent Measurements in a Moving Frame

The measurements from which the results discussed above have been obtained have all used fixed-frame analysis. In general, this gives an impression that there is a random distribution of eddies crossing the observation window and that events relatively remote from each other are statistically independent. However, flow visualization and space-time measurements at laboratory Reynolds numbers show that much of the structure, especially in the larger scale eddies, is ordered and has a longer characteristic decay time than would be apparent from a statistical analysis of the measurements. The experimental measurements of reference 29 for the cross correlation $R_{11}(\xi, \tau)$ are shown in figure 5. In these measurements the average convection speed of the turbulence is almost constant across the mixing region. Its value is related to a group velocity, since the turbulence may be represented by a dispersive wave system, which is a function of frequency. An average value of the convection speed in the initial mixing region of a jet is 0.62 times the mainstream velocity difference between the centerline velocity of the jet U_j and the velocity of the ambient fluid outside U_f. In figure 5 the moving-frame autocorrelation at the speed of convection is the envelope of the cross-correlation curves and has the largest characteristic time scale $L_{\mathcal{T},\max}$. It is found that $L_{\mathcal{T},\max}$ is of the order of the inverse of the mean shear $\partial U_1/\partial x_2$, proof that the eddy distortion is directly related to the mean shear as discussed previously. If u_0 and l_0 are characteristic scales of velocities and eddy sizes, then $L_{\mathcal{T},\max} = O(l_0/u_0)$ also. This simple result is in agreement with Prandtl's mixing length theory, which states $u_0 = l_0 \, \partial U_1/\partial x_2$. The measurements show that with $\omega_0 = 1/L_{\mathcal{T},\max}$, $\omega_0 l_0/u_0 = 1.7$ (or $f_0 l_0/u_0 = 0.27$ approximately), where $\omega_0 = 2\pi f_0$ is the characteristic circular frequency.

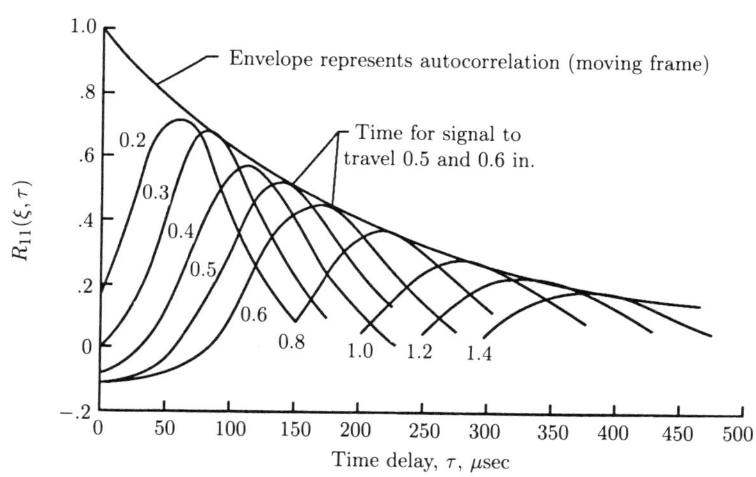

Figure 5. Cross correlation $R_{11}(\xi, \tau)$ in 25-mm-diameter (1.0-in.) jet at $M_j = 0.45$. $y_1 = 1.5D_j$; $r = 0.5D_j$. (From ref. 29.)

We note that $\omega_0 l_0/u_0$ is the characteristic Strouhal number of the turbulence in a moving frame and we may assume it is nearly constant throughout the entire mixing region.

The integral length scale L_{11} of the turbulence is independent of the convection speed and has a value of about $0.12y_1$ near the center of the initial mixing region. The isocorrelation contours as measured in reference 29 in a 25-mm jet at $M = 0.45$ are shown in figure 6 and clearly show the frame of reference in which the correlation falls most slowly. In this case it is 0.62 times the jet exit speed. The variation of the convection speed across the initial mixing region of a jet at two stations downstream of the jet exit is shown in figure 7.

The Large-Scale Structure

The large-scale structure of the turbulence in the mixing region of a jet has been shown experimentally, as observed through flow visualization and methods involving conditional sampling, to possess self-similar structures that are coherent and extend in the direction of their convection. These are discussed by many authors, including Yule (ref. 57) and Browand and Weidman (ref. 58). These structures may similarly be described in terms of their wave number and frequency structure and are amenable to theoretical description. They have been termed wave models or wavelike (refs. 59 and 60). The recent work of reference 61 provides a suitable model for the structure of the two-dimensional mixing region of a jet based on this wave theory of turbulence. With this weakly nonlinear, finite-amplitude model reference 61 finds that the primitive large-scale structure of the mixing region, as shown in figure 8, is the result of the instability of the basic turbulent flow to small disturbances. Corresponding experimental results (ref. 62) are shown in figure 9. The amplitude of the unstable disturbances and their subharmonics grow initially exponentially with both time and space and are convected downstream with a phase speed of about $0.6U_1$. Eventually, though, strong nonlinear and three-dimensional distortion sets in and the simple waveform of the most unstable wave becomes more broadband, with the result that the local flow develops into a complex eddy structure of many different sizes, as discussed above, and the turbulence possesses a near continuous spectrum. As a result the width of the local mean flow grows with downstream distance, as shown in figure 10. Accordingly the properties of the most unstable wave change and largest eddy structures dominate. This condition is accommodated by a "pairing," or some related interaction, between subsequent yet randomly formed upstream disturbances as they are convected downstream. Some irregularity in the structures develops, and overall the new structures suffer a pronounced jitter due to the irregular, turbulent flow developing downstream. Yet on average, as confirmed by the conditionally sampled results, the large-scale structures possess a remarkably coherent structure convecting downstream. The main turbulence is smaller in scale but is also convected downstream along with the large-scale turbulence. The smaller scale turbulence eventually decays through a Kolmogoroff cascade process down to the smallest scales of turbulence at which dissipation occurs. The irregularity in the large-scale structures becomes more marked in the mixing region downstream of the potential core, but nevertheless such a large-scale structure appears to exist and acts to control both the mean flow local growth and the entrainment of the external irrotational fluid into the jet. Different modal structures, reflected in different large-scale structures, develop when the jet is induced to spin about its axis and when the jet is nonuniform and highly disturbed at the exit plane.

All the results discussed above relate to the case when the jet is devoid of both internal and external excitation. Our description of the large-scale structure makes no mention of the sound field generated by it. The randomness of this sound field and its low amplitude compared with the kinetic energy of the eddy structures from which it is generated suggest that the large-scale eddy structure is unchanged as a result of the presence of this sound field, even though the sound field suffers scattering,

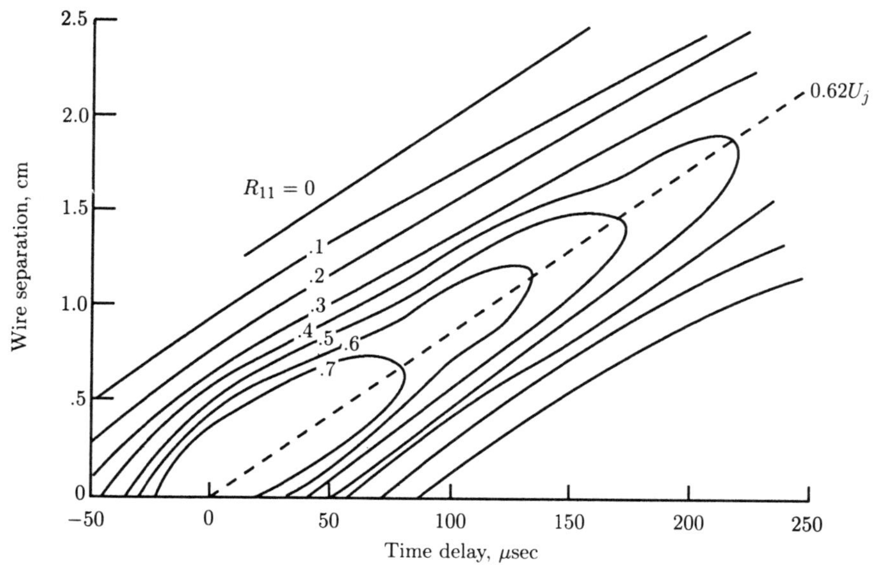

Figure 6. Isocorrelation curves of $R_{11}(\xi, \tau)$ in 25-mm-diameter jet at M_j = 0.45. $y_1 = 1.5D_j$; $r = 0.5D_j$. (From ref. 29.)

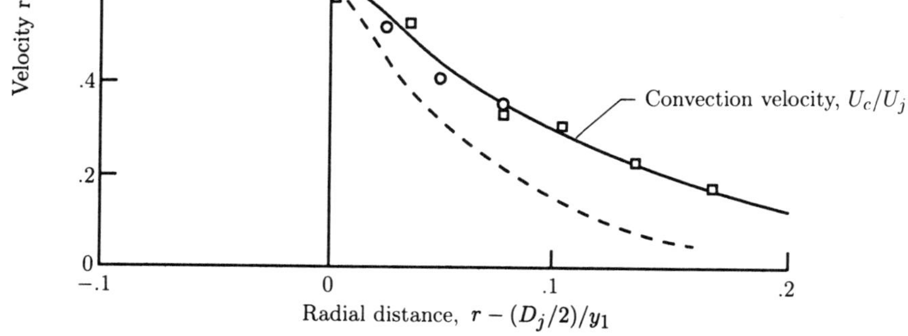

Figure 7. Variation of eddy convection speed across mixing region. (From ref. 29.)

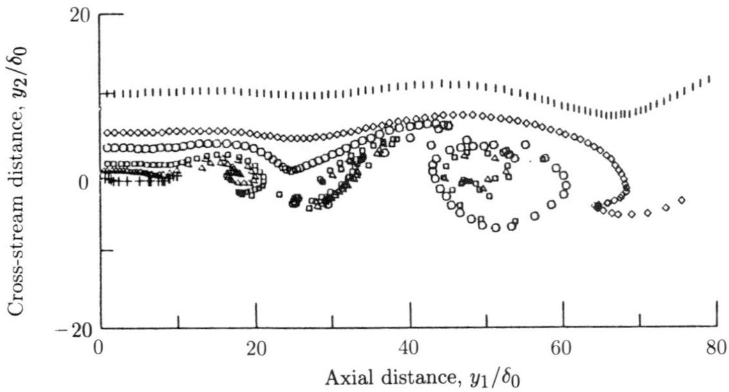

Figure 8. Calculated streaklines for two-dimensional shear layer at nondimensional time of 90 units. (From ref. 61.)

diffraction, and refraction as it traverses the turbulent flow before issuing into the external irrotational flow and propagating toward the distant observer.

Discrete frequency aerodynamic or acoustic forced excitation of the jet generated internally or externally has, on the other hand, a marked effect on the jet structure, at least in the initial mixing region downstream of the jet exit. Provided such disturbance is of sufficient amplitude, the most unstable waves are now closely related to the excitation frequency and its harmonics. Violent changes in the structure of the jet mixing region occur and in extreme cases result in the rapid spreading of the jet in one plane to the splitting of the jet into two or more separate jets. The sound field from an excited jet is treated in another chapter. Further work on excited jets may be found in references 63 to 66.

The Self-Preserving Properties in Jets

For both plane and circular jets at high Reynolds numbers and low Mach numbers, experiments show that throughout the mixing region the mean flow is self-preserving and depends on u_0^* and l_0^*, which are functions of the axial coordinate y_1, and the flow is geometrically similar at all sections. As discussed in references 30 and 31, self-preserving flow is limited to either axisymmetric flows or flows in which the width in one direction is effectively infinite, such as the plane jet or plane mixing layer. Here u_0^* is the scale of the mean velocity variation, and l_0^* is the length scale of the flow; u_0 is the scale of the turbulent velocities and is proportional to u_0^*. However, the complete turbulent structure has a response time which is, in general, long compared with the time for the mean flow development. Thus we find for the jet that the conditions for self-preserving flow are broadly met for scales of turbulence of the order of l_0^* in length, where the mean width b of the mixing zone is of the order of $2l_0^*$ to $3l_0^*$ and $b = 0.32y_1$ for the jet issuing into fluid at rest. However, for the larger eddies in the mixing zone the response time is longer than for the mean flow development and the large eddy structures persist for many jet diameters downstream, as shown in many flow visualization photographs discussed previously. The differences between

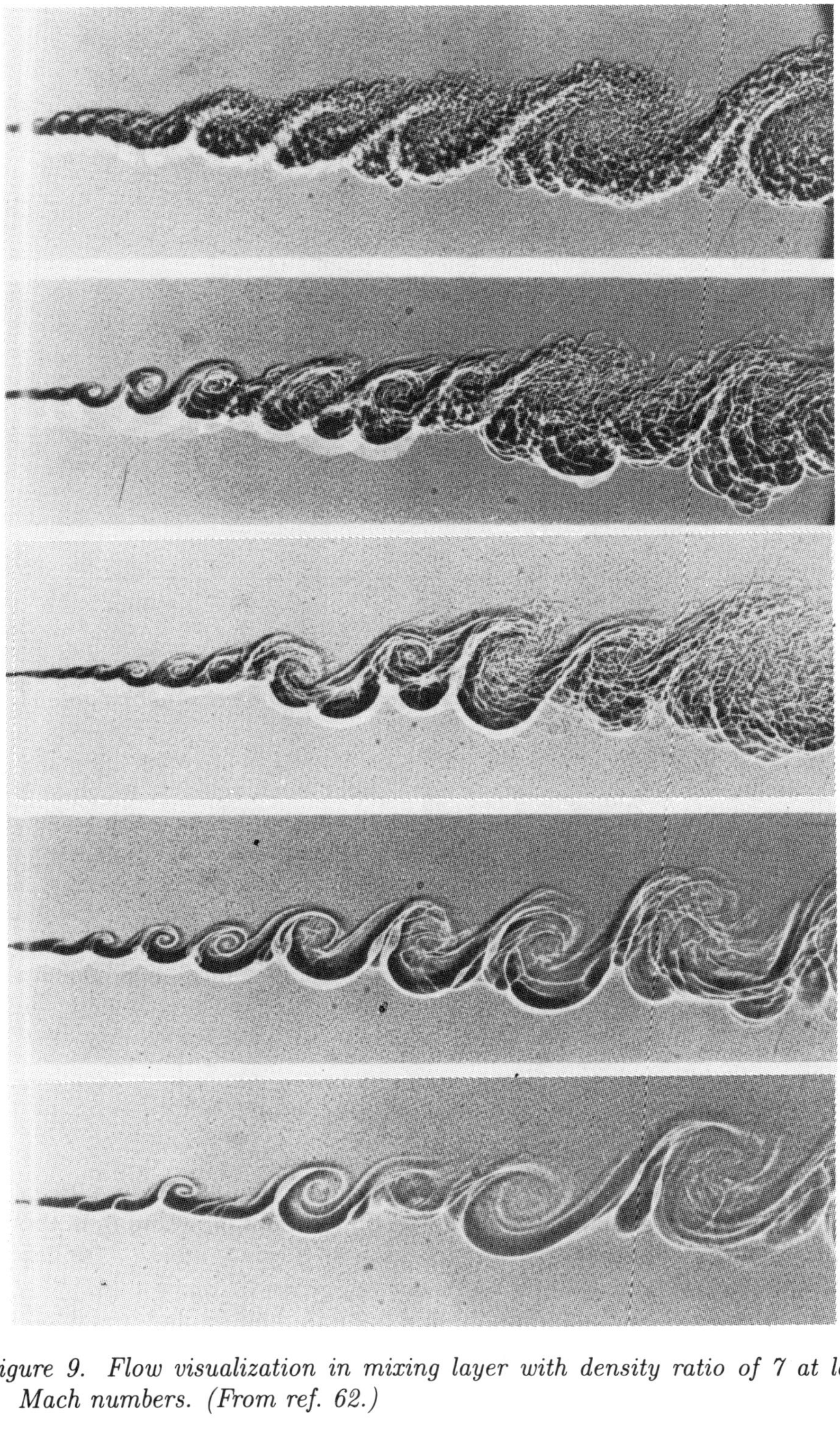

Figure 9. Flow visualization in mixing layer with density ratio of 7 at low Mach numbers. (From ref. 62.)

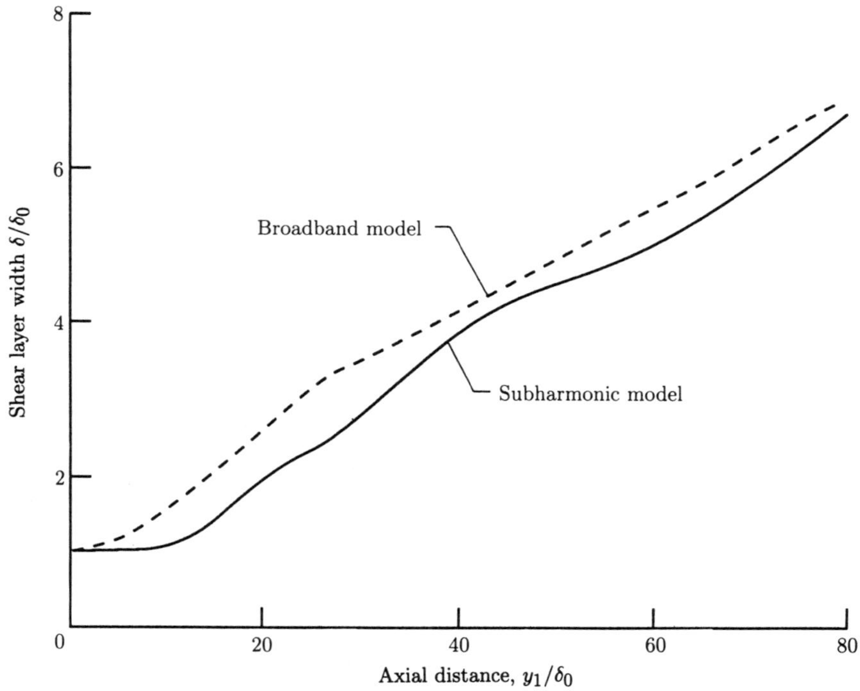

Figure 10. Calculated growth of shear layer at low speeds. (From ref. 61.)

the structure of a jet and a wake far downstream from their respective origins have been noted in reference 31. In the jet the lateral mean velocity, which is directed radially inward, is much greater than that for the wake, and according to Townsend, it restricts the growth of the large eddies. The intermittency of turbulence in a jet is less than it is in a wake. For a two-dimensional high-speed jet issuing into a medium at rest, the spread of the mixing region into the quiescent medium occurs at a faster rate than it does into the high-speed flow (ref. 67).

For the circular jet we can describe certain basic flow properties. Following the work of reference 31, we find that if the Reynolds number is sufficiently high, the initial mixing region may be assumed to be planar, with a mean velocity distribution

$$\frac{U}{U_j} - \lambda = \frac{1-\lambda}{\sqrt{2\pi}} \int_{-\infty}^{\eta+\eta_0} \exp(-x^2/2)\, dx \tag{55}$$

where $\lambda = U_f/U_j$, $\eta_0 \approx 0.33(1-\lambda)/(1+\lambda)$, and $\eta = y_2/l_0^*$. If we assume that the energy-containing eddies have a scale of the order of l_0^*, where l_0^* is about half the width of the mixing region, then the dissipation length scale L_ϵ (as defined by Townsend) is about $3l_0^*$, where the turbulent energy dissipation $\epsilon = (u_0^*)^{3/2}/L_\epsilon$. For the plane mixing region l_0^* increases linearly with y_1, and similar growth occurs for the circular jet issuing into the ambient fluid. For the jet issuing into a moving fluid, with the external speed small compared with the jet exit speed, the growth of the jet is also linear with y_1. When the two speeds become nearly equal the growth is $(y_1 - y_0)^{1/3}$, and such a case occurs asymptotically far downstream when the jet centerline velocity approaches the external velocity. In all these flow cases the flow is self-preserving.

References 68 and 69 show that the length of the potential core L increases with the ratio $\lambda = U_f/U_j$. Similarly, reference 70 shows

$$\frac{L}{D_j} = \frac{4.39}{1 - 0.92\lambda} \tag{56}$$

where D_j is the jet exit diameter. This result was obtained for low speeds, but further experiments show L/D_j increases slowly with increasing Mach number. In addition, the growth of the mixing region can be expressed by

$$b(y_1) = 0.32 y_1 \frac{1 - \lambda}{1 + \lambda} \tag{57}$$

with $l_0^* \approx b(y_1)/2.4$. Since the overall width of the mixing region is not defined with any precision, we will assume in the following applications relating to the determination of the strength of the effective noise sources that to a sufficient approximation, the width of the mixing region at the end of the potential core is equal to the jet exit diameter. The overall growth of the mixing region with downstream distance varies from a model circular jet to a full-scale, straight-jet engine, and the limited experimental data suggest a variation similar to that shown in figure 11.

The intensity of turbulence varies considerably throughout a jet. Typical results from experiments are shown in figure 12 (from ref. 71). These results are for the overall turbulence intensities, which include both the fully turbulent and the nonturbulent components. These components differ markedly from the separate rotational and irrotational components, which arise as a result of the turbulence intermittency. Thus much of the variation of the mean turbulence properties across the jet, as shown in figure 13, arises from the turbulence intermittency, with the result being that in the periods when the flow is fully turbulent, the turbulence intensity distribution is more uniform across the mixing region.

Outside the mixing region in the irrotational fluid, experiments confirm that the fluctuating velocity components decrease as y_2^{-2} at large values of y_2, where y_2 is the distance normal to the boundary of the mixing region.

The Flow Properties Downstream of the Potential Core

The average turbulence properties of the mixing region of a circular jet of diameter D_j in the regions upstream of and downstream of the end of the potential core L are shown in figure 12. Reference 31 shows that these properties depend on the flow constant R_s, which is defined as $|u_0^*| l_0^* / \nu_T$, where ν_T is the eddy viscosity and u_0^* and l_0^* are, respectively, a characteristic velocity and length scale of the mean flow. The mean velocity distribution for the jet downstream of the potential core is given by

$$U = U_f + u_0^* \exp\left(-\frac{y_2^2}{2l_0^{*2}}\right) \tag{58}$$

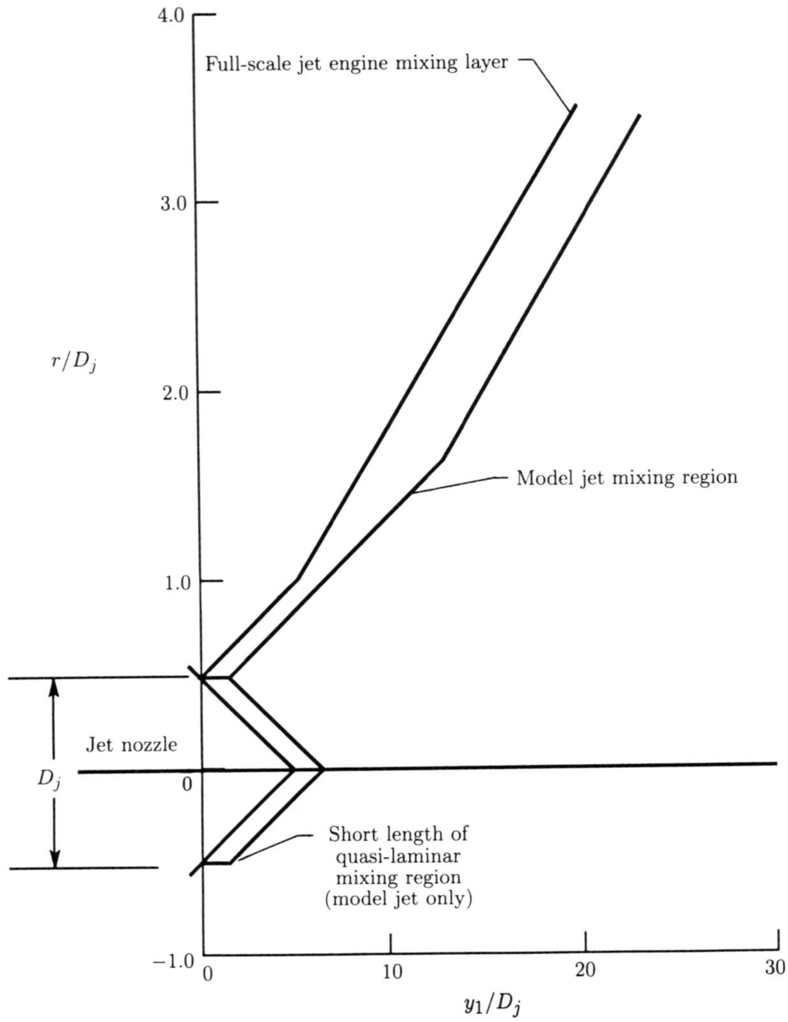

Figure 11. Diagram of composite mean width of mixing region.

where U_f is the velocity of the outer flow and $U_1 > U_f$ is the velocity on the jet centerline. The centerline of the jet is $y_2 = 0$, so that $u_0^* = U_1 - U_f$. If the vorticity thickness is defined as

$$\delta = \frac{U_1 - U_f}{(dU/dy_2)_{\max}} \tag{59}$$

then we find $l_0^*/\delta = 0.520$. According to reference 68, $U_1/(U_j - U_f) = L/y_1$ when $y_1 > L$. When $\lambda = 0$ we obtain $l_0^* = 0.068y_1$.

The Flow Properties in the Initial Mixing Region

For the initial mixing region the transverse distribution of the three normal components of the turbulence velocity is shown in figure 14. In this region the

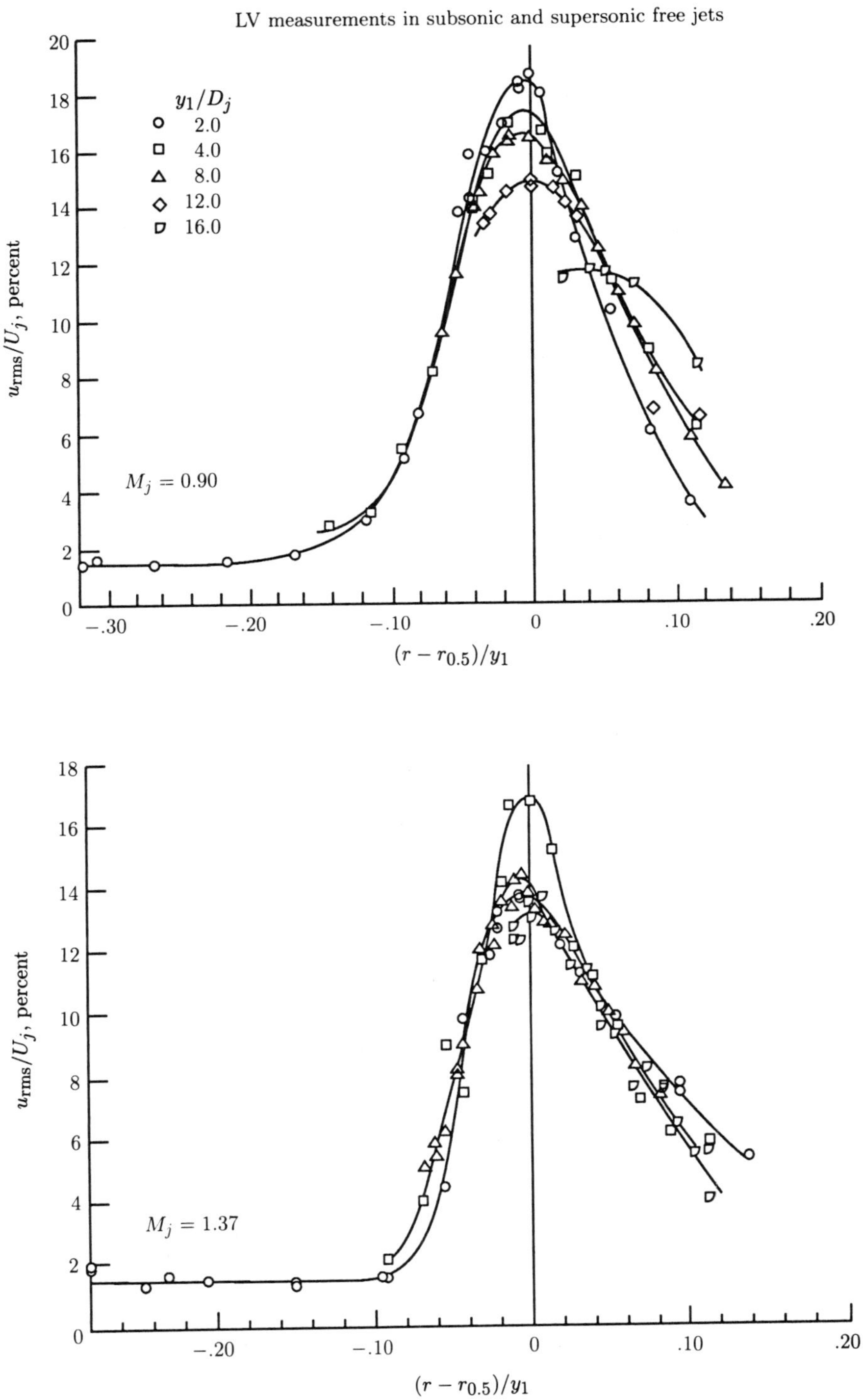

Figure 12. Variation of u_{rms}/U_j across jet at different distances from jet exit at jet Mach numbers of 0.90 and 1.37. (From ref. 71.)

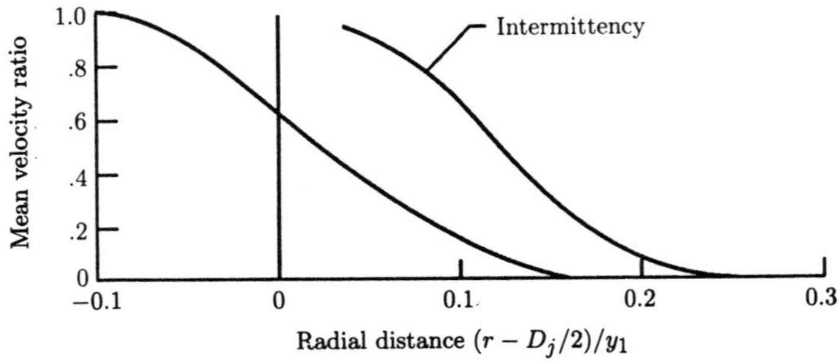

Figure 13. Variation of intermittency factor across mixing region of jet. (From ref. 31.)

mean velocity distribution is given by (see eq. (55))

$$U = U_f + \frac{u_0^*}{\sqrt{2\pi}} \int_{-\infty}^{\eta+\eta_0} \exp(-x^2/2)\, dx \tag{60}$$

where $\eta = y_2/l_0^*$. In this region $u_0^* = U_j - U_f$, and in both regions U_f is the external velocity. For a jet that issues into the ambient medium at rest, $U_f = 0$. We find that $l_0^*/\delta = 0.40$, where again δ is the vorticity thickness. From reference 62,

$$\delta = 0.18 \frac{1-\lambda}{1+\lambda} y_1 \tag{61}$$

with $l_0^* = 0.4\delta$, if we assume the thickness of the shear layer is zero at the jet exit.

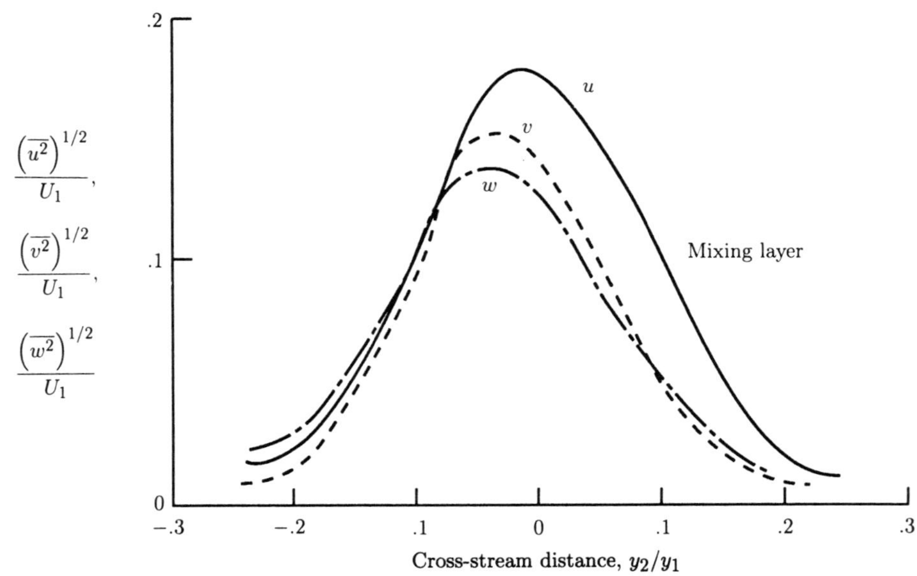

Figure 14. Variation of turbulence velocity components across mixing region of jet. (From ref. 31.)

Reference 31 indicates that $R_s \approx 35.7$ for the plane mixing layer when $\lambda \ll 1$. Therefore,

$$l_0^* = \frac{2}{R_s}\frac{1-\lambda}{1+\lambda}y_1 = 0.056\frac{1-\lambda}{1+\lambda}y_1 \tag{62}$$

However, if we adjust the value of R_s to 30 and make a similar adjustment to the experimental value stated above, the agreement is satisfactory between the results given in reference 31 and the experimental data of reference 62. When $\lambda = 0$, the rate of growth of the initial mixing layer is similar to that of the jet far downstream of the end of the potential core. The overall width of the mixing region is given by $b(y_1) \approx 0.3y_1$ and an average value of $l_0^* \approx 0.1y_1$. For the region downstream of the potential core and considering only the case $\lambda \ll 1$, we find

$$b(y_1) \approx 0.24y_1 \tag{63}$$

The Entrainment Into the Jet

The growth of the width of a jet depends on the entrainment, although both quantities are part of the equilibrium balance imposed on the jet structure by the conservation integral properties of the jet flow and its boundary conditions. If we first consider the special case of incompressible flow and let U_E be the effective average entrainment velocity at the jet boundary, then for self-preserving flow in the jet far downstream of the end of the potential core it follows that U_E is inversely proportional to downstream distance and is only weakly dependent on the velocity distribution across the jet for a given jet and jet thrust.

In the initial mixing layer, which we assume is planar, the entrainment from the high-speed stream differs from that from the ambient medium. Both entrainments are directed toward the mixing layer. If the effective entrainment velocities from the high-speed fluid and ambient fluid are U_+ and U_-, respectively, then we find

$$\frac{U_+}{U_j} = -I_2\frac{dl_0}{dy_1} \qquad \frac{U_-}{U_j} = (I_1 - I_2)\frac{dl_0}{dy_1} \tag{64}$$

where I_1 and I_2 are, respectively, $\int f(\eta)\, d\eta$ and $\int f(\eta)^2\, d\eta$ and the mean velocity distribution is $f(\eta) = (U - U_f)/(U_j - U_f)$. Since η_0 is found from experiment to be -0.03, it follows that the turbulent diffusion into the ambient medium is greater than that into the high-speed flow, a condition that agrees with the results of reference 67.

Further study shows that in all regions of the jet the entrainment is a strong function of the velocity ratio λ, the density ratio ρ_j/ρ_∞, and the jet Mach number M_j. Some typical results for the ratio of mass flux in the jet m_{jet} to mass flux at the jet exit m_j are shown in figure 15 (from ref. 72). The values of $\lambda, \rho_j/\rho_\infty$, and M_j therefore influence the structural parameters of the jet, such as the spreading rate of the jet, the centerline velocity decay downstream of the potential core, and the local turbulent intensity. Thus, as might have been expected, the flow structure of the jet on a full-scale aircraft jet engine in flight may differ dramatically from that of a static model jet tested in the laboratory at ambient temperature.

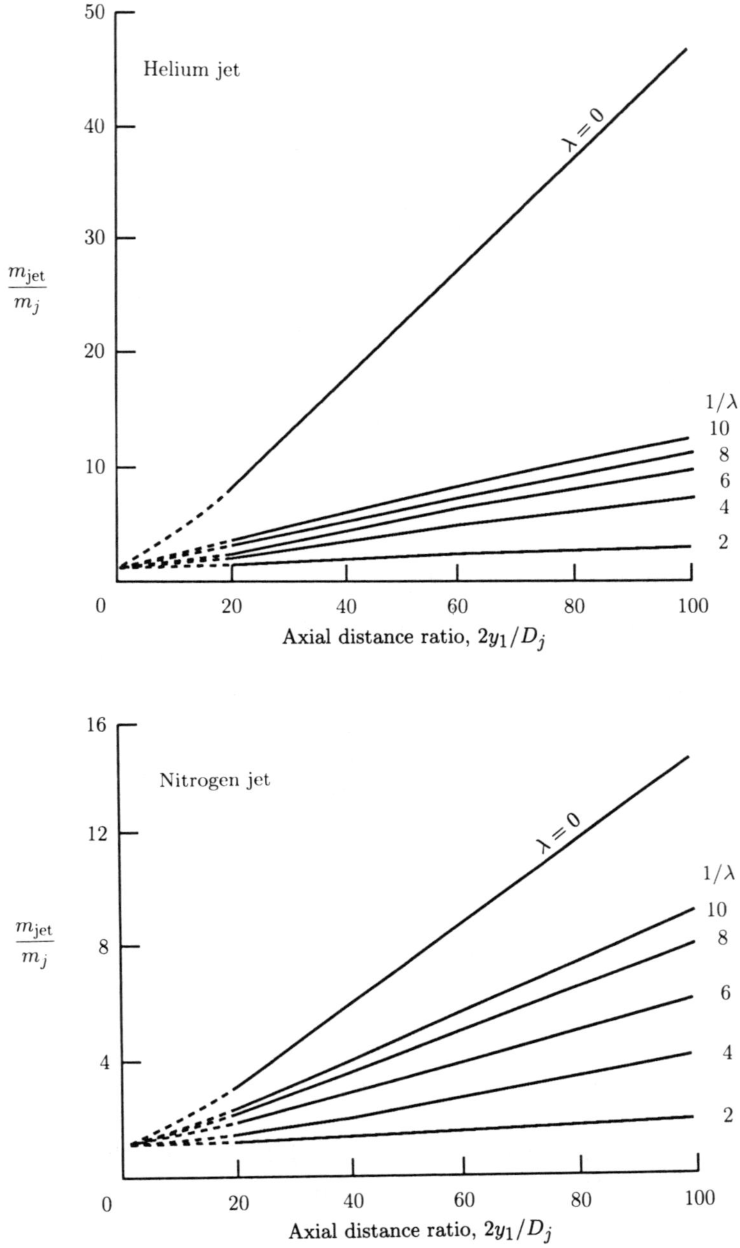

Figure 15. Ratio of mass flux in jet to exit mass flux. (From ref. 72.)

The Properties of the Mixing Region at High Speeds

We have discussed the properties of the mixing region in a jet at low subsonic Mach numbers. References 70 and 71 give information on the changes that occur with an increase in Mach number from subsonic to supersonic shock-free flows. The main conclusions from their results are that, with increasing Mach number, the

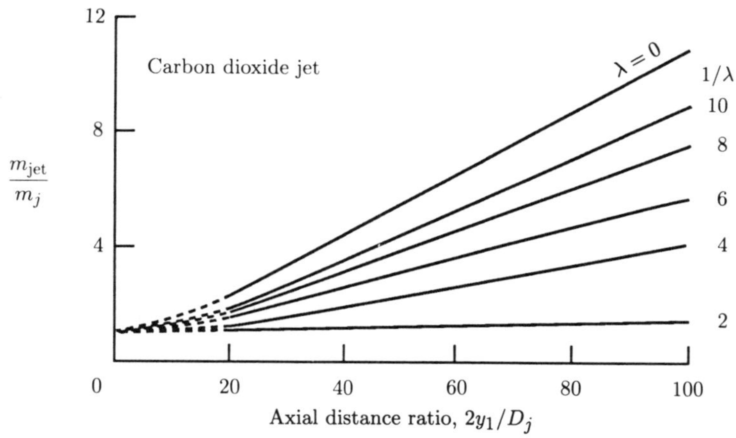

Figure 15. Concluded.

growth of the mixing regions decreases, the length of the potential core increases, and the intensities of the longitudinal and the lateral velocity components decrease. Figure 16 shows the variation of u_{rms}/U_j with distance along the centerline y_1/x_c (x_c is length of the potential core) of a jet at three Mach numbers as measured in reference 71. They obtained similar results for v_{rms}/U_j. The intensity does not fall to zero throughout the potential core, although the level is small compared with the maximum intensity as shown in figure 12. The value of u_{rms}/U_j reaches its peak at nearly twice the length of the potential core and then decays at a rate similar to that found in the self-preserving region farther downstream. The growth of the length of the potential core is shown in figure 17(a), which includes results from reference 71 as well as from other experiments. Figure 17(b) shows the corresponding decrease in the nondimensionalized vorticity thickness δ_η with increasing Mach number. At high Mach numbers, δ_η decreases as $1/M_j$ as M_j tends to infinity (ref. 62).

Concluding Remarks

We have shown in this section that the flow structure in a turbulent circular jet defies simple description even in low-speed flow. At subsonic speeds the jet structure is broadly divided into the initial mixing region, covering the length of the potential core, and a more extensive region downstream. Between these regions is an intermediate region that, although continuous with the upstream and downstream regions, has a non-self-preserving structure, and that structure is not well documented. It is possibly the region contributing most to the radiated acoustic power.

The information we require as input to our model for the noise generation from the turbulent flow includes the mean flow properties of the jet; the instability of the mixing region close to the nozzle exit and its breakup into large-scale vortical structures and, eventually, into fully turbulent flow; the structure of the turbulent flow in all regions of the jet, including its amplitude, length and time scales, and mean speed of convection; the influence of large eddy structures on the growth of the mixing region and the intermittency of the turbulent flow; and the structure of the turbulent flow close to the flow boundaries, its relation to the irrotational flow outside, and the entrainment of that irrotational fluid.

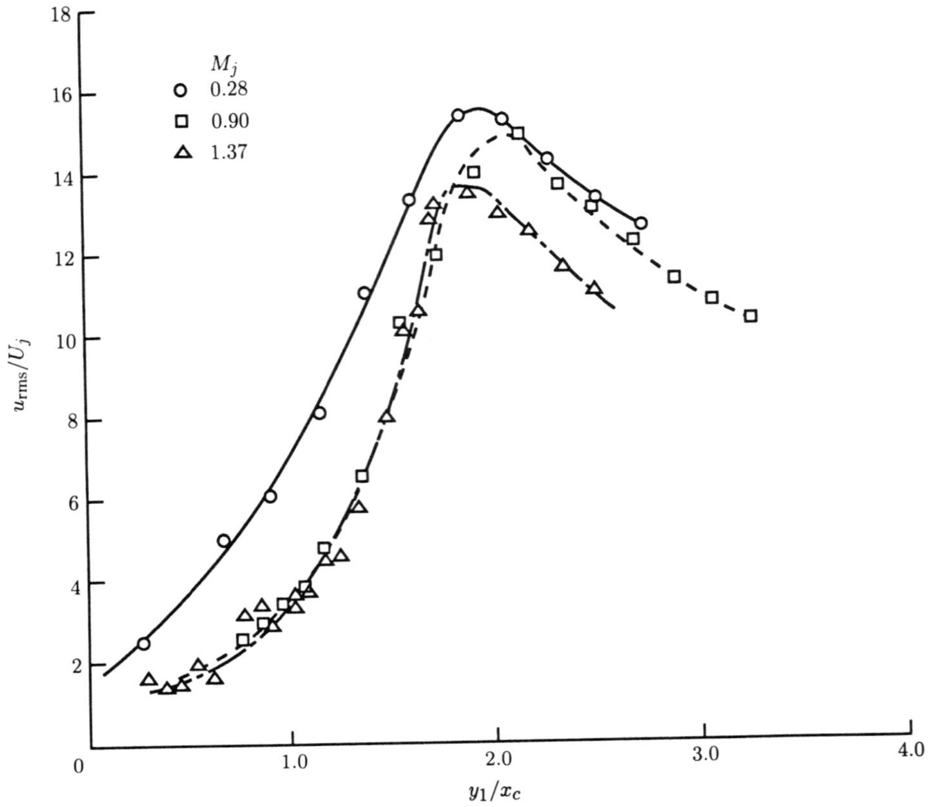

Figure 16. Variation of u_{rms}/U_j along centerline of jet at three Mach numbers. (From ref. 71.)

However, even with this vast experimental data bank, we still need to make a judgement on those regions of the turbulent shear flow that generate the greatest contribution to the radiated noise and their contribution to the amplitude and the length and time scales of the corresponding effective acoustic source function in a moving frame, as required in the Lighthill acoustic analogy. All the information we have included in this section is relevant to the understanding and justification for the parameters we use in the model for the jet noise source function and its distribution. It is this source function that must contain all the details of the convecting turbulent flow, since in Lighthill's acoustic analogy this source function replaces the entire flow. But here we issue a word of caution. The source function involves a moving-frame, fourth-order covariance with spatial and corresponding retarded-time separations with respect to a fixed far-field observer. The experimental data on this covariance are almost nonexistent, and the best we can do is to infer its properties from the experimental data we have already briefly reviewed. The success or failure of our attempts to find a suitable approximation to the source function and its distribution for insertion in Lighthill's acoustic analogy, based on the turbulent structure information, depends on the agreement we finally obtain

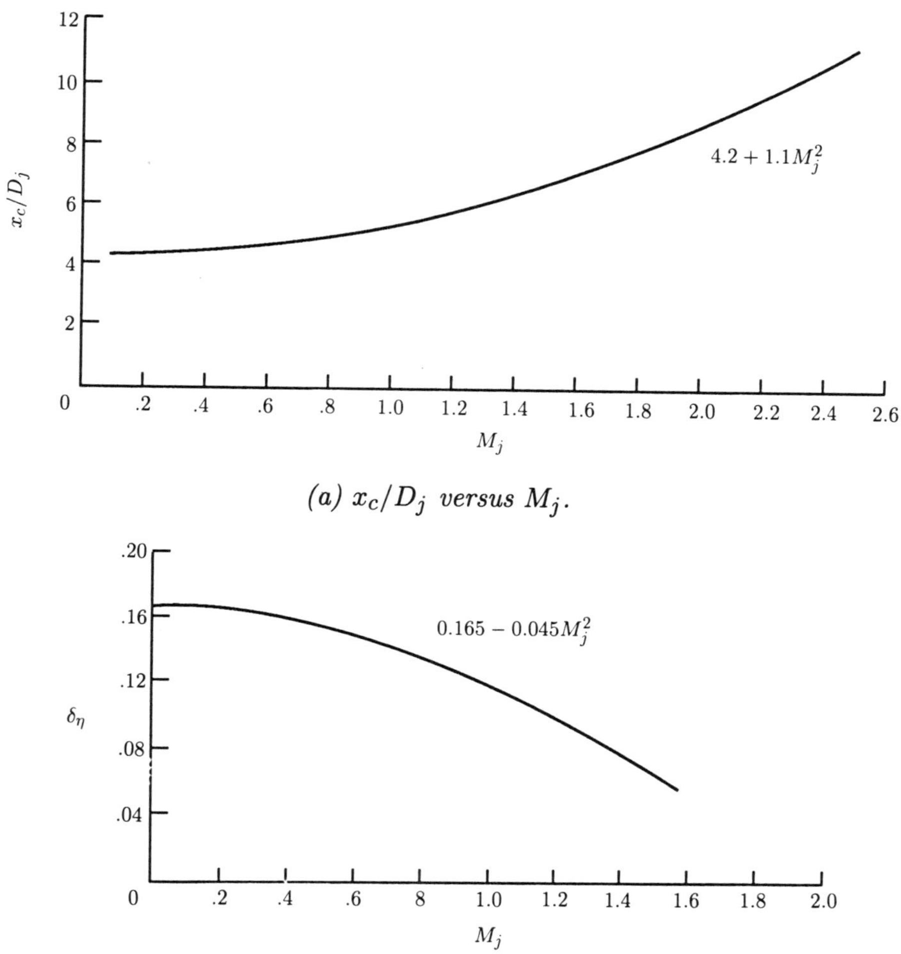

(b) δ_η versus M_j.

Figure 17. Growth of length of potential core and variation of mixing region vorticity thickness with jet Mach number. (From ref. 71.)

between the calculated and experimental characteristics of the radiated noise field. At best we hope to uncover the sources of error in the values of the parameters used in our model, as well as those aspects of the application of the Lighthill acoustic analogy that require further study, through the introduction of the flow-acoustic interaction theories.

The Acoustic Analogy Source Model in Jet Noise

The Acoustic Analogy Equations

In previous sections it has been found that Lighthill's acoustic analogy leads to the following result (eq. (45)) for the autocorrelation of the sound intensity in the far field of a stationary jet when the convection velocity of the bulk of the turbulent flow at any cross-section of the jet is subsonic and the jet, if supersonic, is shock free:

$$I(\mathbf{x}, t^*) \approx \frac{1}{16\pi^2 \rho_\infty c_\infty^5 x^2 |1 - M_c \cos\theta|^5} \iiint d\mathbf{y} \iiint \frac{\partial^4}{\partial \tau^4} P_\theta(\mathbf{y}, \boldsymbol{\delta}, \tau)\, d\boldsymbol{\delta}$$

where P_θ is the source function evaluated in a frame moving at the velocity $U_c(\mathbf{y})$, $\boldsymbol{\delta}$ is the space separation with respect to moving coordinates, and τ is the corresponding retarded-time difference. The corresponding power spectral density (eq.(46)) is

$$\overline{I}(\mathbf{x}, \omega) \approx \frac{\pi}{2\rho_\infty c_\infty^5 x^2} \iiint \omega^4 \overline{P}_\theta(\mathbf{y}, \mathbf{k}, \omega_D)\, d\mathbf{y}$$

where

$$\overline{P}_\theta(\mathbf{y}, \mathbf{k}, \omega_D) = \frac{1}{(2\pi)^4} \iiint \exp(-i\mathbf{k} \cdot \boldsymbol{\delta})\, d\boldsymbol{\delta} \int_{-\infty}^{\infty} \exp(-i\omega_D \tau) P_\theta(\mathbf{y}, \boldsymbol{\delta}, \tau)\, d\tau$$

and ω is the frequency of the radiated noise and ω_D is the frequency of the turbulence in the moving frame.

The term $\overline{P}_\theta(\mathbf{y}, \mathbf{k}, \omega_D)$ is the four-dimensional Fourier transform of $P_\theta(\mathbf{y}, \mathbf{k}, \tau)$, which is assumed to be a symmetric function about the origins of $\boldsymbol{\delta}$ and τ. It is a real function of $\mathbf{k}$ and ω_D, and for each value of ω it is expected to have a maximum amplitude at some position $\mathbf{y}$ within the source volume. This is a reflection of the self-preserving structure of the mixing region of a jet, whereby the dominant frequencies in the turbulence at any station downstream from the jet exit are inversely proportional to the growth of the width of the mixing region. Since the far-field noise spectrum involves an integration of P_θ over the entire source region, it is apparent that in general the contribution to that spectrum within a given frequency band involves only sources located within a small section of the jet, and in particular it involves only that part of their wave-number–frequency spectrum function surrounding $\mathbf{k}$ and ω_D corresponding to ω. Thus the complete determination of the wave-number–frequency spectrum function at each source location in the jet is unnecessary, since it is only the region of the spectrum around the matched values of ω_D and $\mathbf{k}$ that contributes to the far-field noise.

The Model for the Space-Time Correlation Function

In a previous section we also refer to the variation of the turbulent structure across any section of the mixing region, including its intermittency near the jet boundaries. The detailed analysis of reference 43, in which the jet intermittency was neglected, showed the amplitude of P_θ to be distributed across the jet in a Gaussian distribution. However, if we extend that method to include the jet intermittency and, moreover, take account of the large eddy structure in the jet, it appears the source function is likely to be approximately uniform, on average, across the mixing region at any station. On the assumption that the source function distribution is uniform at all stations of the jet at high Reynolds numbers and the length and velocity scales of the turbulence, which determine the properties of the source function, have the self-preserving properties as determined for the incompressible jet, a physically possible form for the source function can be proposed that in turn can be reduced to an effective source function that is a function of the axial coordinate y_1 only for each value of the far-field frequency ω. We refer to this function as the axial source function $R(y_1, \omega)$, where

$$R(y_1, \omega) = \iint \overline{P}_\theta(\mathbf{y}, \mathbf{k}, \omega_D)\, dy_2\, dy_3 \tag{65}$$

Thus, within the elongated source region of a jet we have reduced our problem to the determination of the ensemble average of the turbulent structures that contribute to the space–retarded-time correlation function $P_\theta(\mathbf{y}, \boldsymbol{\delta}, \tau)$ and its integration throughout the source region. Our model, which at best is a gross approximation to the average properties of P_θ, is unlikely to be equal to its value at any one realization of the jet mixing region flow, but then neither should it be so. Few experiments have been performed that relate to the fourth-order covariance $P_\theta(\mathbf{y}, \boldsymbol{\delta}, \omega_D)$, even for zero time delay. Hence, a detailed comparison with experiment with respect to the source structure is not possible. However, indirect comparisons are possible through the far-field noise results and use of the polar correlation technique, as discussed subsequently.

From reference 43, following references 45 and 73, we find typical curves for the pressure and time-gradient pressure space correlations in a free shear layer, and these are shown in figure 18. The longitudinal correlation has large negative values for large separation distances σ, whereas the transverse correlations are positive for all separations. Similar curves might be expected for the space separation properties for the covariance P_θ. But the moving-axis retarded-time curves of the covariance P_θ are more likely to have a shape similar to the envelope of the space-time correlation curves for the turbulent velocity as discussed in the section *The Structure of a Turbulent Jet*, so that P_θ is predominantly positive except at very long separation distances. Even allowing that the true space-time properties of P_θ have positive and negative regions, the sextuple-weighted integral of its fourth time derivative smears out most of these complex details, as found in reference 43, and leaves the function $R(y_1, \omega)$ heavily weighted in terms of the properties of the characteristic values of the turbulence velocity and length scales.

We accordingly define $P(\mathbf{y}, \boldsymbol{\delta}, \tau)$ in terms of the moving-frame turbulence quantities $\rho_0, u_0, l_1, l_2, l_3$, and ω_0, all of which vary with y_1 only. These quantities are, respectively, the characteristic mean density, the root-mean-square turbulent velocity, the turbulence length scales in directions y_1, y_2, and y_3, and the moving-frame frequency. We write

$$P_\theta(\mathbf{y}, \boldsymbol{\delta}, \tau) = \rho_0^2 u_0^4 f\left(\mathbf{y}, \frac{\delta_i}{l_i}, \omega_0 \tau\right) \tag{66}$$

If $f(\mathbf{y}, \boldsymbol{\delta}, \tau)$ is assumed to be equal to a Gaussian distribution for the space–retarded-time separations,

$$R(y_1, \omega) = \iint \frac{\rho_0^2 u_0^4 l_0^3}{16\pi^2 \omega_0} \frac{l_\perp^2}{l_0^2} \exp\left(-\frac{\omega^2 l_0^2 C_0}{4 N_{\mathrm{St},t}^2 u_0^2}\right) dy_2\, dy_3 \tag{67}$$

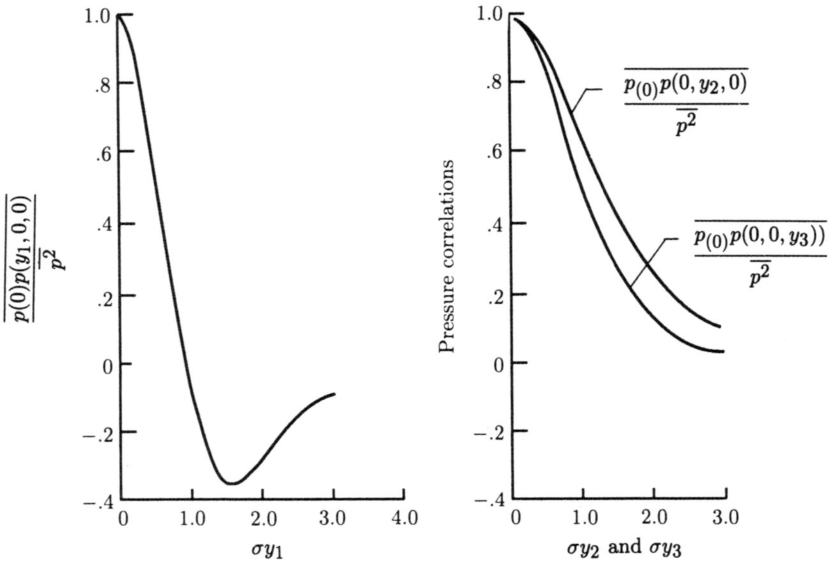

(a) Pressure correlation.

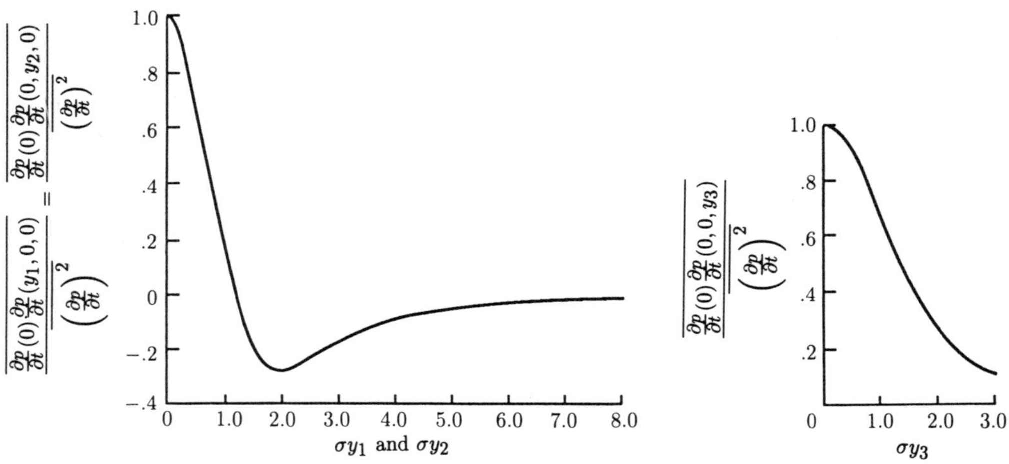

(b) Time-gradient pressure correlation.

Figure 18. Pressure and time-gradient pressure correlations in isotropic turbulence in presence of mean shear. $f(x) = \exp(-\sigma^2 x^2)$.

where the effective Doppler factor is $\sqrt{C_0}$, with

$$C_0 = (1 - M_c \cos\theta)^2 + \frac{N_{\mathrm{St},t}^2 u_0^2}{c_\infty^2}\left(\cos^2\theta + \frac{l_\perp^2}{l_0^2}\sin^2\theta\right) \tag{68}$$

and is finite when $M_c \cos\theta = 1$. The transverse scales of turbulence l_2 and l_3 have been assumed to be equal and replaced by $l_\perp$. The longitudinal scale l_1 has been replaced by l_0. The Strouhal number of the turbulence in a moving frame $N_{\mathrm{St},t} = \omega_0 l_0 / u_0$ is assumed to be a constant throughout the entire jet flow. The angle θ between the observer, relative to the jet exit, and the jet axis is positive when measured in the downstream direction.

The effective cross-sectional area of the jet, over which we assume P_θ to be nearly constant, is given by

$$\iint dy_2\, dy_3 = \left\{ \begin{array}{ll} \pi D_j b & (0 < y_1 < L) \\ \pi b^2 & (L \leq y_1 < \infty) \end{array} \right\} \tag{69}$$

where the first region covers the initial mixing region, where the mixing layer is almost planar since its width is small compared with the downstream distance, and the second region covers the entire jet downstream of the potential core $y_1 = L$. The width of the mixing region b is taken as the mean overall width, as described in the previous main section. From those results we find b/l_0 is a constant throughout the entire jet.

The Strouhal number of the radiated noise is $N_{\mathrm{St},r} = \omega D_j / U_j$, where U_j is the mean jet exit velocity and D_j is the jet exit diameter. The (acoustic) Mach number of the jet is $M_j = U_j / c_\infty$, and M_c is the (acoustic) convection Mach number, which is a function of y_1.

The Model Equation for the Power Spectral Density

The power spectral density of the far-field noise is found by substitution of these results into equation (46), giving

$$\begin{aligned} \overline{I}(x,\theta,\omega) \approx \rho_\infty c_\infty^3 \frac{D_j}{U_j} \frac{D_j^2 M_j^8 N_{\mathrm{St},r}^4}{32x^2} & \left[\int_0^{L/D_j} \frac{\overline{\rho}_0^2\, \overline{u}_0^3\, \overline{l}_0^5}{N_{\mathrm{St},t}} \frac{b}{l_0} \exp\left(-\frac{N_{\mathrm{St},r}^2 \overline{l}_0^2 C_0}{4\overline{u}_0^2 N_{\mathrm{St},t}^2} \right) d\overline{y}_1 \right. \\ & \left. + \int_{L/D_j}^{\infty} \frac{\overline{\rho}_0^2\, \overline{u}_0^3\, \overline{l}_0^6}{N_{\mathrm{St},t}} \frac{b^2}{l_0^2} \exp\left(-\frac{N_{\mathrm{St},r}^2 \overline{l}_0^2 C_0}{4\overline{u}_0^2 N_{\mathrm{St},t}^2} \right) d\overline{y}_1 \right] \end{aligned} \tag{70}$$

where $\overline{y}_1 = y_1/D_j, \overline{u}_0 = u_0/U_j, \overline{l}_0 = l_0/D_j$, and $N_{\mathrm{St},t}$, b/l_0, and $l_\perp / l_0$ are constants.

The Model Equation for the Intensity

The intensity is found by integrating over all frequencies:

$$\begin{aligned} I(x,\theta) \approx \frac{3\sqrt{\pi}}{4} \rho_\infty c_\infty^3 D_j^2 M_j^8 & \left(\int_0^{L/D_j} \frac{\overline{\rho}_0^2 \overline{u}_0^8 N_{\mathrm{St},t}^4}{C_0^{5/2}} \frac{b}{l_0} \frac{l_\perp^2}{l_0^2}\, d\overline{y}_1 \right. \\ & \left. + \int_{L/D_j}^{\infty} \frac{\rho_0^2 u_0^8 N_{\mathrm{St},t}^4}{C_0^{5/2}} \frac{b^2}{l_0^2} \frac{l_\perp^2}{l_0^2} \overline{l}_0\, d\overline{y}_1 \right) \end{aligned} \tag{71}$$

where the integrands have the same value at $y_1 = L$. In both regions the turbulence longitudinal length scale l_0 increases linearly with y_1. In the initial mixing layer u_0 is constant, but downstream of the potential core ($y_1 = L$), u_0 decays inversely proportional to y_1. All these assumptions are consistent with the assumptions of self-preserving flow in both regions and with the low-speed experimental data referred to in the next main section. We find, accordingly,

$$I(x,\theta) \approx \frac{3\sqrt{\pi}}{4}\rho_\infty c_\infty^3 D_j^2 M_j^8 \left[\frac{\bar{\rho}_0^2 \bar{u}_0(L)^8 N_{\mathrm{St},t}^4 (b/l_0)(l_\perp/l_0)^2(L/D_j)}{C_0^{5/2}}\right]\left[1 + \frac{(b/l_0)\bar{l}_0(L)}{6}\right] \tag{72}$$

where, for convenience only to keep the final result as simple as possible, we have assumed ρ_0 and C_0 retain their values at the end of the potential core throughout the downstream jet mixing region. This is justified because the region between the initial mixing region and that downstream of the potential core is continuous and the downstream integral is heavily weighted to the properties of the flow in this "intermediate" region. The quantity $(b/l_0)(l_0/D_j)/6$ represents the ratio of the total acoustic power generated in the region downstream of the potential core to that generated upstream. The ratio is of the order 1/6. Thus we conclude that the initial mixing region is the dominant noise-generating region in a jet when the jet exit velocity is subsonic, and possibly when it is supersonic, in the absence of "shock cells," and provided the average convection velocity is subsonic also.

The Changes in the Model for Supersonic Flow

When the jet is supersonic the structure of the initial mixing region changes, although a potential core still exists if the jet is shock free. The length of the potential core, however, is increased compared with its value in a subsonic flow. When the jet is underchoked or overchoked, the potential core is transformed from a uniform flow at the jet exit, as in subsonic flow, to a flow containing the shock-wave expansion system and extending for a distance from the nozzle exit until the velocity on the axis becomes subsonic. The initial mixing region, as shown in figure 2(b), grows at a slower rate and reflects the structure of shock and expansion waves. Experiments suggest its length increases as a function of the "fully expanded" Mach number of the jet at the exit when the jet is underchoked or overchoked. In these flow cases the large-scale structure of the jet dominates the mixing region and interacts with the shock cell structure.

The Lighthill theory, as applied in equation (70), continues to provide an input to the estimate of order of magnitude for the total acoustic power radiated from a jet, even when the jet is supersonic, provided the T_{ij} covariance reflects, to some approximation, the true flow properties. Thus, in principle, the Lighthill theory can include shock-associated noise and screech tones, although alternative theories presented in another chapter are better adapted to that purpose since they are based on flow-acoustic interaction. We can argue here that if the T_{ij} covariance is a continuous function everywhere in the mixing region, even when shock and expansion waves are incident to it, then the analysis above can be used with only minor changes to the properties of the flow quantities. We introduce L_s, the length of the supersonic region, to replace L, the length of the potential core. We further assume that the characteristic frequency of the turbulence ω_0 is approximately constant over the entire length of the initial mixing region, a reflection of the presence of a large-scale coherent motion. The mean speed and density at the commencement of the subsonic region are those satisfying adiabatic conditions between the nozzle exit and the termination of the supersonic region. With these simple modifications incorporated into the acoustic analogy theory we find the initial mixing region no longer dominates the generation of acoustic power, as it does in the case of the subsonic jet. In the supersonic case approximately half the overall acoustic power is generated by the initial, "supersonic" mixing region, and the remainder comes from the wholly subsonic region downstream of the supersonic region and terminated at $y_1 = L_s$. In both cases the region of maximum acoustic power generation, and in particular the peak in the spectrum at angles near 90°, is that region lying between the upstream and downstream regions. Such a model was first proposed in reference 74, and provided the jet gas properties and the jet exit velocity and temperature are included, the overall acoustic radiated power can be predicted satisfactorily over a very wide range of jet Mach numbers.

Limitations of Model

The results given in equations (72) and (70) for the far-field noise intensity and power spectral density at an observer based at $Q(x, \theta)$ are derived entirely from Lighthill's acoustic analogy with a physically plausible model used for the source function and based as far as possible on relevant experimental data on the structure of turbulence in the mixing region of a jet. The source function used in this section is based on a volume distribution of moving quadrupoles representing the unsteady flow field in a turbulent jet at high Reynolds numbers. These results need to be modified to account for the presence of additional dipole sources at low Mach numbers, with a noise intensity proportional to M_j^6, when the jet is heated to well above ambient temperatures as would be the case for the full-scale jet engine.

The results as given in this section apply only to the static jet and need modification when applied to the jet in flight. The first modification concerns the changes in the structure of the turbulent mixing region, both the intensity of the turbulence and its scale, when the jet mounted on an aircraft is in motion with a velocity U_f in the opposite direction to the jet efflux at a velocity U_j. These velocities have been discussed previously, and it was shown that the turbulent structure depends on $\lambda = U_f/U_j$. The reference density ρ_0 of the fluid within the moving eddy structures responsible for noise generation is also a function of λ as well as of the ratio of the jet to ambient temperature. This is discussed in the next main section. The second modification concerns the additional Doppler effect experienced by the observer because of the motions of the downstream convecting eddies and the bodily motion of the entire jet in a direction upstream as observed by the observer. The result, as first presented in reference 39, requires the additional term $|1 + M_f \cos\theta|^{-1}$ in both the intensity and the power spectral density.

Concluding Remarks

We can draw some interesting conclusions from the results given by equations (70) and (72). The first concerns the effective source distribution along the axis of a low-speed static jet, as shown in figure 19. In the initial mixing region the overall effective source strength is constant, but downstream of the end of the potential core the strength falls as $1/y_1^7$. This result was found by Lilley (ref. 43) and independently by Ribner (ref. 75).

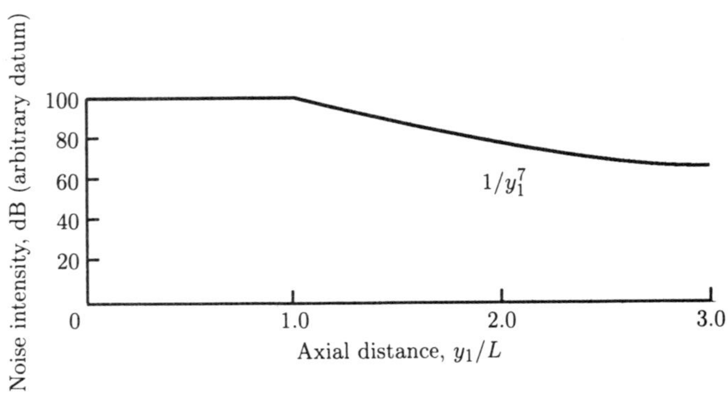

Figure 19. Jet noise effective source distribution at low Mach numbers for static jet.

In figure 20 the corresponding results are given for the effective source distribution along the jet axis for a series of constant values of the far-field Strouhal number. These results show that the high-frequency end of the far-field spectrum is generated almost entirely from turbulence in the initial mixing region, whereas the low frequencies are generated over a very large region of the jet extending far downstream. The region of most intense radiation is near the end of the potential core and is centered at $N_{\mathrm{St},r} = 2.0$. (The Strouhal number here is $N_{\mathrm{St},r} = \omega D_j/U_j$, where $\omega = 2\pi f$.) In summary, we see that the main contribution to the power spectral density for Strouhal numbers from 0.1 to 2.0 comes from the region $y_1/D_j = 5$ to 20, while for Strouhal numbers greater than 2.0 the region of greatest contribution stretches from $y_1/D_j = 0$ to 5. In the region near the end of the potential core the dominant frequency has values of $\omega D_j/U_j = 0.3$ to 0.5. Although the low-frequency noise-generating region is spread over a very large region of the jet downstream of the potential core, its contribution to the total far-field noise power is small.

We see from figure 20 that although the shape of the source distribution curves depends on the choice of the Gaussian distribution for P_θ, the envelope through the peaks is more or less independent of the function approximating P_θ. Moreover it is the envelope through the peaks that determines the power spectral density. Thus we need only choose, or derive, a form for P_θ that includes all the physical variables of the turbulent flow and satisfies certain simple boundary conditions with respect to its variation over $\boldsymbol{\delta}$ and τ. Our answer will then be qualitatively correct and the quantitative error in terms of the far-field noise prediction will be almost negligible. However, it would not be permissible to replace the distributed acoustic sources by a single effective source. If this were done gross errors are likely to be present, since it has been shown that the properties of the far-field noise are highly dependent on the spatial properties of the characteristic length and time scales within the entire mixing region.

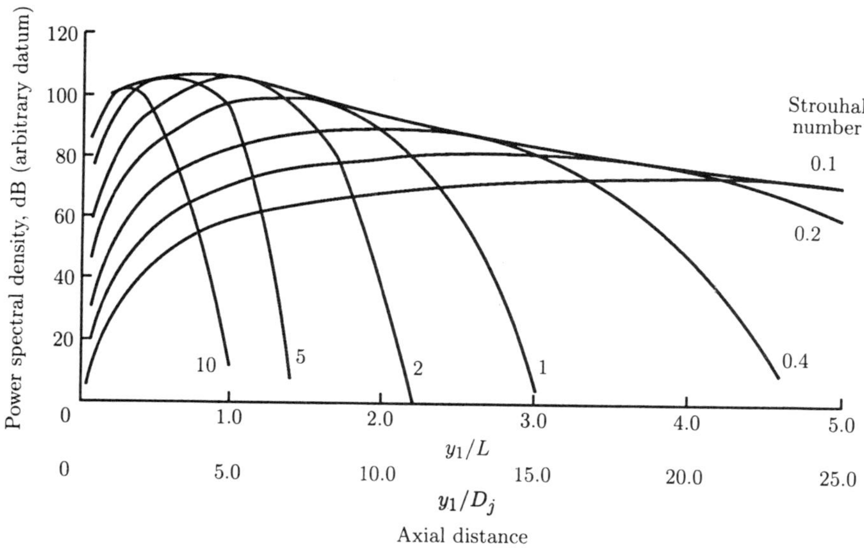

Figure 20. Jet noise effective source distribution at low Mach numbers for constant Strouhal number.

Jet Noise at Subsonic Speeds

Introduction

The results obtained in the previous section are used in this section to determine the far-field noise from a jet at subsonic speeds. (As discussed previously, the subsonic model, with some modifications, may also be used to provide an estimate of order of magnitude for the overall radiated acoustic power from a jet at supersonic speeds. However, since the model does not include details of the jet Mach wave and shock-wave structures, it is expected that the results would become less and less reliable as the jet Mach number is increased, especially where the convection Mach number is sufficiently above unity for Mach wave radiation to persist in regions well outside the jet.) However, our simple formula can give results over a very wide speed range and for different jet gases, and when these results are compared with the few available experimental data the agreement is surprisingly, and perhaps fortuitously, good. As stated previously, our results for the jet at subsonic speeds are not applicable to the heated jet at low Mach numbers, since the additional dipole source has not been included. The necessary extensions to include this case can easily be made with the information on the dipole term contained in the section *Lighthill's Theory of Aerodynamic Noise.* More accurate prediction methods are available, but these are based on applications of the flow-acoustic interaction theory.

The prediction of the characteristics of the far-field noise from a jet based on Lighthill's acoustic analogy and using the particular source function derived in the previous main section depends on the specification of a number of quantities concerning the properties of the jet and the surrounding medium. These are as follows:

Jet:

ρ_j	density at exit, kg/m^3
U_j	velocity at exit, m/sec
A_j	area at exit, $\frac{\pi D_j^2}{4}$, m
γ_j	ratio of specific heats
T_j	temperature at exit, K
m_j	molecular weight
h_j	enthalpy at exit, $C_{p,j}T_j$
c_j	speed of sound, m/sec
M_j	Mach number, U_j/c_∞

Ambient medium:

ρ_f	density, kg/m^3
U_f	velocity of flight, m/sec
T_f	temperature, K
γ_f	ratio of specific heats
m_f	molecular weight
c_f	$= c_\infty$, speed of sound, m/sec
h_f	enthalpy
M_f	Mach number, U_f/c_∞

Flow-Acoustic Interaction at High Frequencies

A simple result from a study of flow-acoustic interaction is that at high frequencies sound generated within the flow field is refracted according to simple acoustic ray theory (Snell's Law), so that in the real flow, for an angle of emission θ_c, sources convected with velocity U_c generate sound rays that are refracted by the flow. The result is that the directivity of the radiated sound θ_f is obtained from

$$\frac{c_f}{\cos\theta_f} + c_\infty M_f = \frac{c_c}{\cos\theta_c} + c_\infty M_c \tag{73}$$

Hence, sound directed at emission along the jet axis ($\theta_c = 0°$) is refracted to θ_{crit}, so at high frequencies a "zone of silence" forms because no high-frequency sound enters the far field in the range $0 < \theta_f < \theta_{\text{crit}}$. Strictly in applications of Lighthill's acoustic analogy it is wrong to apply any correction to account for refraction, since this phenomenon is already included in the definition of the source strength T_{ij}.

However, in our description of the source strength distribution P_θ we have only included the contributions from the unsteady flow field and not the effects of the sound waves generated by it. The amplitude of P_θ would not have been changed significantly by their inclusion, but a change in directivity would have resulted, since the wave-number vector of the turbulence would no longer equal the wave-number vector of the sound in the far field. Thus the directivities $I(x,\theta)$ and $\overline{I}(x,\theta,\omega)$, as calculated in equations (70) and (72), need some correction to allow for this resulting change in the sound wave-number vector from the turbulence wave-number vector. This is part of the analysis in the study of flow-acoustic interaction, but it is not our purpose to infer results from that theory in the results we present here. Let us simply present the results obtained with Lighthill's acoustic analogy uncorrected for flow-acoustic interaction, except that we will assume a zone of silence exists for values of $\theta_f < \theta_{\text{crit}}$. From equation (73) we find θ_{crit} corresponding to $\theta_c = 0°$, with

$$\cos\theta_{\text{crit}} = \frac{1}{M_c - M_f + \sqrt{\gamma_j T_c m_f / \gamma_f T_f m_j}} \tag{74}$$

on the assumption $\gamma_c = \gamma_j$ and $m_c = m_j$.

The convection velocity U_c of the effective sources of sound relative to an observer at rest, when the jet is in motion at the flight velocity U_f and the jet exit velocity U_j, is given by

$$M_c = \frac{U_c}{c_\infty} = K(M_j - M_f) \tag{75}$$

where K is a constant that we will set equal to 0.62, a suitable average value based on reference 29. (The value K is strictly a function of the frequency of the turbulence.) The ambient speed of sound c_∞ is equal here to c_f. When $U_f = 0$ we find that the convection Mach number $M_c = 1.00$ when $M_j = 1.61$, equivalent to a true exit Mach number of 2.32 for an unheated jet.

Specification of the Flow Properties

The results of turbulence intensity measurements in the mixing regions of a jet suggest a strong dependence on $\lambda = U_f/U_j$. An average result for the characteristic turbulence velocity u_0 follows reference 70:

$$\frac{u_0}{U_{\text{axis}}} = \left(\frac{u_0}{U_{\text{axis}}}\right)_{\lambda=0} (1-\lambda)^{0.7} \tag{76}$$

We assume this result holds for all λ and M_j. The value of u_0/U_j when $U_f = 0$ is assumed equal to 0.275 at subsonic Mach numbers. The experimental evidence, reviewed previously, shows that u_0/U_j decreases with increasing Mach number, but the data are sparse, especially for the heated jet. The value of u_0/U_j, when $\lambda = 0$, must be selected from the available experimental evidence for the prescribed test conditions.

The value of ρ_0 is defined as the mean density in the mixing region corresponding to the position where the mean velocity is equal to U_c. Thus, ρ_0 is linked with h_0

and U_c. The mean flow equations for a gas having a Prandtl number of unity lead to

$$h_0 + \frac{U_c^2}{2} + \chi U_c = h_f + \frac{U_f^2}{2} + \chi U_f = h_j + \frac{U_j^2}{2} + \chi U_j \tag{77}$$

where χ is a constant. From equation (77) we find

$$\chi = -\frac{c_\infty \left[\frac{(h_j/h_f)-1}{\gamma_f - 1} + \frac{M_j^2 - M_f^2}{2}\right]}{M_j - M_f} \tag{78}$$

and

$$\frac{h_0}{h_\infty} = \frac{\left\{1 + \left[(\gamma_f - 1)/2\right] M_j^2 (1-\lambda)(K - K\lambda - \lambda)\right\} + (h_j/h_f)(K - K\lambda - \lambda)/(1 - K + K\lambda)}{(1-\lambda)/(1 - K + K\lambda)} \tag{79}$$

with $\rho_0/\rho_\infty = (h_f/h_0)(\gamma_0/\gamma_f)(\gamma_f - 1)/(\gamma_0 - 1), \lambda = M_f/M_j,\ M_c = K(M_j - M_f)$, and $\gamma_0 = (\gamma_f + \gamma_j)/2$. As stated previously, we assume $l_1 = l_0$ and $l_2 = l_3 = l_\perp$, and we put $l_\perp/l_0 = 0.3$.

The length of the potential core is found from

$$\frac{L}{D_j} = \frac{(L/D_j)_{\lambda=0}}{1 - 0.92\lambda} \tag{80}$$

as given in reference 70.

The width of the mixing region is given by

$$\frac{b}{l_0} = \left(\frac{b}{l_0}\right)_{\lambda=0} \tag{81}$$

where $(b/l_0)_{\lambda=0} = 3.0$.

All the turbulent parameters in our source model have been based on the low-speed turbulent properties of the jet, although the changes with Mach number can be included based on the results given in the section entitled *The Structure of a Turbulent Jet.* Most of the results we present below are based on the model of the low-speed properties of the source in order to present the Lighthill acoustic analogy in its simplest possible form for comparison with experimental data. At subsonic convection velocities the changes with Mach number in $u_0/U_j, b/l_0, l_\perp/l_0$, and L/D_j in the initial mixing region result in small changes to the values of the intensity and the power spectral density as obtained from equations (70) and (72). Unfortunately, we have no information on whether or not the measured changes in the properties of the turbulent flow with Mach number apply equally to the space–retarded-time covariance of T_{ij}. We prefer to leave these possible refinements for future study, noting that without more accurate data, our model for the covariance is at best a very crude approximation.

Discussion of Results and Comparison With Experiment

Figure 21 shows the change in overall acoustic radiated power with increase in jet exit Mach number. We see that at subsonic Mach numbers the Lighthill jet velocity eighth power law is in fair agreement with reference 76. At supersonic jet exit Mach numbers the results conform to the Ffowcs Williams-Lighthill convection theory, which asymptotically reduces to the jet velocity third power law. Again the agreement with experiment is fair.

An examination of the spectrum, shown in figure 22, predicted with the subsonic theory using the Gaussian approximation for the fourth-order covariance P_θ and $\theta = 90°$, where convective amplification is absent, shows that its slopes at high and low frequencies are $1/f^2$ and f^2, respectively, and agree with the experimental results of reference 76. However, the peak in the spectrum is slightly displaced and is more prominent than that found experimentally. The reasons for these discrepancies are not difficult to find. The model shows that the $1/f^2$ condition arises from the upstream mixing region and the f^2 condition arises from that mixing region downstream of the potential core. The strengths of the resulting sources in the acoustic analogy theory depend on the turbulent properties prescribed in these two domains. But the turbulence is continuous in structure throughout the intermediate region between these two major mixing regions, and the characteristic turbulence velocity does not decrease discontinuously as in the model. In addition, the rate of growth of the mixing region changes continuously from upstream to downstream, and this variation has not been included adequately in the model. The downstream region is perhaps better modeled, whereas the upstream region is more variable and depends critically on the flow conditions at exit and on the thickness at the commencement of the initial mixing region, including the region occupied by transition from laminar to turbulent flow. In extreme cases the contributions from the two regions to the noise spectrum could become more separated, as shown in figure 23, where the spectrum peak is not only broader but also has a pronounced depression, a reflection of the decay in the strengths of the downstream sources toward high frequency and in the upstream sources toward the low frequencies. Some experiments, such as those of reference 77, confirm this type of behavior under certain jet conditions, although the dips in the spectra are only just outside the limits of experimental accuracy. (The spectra shown here are those for the spectral density and not those for the 1/3-octave or octave band levels, which obviously display different characteristics.)

Figure 24, taken from reference 76, shows the changes that occur in the spectrum at smaller angles to the jet axis. At $\theta = 15°$ the high-frequency content is reduced and no longer displays the $1/f^2$ dependence. The frequency for peak intensity is no longer dependent on the Strouhal number and is almost independent of jet velocity. At larger angles to the jet axis, such as $\theta = 90°$, the Strouhal number dependence for the peak-intensity frequency is regained. (See fig. 25.) The loss of high-frequency sound at small angles to the jet axis is a result of strong flow-acoustic interaction in the initial mixing region, with the result that this contribution to the far-field noise is preferentially radiated at larger angles to the jet axis, and a zone of silence in the higher frequencies is generated near the jet axis. The remaining contribution to the high-frequency sound is generated farther downstream, where its source strength is smaller. However, the overall acoustic power is not affected by this refraction of the high-frequency sound, since little sound is lost by absorption within the flow

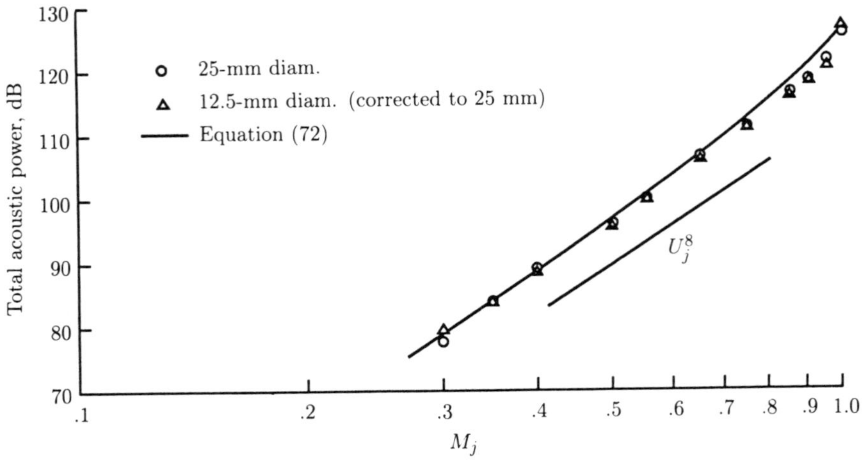

(a) Data from reference 76.

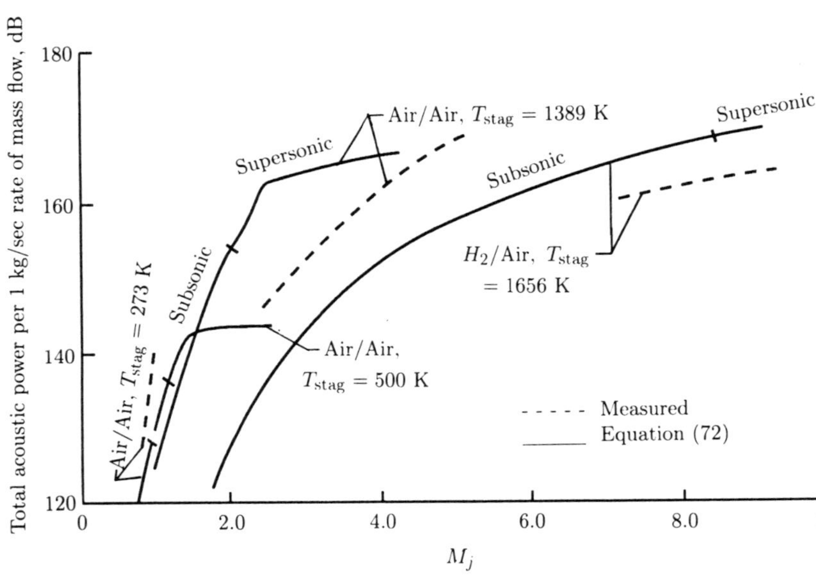

(b) Data from reference 74.

Figure 21. Variation of calculated and experimental acoustic power with jet exit Mach number.

field. The overall changes in sound generation and propagation within the flow field resulting from flow-acoustic interaction are discussed in another chapter.

Figure 26 shows the results for a velocity external to the jet, analogous to the case of a jet in flight. The figure for the simple model displays qualitatively the effects of varying the ratio $\lambda = M_f/M_j$. The amplitude of the sound intensity is decreased, according to this model, at $\theta = 90°$ by $(1 - \lambda)^5$. Others, such as Buckley

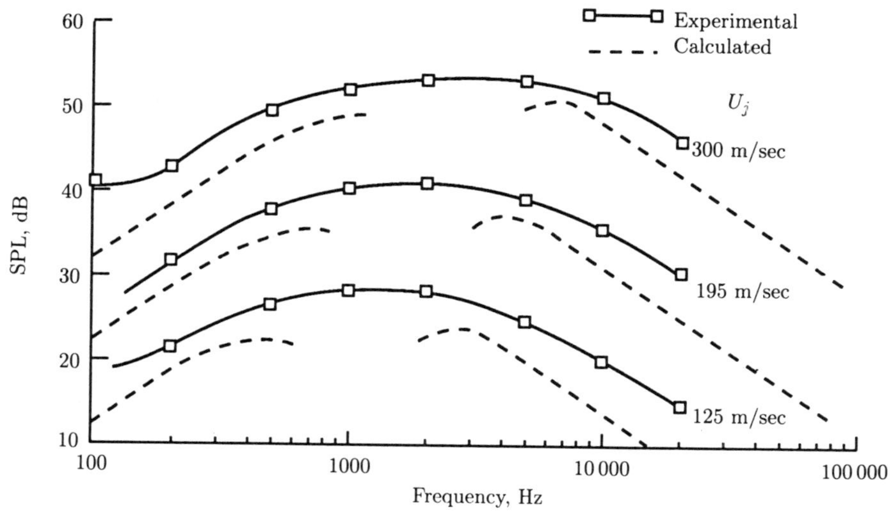

Figure 22. Calculated and experimental (ref. 76) jet noise spectra at $\theta = 90°$ and $y_1/D_j = 120$ ($D_j = 25$ mm).

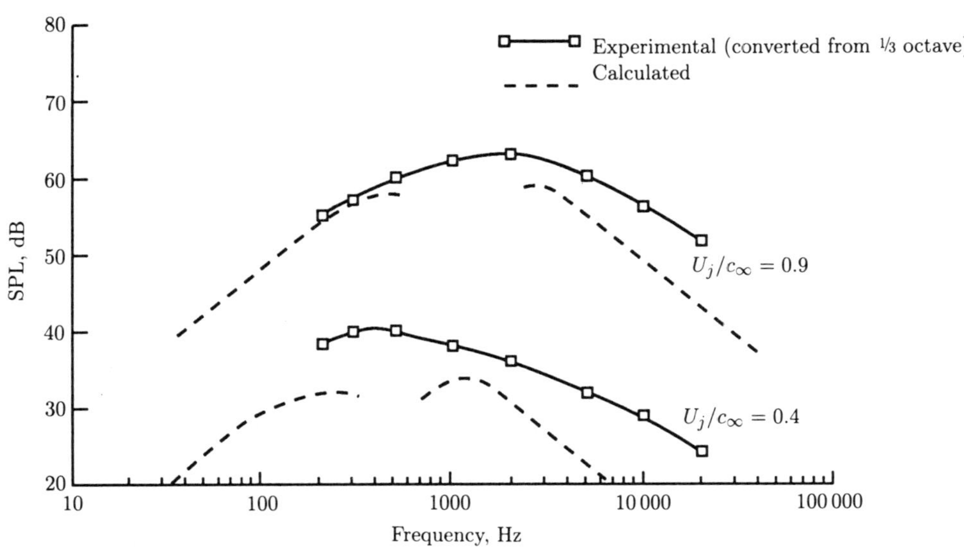

Figure 23. Calculated and experimental (ref. 77) jet noise spectra. D_j = 51 mm; $\theta = 82.5°$; $y_1/D_j = 54$; $M_f/M_j = 0.05$.

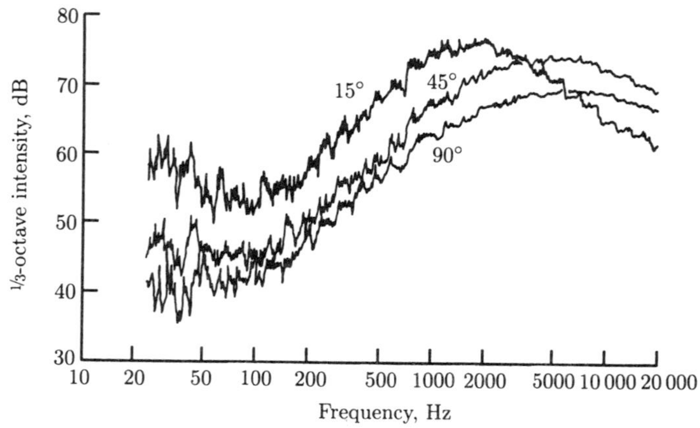

Figure 24. Jet noise spectra at $\theta = 15°, 45°$, *and* $90°$. $U_j = 195$ *m/sec. (From ref. 76.)*

and Morfey (ref. 78), who used flow-acoustic interaction theory, and Michalke and Michel (ref. 79), who used a modified source function, have obtained results that agree better with references 77 and 80 to 82.

The results from this simple model show that Lighthill's acoustic analogy theory is capable of providing a satisfactory baseline for the acoustic radiation from a jet when the main source of sound is due to turbulent mixing. The gross changes to these results, especially with respect to sound directivity and spectra, when strong flow-acoustic interaction exists, are discussed in another chapter. *The directivity of the radiated noise from a jet can only be satisfactorily established by application of flow-acoustic interaction theory.*

In this section we have discussed the application of Lighthill's acoustic analogy to the prediction of the far-field noise radiated from a single, isolated circular jet. The application of the theory to more complex situations is possible provided all the relevant flow-field data are available, which include both the mean and turbulent velocity distributions and all the requisite flow-field scales and flow dimensions. These situations include the noise from noncircular jets, the noise interference between two or more similar jets in proximity to each other, the noise from coaxial jets in which the core jet is at the higher speed, and the noise from coannular jets where the outer jet speed is both less than and greater than the core speed; these cases include both static and in-flight jets. However, in each complex jet problem, a flow-acoustic interaction exists that is far more dominant than in the case of the single, isolated static jet. Thus it is more profitable to explore the sound fields from these complex jet flows in terms of the flow-acoustic interaction theory described in another chapter.

Experimental Considerations

Flow Uniformity and "Excess Noise"

The determination of the far-field noise characteristics of model and full-scale jets from experiments involves elaborate test rigs and extensive instrumentation.

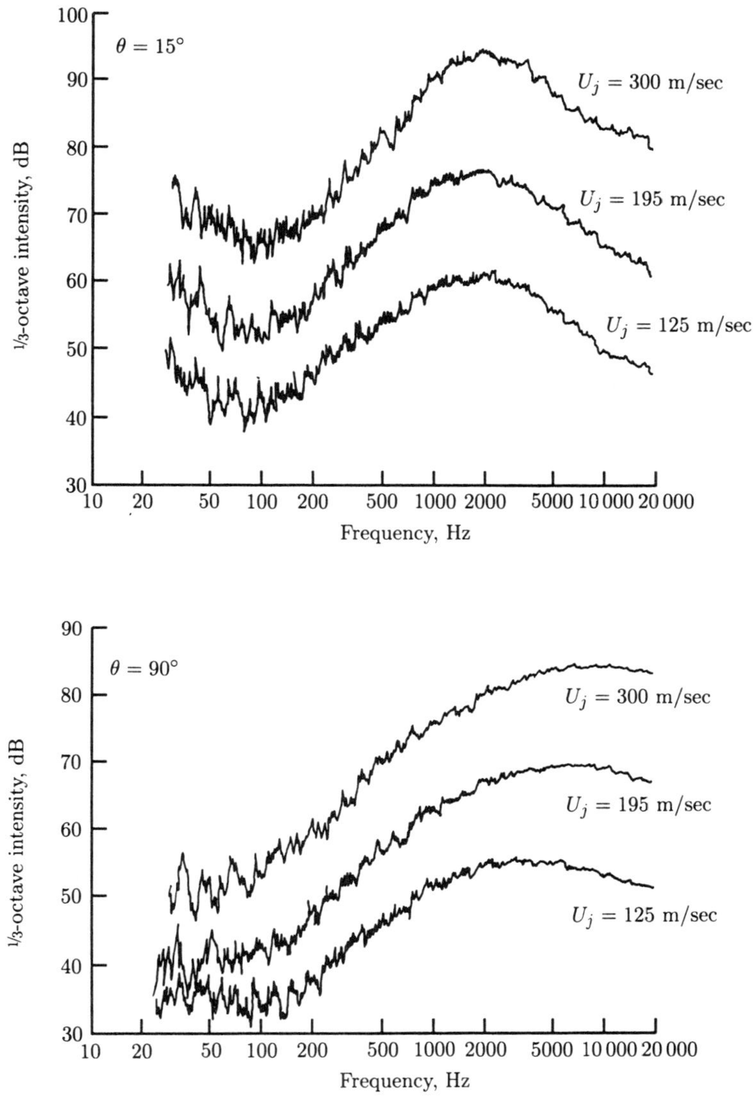

Figure 25. Jet noise spectra at $\theta = 15^\circ$ and 90° for various jet velocities. (From ref. 76).

The flow conditions at the jet exit must be accurately measured. Normally the aim must be to obtain jet exit conditions as nearly uniform as possible.

The result of flow nonuniformity at the jet exit is an increase in noise intensity arising from additional noise sources within the jet pipe and close to the jet exit plane. It is usual to classify this noise as "excess noise." In a jet engine under test conditions, the flow downstream of the combustion chamber and turbine is normally far from uniform and possesses some unsteadiness and swirl, with the result that the flow at the jet exit is nonuniform.

In addition, the presence of solid surfaces forming the jet pipe and its supporting structure in laboratory experiments, and the wing, fuselage, and tail sections in flight

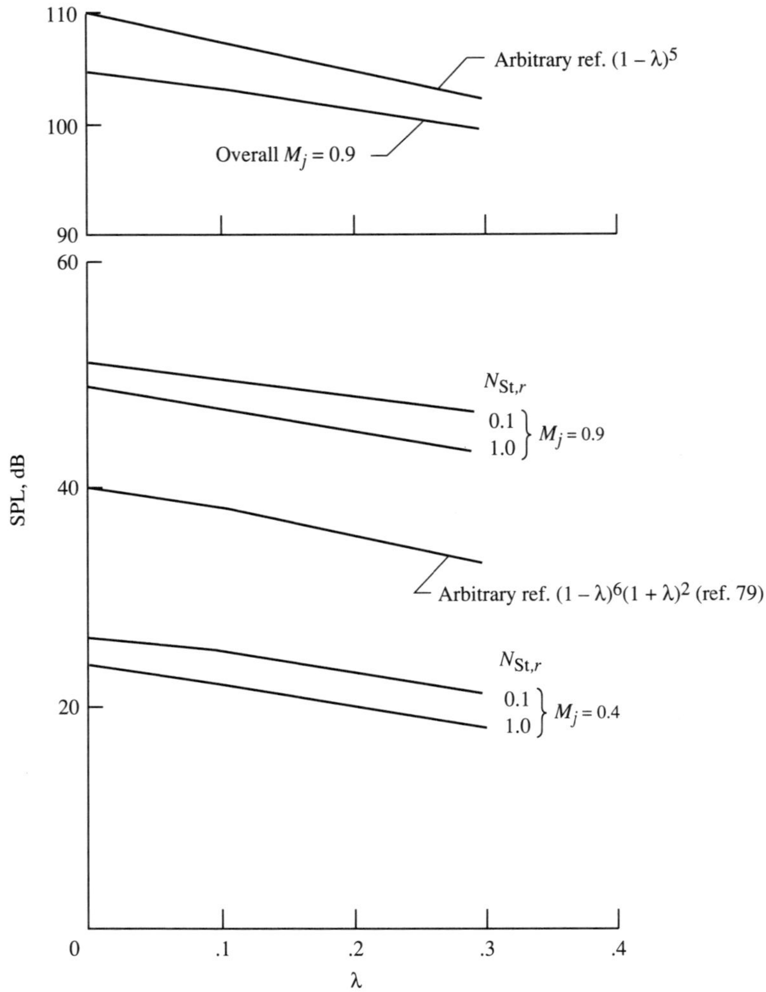

Figure 26. Variation in SPL due to flight effects.

test, all provide interference surfaces for the radiated sound and result in a change in the far-field noise directivity and amplitude. These combined effects, which are normally classified as excess noise, are summarized in reference 83.

Excess noise is important as a source of noise, especially at low jet Mach numbers, since in general it includes monopole and dipole excess noise sources that have dependences of U^4 and U^6, respectively. Hence, at low jet Mach numbers, excess noise has a greater sound intensity than the noise from the mixing region of the jet, which normally has a dependence of U^8. The low Mach number heated jet presents a more complex case, since its effective source is dipole with a dependence of U^6. For any jet engine installation, excess noise is difficult to quantify and invariably is specific to the given installation.

Experimental Conditions

In many of the early experiments on the noise from air jets external rigs were used, with jets blowing horizontally at over 100 D_j from the ground and over prepared surfaces, as in reference 6, or over grass, as in reference 10. In most of these tests it was impossible to avoid the effects of ground reflection and, in some cases, the reflection from adjacent buildings or from the jet supporting structure. Thus it was not possible to obtain reliable free-field measurements of noise intensity from the jet axis to 180°. Also, since the noise radiated in the upstream arc had an intensity well below that radiated downstream, the acoustic radiated power from a jet could be obtained from measurements between 15° and about 135° only. In corresponding noise measurements of jet engines in references 17 and 84, the jet was mounted closer to the ground and distances to the far field were correspondingly greater. The effects of ground absorption and ground reflection were also greater and required separate investigations. Thus, each jet measuring site has its own set of ground corrections, which must be applied irrespective of whether the jet being tested is a model or full scale.

Many of the problems discussed above, which were typical of the early studies on jet noise (at least on model jets), can be avoided by mounting the jet in an anechoic chamber (as shown in fig. 27), which allows "clean" measurements to be made at distances well beyond 100 D_j. (See ref. 76.) These experiments, under near "ideal" conditions, were the first measurements to show quantitatively the effects of flow-acoustic interaction and the loss of convective amplification in the downstream direction at angles close to the jet axis. These effects had been the subject of debate since the Lighthill theory of aerodynamic noise was first published in 1952, but not until reference 76 was published in 1971 was it made clear that the theory of aerodynamic noise involved significant interaction between the flow and the sound generated by it.

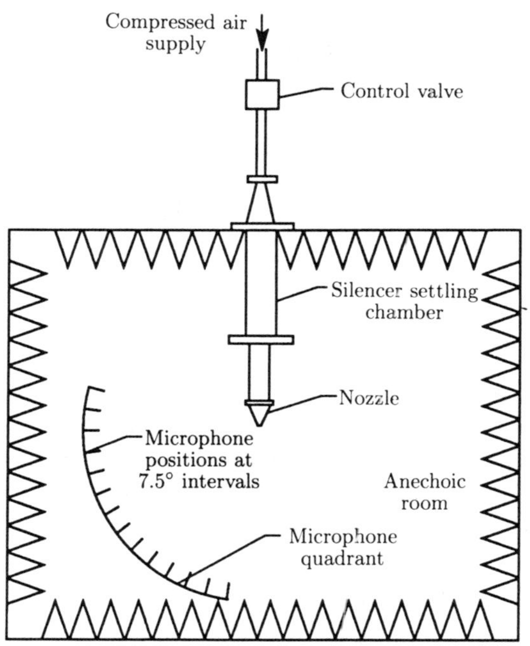

Figure 27. Anechoic jet noise facility. (From ref. 76.)

Simulation of Flight Effects

The simulation of flight effects presents even more complex installation problems. Jet engines of limited size can be mounted on a tracked vehicle (ref. 82) and the noise is measured from a stationary set of microphones during the flyby. Another moving model jet facility is the "spinning rig," in which the jet is mounted at the tip of a rotating arm similar to the blade of a helicopter. (See ref. 85.) For model jet experiments in which the jet is static and mounted in an anechoic chamber, a large-diameter secondary airflow is discharged around the jet into the chamber and both the primary and the secondary air are exhausted to the atmosphere to create as little disturbance as possible to the ambient air in the remainder of the chamber. (See ref. 86.) The noise radiated from the primary jet therefore propagates across the entire secondary jet and across its mixing region with the ambient air, toward the far-field microphones at rest in the ambient fluid. The measurements made under such simulated flight conditions, provided the secondary flow jet is sufficiently large in diameter compared with the primary jet to provide adequate resolution in the lower range of frequencies, are easy to obtain but difficult to translate into corresponding free-field data. At high frequencies acoustic ray theory has been used to convert the measured data to equivalent free-field data (see refs. 77 and 87) based on the flow-acoustic interaction between the primary jet noise field and the various structures of the secondary jet. Attempts to improve the free-field corrections have shown that ray theory is satisfactory in most practical situations.

A more satisfactory simulation of flight effects on jet noise is obtained by mounting the model jet in a specially designed wind tunnel. If the tunnel is of the open type, then surrounding the working section with a large anechoic chamber gives a facility resembling that described above. The corrections of the measured data to equivalent free-field conditions follow by the use of ray theory. The advantages of the wind tunnel are that higher secondary jet speeds can be obtained and the ratio of wind tunnel diameter to primary jet diameter is greater, so data can be obtained at lower frequencies, provided the wind tunnel is carefully designed to give a low background noise level. The wind tunnel may be of the closed type, provided it is made as near anechoic as possible. Noise measurements are now made in the wind tunnel working section in the moving flow. Both types of facilities have been successfully used and are described in reference 86.

Jet Noise Measurement Instrumentation

The instrumentation required for jet noise measurements and their analysis, including instrument corrections for wind speed, ground reflection, and ground and air absorption, is given in references 88 and 89. In references 89 and 90 details are also given of flyover measurements and particularly the type of data collection necessary for aircraft noise certification.

Source Location Techniques

Lighthill's acoustic analogy of aerodynamic noise is based on a distribution of equivalent acoustic sources of density, which replace the flow field and move in the region defined by the flow and its boundaries. In practical applications of Lighthill's acoustic analogy, the details of the jet flow field are rarely known apart from certain gross features such as the magnitudes and distributions of the mean velocity and temperature over the exit plane. In the case of most model jet rigs and those for full-scale jet engines, the jet flow field is installation specific. Thus it is a requirement to readily identify both qualitatively and quantitatively those regions of the jet generating the greatest contributions to the far-field sound intensity.

A number of source location methods have been introduced, such as those of references 91 to 93, which are mainly for use on small-scale jet rigs only. Of wider application to both model and full-scale jet engines is the "acoustic telescope" described in reference 94, and the "polar correlation technique" of reference 95. In the acoustic telescope the far field is surveyed with a linear array of microphones, and from a digital data processing of the cross correlation of the outputs from the microphones in the array, the strength distribution of an equivalent line source distribution on the jet axis can be derived. In the polar correlation method the microphones are distributed around an arc in the far field with its origin at the jet exit, and a Fourier transform of the microphone signals is employed whereby the variations in phase can be interpreted in terms of an equivalent acoustic line source distribution along the axis of the jet. The problem of the lack of uniqueness in the definition of such an axial line source distribution of equivalent acoustic sources is discussed in reference 96.

The Polar Correlation Technique

The underlying theory behind the polar correlation method and, with suitable modifications, all the acoustic source location techniques can be derived from Lighthill's acoustic analogy. If we take two points $P(\mathbf{x}, t)$ and $Q(\mathbf{x}', t')$ in the far field, where the fluctuating densities are ρ_P and ρ_Q, their cross correlation is

$$\overline{\rho_P(x,\theta,t)\rho_Q(x,\theta',t')} \equiv B(x,\theta'-\theta,t'-t) \tag{82}$$

where, in the polar correlation method, the microphone at $Q(x,\theta')$ is fixed and set at $\theta' = 90°$. We will assume that B is equal to its value with $\theta' = 90°$. It is convenient to let $\alpha = \theta' - \theta$, the polar separation angle, and t replaces $t' - t$, the time difference between the microphone signals received at Q and P. We can consider a polar array of equally spaced microphones, an array of arbitrarily spaced microphones in the range $-\alpha_m < \alpha < \alpha_m$, or a fixed and a traversable microphone over the same range of angles. It is convenient to consider just two microphones spaced α degrees apart. The cross-power spectral density corresponding to $B(\alpha, t)$ is

$$\overline{B}(\alpha,\omega) = \frac{1}{2\pi}\int_{-\infty}^{\infty} \exp(i\omega t)B(\alpha,t)\, dt \tag{83}$$

Let us write

$$\overline{B}(\alpha,\omega) = |\overline{B}(\alpha,\omega)| \exp[i\phi(\alpha,\omega)] \tag{84}$$

where the phase is $\phi(\alpha, \omega)$, so that

$$B(\alpha,t) = \int_{-\infty}^{\infty} \exp[i(\phi - \omega t)]|\overline{B}(\alpha,\omega)|d\omega \tag{85}$$

is a real quantity. The maximum value of $B(\alpha, t)$ occurs for a given α when $\phi = \omega t$ at $t = t_m$, where t_m is a function of α. We denote this value of $B(\alpha, t)$ as $B(\alpha, t_m)$, and

$$B(\alpha, t_m) = \int_{-\infty}^{\infty} |\overline{B}(\alpha, \omega)| d\omega \tag{86}$$

Figure 28 shows typical values of $B(\alpha, t)$, $B(\alpha, t_m)$, correlation amplitude, phase, and source strength measured on a jet with a diameter of 25 mm when $M_j = 0.8$.

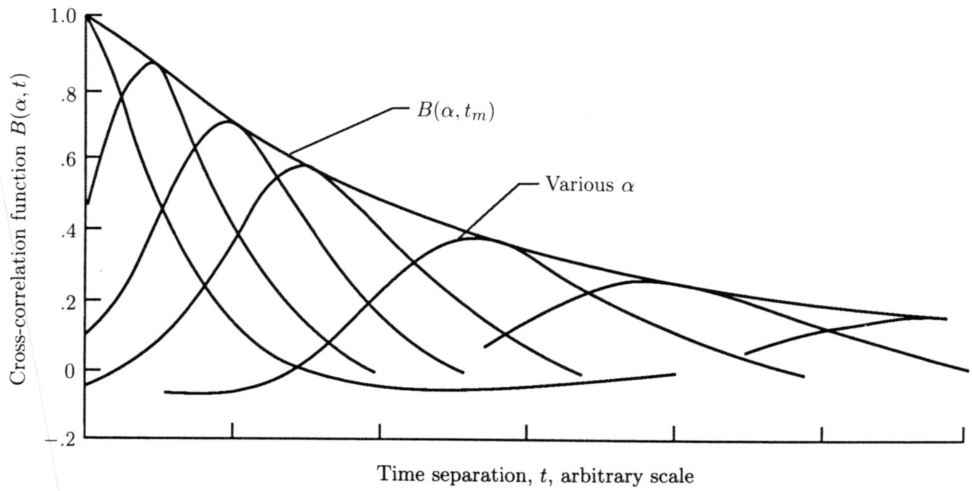

(a) Measurements of cross-correlation function.

Figure 28. Polar correlation technique.

The polar correlation technique attempts to find the position on the jet axis that generates the maximum contribution to $B(\alpha, t)$ and, at a given frequency ω, the position of an equivalent source on the jet axis that makes the greatest contribution to the corresponding cross-power spectral density. For a subsonic jet that is known to be free of excess noise and whose acoustic power approximately follows the U_j^8 law, the equivalent stationary source distribution on the jet axis may be assumed to be given from the results of the Lighthill acoustic analogy in the form

$$16\pi^2 c_\infty^8 x^2 B(x, \alpha, t) = \iiint d\mathbf{y} \iiint \frac{\partial^4}{\partial \tau^4} P_\theta(\mathbf{y}, \boldsymbol{\delta}, \tau)\, d\boldsymbol{\delta} \tag{87}$$

from which we derive the wave-number–frequency spectrum function of the source distribution P_θ:

$$\widehat{P}(y_1, \mathbf{k}, \omega) = \iint \overline{P}(\mathbf{y}, \mathbf{k}, \omega)\ \exp[(-i\omega/c_\infty x)(\sigma_2 y_2 + \sigma_3 y_3)]\ dy_2\, dy_3 \tag{88}$$

With

$$|C(\alpha, \omega)| = \frac{2c_\infty^7 x^2}{\pi \omega^3}\ |\overline{B}(\alpha, \omega)\ | \tag{89}$$

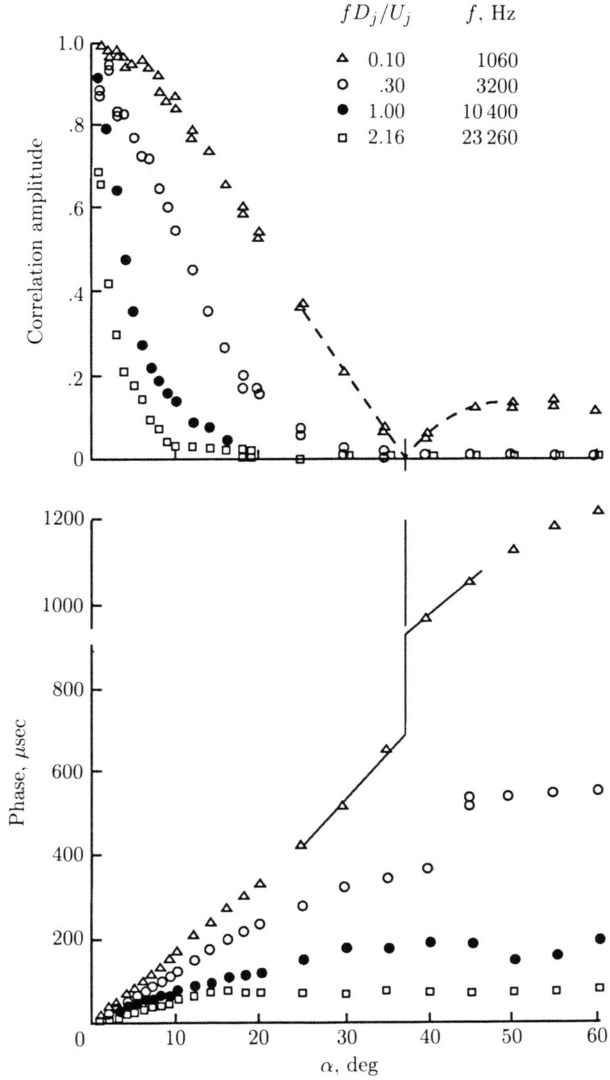

(b) Polar data for 25-mm-diameter jet; $U_j/c_\infty = 0.8$. (From ref. 95.)

Figure 28. Continued.

we find

$$|C(\alpha,\omega)| \ \exp(i\phi) = \int_0^\infty \exp(-i\mu p)\widehat{P}(\mu,\mathbf{k},\omega)\ d\mu \tag{90}$$

where $\mu = \omega y_1/c_\infty$ and $p = \cos\theta' - \cos\theta$. We have assumed for this source distribution that $p = 0$ when $y_1 < 0$, upstream of the nozzle exit. Since we have specified a stationary source distribution the effects of eddy convection at a speed $c_\infty M_c$, relative to the observer, must be included in the description of the source distribution function. In equation (90) the wave-number vector $\mathbf{k} = -\mathbf{x}'\omega/(xc_\infty)$, the spatial separation at points in the array $\boldsymbol{\sigma} = \mathbf{x}' - \mathbf{x}$, and the phase $\phi = \omega t_m$.

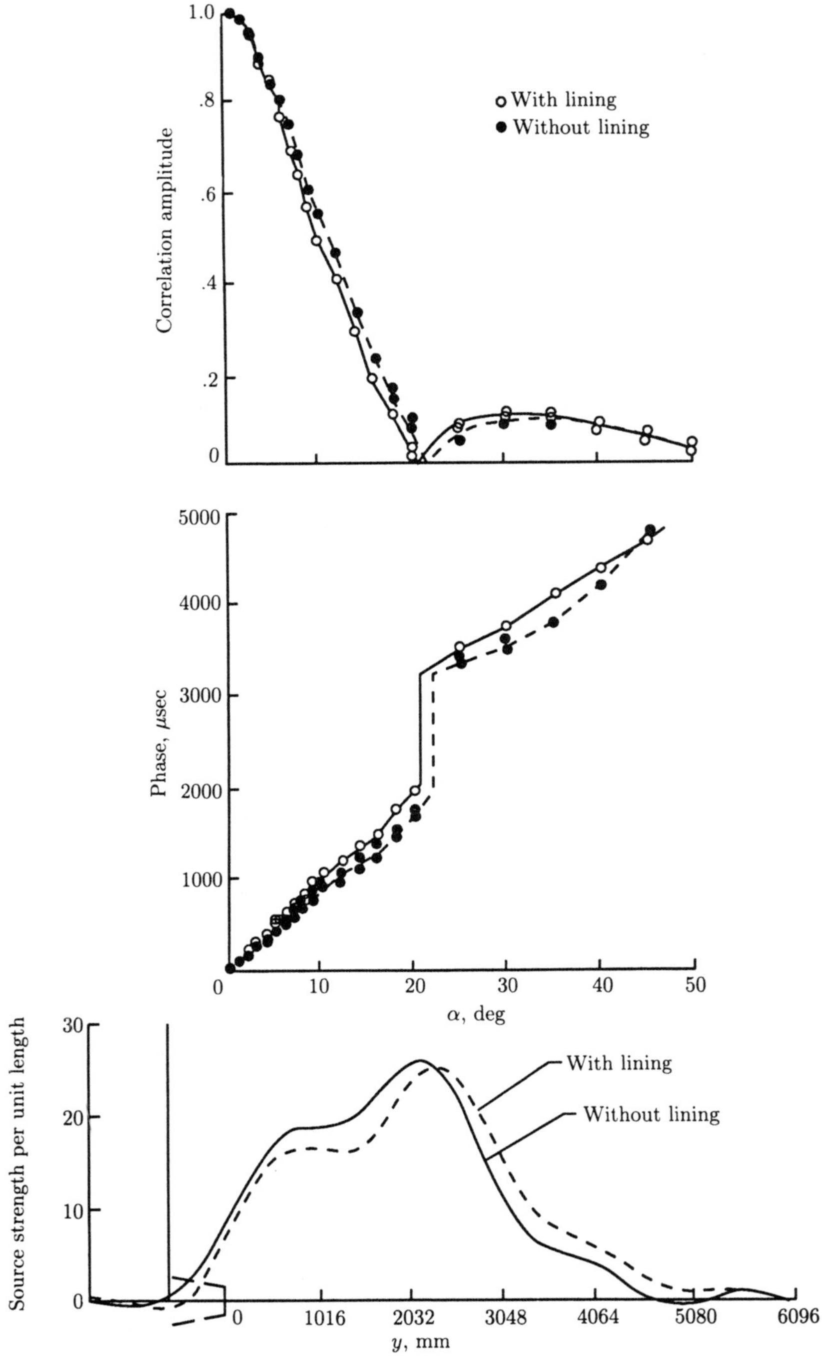

(c) Polar data for full-scale engine; 80% max rpm; 303 Hz. (From ref. 95.)

Figure 28. Concluded.

The Fourier transform of equation (90) is

$$\widehat{P}(\mu, \mathbf{k}, \omega) = \frac{1}{2\pi} \int_{-\infty}^{\infty} \exp[i(\mu p + \omega t_m)] |C(\alpha, \omega)| \, dp \tag{91}$$

and gives the required relation between the line source, the wave-number–frequency spectrum function, and the cross-power spectral density in the far field. In equation (91) the integration over p is from $-\infty$ to ∞, whereas in the practical method p is limited to, at most, a range of -1 to 1. However, if we define the Strouhal number of the jet as $N_{\mathrm{St},r} = \omega D_j/U_j$ and the jet acoustic Mach number $M_j = U_j/c_\infty$, then $N_{\mathrm{St},r}M_j$ is the Helmholtz number. In practical applications concerning jet noise, $N_{\mathrm{St},r}M_j$ may vary from just less than 1 to about 60. Since $\mu = \omega y_1/c_\infty = N_{\mathrm{St},r}M_j y_1/D_j$, we see that μ may be regarded as a large quantity for $y_1/D_j > 1$, and since α is a function of p only, $|C(\alpha, \omega)|$ is a real function of p for given values of ω.

Thus to find a value for the integral in equation (91) we can use the Stokes-Kelvin method of stationary phase, which states that the major contribution to the integral comes from the vicinity of the stationary points $h(p)$. In our case, $h(p) \equiv \mu p + \omega t_m$. Hence we find the value $m = p$ where $h'(p) = 0$ and then

$$2\pi \widehat{P}(\mu, \mathbf{k}, \omega) \approx \int_{m-\epsilon}^{m+\epsilon} |C(\alpha(p), \omega)| \, \exp[i(\mu p + \omega t_m)] \, dp \tag{92}$$

$$\approx \sqrt{2/|\phi''(m)|} \, |C(\alpha(m), \omega)| \, \cos\left[\mu m + \phi(m) + \frac{\pi}{4}\right] \tag{93}$$

But m is the value of p for which $B(\alpha, t)$ is a maximum, and thereat $t = t_m$. Therefore when $p = m$, we find

$$\mu = -\omega t'_m = -\phi'(\alpha(m), \omega) \tag{94}$$

or

$$\frac{y_1}{D_j} = -\frac{\phi'(\alpha(m), \omega)}{N_{\mathrm{St},r}M_j} \tag{95}$$

where α is a function of p only and ϕ' is evaluated at $p = m$. From equation (95) we see that for a given frequency the position of the equivalent source that makes the greatest contribution to the far-field noise intensity is inversely proportional to the frequency. Although in practical applications of the polar correlation technique the range of α is limited to $-\alpha_m < \alpha < \alpha_m$, we see a "good" value for the effective axial line source strength can be obtained from the measured value $|C(\alpha(m), \omega)|$.

In summary, we select points on a polar arc in the far field centered on the jet exit and we measure the cross correlation $B(\alpha, t)$ between a fixed microphone (e.g., at 90°) and each of the other microphones at points on the arc. For each value of α, the angular separation, we find the time delay $t = t_m$ for which $B(\alpha, t)$ is a maximum. We also find the cross-power spectral density $B(\alpha, \omega)$, which is complex and has real and imaginary parts B_R and B_I, with the phase $\phi(\alpha, \omega) = \tan^{-1}(B_I/B_R)$. Since $\phi(\alpha, \omega) = \omega t_m$, we have a check on the value of t_m. Noting that α is a function of

p only, we find $\partial\phi/\partial p$ at $p = m$, and then the position along the jet axis with the greatest contribution to $B(\alpha, \omega)$, for the given frequency, is found from

$$\frac{y_1}{D_j} = -\frac{(\partial\phi/\partial p)_{p=m}}{N_{\mathrm{St},r} M_j} \tag{96}$$

For a uniform jet we would expect one source position for each frequency. From reference 95 we find the following results:

fD_j/V_j	$N_{\mathrm{St},r}$	$N_{\mathrm{St},r}M_j$	ϕ'	y_1/D_j Calculated, eq. (96)	y_1/D_j Measured, ref. 95
0.1	0.628	0.50	−6.9	13.8	13.0
.3	1.885	1.51	−13.8	9.1	8.5
1.0	6.283	5.03	−30.0	6.0	5.5
2.16	13.572	10.86	−42.0	3.9	4.0

The method of data reduction used in reference 95 differs from that used above, but similar results are still obtained. Thus the method of stationary phase provides an adequate approximation for the determination of the effective source location from use of the polar correlation technique.

Reference 95 also discusses applications of the polar correlation technique when multiple sources are present at a given frequency. Thus the polar correlation technique has application to such cases where excess noise is present and typically results in a further effective source located at or upstream of the nozzle exit. The polar correlation technique is also applied to the case of coaxial jets, for which it can distinguish the effective location of the dominant sources in the inner and outer streams. For details of the application of the polar correlation technique to these jet configurations, see reference 95.

Comparison With the Lighthill Acoustic Analogy Model

A particular application of the polar correlation technique is to provide experimental verification of the assumptions used in the simple acoustic analogy model, particularly the values introduced to define the T_{ij} covariance in terms of specified local, average, and characteristic values of the flow quantities. Remember that these quantities are introduced to define the Gaussian approximation used to describe the T_{ij} covariance in the turbulent mixing regions of the jet. A justification for the use of such a crude approximation has already been partially given, but here we will concentrate on providing experimental support for our model.

We can determine $B(\alpha, \omega)$ and $|C(\alpha, \omega)|$ as required in the polar correlation technique by the use of methods similar to those described in the previous section for the estimation of the far-field noise intensity and the corresponding far-field spectral density. Thus the position of the maximum contribution to the far-field intensity, in a given band of frequencies, can be calculated for a jet configuration similar to that tested in references 95 and 97. The results are given in figure 29 and in the table below for a model at $M_j = 0.8$ with a jet diameter of 25 mm and a full-scale engine running at 80 percent of maximum rpm and using an unlined tail pipe.

Model jet

$[(y_1)_0 = 1.25D_j]$

$(y_1/D_j)_m$	$N_{St,r}$
1.40	100.0
1.54	50.0
1.98	20.0
2.22	15.0
2.70	10.0
4.16	5.0
4.88	4.0
5.09	3.0
7.51	2.0
6.42	5.0
7.24	3.0
8.06	2.0
9.91	1.0
12.52	.5
17.70	.2
23.55	.1

Full-scale engine

$[(y_1)_0 = 0]$

$(y_1/D_j)_m$	$N_{St,r}$
0.15	100.0
.29	50.0
.73	20.0
.97	15.0
1.45	10.0
1.94	7.5
2.91	5.0
3.63	4.0
4.84	3.0
3.64	3.0
4.46	2.0
6.31	1.0
8.92	.5
14.10	.2
19.95	.1

In order to improve agreement, it was first determined that the discrepancy was the result of the assumption that in both trial cases the nozzle exit conditions were similar and the initial thickness of the mixing region downstream of the nozzle exit was zero. The comparisons between the calculated and measured results suggest this is a good assumption for the full-scale engine. However, for the model jet this appears to be a poor approximation. For a model jet, the early mixing region is unlikely to be fully turbulent unless special measures are taken to disturb it sufficiently to force transition at or near the origin of the mixing region. The results of reference 95 suggest that transition was free, so we can expect that a certain length downstream of the jet exit the mixing layer is in a transitional state, and even though this region may generate noise its characteristics will be very different from those associated with a fully turbulent mixing region. Accordingly, in the results presented in figure 29 and in the above table, an artificial origin at $1.25D_j$ downstream of the exit has been introduced for the fully turbulent mixing layer.

The overlap in the results around the end of the potential core, which we took as $y_1 = 5D_j$, is the result of the assumption that the growth of the mixing region has the same value upstream and downstream of the potential core, but the characteristic turbulence velocity discontinuously decays inversely proportional to the axial distance, beginning at the end of the potential core following its constant value throughout the initial mixing region. An improved model would be one in which the flow properties were made continuous in the three regions covering the initial

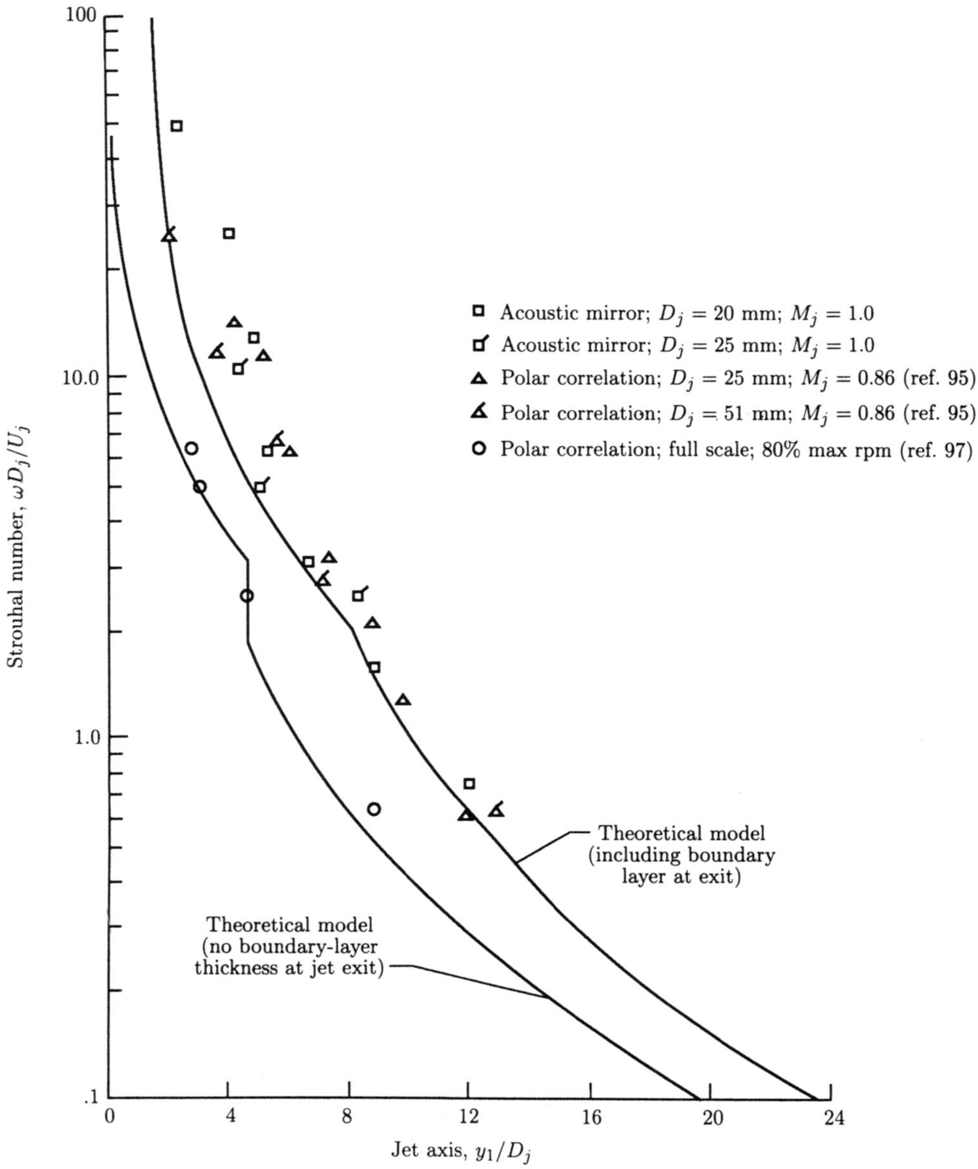

Figure 29. Calculated and measured peak source strength locations.

mixing region, the region around the end of the potential core, and the downstream mixing region.

The conclusion reached is that a simple model for the T_{ij} covariance for use in the Lighthill acoustic analogy is satisfactory and models the flow in the mixing region of a jet. Any arbitrariness in the chosen values of the constants representing the values of the characteristic flow quantities is a reflection on the likely differences that could exist in jets having different flow properties at exit.

It should be emphasized here that the polar correlation technique itself neither employs nor depends on the Lighthill acoustic analogy for an experimental determination of the effective acoustic source distribution. The comparison that we have made between the results we obtained using the Lighthill acoustic analogy and the data obtained experimentally with the polar correlation technique is confirmation that our simple model for the effective acoustic source strength provides a fair approximation of the flow-field characteristics required in the application of the Lighthill acoustic analogy to the estimation of the characteristics of the far-field noise from a subsonic jet, but only in those regions of the jet and certain angular regions within the radiated field where flow-acoustic interaction can be ignored. The comparison we made was taken at 90° only, and at this angle in the far-field convective amplification effects are zero.

Jet Noise Reduction Techniques

One of the goals of the early experimental studies on jet noise was the exploitation of the knowledge gained, with respect to the characteristics of the sources of noise generation in the turbulent mixing region of a jet, to find means by which the radiated noise intensity could be reduced with no loss in nozzle performance (i.e., jet thrust). The work of reference 6 on simple modifications to the shape of the nozzle exit with a number of different nozzle extensions, shown in figure 30, showed that changes to the initial mixing region of the jet changed the flow structure of the entire jet. With some of the devices the noise was reduced by 8 dB in certain directions with consequent changes in spectral shape, and this reduction was achieved with a relatively small loss in nozzle performance. The noise reduction was even greater when the nozzle was choked, whereas the unchoked nozzle exhibited the characteristics of screech and shock-associated noise. An analysis of all the model experimental data indicated that for the reduction of jet noise on full-scale jet aircraft an aerodynamically smooth transition was required between the upstream circular tail pipe and the "fluted" circumference at the nozzle exit. The result was the "corrugated nozzle," designed by R. Westley, G. M. Lilley, and A. D. Young and developed by Greatrex (ref. 17). It was fitted to many of the civil aircraft flying between 1955 and 1980. Two examples are shown in figure 31. Derivatives of the corrugated nozzle are used on many modern aircraft, as discussed in reference 89. Although the noise reduction obtained with the corrugated nozzle may be considered modest, it nevertheless is accepted as the one major practical device that has reduced jet noise for minimum loss in performance. Apart from its performance loss a further disadvantage in the use of the corrugated nozzle was its additional weight, which when combined with the thrust loss produced a significant increase in fuel consumption for a modest reduction in noise.

The original application of the corrugated nozzle to jet noise reduction was on the straight-jet engine operating at or just above choking for takeoff. Once far greater noise reductions were required on civil aircraft power plants than could be successfully achieved with the corrugated nozzle, it was realized that a major change in aircraft engine design was required. Substantial noise reductions could only be obtained by a large reduction in the final jet velocity, and this was accomplished with the bypass jet engine and later with the turbofan engine. For a jet whose overall radiated acoustic power was proportional to U_j^8, the potential noise reduction with a halving of the final exhaust velocity was 24 dB, which was far greater than

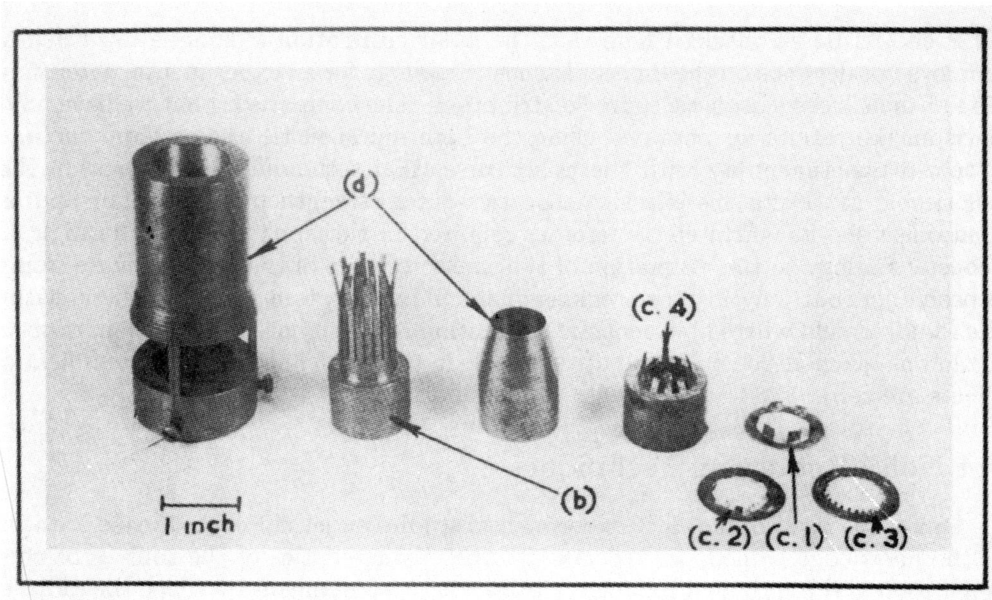

Figure 30. Noise reduction devices. (From ref. 6.)

Figure 31. Types of aircraft engine noise suppressors. (Courtesy of Rolls-Royce Limited.)

could be achieved with the corrugated nozzle. Since it was found that the different thermodynamic cycle used on the bypass engine produced a smaller specific fuel consumption, this type of power plant quickly replaced the straight-jet engine as the basic civil aircraft power plant for all the airlines of the world since it was technically more efficient and environmentally more acceptable.

The early reasoning for why noise reduction was achieved with the corrugated nozzle was that the initial mixing region structure had changed at subsonic speeds to produce an increased mixing rate, a reduction in the mean shear, and a consequent reduction in the measured length of the potential core. Thus it was argued the effective acoustic source volume was reduced, with a consequent reduction in the sound power. In the over-choked case the corrugated nozzle prevented an ordered eddy structure from developing, and this interacting with the regular spacing of the shock cells generated screech and shock-associated noise. The results of flow visualization broadly confirmed these conclusions. A further feature of the corrugated nozzle was that the circumference of the jet at exit was now broken up into a series of smaller jets, each jet being associated with each corrugation, and these structures persisted along a substantial length of the jet mixing region. It had been established that decreasing the diameter of a jet proportionately increased the peak frequency, and hence a further change in the noise reduction characteristics of the corrugated nozzle was the shift in the noise spectrum to higher frequencies dependent on the number of corrugations. A further derivative of the corrugated nozzle was the multitube nozzle, which operated on the same principle but possessed the disadvantages of increased weight and internal losses. The combination of these devices with an ejector gave increased noise reduction, but again at the expense of increased weight and increased drag in flight. The noise reduction characteristics of many of these devices are discussed in another chapter and in reference 89.

It would be wrong to argue that the introduction of the Lighthill acoustic analogy had little influence on the design and development of the corrugated nozzle and its derivatives. Nevertheless it has to be accepted that all the noise reduction devices discussed above were developed experimentally, and even today their performance cannot be satisfactorily predicted theoretically. However, once it had been established that flow-acoustic interaction played a significant role in the radiated noise characteristics of a jet, it became clear that any device added to a nozzle-exhaust system that modified the jet mixing region and the surrounding flow field would result in a change, and almost certainly an increase, in flow-acoustic interaction. Thus, it is suggested in reference 36 that the reduction in noise arising from the corrugated nozzle and its derivatives occurs within the zone of silence and is negligible outside it, especially at large angles to the jet axis. Hence a necessary condition in a device to reduce jet noise at subsonic speeds is to provide a gaseous shield around the jet and between the fast-moving turbulent structures and the far-field observer. The application of flow-acoustic interaction theory, as performed in reference 36, provides a satisfactory qualitative explanation of the noise reduction properties of the corrugated nozzle and its derivatives.

Alternative Theories of Aerodynamic Noise

The Determination of T_{ij}

Following the publication of Lighthill's theory of aerodynamic noise many scientists and engineers adapted the theory to provide prediction methods for jet noise covering a wide range of jet conditions, such as jet exit temperature and speed. An early stumbling block was the modeling of the space–retarded-time covariance of T_{ij} in terms of readily measurable turbulence quantities, such as second-order turbulence velocity covariances, their energy spectra, and their integral scales. Some researchers, such as Jones (ref. 98), attempted the difficult measurement of the fourth-order velocity covariances, but the complexity of the problem (T_{ij} has six independent components) has meant that more attention has been placed on theoretical, rather than further experimental, developments in the determination of T_{ij}. Lighthill (ref. 3) noted that a major contribution to the T_{ij} covariance in a turbulent shear flow came from $e_{ij}\,\partial p/\partial t$, where e_{ij} is the mean rate of strain $\partial v_i/\partial x_j + \partial v_j/\partial x_i$. Thus, in the presence of an intense mean shear, the fluctuations in pressure would be highly amplified and the noise radiation would be enhanced. A theory of subsonic jet noise based on this model was attempted in reference 43, and many of the results obtained were shown to be in broad agreement with experiment. The method, however, was unsatisfactory at and near $M_c \cos\theta = 1$. It was later that Lighthill's theory was extended both by Ffowcs Williams (ref. 39) and Lighthill (refs. 4 and 5) to cover convection speeds at all subsonic and supersonic Mach numbers, including $M_c \cos\theta = 1$, at least up to those Mach numbers for which density fluctuations inside the flow were considered to have little influence on the turbulence.

The Theories of Ribner and Michalke

Ribner (refs. 99 and 100) noted that inside a turbulent shear flow, when the fluid is incompressible, the double divergence of the Lighthill stress tensor, with $T_{ij} = \rho_\infty v_i v_j$, is exactly equal to $\nabla^2 p$ and therefore can be written

$$\frac{\partial^2 T_{ij}}{\partial x_i\,\partial x_j} = \nabla^2 \phi \tag{97}$$

where ϕ is a scalar function and is termed a "pseudo-incompressible pressure." Ribner discarded the notion of a quadrupole source and referred to his theory as the simple source theory of aerodynamic noise. The theory was criticized by Lighthill (ref. 5) on the grounds that it suppressed the tensorial properties of T_{ij} and that it was wrong to imply that T_{ij} decayed near, and beyond, the boundaries of an incompressible flow at the same rate as the pressure. Therefore, the neglect of T_{ij} outside a flow field would not apply to the pressure. However, as an approximation to the Lighthill stress tensor and as used to derive a model containing empirical constants for the prediction of the radiated noise from a jet, Ribner's results were shown to be satisfactory. Ribner found it necessary to include the effects of refraction due to the mean flow-acoustic interaction, and since this was a high-frequency phenomenon, he found it was satisfactory to use ray acoustics. Ribner's results were qualitatively in agreement with experiment and helped to explain the so-called zone of silence near the jet axis. Further developments of Ribner's theory are given in references 44 and 101.

It was proposed in references 102 and 103 that the radiated noise from a turbulent flow, such as a jet, could be obtained from the method of matched asymptotic expansions, whereby the inner region would be the field of turbulent flow and the outer region the radiated noise field. This suggested that the outer solution could be represented by a distribution of axisymmetric emitting noise sources ($m = 0$) and nonaxisymmetric emitters ($m \neq 0$). It was found that a relatively small number of azimuthal modes were needed to provide a good representation of the sound field of a circular jet found from experiment. Of course, a major problem was to relate the external sound field to the characteristics of the turbulence in the jet mixing region, which in Lighthill's theory is given by $\partial^2 T_{ij}/\partial t^2$. In reference 103 a single parameter of the jet turbulence $\delta = kr\sin\theta$ was used, where k is the sound wave number ω/c_∞ and $2r = D_j$, the jet exit diameter. The length δ is related to D_j, but as it does not vary with position along the jet it was assumed to be some suitably averaged length scale of the turbulence, possibly related to the region of the jet concerned with the sound generation of greatest amplitude. In spite of this apparent oversimplification, the theory showed agreement with a wide range of experimental data, including that of references 76, 104, and 105. However the complete theory required the effective surface source distribution on, for example, a cylinder of radius r to be matched with the near-field radiation from the true vortical sources within r, whose length scale and amplitude depend on the local turbulence parameters and not on just a single parameter δ. Related work on that problem was undertaken in reference 106, extending earlier work (ref. 99) and unpublished work by Csanady. A further extension of that work led to the study of flow-acoustic interaction, as described in reference 18, and to its practical application as a jet-noise prediction scheme (e.g., ref. 107). The flow-acoustic interaction problem has been investigated by many researchers (e.g., refs. 108 to 113). Further discussion on the choice of acoustic analogy is given in references 114 and 115.

The Neglect of the Fluctuations in Density at Source

An aspect of the Lighthill acoustic analogy theory that has caused much discussion is that Lighthill's equation is strictly an integro-differential equation for the density, since ρ appears as the independent variable and also in the source function through the stress tensor T_{ij}. (See ref. 52.) In low Mach number flows, the strength of the acoustic sources is such a small fraction of the flow kinetic energy that it is a good approximation to replace ρ in T_{ij} with the ambient density ρ_∞. We can assume that in such flows sound waves present in the flow do not modify the turbulent flow. If we argue that pressure fluctuations inside a turbulent flow are of the order of $\rho_\infty u_0^2$ and fluctuations in density are of the order of p'/c_∞^2, then since for plane sound waves $\rho'/\rho_\infty = u'/c_\infty$ (where u' is the particle velocity), it follows that for a circular frequency $\omega = 2\pi f$ and a sound wavelength λ, $u'/u_0 \ll 1$ and $l_0/\lambda \ll 1$, where $\omega l_0/u_0$ is of the order of unity and l_0 is a characteristic eddy length. Since the particle velocity is very small compared with the turbulence velocity we see that the influence of the sound on the turbulence can be neglected. However the opposite is not true and the result is embodied in flow-acoustic interaction.

Introduction to Flow-Acoustic Interaction

Consider the disturbance created in a turbulent flow, or indeed in any unsteady flow, that results in alternate compressions and expansions of a fluid element as it is convected by the flow. The time rate of change in the volume of this fluid element δV per unit volume of fluid, following the flow, is given by

$$\lim_{\delta V \to 0} \frac{1}{\delta V}\frac{D\delta V}{Dt} = -\frac{D \ln \rho}{Dt} = \operatorname{div} \mathbf{v} \tag{98}$$

and it follows that in a compressible flow the sound generation is directly related to the value of the time-dependent part of div $\mathbf{v}$ inside the flow. The value of div $\mathbf{v}$ inside a flow is negligibly small and would be almost impossible to measure experimentally. Nevertheless, it is only when the time-dependent part of div $\mathbf{v}$ is finite inside an unsteady compressible flow that sound generation can occur.

Pressure fluctuations exist in any turbulent flow. These are present in an incompressible flow and are of similar amplitude in a compressible flow. In an incompressible turbulent flow, where the vorticity $\mathbf{\Omega} = \nabla \times \mathbf{v}$, it is easily shown from the equations of motion that

$$\nabla^2 p = -\rho_\infty \nabla^2(v^2/2) + \rho_\infty \mathrm{div}(\mathbf{v} \times \mathbf{\Omega}) = -\rho_\infty\, \partial^2 v_i v_j / \partial x_i\, \partial x_j \tag{99}$$

where the constant density is set equal to ρ_∞. In a compressible flow the positive and negative fluctuations in pressure give rise to density fluctuations, which are then propagated outward at the local speed of sound relative to the local flow velocity. The pressure fluctuations in the flow are barely modified by the resulting sound field, except at high Mach numbers when shocklets are generated. Let the fluid velocity $\mathbf{v} = \mathbf{u} + \nabla\phi$, where $\mathbf{u}$ is the instantaneous velocity of the vortical field. Since the sound field is irrotational, we can assume it is defined in terms of the time-dependent part of the velocity potential ϕ. For a turbulent shear flow, $\nabla \cdot \mathbf{u} = 0$ by definition of the vortical field, so that $\nabla^2\phi = \mathrm{div}\ \mathbf{v}$ and in compressible flow is not zero. If div $\mathbf{v}$ is identically zero everywhere, as in incompressible flow, there can be no sound. Inside the flow we set $|\nabla\phi| \ll |\mathbf{u}|$, and hence to a good approximation,

$$\frac{D}{Dt} = \frac{\partial}{\partial t} + \mathbf{v} \cdot \nabla \approx \frac{\partial}{\partial t} + \mathbf{u} \cdot \nabla \equiv \frac{\widetilde{D}}{Dt} \tag{100}$$

Let us assume that in a given flow the vorticity and the enthalpy distributions are known. Then in such a compressible flow we find

$$\frac{D \ln \rho}{Dt} \approx \frac{\widetilde{D} \ln \rho}{Dt} \tag{101}$$

to a good approximation, and from equation (98)

$$\frac{\widetilde{D} \ln \rho}{Dt} = -\nabla^2\phi \tag{102}$$

when sound is generated by the flow. The sound field is given exactly by the time-dependent part of $\nabla^2\phi$. If we define a new variable $r = \ln\left(p^{1/\gamma}\right)$ such that $\nabla r = \nabla p/\rho c^2$, then from the energy equation we find, to the same order of approximation,

$$\frac{\widetilde{D} r}{Dt} + \nabla^2\phi = \frac{\widetilde{D}}{Dt}\frac{s}{C_p} \tag{103}$$

where s is the specific entropy and C_p is the specific heat at constant pressure. If we omit the diffusive terms in the equation of motion and again approximate to the convective operator only, we find that

$$\frac{\widetilde{D}\mathbf{v}}{Dt} = -c^2\, \nabla r \tag{104}$$

leading to

$$\frac{\widetilde{D}}{Dt}\nabla^2\phi = -\nabla\cdot(c^2\ \nabla r) - \nabla\mathbf{u}\ :\ \mathbf{u}\nabla \tag{105}$$

where : is the double dot product. Since $c^2 = (\gamma - 1)h$, we can write equation (105) in the form

$$\frac{\widetilde{D}}{Dt}\nabla^2\phi = -\nabla\mathbf{u}\ :\ \mathbf{u}\nabla - c^2\ \nabla^2 r - c^2\ \nabla\ln h\cdot\nabla r \tag{106}$$

In a given flow devoid of all irrotational components and with $\mathbf{u}, h$, and s known everywhere, all sound is absent. Yet when such a flow is compressible it generates a sound field described by $\nabla\phi$. In the absence of flow,

$$\left(\frac{\partial^2}{\partial t^2} - c_\infty^2\ \nabla^2\right)\nabla^2\phi = 0 \tag{107}$$

but in the presence of flow, after we eliminate terms in r between equations (105) and (106), the corresponding convective wave equation for $\nabla^2\phi$ is found with a forcing function that is a unique function of $\mathbf{u}, h$, and s only. It is easy to deduce by reference to equation (102) that this forcing function is equal to

$$\nabla^2\frac{\widetilde{D}}{Dt}\ln\rho \tag{108}$$

and, as proposed in reference 116, may be regarded as the source of sound in an unsteady shear flow. It is equal to $-\nabla^4\phi$ inside the flow to the same order of approximation. Not surprisingly, one of the dominant terms in this forcing function is simply $\partial^2 u_i u_j/\partial x_i\ \partial x_j$ as in Lighthill's source function. The comparison between the source functions in the two theories, if all diffusive terms are omitted, is as follows:

Lighthill:

$$\frac{\partial^2 T_{ij}}{\partial x_i\ \partial x_j},\quad \text{where } T_{ij} \approx \rho u_i u_j + (p - \rho c_\infty^2)\delta_{ij}$$

Legendre:

$$\nabla^2\frac{\widetilde{D}}{Dt}\ln\rho = -\frac{\widetilde{D}}{Dt}\left(\frac{\nabla\mathbf{u}:\mathbf{u}\nabla}{c^2} + \frac{\nabla\ln h}{c^2}\cdot\frac{\widetilde{D}\mathbf{u}}{Dt}\right)$$
$$-\frac{\nabla^2\mathbf{u}}{c^2}\cdot\frac{\widetilde{D}\mathbf{u}}{Dt} - 2\nabla\mathbf{u}:\nabla\left(\frac{1}{c^2}\frac{\widetilde{D}\mathbf{u}}{Dt}\right)$$

In a weakly nonisentropic flow we see from equations (102) and (103) that $\widetilde{D}r/Dt$ is also equal to $-\nabla^2\phi$, and hence $\ln\rho$ and r are interchangeable. Thus with the elimination of $\nabla^2\phi$ between equations (103) and (104),

$$\frac{D^2 r}{Dt^2} - \nabla\cdot(c^2\ \nabla r) = \nabla\mathbf{v}:\mathbf{v}\nabla \tag{109}$$

Equation (109) is known as Phillips' equation (ref. 117), where the entropy and viscous diffusion terms are omitted because they are small compared with $\nabla \mathbf{v} : \mathbf{v}\nabla$. Derivatives of this equation led Lilley (ref. 106) to investigate not only those flow quantities responsible for noise generation but also the flow variables associated with the propagation of sound out of the flow and the interaction between the flow field and the sound waves within the flow. The resulting interaction is referred to as flow-acoustic interaction.

Concluding Remarks

In conclusion, we should point out that Lighthill's acoustic analogy theory of aerodynamic noise is exact. All other theories, at best, involve approximations of the real flow. Lighthill's theory includes all flow-acoustic interactions as well as the scattering of noise by the turbulence. (See ref. 118.) The underlying difficulty in applications of Lighthill's acoustic analogy is that the full space-time history of T_{ij} cannot readily be evaluated in any given flow with specified initial and boundary conditions. In order to unravel the effects of noise generation in a flow from those of flow-acoustic interaction it is necessary to consider the true unsteady-flow equations, rather than Lighthill's acoustic analogy equation involving the forcing term $\partial^2 T_{ij}/\partial x_i \, \partial x_j$, which represent the quadrupole distribution of equivalent acoustic sources moving in a uniform medium at rest. The beauty of Lighthill's approach, however, is that it is an analogy that provides a good approximation of the order of magnitude of the radiated sound from a turbulent shear flow, even when the true unsteady flow field can only itself be described very approximately. However, the more accurately T_{ij} is known, the more accurate the estimate of the radiated noise is.

In the alternative theories of aerodynamic noise based on the convected flow equations, which form the basis for the methods of flow-acoustic interaction, the aim is to find suitable approximations to the space-time covariance of Dp/Dt throughout the flow. The various attempts to achieve this are included in the works of references 44, 106, 110, 113, 116, 117, and 119 to 124. All these studies are based on the exact equations for unsteady viscous compressible flow, just as in Lighthill's theory, but differ from it in that the acoustic sources are now required to move relative to the real flow, rather than being embedded in a uniform medium at rest. This requirement is only achieved at the expense of the introduction of a modified wave equation of greater complexity, and the simplicity of Lighthill's approach is lost.

The various attempts to achieve this goal of approximating the space-time covariance differ essentially in the choice of the independent variable used. In essence all are equally valid, although their results reflect the further assumptions introduced and, in particular, the flow quantities specified as known in a given flow. It is interesting to note that Lilley et al. (ref. 106) (see Goldstein (ref. 18)) used ϕ and r as independent variables, Howe (ref. 122) used ϕ and h_s, Yates and Sandri (ref. 124) used ϕ and $h + (\partial\phi/\partial t) + (\nabla\phi)^2/2$, and Legendre (ref. 116) used ϕ and $\ln \rho$. In each approach one of the variables was eliminated so that a single equation could be obtained. Thus Lilley derived a single equation in r, Howe derived a single equation in h_s, and Legendre used the single equation in ϕ. All these methods lead to the determination of the sound field generated by an unsteady flow. The source of noise is unaffected by the choice of the independent variable. Figure 32 shows pictorially the differences between flow-acoustic interaction theories and Lighthill's acoustic analogy.

We conclude that the theories discussed above may prove useful as a guide to further experimental studies or to studies based on computer simulations, such as LES (large-eddy simulation) or direct numerical simulation, wherein the need to introduce approximations into the system of equations may no longer be necessary. In most practical situations the basic unsteady flow field is not known in sufficient detail to use any of the above theoretical methods to obtain a good quantitative and accurate assessment of the properties of the radiated noise from the flow. At best the theoretical methods can give a physical insight into the properties of the noise-generation processes and a qualitative picture of the characteristics of the radiated noise. Quantitative prediction methods, in general, have to be based on good, reliable experimental data within the framework of the theories discussed above.

Conclusions

This chapter has presented Lighthill's theory of aerodynamic noise as the foundation on which to build all other theories of aerodynamic noise. The application of the Lighthill acoustic analogy to the estimation of the characteristics of the noise radiated from jets is central to this chapter.

Attention is given to the assumptions on which the Lighthill acoustic analogy is based and it is shown why the theory gives results different from experiment when flow-acoustic interaction occurs. The details of flow-acoustic interaction are invariably unavailable to provide the necessary fine adjustments to the Lighthill source function to render it such that the noise radiation as calculated is exact. The alternative approaches to the understanding of aerodynamic noise theory are discussed, wherein the emphasis is placed on the flow-acoustic interaction and such theories are required to complement the results obtained by application of Lighthill's acoustic analogy.

The application of the Lighthill acoustic analogy to any aerodynamic noise problem involves a detailed knowledge of the time-dependent flow to an extent that is rarely available, especially when the flow is turbulent. We discuss some of the dominant features of the mean flow and turbulent structure of a jet to guide us in modeling the T_{ij} fourth-order covariance, which is central to applications involving Lighthill's acoustic analogy. We avoid discussion of the structure of more complex jet configurations since we need to retain a gross simplicity in our model in order to establish whether qualitative and possibly quantitative agreement can be obtained when comparison is made with experiment. The modeling assumptions are severe, and yet we are able to establish an agreement with experiment better than an order of magnitude. This in itself is surprising when we consider that the acoustic source function based on T_{ij} is related to the kinetic energy of the turbulence, whereas the overall radiated acoustic power is of the order of 10^{-4} smaller.

The results obtained from the acoustic analogy model are compared with experimental data obtained by application of the polar correlation technique to both a model-scale jet and a full-scale jet engine. The relatively close agreement is evidence that the flow-field data are pertinent to the description of the acoustic analogy model.

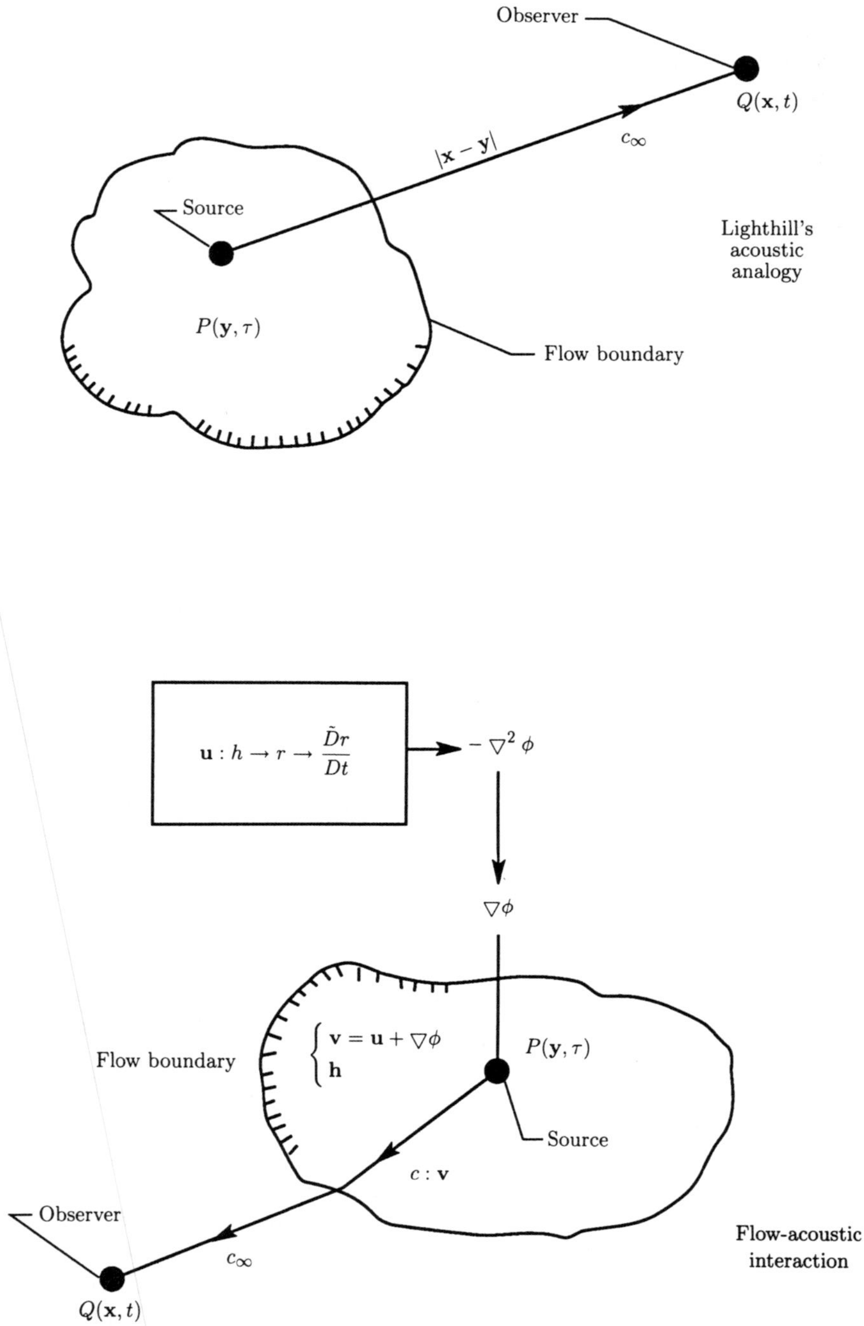

Figure 32. Differences between Lighthill's acoustic analogy theories of aerodynamic noise related to flow-acoustic interaction.

The wider applications of the use of the polar correlation technique are mentioned, especially the determination of the location of additional noise sources, such as the source of excess noise, from a full-scale engine.

A brief description is given of attempts to reduce jet noise without incurring an undue penalty in the loss of nozzle efficiency. Attempts to understand their noise reduction characteristics on the basis of the Lighthill acoustic analogy are shown to be relatively unsuccessful. However, the main feature of all noise reduction schemes is shown to be the large changes in the jet flow structure that result, notably a shielding of the high-speed flow near the jet boundary. It is shown that flow-acoustic interaction theory gives a more satisfactory explanation of the main changes to the radiated noise characteristics, especially within the zone of silence and an almost negligible change outside.

Finally, the importance of good, reliable, and accurate experimental data in all studies on aerodynamic noise is stressed. At best the theoretical work can only assist in providing a suitable framework in which to analyze the results and the presentation of the experimental data for prediction purposes.

References

1. Morley, A. W.: Estimation of Aeroplane Noise Level: Some Empirical Laws With an Account of the Present Experiments on Which They Are Based. *Aircr. Eng.*, vol. 11, no. 123, May 1939, pp. 187–189.
2. Lighthill, M. J.: On Sound Generated Aerodynamically: I. General Theory. *Proc. Royal Soc. London*, ser. A, vol. 211, no. 1107, Mar. 20, 1952, pp. 564–587.
3. Lighthill, M. J.: On Sound Generated Aerodynamically: II. Turbulence as a Source of Sound. *Proc. Royal Soc. London*, ser. A, vol. 222, no. 1148, Feb. 23, 1954, pp. 1–32.
4. Lighthill, M. J.: The Bakerian Lecture, 1961: Sound Generated Aerodynamically. *Proc. Royal Soc. London*, ser. A, vol. 267, no. 1329, May 8, 1962, pp. 147–182.
5. Lighthill, M. J.: Jet Noise. *AIAA J.*, vol. 1, no. 7, July 1963, pp. 1507–1517.
6. Westley, R.; and Lilley, G. M.: *An Investigation of the Noise Field From a Small Jet and Methods for Its Reduction.* Rep. No. 53, College of Aeronautics, Cranfield (England), Jan. 1952.
7. Powell, Alan: *Some Experimental Observations on the Behaviour of Free Air Jets.* lst Noise Research Interim Note, Dep. of Aeronautical Engineering, Univ. College, Southampton (England), May 30, 1951.
8. Powell, Alan: The Noise of Choked Jets. *J. Acoust. Soc. America*, vol. 25, no. 3, May 1953, pp. 385–389.
9. Powell, Alan: A Survey of Experiments on Jet Noise. *Aircr. Eng.*, vol. XXVI, no. 299, Jan. 1954, pp. 2–9.
10. Lassiter, Leslie W.; and Hubbard, Harvey H.: *Experimental Studies of Noise From Subsonic Jets in Still Air.* NACA TN 2757, 1952.
11. Lassiter, Leslie W.; and Hubbard, Harvey H.: *The Near Noise Field of Static Jets and Some Model Studies of Devices for Noise Reduction.* NACA Rep. 1261, 1956. (Supersedes NACA TN 3187.)
12. Lassiter, L. W.; and Hubbard, H. H.: Some Results of Experiments Relating to the Generation of Noise in Jets. *J. Acoust. Soc. America*, vol. 27, no. 3, May 1955, pp. 431–437.
13. Fitzpatrick, H. M.; and Lee, Robert: *Measurements of Noise Radiated by Subsonic Air Jets.* Rep. 835, David W. Taylor Model Basin, Navy Dep., Nov. 1952.
14. Richards, E. J.: Research on Aerodynamic Noise From Jets and Associated Problems. *J. Royal Aeronaut. Soc.*, vol. 57, no. 509, May 1953, pp. 318–342.
15. Gerrard, J. H.: An Investigation of the Noise Produced by a Subsonic Air Jet. *J. Aeronaut. Sci.*, vol. 23, no. 9, Sept. 1956, pp. 855–866.
16. Greatrex, F. B.: Engine Noise. *J. Royal Aeronaut. Soc.*, vol. 58, no. 250, Apr. 1954, pp. 223–231.
17. Greatrex, F. B.: Jet Noise. Preprint No. 559, S.M.F. Publ. Fund Preprint, Inst. of Aeronautical Science, Inc., June 1955.
18. Goldstein, Marvin E.: *Aeroacoustics.* McGraw-Hill Book Co., c.1976.
19. Curle, N.: The Influence of Solid Boundaries Upon Aerodynamic Sound. *Proc. Royal Soc. London*, ser. A, vol. 231, no. 1187, Sept. 20, 1955, pp. 505–514.
20. Phillips, O. M.: On the Aerodynamic Surface Sound From a Plane Turbulent Boundary Layer. *Proc. Royal Soc. London*, ser. A, vol. 234, no. 1198, Feb. 21, 1956, pp. 327–335.

21. Kraichnan, Robert H.: Noise Transmission From Boundary Layer Pressure Fluctuations. *J. Acoust. Soc. America,* vol. 29, no. 1, Jan. 1957, pp. 65–80.
22. Doak, P. E.: Acoustic Radiation From a Turbulent Fluid Containing Foreign Bodies. *Proc. Royal Soc. London,* ser. A, vol. 254, no. 1276, Jan. 19, 1960, pp. 129–145.
23. Hodgson, Thomas H.: Pressure Fluctuations in Shear Flow Turbulence. Ph.D. Thesis, Univ. of London, 1962.
24. Ffowcs Williams, J. E.; and Hawkings, D. L.: Sound Generation by Turbulence and Surfaces in Arbitrary Motion. *Philos. Trans. Royal Soc. London,* ser. A, vol. 264, no. 1151, May 8, 1969, pp. 321–342.
25. Tam, Christopher K. W.: Intensity, Spectrum, and Directivity of Turbulent Boundary Layer Noise. *J. Acoust. Soc. America,* vol. 57, no. 1, Jan. 1975, pp. 25–34.
26. Landahl, Marten T.: Wave Mechanics of Boundary Layer Turbulence and Noise. *J. Acoust. Soc. America,* vol. 57, no. 4, Apr. 1975, pp. 824–831.
27. Whitham, G. B.: The Flow Pattern of a Supersonic Projectile. *Commun. Pure & Appl. Math.,* vol. 5, no. 3, Aug. 1952, pp. 301–348.
28. Guiraud, Jean-Piere: *Theorie du Bruit Ballistique Provoque en Atmosphere non Homogene par le Vol d'un Avion Supersonique.* ONERA Tech. Note No. 79, 1964.
29. Davies, P. O. A. L.; Fisher, M. J.; and Barratt, M. J.: The Characteristics of the Turbulence in the Mixing Region of a Round Jet *J. Fluid Mech.,* vol. 15, pt. 3, Mar. 1963, pp. 337–367.
30. Townsend, A. A.: *The Structure of Turbulent Shear Flow.* Cambridge Univ. Press, 1956.
31. Townsend, A. A.: *The Structure of Turbulent Shear Flow, Second ed.* Cambridge Univ. Press, c.1976.
32. Tanna, H. K.; Dean, P. D.; and Fisher, M. J.: The Influence of Temperature on Shock-Free Supersonic Jet Noise. *J. Sound & Vibration,* vol. 39, no. 4, Apr. 22, 1975, pp. 429–460.
33. Hoch, R. G.; Duponchel, J. P.; Cocking, B. J.; and Bryce, W. D.: Studies of the Influence of Density on Jet Noise. *J. Sound & Vibration,* vol. 28, no. 4, June 22, 1973, pp. 649–668.
34. Tanna, H. K.: An Experimental Study of Jet Noise. Part I: Turbulent Mixing Noise. *J. Sound & Vibration,* vol. 50, no. 3, Feb. 8, 1977, pp. 405–428.
35. Morfey, C. L.: Amplification of Aerodynamic Noise by Convected Flow Inhomogeneities. *J. Sound & Vibration,* vol. 31, no. 4, Dec. 22, 1973, pp. 391–397.
36. Fisher, M. J.; and Morfey, C. L.: Jet Noise. *Aerodynamic Noise,* AGARD-LS-80, Jan. 1977, pp. 3-1–3-23.
37. Morfey, C. L.; and Szewczyk, V. M.: *Jet Noise Modelling by Geometric Acoustics. Part I: Theory and Prediction Outside the Cone of Silence.* ISVR-TR-91-Pt. 1, Southampton Univ. (England), Sept. 1977.
38. Krasil'nikova, T. N.: Dipole Nature of Sound Radiation by Free Turbulence With Shear. *Fluid Dyn.,* vol. 10, no. 1, Jan.–Feb. 1975, pp. 94–98.
39. Ffowcs Williams, J. E.: The Noise From Turbulence Convected at High Speed. *Philos. Trans. Royal Soc. London,* ser. A, vol. 255, no. 1061, Apr. 18, 1963, pp. 469–503.
40. Seiner, John M.: Advances in High Speed Jet Aeroacoustics. AIAA-84-2275, Oct. 1984.
41. Morkovin, Mark V.: Effects of Compressibility on Turbulent Flows. *The Mechanics of Turbulence,* Gordon & Breach Scientific Publ., Inc., c.1964, pp. 367–380.
42. Bradshaw, P.: Compressible Turbulent Shear Layers. *Annual Review of Fluid Mechanics,* Volume 9, Milton Van Dyke, J. V. Wehausen, and John L. Lumley, eds., Annual Reviews Inc., 1977, pp. 33–54.
43. Lilley, G. M.: *On the Noise From Air Jets.* Rep. No. 20,376, British Aeronautical Research Council, Sept. 8, 1958.
44. Ribner, H. S.: The Generation of Sound by Turbulent Jets. *Volume 8 of Advances in Applied Mechanics,* H. L. Dryden and Th. von Kármán, eds., Academic Press, Inc., 1964, pp. 103–182.
45. Proudman, I.: The Generation of Noise by Isotropic Turbulence. *Proc. Royal Soc. London,* ser. A, vol. 214, no. 1116, Aug. 7, 1952, pp. 119–132.
46. Liepmann, Hans Wolfgang; and Laufer, John: *Investigations of Free Turbulent Mixing.* NACA TN 1257, 1947.
47. Crow, S. C.; and Champagne, F. H.: Orderly Structure in Jet Turbulence. *J. Fluid Mech.,* vol. 48, pt. 3, Aug. 16, 1971, pp. 547–591.
48. Michalke, A.: On the Inviscid Instability of the Hyperbolic-Tangent Velocity Profile. *J. Fluid Mech.,* vol. 19, pt. 4, Aug. 1964, pp. 543–556.
49. Michalke, A.: On Spatially Growing Disturbances in an Inviscid Shear Layer. *J. Fluid Mech.,* vol. 23, pt. 3, Nov. 1965, pp. 521–544.
50. Bradshaw, P.; Ferriss, D. H.; and Johnson, R. F.: Turbulence in the Noise-Producing Region of a Circular Jet. *J. Fluid Mech.,* vol. 19, pt. 4, Aug. 1964, pp. 591–624.

51. Morkovin, Mark V.: *Recent Insights Into Instability and Transition to Turbulence in Open-Flow Systems.* NASA CR-181693, ICASE Rep. No. 88-44, 1988.
52. Crow, S. C.: Aerodynamic Sound Emission as a Singular Perturbation Problem. *Stud. Appl. Math.*, vol. XLIX, no. 1, Mar. 1970, pp. 60–83.
53. Hussain, A. K. M. F.: Coherent Structures—Reality and Myth. *Phys. Fluids,* vol. 26, no. 10, Oct. 1983, pp. 2816–2850.
54. Zaman, K. B. M. Q.; and Hussain, A. K. M. F.: Natural Large-Scale Structures in the Axisymmetric Mixing Layer. *J. Fluid Mech.,* vol. 138, Jan. 1984, pp. 325–351.
55. Laufer, J.; Kaplan, R. E.; and Chu, W. T.: On the Generation of Jet Noise. *Noise Mechanisms,* AGARD-CP-131, Mar. 1974, pp. 21-1–21-6.
56. Grant, H. L.: The Large Eddies of Turbulent Motion. *J. Fluid Mech.,* vol. 4, pt. 2, June 1958, pp. 149–190.
57. Yule, A. J.: Large-Scale Structure in the Mixing Layer of a Round Jet. *J. Fluid Mech.*, vol. 89, pt. 3, Dec. 13, 1978, pp. 413–432.
58. Browand, F. K.; and Weidman, P. D.: Large Scales in the Developing Mixing Layer. *J. Fluid Mech.,* vol. 76, pt. 1, July 14, 1976, pp. 127–144.
59. Liu, J. T. C.: Developing Large-Scale Wavelike Eddies and the Near Jet Noise Field. *J. Fluid Mech.,* vol. 62, pt. 3, Feb. 11, 1974, pp. 437–464.
60. Landahl, Marten T.: A Wave-Guide Model for Turbulent Shear Flow. *J. Fluid Mech.,* vol. 29, pt. 3, Sept. 6, 1967, pp. 441–459.
61. Morris, Philip J.; Giridharan, Manampathy G.; and Lilley, Geoffrey M.: On the Turbulent Mixing of Compressible Free Shear Layers. *Proc. Royal Soc. London,* ser. A, vol. 431, no. 1882, Nov. 8, 1990, pp. 219–243.
62. Brown, Garry L.; and Roshko, Anatol: On Density Effects and Large Structure in Turbulent Mixing Layers. *J. Fluid. Mech.,* vol. 64, pt. 4, July 24, 1974, pp. 775–816.
63. Moore, C. J.: The Role of Shear-Layer Instability Waves in Jet Exhaust Noise. *J. Fluid Mech.,* vol. 80, pt. 2, Apr. 25, 1977, pp. 321–367.
64. Ho, Chih-Ming; and Huang, Lein-Saing: Subharmonics and Vortex Merging in Mixing Layers. *J. Fluid Mech.,* vol. 119, June 1982, pp. 443–473.
65. Gaster, M.; Kit, E.; and Wygnanski, I.: Large-Scale Structures in a Forced Turbulent Mixing Layer. *J. Fluid Mech.,* vol. 150, Jan. 1985, pp. 23–39.
66. Tam, C. K. W.; and Morris, P. J.: Tone Excited Jets, Part V: A Theoretical Model and Comparison With Experiment. *J. Sound & Vibration,* vol. 102, no. 1, Sept. 8, 1985, pp. 119–151.
67. Wygnanski, I.; and Fiedler, H. E.: The Two-Dimensional Mixing Region. *J. Fluid Mech.,* vol. 41, pt. 2, Apr. 13, 1970, pp. 327–361.
68. Forstall, Walton, Jr.; and Shapiro, Ascher H.: Momentum and Mass Transfer in Coaxial Gas Jets. *J. Appl. Mech.,* vol. 17, no. 4, Dec. 1950, pp. 399–408. (Discussion in *J. Appl. Mech.,* vol. 18, no. 2, June 1951, pp. 219–220.)
69. Abramovich, G. N.: *The Theory of Turbulent Jets.* M.I.T. Press, c.1963.
70. Morris, Phillip J.: Turbulence Measurements in Subsonic and Supersonic Axisymmetric Jets in a Parallel Stream. *AIAA J.*, vol. 14, no. 10, Oct. 1976, pp. 1468–1475.
71. Lau, Jark C.; Morris, Phillip J.; and Fisher, Michael J.: Measurements in Subsonic and Supersonic Free Jets Using a Laser Velocimeter. *J. Fluid Mech.,* vol. 93, pt. 1, July 12, 1979, pp. 1–27.
72. Craven, A. H.: *The Effect of Density on Jet Flow at Subsonic Speeds.* Rep. No. 120, College of Aeronautics, Cranfield (England), July 1959.
73. Batchelor, G. K.: Pressure Fluctuations in Isotropic Turbulence. *Proc. Cambridge Philos. Soc.,* vol. 47, pt. 2, Apr. 1951, pp. 359–374.
74. Nagamatsu, H. T.; Sheer, R. E., Jr.; and Horvay, G.: Supersonic Jet Noise Theory and Experiments. *Basic Aerodynamic Noise Research,* Ira R. Schwartz, ed., NASA SP-207, 1969, pp. 17–51.
75. Ribner, H. S.: Strength Distribution of Noise Sources Along a Jet. *J. Acoust. Soc. America,* vol. 30, no. 9, Sept. 1958, pp. 876–877.
76. Lush, P. A.: Measurements of Subsonic Jet Noise and Comparison With Theory. *J. Fluid Mech.,* vol. 46, pt. 3, Apr. 1971, pp. 477–500.
77. Lockheed-Georgia Co.: *Effects of Forward Velocity on Turbulent Jet Mixing Noise.* NASA CR-2702, 1976.
78. Buckley, R.; and Morfey, C. L.: Effects on Jet Mixing Noise: Scaling Laws Predicted for Single Jets From Flight Simulation Data. AIAA-83-0748, Apr. 1983.
79. Michalke, A.; and Michel, U.: Prediction of Jet Noise in Flight From Static Tests. *J. Sound & Vibration,* vol. 67, no. 3, Dec. 8, 1979, pp. 341–367.

80. Bushell, K. W.: Measurement and Prediction of Jet Noise in Flight. AIAA Paper 75-461, Mar. 1975.
81. Cocking, B. J.; and Bryce, W. D.: Subsonic Jet Noise in Flight Based on Some Recent Wind-Tunnel Tests. AIAA Paper 75-462, Mar. 1975.
82. Drevet, P.; Duponchel, J. P.; and Jacques, J. R.: The Effect of Flight on Jet Noise as Observed on the Bertin Aérotrain. *J. Sound & Vibration,* vol. 54, no. 2, Sept. 22, 1977, pp. 173–201.
83. Crighton, D. G.: Mechanisms of Excess Jet Noise. *Noise Mechanisms,* AGARD-CP-131, Mar. 1974, pp. 14-1–14-7.
84. Callaghan, Edmund E.; and Coles, Willard D.: *Far Noise Field of Air Jets and Jet Engines.* NACA Rep. 1329, 1957. (Supersedes NACA TN 3590 by Callaghan and Coles and TN 3591 by Coles and Callaghan.)
85. Smith, W.: The Use of a Rotating Arm Facility To Study Flight Effects on Jet Noise. *Proceedings of the Second International Symposium on Air Breathing Engines,* Royal Aeronautical Soc., 1974.
86. Williams, John: Ground-Based Facilities With Forward-Speed Representation for Aircraft Noise Research. *Aerodynamic Noise,* AGARD-LS-80, Jan. 1977, pp. 11-1–11-43.
87. Morfey, C. L.; and Tester, B. J.: Noise Measurements in a Free Jet Flight Simulation Facility: Shear Layer Refraction and Facility-to-Flight Corrections. *J. Sound & Vibration,* vol. 54, no. 1, Sept. 8, 1977, pp. 83–106.
88. House, Michael E.: Aero-Acoustic Measurement and Analysis Techniques. *Aerodynamic Noise,* AGARD-LS-80, Jan. 1977, pp. 8-1–8-48.
89. Smith, Michael J. T.: *Aircraft Noise.* Cambridge Univ. Press, c.1989.
90. House, Michael E.: Aircraft Flyover Measurements. *Aerodynamic Noise,* AGARD-LS-80, Jan. 1977, pp. 9-1–9-14.
91. Siddon, Thomas E.: Surface Dipole Strength by Cross-Correlation Method. *J. Acoust. Soc. America,* vol. 53, no. 2, Feb. 1973, pp. 619–633.
92. Chu, W. T.; Laufer, J.; and Kao, K.: Noise Source Distribution in Subsonic Jets. *INTER-NOISE 72 Proceedings—International Conference on Noise Control Engineering,* Malcolm J. Crocker, ed., Inst. of Noise Control Engineering, 1972, pp. 472–476.
93. Grosche, F.-R.: Distributions of Sound Source Intensities in Subsonic and Supersonic Jets. *Noise Mechanisms,* AGARD-CP-131, Mar. 1974, pp. 4-1–4-10.
94. Billingsley, J.; and Kinns, R.: The Acoustic Telescope. *J. Sound & Vibration,* vol. 48, no. 4, Oct. 22, 1976, pp. 485–510.
95. Fisher, M. J.; Harper-Bourne, M.; and Glegg, S. A. L.: Jet Engine Noise Source Location: The Polar Correlation Technique. *J. Sound & Vibration,* vol. 51, no. 1, Mar. 8, 1977, pp. 23–54.
96. Ffowcs Williams, J. E.: Technical Evaluation Report. *Noise Mechanisms,* AGARD-CP-131, Mar. 1974, pp. vii–xx.
97. Tester, B. J.; and Fisher, M. J.: Engine Noise Source Breakdown: Theory, Simulation and Results. AIAA-81-2040, Oct. 1981.
98. Jones, Ian S. F.: Fluctuating Turbulent Stresses in the Noise-Producing Region of a Jet. *J. Fluid Mech.,* vol. 36, pt. 3, May 1, 1969, pp. 529–543.
99. Ribner, H. S.: New Theory of Jet-Noise Generation, Directionality, and Spectra. *J. Acoust. Soc. America,* vol. 31, no. 2, Feb. 1959, pp. 245–246.
100. Ribner, Herbert S.: Perspectives on Jet Noise. AIAA-81-0428, Jan. 1981.
101. Ribner, H. S.: Quadrupole Correlations Governing the Pattern of Jet Noise. *J. Fluid Mech.,* vol. 38, pt. 1, Aug. 14, 1969, pp. 1–24.
102. Michalke, Alfons: An Expansion Scheme for the Noise From Circular Jets. *Z. Flugwiss.,* Jahrg. 20, Heft 6, June 1972, pp. 229–237.
103. Michalke, A.; and Fuchs, H. V.: On Turbulence and Noise of an Axisymmetric Shear Flow. *J. Fluid Mech.,* vol. 70, pt. 1, July 15, 1975, pp. 179–205.
104. Mollo-Christensen, Erik; Kolpin, Marc A.; and Martuccelli, John R.: Experiments on Jet Flows and Jet Noise Far-Field Spectra and Directivity Patterns. *J. Fluid Mech.,* vol. 18, pt. 2, Feb. 1964, pp. 285–301.
105. Krishnappa, G.; and Csanady, G. T.: An Experimental Investigation of the Composition of Jet Noise. *J. Fluid Mech.,* vol. 37, pt. 1, June 5, 1969, pp. 149–159.
106. Lilley, Geoffrey M.; Plumblee, Harry E.; Strahle, Warren C.; Ruo, Song-Yeong; and Doak, Philip E.: *The Generation and Radiation of Supersonic Jet Noise. Volume IV—Theory of Turbulence Generated Jet Noise, Noise Radiation From Upstream Sources, and Combustion Noise.* AFAPL-TR-72-53, Vol. IV, U.S. Air Force, July 1972. (Available from DTIC as AD 749 139.)

107. Tester, B. J.; and Morfey, C. L.: Developments in Jet Noise Modelling—Theoretical Predictions and Comparisons With Measured Data. *J. Sound & Vibration,* vol. 46, no. 1, May 8, 1976, pp. 79–103.
108. Mani, R.: A Moving Source Problem Relevant to Jet Noise. *J. Sound & Vibration,* vol. 25, no. 2, 1972, pp. 337–347.
109. Mani, Ramani: The Issue of Convective Amplification in Jet Noise. *Noise Mechanisms,* AGARD-CP-131, Mar. 1974, pp. 10-1–10-12.
110. Mani, R.: The Influence of Jet Flow on Jet Noise. *J. Fluid Mech.*, vol. 73, pt. 4, 1976.
Part 1. The Noise of Unheated Jets, pp. 753–778.
Part 2. The Noise of Heated Jets, pp. 779–793.
111. Berman, C. H.: Some Analytical Consideration in Jet Noise Prediction. *Mechanics of Sound Generation in Flows,* E.-A. Müller, ed., Springer-Verlag, 1979, pp. 160–166.
112. Balsa, T. F.: The Acoustic Field of Sources in Shear Flow With Application to Jet Noise: Convective Amplification. *J. Fluid Mech.,* vol. 79, pt. 1, Jan. 1977, pp. 33–47.
113. Dowling, A. P.; Ffowcs Williams, J. E.; and Goldstein, M. E.: Sound Production in a Moving Stream. *Philos. Trans. Royal Soc. London,* ser. A, vol. 288, no. 1353, Mar. 23, 1978, pp. 321–349.
114. Crighton, D. G.: Why Do the Acoustics and the Dynamics of a Hypothetical Mean Flow Bear on the Issue of Sound Generation by Turbulence? *Mechanics of Sound Generation in Flows,* E.-A. Müller, ed., Springer-Verlag, 1979, pp. 1–11.
115. Crighton, D. G.: Acoustics as a Branch of Fluid Mechanics. *J. Fluid Mech.,* vol. 106, May 1981, pp. 261–298.
116. Legendre, Robert: *Bruits Emis par la Turbulence.* ONERA Publ. 1981-3, 1981.
117. Phillips, O. M.: On the Generation of Sound by Supersonic Turbulent Shear Layers. *J. Fluid Mech.,* vol. 9, pt. 1, Sept. 1960, pp. 1–28.
118. Lighthill, M. J.: On the Energy Scattered From the Interaction of Turbulence With Sound or Shock Waves. *Proc. Cambridge Philos. Soc.,* vol. 49, pt. 3, Cambridge Univ. Press, July 1953, pp. 531–551.
119. Lilley, G. M.: Aerodynamic Noise. *J. Royal Aeronaut. Soc.,* vol. 58, Apr. 1954, pp. 235–239.
120. Lilley, G. M.: On the Noise From Jets. *Noise Mechanisms,* AGARD-CP-131, Mar. 1974, pp. 13.1–13.12.
121. Pao, S. P.: Aerodynamic Noise Emission From Turbulent Shear Layers. *J. Fluid Mech.,* vol. 59, pt. 3, July 1973, pp. 451–479.
122. Howe, M. S.: Contributions to the Theory of Aerodynamic Sound, With Application to Excess Jet Noise and the Theory of the Flute. *J. Fluid Mech.,* vol. 71, pt. 4, Oct. 28, 1975, pp. 625–673.
123. Powell, Alan: Theory of Vortex Sound. *J. Acoust. Soc. America,* vol. 36, no. 1, Jan. 1964, pp. 177–195.
124. Yates, John E.; and Sandri, Guido: Bernoulli Enthalpy: A Fundamental Concept in the Theory of Sound. AIAA Paper 75-439, Mar. 1975.

5 Noise From Turbulent Shear Flows

Lead author

M. E. Goldstein
NASA Lewis Research Center
Cleveland, Ohio

Introduction

This chapter is primarily concerned with the generation of sound in turbulent shear flows with high Reynolds numbers. The subject became a serious scientific discipline in the early 1950's when Lighthill (refs. 1 and 2) published his acoustic analogy theory of jet noise. That work has more or less dominated the subsequent development of this field, which is still somewhat incomplete and has undergone little change in the past several years.

Lighthill achieved considerable success in explaining some of the most prominent features of the experimentally observed jet sound field (such as the directivity patterns of the overall sound pressure levels), but when more detailed experiments were conducted (refs. 3 and 4, for example) it became clear that there were other, more detailed features (such as the directivity patterns of the acoustic radiation in individual frequency bands) that could not be explained by Lighthill's analogy. Reference 5 extended the analogy to account for such features, but attempts to explain the new observations were mainly based on more complex analogies such as those of references 6 to 9. All these analogies involve, in one form or another, a nonlinear wave operator that eventually must be linearized before meaningful calculations can be carried out.

Lighthill's approach is discussed in considerable detail in chapter 2 of Goldstein (ref. 10). This chapter therefore places little emphasis on the acoustic analogy, but rather concentrates on an alternative approach which may be more readily adapted for use on large-scale computers to obtain more detailed information about the sound field than would be possible from the acoustic analogy. This approach amounts to little more than calculating the unsteady flow that produces the sound simultaneously with the resulting sound field. One starts from some prescribed upstream state that is ideally specified just ahead of this region where the sound generation takes place. To make progress without resorting to full-scale numerical computation requires that the governing equations be linearized about some appropriate mean flow. But that ultimately has to be done, either implicitly or explicitly, even with the acoustic analogy approach. I have no doubt that the day will come when turbulence-generated sound is calculated directly from the Navier-Stokes equation, but, to my knowledge, that has yet to be done. There is still much to be learned from the existing work, which has often led to relatively simple formulas that show encouraging agreement with experiments and produce a great deal of insight into and physical understanding of the sound generation process. This chapter is a somewhat selective review of that work.

The use of linearized theory to calculate turbulent flows or, better yet, changes in turbulent flows is a branch of turbulence theory now known as "rapid distortion theory." (See ref. 11.) It assumes that the following conditions are satisfied (ref. 12): (1) $u'/U << 1$, where u' is the rms turbulence velocity and U is the local mean-flow velocity, and (2) the interaction or change being calculated is completed in a time, say τ_I, that is short compared with τ_{decay}, where τ_{decay} is the decay time or lifetime of a typical turbulent eddy $O(\ell/u')$, ℓ being the characteristic size of turbulent eddies. Rapid-distortion calculations are usually based on the inviscid equations—an approximation that is justified when both the mean-flow and the turbulence Reynolds numbers are large. The important point here is that *the radiated sound field can be determined as a by-product of any such rapid-distortion calculation, as long as compressibility effects are retained.*

Solid-Surface Effects

While it might seem most logical to begin by omitting solid-surface effects and to include them only after the turbulence self-noise problem has been appropriately dealt with, it turns out that the solid boundaries actually simplify the problem and allow a more rigorous treatment in at least some cases. Consider then a high Reynolds number turbulent air jet such as that shown schematically in figure 1, where U_j is the jet velocity. The maximum turbulence occurs along the centerline of the initial mixing layer, indicated by the dashed line in the figure. Here the ratio of the rms turbulence velocity to the local mean-flow velocity is roughly 0.24 (ref. 13), which is not all that small but would probably still be considered to be an acceptable "small parameter" to many classic applied mathematicians. Condition (1) is therefore reasonably well satisfied.

Now suppose that a semi-infinite, but infinitesimally thin, flat plate is inserted into the flow as shown in figure 1. Then the interaction between the turbulence and the leading edge will be completed in a time $\tau_I = O(\ell/U)$, which is fairly small compared with $\tau_{\text{decay}} = O(\ell/u')$, considering the smallness of the turbulence intensity. Thus, inviscid rapid distortion theory applies, and the interaction between the turbulence and the edge can be calculated by linearizing the inviscid equations (the Euler equations) about the mean flow.

Since the ratio of the cross-stream to streamwise components of the mean-flow velocity is of the order of $(u'/U)^2$ (ref. 14), the order of approximation will certainly not be diminished if this flow is taken to be a unidirectional transversely sheared flow. The important advantage of using this flow is that it is itself a solution of the inviscid equations (for any velocity profile). The resulting expansion is then a rational perturbation that can, in principle, be carried to arbitrary order without internal inconsistency. The lowest order equations are now the same as those used

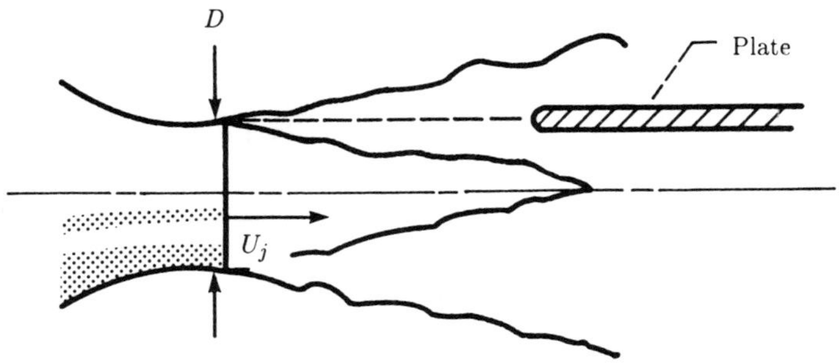

Figure 1. Plate embedded in turbulent air jet.

in inviscid stability theory (i.e., the Rayleigh equations; see ref. 15), and as already indicated, the radiated sound field can be determined as part of the solution to these equations—provided, of course, that compressibility effects are retained.

Representation of Incident Turbulence

To determine the radiated sound field, one must first decide on an appropriate representation of the incident turbulence. This representation would be rather easy if the mean flow were completely uniform, since any solution for the unsteady velocity-pressure fluctuations would then be decomposed into the sum of an "acoustic" solution that carries no vorticity and a "vortical" solution that produces no pressure fluctuations and is often referred to as the "gust" or "hydrodynamic" solution. The vortical solution is used to represent the incident turbulence in most problems that involve the interaction of turbulence with solid surfaces embedded in uniform mean flow. Its suitability for this purpose is largely due to the following reasons:

1. It does not become infinite anywhere in space, even in the absence of solid surfaces, so it can describe the turbulence field that would exist if the surfaces were not present.
2. It involves two arbitrary "convected" quantities that can be specified as upstream boundary conditions to describe the turbulence entering the interaction zone in any given problem. This seems to be the appropriate degree of generality, because the vorticity is a convected quantity that has only two independent components (since its divergence must vanish).
3. It has no acoustic radiation field at subsonic speeds and will, in fact, vanish exponentially fast at transverse infinity if the mean and unsteady vorticity fields are sufficiently compact.

Decomposition of the solution into completely decoupled acoustic and vortical parts is no longer possible when the mean flow is nonuniform, but the compressible Rayleigh equations still possess a solution that has the three properties listed above and, in fact, approach the vortical solution on a uniform mean flow in the limit as the mean flow approaches a uniform flow (refs. 16 to 18). This would then seem to be the natural generalization of the vortical solution to nonuniform flows, and it would therefore seem appropriate to refer to it as the gust, or hydrodynamic, solution and, more importantly, to use it to represent the incidence turbulence.

In the general case, this solution can be written as follows (refs. 16 and 17):

$$u_\sigma = A_\sigma\left(\frac{x_1}{U(\mathbf{x}_t)} - t, \mathbf{x}_t\right) + \int_{-\infty}^{\infty}\int_{-\infty}^{\infty} G_\sigma(\mathbf{x}, t|\mathbf{y}, \tau)\omega_c\left(\frac{y_1}{U(\mathbf{y}_t)} - \tau, \mathbf{y}_t\right) d\mathbf{y}\, d\tau \qquad (\sigma = 1, 2, 3, 4) \tag{1}$$

where u_σ denotes one of the perturbation velocity components when $\sigma = 1, 2, 3$ and u_4 denotes the associated normalized pressure fluctuation. For this equation, t is the time; (x_1, x_2, x_3) are Cartesian coordinates, with x_1 in the mean-flow direction; $\mathbf{x}_t = \{x_2, x_3\}$ in the transverse direction; G_σ is a free-space vector Green's function for the compressible Rayleigh equations, a slightly unusual case in that it is defined by placing the convective derivative (based on the mean-flow velocity) of the delta function on the right side of the Rayleigh equations, rather than being the delta function itself; ω_c is a convected quantity that can be arbitrarily specified as an upstream boundary condition; and A_σ is another convected quantity. The fourth component of A_σ is identically zero, and the remaining three components form a three-dimensional vector that has zero divergence and is perpendicular to the gradient of the mean-flow velocity. Therefore, A_σ has one independent component, and this component can be an arbitrary function of its argument.

Equation (1) thus involves two arbitrary convected quantities. It is certainly defined over all space, since G_σ is the free-space Green's function. That it is a homogeneous solution of the linearized Rayleigh equations can be seen by inspection. For example, substituting the second term into the Rayleigh equations will, in view of the definition of G_σ, transform the integrand into the convective derivative of $\delta(x - y)\ \sigma(t - \tau)$ times ω_c. Integration by parts produces a convective derivative of ω_c, which by construction is identically zero. This inspection shows that the second member of equation (1) is indeed a homogeneous solution of the Rayleigh equations. It is easy to show that the first member also has this property.

Sound Generation and the Role of Instability Waves

Returning now to the problem of a large, flat plate embedded in a turbulent shear flow, we can, as argued above, use the gust solution to represent the incident turbulence. Since this gust solution does not satisfy the boundary condition of zero normal velocity at the plate, it is necessary to add another solution to cancel this component of velocity. Unlike the gust solution, this latter solution does not vanish exponentially fast at infinity, but rather behaves like a propagating acoustic wave there (ref. 17). In other words, the plate is able to "scatter" the nonpropagating motion associated with the gust into a propagating acoustic wave (ref. 19).

The problem also possesses an eigenfunction solution associated with the spatially growing instability wave that can propagate downstream from the edge of the inflectional mean-velocity profile (refs. 20 and 21). The solution is therefore not unique! It could be made unique if we required that it remain bounded at infinity (since that would eliminate the eigenfunction solution that grows without bound there). But since the linearization is only valid in the vicinity of the leading edge, it is probably not appropriate to impose a "boundary" condition far downstream in the flow where all sorts of nonlinear effects will have had a chance to intervene (ref. 22).

One can therefore look for an alternative way to make the solution unique. This can be done by treating the steady-state solution, which is, of course, the one of interest here, as the long time limit of the solution to an initial-value problem. A "causality" condition is then imposed in the sense that the solution is required to be identically zero before the initial time when the incident disturbance is "turned on" (ref. 20).

But in reference 22 it is argued that an initial condition imposed in the distant past may not be relevant to the steady-state solution, since the linearization might only be valid over a relatively short interval of time. One might therefore consider a third way of making the solution unique. This amounts to using the eigenfunction solution to eliminate the leading-edge singularity that appears in both the bounded and the causal solution by satisfying a leading-edge "Kutta" condition (ref. 21). This procedure may be rationalized by noting that the instability wave represents downstream vortex shedding that could adjust itself to eliminate the singularity in the inviscid solution and thereby prevent any flow separation that would otherwise occur at a very sharp edge.

Comparison With Data

It is not entirely clear which of these three solutions is correct, but I suspect the argument in reference 22 is invalid and that imposition of causality is probably appropriate. In reference 17, I compared the theory with the data of reference 23, in which the sound radiated was measured in 1/3-octave frequency bands as a function of the angle from the jet axis in a plane perpendicular to that of the plate. Comparison of the experiment and theory is shown in figure 2. The top part of the figure corresponds to the high-frequency limit where the instability waves are "cut-off" and the issues of causality and Kutta conditions are irrelevant. However, at low frequencies the causal solution, which is shown at the bottom, is strongly affected by the instability wave. The agreement between experiment and theory is good, but the causal and leading-edge Kutta conditions have the same low-frequency limit, and one cannot conclude from this comparison which is correct. However, the bounded solution behaves quite differently in this limit and consequently does not agree with the data.

Sound Generated by Turbulence Interacting With Itself: The Jet Noise Problem

Having achieved some success in using linear theory for the turbulence–leading-edge interaction, it is natural to try using it to calculate the sound generated by turbulence interacting with itself (i.e., to deal with the problem of jet noise). I have already pointed out that the ratio of rms turbulence velocity to local mean-flow velocity is reasonably small in the region of maximum turbulence, so that the first requirement for the validity of the rapid distortion theory is satisfied. (See *Introduction.*) However, the interaction time τ_I, which in the present context should be taken as the time for the sound generation to occur, is now equal to the decay time τ_{decay} of the turbulence, and thus the second requirement of the theory is not satisfied. But with no better alternative at hand, we might still attempt to introduce the same small parameter as before (i.e., u'/U) and carry the corresponding asymptotic expansion to its logical conclusion. Like the more ad hoc acoustic analogy approach, this systematic procedure assures that all appropriate conservation laws

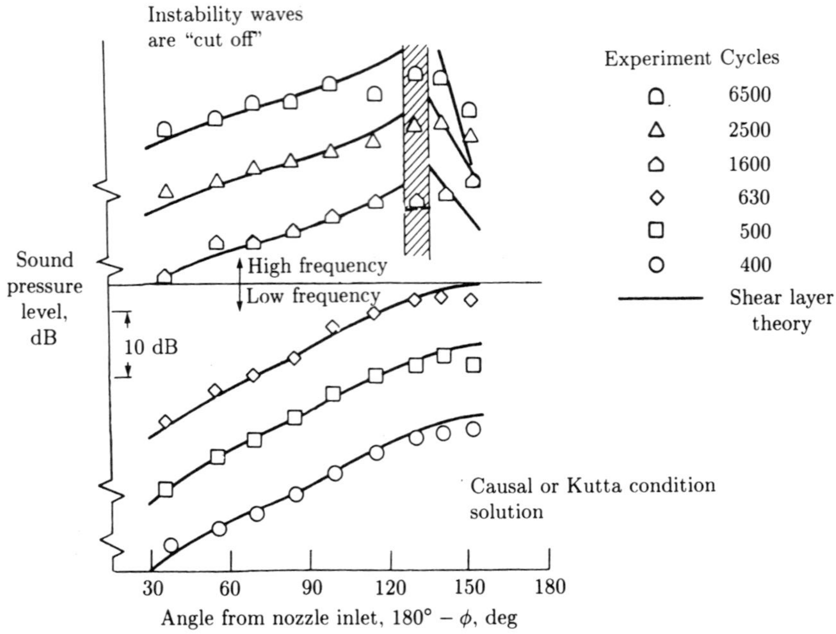

Figure 2. Comparison of causal or leading-edge Kutta condition solution with data of ref. 23 for $U_j = 213$ m/sec.

are satisfied and that the acoustic sources are of the appropriate multipole order. But it seems to have certain advantages over the acoustic analogy in that it provides a "rational" framework for assessing the internal consistency of the various jet noise analyses. It may also apply to some physically realizable flow, which we hope is not too different from the real turbulent flow of interest, and finally, it provides a method for identifying acoustic sources and distinguishing acoustic and nonacoustic components of the unsteady motion.

The Basic Equation

The lowest order equations are, on the face of it, the same as before, that is, they are the compressible Rayleigh's equations. It is well-known (ref. 15) that the velocity components can be eliminated between these equations to obtain a single equation for the normalized first-order pressure fluctuation $\Pi_1 \equiv p_1/\rho_o c_o^2$, where p_1 is the actual first-order pressure fluctuation, $\rho_o(\mathbf{x}_t)$ is the mean-flow density, and $c_o(\mathbf{x}_t)$ is the mean-flow sound speed, where the latter two quantities depend only on the cross-stream coordinate $\mathbf{x}_t = \{x_2, x_3\}$, with (x_1, x_2, x_3) denoting Cartesian coordinates and x_1 in the mean-flow direction. This equation can be written symbolically as

$$L\Pi_1 = 0 \tag{2}$$

where L denotes the third-order linear wave operator:

$$L \equiv \frac{D}{Dt}\left(\frac{D^2}{Dt^2} - \nabla \cdot c_o^2 \nabla\right) + 2c_o^2(\nabla U) \cdot \nabla \frac{\partial}{\partial x_1} \tag{3}$$

where t denotes the time and $D/Dt = \partial/\partial t + U\partial/\partial x_1$ is the convective derivative based on the mean-flow velocity $U(\mathbf{x}_t)$.

Since solid boundaries are acoustically irrelevant for the turbulence self-noise problem, it is appropriate to suppose that the flow is defined over all space. Then (for reasons given in the section on solid surface interactions) the gust (hydrodynamic) solution (i.e., eq. (1)) is a relevant solution of equation (2). But equation (2) also has (spatially growing) instability-wave solutions which can exist whenever the mean flow is inflectional (ref. 15). Since many investigators (e.g., refs. 24 to 29) have argued that these latter solutions correspond to the experimentally observed large-scale turbulent structures, it would seem appropriate to identify the gust solution with the "fine-grained" (or relatively fine-grained) turbulent motions.

However, there are experimentally observed motions that, on a global basis, seem to bear little resemblance to any motion that can be represented by either the gust or the linear instability-wave solution. This should come as no surprise, since we have already noted that the linearized solution can at best remain valid over relatively small streamwise distances.

We have seen that the gust solution produces no acoustic radiation at subsonic speeds, and the same can be said for the instability waves. (However, see below.) The asymptotic expansion must therefore be carried to the next order if it is to be used to calculate radiated sound. The normalized second-order pressure fluctuation Π_2 again satisfies a third-order wave equation, but it is more convenient to work with the isentropic density fluctuation:

$$\Pi \equiv \Pi_2 - \frac{\gamma - 1}{2}\Pi_1^2 \tag{4}$$

where γ is the specific heat ratio. Then Π satisfies

$$L\Pi = \Gamma \tag{5}$$

which, except for the inhomogeneous source term

$$\Gamma \equiv \frac{D}{Dt}\nabla \cdot \mathbf{f} - 2\frac{\partial \mathbf{f}}{\partial x_1} \cdot \nabla U \tag{6}$$

is the same as equation (2) for the first-order normalized pressure fluctuation.

Equation (6) is identical to the source term that would be produced by an externally applied fluctuating force per unit mass $\mathbf{f} = \{f_1, f_2, f_3\}$ and might therefore be thought of as a dipole-type source, since a fluctuating force produces such a source when there is no mean flow. The force $\mathbf{f}$ is not arbitrary, of course, but is now given as a quadratic function of the first-order solutions, namely,

$$f_1 \equiv \frac{\partial}{\partial x_j} u_i^{(1)} u_j^{(1)} + c_1^2 \frac{\partial}{\partial x_1}\Pi_1 \qquad (i, j = 1, 2, 3) \tag{7}$$

where $u_i^{(1)}$ denotes the first-order velocity fluctuation, and $c_1^2 = \gamma R T_1$ is the squared first-order sound-speed fluctuation, R being the gas constant and T_1 being the first-order temperature fluctuation.

Equations (5) to (7), with some relatively minor differences, were first derived by Lilley (ref. 7), who used a generalization of the acoustic analogy approach. The result is now commonly referred to as Lilley's equation. In the present approach, it arises as the equation for the composite second-order pressure fluctuation with a source term Γ that involves only first-order solutions. Since these solutions, which satisfy the homogeneous equation (2), have no acoustic fields at subsonic speeds while the second-order solution does, our expansion provides a conceptual if not experimental mechanism for identifying acoustic and nonacoustic parts of the unsteady motion. But there are some complications.

The Sources of Sound

The second term in equation (7) represents a dipole-type source due to the temperature fluctuations in the flow (ref. 30). Although this source is of real significance in actual high-temperature jet exhausts, I will not discuss it in this chapter. I will concentrate instead on the first term, which, being the divergence of the fluctuating (first-order) Reynolds stress $u_i^{(1)}u_j^{(1)}$, corresponds to the source that would be produced by an externally applied fluctuating stress field. It might therefore be interpreted, by analogy with the zero mean-flow case, as a quadrupole-type source.

This latter term can be further decomposed into a number of subsources by separating the first-order solution $u_i^{(1)}$ into its gust and linear (spatially growing) instability-wave components and, as before, identifying the gust with the fine-grained turbulent motion. Unfortunately, this procedure cannot be carried to its logical conclusion because the linear instability waves, which grow without bound in a parallel mean flow, ultimately produce an unbounded source term in equation (5). It would then be inappropriate to use this equation to calculate the acoustic field, since it is its global, and not its local, solutions that must be used in such a calculation. However, the real flow is only locally parallel, and the slowly varying (rather than the parallel-flow) approximation should be used to represent the instability waves, as was done in references 31 and 32, for example. Then the source term in equation (5) remains bounded, since the local growth rate of the instability wave varies with the thickness of the jet or shear layer. (It first increases, reaches a maximum, and then becomes negative as the thickness increases.)

However, supersonically traveling waves can be produced as a by-product of this approximation, and these latter waves couple to the radiation field (ref. 32) when the first-order solution is rendered uniformly valid (through use of an appropriate singular perturbation procedure such as the method of multiple scales, ref. 33). The previous comment that the first-order solution has no radiation field therefore needs to be qualified. We return to this subsequently, but for now the important point is that the first term in **f** should then describe the sound generation due to the following types of interactions:

1. Linear instability wave and fine-grained turbulence
2. Linear instability wave and linear instability wave
3. Fine-grained turbulence and fine-grained turbulence

Of course, this list may be incomplete or even inappropriate since, as I already indicated, there are other types of large-scale motions in the jet that do not seem to be globally representable by either the gust or the instability wave. In any case, it is clear that this list should only be taken as an indication of the types of interactions that can occur and should not be considered to be the result of a rigorous analysis. In fact, we shall eventually show that nonuniformities in the asymptotic expansion cause these interactions to occur at different asymptotic orders than the present formal expansion would suggest. One might then choose to ignore this list entirely and argue that the experimentally observed turbulent motions should be used in place of the first-order solutions that appear in the source term (eq. (6)). This use of observed motions in place of the first-order solutions is, in effect, what is done in the acoustic analogy approach. However, I do not think that it should be dismissed entirely and therefore consider it in some detail.

The first item (i.e., the instability-wave–fine-grained turbulence interaction) has only been considered very briefly in the literature (e.g., refs. 26 and 34) and only limited quantitative results have been obtained for this interaction. It can be thought of as the sound generated by the fine-grained turbulence shaking the instability waves and is likely to emerge as an important source mechanism in flows at relatively low Reynolds numbers.

The instability-wave–instability-wave interaction may be related to the vortex-pairing events that occur in the initial mixing region of a high-speed jet. These events can be experimentally enhanced by exciting the jet with an external acoustic source tuned to the most unstable frequency of the shear layer at the nozzle lip, as was done by Kibens (ref. 35). He found that this enhancement caused the natural broadband noise of the jet to be suppressed and that most of the sound was then generated at subharmonics of the excitation frequency. By taking measurements in the near and far fields, Kibens showed that there was no Doppler shift in frequency, an indication that the sound was generated by nonconvecting sources within the jet. He subsequently identified the locations of these sources with the vortex-pairing locations.

However, quadratic interactions between two-dimensional (or between axisymmetric) instability waves produce only subsonically traveling waves in a subsonic parallel flow, and these waves do not radiate sound. But the straightforward perturbation analysis of these interactions is (as in the straightforward nonparallel-mean-flow analysis) nonuniformly valid in the streamwise direction (leading to the so-called Kelly resonance, ref. 36), and supersonically traveling waves are produced when the straightforward asymptotic solution is rendered uniformly valid in that direction. The sound field can then be calculated with a procedure similar to the one used in reference 37 for the sound generated by the nonlinear saturation of a single instability wave. A more systematic approach might be to adapt the nonparallel-flow analysis of reference 32 to this case.

The Lighthill Result

The interaction of fine-grained turbulence and fine-grained turbulence is essentially the mechanism originally considered by Lighthill (refs. 1 and 2). Difficulties such as those discussed in conjunction with instability-wave–instability-wave interaction may also occur when the present perturbation approach is applied to this case. Lighthill's acoustic analogy theory leads to a stationary medium (i.e., classic) wave equation. He suggested that it should be possible to neglect variations in retarded time across the turbulent eddies (or correlation volumes) in this case, since the time

$\ell/c_o(1 - M_c \cos\theta)$ for a sound wave to cross a turbulent eddy is small compared with the characteristic time ℓ/u' of the sound source at subsonic jet velocities (ref. 10). Here M_c is the convection Mach number of the turbulence and θ is the angle between the downstream jet axis and the line connecting the source point and the observation point.

It is a consequence of the linearity of the wave operator on the left side of equation (5) that the sound radiated by any given turbulent eddy is then independent of that radiated by any other eddy. Lighthill therefore argued that each eddy should behave like a point quadrupole source moving downstream with the "convection velocity" of the turbulence and that the entire sound field of the jet could then be estimated by calculating the sound radiated by a "typical" turbulent eddy. This picture turns out to be a slight oversimplification and was later corrected by Ffowcs Williams (ref. 38).

In the Lighthill-Ffowcs Williams result (refs. 1 and 38), the mean-square pressure p^2 radiated in any proportional frequency band at a fixed source frequency Ω, where $\omega = \Omega/(1 - M_c \cos\theta)$ is the actual frequency of the sound, behaves as follows:

$$p^2 \sim \frac{f(\Omega)}{(1 - M_c \cos\theta)^5} \tag{8}$$

so that its "directivity pattern" is primarily determined by the Doppler factor $(1 - M_c \cos\theta)$ raised to the -5 power.[1] These five inverse Doppler factors produce a highly directional radiation pattern at high subsonic Mach numbers—a result which is remarkably similar to experimental observation.

Solutions of Lilley's Equation

Solutions of equation (5) and Γ treated as a moving point source (eq. (7)) can be interpreted as corrections to the Lighthill-Ffowcs Williams result that account for the effects of the nonuniform surrounding mean flow. A number of researchers (e.g., refs. 7, 30, 34, and 39 to 45) therefore decided to calculate the acoustic radiation from point quadrupole sources moving through transversely sheared mean flows. The relevant solutions usually had to be obtained numerically, but relatively simple closed-form (or nearly closed-form) solutions were obtained in the low- and high-frequency limits $\omega D/U_j << 1$ and $\omega D/U_j >> 1$, respectively, where D denotes the jet diameter (see fig. 1) and U_j denotes the jet velocity.

Low-frequency solutions were obtained for a round jet with an arbitrary mean-velocity profile in references 40 to 42. All components of an idealized quadrupole convecting through a stationary medium exhibit directivity patterns given by inverse Doppler factors times sines and cosines of the observation angle. The low-frequency analyses show that only the $x_1 - x_1$ and $x_1 - r$ quadrupole components (where r is the radial coordinate) retain this property in the presence of a parallel but nonuniform mean flow. The remaining quadrupole components exhibit directivity patterns given by more complex formulas involving the complete mean-velocity profile and the location of the sources within the jet (ref. 42).

---

[1] The Doppler factor can be corrected to avoid the singularity at $M_c \cos\theta = 1$ by accounting for finite source volume.

However, the low-frequency analyses uncovered the very surprising result that the mean flow causes certain quadrupole components to emit sound much more efficiently than they otherwise would; the mean-square pressures in the absence and presence of the mean flow are, respectively, $O(\Omega^4)$ and $O(\Omega^2)$ as $\Omega \to 0$. The acoustic field of the $x_1 - r$ quadrupole, which is the only one of these more efficient sources that can be expressed in simple Doppler factor form, is proportional to the local mean-velocity gradient. It is worth noting that this source arises as much from the first member of the source term (eq. (6)) as from the second, even though the first does not explicitly involve the mean-velocity gradient (ref. 40).

Observed low-frequency jet noise directivity patterns therefore depend on complex properties of the jet turbulence and mean flow that are difficult to estimate. But the mathematical results are consistent with the experimentally observed results that the patterns are always more directional than Lighthill's five inverse Doppler factors would indicate. To be more specific, the analytical and experimental results show that the low-frequency sound should be more concentrated on the downstream axis than Lighthill's result implies, with the on-axis sound being produced by the quadrupoles with one axis in the streamwise direction.

The high-frequency solutions, which were obtained in references 7, 10, 34, 39, 40, 43, and 46, exhibit a "zone of silence" on the downstream jet axis. The acoustic field is exponentially small in that region, which is circumferentially asymmetric when the jet is nonaxisymmetric or the sound source is located off axis. It fills the entire range of circumferential angles ϕ when θ is sufficiently close to the downstream jet axis (see fig. 3), but it only occupies a limited range of angles (say, $\phi_{\text{min}} < \phi < \phi_{\text{max}}$) at larger values of θ (say, $\theta_{\text{min}} < \theta < \theta_{\text{max}}$), and finally, it will disappear completely when $\theta > \theta_{\text{max}}$ (see fig. 4). These remarks only apply to subsonic isothermal jets with monotonic or nearly monotonic mean-velocity profiles. A host of complex interference effects may occur when these restrictions are relaxed.

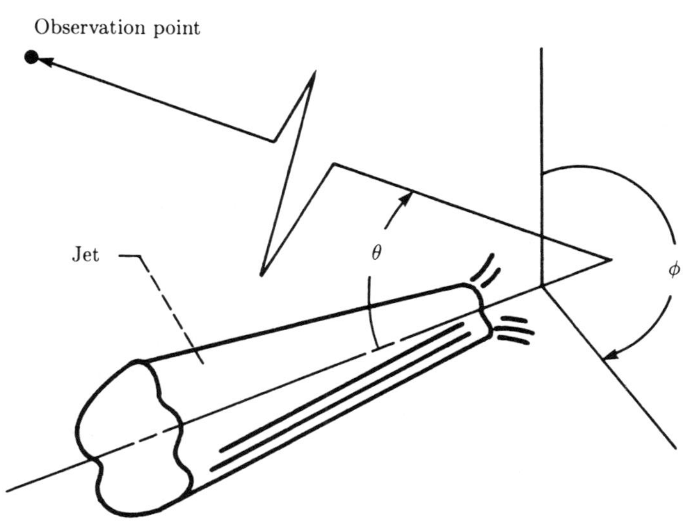

Figure 3. Coordinates for observation point.

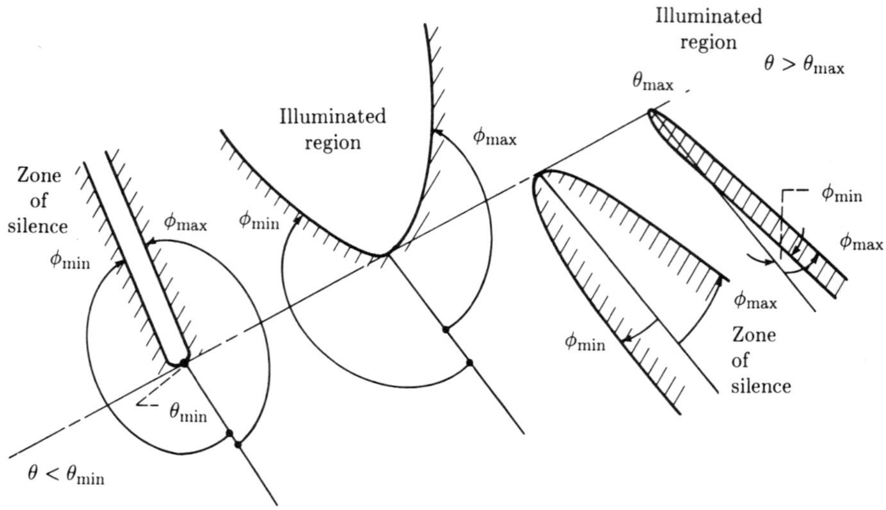

Figure 4. Asymmetric zone of silence.

As in geometric optics, the sound propagates along distinct rays in the high-frequency approximation. Only one ray can reach the observer when $\theta > \theta_{max}$, but there will be at least two rays reaching the observer when $\theta_{min} < \theta < \theta_{max}$—a direct ray and a ray reflected from the boundary of the zone of silence (ref. 43). The corresponding sound waves can then interfere, but the interference term is a rapidly oscillating function of angle and, since all acoustic measurements involve some form of spatial averaging, may not be experimentally observable.

The mean-square pressure radiated in any proportional frequency band of fixed-source frequency Ω is thus the sum of the mean-square pressures for each ray reaching the observer. The result for a convecting quadrupole source, corrected as in reference 38, is given by (ref. 43)

$$\overline{p^2} \propto \frac{\rho_\infty^2 \Omega^5 \left| \sum_{i,j=1}^{3} Q_{ij} v_i v_j \right|^2}{(4\pi r c_\infty c_o)^2 (1 - M\cos\theta)^2 (1 - M_c \cos\theta)^5} \Delta\phi \tag{9}$$

where r is the distance between the source point and the observation point, ρ_∞ and c_∞ are the density and the sound speed at infinity, M is the Mach number based on the mean flow at the source location and the speed of sound at infinity, and Δ denotes a circumferential directivity factor that depends on the circumferential observation angle ϕ, the location of the sound source within the jet, and the mean-velocity and temperature profiles of the jet. The term Q_{ij} denotes the relative quadrupole strengths:

$$\nu_1 = q_o \cos\Lambda \qquad \nu_2 = q_o \sin\Lambda \qquad \nu_3 = \cos\theta \tag{10}$$

where

$$q_o = \sqrt{\left(\frac{1 - M\cos\theta}{c_o/c_\infty}\right)^2 - \cos^2\theta} \tag{11}$$

and Λ is the initial circumferential angle made by the acoustic ray associated with equation (9).

Equation (9) is an *exact* high-frequency result that applies to jets of any cross section and with any transverse mean-velocity and temperature profiles, but Λ and Δ must be calculated by solving a second-order, ordinary differential equation in the general case. They are, however, given by relatively simple analytic formulas for off-axis sources at arbitrary locations in a circular jet with arbitrary velocity and temperature profiles (ref. 43).

The circumferential directivity factor can be used to study the effect of non-axisymmetric jet velocity and temperature profiles in reducing jet noise below the flight path of a jet aircraft, this reduction being of considerable interest for technological applications (ref. 47). But for the present purpose, it is appropriate to concentrate on the azimuthal directivity pattern, which is relatively unaffected by this factor.

Equation (9) shows that the inverse Doppler factor exponent is increased from five to seven in the high-frequency limit, since the local mean-flow Mach number and the turbulence convection Mach number are usually not very different in the regions of peak turbulence intensity. This taken by itself would cause the high-frequency sound (like the low-frequency sound) to be more directional than Lighthill's equation (eq. (8)) would predict. If, however, the quadrupole is assumed to be isotropic (ref. 40) so that

$$Q_{ij} = \delta_{ij} Q_o \qquad (i,j = 1,2,3) \tag{12}$$

where δ_{ij} is the Kronecker delta, it follows from equations (10) and (11) that

$$\left| \sum_{i,j=1}^{3} Q_{ij} \nu_i \nu_j \right|^2 = \left(\frac{1 - M\cos\theta}{c_o/c_\infty} \right)^4 Q_o^2$$

This equation more than compensates for the additional two Doppler factors in the denominator of equation (9) and produces a net azimuthal directivity pattern that is given by three inverse Doppler factors and results in excellent agreement with the experimentally observed 1/3-octave directivity patterns.

The interpretation of this result is that the reduced directivity of the high-frequency sound is due to interference between the various quadrupole components. The mean-square pressure is now the product of an azimuthal directivity factor and a circumferential directivity factor that depends only on ϕ. It is, of course, highly unlikely that the actual quadrupoles will precisely satisfy equation (12), but the analysis strongly suggests that the quadrupole component interference effects can greatly reduce jet noise directivity.

Sound Generation From Streamwise Variations in Mean Flow

The formal asymptotic expansion in powers of u'/U can be continued to the third order. At this stage, interactions between the first-order perturbation solution and the streamwise variations in the mean flow appear in the source term. (Recall that the ratio of the cross-stream to streamwise components of the mean-flow velocity is $O(u'/U)^2$, while the first-order solution is $O(u'/U)$.) Then by decomposing the first-order solution into its gust and instability-wave components and making the connection between the gust and fine-grained turbulence that we discussed above, we infer that the source term now describes sound generation because of (1) the fine-grained turbulence interacting with streamwise variations (i.e., spreading) of the mean flow and (2) the instability waves interacting with the streamwise mean-flow variations. To my knowledge, the first mechanism has not yet been considered in the literature. The second is already accounted for in the first-order analysis if the slowly varying approximation is used to described the instability waves. This mechanism has been analyzed in a more ad hoc fashion in references 26, 37, 48, and 49 and in a systematic way in reference 32. However, reference 32 ultimately concludes that this source is not important at subsonic speeds, and this conclusion is consistent with the findings of reference 50, in which the phenomenon was studied experimentally by artificially exciting a jet under conditions that tended to minimize vortex pairing. Unlike reference 35, it was found in reference 50 that the broadband noise was usually increased rather than suppressed by the external excitation. It was concluded that the instability wave, while not radiating noise directly, acted as a conduit through which energy could be transferred to the small-scale turbulent motion.

Comparison of Experiment and Theory

Figure 5 is a plot of the sound radiated in 1/3-octave frequency bands at fixed source frequencies by a high-Reynolds-number turbulent air jet as a function of azimuthal angle measured from the downstream jet axis (ref. 51). The jet velocity was 994 ft/sec. The data indicated by the open circles coincide with the peak frequency f_p of the jet noise at 90° from the jet axis; this frequency corresponds to a Strouhal number $\Omega D/2\pi U_j$ of unity. The squares represent low-frequency data corresponding to one-tenth of this value, and the triangles represent relatively high-frequency data corresponding to about three times the peak frequency.

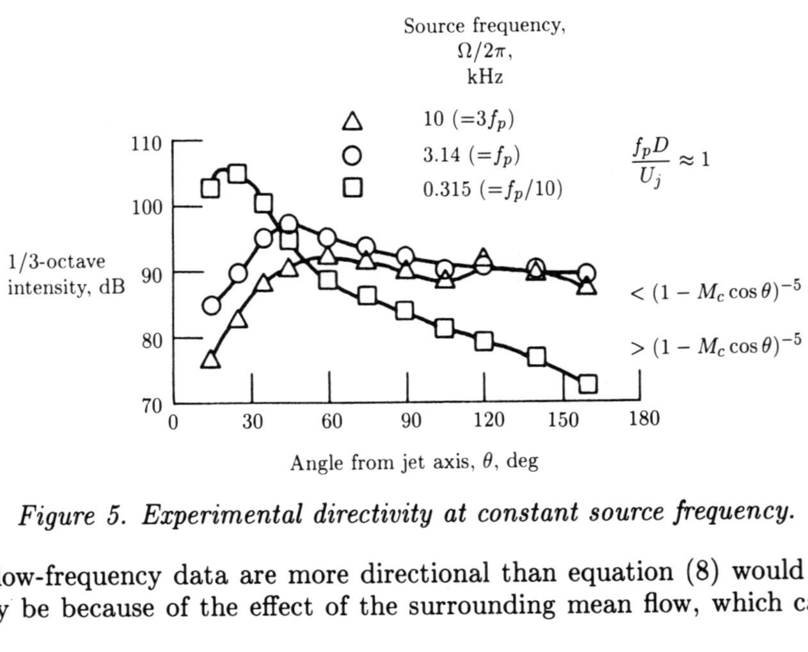

Figure 5. Experimental directivity at constant source frequency.

The low-frequency data are more directional than equation (8) would indicate. This may be because of the effect of the surrounding mean flow, which causes the low-frequency sound to be more concentrated on the downstream jet axis. However, as already noted, it is necessary to make specific assumptions about source locations, relative quadrupole strengths, and mean profile shapes before explicit calculations can be made. This was, in fact, done by Mani (ref. 44)[2], who assumed the sound source to be located on the axis in a slug flow velocity profile to simplify his calculation. However, this model effectively precludes the exceptionally efficient low-frequency sources discussed above. Mani also introduced a specific assumption about the relative quadrupole strengths which he attempted to justify at least partially by evoking an analysis presented in reference 52.

Mani's calculations are compared with data taken from reference 3 in figure 6, which is a plot of the sound radiated in a 1/3-octave frequency band at a very low, constant source Strouhal number of 0.03 for three different jet velocities. The theoretical curves are adjusted to pass through the data at 60° from the downstream jet axis rather than at 90°, as is usually done when overall sound pressure levels are not predicted by the theory.[3] The agreement appears to be fairly good, but it is probably necessary to test the sensitivity of the analysis to its numerous assumptions before definite conclusions can be drawn.

The remaining curves in figure 5 (f_p and $3f_p$) exhibit a zone of silence on the downstream jet axis, as predicted by the high-frequency solution, and are less directional outside this zone than equation (8) would indicate. The latter effect could be the result of the cancellation between the various components of the quadrupole source that occurs in the high-frequency solution (eq. (9)) when this source is assumed to be isotropic. Then as we have seen, the 1/3-octave band pressure fluctuations vary like three inverse Doppler factors.

Figure 7 is a plot of the jet noise radiated in 1/3-octave frequency bands at constant source frequency as a function of the angle measured from the downstream jet axis. The measurements, which are indicated by the open symbols, are from reference 51. They correspond to the peak frequency of the sound radiated at 90° to the jet axis (i.e., to a source Strouhal number of unity) but are taken at three different subsonic jet velocities. It is worth noting that the Reynolds number is quite high in these experiments.

The solid curves are obtained by using three inverse Doppler factors for the data at 90° to the jet axis. The turbulence convection Mach number M_c is taken to be 0.62 times the jet exit Mach number based on the speed of sound at infinity and is the value usually recommended by experimentalists. (The dashed curve is the result of using the five inverse Doppler factors implied by eq. (8).)

Although the agreement between experiment and theory is good, one might feel somewhat uncomfortable about extending the high-frequency solution to such low Strouhal numbers. However, reference 45 compared the exact and high-frequency solutions for fixed quadrupoles in a linear shear flow and found the results to be in close agreement, even at a Strouhal number of unity. Moreover, the peak frequency

[2] Mani's analysis is not restricted to low frequencies, and he did not explicitly take the low-frequency limit of his result. But he did carry out numerical calculations at a very low frequency, and we consider only these results.

[3] The usual argument is that the sound is unaffected by both source convection and mean-flow effects at 90°.

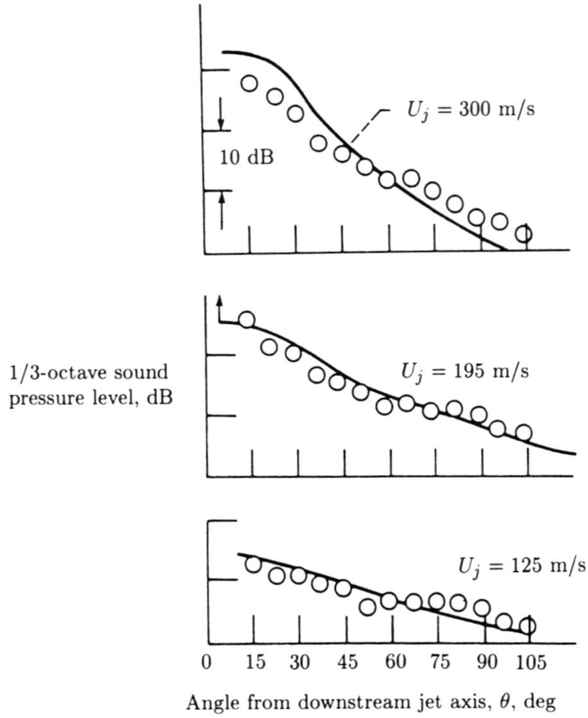

Figure 6. Comparisons of Mani's (ref. 44) analysis with the 1/3-octave directivity data of Lush (ref. 3) for source Strouhal number $\Omega D/2\pi U_j$ = 0.03.

sound is believed to be generated at the end of the potential core, where the relevant length scale for computing the Strouhal number is nearer 2D than D. (See fig. 1.)

There are, of course, other effects that might explain the disagreement between Lighthill's original theory and the high Reynolds number experiments. First, the decreased directivity at the higher frequencies could be due to the scattering of the sound by the turbulence, an effect that is certainly more important at the higher frequencies and that tends to make the sound field less directional. It could also be due to variations in retarded time across the source (i.e., source-coherence effects), which were neglected in the point-source models described above and which, for a fixed source size, would also be more important at the higher frequencies. This phenomenon was analyzed in references 38 and 53 and it was shown that it tends to diminish the increased directionality that results from source convection. It is important, however, to note that these analyses are based on an assumed source model.

Extensions to More Complex Turbulent Flows

The results of the preceding section suggest that the high-frequency solution may remain valid at frequencies that are low enough to include the most energetic portion

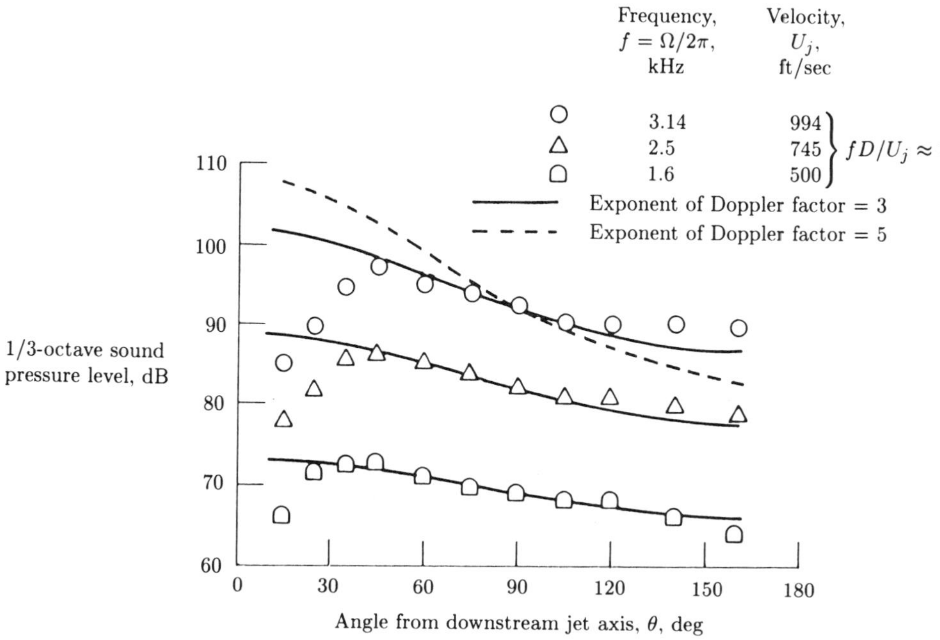

Figure 7. Experimental 1/3-octave directivity data of Olsen and Friedman (ref. 51) plotted at constant source frequency Ω, where $\Omega/2\pi$ is equal to peak frequency of spectrum at 90° from jet axis.

of the jet noise spectrum. This might remain true for other more complex turbulent flows. We may therefore be able to calculate the sound generated in such flows by finding the high-frequency solution for a point quadrupole source moving through the appropriate mean flow. Recently such a solution was obtained for a completely general mean flow (ref. 54). The final equation in that reference is somewhat formal in that (1) it involves quantities that depend on the solution of a system of ordinary differential equations (which must, in general, be obtained numerically), and (2) it does not explicitly account for the presence of the caustics that, as we have seen, can occur in the high-frequency limit. However, the caustics can easily be incorporated into the theory of reference 43, which is explicit enough to provide considerable information about the high-frequency sound generation in these more complex flows.

Reference 55 used the analysis to study the effect of mean-flow divergence on subsonic jet noise. (See fig. 3.) It showed that even small jet spreading eliminates the zone of silence that occurs in the axisymmetric parallel-flow model in the sense that the radiated sound is no longer exponentially small in that region but is merely greatly reduced because of a strong divergence of the acoustic rays. Unlike the parallel-flow calculation, these results indicate that the shallow-angle acoustic field is nonzero and exhibits a directivity that is independent of frequency in the limit as $\omega \to \infty$. It is worth noting that the experimental zone of silence directivity pattern only becomes independent of frequency at very high frequencies.

Supersonic Flows

Up to now our remarks have been confined to subsonic flows, though much of what has been said also applies to supersonic flows. However, some new phenomena also come into play with supersonic flows. First, the linear gust solution (eq. (1)) no longer decays exponentially at large distances from a jet but involves propagating wave components. These waves probably correspond to the so-called Mach wave radiation observed in the initial mixing region of supersonic jets in references 56 and 57, for example. They are the leading-edge shock waves (or bow waves) of the supersonically moving eddies. It is generally agreed that this phenomenon is not very important at moderate supersonic Mach numbers.

Second, the instability waves can achieve supersonic phase speed and generate sound directly at very high Mach numbers. But reference 32 argues that the instability waves in a slowly varying mean flow involve supersonically traveling components that can radiate significant sound even at moderate supersonic Mach numbers. This is consistent with the low Reynolds number supersonic jet noise experiments of references 58 and 59, which imply that the majority of the noise is generically related to the relatively slow growth and decay of organized wavelike structures. This behavior seems to persist even at high Reynolds numbers.

Finally, both the fine-grained turbulence and the instability waves can interact with the shock waves, which can now appear in the flow, to generate noise in a very efficient manner. This is usually referred to as the "shock-associated noise" (or shock "screech" if there is feedback to the nozzle lip). Reference 60 suggested that turbulent eddies can retain their identities long enough to pass through several shock waves and that the resulting coherence between the noise sources has a dominant influence on the radiated sound. This is even more true if the sound is generated by interactions between instability waves and the shock structure (ref. 61).

References

1. Lighthill, M. J.: On Sound Generated Aerodynamically: I. General Theory. *Proc. Royal Soc. London,* ser. A, vol. 211, no. 1107, Mar. 20, 1952, pp. 564–587.
2. Lighthill, M. J.: On Sound Generated Aerodynamically: II. Turbulence as a Source of Sound. *Proc. Royal Soc. London*, ser. A, vol. 222, no. 1148, Feb. 23, 1954, pp. 1–32.
3. Lush, P. A.: Measurements of Subsonic Jet Noise and Comparison With Theory. *J. Fluid Mech.*, vol. 46, pt. 3, Apr. 13, 1971, pp. 477–500.
4. Ahuja, K. K.; and Bushell, K. W.: An Experimental Study of Subsonic Jet Noise and Comparison With Theory. *J. Sound & Vibration,* vol. 30, no. 3, Oct. 8, 1973, pp. 317–341.
5. Dowling, A. P.; Ffowcs Williams, J. E.; and Goldstein, M. E.: Sound Production in a Moving Stream. *Philos. Trans. Royal Soc. London,* ser. A, vol. 288, no. 1353, Mar. 23, 1978, pp. 321–349.
6. Phillips, O. M.: On the Generation of Sound by Supersonic Turbulent Shear Layers. *J. Fluid Mech.*, vol. 9, pt. 1, Sept. 1960, pp. 1–28.
7. Lilley, G. M.: On the Noise From Jets. *Noise Mechanisms,* AGARD-CP-131, Mar. 1974, pp. 13.1–13.12.
8. Howe, M. S.: Contributions to the Theory of Aerodynamic Sound, With Application to Excess Jet Noise and the Theory of the Flute. *J. Fluid Mech.*, vol. 71, pt. 4, Oct. 28, 1975, pp. 625–673.
9. Yates, John E.; and Sandri, Guido: Bernoulli Enthalpy: A Fundamental Concept in the Theory of Sound. AIAA Paper 75-439, Mar. 1975.
10. Goldstein, Marvin E.: *Aeroacoustics.* McGraw-Hill Book Co., c.1976.
11. Moffatt, H. K.: Some Developments in the Theory of Turbulence. *J. Fluid Mech.*, vol. 106, May 1981, pp. 27–47.
12. Hunt, J. C. R.: A Theory of Turbulent Flow Round Two-Dimensional Bluff Bodies. *J. Fluid Mech.*, vol. 61, pt. 4, Dec. 18, 1973, pp. 625–706.

13. Bradshaw, P.; Ferriss, D. H.; and Johnson, R. F.: Turbulence in the Noise-Producing Region of a Circular Jet. *J. Fluid Mech.,* vol. 19, pt. 4, Aug. 1964, pp. 591–624.
14. Tennekes, H.; and Lumley, J. L.: *A First Course in Turbulence.* M.I.T. Press, c.1972.
15. Betchov, Robert; and Criminale, William O., Jr.: *Stability of Parallel Flows.* Academic Press, Inc., 1967.
16. Goldstein, M. E.: Characteristics of the Unsteady Motion on Transversely Sheared Mean Flows. *J. Fluid Mech.,* vol. 84, pt. 2, Jan. 30, 1978, pp. 305–329.
17. Goldstein, M. E.: Scattering and Distortion of the Unsteady Motion on Transversely Sheared Mean Flows. *J. Fluid Mech.,* vol. 91, pt. 4, Apr. 27, 1979, pp. 601–632.
18. Möhring, W.: Über Schallwellen in Scherströmungen. *Forschritte der Akustik,* DAGA 1976, pp. 543–546.
19. Ffowcs Williams, J. E.; and Hall L. H.: Aerodynamic Sound Generation by Turbulent Flow in the Vicinity of a Scattering Half Plane. *J. Fluid Mech.,* vol. 40, pt. 4, Mar. 1970, pp. 657–670.
20. Crighton, D. G.; and Leppington, F. G.: Radiation Properties of the Semi-Infinite Vortex Sheet: The Initial Value Problem. *J. Fluid Mech.,* vol. 64, pt. 2, June 19, 1974, pp. 393–414.
21. Goldstein, M. E.: The Coupling Between Flow Instabilities and Incident Disturbances at a Leading Edge. *J. Fluid Mech.,* vol. 104, Mar. 1981, pp. 217–246.
22. Rienstra, S. W.: *Edge Influence on the Response of Layers to Acoustic Forcing.* Ph.D. Thesis, Tech. Hogeschool, Eindhoven, 1979.
23. Olsen, W. A.: Noise Generated by Impingement of Turbulent Flow on Airfoils of Varied Chord, Cylinders, and Other Flow Obstructions. AIAA Paper No. 76-504, July 1976. (Available as NASA TM X-73464.)
24. Crighton, D. G.: Why Do the Acoustics and the Dynamics of a Hypothetical Mean Flow Bear on the Issue of Sound Generation by Turbulence? *Mechanics of Sound Generation in Flows,* E.-A. Müller, ed., Springer-Verlag, 1979, pp. 1–11.
25. Crighton, D. G.: Acoustics as a Branch of Fluid Mechanics. *J. Fluid Mech.,* vol. 106, May 1981, pp. 261–298.
26. Liu, J. T. C.: Developing Large-Scale Wavelike Eddies and the Near Jet Noise Field. *J. Fluid Mech.,* vol. 62, pt. 3, Feb. 11, 1974, pp. 437–464.
27. Tam, Christopher K. W.; and Chen, K. C.: A Statistical Model of Turbulence in Two-Dimensional Mixing Layers. *J. Fluid Mech.,* vol. 92, pt. 2, May 28, 1979, pp. 303–326.
28. Haertig, J.: Theoretical and Experimental Study of Wavelike Disturbances on Round Free Jet With Emphasis Being Placed on Orderly Structures. *Mechanics of Sound Generation in Flows,* E.-A. Müller, ed., Springer-Verlag, 1979, pp. 167–173.
29. Gaster, M.; Kit, E.; and Wygnanski, I.: Large-Scale Structures in a Forced Turbulent Mixing Layer. *J. Fluid Mech.,* vol. 150, Jan. 1985, pp. 23–39.
30. Tester, B. J.; and Morfey, C. L.: Developments in Jet Noise Modelling—Theoretical Predictions and Comparisons With Measured Data. *J. Sound & Vibration,* vol. 46, no. 1, May 8, 1976, pp. 79–103.
31. Crighton, D. G.; and Gaster, M.: Stability of Slowly Diverging Jet Flow. *J. Fluid Mech.,* vol. 77, pt. 2, Sept. 24, 1976, pp. 397–413.
32. Tam, Christopher K. W.; and Morris, Philip J.: The Radiation of Sound by the Instability Waves of a Compressible Plane Turbulent Shear Layer. *J. Fluid Mech.,* vol. 98, pt. 2, May 29, 1980, pp. 349–381.
33. Nayfeh, Ali Hasan: *Perturbation Methods.* John Wiley & Sons, Inc., c.1973.
34. Berman, C. H.: Noise From Nonuniform Turbulent Flows. AIAA Paper No. 74-2, Jan.–Feb. 1974.
35. Kibens, Valdis: Discrete Noise Spectrum Generated by an Acoustically Excited Jet. *AIAA J.,* vol. 18, no. 4, Apr. 1980, pp. 434–441.
36. Kelly, R. E.: On the Stability of an Inviscid Shear Layer Which Is Periodic in Space and Time. *J. Fluid Mech.,* vol. 27, pt. 4, Mar. 20, 1967, pp. 657–689.
37. Huerre, P.; and Crighton, D. G.: Sound Generation by Instability Waves in a Low Mach Number Jet. AIAA-83-0661, Apr. 1983.
38. Ffowcs Williams, J. E.: The Noise From Turbulence Convected at High Speed. *Philos. Trans. Royal Soc. London,* ser. A, vol. 255, no. 1061, Apr. 18, 1963, pp. 469–503.
39. Balsa, Thomas F.: The Far Field of High Frequency Convected Singularities in Sheared Flows, With an Application to Jet-Noise Prediction. *J. Fluid Mech.,* vol. 74, pt. 2, Mar. 23, 1976, pp. 193–208.
40. Balsa, T. F.: The Acoustic Field of Sources in Shear Flow With Application to Jet Noise: Convective Amplification. *J. Fluid Mech.,* vol. 79, pt. 1, Jan. 1977, pp. 33–47.
41. Goldstein, M. E.: The Low Frequency Sound From Multipole Sources in Axisymmetric Shear Flows, With Applications to Jet Noise. *J. Fluid Mech.,* vol. 70, pt. 3, Aug. 12, 1975, pp. 595–604.
42. Goldstein, M. E.: The Low Frequency Sound From Multipole Sources in Axisymmetric Shear Flows, Part 2. *J. Fluid Mech.,* vol. 75, pt. 1, May 13, 1976, pp. 17–28.

43. Goldstein, M. E.: High Frequency Sound Emission From Moving Point Multipole Sources Embedded in Arbitrary Transversely Sheared Mean Flows. *J. Sound & Vibration,* vol. 80, no. 4, Feb. 22, 1982, pp. 499–522.
44. Mani, R.: The Influence of Jet Flow on Jet Noise. *J. Fluid Mech.,* vol. 73, pt. 4, Feb. 24, 1976.
    Part 1: The Noise of Unheated Jets, pp. 753–778.
    Part 2: The Noise of Heated Jets, pp. 779–793.
45. Scott, James N.: Propagation of Sound Waves Through a Linear Shear Layer. *AIAA J.,* vol. 17, no. 3, Mar. 1979, pp. 237–244.
46. Tester, Brian J.; and Burrin, Robert H.: On Sound Radiation From Sources in Parallel Sheared Jet Flows. *The Generation and Radiation of Supersonic Jet Exhaust Noise—A Progress Report on Studies of Jet Noise Generation and Radiation, Turbulence Structure and Laser Velocimetry,* Harry E. Plumblee, Jr., ed., AFAPL-TR-74-24, U.S. Air Force, June 1974, pp. 59–87. (Available from DTIC as AD 787 192.)
47. von Glahn, U.; and Goodykoontz, J.: *Noise Suppression Due to Annulus Shaping of a Conventional Coaxial Nozzle.* NASA TM-81461, 1980.
48. Crow, S. C.; and Champagne, F. H.: Orderly Structure in Jet Turbulence. *J. Fluid Mech.,* vol. 48, pt. 3, Aug. 16, 1971, pp. 547–591.
49. Ffowcs Williams, J. E.; and Kempton, A. J.: The Noise From the Large-Scale Structure of a Jet. *J. Fluid Mech.,* vol. 84, pt. 4, Feb. 27, 1978, pp. 673–694.
50. Moore, C. J.: The Role of Shear-Layer Instability Waves in Jet Exhaust Noise. *J. Fluid Mech.,* vol. 80, pt. 2, Apr. 25, 1977, pp. 321–367.
51. Olsen, W.; and Friedman, R.: Jet Noise From Co-axial Nozzles Over a Wide Range of Geometric and Flow Parameters. AIAA Paper No. 74-43, Jan.–Feb. 1974.
52. Ribner, H. S.: Quadrupole Correlations Governing the Pattern of Jet Noise. *J. Fluid Mech.,* vol. 38, pt. 1, Aug. 14, 1969, pp. 1–24.
53. Ribner, H. S.: *Aerodynamic Sound From Fluid Dilatations—A Theory of the Sound From Jets and Other Flows.* UTIA Rep. No. 86 (AFOSR TN 3430), Inst. of Aerophysics, Univ. of Toronto, July 1962.
54. Durbin, P. A.: High Frequency Green Function for Aerodynamic Noise in Moving Media, Part I: General Theory. *J. Sound & Vibration,* vol. 91, no. 4, Dec. 22, 1983, pp. 519–525.
55. Durbin, P. A.: High Frequency Green Function for Aerodynamic Noise in Moving Media, Part II: Noise From a Spreading Jet. *J. Sound & Vibration,* vol. 91, no. 4, Dec. 22, 1983, pp. 527–538.
56. Dosanjh, Darshan S.; and Yu, James C.: Noise From Underexpanded Axisymmetric Jet Flows Using Radial Jet Flow Impingement. *Aerodynamic Noise,* Univ. of Toronto Press, c.1969, pp. 169–188.
57. Lowson, M. V.; and Ollerhead, J. B.: Visualization of Noise From Cold Supersonic Jets. *J. Acoust. Soc. America,* vol. 44, no. 2, Aug. 1968, pp. 624–630.
58. McLaughlin, Dennis K.; Morrison, Gerald L.; and Troutt, Timothy R.: Experiments on the Instability Waves in a Supersonic Jet and Their Acoustic Radiation. *J. Fluid Mech.,* vol. 69, pt. 1, May 13, 1975, pp. 73–95.
59. Troutt, T. R.; and McLaughlin, D. K.: Experiments on the Flow and Acoustic Properties of a Moderate-Reynolds-Number Supersonic Jet. *J. Fluid Mech.,* vol. 116, Mar. 1982, pp. 123–156.
60. Harper-Bourne, M.; and Fisher, J. J.: The Noise From Shock Waves in Supersonic Jets. *Noise Mechanisms,* AGARD-CP-131, Mar. 1974, pp. 11.1–11.13.
61. Tam, C. K. W.; and Tanna, H. K.: Shock Associated Noise of Supersonic Jets From Convergent-Divergent Nozzles. *J. Sound & Vibration,* vol. 81, no. 3, Apr. 8, 1982, pp. 337–358.

6 Jet Noise Generated by Large-Scale Coherent Motion

Lead author

Christopher K. W. Tam
Florida State University
Tallahassee, Florida

Introduction

In this chapter the noise generated by large-scale turbulence structures and instability waves of jets is discussed. The primary emphasis is on supersonic jets with moderate to high Reynolds numbers (i.e., $N_{\mathrm{Re}} > 10^5$). This is because it is in these jets that unambiguous experimental and theoretical evidence is found indicating that large turbulence structures and instability waves are directly responsible for generating the dominant part of the noise. For subsonic jets similar large turbulence structures and instability waves do play a crucial role in the dynamics, spread, and mixing of the jet fluid. However, at subsonic convection speeds they do not appear to be efficient noise generators. Many investigators believe that the dominant noise source of subsonic jets is, in fact, the small-scale turbulence. As yet this belief has not received universal acceptance. The issues involved are complicated and are not easy to resolve. In order not to divert attention from the main theme, they are left to the end of the chapter where they are examined in greater detail.

Large Turbulence Structures and Instability Waves

Experimental Observations of Large Turbulence Structures and Instability Waves in Jet Flows at Moderate to High Reynolds Numbers

One of the most significant developments in turbulence research in recent years is the recognition that turbulence in free shear flows is far more coherent and orderly than previously believed. In a study of a moderate Reynolds number low-speed jet, it was discovered that the shear layer could support large, orderly vortical toroidal structures (ref. 1). It was independently found in reference 2 that large-scale coherent vortical structures (see fig. 1) are intrinsic features of two-dimensional turbulent mixing layers at high Reynolds numbers. Since these pioneering works there have been many investigations on large structures at various Reynolds numbers, Mach numbers, densities, and temperature ratios. These works provide the foundation of our present understanding of the dynamics of turbulent free shear flows.

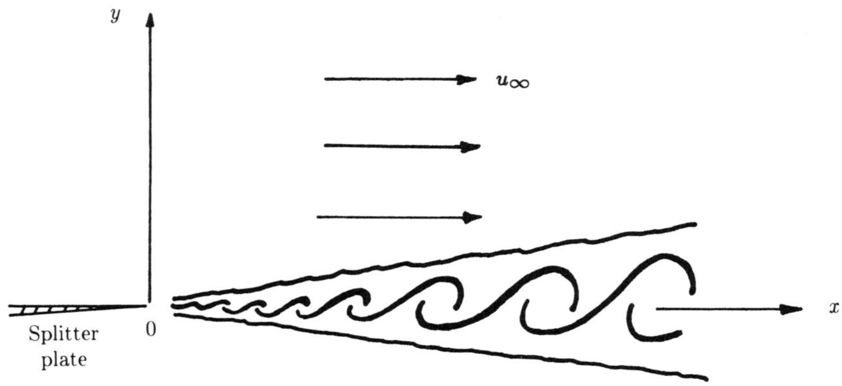

Figure 1. Large turbulence structures in high Reynolds number two-dimensional mixing layer.

In order to study the space-time evolution of large turbulence structures in two-dimensional mixing layers, several researchers (e.g., refs. 2 and 3) took high-speed motion pictures of these flows. These pictures reveal that the large structures are initiated near the trailing edge of the splitter plate, which marks the beginning of the mixing layer. These structures grow in size as they are convected downstream. To accommodate this growth the spacing between neighboring structures undergoes constant changes. Every now and then two (or three) of these vortex-like structures coalesce to form a single large structure. This process, which was observed to occur more prominently at low to moderate Reynolds numbers (ref. 4), is generally referred to as "vortex pairing" (or tripling). In high Reynolds number flows the pairing process, once started, is usually completed in a very short time. Contrary to early expectations, Hernan and Jimenez (ref. 5), using digital image analysis of a flow visualization film, found that most of the entrainment of ambient fluid into the mixing layer takes place during the growth stage of the large vortical structures and not during pairing. In addition to the pairing phenomenon, it was observed in reference 3 that a large structure may abruptly disintegrate in the straining field of the adjacent large structure (or structures). This process is called "tearing." Acting together, the mechanisms of vortex pairing and tearing are instrumental in randomizing the space-time trajectories of the large turbulence structures. Thus, although a single large structure may appear as quasi-deterministic, the sum total of all the large structures in the mixing layer amalgamating randomly in space and time gives the overall phenomenon a stochastic and chaotic character typical of turbulent flows. As far as it is known, the surviving large structures have extremely long lifetimes. In all the experiments mentioned above they seemed to persist all the way to the end of the test sections.

In the case of jets, the circular geometry and the limited length of the potential more impose severe constraints on the possible geometry and time history of the large, coherent structures. Yule (ref. 6), who carried out extensive visual as well as conditional hot-wire measurements of these structures in moderate Reynolds number low-speed jets, suggested that the jet mixing layer be divided into two regions, as shown in figure 2. In the transitional-flow region closest to the nozzle exit, the shear layer rolls up into toroidal vortices. These toroidal vortices are the equivalent of the large vortical structures of the two-dimensional mixing layer described above. The vortices undergo pairing and tearing. Further downstream in the turbulent-flow region they break up into three-dimensional large turbulent structures. These structures, which are made up of rotational fluid, have highly irregular boundaries and bear no resemblance to the strongly axisymmetric toroidal vortices. The axial length of the transitional-flow region where toroidal vortices are found depends on the Reynolds number. This region shortens as Reynolds number increases. Thus, for high Reynolds number jets, vortex pairing in the sense of references 2 and 4 is, in fact, an infrequent event. The pairing process, therefore, is not expected to be dynamically very significant for high Reynolds number jets.

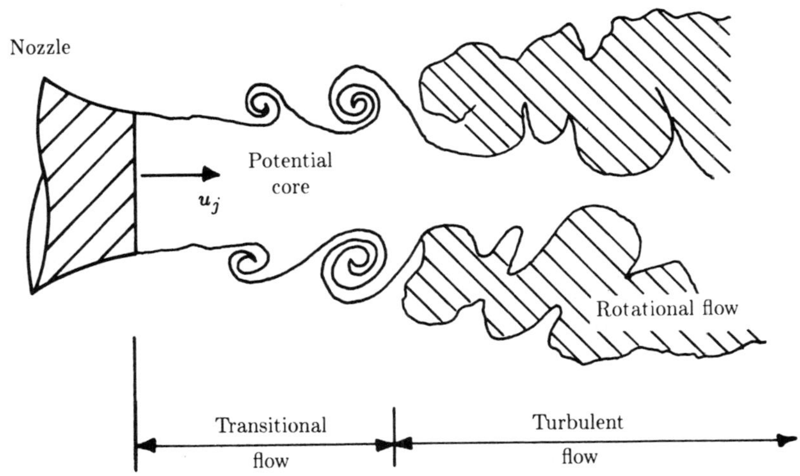

Figure 2. Schematic drawing of large turbulence structures in mixing layer of round jet at moderate Reynolds number. (Based on ref. 6.)

When a jet at a sufficiently high Reynolds number is slightly excited by upstream tones or other external means, it has been observed (refs. 7 to 13) that, unlike natural jets, the large turbulence structures in the mixing layer of the jets are more orderly and azimuthally correlated. Under excitation these large structures can be found throughout the entire potential core of the jet and in the region immediately downstream. A conditional sampling technique was used in reference 14 to study both the naturally occurring large turbulence structures in an unexcited jet and the phase-locked large turbulence structures in a sinusoidally excited jet. Upon comparison of the characteristics of the educed natural structures and the induced structures, excellent agreements were found. The implication appears to be that on a statistical-average basis the natural large turbulence structures of a jet are nearly as coherent and orderly as those in a forced jet, although instantaneously they might appear to be quite disorganized.

In recent years numerous experiments have been carried out to investigate the space-time evolution and fine details of the large turbulence structures in low-speed free shear flows. The amount of literature on this subject has grown so much that it has become, more or less, an entire research area by itself. Herein only the salient features of these structures that are directly relevant to jet noise generation are discussed. Readers who wish to pursue the subject of large structures in low Mach number flows beyond this brief discussion should consult some of the more in-depth review articles (refs. 15 to 20).

For high subsonic and supersonic axisymmetric jets, visual observations of possible coherent flow structures have been carried out over an extended period of time (refs. 21 to 28). These observations have provided clear-cut evidence of the existence of similar large turbulence structures in the mixing layers of these jets. More recently, a systematic visual study of these structures in axisymmetric supersonic jets was presented in reference 29. At Reynolds numbers exceeding 10^6, the naturally occurring large structures are not as well organized and are, therefore, difficult to observe. To enhance their visibility a low-level acoustic excitation of the jet upstream of the nozzle exit was introduced. The level of excitation was kept low enough so that no detectable modification of the mean flow was measured. Figure 3 is a sequence of ensemble-averaged photographs taken from reference 29. The large-scale structures in the mixing layer of the supersonic jet can clearly be seen. The strobing light was triggered at different phase or time delays relative to the excitation signal, and thus the downstream movement of the large structures was recorded (indicated by an arrow). A shock cell structure inside the jet is also evident in these pictures, an indication the jet is not perfectly expanded. Thus the visual observations in reference 29 provide unambiguous evidence that large turbulence structures exist in shock-containing high Reynolds number supersonic jets as well.

Quantitative information on the dynamic and space-time properties of the large turbulence structures in high-speed flows is difficult to obtain because of the well-known hot-wire breakage problem. To avoid this inherent difficulty, a very detailed sequence of experimental studies (refs. 30 to 32) was performed in a low-density supersonic jet facility. It was found that for low Reynolds number supersonic jets, the large-scale structures took the form of instability waves of discrete frequencies. These waves were coherent over an axial distance of several jet diameters. In addition, these large-scale instability waves generated an intense acoustic field which extended from the jet all the way to the boundary of the anechoic jet flow facility. To assess the effects of Reynolds number, the earlier experiment was repeated at a moderately high Reynolds number (ref. 33). It was found that the unsteady motion of the supersonic jet was dominated by a band of large-scale instability waves. The characteristics of these instability waves were similar to those of the low Reynolds number experiments. Furthermore, at moderate Reynolds numbers the acoustic field associated with the large-scale instability waves also dominated the total noise field of the jet. The near-field noise contours and the far-field noise directivity and spectral characteristics of these jets were carefully compared with those at high Reynolds number in references 33 and 34. Remarkable resemblance was found. The findings suggest that, at least in a statistical-average sense, the large-scale turbulence

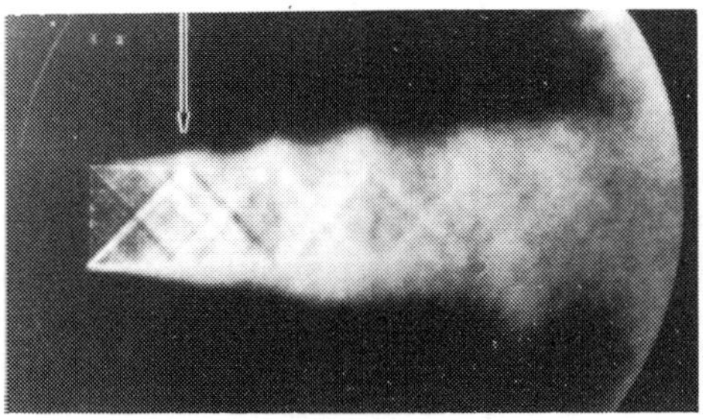

(*a*) $\phi = 0°$.

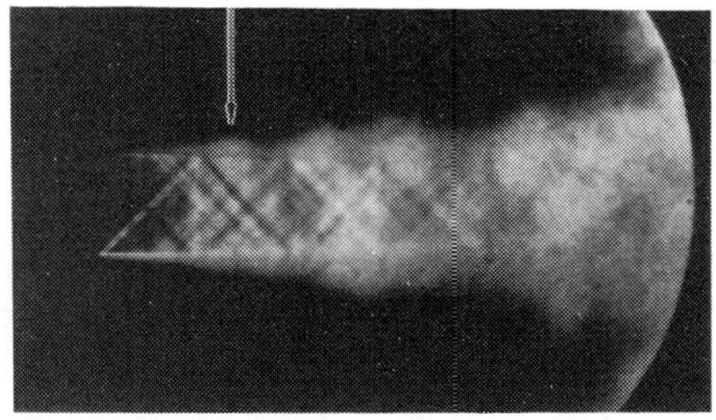

(*b*) $\phi = 60°$.

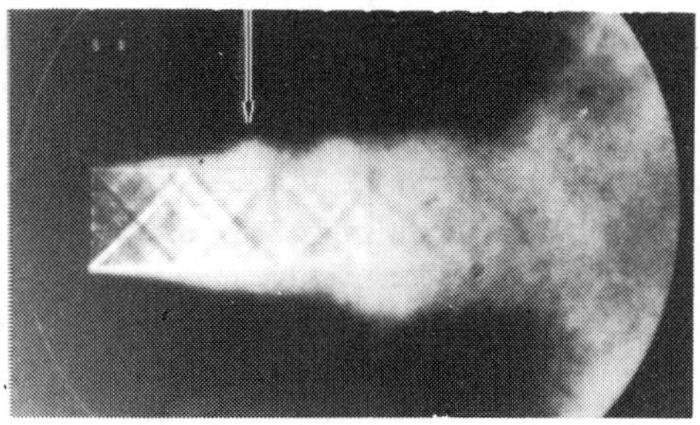

(*c*) $\phi = 120°$.

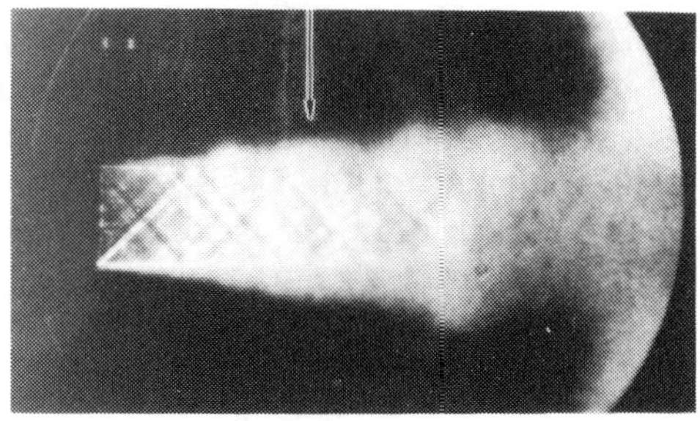

(*d*) $\phi = 180°$.

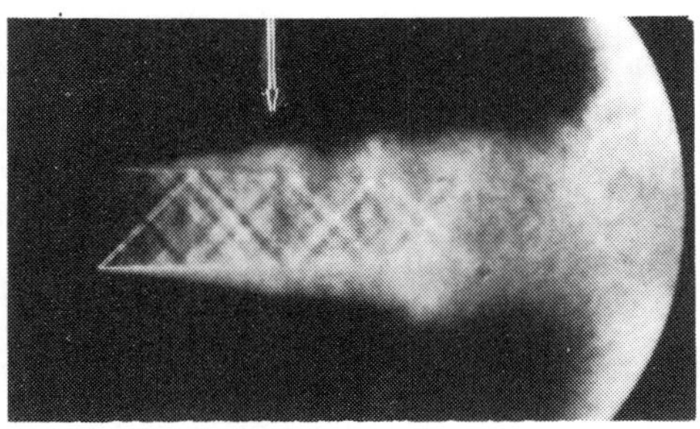

(*e*) $\phi = 240°$.

Figure 3. Ensemble-averaged photographs of large-scale structures in high Reynolds number supersonic jet under low-level excitation at various phase angles ϕ relative to triggering signal. Arrow indicates position of same large structure turbulence as function of time. Excitation Strouhal number = 0.4; $M_j = 1.37$. (From ref. 29.)

structures in high Reynolds number supersonic jets most probably have dynamic characteristics similar to those of the large-scale instability waves observed at low and moderate Reynolds numbers.

Models of Large Turbulence Structures in Free Shear Layers

During the last 15 years many attempts have been made to develop ways to calculate or predict the dynamic behavior of the large turbulence structures in free shear layers. Because of the great complexity of turbulent flow it becomes immediately clear that a complete deterministic approach to the problem by solving the full Navier-Stokes equation is both difficult and fruitless. A good deal of the past effort was, therefore, devoted to modeling or simulating the main features of the large structures. Most of these works can be classified into one of the following categories: discrete vortex models, discrete wave models, direct numerical simulations (with small-scale turbulence closure), and stochastic wave models.

Discrete Vortex Model

Shadowgraphic observations of two-dimensional mixing layers (e.g., ref. 2) reveal that the large turbulence structures appear to take the form of two-dimensional vortices. This motivated a number of investigators to model the large structures by vortices (refs. 35 to 38). It was reported in references 35 and 38 that if discrete line vortices were used to simulate the mixing layer, then the model-produced flow entities would not be consistent with experimental observations. Vorticity elements of finite sizes were needed in the model if many of the prominent features of the large structures were to be reproduced. For axisymmetric jets, with the same line of reasoning, vortex rings were used in references 39 and 40 to model the large structures. However, unlike two-dimensional mixing layers, the large structures in jets do not persist beyond the potential core. To simulate this fact, the vortices in reference 40 were artificially forced to decay once they had travelled a certain distance downstream from the nozzle lip. The location and the rate of decay were determined so that the calculated near pressure field fit closely to that of the measurements of reference 41. Unfortunately empiricism of this kind compromises any real value of such models. Aside from this, vortex-ring models suffer two other major drawbacks. Because of their restricted geometry and the requirements of conservation of vorticity, they cannot readily simulate helical and other nonaxisymmetric large-scale motions of the jet flow. Moreover, for supersonic jets they appear to be inappropriate.

Discrete Wave Model

Free shear layers with inflection point velocity profiles are known to be intrinsically unstable. These instability waves have wavelengths which scale according to the mixing layer thickness consistent with the observed length scale of the large turbulence structures. Several researchers (refs. 20, 23, and 42 to 48) modeled the large turbulence structures in jets, mixing layers, and wakes by the intrinsic instability waves of the flows. In most of these works a single-frequency wave was used. As such they are more appropriate as models for excited large structures. The excited large structures are more or less completely deterministic. In references 20, 47, and 48, ample experimental evidence is provided to show that the single instability wave model can indeed offer an excellent quantitative description of the excited large structures. Use of discrete waves to model naturally occurring large turbulence structures is another matter. Since natural large turbulence structures have characteristics which are random while discrete instability waves are completely deterministic, an obvious problem arises in the interpretation of such models. The discrete wave models are meaningful only if they are used to represent the statistical mean value of the natural large structures. Even in this context the spectral representation is definitely incorrect. The best that can be done is to pick the frequency of the instability wave to coincide with the dominant frequency of the large structures. This approximation is reasonable provided the naturally occurring large structures are confined to a relatively narrow frequency band.

Direct Numerical Simulation

A very attractive way to model the large turbulence structures is to perform direct numerical simulations. Such a direct simulation of a two-dimensional mixing layer was performed in reference 49 by solving the three-dimensional Navier-Stokes equation. However, because of computer storage limitations, spatially periodic boundary conditions in both the streamwise and spanwise directions had to be imposed. That is, growth of the mixing layer was partly suppressed in the calculation. For axisymmetric jets, a computer code capable of calculating the large-scale instability waves in subsonic jets was developed in references 50 and 51. In this code measured mean velocity profiles of the jet are a part of the input. Three-dimensional flow structures are, however, not simulated by the code. Only the linearized axisymmetric small-amplitude inviscid disturbances are calculated. The direct simulation model of reference 51 exhibited a very interesting vortex shedding phenomenon. Although at the present time no useful information related to jet noise generation has been obtained through direct simulation of the large turbulence structures, future development may yet prove this to be a very fruitful approach.

Stochastic Wave Model

For the purpose of modeling naturally occurring large turbulence structures in both subsonic and supersonic flows, a relatively well-developed stochastic wave model is currently available. This model has been applied to the study of these structures in two-dimensional mixing layers and subsonic jets in references 52 to 54 and in supersonic jets for broadband shock associated noise calculations in reference 55. The calculated results in each of these instances agree favorably with experimental measurements. Compared with the discrete vortex model and the direct numerical simulation, the stochastic wave model appears to be the most successful up to this time. In view of its importance in shock-associated noise theory, the physics and formulation of this model are now briefly examined.

In the shear layer of a jet or in a two-dimensional mixing layer, the thickness and other statistics of turbulence dynamics change very slowly in the downstream direction. If these slow changes are neglected, then the turbulence statistics are invariant to translation in the flow direction. In other words, statistically the flow is nearly unchanged in the downstream direction. That this is true implies the flow is, at least, locally in a state of dynamic equilibrium. According to statistical mechanics for a system which is in dynamic equilibrium, the large-scale fluctuations (the large turbulence structures in the present consideration) can be represented by a superposition of the normal modes of the system. For jet flows or two-dimensional mixing layers, the normal modes are the instability wave modes (damped waves included). Thus it is statistically meaningful to represent the large turbulence structures by a superposition of the instability wave modes. The amplitudes of the instability wave modes are unknown. Since the large turbulence structures are random in space and time the amplitudes are, therefore, taken to be stochastic random functions.

Instability wave solutions or modes in a slightly divergent flow have been constructed and studied in references 56 to 62 using the method of multiple scales. (See ref. 63.) For supersonic jets, the fact that the problem has two intrinsic length scales was used in reference 62. One scale is imposed by the jet diameter. The other scale characterizes the slow variation of the mean flow in the downstream direction and is given by the potential core length of the jet. The ratio of the two scales ϵ is a very small number. Suppose (r, ϕ, x) are the cylindrical coordinates of a coordinate system centered at the nozzle exit of the supersonic jet. The x-axis is chosen to coincide with the direction of flow. We can now take advantage of the existence of the small parameter ϵ to introduce a slow variable $s = \epsilon x$ in a multiple-scales expansion. Let $p_t(r, \phi, x, t)$ and $\mathbf{v}_t\ (r, \phi, x, t)$ be the pressure and velocity fluctuations associated with the large turbulence structures. According to the stochastic wave model they can be represented by a superposition of normal modes in the form

$$\begin{bmatrix} p_t(r,\phi,x,t) \\ \mathbf{v}_t(r,\phi,x,t) \end{bmatrix} = \int_{-\infty}^{\infty} \sum_{n=-\infty}^{\infty} a_n(\omega) \begin{bmatrix} \tilde{p}_n(r,s,\omega) \\ \tilde{\mathbf{v}}_n(r,s,\omega) \end{bmatrix} \exp\Big\{ i[\theta_n(s,\omega)/\epsilon + n\phi - \omega t] \Big\} d\omega \quad (1)$$

In equation (1), $\theta_n(s, \omega)$ is the slowly varying phase function which is related to the local eigenvalue or wave number α_n of the inviscid Orr-Sommerfeld or Rayleigh equation by

$$\theta_n(s,\omega) = \int_0^s \alpha_n(s,\omega)\, ds \quad (2)$$

and $\tilde{p}_n(r, s, \omega)$ and $\tilde{\mathbf{v}}_n(r, s, \omega)$ are the corresponding eigenfunctions. In particular, $\tilde{p}_n(r, s, \omega)$ satisfies the equation

$$\frac{\partial^2 \tilde{p}_n}{\partial r^2} + \left[\frac{1}{r} - \frac{1}{\bar{\rho}}\frac{\partial \bar{\rho}}{\partial r} + \frac{2\alpha_n}{(\omega - \alpha_n \bar{u})}\frac{\partial \bar{u}}{\partial r} \right] \frac{\partial \tilde{p}_n}{\partial r} + \left[\frac{(\omega - \alpha_n \bar{u})^2}{\bar{c}^2} - \frac{n^2}{r^2} - \alpha_n^2 \right] \tilde{p}_n = 0 \quad (3)$$

where $\bar{u}(r, s)$, $\bar{\rho}(r, s)$, and $\bar{c}(r, s)$ are the mean velocity, density, and speed of sound. The amplitude $a_n(\omega)$ of equation (1) is a stochastic random function. In reference 52, for two-dimensional mixing layers a similarity argument is used to determine its stochastic properties. Here, in the initial mixing layer of the jet, self-similarity applies and so a similar argument holds. The similarity argument suggests that the instability wave modes of the large turbulence structures may be regarded as having been initiated by excitation that has no intrinsic characteristic length and time scales, namely, white noise. This condition leads to the following stochastic property for the amplitude function $a_n(\omega)$ (see, e.g., ref. 52):

$$\langle a_n(\omega) a_{n'}(\omega') \rangle = \frac{1}{2} \tilde{D} \frac{R_j}{\bar{u}_j} \delta(\omega + \omega') \delta_{n,-n'} \quad (4)$$

where $\delta_{n,-n'}$ is the Kronecker delta, R_j is the fully expanded jet radius, and $\langle\ \rangle$ indicates ensemble average.

Equations (1) to (4) provide a complete stochastic description of the large turbulence structures in a jet. Clearly the characteristics of the large structures are directly related to the instability waves of the mean flow through the eigenfunctions

$\tilde{p}_n$ and $\tilde{\mathbf{v}}_n$ and eigenvalues α_n. Because of this, in the rest of this chapter the terms "large turbulence structures" and "instability waves" are used interchangeably to refer to the same physical entities. This stochastic model may be used to calculate the second-order statistics such as Reynolds stresses and root-mean-squared pressure and velocity fluctuations associated with the large structures as in references 52 to 54. In a subsequent section this stochastic model is applied to the computation of the noise spectrum and directivity of broadband shock-associated noise of supersonic jets.

Introduction to Jet Noise Generated by Instability Waves

Prior to and at about the same time as the discovery of large turbulence structures in free shear flows, a number of investigators (refs. 22 to 24, 42, 45, and 64 to 66) suggested that instability waves might play an important role in the generation of jet noise. However, some of the suggestions were mere intuition and speculation with little experimental or theoretical support. Among these early works the first successful theory of noise generated by instability waves was developed in an investigation of the strong directional acoustic radiation emitted from the shear layer close to the nozzle exit of supersonic jets. (See fig. 4; also see ref. 22.) The theory was subsequently extended to be applicable to lower Strouhal number instability waves (ref. 24). This reference also provided further and more complete experimental verification of the calculated results. One of the goals of these early works was to predict the observed acoustic radiation pattern by seeking appropriate solutions of the equations of motion of compressible flows. The theory is self-contained without requiring external theoretical or empirical input. As such it represents a radical departure from the then prevalent acoustic analogy theory of references 67 to 70. In the acoustic analogy theory the source terms are regarded as known, making it dependent on separate theoretical calculations or experimental measurements for the provision of these quantities. In addition to identifying instability waves as a noise source, the mathematical analysis of these early works provides a firm basis for the subsequent development of a more comprehensive mathematical theory of supersonic jet noise.

Figure 4 (see ref. 22) and shadowgraphs taken of the sound field pattern of supersonic jets (refs. 21, 24, and 71) show that strong directional acoustic waves are emitted from the shear layer close to the nozzle exit. On the shadowgraphs, these waves appear more or less as parallel straight lines radiating in the downstream direction. In a rather puzzling manner they exist only in a limited region downstream of the nozzle. They are never found beyond a certain acute angle measured from the jet boundary. It is suggested (ref. 22) that this complex directional acoustic wave pattern was generated by the instability waves in the mixing layer of the jet. Through use of a simple top-hat velocity profile to model the jet flow, it was shown that the spatially amplifying Kelvin-Helmholtz instability waves possessed near acoustic fields which exhibited features identical to those of figure 4. Numerical results on the geometry of the wave pattern computed in references 22 and 24 were found to agree very favorably with experimental measurements for nitrogen as well as helium supersonic jets. In addition, the theory predicted that the parallel wave fronts in the near field propagated with a speed less than the ambient speed of sound. This

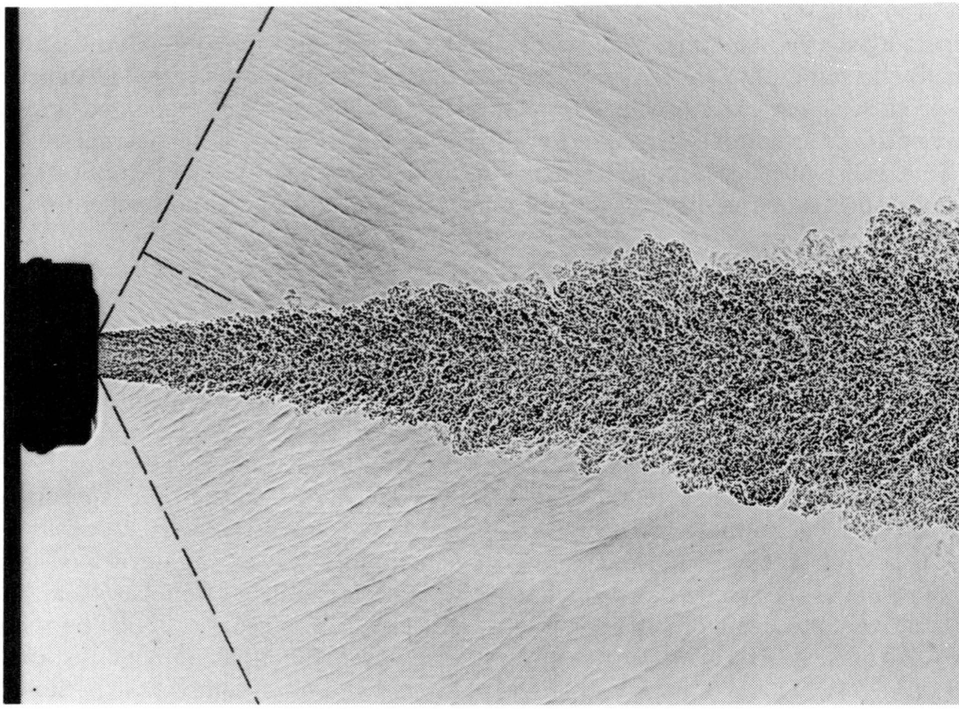

Figure 4. Directional acoustic radiation from supersonic jet generated by instability waves. Helium jet with reservoir pressure of 54.7 psia. (From ref. 22.)

totally unexpected result was confirmed by the measurements in reference 24. The predicted velocities agreed closely with the measured values over a substantial range of supersonic Mach numbers.

Characteristics of Supersonic Jet Noise

Unless a supersonic jet issuing from a convergent-divergent nozzle is operated very close to the nozzle design condition, its noise spectrum invariably consists of discrete and broadband components. (See fig. 5, which is based on ref. 72.) The discrete components are generally referred to as screech tones. For imperfectly expanded supersonic jets with rather strong shock cells, the screech tone is often accompanied by its harmonics. In some cases, such as underexpanded jets from convergent nozzles, as many as four harmonics have been observed. The screech component disappears when the jet is perfectly expanded. For perfectly expanded jets the noise spectrum is made up of a broad, smooth peak as illustrated by the lower spectra of figure 5. This broadband noise component is generated by the turbulence in the mixing layer of the jet. For this reason it is called the turbulent mixing noise. If the ratio of reservoir to ambient pressure of the jet is changed so that the jet is operating in an off-design condition, then experiments show that additional broadband noise would be emitted. This noise component, which owes its origin to the presence of a quasi-periodic shock cell structure inside the jet plume, is known as broadband

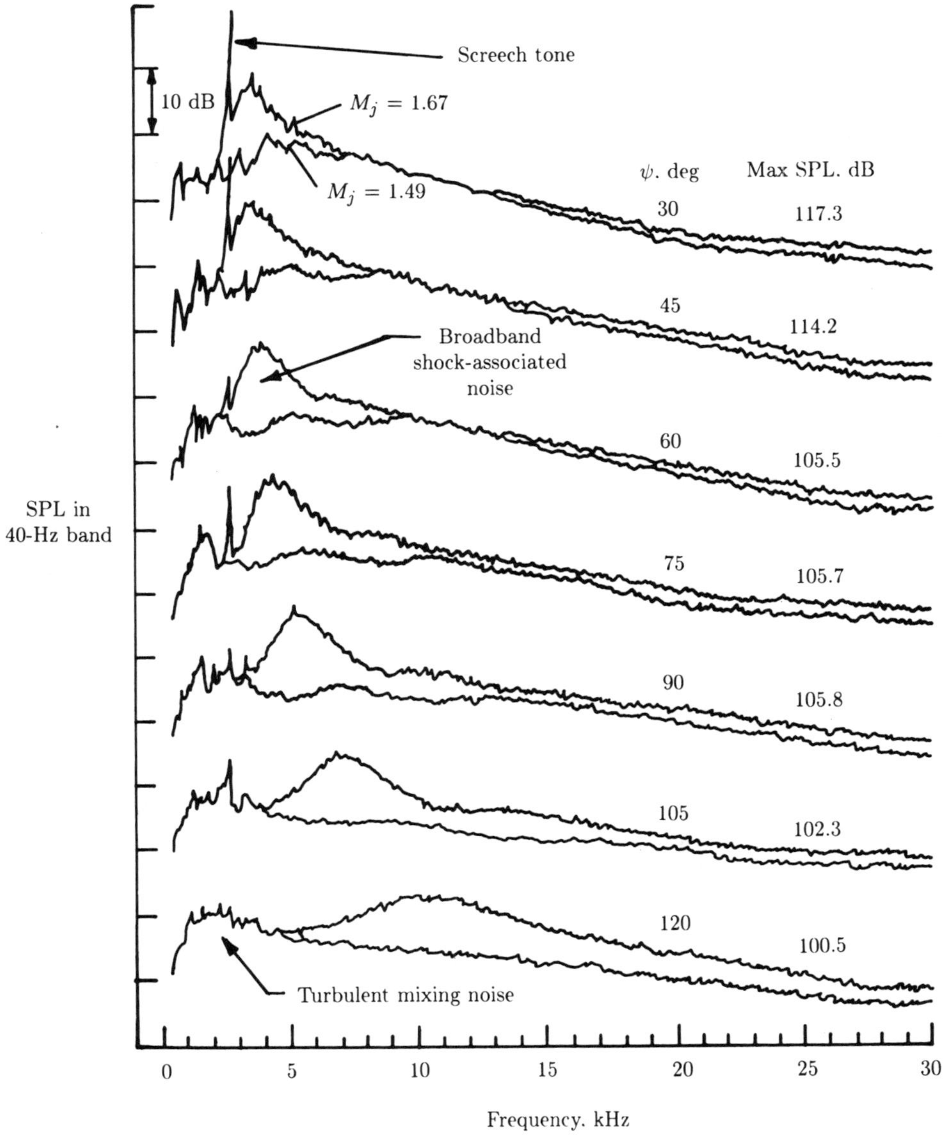

Figure 5. Narrow-band noise spectra of supersonic jet for fully expanded Mach numbers M_j of 1.49 and 1.67 and for jet inlet angles ψ of 30° to 120°. Nozzle design Mach number $M_d = 1.5$. (Based on ref. 72.)

shock-associated noise, or simply shock-associated noise. The dominant part of the broadband shock-associated noise is comprised of a spectral peak with a relatively narrow half-width. (See fig. 5.) One of the most peculiar characteristics of broadband shock-associated noise is that the frequency of the spectral peak is a function of the direction of radiation. The spectral peak frequency is lowest near the jet inlet direction and increases monotonically towards the jet flow direction. Recently, after a careful examination of all available narrow-band shock-associated noise spectra, it was pointed out that the fundamental screech tone frequency is always smaller than the frequencies of broadband shock-associated noise (ref. 73). As a matter of fact, the screech tone frequency is a very reliable indicator of the lower bound of the broadband shock-associated noise spectrum. We now briefly examine some of the most prominent directional and spectral characteristics of turbulent mixing noise, broadband shock-associated noise, and screech tones of axisymmetric supersonic jets.

Turbulent Mixing Noise

Good quality noise measurements of perfectly expanded high Reynolds number supersonic jets are available in references 72 and 74 to 77. These studies indicate clearly that turbulent mixing noise is highly directional. Figure 6(a) shows a typical directional distribution of the SPL of turbulent mixing noise for cold supersonic jets (ratio of jet temperature to ambient temperature T_j/T_0 of 0.73). As shown, the predominant part of the noise is radiated in the downstream direction. The peak value is about 25° from the jet flow direction. This maximum directivity angle θ varies as a function of Mach number and jet temperature. Normally for jets with fully expanded Mach number M_j less than 2.0, θ is between 25° to 45°. Figure 6(b) shows the noise directivity for a hot jet with temperature ratio T_j/T_0 of 2.27 and the same fully expanded jet Mach number as in figure 6(a). In this case the jet velocity is higher and the SPL becomes even more directional.

As has been pointed out, the power spectrum of turbulent mixing noise of supersonic jets is characterized by a smooth broad peak. Figure 7 shows a typical noise power spectrum for a cold supersonic jet from reference 74. To the left of the spectral peak it was found that the spectral density varies as the square of the frequency (refs. 74 and 75). On the other hand, in the mid-frequency range to the right of the spectral peak, the spectral density decreases as the inverse of the frequency. At still higher frequencies the noise spectral density drops off rather abruptly. This dependence of the noise power spectrum on frequency was also observed in the peak noise radiation direction (ref. 78). The variation of the noise spectral distribution as a function of jet temperature was presented in reference 76. Generally speaking, for supersonic jets, as the jet temperature increases and the jet velocity stays the same, there is a reduction in the radiated noise across all frequencies. However, the reduction in high-frequency noise is significantly more dramatic.

Another way of seeing that turbulent mixing noise is highly directional is to examine the near acoustic field of the jet. Figure 8 is a near-field noise contour plot from reference 74. This figure represents a plane passing through the centerline of the jet. The x-axis is in the direction of the jet flow and the r-axis points in the radial direction. The value $x = 0$ is at the nozzle exit. Plotted in this figure are contours of equal 1/3-octave-band noise at a center frequency of 10 kHz. This frequency corresponds closely to the broad peak of the noise power spectrum of figure 7. In the lower part of figure 8 the space with no contour curves is occupied by the jet flow. It is evident that the contours in this figure form a highly directional lobe pointing in a direction approximately 30° from the x-axis. This implies that for this jet the direction of maximum noise radiation at a 10-kHz frequency makes an angle of about 30° with the jet axis. This is consistent with far-field noise

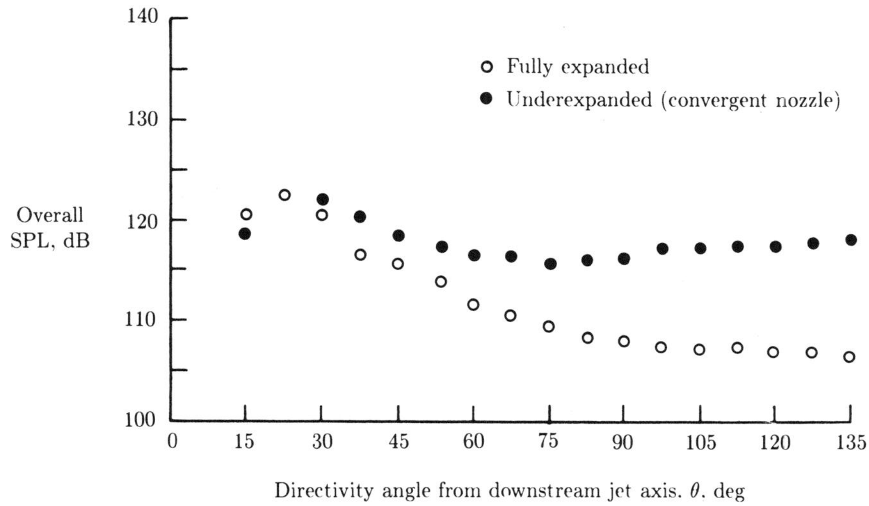

(a) $T_j/T_0 = 0.73$ *(cold jet).*

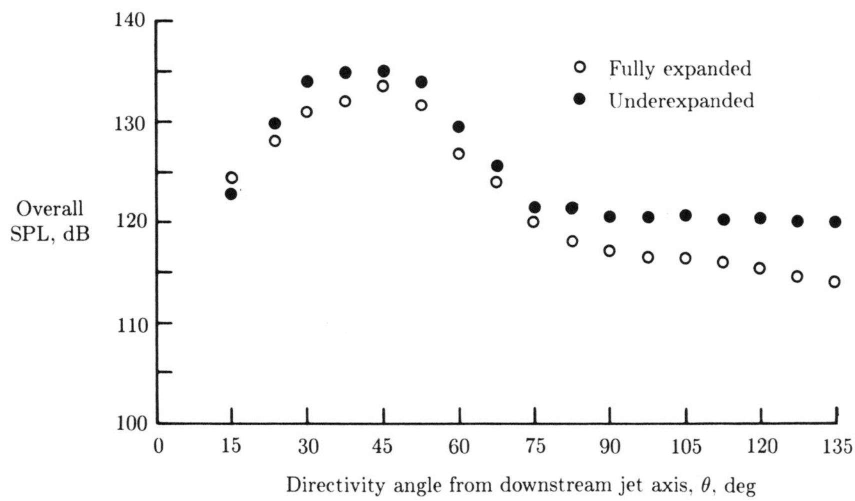

(b) $T_j/T_0 = 2.27$ *(hot jet).*

Figure 6. Directivity of overall sound pressure level. $M_j = 1.372$. *(From part II of ref. 76.)*

measurements. (See ref. 74.) A large body of near-field noise measurements of this kind can be found in references 74 and 77. An examination of these data reveals that at low frequencies the near-field 1/3-octave-band noise contours do not exhibit a prominent lobe as in figure 8. Instead they form flat curves more or less parallel to the jet flow boundary. (See the 138-dB contour of fig. 8.) This contour pattern suggests that the peak noise radiation directions at these frequencies are below the direction of the jet boundary; that is, they are very close to the jet axis. At very high

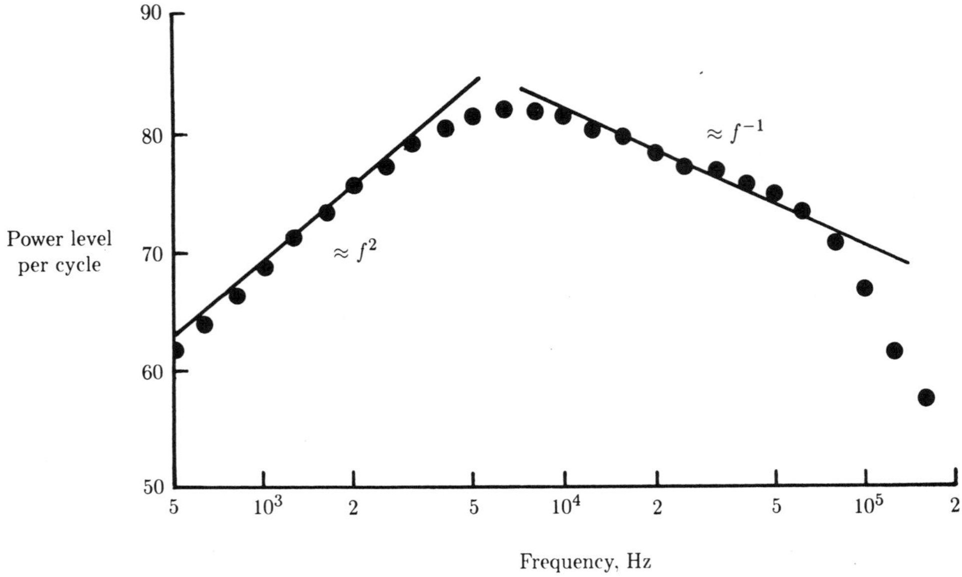

Figure 7. Noise power spectrum of $M_j = 1.5$. (From ref. 74.)

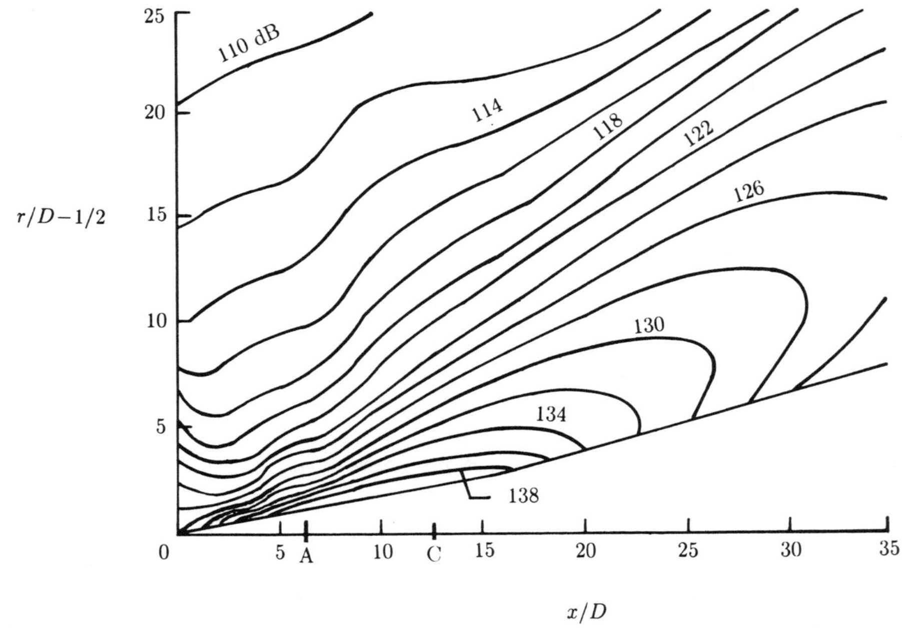

Figure 8. Near-field 1/3-octave-band sound-pressure-level contours of Mach 1.50 supersonic jet at 10-kHz center frequency. (From ref. 74.)

frequencies the near-field noise contours display an entirely different pattern. They are roughly in the form of concentric circles. This means that the high-frequency noise radiation is more or less uniform in all directions.

Broadband Shock-Associated Noise

Broadband shock-associated noise possesses many characteristics which are distinctly different from those of turbulent mixing noise. For instance, as noted before, turbulent mixing noise is most intense in the downstream direction and peaks around $\theta = 30°$, where θ is the polar angle measured from the jet flow direction. Broadband shock-associated noise, on the other hand, is most prominent in the forward arc or upstream direction. Figure 6(a) shows the directivity of the overall SPL of a fully expanded jet at a Mach number of 1.37 and a temperature ratio of 0.73 (from ref. 76). The jet is practically shock-free so that this is the directivity of turbulent mixing noise. Also plotted in this figure is the directivity of overall SPL of an underexpanded supersonic jet with the same fully expanded Mach number and jet operating conditions. The noise from this underexpanded jet contains both turbulent mixing noise and broadband shock-associated noise. The differences in these two directivities, therefore, would provide a good estimate of the distribution of shock-associated noise. It is clear from this figure that shock-associated noise is important mainly in the forward direction of the jet. Figure 6(b) shows a similar directivity plot for a hot jet with a temperature ratio of 2.27. Although quantitatively shock-associated noise appears to be relatively less intense for hot jets, it still predominates in the forward arc. A somewhat different way of showing that the importance of shock-associated noise is confined primarily to the forward directions of the jet is to compare the overall SPL's between perfectly expanded and imperfectly expanded supersonic jets at different observation angles as the velocity of the jet increases from subsonic to supersonic. Data of this kind were presented in reference 76 and are shown in figure 9. Figure 9(a) shows that in the downstream direction the noise from a perfectly expanded jet and that from an underexpanded jet are nearly identical. In other words, shock-associated noise makes no significant contribution to the overall noise level in this downstream direction. On the other hand, figures 9(b) and 9(c) indicate clearly that at 90° and in the forward arc, the underexpanded jet is noisier. Broadband shock-associated noise is responsible for the difference in radiated sound pressure levels. The fact that shock-associated noise can be easily identified in the forward arc does not mean that there is no broadband shock-associated noise in the downstream direction. Indeed, by analyzing the noise data carefully it is not difficult to identify the presence of this noise component in all directions. It is, however, a weaker contributor to the overall noise in the downstream direction than the turbulent mixing noise and hence is not as important.

For a given far-field direction, the dominant part of broadband shock-associated noise is always concentrated in a relatively narrow frequency band compared with, say, turbulent mixing noise. The noise spectrum consists essentially of a dominant characteristic peak, as illustrated in figure 5. A careful examination of the measured data, however, reveals that the noise spectrum actually contains one or more secondary peaks. For instance, a secondary peak centered at approximately 10 kHz can be easily identified in figure 5 for the $\psi = 90°$ spectrum. In reference 73 some spectra of broadband shock-associated noise with very well-defined secondary

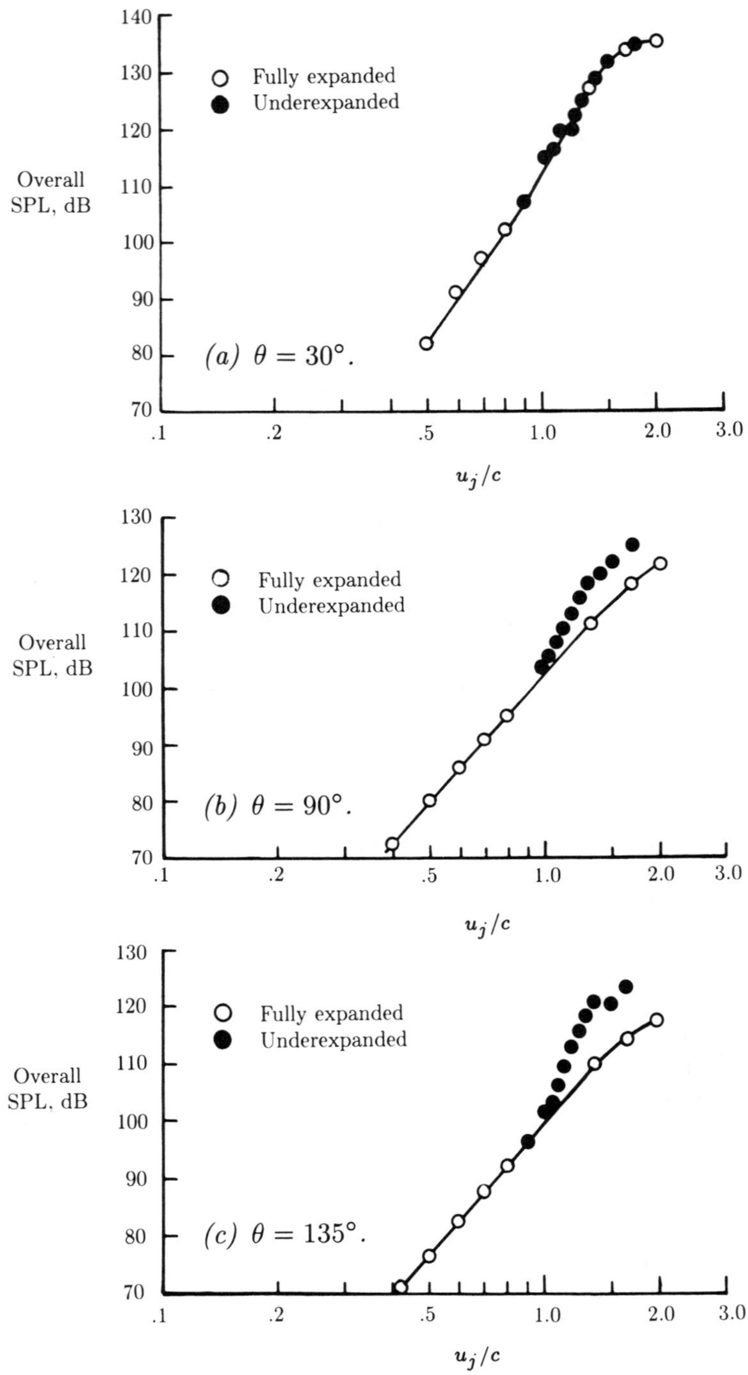

Figure 9. Overall sound pressure level of broadband shock-associated noise. $T_j/T_0 = 1.0$. (From part II of ref. 76.)

peaks are provided. One distinct characteristic of broadband shock-associated noise, first identified in reference 79, is that the peak frequency of the noise spectrum is a function of the observation angle. This can readily be seen in figure 5. The peak frequency decreases as the angle of observation increases toward the forward direction. Recently it was pointed out in reference 73 that the half-width of the dominant peak was also a function of the direction of radiation. The half-width decreases as the radiation direction approaches the inlet direction of the supersonic jet.

Instead of relying on a comparison of the far-field noise intensities of a perfectly expanded jet with those of a similar imperfectly expanded jet to show that broadband shock-associated noise and turbulent mixing noise are indeed two different supersonic jet noise components, it is possible to achieve the same goal by studying only the near-field sound pressure distribution of a shock-containing jet. Figure 10 shows a

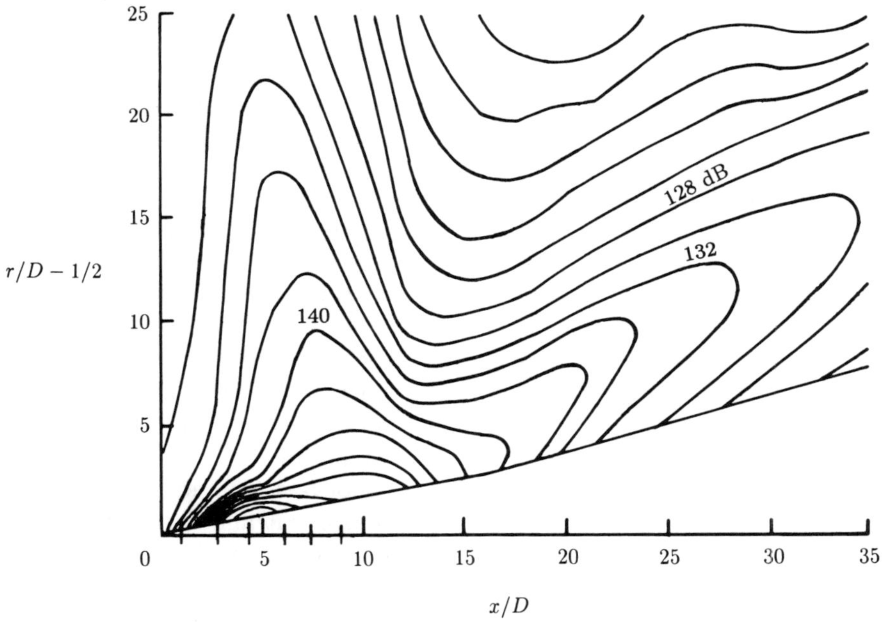

Figure 10. Near-field 1/3-octave-band sound-pressure-level contours of Mach 1.50 supersonic jet at 20-kHz center frequency. $M_j = 1.67$. (From ref. 74.)

sound-pressure-level contour map of an underexpanded jet from reference 74. The fully expanded Mach number of the jet is 1.67 and the nozzle design Mach number is 1.50. In this figure it is easy to see that the contours form a pattern dominated by two distinct lobes. One lobe points in the downstream direction at approximately 30° to the jet axis, while the other lobe points upstream at approximately 80° to the jet inlet direction. The downstream-pointing lobe is very similar to that of figure 8. Clearly it is associated with the turbulent mixing noise of the jet. The upstream-pointing lobe does not exist for a perfectly expanded jet. It is associated with the broadband shock-associated noise of the jet. On examining near-field sound-pressure-level contour maps at different 1/3-octave-band center frequencies, it is found that the direction of this lobe changes with frequency. For low-frequency noise the direction of this lobe makes a small angle with the jet inlet direction. As the frequency increases this angle increases also. This variation of the direction of noise radiation with frequency is consistent with the far-field narrow-band spectral and directional measurements shown in figure 5. For that figure it was found that the peak frequency is a function of the direction of radiation, with the lowest frequency component radiating in the jet inlet direction. It is interesting to point out that at a sufficiently high frequency the near-field sound-pressure-level contours exhibit secondary lobes. The weak secondary lobe is the counterpart of the first secondary peak of the sound power spectrum discussed previously. The secondary lobes in the near-field sound-pressure-level contour maps and the secondary peaks in the sound power spectra are some of the many fine, distinctive features of broadband shock-associated noise. This is in sharp contrast to the turbulent mixing noise, which is almost totally devoid of any similar structural characteristics.

When a supersonic jet from a convergent-divergent nozzle is operated at the design pressure ratio, the jet is shock free and hence the radiated sound is entirely turbulent mixing noise. If the pressure ratio of the jet is changed, a quasi-periodic shock cell structure forms in the jet. This is so regardless of whether the jet is overexpanded or underexpanded. In both cases shock-associated noise is produced, increasing the SPL radiated forward. The level of shock-associated noise is naturally a function of the fully expanded jet Mach number M_j. The dependence of shock-associated SPL on M_j is illustrated in figure 11. This figure shows the measured overall SPL at a 150° angle to the jet axis as a function of jet Mach number obtained with a convergent-divergent nozzle with design Mach number of 1.5 (ref. 80). At the design Mach number, there is no shock-associated noise and the curve attains a local minimum. This is denoted in the figure by point A. With decreasing Mach number the overall noise level increases, following the curve AC. This occurs even though the level of the turbulent mixing noise, given by the curve of the solid symbols, decreases. The reason for the increase in overall noise level is that as the operating Mach number deviates more and more from the design Mach number, the shock cell strength increases very rapidly and gives rise to intense shock-associated noise. At point C the mismatch between the pressure at the jet nozzle exit and the ambient pressure is so large that the oblique shocks of the jet can no longer form a quasi-periodic shock cell structure. A Mach disk forms near the end of the first shock cell. This in turn reduces the shock cell strength farther downstream. As a result the level of shock-associated noise is reduced. If the jet is operated in the underexpanded mode, the overall noise level follows the curve AB as the operating fully expanded Mach number increases. Point B, at which the curve reaches the peak value, corresponds approximately again to the condition for the appearance of a Mach disk in the jet. Beyond this point the shock structure is highly complicated, with mixed subsonic and supersonic flows. Qualitatively figure 11 is typical of all overall noise level curves for an observation angle θ in the forward direction. The dip in the noise level curve at the nozzle design point A is, of course, a strong function of θ. This is illustrated in figure 12. The data are taken from the measurements of reference 81. The design Mach number of the convergent-divergent nozzle used in this series of experiments is 1.67. Figure 12(c) indicates that the characteristic dip at the nozzle design Mach number can be seen even at rear arc angles as small as $\theta = 60°$.

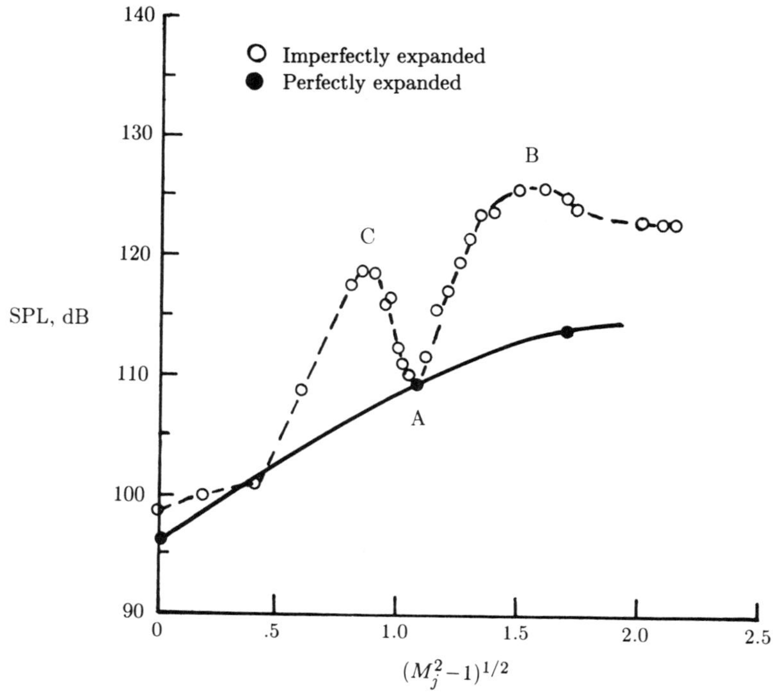

Figure 11. Overall sound pressure level at 150° angle to jet axis. $M_d = 1.5$. *(Based on ref. 80.)*

Screech Tones

The flow and acoustic fields associated with the jet screech phenomenon are highly complicated and sensitive to the presence of reflecting surfaces in the vicinity of the jet. For simplicity, it is assumed throughout this chapter that such surfaces are absent in the near field of the jet. Extensive visual observations (e.g., refs. 82 and 83) of the motion of screeching jets indicate that during a screech cycle the jet undergoes principally two types of large-scale motion. These motions are associated with the propagation of toroidal and helical mode disturbances along the jet column. More recent observations (refs. 84 and 85), however, reveal that the left- and right-hand helical disturbances are usually excited simultaneously so that the overall motion of the jet is, in fact, a flapping mode. The flapping motion is highly reproducible relative to the flapping plane. The orientation of the flapping plane, however, can change over a period of time even in the same experimental facility. At about three to five shock cells downstream of the nozzle exit, strong acoustic waves are generated. These waves propagate outside the jet flow predominantly in the upstream direction, as shown in figure 13 (from ref. 86). Screech tone radiation being strongest in the upstream direction has been confirmed by acoustic far-field measurements.

It has been reported by a number of investigators (e.g., refs. 87 to 91), who studied the screech phenomenon using axisymmetric sonic nozzles, that the jets exhibit a staging behavior. As the operating pressure ratio of the jet is increased, the screech frequency decreases so that the acoustic wavelength increases. When

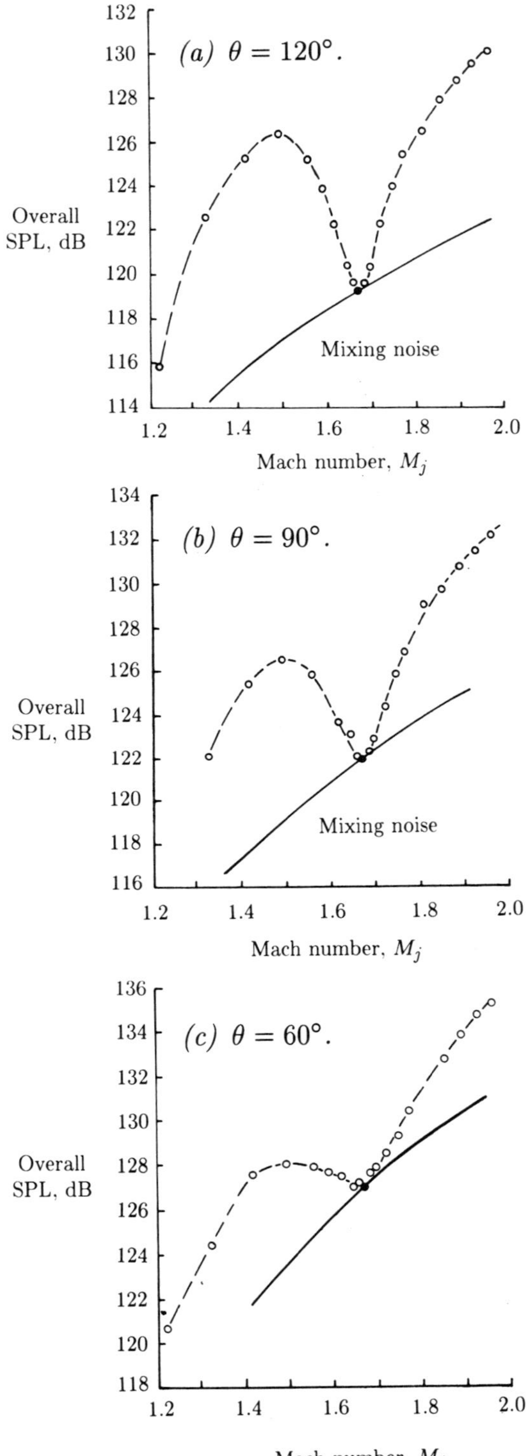

Figure 12. Shock-associated and turbulent mixing noise components for convergent-divergent nozzle operated with $T_j/T_0 = 1.0$. (Based on ref. 81.)

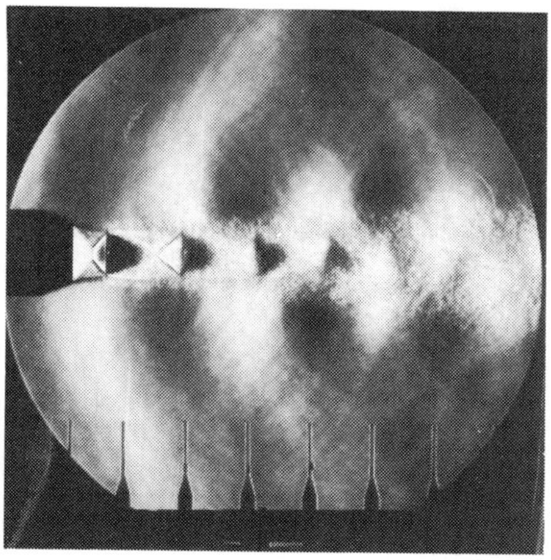

(a) $\phi = 0$.

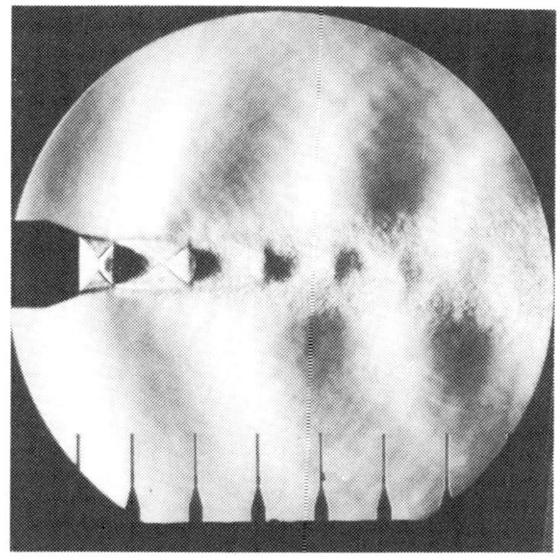

(b) $\phi = \pi/2$.

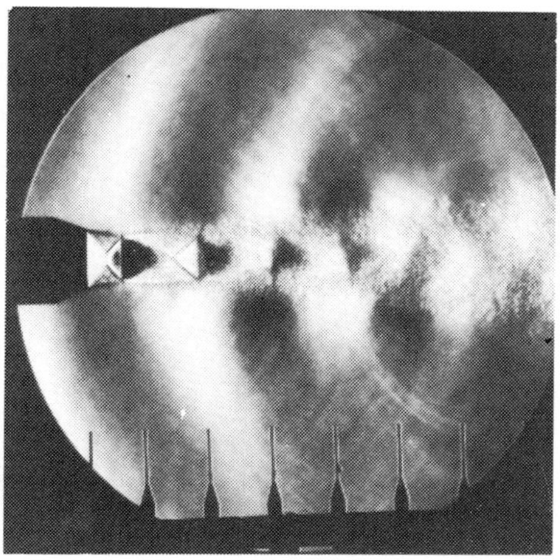

(c) $\phi = \pi$.

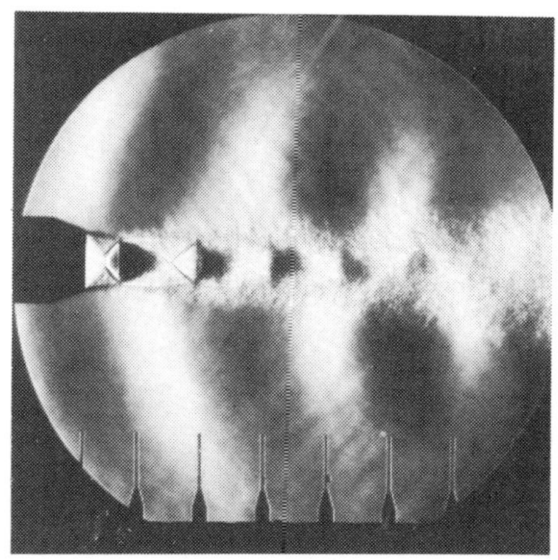

(d) $\phi = 3\pi/2$.

Figure 13. Sequence of phase-averaged schlieren records showing helical screech mode for $M_d = 2.00$ *nozzle operating overexpanded at* $M_j = 1.58$. $f_s = 2810$ *Hz. (From ref. 86.)*

the wavelength reaches a certain critical value, a sudden jump in the wavelength occurs. This discontinuous change in the acoustic wavelength, and hence in the screech frequency, is not an isolated event but repeats itself as the pressure ratio keeps increasing. As many as five such separate transitions have been identified. In reference 92, the measured data of the previous investigators were combined into figure 14(a) and the five stages (or modes) were labelled as A_1, A_2, B, C, and D. The measured wavelengths from different investigators are not identical but are sufficiently close for modal identification. Figure 14(b) shows the A_1, A_2, B, and

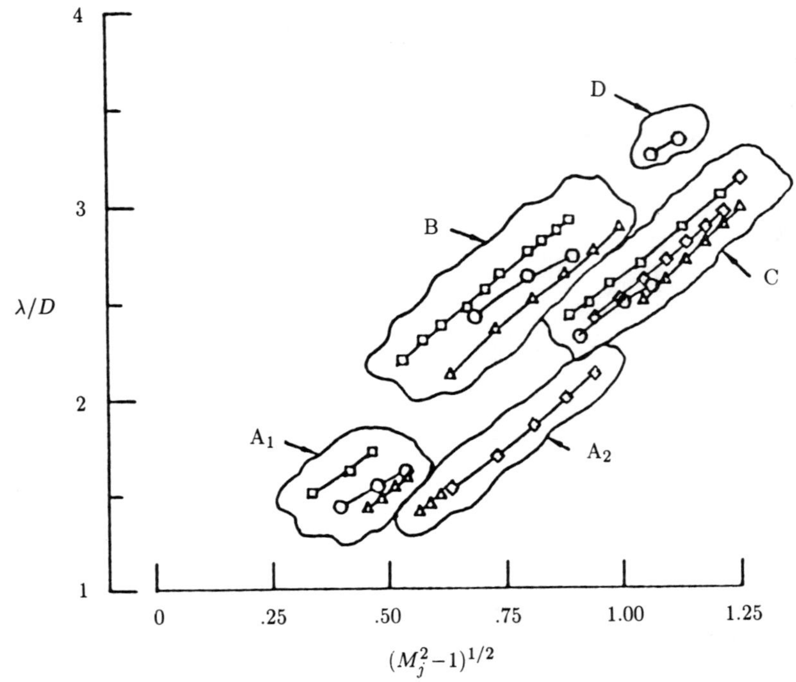

(a) Axisymmetrtic sonic nozzles.

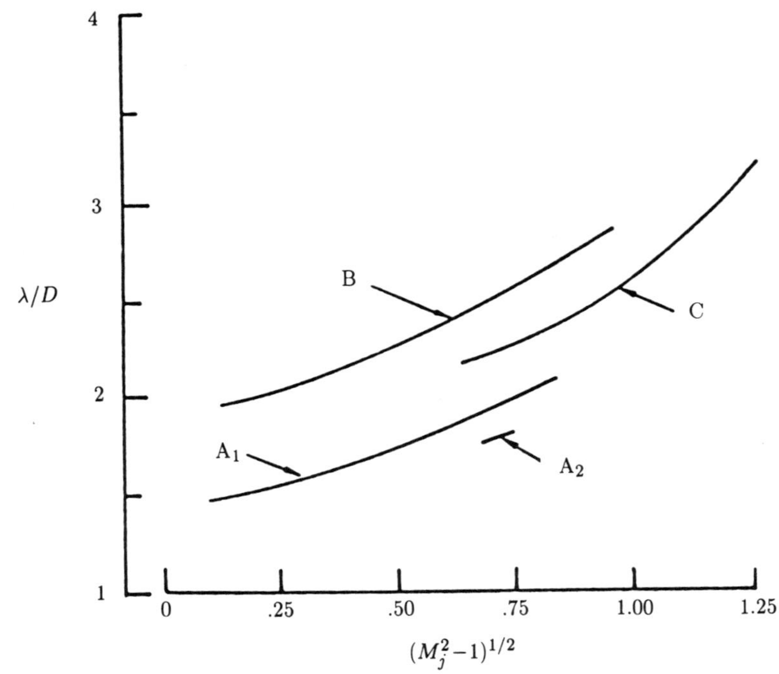

(b) Baseline thick-lipped nozzle.

Figure 14. Wavelength variations of different screech stages. (From ref. 92.)

C modes measured for a thick-lipped nozzle. Not all five modes were identified in each experiment. However, over the range of the fully expanded jet Mach numbers shown, at least two modes of screech existed at a given pressure ratio. In addition to the basic five modes, harmonics of the fundamental frequency of each mode were present whenever the amplitude of the fundamental was sufficiently large.

When transition from one screech mode to the other takes place, the new screech condition is sometimes not stable. References 83, 90, and 92 reported jet screech switching back and forth from one mode to the other every few seconds. This mode switching phenomenon could, however, be stabilized by placing a reflection surface nearby. Perhaps because of this, the mode transition points were never found to be identical by different experimenters. Apparently for jets issuing from convergent-divergent nozzles, mode switching of screech tones occurs only infrequently. On the other hand, even for relatively stable screech tones some unsteadiness in the acoustic wave amplitude is often inevitable. Real-time screech amplitude measurements (ref. 84) provide the best illustration of this type of unsteadiness. Instead of having a constant amplitude, the measurements show a nearly quasi-periodic amplitude modulation. Sometimes the amplitude modulation is quite large so that the screech tone arrives at an observer almost in bursts.

The staging of screech tones appears to be unique to sonic nozzles. Supersonic jets from convergent-divergent nozzles do not show similar behavior (refs. 83 and 85). Figure 15(a) shows the screech wavelengths as a function of fully expanded jet Mach number M_j for a supersonic jet issued from an axisymmetric convergent-divergent nozzle with a design Mach number of 1.41 (ref. 85). Figure 15(b) shows the corresponding sound pressure levels of the screech tones in the nozzle lip region of the jet. When the jet is operating in the overexpanded condition, the screech is generated by toroidal disturbances in the jet flow. When the jet is underexpanded, the helical mode screech dominates. For a jet operating at a Mach number close to the nozzle design value the screech tones are weak and both the toroidal and helical mode disturbances are detected. The reason for the change from toroidal to helical mode screech as the jet Mach number increases is not clear. Most probably it is related to the changes in both the shock cell structure inside the jet plume and the instability characteristics of the jet. Unlike the sonic nozzle case, even though several modes can be observed at a given value of M_j, only one dominates within a range of M_j.

Over the years a large collection of the dominant screech tone frequencies f_s of supersonic jets from convergent-divergent nozzles and from sonic nozzles have been measured. Since the screech tones are generated while the jet is operated at an off-design condition, the nozzle exit diameter D is not the best length scale for correlating these data. It is pointed out in reference 81 that the fully expanded jet diameter D_j, obtained by imposing the condition of conservation of mass flux, is the more appropriate length scale for shock cell structure and hence screech tone considerations. The D_j value is related to D by

$$\frac{D_j}{D} = \left\{ \frac{1 + \left[(\gamma - 1)M_j^2/2\right]}{1 + \left[(\gamma - 1)M_d^2/2\right]} \right\}^{\frac{\gamma+1}{4(\gamma-1)}} \left(\frac{M_d}{M_j} \right)^{1/2}$$

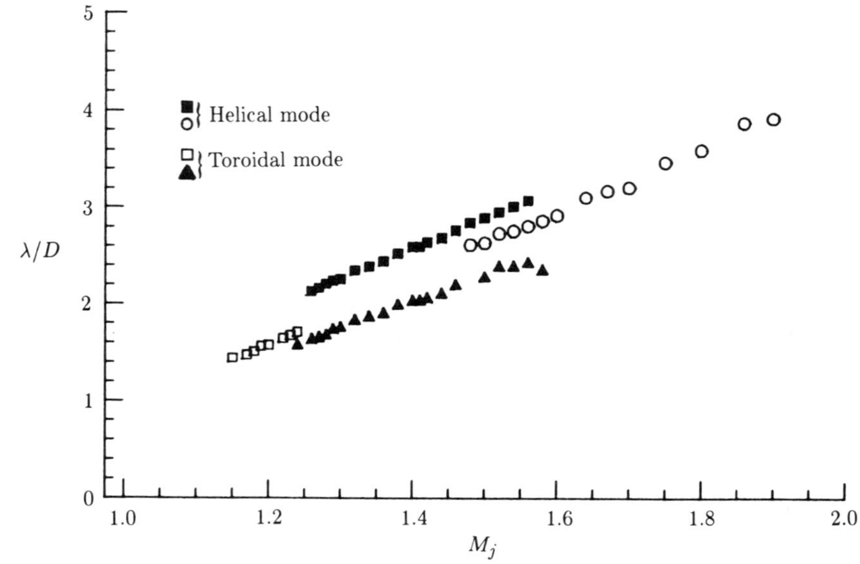

(a) Wavelength.

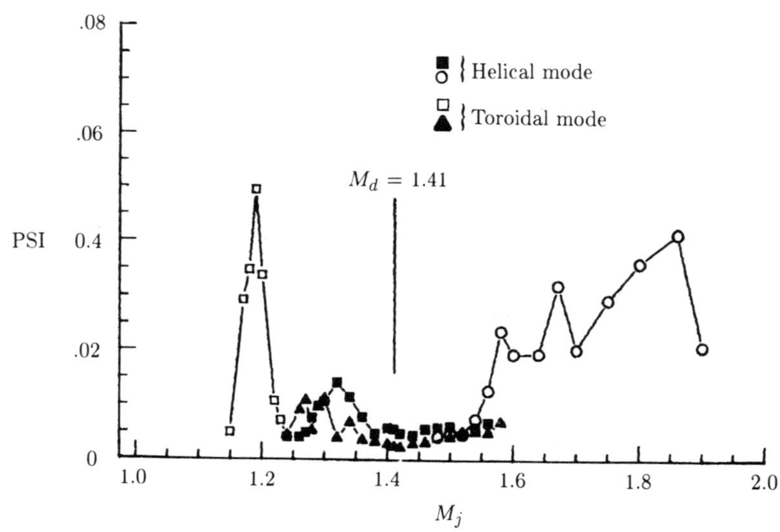

(b) Sound pressure.

Figure 15. Screech modes of axisymmetric supersonic jet from convergent-divergent nozzle with design Mach number of 1.41. (From ref. 85.)

where M_d and M_j are the nozzle design and fully expanded jet Mach numbers and γ is the ratio of specific heats of the gas. Through use of D_j and the fully expanded jet velocity u_j, it was found that there is a reasonable collapse of the measured screech tone Strouhal number $f_s D_j/u_j$ as a function of M_j. Figure 16 shows such a nearly universal curve for cold jets. The experimental measurements are from references 83, 85, 87, 90, 93, and 94. In this figure only the helical screech mode Strouhal numbers are included.

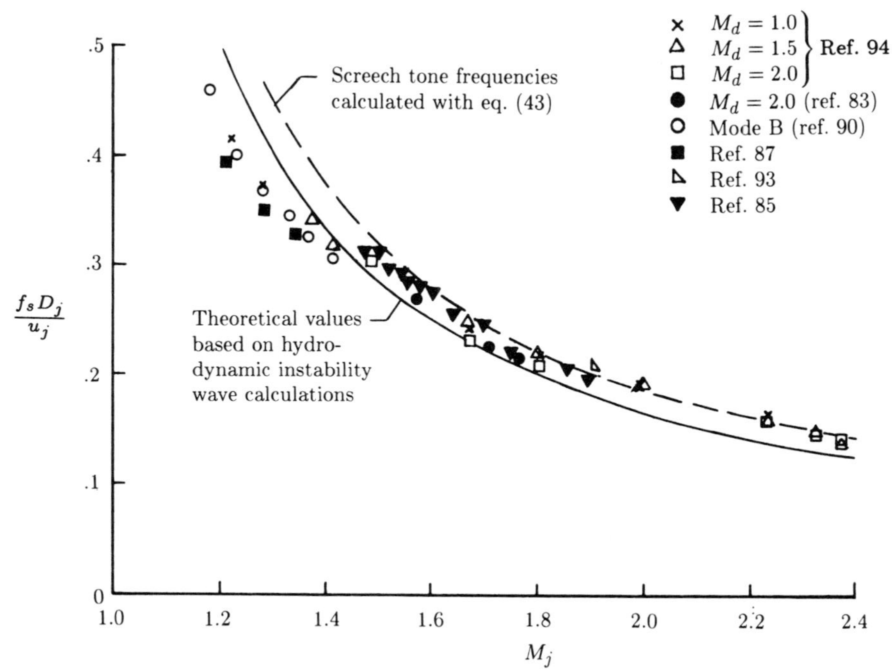

Figure 16. Strouhal number of dominant screech tone versus fully expanded jet Mach number for cold jets.

Detailed far-field directivities of screech tones and their harmonics have been measured in reference 92. Figure 17 shows a typical directivity pattern for the first four harmonics. The harmonics are produced by nonlinear effects. Two types of nonlinearities are involved. The first is the source nonlinearity. This is the nonlinearity of the downstream-propagating large-scale disturbances (instability waves) inside the jet plume which generate the screech tones in the first place. The second is the nonlinear propagation effect. This is similar to the sonic boom phenomenon. Because the screech intensity is fairly high, different parts of the acoustic wave which form the tone would propagate with slightly different speeds. The nonuniformity in the propagation velocity arises from the slight change in sound speed because of compression or rarefaction and the fluid particle velocity. This nonuniformity causes the waveform to become distorted as the acoustic waves propagate away from the jet. The distortion of the waveform creates higher harmonics at the expense of the fundamental. Thus higher harmonics can be observed in the far field even if only the fundamental screech tone is generated by the jet. In figure 17 all the harmonics show a peak direction identical to that of

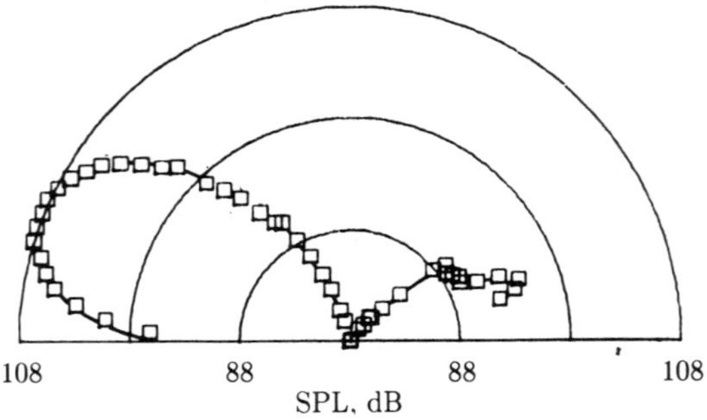

(a) Fundamental.

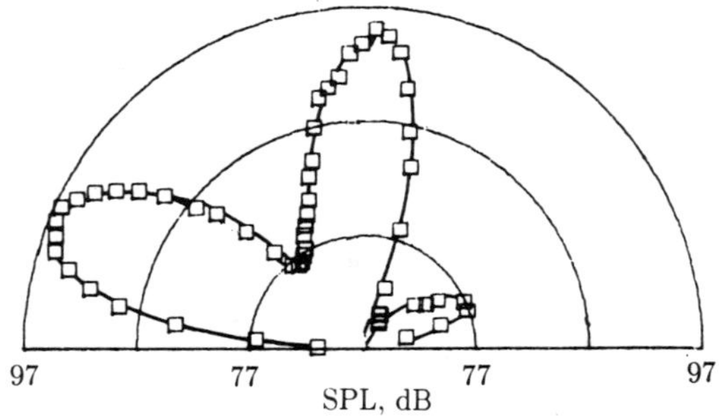

(b) Second harmonic.

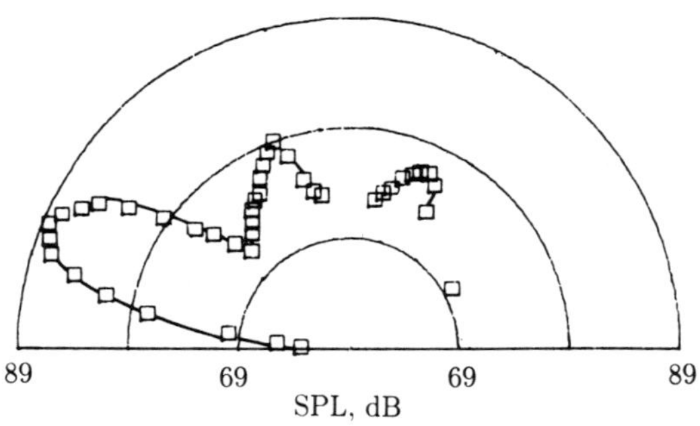

(c) Third harmonic.

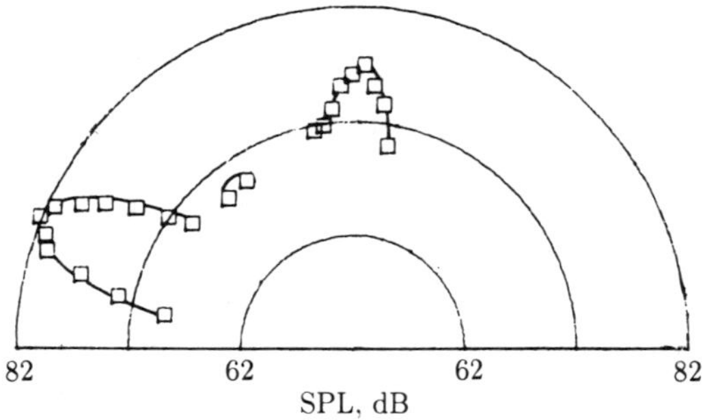

(d) Fourth harmonic.

Figure 17. Far-field directivity of stage C screech at $M_j = 1.49$. Jet exhausts to right from sonic nozzle. (From ref. 92.)

the fundamental in the upstream direction. This coincidence suggests that nonlinear propagation effects or distortion of the fundamental waveform may be the cause of the higher harmonics in this particular direction. In figure 17(b) the second harmonic shows another strong radiation at approximately 90° from the jet flow. In this direction there is almost a complete absence of the fundamental component. Because of this, it is believed that the second harmonic in this direction is produced by the nonlinearity of the noise source. Similar consideration suggests that the two weaker radiations of the third harmonic of figure 17(c) are also generated by nonlinear source effects while the weak radiation of the fourth harmonic may be because of nonlinear propagation effects associated with the second harmonic. At this time the relative importance of nonlinearity in the noise source versus nonlinear propagation effects in the generation of higher screech harmonics is still not fully understood.

The directivity of screech tones is not always axisymmetric, although it is normally so. When the jet screech phenomenon is associated with the flapping motion of the jet, the directivity is three-dimensional. This three-dimensional radiation pattern has been demonstrated recently in reference 84. In this reference a circular array of microphones was mounted at the nozzle exit plane centered at the axis of the jet. The intensities measured by the microphones provided an azimuthal distribution of the screech tones. Figure 18 shows the normalized azimuthal pressure pattern at the fundamental screech frequency. In figures 18(a), 18(b), and 18(d) the radiation pattern is associated with the flapping motion of the jet. Figures 18(a) and 18(d) refer to the same jet operating conditions for the same nozzle in the same experimental facility except they were measured at substantially different times. The field shapes of the pressure patterns are nearly the same. However, the orientations of the major lobes (the flapping planes) are quite different. The reason for this change in orientation over time is unknown. It is suspected that the pattern is sensitive to subtle boundary conditions which are difficult to control experimentally.

The near pressure fields of screeching supersonic jets have been carefully measured in references 95 to 97. This work covered choked jets undergoing toroidal as well as helical screeching motion. Figure 19 shows a typical (1/10-octave-band) near-field SPL contour map centered at the screech frequency. By comparison with figures 8 and 10 it is evident that the pattern of the near pressure field associated with a screech tone is totally different from that of the turbulent mixing noise or broadband shock associated noise. The SPL contours form almost equally spaced peaks and valleys. This field shape represents virtually a standing wave pattern wrapped around the jet. This standing wave pattern is the result of superimposing the near pressure field associated with the downstream-propagating large-scale instability waves inside the jet onto the observed upstream-propagating acoustic waves just outside the jet. The phases of these two wave fields are highly correlated, the fields being generated (as discussed subsequently) by the same feedback cycle. In references 95 to 97 the nozzle had a fairly large flange. Part of the upstream-propagating sound waves were, therefore, reflected back, partly contributing to the formation of the standing wave pattern. However, the presence of the flange is not crucial to the development of the standing wave pattern.

In addition to the fully expanded jet Mach number, the intensities and frequencies of screech tones are also affected by a number of nozzle design and jet operating parameters. It turns out that the thickness of the nozzle lip has a major influence on the radiated screech tone SPL but not so much influence on the tone frequency.

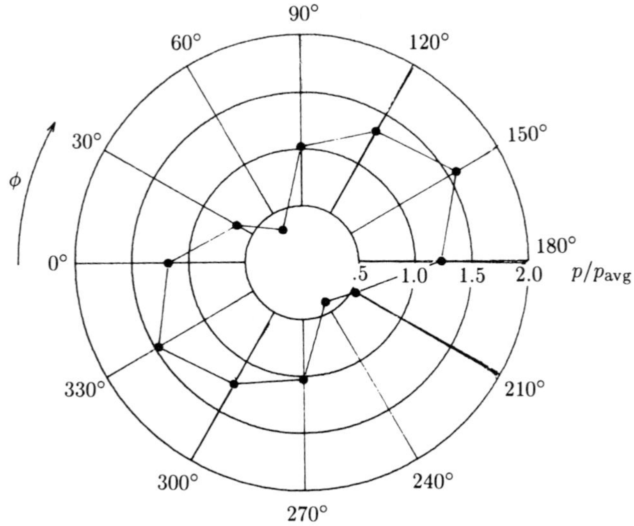

(a) $M_j = 1.56$; $N_{St} = 0.27$ *(Mar. 1985).*

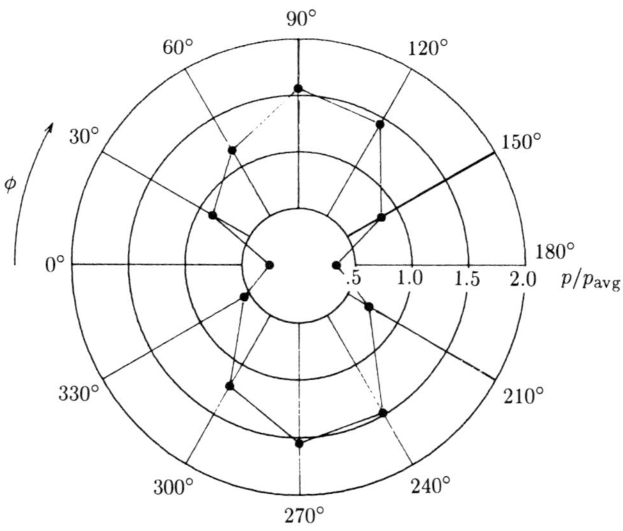

(b) $M_j = 1.63$; $N_{St} = 0.25$ *(Mar. 1985).*

Figure 18. Normalized azimuthal pressure pattern of time-averaged pressure band-pass filtered around fundamental screech frequency. Measurements made at nozzle exit plane at radial distance of 2.96 jet diameters. (From ref. 84.)

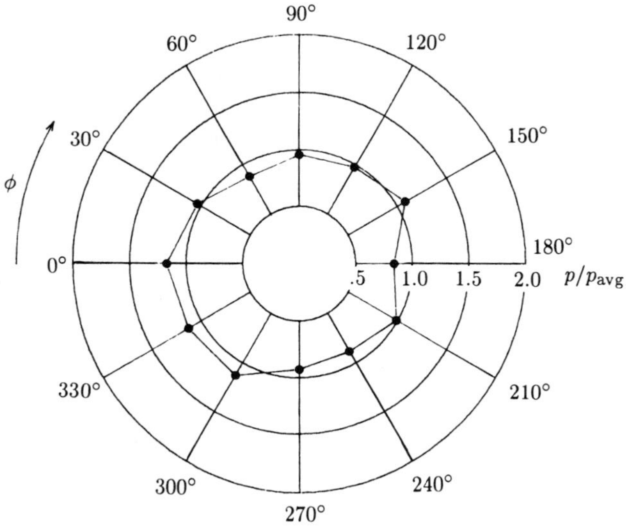

(c) $M_j = 1.80$; $N_{St} = 0.21$ (Mar. 1985).

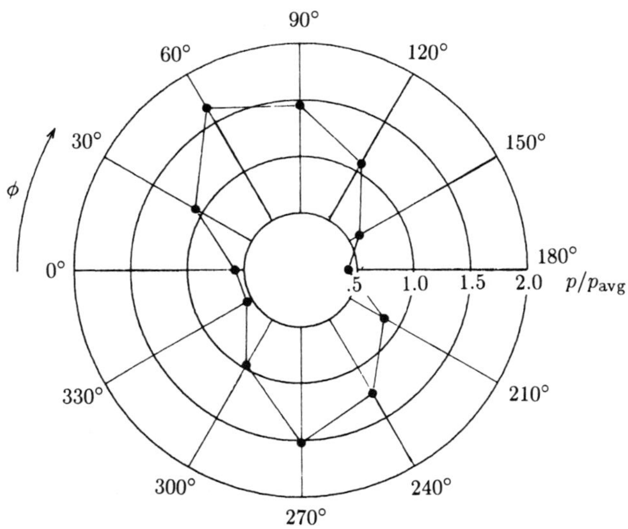

(d) $M_j = 1.56$; $N_{St} = 0.27$ (Apr. 1985).

Figure 18. Concluded.

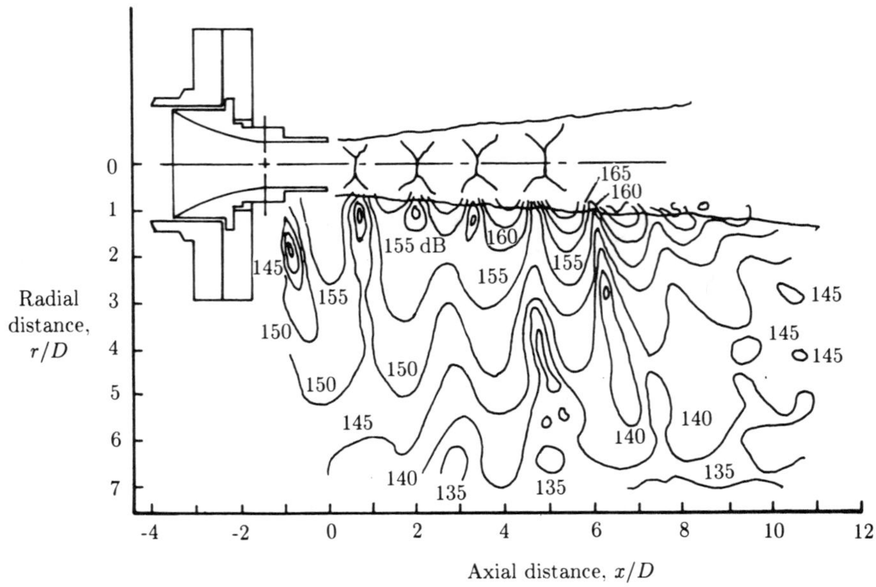

Figure 19. One-tenth-octave-band near-field SPL contours of screeching jet. $f_s = 2$ kHz; $M_j = 1.67$; helical screech mode. (From ref. 97.)

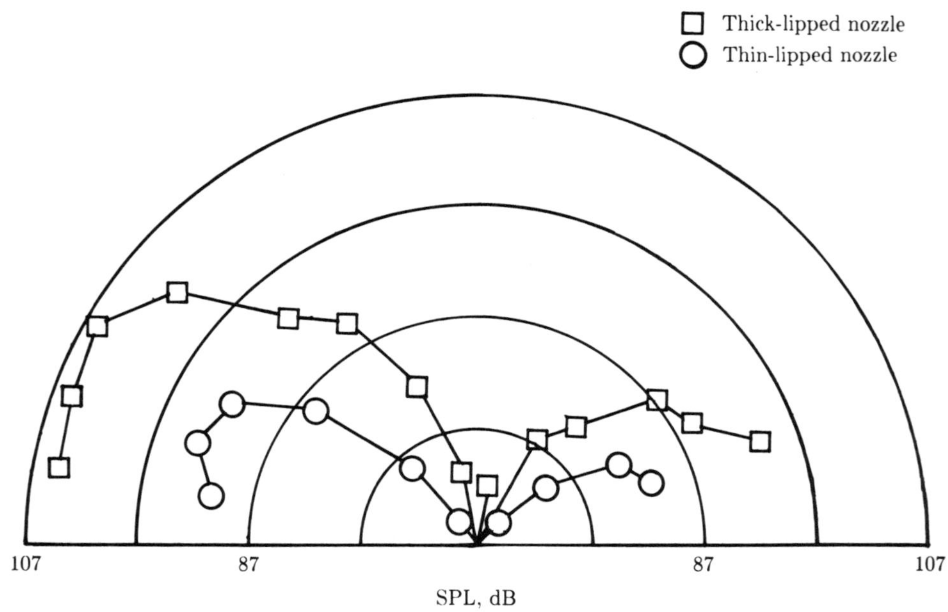

Figure 20. Comparison of far-field directivities of fundamental screech tones of two convergent nozzles at $M_j = 1.49$. (From ref. 92.)

Figure 20, taken from reference 92, shows the directivities of the fundamental screech tones at $M_j = 1.49$ issued from two choked nozzles with the same internal dimensions. The directivity patterns of the jets are very similar. However, the screech tone level associated with the thick-lipped nozzle is 10 dB higher than that of the thin-lipped nozzle. This fact is consistent with the belief that sound waves reflected off the thick nozzle lip are instrumental in exciting stronger instability waves in the jet flow which, in turn, generate louder screech tones. The jet temperature and forward flight, on the other hand, influence both the screech tone frequency and intensity. At higher jet temperatures with the same jet Mach number, the fundamental screech frequency increases. (See ref. 98.) The effect of forward flight has been investigated in references 99 to 103. By performing the jet screech experiment in an open wind tunnel to simulate forward flight, Norum and Shearin (refs. 102 and 103) found that, for an observer fixed relative to the jet, the tone frequency decreases as the forward-flight Mach number increases. Generally speaking the tone intensity also decreases. However, it is possible for tones that are nonexistent in the static case to be excited by forward flight. This last aspect, although important for application to aircraft structural fatigue problems, has unfortunately only been studied rather superficially.

Turbulent Mixing Noise Generated by Large Turbulence Structures and Instability Waves of Supersonic Jets

In a series of experiments on supersonic jets at low to moderate Reynolds numbers (refs. 30 to 33), it was found that the large coherent structures of these flows were in the form of instability waves. Outside the jet, near- and far-field microphone measurements indicated that these instability waves were directly responsible for the generation of the dominant part of the noise of these jets. To ascertain if the findings could shed light on the noise generation processes of high Reynolds number supersonic jets, the near- and far-field noise characteristics of the jet were compared with those at high Reynolds number in references 33 and 34. Overall, many extraordinary similarities were found. Figure 21 shows the acoustic noise spectra in the peak noise radiation direction of three supersonic jets of comparable Mach numbers but drastically different Reynolds numbers. Despite the several orders of magnitude differences in Reynolds numbers the Strouhal frequencies of the spectral peaks are nearly the same. At the low Reynolds number the radiated noise consists of an almost discrete spectrum corresponding to that of a single instability wave. At the moderate and high Reynolds numbers the spectra are broadband. Most important of all, however, is that as a function of Strouhal number the spectral distributions of the moderate and high Reynolds number jets are almost identical. Figure 22(a) shows the near-field SPL contour map of the Mach 2.1 jet with a Reynolds number of 7×10^4 (ref. 33). Figure 22(b) shows a similar map of a Mach 2.0 jet with a high Reynolds number ($N_{\mathrm{Re}} = 5.19 \times 10^6$, ref. 77). The two maps are very much alike. This is especially true in terms of the direction of the principal lobe, the general field shape of the contours, and the spacing separating neighboring contours. These and other remarkable resemblances between the near- and far-field noise characteristics of moderate and high Reynolds number supersonic jets strongly suggest that the basic noise generation mechanism of these jets is probably the same regardless of Reynolds number. In other words, the dominant part of the turbulent mixing noise of high

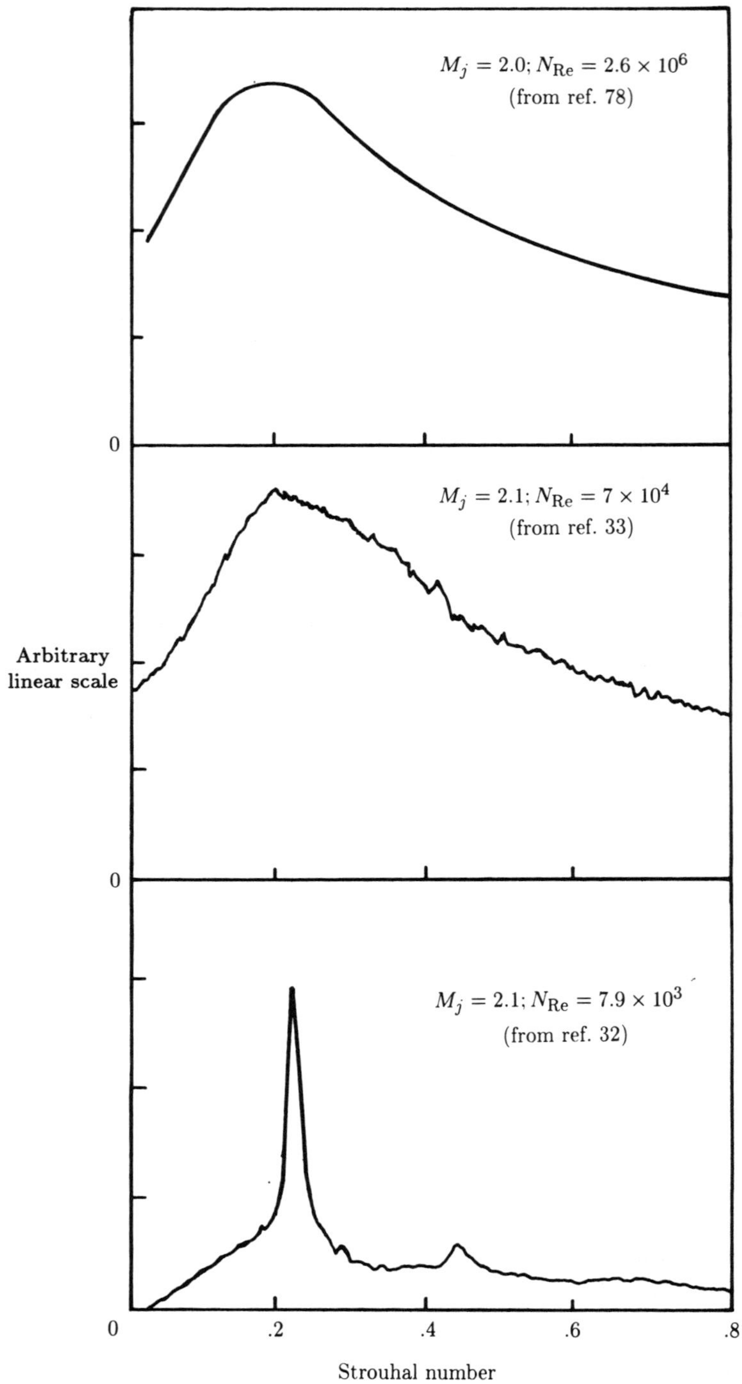

Figure 21. Acoustic spectra in direction of maximum noise radiation.

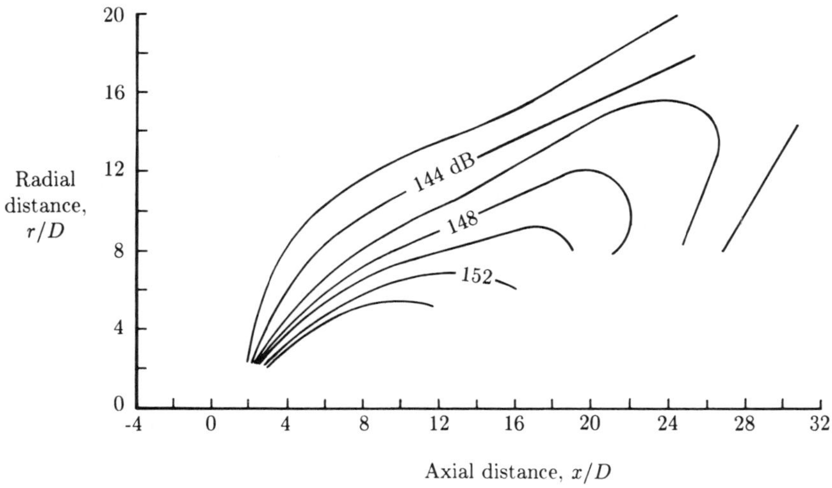

(a) $M_j = 2.1$; $N_{\mathrm{Re}} = 7 \times 10^4$. *(From ref. 33.)*

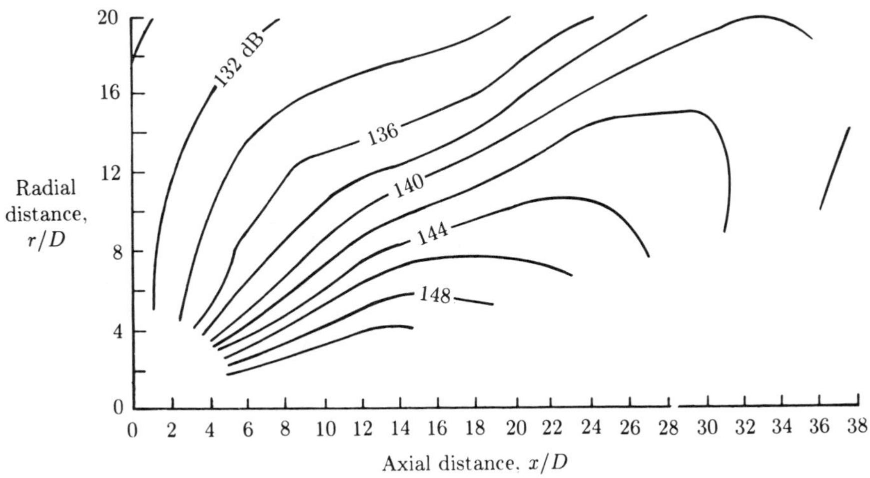

(b) $M_j = 2.0$; $N_{\mathrm{Re}} = 5.19 \times 10^6$. *(From ref. 77.)*

Figure 22. Overall SPL contours.

Reynolds number supersonic jets may be considered to be generated by a random superposition of the intrinsic large-scale instability waves of the jet flow, as is the case for moderate Reynolds number jets. These instability waves, in accordance with the stochastic random waves model described in a previous section, are synonymous with the large turbulence structures of these jets, as observed by researchers (e.g., ref. 29).

Noise Generation Processes

We will now examine the processes by which instability waves which propagate down a jet column produce acoustic radiation. In free shear flows such as mixing layers or jets, the mean flow diverges slowly in the downstream direction because of the entrainment of ambient fluid. Over the initial region, where the shear layer is thin and the mean velocity gradient is large, the amplitude of an excited instability wave grows very rapidly. As the wave propagates downstream the growth rate reduces. The growth rate is smaller because as the flow slowly diverges the transverse velocity gradient is gradually reduced. Eventually the instability wave of a given frequency will reach a point at which its growth rate becomes zero. On propagating farther downstream, the wave becomes damped. Its amplitude decreases as it continues to propagate until it becomes vanishingly small. The growth and decay of the wave amplitude are extremely important to the sound radiation process. This is especially true for instability waves with subsonic phase velocities. It is well-known that a subsonic wave of constant amplitude does not generate sound in a compressible medium. Such a wave has a discrete wave number spectrum. However, for a fixed-frequency instability wave for which amplitude undergoes growth and decay spatially, its wave number spectrum is no longer discrete. Instead it is broadband. Some of the broadband wave components, especially those of the small wave numbers, would actually be moving with supersonic phase velocities. These supersonic phase velocity disturbances, by the wavy wall analogy, immediately lead to acoustic radiation.

The Acoustic Field of Instability Waves as an Outer Solution

Classic hydrodynamic stability theory of a compressible flow (see refs. 104 to 110) does not predict acoustic radiation by instability waves. In fact, the whole question had been completely ignored until recently. (See ref. 60.) The point of departure of that analysis from classic hydrodynamic stability theory lies in the recognition that to determine sound radiation, a global solution of the entire wave propagation phenomenon is necessary. To describe the growth and decay of the excited instability waves resulting from the slight mean flow divergence, it is most convenient to use the method of multiple-scales asymptotic expansion. (See, e.g., refs. 56, 57, and 63.) This method exploits the fact that there are two disparate length scales in the problem. The ratio of these two length scales is quite small and is chosen to be the small expansion parameter. However, the multiple-scales instability wave solution predicts no sound radiation, as is the case for the classic locally parallel flow normal mode solution. As a matter of fact, all these solutions are constructed with the boundary condition that the wave disturbances decay to zero far away from the jet or mixing layer. Thus, by itself the multiple-scales solution could never yield any possible acoustic field associated with the instability wave. This inadequacy of the multiple-scales asymptotic expansion solution was recognized in reference 60, which showed that the asymptotic expansion is actually nonuniformly valid outside the flow. Away from the jet, acoustic disturbances propagate in all directions, so that all spatial coordinates must be treated on an equal basis. Solutions obtained by the multiple-scales asymptotic expansion method, which purposely scales different spatial coordinates unevenly, are therefore inappropriate. They should not be used in the far-field region. Based on this reasoning, it was proposed in reference 62 to apply the method of matched asymptotic expansions (e.g., refs. 63 and 111) to the problem of acoustic radiation by instability waves. In this approach two separate asymptotic expansions of the solutions are constructed. One, called the inner expansion, is valid inside the jet flow and in the region immediately outside. The lowest order term of this expansion is the multiple-scales instability wave solution. The other expansion, referred to as the outer expansion, is valid outside the jet all the way to the acoustic far field. The two expansions are related to each other by the process of matching. The matching process is crucial to the success of the technique. It ensures that one expansion is the proper analytic continuation (at least in an asymptotic sense) of the other in different parts of the physical space.

Inner and Outer Solutions of Instability Waves of an Axisymmetric Jet

To apply the method of matched asymptotic expansions, the first important step is the choice of the appropriate inner and outer variables. For supersonic jet flows the rate of spread of the mean flow ϵ is a small parameter. Let (r, ϕ, x) be the cylindrical coordinates of a coordinate system centered at the nozzle exit with the x-axis pointing in the jet flow direction. The mean flow of the jet is a function of r and the slow variable $s = \epsilon x$. It may be represented analytically in the form

$$\overline{\mathbf{V}} = (\bar{u}(r,s), \epsilon\bar{v}_1(r,s), 0) \tag{5}$$

where

$$\bar{u} = 0 \tag{6a}$$

$$\epsilon\bar{v}_1 = \frac{\epsilon\bar{v}_\infty}{r} \qquad (r \geq r_m(s)) \tag{6b}$$

The set of inner variables suitable for the description of the excited instability waves in the jet is the same as that for the mean flow, namely (r, ϕ, s). Before choosing the outer variables it would be helpful to recall that the overall spatial growth and decay of the wave amplitude are crucial to sound radiation. Clearly, this gradual amplitude variation in the flow direction is a function of the slow variable s. This suggests that the appropriate outer variable in the flow direction is s. Further, since sound propagates without preferred direction in the far field, the spatial variables in this region must be scaled in the same manner in all directions. Hence, a suitable set of outer variables is $(\bar{r}, \phi, s)$, where $\bar{r} = \epsilon r$.

Let us consider the spatial evolution of a small-amplitude instability wave of angular frequency ω in an axisymmetric jet (as shown in fig. 23). The instability waves and their associated sound field are governed by the linearized equations of motion for an inviscid compressible fluid. In the following equations, dimensionless variables are used. The respective length, velocity, time, density, and pressure scales are $D/2$ (where D is the jet diameter at the nozzle exit), u_j (the jet exit velocity), $(D/2)/u_j$, ρ_j (the jet exit density), and $\rho_j u_j^2$. Since the jet flow is axisymmetric, the instability waves can be Fourier decomposed into azimuthal modes. All the physical variables can be represented in the form $p(r, x, \phi, t) = \mathrm{Re}[p(r, x) \exp(in\phi - i\omega t)]$ and so forth, where n is an integer. After the

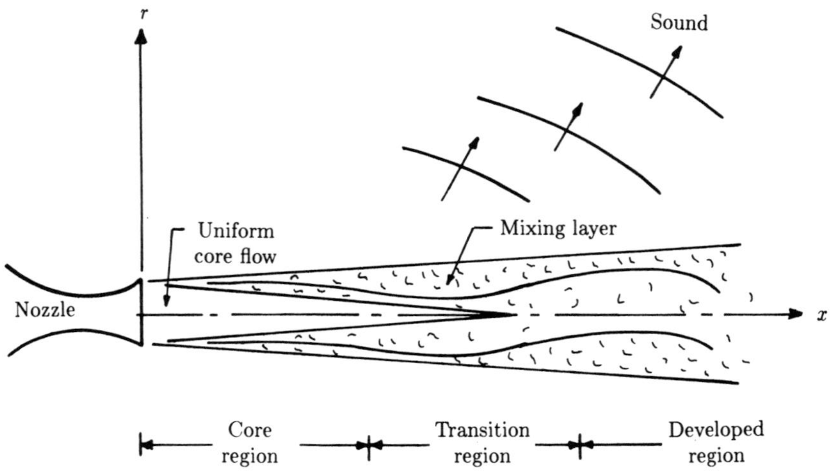

Figure 23. Instability waves and their sound field for axisymmetric jet.

exponential dependence on ϕ and t is factored out, the governing equations for the spatial part of the solution written in cylindrical coordinates are

$$\left.\begin{aligned}
&-i\omega v + \bar{v}\frac{\partial v}{\partial r} + v\frac{\partial \bar{v}}{\partial r} + \bar{u}\frac{\partial v}{\partial x} + u\frac{\partial \bar{v}}{\partial x} = -\frac{1}{\bar{\rho}}\frac{\partial p}{\partial r} \\
&-i\omega w + \bar{v}\frac{\partial w}{\partial r} + \frac{\bar{v}w}{r} + \bar{u}\frac{\partial w}{\partial x} = -\frac{in}{\bar{\rho}r}p \\
&-i\omega u + \bar{v}\frac{\partial u}{\partial r} + v\frac{\partial \bar{u}}{\partial r} + u\frac{\partial \bar{u}}{\partial x} + \bar{u}\frac{\partial u}{\partial x} = -\frac{1}{\bar{\rho}}\frac{\partial p}{\partial x} \\
&-i\omega p + \bar{v}\frac{\partial p}{\partial r} + \bar{u}\frac{\partial p}{\partial r} + \frac{1}{M_j^2}\left(\frac{1}{r}\frac{\partial vr}{\partial r} + \frac{in}{r}w + \frac{\partial u}{\partial x}\right) = 0
\end{aligned}\right\} \tag{7}$$

where M_j is the Mach number of the jet.

The inner solution represents a wave propagating in an inhomogeneous medium formed by the mean flow of the jet. Such a wave may be written in the form (see chapter 11 of ref. 112)

$$\begin{bmatrix} u(r,x) \\ v(r,x) \\ w(r,x) \\ p(r,x) \end{bmatrix} = \sum_{m=0}^{\infty} \delta_m(\epsilon) \begin{bmatrix} u_m(r,s) \\ v_m(r,s) \\ w_m(r,s) \\ p_m(r,s) \end{bmatrix} \exp[i\theta(s)/\epsilon] \tag{8}$$

In equation (8), $\delta_m(\epsilon)$ ($m = 0, 1, 2, \ldots,$ with $\delta_0 = 1$) represents the gauge functions of the asymptotic expansion. These functions are to be determined by the process of matching inner and outer solutions.

Through substitution of equation (8) into equation (7) and partitioning terms according to $\delta_m(\epsilon)$ ($m = 0, 1, 2, \ldots$), a set of equations are found for the amplitude functions (u_m, v_m, w_m, p_m). For noise prediction purposes, only the lower order solutions are needed. It turns out that the lowest order solution is the multiple-scales instability wave solution (see the *Stochastic Wave Model* section and ref. 62). The outer solution of equation (7), which describes the acoustic near and far fields, is to be valid in the region $r > r_m$. In terms of the outer variables, equation (7), for small ϵ, becomes

$$\left.\begin{aligned}
&-i\frac{\omega}{\epsilon}v + \frac{\epsilon^2\bar{v}_\infty}{\bar{r}}\frac{\partial v}{\partial \bar{r}} + \epsilon^2 v\frac{\partial}{\partial \bar{r}}\left(\frac{\bar{v}_\infty}{\bar{r}}\right) = -\frac{1}{\bar{\rho}_\infty}\frac{\partial p}{\partial \bar{r}} \\
&-i\frac{\omega}{\epsilon}w + \frac{\epsilon^2\bar{v}_\infty}{\bar{r}}\frac{\partial w}{\partial \bar{r}} + \epsilon^2\frac{\bar{v}_\infty w}{\bar{r}^2} = -\frac{in}{\bar{\rho}_\infty \bar{r}}p \\
&-i\frac{\omega}{\epsilon}u + \frac{\epsilon^2\bar{v}_\infty}{\bar{r}}\frac{\partial u}{\partial \bar{r}} = -\frac{1}{\bar{\rho}_\infty}\frac{\partial p}{\partial s} \\
&-i\frac{\omega}{\epsilon}p + \frac{\epsilon^2\bar{v}_\infty}{\bar{r}}\frac{\partial p}{\partial \bar{r}} + \frac{1}{M_j^2}\left(\frac{1}{\bar{r}}\frac{\partial v\bar{r}}{\partial \bar{r}} + \frac{in}{\bar{r}}w + \frac{\partial u}{\partial s}\right) = 0
\end{aligned}\right\} \tag{9}$$

A general solution of equation (9) satisfying the boundedness or radiation condition for large $\bar{r}$ may be constructed by first applying a Fourier transform to the variable s. A solution of equation (9) to order ϵ^2, which satisfies the outgoing wave or boundedness condition as $\bar{r} \to \infty$, can be found in terms of a Hankel function of the first kind. When the Fourier transform is inverted, the explicit solution for the pressure p to the lowest order is

$$p(r, x, \phi, t) = \int_{-\infty}^{\infty} g(\eta) H_n^{(1)}[i\lambda(\eta)r] \exp[i(\eta x + n\phi - \omega t)]\, d\eta \tag{10}$$

where

$$g(\eta) = \frac{1}{2\pi}\int_{-\infty}^{\infty} A_0(\epsilon x) \exp[i\theta(\epsilon x)/\epsilon - i\eta x]\, dx \tag{11}$$

To ensure that the inner and outer solutions are solutions of the same physical problem, although valid in different parts of the physical space, they have to be properly matched. For the present problem the intermediate matching principle of references 111 and 113 is to be followed. The matching processes to orders 1, $\epsilon \ln(\epsilon)$, and ϵ were implemented in reference 62. To order 1, matching requires the inner expansion to be an eigenvalue solution and the amplitude functions of the inner and outer solutions (A_0 of eq. (11)) to be the same. To orders $\epsilon \ln(\epsilon)$ and ϵ, matching determines the nonparallel flow correction to the instability wave amplitude. In this way we can find a complete instability wave solution and its associated sound field in an axisymmetric supersonic jet to the lowest order.

Comparisons With Experiments

In reference 62 the above instability wave theory was applied to the experiments of reference 33. In this reference the amplitude distributions of the instability wave along the centerline of a Mach 2.1 jet, slightly excited at Strouhal numbers of 0.2 and 0.4, were measured. The near-field SPL contours were also determined. In calculating the instability wave amplitude and its associated acoustic field outside the jet, the measured mean velocity profile of the jet was used, so that the calculated results contained only one single unknown (namely, the initial amplitude of the instability wave at the nozzle exit). This constant was chosen by fitting the computed results to the measured data at one point.

Figure 24 shows a comparison of the measured and calculated axial distributions of the centerline mass-velocity fluctuation of the Mach 2.1 jet excited at a Strouhal

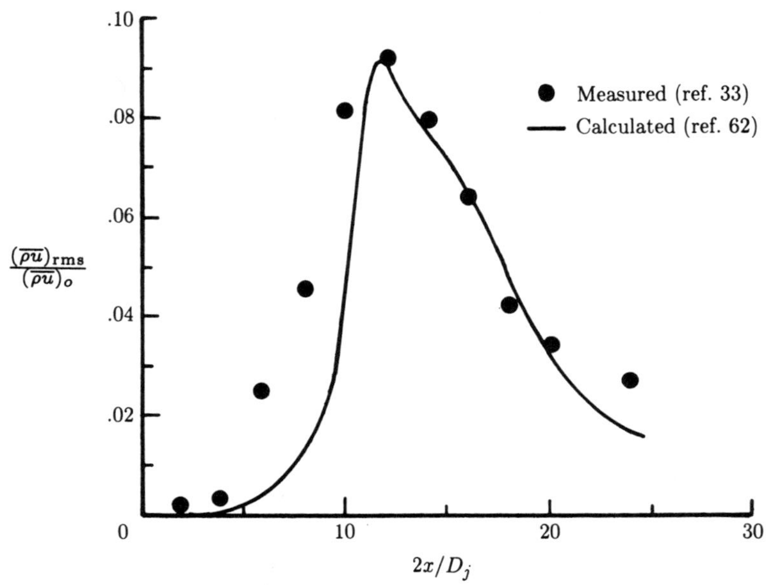

Figure 24. Measured and calculated axial distributions of centerline mass-velocity fluctuation of Mach 2.1 jet excited at $N_{\mathrm{St}} = 0.4$.

number of 0.4. The initial amplitude of the theory has been adjusted so that both the measured and the calculated results have the same maximum value. Overall, there is favorable agreement. This is especially so in the location of the maximum fluctuation and in the half-width of the amplitude distribution. Figure 25(a) shows the calculated near-field SPL in dB associated with the excited instability wave at a Strouhal number of 0.4. Figure 25(b) is the corresponding contour plot measured in reference 33. The unknown constant of the calculated field has been adjusted so that the pressure level at the point marked by a black circle is 148 dB, the same as the corresponding point in figure 25(b). A comparison of these two figures shows excellent agreement. The agreement between the calculated and the measured 150- and 148-dB contours is nearly perfect. The lobed nature of the contours, the direction of the lobe, and the spacings of the contours are correctly predicted. A similar comparison was carried out in reference 62 for the instability waves with a Strouhal number of 0.2. Again very favorable agreements are found between theoretical calculations based

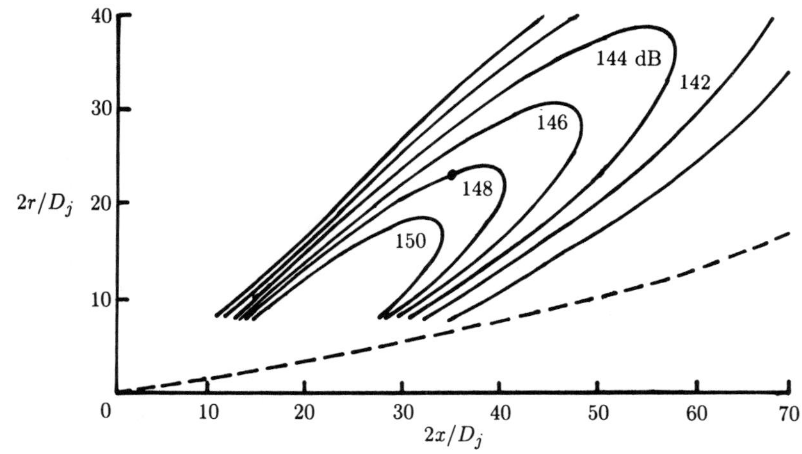

(a) Calculated (ref. 62).

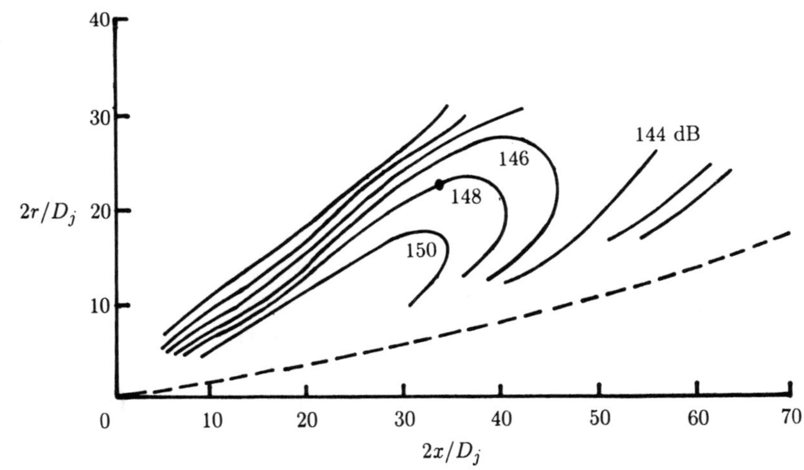

(b) Measured (ref. 33).

Figure 25. Contours of near-field SPL's for jet excited at $N_{St} = 0.4$.

on the method of matched asymptotic expansions and experimental measurements. These agreements provide further support for the proposition that instability waves are the dominant mixing noise sources in supersonic jets.

Theories of Broadband Shock-Associated Noise

The possibility that acoustic waves could be generated by shock-turbulence interaction in imperfectly expanded supersonic jets was recognized many years ago. (See, e.g., refs. 114 and 115.) In these early works the turbulence was pictured as a random array of small eddies resembling what is now referred to as fine-scale turbulence. To reduce the problem to a manageable form the quasi-periodic shock cell structure was all but discarded. It was usually replaced by a single shock wave of infinite spatial extent in the analysis. Basically the noise generation process as envisaged by these early investigators consisted essentially of the random scattering of blobs of fine-scale turbulence by plane shock waves. In subsequent years no effort was ever made to correlate the predictions of this simple random scattering model to the measured spectral, directional, and intensity characteristics of the broadband noise of supersonic jets. Thus these early theoretical results were largely academic and conceptual in nature. It turns out that the observed shock-associated noise is not generated by random incoherent interaction between fine-scale turbulence and individual plane shock waves of a supersonic jet. Instead, the source of this noise component is spatially coherent over an extended length of the jet. Furthermore, the quasi-periodicity of the shock cell structure also plays a crucial role in the noise generation processes.

Current interest in broadband shock-associated noise appears to begin with the work of Harper-Bourne and Fisher (ref. 79), who carried out extensive experimental measurements of the noise of choked jets. By analyzing these data carefully they were able to identify a broadband noise component having spectral and directional characteristics which differed completely from those of the turbulent mixing noise. To model how this noise component was generated they adopted a simple model comprised of an array of phased simple sources spaced at regular intervals. The distance between the point sources was equal to that of the shock cell spacing of the supersonic jet. By using this model they were able to show that this new component of broadband noise was in all likelihood generated by the interaction between the components of the turbulence that are spatially quite coherent and the quasi-periodic shock cell structure of the jet. The crucial discovery, namely, that it requires a noise source which is spatially coherent and quasi-periodic over an extended length of the jet to account for the observed far-field noise characteristics, set their work completely apart from all the previous works. Since this pioneering study a good number of experimental and theoretical investigations on this subject have been carried out (e.g., refs. 55, 72, 76, 80, 81, 86, 94, 102, 103, and 116 to 121). They provide the basis of our present-day view of how broadband shock-associated noise is generated. Herein attention is confined exclusively to the findings and theories of these more recent studies.

Shock Cell Structure of Supersonic Jets

Shock cells in underexpanded and overexpanded axisymmetric jets are quasi-periodic. They are formed by oblique shocks and expansion fans. These shocks and expansion fans are generated at the nozzle lip because of the mismatch of the static pressure inside and outside the jet. The reason the shock cell structure is quasi-periodic is that when the oblique shocks or expansion fans impinge on the jet boundary they are reflected back into the jet. In a sense the shocks and expansion fans are trapped inside the jet, bouncing from one side to the other and forming more or less a standing wave pattern.

To estimate the shock cell spacing, Prandtl (ref. 122) employed a linear vortex sheet model. In this model the jet flow is taken to be uniform and bounded by a vortex sheet. The oblique shocks and expansion fans which make up the shock cell structure are assumed to be weak so they can be represented by small amplitude disturbances superimposed on the mean flow. The complete solution of the vortex sheet model was first found in reference 123. It can be expressed in the following simple form:

$$p_s(r,x) = \sum_{i=1}^{\infty} A_i \Phi_i(r) \cos(k_i x) \tag{12}$$

where

$$\left.\begin{aligned} A_i &= 2\Delta p/\sigma_i \\ k_i &= \frac{2\sigma_i}{D_j(M_j^2-1)^{1/2}} \\ J_0(\sigma_i) &= 0 \end{aligned}\right\} \qquad (i = 1, 2, \ldots)$$

and p_s is the pressure disturbance associated with the shock cells and Δp is the difference between the static pressure inside and outside the jet at the nozzle exit. In equation (12), $\Phi_i(r)$ is a set of orthonormalized eigenfunctions $J_0(2\sigma_i r/D_j)/J_1(\sigma_i)$. The terms J_0 and J_1 are the zeroth and first-order Bessel functions, respectively. To a good approximation the shock cell spacing L_s is given by the wavelength of the first term of the series in equation (12). That is,

$$L_s \approx 2\pi/k_1 = \pi(M_j^2-1)^{1/2}\, D_j/\sigma_1 \qquad (\sigma_1 = 2.405) \tag{13}$$

Since the thickness of the mixing layer of the jet increases in the downstream direction, the vortex sheet model is valid only in the initial region of the supersonic jet immediately downstream of the nozzle. To provide a more accurate description of the shock cells, a number of investigators (e.g., refs. 124 and 125) used inviscid flow models and the method of characteristics to determine the structural details of the shock cells close to the nozzle exit. No attempt, however, was made by these investigators to extend their studies beyond the first two to three shock cells. More recently, inviscid Euler codes were developed to calculate the shock cells numerically (refs. 126 and 127). Comparisons of these inviscid numerical methods with experimental data reported in references 94 and 117 showed that these models do not provide acceptable results for the prediction of broadband shock-associated noise. The method has since been improved to include the effect of turbulent mixing through the use of turbulence closure equations (ref. 128). The numerical code was tested in reference 129 and found to provide results that agree reasonably well with experimental measurements.

For supersonic jets that are not severely underexpanded or overexpanded, the shock cells are relatively weak. In this case we can develop a simple but reasonably accurate analytical model of the shock cells by exploiting the fact that the shock cell structure is characterized by two basic disparate length scales. One length scale is the jet diameter. The other is the potential core length of the jet, which controls the slow rate of change of the jet mean flow. Such a multiple-scales shock cell model has recently been developed in reference 130. In this model the flow quantities associated with the shock cells are decomposed into the time-independent waveguide or Fourier modes of the mean flow. These modes are calculated by the method of multiple-scales asymptotic expansion. If only the lowest order terms of the expansion are kept, the flow quantities are given analytically in the following form (a subscript s is used to denote their association with the shock cell structure):

$$\begin{pmatrix} u_s \\ v_s \\ p_s \end{pmatrix} = \sum_{m=1}^{\infty} \left\{ \begin{pmatrix} u_m(r,s) \\ v_m(r,s) \\ p_m(r,s) \end{pmatrix} \exp[i\Phi_m(s)/\epsilon] + \text{Complex conjugate} \right\} \tag{14}$$

where $\frac{d\Phi_m}{ds} = \beta_m(s)$ is the local wave number (eignevalue) of the mth waveguide or Fourier mode and $s = \epsilon x$ is again the slow variable. The terms u_m, v_m, and p_m are the shock cell structure functions (eigenfunctions). They provide the radial distribution of the flow variables associated with each mode. To determine the starting amplitudes of the solution, we impose an initial condition that each mode of equation (14) must match the corresponding mode of the vortex sheet shock cell solution of references 122 and 123 at the nozzle exit. It was pointed out in reference 81 that within the framework of the vortex sheet model the amplitudes of the waveguide or Fourier modes are proportional to $\Delta p/\sigma_i$ ($i = 1, 2, \ldots$; see eq. (12)). It was further shown that, to a good approximation, $\Delta p/p_\infty$ is equal to $0.5\gamma(M_j^2 - M_d^2)\Big/\Big[1 + (\gamma - 1)M_d^2/2\Big]$, where p_∞ is ambient pressure. Thus the amplitudes of the different waveguide modes are approximately proportional to

$$\left.\begin{array}{c} \dfrac{p_\infty\left(M_j^2 - M_d^2\right)}{\left[1 + (\gamma-1)M_d^2/2\right]\sigma_i} \\[2em] \text{or} \\[1em] \dfrac{\rho_\infty c^2\left(M_j^2 - M_d^2\right)}{\left[1 + (\gamma-1)M_d^2/2\right]\sigma_i} \end{array}\right\} \qquad (i = 1, 2, \ldots) \tag{15}$$

where ρ_∞ and c are the ambient gas density and speed of sound.

Carried out in reference 130 are extensive comparisons of the calculated pressure distributions of the shock cell structures based on the multiple-scales model with the measurements of reference 72. The test cases included both underexpanded and overexpanded jets issued from nozzles of three design Mach numbers of 1.0, 1.5, and 2.0. The calculated shock cell spacings and amplitudes compared very favorably with the experimental measurements. In addition, many of the fine structures observed in the first three or four shock cells were reproduced by their calculations.

Phased Point-Source Array Model

It was proposed in reference 79 that the source of broadband shock-associated noise is in the form of a synchronized array of periodic point monopoles. These monopoles are of equal strength and are spaced evenly at regular intervals. The researchers believed that the shock-associated noise is emitted from the endpoints of the shock cells of imperfectly expanded supersonic jets. Therefore, point monopoles are used to simulate these very localized noise sources. A similar model was developed previously in reference 87 to describe the generation of screech tones in choked jets. In the reference 79 model the phases of adjacent monopoles are assumed to be correlated by the time taken for turbulence to be convected from one point source to the next, as shown in figure 26. Let L_s be the shock cell spacing and u_c be the turbulence convection velocity. Then the time taken for turbulence to be convected from point-source A to point-source B is $T_{\mathrm{AB}} = L_s/u_c$. Thus the time origin of point-source B is assumed to be delayed by an amount equal to T_{AB} relative to point-source A. A similar time delay is applied to successive point sources of the array. Now consider the sound radiation in the direction θ. The sound from source B follows path BB′ and that from source A follows AA′. Clearly, path length AA′ is longer than BB′ by the length AC $= L_s \cos\theta$. The time needed for a sound wave to travel the distance AC is $T_{\mathrm{AC}} = L_s \cos\theta/c$, where c is the ambient speed of sound. Because of the difference in the propagation path length and in the origin of time, the sound waves emitted from the two point sources A and B are out of phase when they reach the observer far away. Thus there is a tendency for them to partially cancel each other. It is possible, however, for the sound waves to arrive at the far-field observer exactly in phase if certain conditions are met. When this happens there is maximum constructive reinforcement of the sound intensity. For a given direction θ this condition is satisfied only for certain special frequencies f_p. Hence one would expect the broadband shock-associated noise spectra to peak at these frequencies of maximum reinforcement. The condition for maximum wave reinforcement occurs when the difference between the turbulence convection time T_{AB} and the sound propagation time T_{AC} is equal to an integral multiple of the period of oscillation. That is, $T_{\mathrm{AB}} - T_{\mathrm{AC}} = n/f_p$ or

$$\left(\frac{L_s}{u_c} - \frac{L_s \cos\theta}{c}\right) = \frac{n}{f_p} \qquad (n = 1, 2, \ldots)$$

The primary peak frequency corresponds to the case of $n = 1$, so that from the above the peak frequency f_p is given by

$$f_p = \frac{u_c}{L_s(1 - M_c \cos\theta)} \tag{16}$$

In equation (16), $M_c = u_c/c$ is the convection Mach number based on the ambient speed of sound. Comparison of equation (16) with the choked jet noise data of reference 79 resulted in reasonably good agreement. In this way it was possible to offer an explanation to one of the most prominent characteristics of broadband shock-associated noise, namely, the shift of the peak frequency of the noise spectrum with observation angle, as depicted in figure 5.

In addition to the derivation of equation (16), reference 79 found semiempirically (after extensive analysis of the choked jet noise data) that the shock-associated noise intensity I_s varies as β^4, where $\beta = (M_j^2 - 1)^{1/2}$. The parameter β was introduced because it is the parameter which characterizes the pressure jump across a normal shock of upstream Mach number M_j. Subsequent to this work, it was verified in reference 76 that this formula worked not only for cold jets but also for hot jets. The

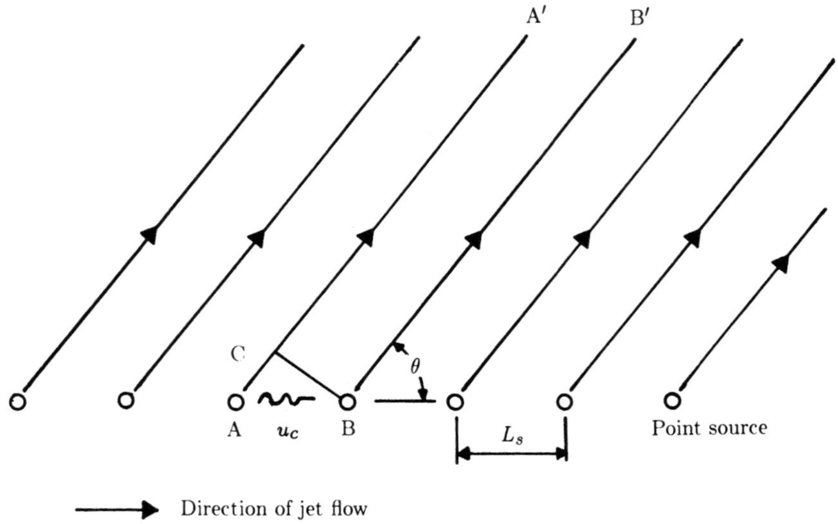

Figure 26. Schematic diagram of phased point-source array model.

only restriction was that they were choked jets, that is, jets issued from convergent nozzles.

Recently a series of experiments were conducted (ref. 102) on the effect of forward flight on broadband shock-associated noise. In an attempt to correlate the data with theory, the phased point-source array model was extended to the case of a jet in forward motion. For noise radiation at 90° the measured data show a downward shift of the frequency of the spectral peak with increase in forward-flight Mach number. However, the model theory predicted an increase in this peak frequency exactly opposite to the measurements. The disagreement between theory and experiment appears to be serious and disturbing. It calls for renewed scrutiny of the validity of the model and its underlying noise generation mechanism.

Large Turbulence Structure and Shock Cell Interaction Theory

Reference 81 proposed that broadband shock-associated noise is produced by the coherent scattering of the large turbulence structures as they pass through the quasi-periodic shock cells. The large turbulence structures are random and consist of wave-like components of a fairly broad range of frequencies. The shock cell structure is spatially quasi-periodic and may be considered as a superposition of the time-independent waveguide or Fourier modes of the mean flow of the jet. The passage of the large turbulence structures through the shock cells, therefore, produces interactions between the broad spectrum of wave-like disturbances which make up the large turbulence structures and each of the individual waveguide modes of the shock cells. Since the wavelengths of different waveguide modes are different, the scattering properties of the modes are not the same. As a result, the principal direction of the noise radiation and the spectral content of the radiated noise associated with different modes are different. In the far field the noise intensity is the sum total of all the noise generated by each of the modes. Hence the noise spectrum at a point is made up of a superposition of many subspectra. Each subspectrum is generated by a waveguide mode of the quasi-periodic shock cell structure. In other words, within the framework of the reference 81 model of broadband shock-associated noise, the noise spectrum could exhibit more than one spectral peak, a characteristic feature consistent with experimental observations.

Formulation

The continuity, momentum, and energy equations of an inviscid compressible fluid are

$$\left.\begin{aligned} &\frac{\partial \rho}{\partial t} + \nabla \cdot (\rho \mathbf{v}) = 0 \\ &\rho\left(\frac{\partial \mathbf{v}}{\partial t} + \mathbf{v} \cdot \nabla \mathbf{v}\right) = -\nabla p \\ &\frac{\partial p}{\partial t} + \nabla \cdot (p\mathbf{v}) + (\gamma - 1)p\, \nabla \cdot \mathbf{v} = 0 \end{aligned}\right\} \tag{17}$$

Let $\bar{p}$, $\bar{\mathbf{v}}$, and $\bar{\rho}$ be the pressure, velocity, and density of the mean flow of the jet and p_t, $\mathbf{v}_t$, ρ_t, p_s, $\mathbf{v}_s$, and ρ_s be the corresponding variables associated with the large turbulence structures (eq. (1)) and the shock cell structures (eq. (14)). The mean flow, the mean flow plus the large turbulence structures, and the mean flow plus the shock cell structures are each solutions of equations (17). However, when a first-order shock cell structure develops inside a turbulent jet, the combined shock cell and large turbulence structure solutions do not satisfy the governing equations of motion. The interaction between the large turbulence structures and the quasi-periodic shock cells gives rise to time-dependent disturbances p', $\mathbf{v}'$, and ρ'. These disturbances, when radiated to the far field, become the broadband shock-associated noise. Hence the flow quantities consist of four main components:

$$\begin{bmatrix} p \\ \mathbf{v} \\ \rho \end{bmatrix} = \begin{bmatrix} \bar{p} + p_s + p_t + p' \\ \bar{\mathbf{v}} + \mathbf{v}_s + \mathbf{v}_t + \mathbf{v}' \\ \bar{\rho} + \rho_s + \rho_t + \rho' \end{bmatrix} \tag{18}$$

By substitution of equations (18) into equations (17), a set of equations for p', $\mathbf{v}'$, and ρ' are found. If only the lowest order interaction terms involving the large turbulence structures and the shock cell solutions are retained, these equations (after dropping the primes from p', $\mathbf{v}'$, and ρ') may be written as

$$\frac{\partial \rho}{\partial t} + \nabla \cdot (\bar{\rho}\mathbf{v} + \rho\bar{\mathbf{v}}) = -\nabla \cdot (\rho_s \mathbf{v}_t + \rho_t \mathbf{v}_s) \tag{19a}$$

$$\begin{aligned} &\frac{\partial \mathbf{v}}{\partial t} + \bar{\mathbf{v}} \cdot \nabla \mathbf{v} + \mathbf{v} \cdot \nabla \bar{\mathbf{v}} + \frac{1}{\bar{\rho}} \nabla p \\ &\qquad = -\mathbf{v}_s \cdot \nabla \mathbf{v}_t - \mathbf{v}_t \cdot \nabla \mathbf{v}_s + \frac{1}{\bar{\rho}^2} (\rho_s \nabla p_t + \rho_t \nabla p_s) \end{aligned} \tag{19b}$$

$$\frac{\partial p}{\partial t} + \nabla \cdot (\bar{p}\mathbf{v} + p\bar{\mathbf{v}}) + (\gamma - 1)(\bar{p}\ \nabla \cdot \mathbf{v} + p\ \nabla \cdot \bar{\mathbf{v}})$$
$$= -\nabla \cdot (p_t \mathbf{v}_s + p_s \mathbf{v}_t) - (\gamma - 1)\,(p_s\ \nabla \cdot \mathbf{v}_t + p_t\ \nabla \cdot \mathbf{v}_s) \tag{19c}$$

The right-hand sides of the above equations represent the source terms of broadband shock-associated noise. Since the shock cell structure solution vanishes outside the jet flow and in the fully developed region of the jet, these source terms are effectively confined to the volume inside the jet plume, extending from the nozzle exit to somewhere slightly downstream of the end of the potential core of the jet.

Solution by the Method of Matched Asymptotic Expansions

To solve the nonhomogeneous equations (19) it is convenient to expand the solution as a Fourier series in ϕ and a Fourier transform in t. For example,

$$p(r, \phi, x, t) = \sum_{n=-\infty}^{\infty} \int_{-\infty}^{\infty} a_n(\omega)\hat{p}_n(r, x, \omega) \exp[i(n\phi - \omega t)]\ d\omega \tag{20}$$

On application of Fourier transform and Fourier series expansions to equations (19), the equations reduce to a system of four nonhomogeneous partial differential equations in r and x. The homogeneous parts of these equations are identical to those of equations (7), and the nonhomogeneous terms consist of sums over the shock cell modes. For instance, the energy equation may be written in the form

$$-i\omega\hat{p}_n + \bar{v}\frac{\partial \hat{p}_n}{\partial r} + \bar{u}\frac{\partial \hat{p}_n}{\partial x} + \gamma\bar{p}\left(\frac{1}{r}\frac{\partial r\hat{v}_n}{\partial r} + \frac{in}{r}\hat{w}_n + \frac{\partial \hat{u}_n}{\partial x}\right)$$
$$= \sum_{m=1}^{\infty} \tilde{I}_{4m} \exp[i\,(\theta_n - \Phi_m^*)\,/\epsilon] \tag{21}$$

where the asterisk denotes the complex conjugate. The full expressions for $\tilde{I}_{1m}$, $\tilde{I}_{2m}$, $\tilde{I}_{3m}$, and $\tilde{I}_{4m}$ can be found in reference 55. The system of nonhomogeneous equations for $\hat{u}_n$, $\hat{v}_n$, $\hat{w}_n$, and $\hat{p}_n$ can be solved by the method of matched asymptotic expansions as outlined in the section entitled *Turbulent Mixing Noise Generated by Large Turbulence Structures and Instability Waves of Supersonic Jets.* The reasons given there for the need to use separate inner and outer solutions also apply here. Now for each nonhomogeneous term with subscript m a separate nonhomogeneous solution can be found. Thus, for easy identification each of these solutions is labelled by a pair of subscripts n and m. The inner solution, which is valid inside the jet flow and in the region immediately outside, may be constructed easily again by the method of multiple-scales expansion. The outer solution is valid from the edge of the jet flow all the way to the far field. The two solutions are to be properly matched according to the intermediate matching principle. This was carried out in reference 55, in which the lower order terms of the inner and outer expansions were constructed and it was shown that they did match in the intermediate limit. To the lowest order, the formula for the pressure field resulting from the interaction between the large turbulence structures and the mth waveguide or Fourier mode of the shock cells may be written as

$$p_m(r,\phi,x,t) = \sum_{n=-\infty}^{\infty} \iint_{-\infty}^{\infty} a_n(\omega) g_{nm}(\eta,\omega) H_n^{(1)}[i\lambda(\eta,\omega)r] \exp[i(n\phi - \omega t)$$

$$+ ix\eta + i(n+1)(\pi/2)]\ d\eta\ d\omega \tag{22}$$

$$g_{nm}(\eta,\omega) = \frac{1}{2\pi}\int_{-\infty}^{\infty} A_{nm}(x,\omega)\exp\{i[\theta_n(\epsilon x,\omega) - \Phi_m^*(\epsilon x,\omega)]/\epsilon - i\eta x\}\,dx \tag{23}$$

where $\lambda(\eta,\omega) = (\eta^2 - \omega^2/c^2)^{1/2}$ (with branch cuts defined by $-\pi/2 \le \arg[\lambda(\eta,\omega)] < \pi/2$) and $H_n^{(1)}$ is the nth order Hankel function of the first kind.

Near- and Far-Field Noise Spectrum and Directivity

Equations (22) and (23) give the pressure field generated by the interaction between downstream-propagating large turbulence structures and the mth waveguide mode of the quasi-periodic shock cells. These equations may be used for noise calculations both in the near field and in the far field. Let us form the autocorrelation function at a point (r,ϕ,x). By means of these formulas it is easy to find

$$\langle p_m(r,\phi,x,t)p_m(r,\phi,x,t+\tau)\rangle = \sum_{n=-\infty}^{\infty}\sum_{n'=-\infty}^{\infty} \iiiint_{-\infty}^{\infty} \langle a_n(\omega)a_{n'}(\omega')\rangle$$

$$\times\, g_{nm}(\eta,\omega)g_{n'm}(\eta',\omega')H_n^{(1)}[i\lambda(\eta,\omega)r]H_{n'}^{(1)}[i\lambda(\eta',\omega')r]$$

$$\times \exp\big\{i\big[(\eta+\eta')x + (n+n')\phi - (\omega+\omega')t - \omega'\tau$$

$$+\ (n+n'+2)(\pi/2)\big]\big\}\ d\eta\ d\eta'\ d\omega\ d\omega' \tag{24}$$

Through use of the stochastic property of the random amplitude function $a_n(\omega)$ (as given in eq. (4)), it is straightforward to obtain, after summing over the Kronecker delta $\delta_{n-n'}$, integrating over the delta function $\delta(\omega+\omega')$, and changing the variable η to $-\eta$, the following:

$$\langle p_m(r,\phi,x,t)p_m(r,\phi,x,t+\tau)\rangle = \int_{-\infty}^{\infty}\sum_{n=-\infty}^{\infty}\frac{1}{2}\,\tilde{D}\,\frac{R_j}{u_j}\,|G_{nm}(r,x,\omega)|^2\exp(-i\omega\tau)\,d\omega \tag{25}$$

where

$$G_{nm}(r,x,\omega) = \int_{-\infty}^{\infty} g_{nm}(\eta,\omega)H_n^{(1)}[i\lambda(\eta,\omega)r]\exp(i\eta x)\,d\eta \tag{26}$$

Since the noise power spectrum $S_m(r,x,\omega)$ is the Fourier transform of the auto-correlation function, it follows from equation (25) that

$$S_m(r,x,\omega) = \sum_{n=-\infty}^{\infty}\tilde{D}\,\frac{R_j}{u_j}\,|G_{nm}(r,x,\omega)|^2 \qquad (\omega > 0) \tag{27}$$

The angular frequency ω in equation (25) may be positive or negative. To restrict our consideration to positive frequencies only, a factor of two has been incorporated into equation (26).

Equations (25) to (27) are the necessary equations for the calculation of the near-field noise power spectrum. In the far field, equation (26) may be further simplified by evaluating the integral asymptotically by the saddle point method. It is straightforward to find

$$\lim_{R\to\infty} G_{nm}(R,\theta,\omega) \approx \frac{2}{R}\exp[i(\omega/c)R - in\pi/2]\, g_{nm}\,(\omega\cos\theta/c,\omega) \tag{28}$$

where (R,θ,ϕ) are spherical coordinates with the x-axis as the polar axis. Herein it is assumed that the acoustic radiation associated with different waveguide modes is statistically independent. Thus, the total noise spectrum $S(r,x,\omega)$ is a direct sum of the spectra of all the modes, that is,

$$S(r,x,\omega) = \sum_{m=1}^{\infty} S_m(r,x,\omega) \tag{29}$$

A Similarity Source Model

To calculate the noise power spectrum of broadband shock-associated noise with the theory developed above, extensive calculations are needed to determine the source function $A_{nm}\exp[i(\theta_n - \Phi_m^*)/\epsilon]$ of equation (23). After this is done, the spectrum function $S(r,x,\omega)$ may be found systematically by performing numerical evaluations of the integrals in equations (23) and (26). To avoid excessive computation, it was suggested in reference 55 to simulate the source function with a realistic physical model.

In the proposed model the phase velocity of the instability waves and the shock cell wavelengths are assumed to be constants, that is,

$$\mathrm{Re}(\theta_n) = k_t x \equiv \frac{\omega}{u_c}x \tag{30}$$

$$\mathrm{Re}(\Phi_m^*) \approx k_m x \qquad (m = 1, 2, \ldots) \tag{31}$$

Here k_t and k_m are the wave numbers of the instability waves and waveguide modes of the shock cell structure at maximum wave amplitude. The term u_c is the phase velocity of the instability wave or the convection velocity of the large turbulence structures.

The source function $A_{nm}\exp[i(\theta_n - \Phi_m^*)/\epsilon]$ involves both axisymmetric ($n = 0$) and nonaxisymmetric ($n > 0$) components. Numerical calculations reveal that only the first few lower order modes are important; the higher order modes decay spatially rather rapidly. For round jets, the noise radiation pattern is axisymmetric. Hence, statistically the source of noise is expected, on the basis of ensemble averaging, to be axisymmetric. For simplicity here the source is treated as axisymmetric; that is, the dependence on n is dropped. Physically this does not mean that the helical and the higher order modes are unimportant. Instead their contributions are, for convenience, replaced by an equivalent simple axisymmetric source which simulates their ensemble-averaged behavior. The adoption of this approximation allows the deletion of the summation over n in equation (27) and of all the dependence on n in equation (28). The value n is set equal to zero whenever it is appropriate.

In the present theory the shock cells are assumed to be stationary in time so that the spectral content of the source function is completely dictated by that of the turbulence. Extensive experimental evidence is available which indicates that the turbulence spectrum in the potential core region of the jet is nearly self-similar. It is, therefore, not unreasonable to assume that the source function of equation (23) also possesses similarity characteristics. On choosing a Gaussian function of half-width L to approximate the spatial distribution of each instability wave component and by invoking a similarity argument, we arrive at the following similarity source model:

$$A_{nm} \exp[i(\theta_n - \Phi_m^*)/\epsilon] \propto \frac{u_j}{c}\, \omega^{1/2} \exp\Big\{ -\ln 2\, \left[(\omega x/u_j) - X_m\right]^2 / L^2 + i(k_t - k_m)x \Big\} \tag{32}$$

In equation (32), $\omega x/u_j$ is the similarity variable, and X_m is the dimensionless location at which the instability waves attain maximum amplitude. To add the dependence of the source function on the shock cell strength, it is assumed that the source function is directly proportional to the relative strength of the waveguide mode as given by equation (15). Thus upon balancing the dimensions appropriately we find

$$A_{nm} \exp[i(\theta_n - \Phi_m^*)/\epsilon] = \tilde{E} \rho_\infty c^2 \frac{\left(M_j^2 - M_d^2\right)}{\{1 + [(\gamma - 1)/2]\, M_d^2\}\, \sigma_m} \left(\frac{u_j}{c}\right) \left(\frac{\omega R_j}{u_j}\right)^{1/2}$$
$$\times \exp\Big\{ -\ln 2\, \left[(\omega x/u_j) - X_m\right]^2 / L^2 + i(k_t - k_m)x \Big\} \tag{33}$$

where $\tilde{E}$ is an unknown proportionality constant. Through the use of equations (33), (23), and (27) to (29), the noise power spectrum at a point ($R, \psi = \pi - \theta$, where ψ is the inlet angle) is given by

$$\lim_{R \to \infty} S(R, \psi, f) = \frac{CL^2 \left(M_j^2 - M_d^2\right)^2 A_j}{\{1 + [(\gamma - 1)/2] M_d^2\}^2 R^2} \frac{\rho_\infty^2 c^2 u_j^2}{f}$$
$$\times \left\{ \sum_{m=1}^{\infty} \frac{1}{\sigma_m^2} \exp\left[-\left(\frac{f_m}{f} - 1\right)^2 \frac{(1 + M_c \cos\psi)^2}{(u_c/u_j)^2} \frac{L^2}{2 \ln 2} \right] \right\} \tag{34}$$

where $f = \omega/2\pi$ is the frequency, $A_j = \frac{1}{4}\pi D_j^2$ is the area of the fully expanded jet, $M_c = u_c/c$ is the convection Mach number, f_m is defined by

$$f_m = \frac{u_c k_m}{2\pi(1 + M_c \cos\psi)} \tag{35}$$

and C is a constant related to $\tilde{E}$ and $\tilde{D}$ of equations (33) and (25) by

$$C = \frac{\tilde{D}\tilde{E}^2}{2\pi^3 \ln 2}$$

If equation (26) is used instead of equation (28), it is easy to find that the noise power spectrum in the near field is given by

$$S(r,x,f) = \frac{1}{4}\, CL^2 \frac{\left(M_j^2 - M_d^2\right)^2 A_j \rho_\infty^2 c^2 u_j^2}{\{1 + [(\gamma-1)/2]\, M_d^2\}^2 f}$$
$$\times \left(\sum_{m=1}^{\infty} \frac{1}{\sigma_m^2} \left| \int_{-\infty}^{\infty} \exp\left[-(\eta - k_t + k_m)^2 (u_j^2 L^2/f^2)\, (16\pi^2 \ln 2)^{-1} \right] \right.\right.$$
$$\left.\left. \times H_0^{(1)} \left\{ i \left[(\eta^2 - \omega^2)/c^2 \right]^{1/2} r \right\} \exp\{ i[x - (u_j X_m / 2\pi f)]\eta \}\, d\eta \right|^2 \right) \qquad (36)$$

Numerical Results and Comparisons With Experiments

To use the far-field noise power spectrum equation (34), the four parameters u_c, k_m, L, and C must first be specified. In reference 55, u_c was taken to be $0.7u_j$, according to experimental observations (e.g., ref. 79), and k_m was taken to be nearly the mth wave number of the vortex sheet model. (See eqs. (30) and (31).) The remaining parameters L and C were chosen by fitting the theoretical spectrum to the data of figure 27 at $\psi = 90°$. These narrow-band data were measured in reference 72. Figure 27 shows a comparison of the measured and calculated spectra for $L = 3.3$ and $C = 7 \times 10^{-4}$. The absolute levels of the calculated spectra shown have been slightly adjusted by ΔdB (indicated in each spectrum) so as to give a best fit to the measurements. As shown, there is generally good agreement over the entire frequency range and for all observed values of ψ. The calculated peak frequencies at different observation angles agree well with the measured values. The half-width of the measured noise spectra decreases as the direction of noise radiation approaches the nozzle inlet. The calculated spectra exhibit this behavior. At 90° to the jet, the noise spectrum appears to show a less prominent but identifiable second peak. This is more or less reproduced by the calculated spectrum. This second peak is produced by the noise generated by the interaction of the large turbulence structures and the second waveguide mode of the quasi-periodic shock cell structure. At low inlet angles the calculated spectra differ from the measured spectra near the peak noise region, where a dip in each calculated spectrum appears. The cause of this discrepancy seems to be that each calculated spectrum is made up of a superposition of many subspectra. The dip in the calculated overall noise spectrum arises when the subspectra of the first and second waveguide modes do not overlap sufficiently. Apparently, at low inlet angles there is a further broadening of the subspectra by certain physical processes which have not been accounted for in the model.

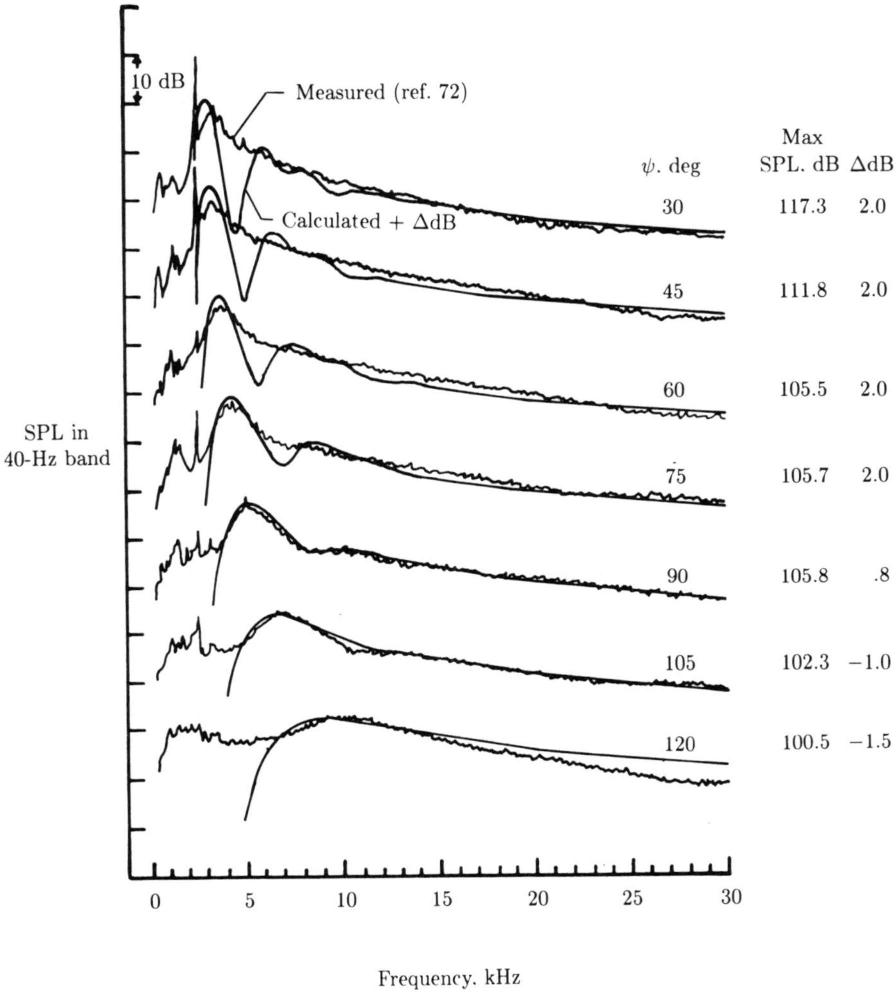

Figure 27. Far-field noise spectra. $M_j = 1.67$; $M_d = 1.50$.

Extensive comparisons between equation (34) and measurements of references 72 and 74 were carried out in reference 55. The cases compared included overexpanded and underexpanded jets from nozzles with design Mach numbers of 1.0, 1.5, and 2.0. Overall, favorable agreements were found in absolute intensity, spectral shape, and directivity.

The near-field sound power spectrum equation (36) contains an extra parameter X_m which characterizes the center of the noise source in the similarity variable. In principle, X_m can be calculated. However, for the purpose of comparison with the measured near-field data of reference 74, the noise source centers for the first two modes were taken in reference 55 to be 8.0 and 7.0 jet diameters downstream of the nozzle exit. In the near field the turbulent mixing noise dominates in the downstream direction. The SPL contours form a strong lobe at around 30° to 40° to the jet flow direction. To reproduce the turbulent mixing noise contribution to the near field, the near-field acoustic pressure distribution formula (ref. 62) is used. The spatial distribution of the noise source is assumed to be in the form of a hyperbolic secant function. The parameters of this function are adjusted to match the overall pattern of the turbulent mixing noise lobe. To evaluate the integrals of equation (36), the method of fast Fourier transforms is extremely useful. Details of this method may be found in reference 62.

Figures 28(a) and 28(b) show the measured (ref. 74) and calculated 1/3-octave-band SPL contours at a center frequency of 16 kHz. The calculated noise contours shown are equal to those calculated with equation (36) plus 1.4 dB. The addition of 1.4 dB is needed to bring the calculated contours in the principal direction of radiation into close agreement with the measured data. This may be regarded as the discrepancy between theory and experiment. Figures 29(a) and 29(b) show the measured and calculated near-field SPL contours at a center frequency of 40 kHz. For this frequency band the dominant direction of broadband shock-associated noise radiation is around $\psi = 100°$. There is also a secondary direction at $\psi \approx 80°$. In figure 29(a) these angles are represented by the lobes labelled "A" and "B," respectively. Figure 29(b) shows the calculated near-field contours with 1.6 dB added for the noise associated with the first and second waveguide modes alone. A comparison of the two figures shows that there are many similarities, especially near the principal direction of noise radiation. For the second principal direction of shock-associated noise radiation, the absolute level is underpredicted. However, the general orientation appears to be correct. Although perfect agreements have not been achieved yet, on taking all the above comparisons between calculated results and measurements together it seems reasonable to conclude that equation (36) is indeed capable of providing a first estimate of the near-field level and spectrum of broadband shock-associated noise.

Peak Frequency and Intensity Scaling Formulas

For practical applications it is sometimes advantageous to have a simple scaling formula for the intensity of broadband shock-associated noise. From equation (34) it is easy to verify that for supersonic jets which are not severely underexpanded or overexpanded, the variation of the noise intensity I_s is controlled by the first factor so that to a good approximation

$$I_s \propto \left(M_j^2 - M_d^2\right)^2 \tag{37}$$

This scaling formula was derived in reference 81, in which it was argued that for a given nozzle, over the range of operating conditions under which broadband shock-associated noise is important, the change in the noise source strength is dictated primarily by the change in the shock cell strength alone. This change in strength is given by equation (15), which leads immediately to equation (37). Figure 30 shows a comparison of calculations from equation (37) with the experimental measurements of reference 81 for a hot jet. Similar comparisons indicate that the scaling formula works for underexpanded as well as overexpanded jets, regardless of whether they are hot or cold. For sonic nozzles M_d is unity. In this case, equation (37) reduces to the intensity scaling formula $I_s \propto \beta^4$, found empirically in reference 79.

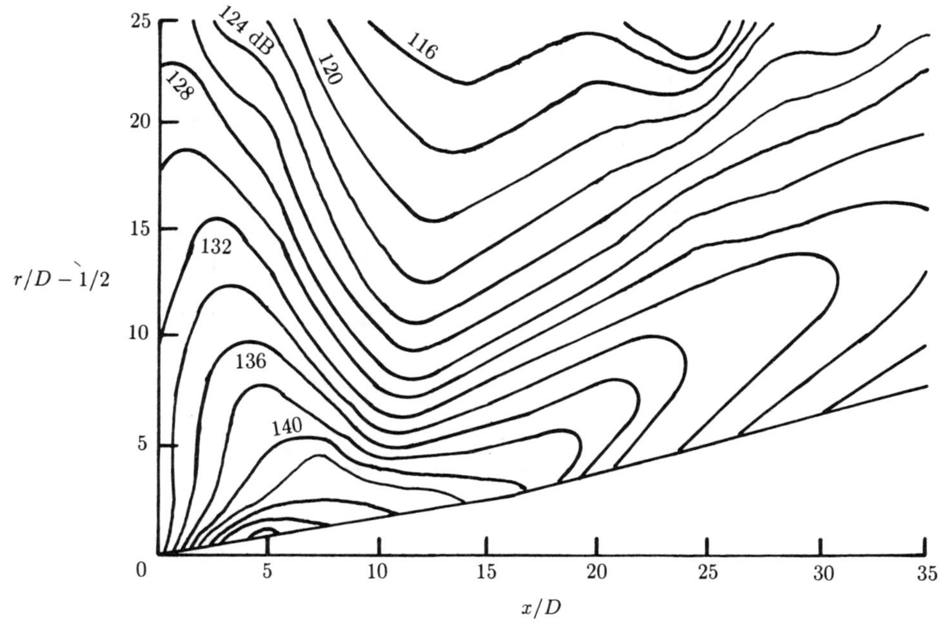

(a) Measured (ref. 74).

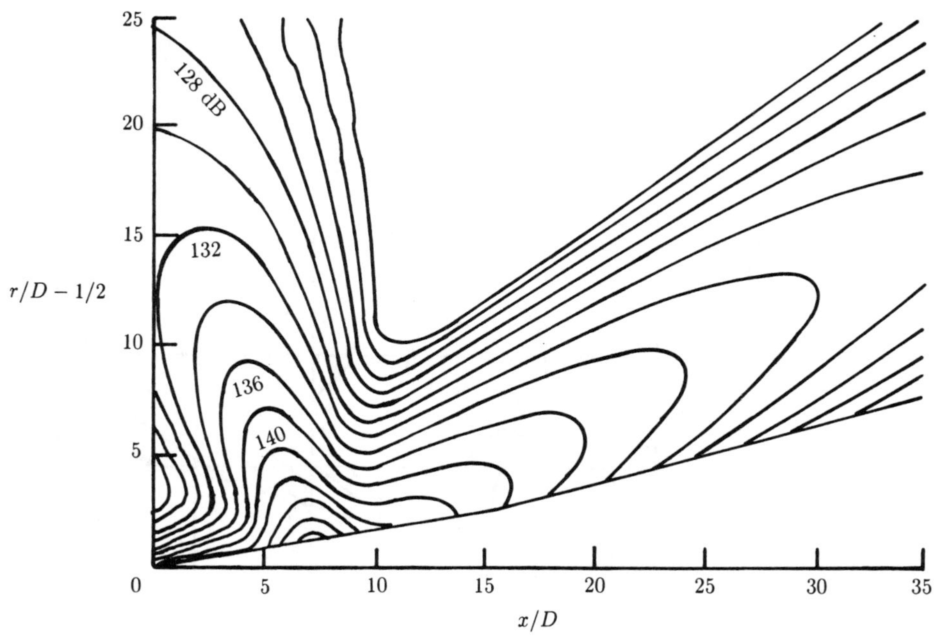

(b) Calculated plus 1.4 dB (first mode only).

Figure 28. Near-field 1/3-octave-band SPL contours at $f = 16$ *kHz.* $M_j = 1.67$; $M_d = 1.50$.

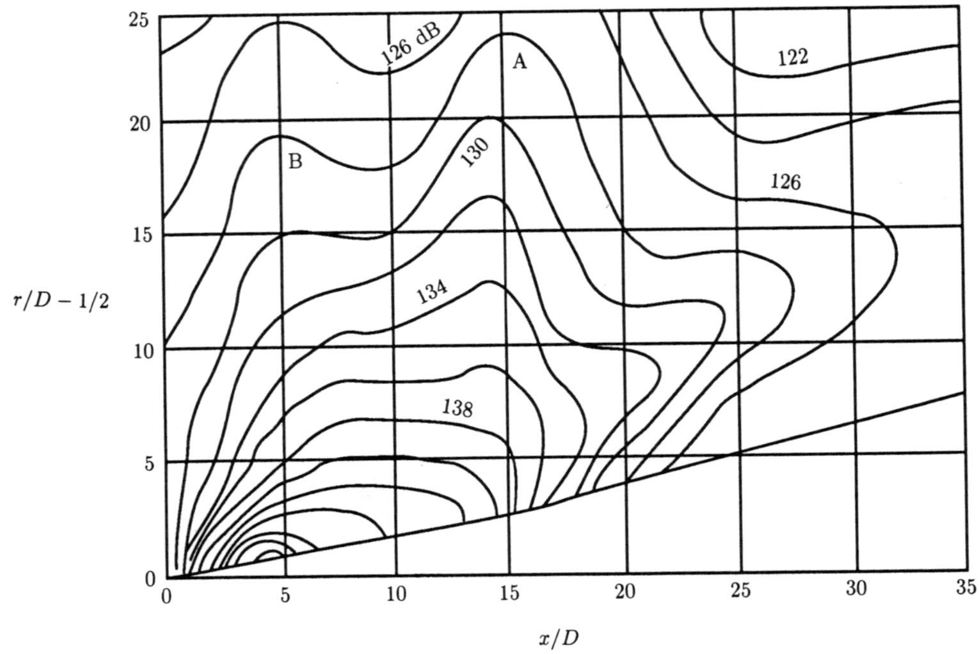

(a) Measured (ref. 74).

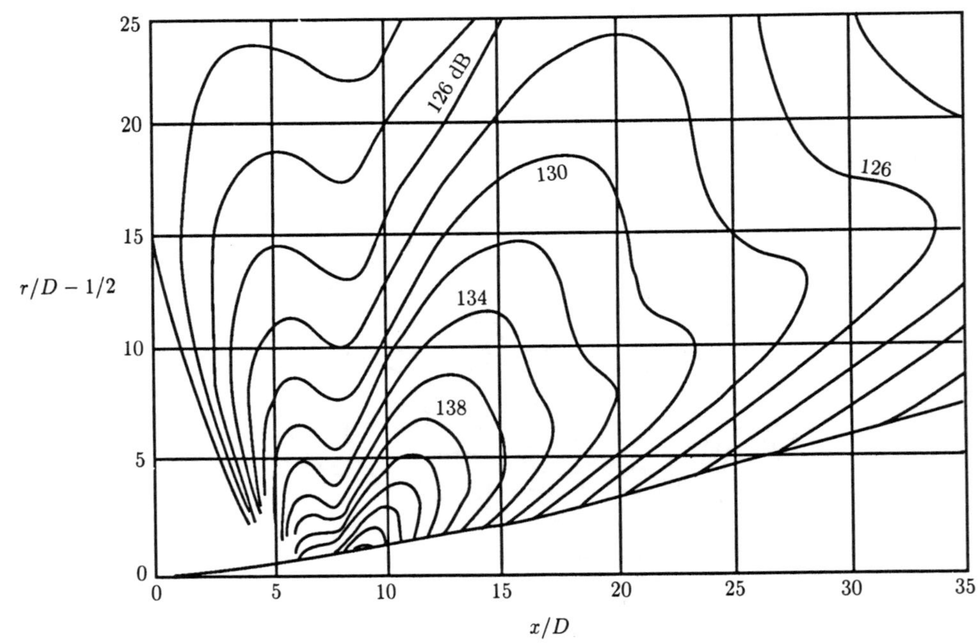

(b) Calculated plus 1.6 dB (first two modes).

Figure 29. Near-field 1/3-octave-band SPL contours at $f = 40\ kHz$. $M_j = 1.67$; $M_d = 1.50$.

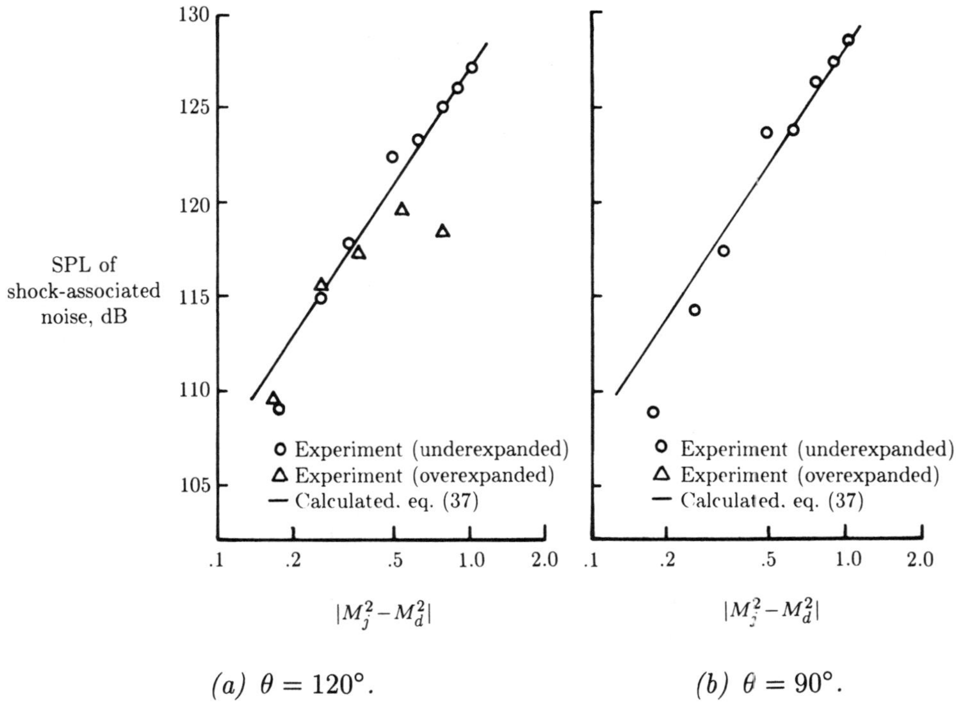

Figure 30. Broadband shock-associated noise SPL scaling. $M_d = 1.67$. $T_r/T_\infty = 2.0$. *(From ref. 81.)*

Equation (34) gives highly peaked spectra. For a given direction the spectral peak is determined essentially by the value of f which maximizes the Gaussian function of the first waveguide mode. It is easy to show that this value, f_p, is equal to f_1 of equation (35); that is,

$$f_p = \frac{u_c k_1}{2\pi(1 + M_c \cos\psi)} = \frac{u_c}{L_s(1 + M_c \cos\psi)} \tag{38}$$

Equation (38) is, for all intents and purposes, the same as equation (16), which was derived in reference 79 from the phased point-source array model. Extensive comparisons of equation (38) with measurements can be found in references 73 and 81. Favorable agreements were obtained in both references.

Effects of Forward Flight

To assess the effects of forward flight on broadband shock-associated noise, an analysis similar to that of the section entitled *Large Turbulence Structures—Shock Cell Interaction Theory* but including a uniform external flow of Mach number M_∞

was carried out. The analysis provided the following formula for the frequency of the spectral peak:

$$f_p = \frac{u_c k_1}{2\pi\left(1 + M_c\left\{\left[M_\infty/\left(1-M_\infty^2\right)\right] + \left[\cos\psi/\left(1-M_\infty^2\right)\left(1-M_\infty^2\sin^2\psi\right)^{1/2}\right]\right\}\right)} \tag{39}$$

For the limit $M_\infty \to 0$, equation (39) reduces to equation (38). In the forward arc and at $\psi = 90°$, for which broadband shock-associated noise is important, equation (39) indicates that the peak frequency generally decreases with increases in forward-flight Mach number. Recently the effects of forward flight on broadband shock-associated noise were studied experimentally up to $M_\infty = 0.4$ (refs. 102 and 103). It was found that over this range of forward-flight Mach numbers the noise intensity remained essentially the same (in the nozzle fixed frame). However, the peak frequency at $\psi = 90°$ decreased substantially at higher forward-flight Mach numbers. Figure 31 shows a comparison between calculations with equation (39) and measured peak frequency at forward Mach numbers of 0, 0.2, and 0.4 for a range of fully expanded jet Mach numbers. As shown, there is good agreement between the calculations and measurements over the entire range of parameters. At high jet Mach numbers the decrease in the peak frequency between $M_\infty = 0$ and $M_\infty = 0.4$ is as large as 30 percent. This unexpectedly large change is correctly predicted by the theory.

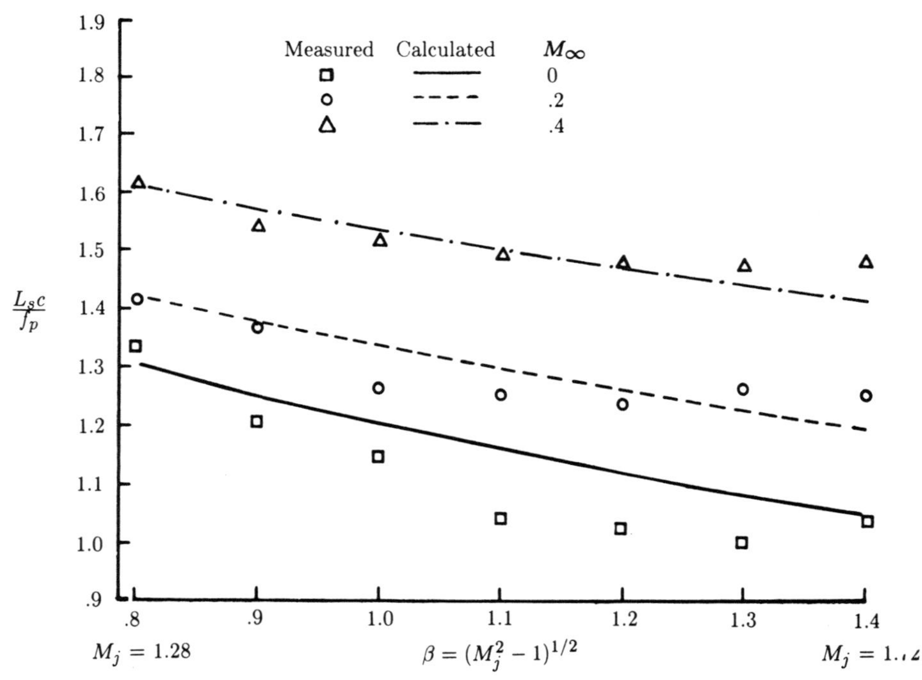

Figure 31. Effects of forward flight on peak frequency of broadband shock-associated noise. $\psi = 90°$. (Based on ref. 103.)

Generation and Prediction of Screech Tones

Although the discovery and study of screech tones of imperfectly expanded supersonic jets predates that of broadband shock-associated noise, current understanding of the screech phenomenon is primarily qualitative in nature. Quantitative understanding or prediction capability is largely restricted to the screech frequency alone. At the present time, there is no screech tone intensity prediction criterion or formula, be it empirical or theoretical. This is not surprising, for the phenomenon is highly complex and sensitive to subtle changes in the boundary conditions of the experimental environment.

Feedback Loop

The screech tone phenomenon associated with choked jets was first investigated systematically in references 87 to 89. Based on the results of this visualization study on two-dimensional jets, it was suggested that screech tones were generated by a feedback cycle. The feedback loop consisted of two halves. The inner part of the loop was made up of vortex-like disturbances which were shed periodically from the nozzle lip. These disturbances were convected downstream by the mean flow in the mixing layer of the jet. As these disturbances traversed the shock cells in the jet plume, the shock structure was set into motion. It was believed that acoustic waves were generated at the shock tips because of this coherent oscillatory motion. It was further shown that because of phase cancellation effects the acoustic waves outside the jet, which formed the outer part of the feedback loop, propagated predominantly in the upstream direction. Upon reaching the nozzle exit the acoustic waves excited the thin shear layer of the jet, leading to the continuous shedding of vortex-like disturbances and thus closing the feedback loop.

A simple phased point-source array model was developed to describe the feedback loop. This model was later adapted in reference 79 to model the generation of broadband shock-associated noise. (See section entitled *Phased Point-Source Array Model.*) In this model the noise sources were assumed to be a linear array of point sources (i.e., tips of the shock cells) spaced L_s apart, as shown in figure 26. For maximum reinforcement of the feedback cycle the time taken for vortex-like disturbances to travel one shock cell inside the jet flow plus the time taken by the acoustic waves to travel one shock cell upstream outside the jet must be equal to the period of oscillation. This condition leads to the following screech tone frequency formula:

$$f_s = \frac{u_c}{L_s(1 + M_c)} \tag{40}$$

where u_c is the convection velocity of the vortices and $M_c = u_c/c$ is the convection Mach number based on the ambient speed of sound.

The point-source array model is highly idealized and oversimplified. As it stands there is little likelihood that it could predict acoustic intensity. In fact, if one wishes to apply the model to screech tones of axisymmetric jets, one immediately faces the dilemma of how to account for the observed fact that there are two types of screech cycles, the toroidal and the helical modes. However, in spite of these difficulties the basic concept of feedback is undoubtedly correct. It forms the central framework of all subsequent theories on screech tones.

Screech Tones Generated by Interaction of Large-Scale Instability Waves and Shock Cells

Based on the results of analytical and experimental investigations, it was proposed in reference 83 that the inner part of the screech tone feedback loop is formed by the large-scale instability waves of the jet flow. According to the proposal, the tones are generated by the weak interaction of these instability waves or large turbulence structures and the shock cell structures. This interaction occurs when the spatially growing instability waves have reached sufficiently large amplitudes, usually near the end of the potential core of the jet. Thus the noise generation processes for screech tones and broadband shock-associated noise are very similar. The only major difference, however, is that in the former case a single, highly excited toroidal or helical wave is involved whereas in the latter case a spectrum of waves is involved. Because of the similarity in the noise generation process it is natural to expect a close relationship between the characteristics of screech tones and broadband shock-associated noise. This is so despite the obvious difference that the frequency of a screech tone is the same regardless of the direction of observation, whereas the dominant frequencies of broadband shock-associated noise have strong directional dependence. Such a relationship was investigated recently in reference 73.

A careful study of all the narrow-band noise data of reference 72 (see fig. 5) suggests that the screech tone, as a first approximation, could be considered as the limiting case of broadband shock-associated noise when the radiation angle θ ($\theta = 180° - \psi$) approaches the limiting value of 180°. This proposition is supported by a reexamination of the behavior of the spectral half-width and the peak frequencies of broadband shock-associated noise as functions of the direction of radiation. First, the spectral half-width of a tone is very narrow. Thus if the screech tone is a member of the broadband shock-associated noise, the half-width of the dominant peak of the broadband shock-associated noise spectrum must approach a reasonably small value as the radiation angle approaches the nozzle inlet axis. This is indeed the case, as has been discussed in the section entitled *Broadband Shock-Associated Noise.* Second, the frequency of the screech tone must have the same value as the limit of the peak frequency of broadband shock-associated noise as θ tends to 180°. It has been demonstrated by various investigations that the dominant peak frequency f_p of broadband shock-associated noise can be correlated or predicted quite accurately by equation (38). This equation may be written in the form

$$\frac{c}{f_p D} = \frac{2\pi}{k_1 D M_c} (1 - M_c \cos\,\theta) \tag{41}$$

where D is the diameter of the jet at the nozzle exit. In equation (41) the phase velocity u_c and the fundamental wave number of the shock cell structure k_1 are not strong functions of frequency. Thus one sees that the inverse of the peak frequency of broadband shock-associated noise is approximately a linear function of cos θ. Figure 32 shows the experimental confirmation of this linear relationship. The experimental measurements shown were obtained from the imperfectly expanded supersonic jets, issued from convergent-divergent nozzles with design Mach numbers of 2.0 and 1.5. The measured data in each case lie close to a straight line. It is clear that the data points of the screech tone frequencies, which lie on the left boundary of the figure, do fall on the end of the straight lines. Thus within experimental accuracy, the spectral characteristics of the fundamental screech tone may be considered, to a good approximation, to be the limit of broadband shock-associated noise as θ tends to 180°.

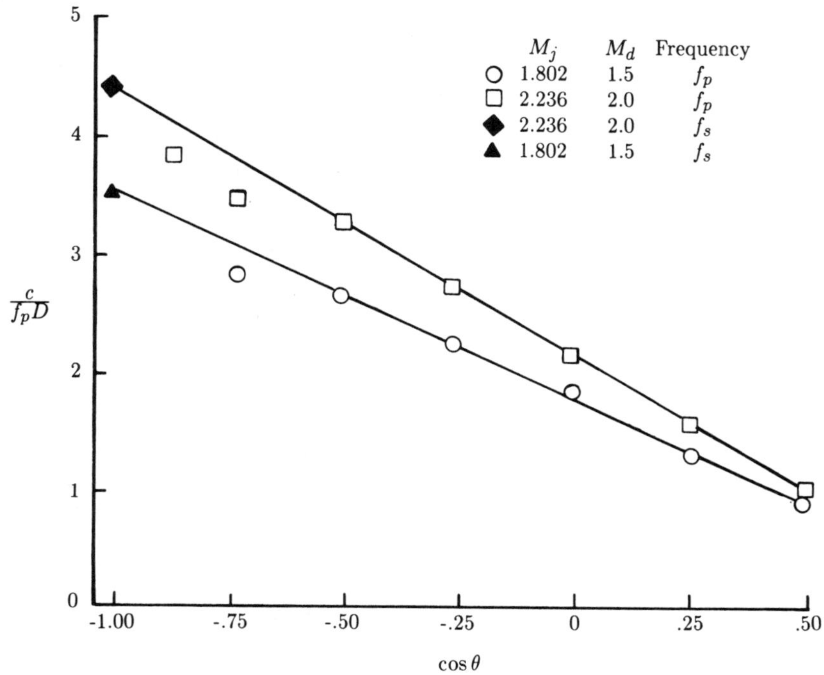

Figure 32. Screech tone frequency as limit of peak frequency of broadband shock-associated noise. (From ref. 73.)

In a feedback loop the phase change taken over the entire loop must be equal to an integral multiple of 2π. In many self-excited oscillation systems, such as cavity tones and edge tones, this phase-integral condition is known to be the controlling factor in selecting the frequencies of oscillations. In the case of screech tones of imperfectly expanded supersonic jets, this does not appear to be so. The main reason it is not so is that in the case of edge or cavity tones the feedback path length is more or less fixed by the geometry of the problem. For jet screech tones the feedback point downstream or the location of the acoustic noise source may vary so that there is no fixed feedback length inherent in the problem. In addition, the feedback loop in this case behaves somewhat nonlinearly in the sense that at a sufficiently large amplitude the instability wave could affect the spread of the mean flow, which in turn affects the instability wave characteristics. Thus the phase-integral condition can be satisfied by a slight adjustment of the wave amplitude, and this adjustment leads to a slight change in the feedback path length without imposing any condition on the oscillation frequency.

What mechanism controls the fundamental screech tone frequency then? Reference 73 proposed that the frequency is determined by the weakest link of the feedback loop. The weakest link of the loop is the joint between the outer loop and the inner loop at the nozzle exit. To avoid breaking up the feedback loop, sound waves of sufficient intensity must reach the nozzle lip region to excite the intrinsic instability wave of the jet. However, if the acoustic waves of the screech tone are generated by the interaction between downstream-propagating instability waves and the shock cell structure of the jet flow as proposed (the same mechanism as for broadband shock-associated noise), then measurements and theory indicate that only sound waves of a narrow frequency band can be radiated to the nozzle lip region with sufficiently large intensity (as discussed in previous sections). Hence, if the feedback loop is to be maintained, the screech tone frequency must be confined to this relatively narrow frequency band. Incidentally, this weakest link mechanism also provides an explanation of why the screech tone frequency is, to a good approximation, given by the limit $\theta \rightarrow 180°$ of the peak frequency of broadband shock-associated noise.

Prior to reference 73, it was noticed in reference 131 that there was numerical agreement between screech tone frequency and peak frequency of shock-associated noise as $\theta \rightarrow 180°$. It was also suggested that if the shock-associated noise was generated by an array of point sources located at the tip of the shock cells, then at the screech tone frequency the shock-associated noise from the array of sources would combine constructively to yield the strong forward radiation needed to maintain the feedback loop. However, one may recognize that this is not sufficient, for the feedback loop can be maintained without maximum constructive reinforcement as long as the intensity of sound is not too low. What is crucial is that the frequency band of the sound wave that can reach the nozzle lip is very narrow, thus effectively limiting the screech tone frequency to this narrow band.

The observations in reference 83 reveal that stable screech tones from convergent-divergent nozzles are generated by the helical instability waves of the jet column. Now from the results above it becomes possible to calculate from first principles the fundamental frequencies of these tones at different jet operating conditions. If $\theta = 180°$ (or $\psi = 0°$ in eq. (38)), the equation by which the screech frequency can be determined is

$$f_s = \frac{u_c k_1}{2\pi\left[1 + (u_c/c)\right]} \tag{42}$$

It was suggested in reference 73 that u_c, the phase velocity of the instability wave, be calculated by the hydrodynamic stability theory for locally parallel flow (see, e.g., ref. 62) and k_1, the fundamental shock cell wave number, be calculated by the multiple-scales shock cell structure model at the location of the jet where the instability wave attains its maximum amplitude (i.e., it becomes neutrally stable). This point is used because it is in this area that maximum interaction between the instability wave and the shock cell structure is expected. By using a mean axial velocity profile of a jet consisting of a uniform core and a mixing layer with a Gaussian profile, it was possible to determine the screech frequencies from equation (42) without any empirical constants. The theoretical values based on hydrodynamic instability wave calculations are shown as the solid curve in figure 16. As shown, over the Mach number range of 1.2 to 2.4 there is generally good agreement between theory and experiment. For Mach numbers below 1.3 experimental observations indicate that there is generally a switch from helical to axisymmetric modes in the screech phenomenon. The values shown in figure 16 calculated with the helical mode only must, therefore, be regarded as an approximate prediction in this low Mach number range. Since there is no empirical or adjustable constant in the calculated results, the favorable agreement must be regarded as suggesting that the "weakest link theory" is essentially correct.

Effects of Jet Temperature

For a fixed pressure ratio the screech tone frequency increases with jet temperature. This is primarily because the jet velocity for hot jets, and hence the phase velocity of the instability wave, is higher. The basic feedback mechanism of the preceding section applies to hot as well as to cold jets. To provide an estimate of the dependence of screech tone frequency on jet temperature, equation (42) was used (ref. 73). For the two unknowns u_c and k_1 of this equation, it was proposed as a first approximation to take $u_c = 0.7u_j$, according to empirical observations, and to take k_1 to be about 0.8^{-1} times the value given by the vortex-sheet shock cell model. The multiplicative factor 0.8^{-1} is to compensate for the finite mixing layer thickness near the end of the potential core of the jet where the screech tone is generated. With these empiricisms and approximations incorporated into equation (42) and the Crocco's relation used to calculate u_j, it is straightforward to find

$$\frac{f_s D_j}{u_j} \approx \frac{0.67}{\left(M_j^2 - 1\right)^{1/2}} \left[1 + \frac{0.7M_j}{\left[1 + (\gamma - 1)M_j^2/2\right]^{1/2}} \left(\frac{T_r}{T_\infty}\right)^{1/2}\right]^{-1} \tag{43}$$

where T_r/T_∞ is the ratio of reservior to ambient temperature of the jet. Equation (43) is valid for hot as well as for cold jets. In figure 16, the dashed curve is the screech tone frequencies calculated with equation (43) for cold jets. The agreement between the semiempirical formula and the measurements is quite good. Figure 33 shows the screech tone frequencies of two hot jets measured in reference 98. The calculated values of equation (43) appear to agree well with these measurements.

Effects of Forward Flight

As discussed in the *Screech Tones* section, the forward motion of a jet can cause a switch in the operating screech mode. Presently, this mode switching phenomenon is not adequately understood. In the absence of mode switching the shift in the screech tone frequency resulting from forward-flight Mach number M_∞ can be calculated by applying the weakest link theory to equation (39). Upon setting $\psi = 0°$, the screech tone frequency formula for a forward-flight Mach number M_∞ is

$$f_s = \frac{u_c k_1}{2\pi\left[1 + M_c/\left(1 - M_\infty\right)\right]} = \frac{u_c}{L_s\left[1 + M_c/\left(1 - M_\infty\right)\right]} \tag{44}$$

Equation (44) has been tested against the limited measurements of reference 102 up to $M_\infty = 0.2$, and satisfactory agreements were found.

At high forward-flight Mach numbers there are considerable changes in the mean flow of the jet as well as of the shock cell structure. The entire screech tone phenomenon becomes even more complicated. So far it has not been adequately investigated.

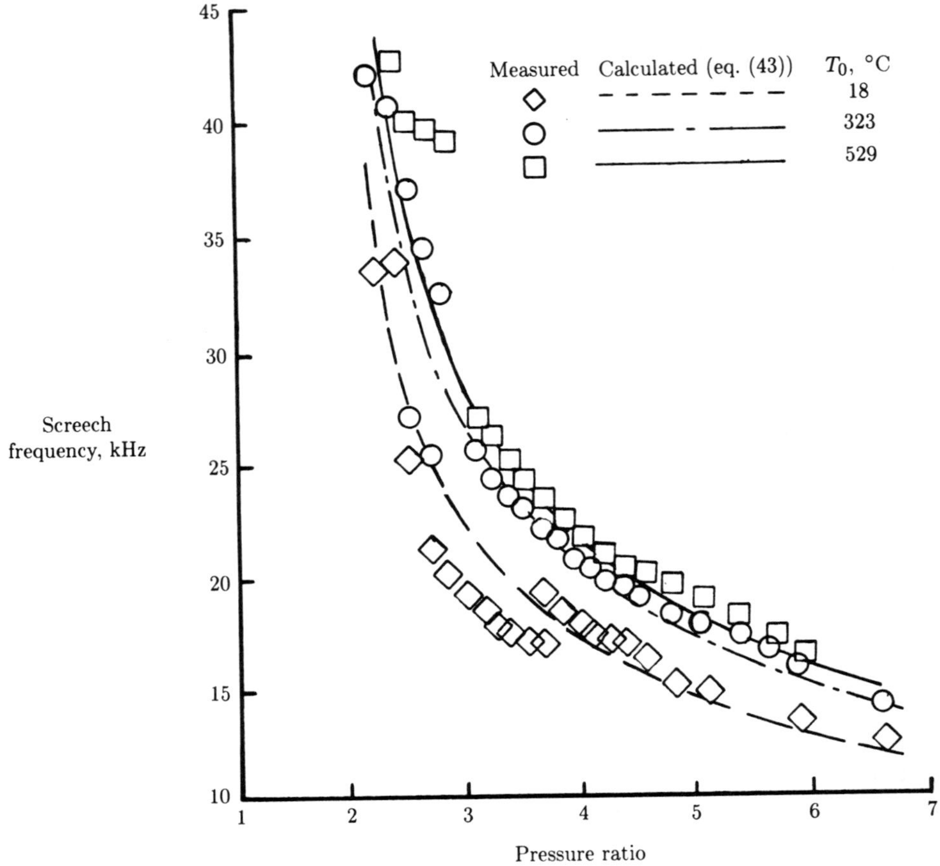

Figure 33. Dependence of screech tone frequency on jet pressure ratio at different total temperatures T_0 for convergent nozzle. (Based on ref. 98.)

Noncircular Supersonic Jets

Currently there is a good deal of interest in the flow characteristics and noise of noncircular supersonic jets. These jets are being considered for potential applications in the propulsive systems of V/STOL aircraft and aircraft requiring thrust vectoring and reversing capabilities.

The mean flow of a supersonic rectangular jet is highly complicated. A detailed mapping of such a flow field is not available in the literature at this time. Reference 132 investigated the gross features of the mean flow of such jets issued from a nozzle with a relatively large aspect ratio of 16.7. It is known from earlier subsonic jet experiments (e.g., ref. 133) that the half-width of the velocity profile of the jet in the centerline plane parallel to the small dimension of the nozzle eventually becomes larger than that parallel to the large dimension at a sufficiently far distance downstream. At the present time the mechanism which is responsible for this crossover phenomenon is not understood. In reference 132 this crossover phenomenon was also observed in supersonic jets. (See fig. 34.)

A perfectly expanded rectangular nozzle is extremely difficult to design and fabricate. Therefore, in almost all cases a shock cell structure inevitably develops in

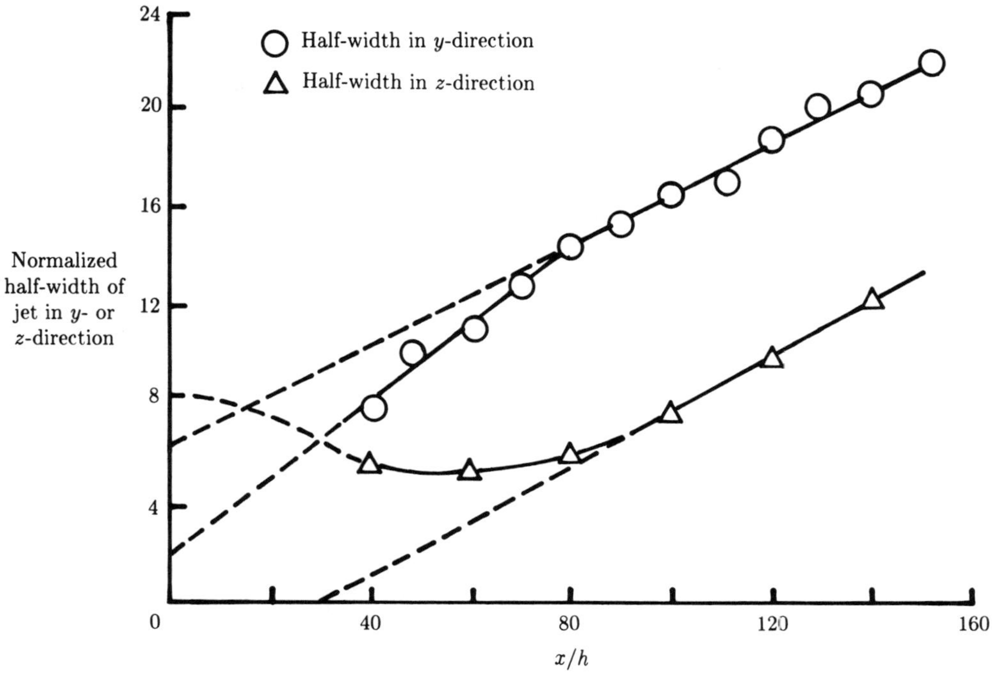

Figure 34. Growth of rectangular jet with downstream distance. Pressure ratio = 3.8; *Aspect ratio* = 16.7. *(From ref. 132.)*

the plume of such a jet. In some applications, the opposite side walls of the nozzle are cut back for aerodynamic purposes. This cutback invariably creates more shock waves and thus leads to the formation of a highly complicated shock cell pattern. (See, e.g., ref. 134.) For jets with large aspect ratios the shock cells are confined mainly to the potential core region of the jet extending approximately the distance of one large nozzle dimension downstream. Immediately at the nozzle exit the shock cells are nearly two-dimensional. As the shock cell structure evolves downstream it becomes increasingly three-dimensional.

To provide a first estimate of the effect of the aspect ratio of a rectangular nozzle on the shock cell spacing, a vortex-sheet model for these jets was developed in reference 135. The model is analogous to the Prandtl-Pack model for axisymmetric supersonic jets. With respect to a Cartesian coordinate system centered at the lower left-hand corner of the nozzle (the x-axis of which points in the direction of the flow while the y- and z-axes are parallel to the sides of the nozzle), the pressure disturbance p_s according to the vortex-sheet shock cell model is given by

$$p_s = \sum_{n=1}^{\infty}\sum_{m=1}^{\infty} \frac{4\Delta p}{nm\pi^2}(1-\cos n\pi)(1-\cos m\pi)\sin\left(\frac{n\pi y}{b_j}\right)\sin\left(\frac{m\pi z}{h_j}\right)\cos k_{nm}x \quad (45)$$

where

$$k_{nm} = \left(\frac{n^2}{b_j^2} + \frac{m^2}{h_j^2} \right)^{1/2} \frac{\pi}{\left(M_j^2 - 1\right)^{1/2}} \qquad \begin{pmatrix} n = 1, 2, \ldots \\ m = 1, 2, \ldots \end{pmatrix} \tag{46}$$

and b_j and h_j are the width and the height of the fully expanded rectangular jet. According to equation (46) the smallest wave number in the flow direction is k_{11}. But the shock cell spacing L_s is approximately equal to the fundamental wavelength. Thus,

$$L_s = \frac{2\pi}{k_{11}} = \frac{2\left(M_j^2 - 1\right)^{1/2} h_j}{\left[1 + \left(h_j^2/b_j^2\right)\right]^{1/2}} \tag{47}$$

For nozzles with aspect ratios greater than 4, equation (47) may be approximated by

$$L_s \approx 2\left(M_j^2 - 1\right)^{1/2} h_j = 2\left(M_j^2 - 1\right)^{1/2} \left\{ \frac{1 + [(\gamma - 1)/2]\, M_j^2}{1 + [(\gamma - 1)/2]\, M_d^2} \right\}^{(\gamma+1)/2(\gamma-1)} \left(\frac{M_d}{M_j} \right) h \tag{48}$$

where h is the small dimension of the rectangular nozzle. In other words, the shock cell spacing of a rectangular jet with aspect ratio greater than 4 may, to a first approximation, be considered to be the same as that of a two-dimensional jet. Numerical results of equation (48) compared very favorably in reference 135 with the measurements of references 89 and 136.

Shadowgraphic and schlieren observations in references 132 and 134 reveal that there are at least two families of large turbulent structures associated with a rectangular supersonic jet of large aspect ratio. One family involves the flapping instability wave mode of the jet near the nozzle exit. These instability waves are instrumental in generating the screech tones of the jet. The other family was tentatively identified in reference 132 in the far-downstream region of the jet beyond the crossover point of the mean flow. On the shadowgraph this family of large turbulence structures appears again in the form of the flapping mode of the jet but in the plane of the long dimension of the nozzle. That is, the two families of large-scale flapping motions of the jet are in planes which are perpendicular to each other. The preliminary results of reference 132 do not shed light on what role this second family of flapping motion of the jet plays in terms of noise generation. Since the flapping motion occurs in the far-downstream region of the jet flow, one possibility is that it might enhance the radiation of low-frequency turbulent mixing noise. Theoretical analyses of the instability wave modes of rectangular and noncircular jets have recently been carried out in references 137 to 142. Earlier the instabilities of an incompressible elliptic vortex sheet jet were studied (ref. 143). The Rayleigh equation incorporating the velocity profiles of noncircular high-speed jets is nonseparable, so that the instability characteristics cannot be determined by the usual instability wave analysis. To solve this class of problems, new and more powerful numerical methods need to be developed. Efforts in this direction are currently under way.

Several sets of high-quality far-field noise data from rectangular supersonic jets have recently been published (refs. 134 and 144). The data indicate that the measured noise can again be divided into turbulent mixing noise, broadband shock-associated noise, and screech tones. Reference 134 reported that for cold supersonic jets the relative importance of the three noise components depended to a great extent on the aspect ratio of the jet. For low-aspect-ratio rectangular jets, the screech component is unimportant. Because of the asymmetry of the flow, both the turbulent mixing noise and the broadband shock-associated noise have nonaxisymmetric radiation patterns. For high-aspect-ratio jets (Aspect ratio > 7), the far-field noise is dominated by screech tones. The tones are accompanied by many harmonics and sometimes even combination tones. This is true in the forward as well as in the rear arc. (See fig. 35.) An examination of the noise spectra shows that there is considerable suppression of mixing and broadband shock-associated noise because of the presence of intense screech tone components. The directivities of the screech tones are highly three-dimensional. They have not, however, been studied systematically at this time.

To obtain an estimate of the fundamental screech tone frequency of a rectangular supersonic jet, one may apply the weakest link argument. According to equation (42) the fundamental screech tone frequency is given by

$$f_s = \frac{u_c k_{11}}{2\pi\left[1 + (u_c/c)\right]}$$

In equation (42), the fundamental wave number k_{11} may be calculated with equation (46) and the phase velocity of the large instability wave u_c may be taken to be $0.7u_j$. Figure 36 shows a comparison of the numerical results calculated with equation (42) and the experimental measurements of references 89 and 132. As shown over the range of fully expanded Mach numbers up to 1.8, there is good agreement between the calculated results and measurements. The accuracy of the prediction is comparable to that for axisymmetric jets.

Acoustically Excited Jets

The Phenomenon of Broadband Noise Amplification

It has been known since the mid-1800's that sound waves can excite and change the flow of laminar jets. However, only in recent years have the effects of acoustic excitation on the noise radiated by high Reynolds number turbulent jets been investigated. References 11 and 145 independently reported a considerable amplification of the broadband noise of a high Reynolds number subsonic jet when the jet was excited by an upstream tone. Since these pioneering works the phenomenon of tone-excited jets has been studied by a number of investigators (refs. 146 to 149). Most recently a coordinated experimental and theoretical investigation on this subject was carried out (refs. 12, 13, 48, 150, and 151). Figure 37 shows a typical result of the noise amplification phenomenon. In this figure the noise power spectra of a high subsonic Mach number jet under unexcited and excited conditions are plotted. These measurements show clearly that when the jet is excited there is

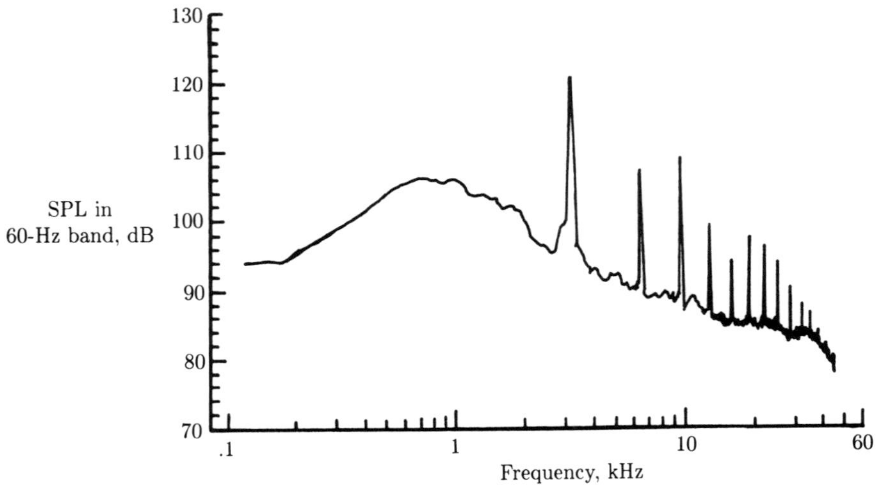

(a) Rear arc spectra.

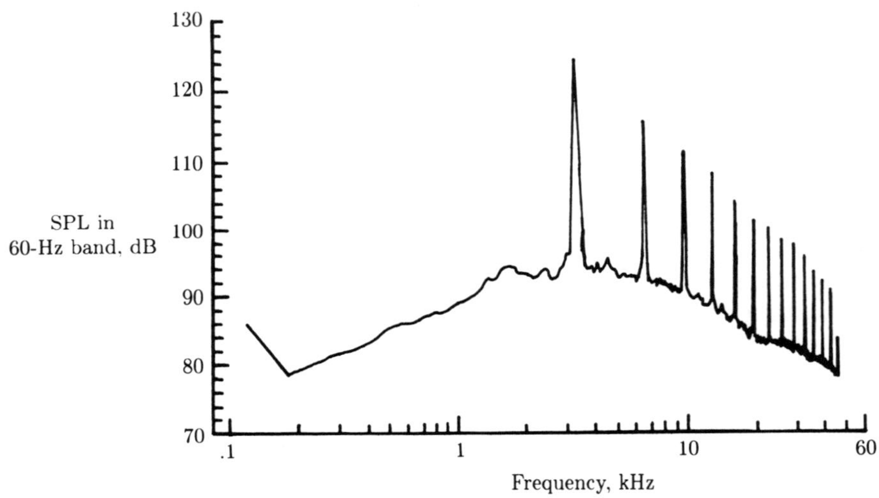

(b) Forward arc spectra.

Figure 35. Noise spectra of underexpanded supersonic jet in plane of symmetry parallel to small dimension of the nozzle. Pressure ratio = 1.45; *Aspect ratio* = 7.6. *(From ref. 134.)*

an increase of 2 to 4 dB in the radiated noise across a broad frequency band. The excitation tone may be a plane wave mode or a higher order azimuthal mode. In reference 151, when the first azimuthal mode was used the excitation tone was not detectable in the far field. In addition to discrete tones, broadband noise excitation was used in reference 146 and similar jet noise amplification was found. The effects of tones on supersonic jet noise were studied in reference 148. These measurements indicated a significant increase in the radiated broadband shock-associated noise as well as in the turbulent mixing noise across a wide spectrum of frequencies.

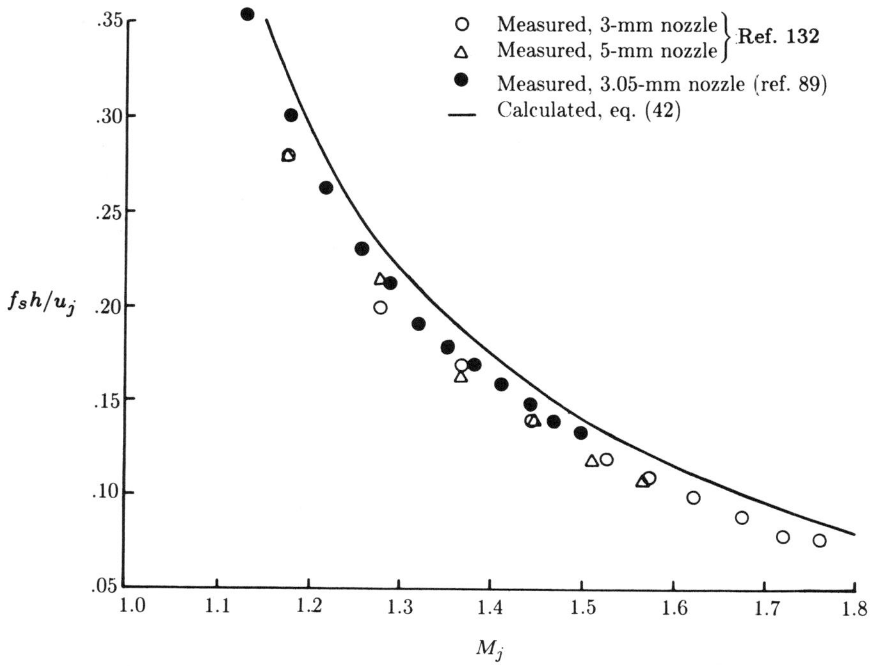

Figure 36. Measured and calculated screech tone frequencies of rectangular supersonic jets.

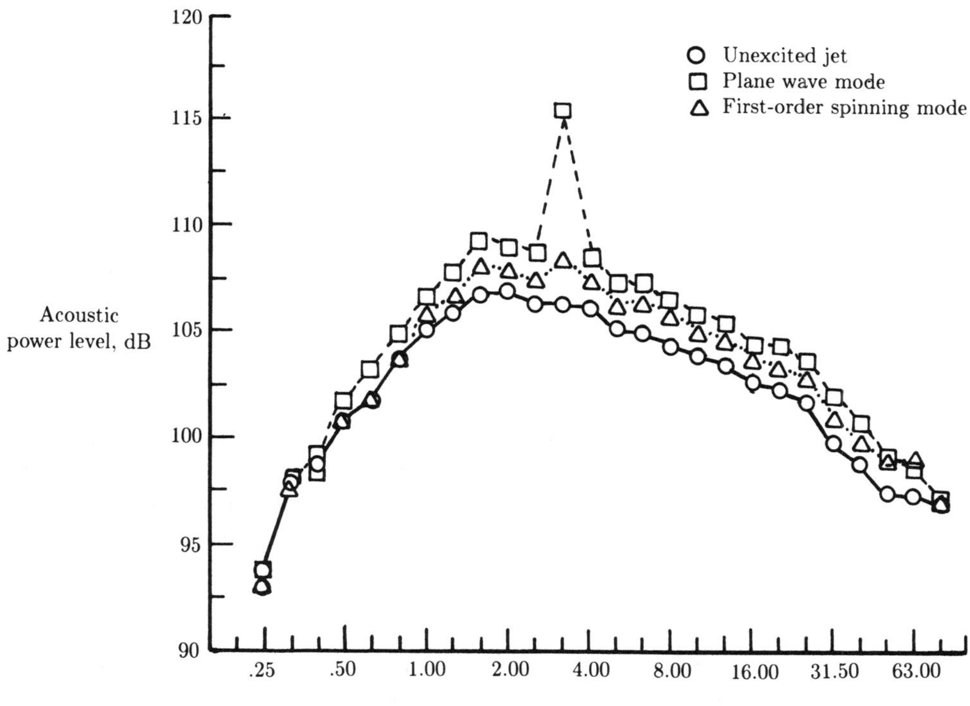

Figure 37. Spectra showing noise amplification by tone excitation. $M_j = 0.78$; $u_j = 250$ *m/sec;* $N_{St} = 0.63$; *Excitation level* $= 136$ *dB. (From ref. 151.)*

The effects of acoustic excitation on the noise of coaxial jets have been investigated in references 148 and 152 to 154. In reference 153 both discrete tones and broadband noise were used to excite the jet. The consensus of all the experimental results appears to be that there is very little change in the noise of a coaxial jet if the excitation is imposed on the primary (inner) jet. On the other hand, if the secondary (outer) jet is excited then there is an appreciable enhancement of the radiated broadband noise, just as in the case of a single jet. For jets issuing from nozzles with complex geometry, such as silencer nozzles, no measurable noise amplification was reported (ref. 152).

Several important characteristic features associated with the broadband noise amplification phenomenon of high Reynolds number subsonic jets are worth noting.

Strouhal Number Effects

Broadband noise amplification is observed only if the Strouhal number of the upstream tone lies within the unstable Strouhal number range of the jet (refs. 11 and 151).

Excitation Level Effects

Through variation of the intensity of the excitation tone (ref. 11), no noise amplification was found unless a certain threshold level was exceeded. The measurements gave a threshold level of 0.08 percent of the dynamic head of the jet. This was confirmed by the experiment in reference 151.

Noise Source Location

It was observed in reference 155, via telescopic source location, that the enhanced broadband noise from the jet appeared to come from a relatively localized region at approximately three to four jet diameters downstream of the nozzle exit. This is supported by the near-field SPL measurements of reference 149.

Temperature Effects

For hot jets a somewhat reduced broadband noise amplification effect was found in references 146 and 148. For cold jets the amplified noise has a nearly omnidirectional radiation pattern. For hot jets the increase in noise is higher in the forward arc. It is also concentrated more in the high-frequency range of the jet noise spectrum. Under similar experimental conditions, however, no noise amplification was reported for hot jets in reference 151. The reason for this anomaly remains unknown.

Forward-Flight Effects

The effects of forward flight on the noise amplification phenomenon were studied in reference 151. For forward-flight Mach numbers up to about 0.25, it was reported that the broadband noise amplification in the forward-flight case was the same as the unexcited jet in the static case. This was true both in intensity as well as in the spectral distribution.

Mechanism of Broadband Noise Amplification

It was recognized at the very beginning that the large-scale instability waves of the jet flow played a crucial role in the noise amplification phenomenon (ref. 11). However, the precise mechanism which led to additional broadband noise radiation was not known for a long period of time. It is now generally accepted that the first link of the entire process is the excitation of the intrinsic large-scale instability wave of the jet flow by the pressure and velocity fluctuations associated with the upstream tone. Because of the large difference in the wavelengths, the coupling between the imposed sound wave and the flow instability wave is effective only in the region near the nozzle exit. (See refs. 48 and 156.) Once excited the instability wave, regardless of whether it is in the form of axisymmetric, helical, or higher order modes, grows rapidly in the downstream direction. When the amplitude of the excited instability wave becomes sufficiently large, the wave interacts with the mean flow and the naturally occurring turbulence. The natural turbulence of the jet consists of both the fine-scale turbulence and the (somewhat random) large turbulence structures. How the three components of the jet flow interact nonlinearly is not fully understood even now. But it is certain that energy from the mean flow is transferred to the excited large-scale instability wave and the natural turbulence. This energy transfer causes a substantial increase in the rate of spread of the mean flow of the jet. At the same time there is also a large increase in the intensities of both the fine-scale turbulence and the large turbulence structures. The increase is most noticeable in the region near the end of the potential core of the jet, where the amplitude of the excited large-scale instability wave attains its maximum value. For subsonic jets the enhanced fine-scale turbulence causes additional broadband noise radiation. For supersonic jets the enhanced random large turbulence structures are responsible primarily for the increased radiation of turbulent mixing noise as well as broadband shock-associated noise.

A Quasi-Linear Theory and Comparisons With Experiments

A mathematical model of tone-excited jets based on the above physical processes has recently been developed (ref. 48). The model consists of two major components. The first component involves a mathematical analysis of the coupling between the excitation tone and the large-scale instability wave of the jet (i.e., receptivity analysis). To determine the coupling constants between the sound waves and the instability waves, a Green's function approach following the work of references 157 and 158 is used. The second component is a nonlinear theory of the interaction between the excited instability waves, the mean flow of the jet, and the natural turbulence (fine-scale turbulence and random large turbulence structures). Here an integral approach using a set of conservation equations is adopted. These conservation equations provide a way to predict the nonlinear spatial development of the instability wave, the mean flow, and the intensity of the natural turbulence. Models of this kind, with different degrees of sophistication, have been employed in references 42, 45, 159, and 160. In the quasi-linear theory of reference 48 the mean flow of the jet is characterized by two parameters. They are the core radius and the half-width of the mixing layer. The natural turbulence is characterized by a peak turbulence intensity function $q(x)$, where x is the axial distance downstream. The radial distribution of the turbulence kinetic energy is assumed to be given by a Gaussian function, as suggested by the measurements of reference 161. The excited large-scale instability wave is represented by the instability wave solution of the locally parallel flow analysis. The amplitude of the instability wave $A(x)$ is left as an unknown in the quasi-linear calculation. It is to be found as the solution of the wave-amplitude equation according to the receptivity analysis. The wave-amplitude equation may be written in the form

$$\frac{dA}{dx} = ik_n A + C_{pn}\hat{p}_n(x) + C_{vn}\hat{v}_n(x) \tag{49}$$

where k_n is the local complex wave number of the excited instability wave, C_{pn} and C_{vn} are the pressure and velocity coupling constants, and $\hat{p}_n(x)$ and $\hat{v}_n(x)$ are the pressure and radial velocity distributions of the incident upstream tone. The appropriate initial condition for $A(x)$ to be used in conjunction with equation (49) is $A(0) = 0$ (see ref. 156); that is, upstream of the nozzle exit the amplitude of the instability wave is zero.

The numerical results of their model were compared in reference 48 with the measurements of references 11 and 13. Figure 38 shows a comparison of the calculated and measured instability wave amplitudes at the centerline of the jet at a Strouhal number of 0.5 and a jet Mach number of 0.575. The jet is excited by a tone of the plane wave mode with an intensity of 141 dB at the nozzle exit. As shown, the calculated wave amplitude compares well with the measurements all the way to the end of the potential core. Figure 39 shows a similar comparison with the measurements of reference 11. The excitation Strouhal number is 0.98 and the jet Mach number is 0.15. At this high Strouhal number the excited instability wave becomes damped in the region close to the nozzle exit so that it does not grow to an appreciable amplitude. Under this condition the excited instability wave amplitude is comparable with that of the acoustic wave which radiates out of the jet nozzle. Since the wavelength of the acoustic wave is long relative to that of the instability wave, the combined pressure distribution of the two waves produces a nearly periodic amplitude modulation pattern as shown. The top part of figure 39 shows the amplitude distribution of the acoustic wave and the calculated amplitude of the instability wave. By combining these two waves and taking into consideration their wavelengths and phases, we can find the characteristic amplitude oscillations shown in this figure. As shown, there is favorable agreement with the reference 11 measurements even in absolute intensity.

Extensive numerical tests of the quasi-linear theory were carried out in reference 48. It was found that the theory predicts very little change in the peak-turbulence kinetic energy in the jet flow unless the excitation acoustic amplitude exceeds a certain level. This nonlinear result of the theory is consistent with the observation of a threshold levels in references 11 and 151. Figure 40 shows the effect of excitation level on the calculated peak-turbulence kinetic energy $\bar{q}$. In this figure the threshold levels of reference 11 are also marked by arrows. As is apparent the measured threshold level falls almost exactly on the threshold level of the theoretical model. The agreement is rather remarkable. This and other agreement suggest that

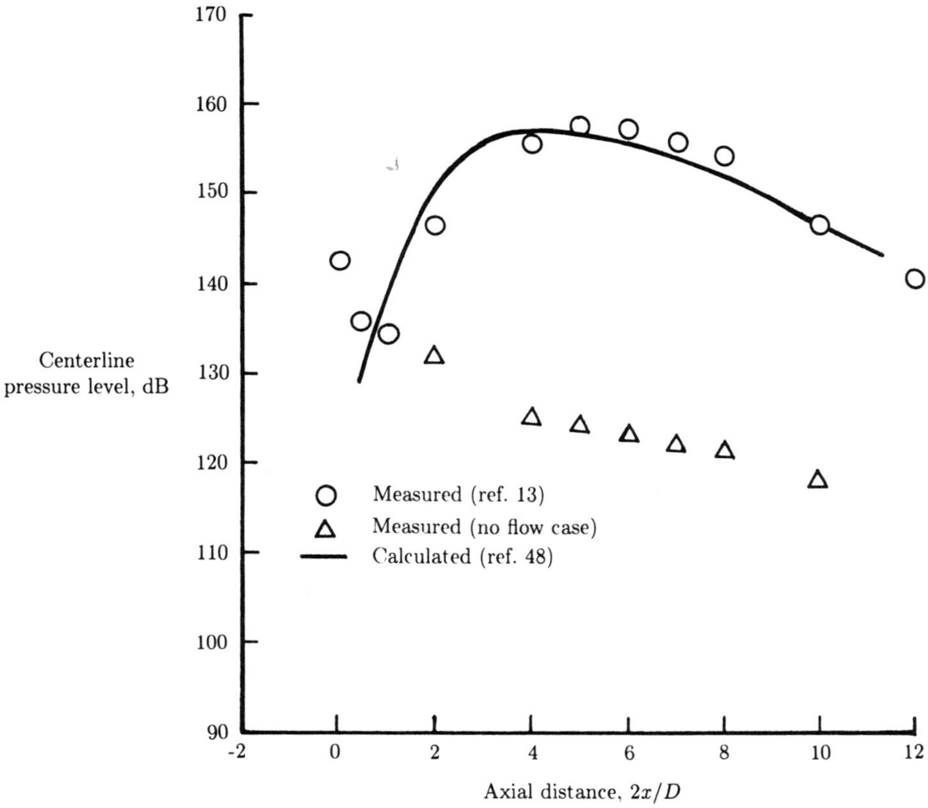

Figure 38. Calculated and measured instability wave amplitudes for cold jet. $M_j = 0.575$; Pressure ratio = 1.25; $N_{\mathrm{St}} = 0.5$; Excitation level at nozzle exit = 141 dB; plane wave mode. (Based on ref. 48.)

the quasi-linear theory does contain all the essential physics of the tone excitation phenomenon.

Additional Topics

We will now turn our attention to the role of large turbulence structures and instability waves in subsonic jet noise generation. Presently there is overwhelming experimental evidence showing that such large structures not only exist in the flow of these jets but also are the controlling factors in the mixing and spreading of the jet flow and in the production of fine-scale turbulence. The pertinent question that needs to be clarified here is whether the noise radiated directly by the large turbulence structures and instability waves forms the dominant part of the noise of subsonic jets (as in the case of supersonic jets) or whether the dominant part of the noise is generated by the fine-scale turbulence. In the latter case the large turbulence structures and instability waves would only play an indirect role in the noise generation processes.

Crow and Champagne (ref. 1) in their pioneering work on large turbulence structures appeared to be the first to suggest that indeed these structures were the

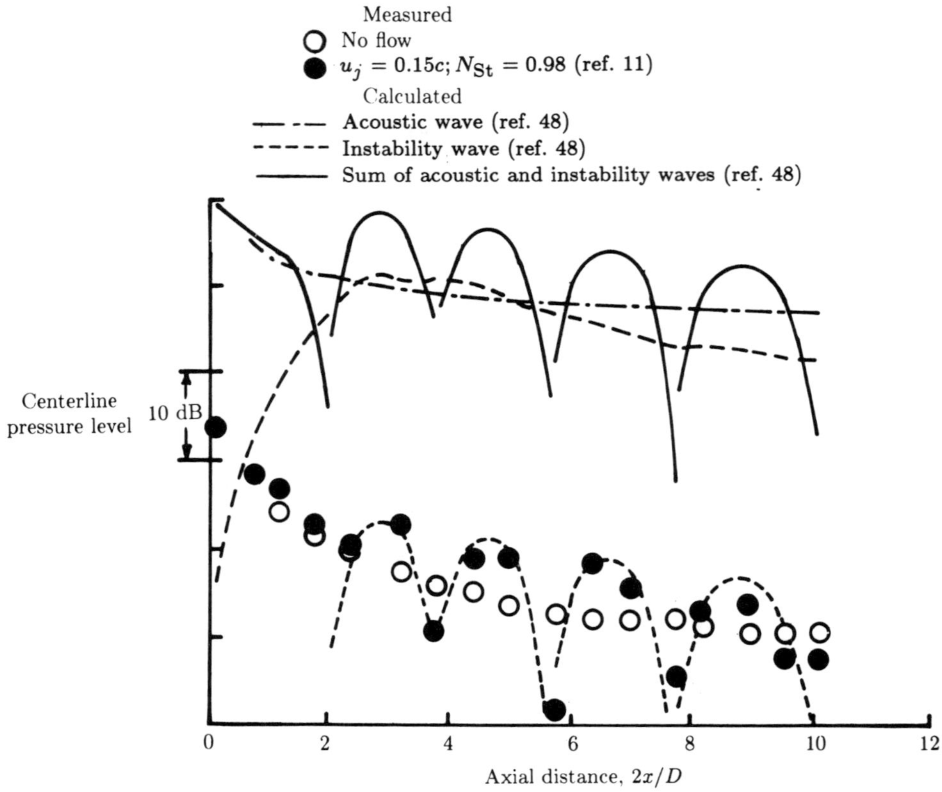

Figure 39. Prediction of centerline pressure fluctuation as sum of acoustic and instability waves and comparisons with experiments. (From ref. 48.)

dominant direct noise sources of subsonic jets. To quantify this idea, Crow developed an oscillating antenna model to describe the noise generation process. This antenna model was later refined and extended in reference 162. Despite all the elaborate analysis involved in the development of this model so far, no numerical results or serious comparisons with experiments were provided to demonstrate its validity. Totally independent of the development of the antenna model, other researchers, inspired by the dominance of the large turbulence structures and instability waves in the dynamics of the flow of subsonic jets, suggested implicitly the existence of a strong direct relationship between subsonic jet noise and these structures (refs. 163 to 165). Unfortunately, after a good deal of effort they were unable to offer definitive experimental evidence to support their contention. At the present time there is no reason to dismiss the possibility that large turbulence structures and instability waves are the dominant direct noise sources of high Reynolds number subsonic jets. However, there is strong experimental and theoretical evidence indicating that the unsteady motions (noise sources) of the oscillating antenna model are inefficient noise generators. It is easy to recognize that the oscillating antenna is nothing more than just a highly simplified form of the instability waves of the jet. The noise generation processes of instability waves have been discussed in the *Noise Generation Processes* section. As noted therein the crucial factor which governs the noise generation efficiency of these physical entities is the effective phase velocity of the waves. For cold

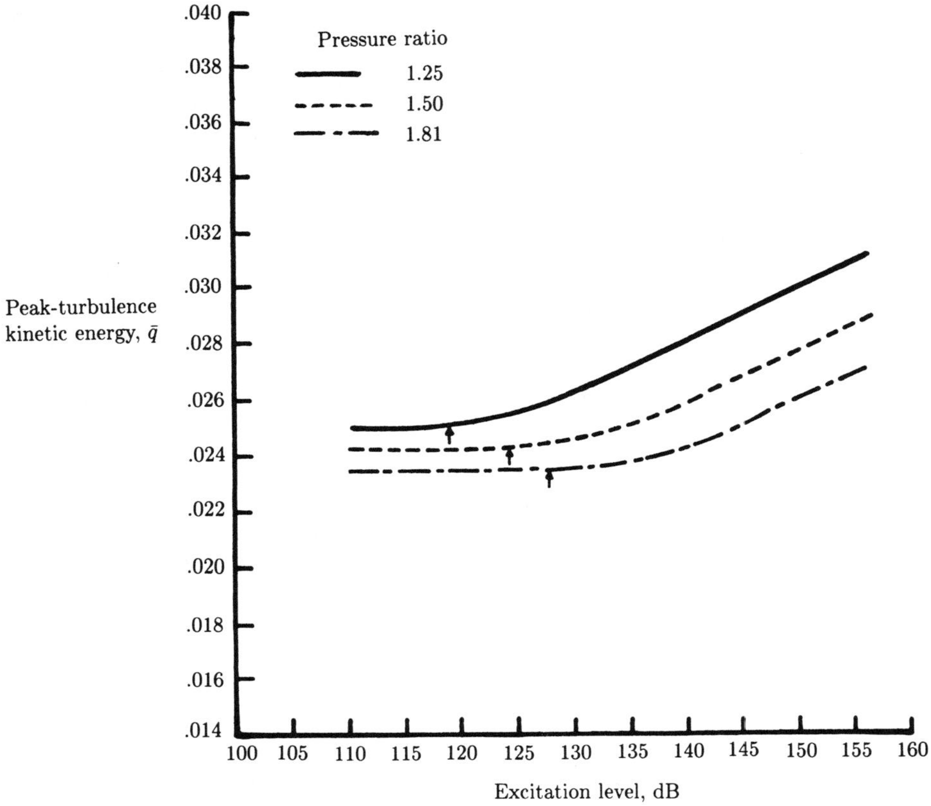

Figure 40. Effect of excitation level and Mach number on peak turbulent kinetic energy for cold jet. Arrows indicate measured threshold levels (ref. 11). $N_{St} = 0.5$*; plane wave mode. (From ref. 48.)*

subsonic jets the phase velocities of the instability waves are subsonic. From the wavy wall analogy, these waves are extremely ineffective in producing acoustic radiation. In reference 60 on sound generated by instability waves, quantitative comparisons of the noise produced by instability waves in supersonic and subsonic flows were carried out. The calculations show that instability waves in subsonic flows produce an insignificant amount of noise when compared with those in supersonic flows. Experimentally this loss of noise radiation efficiency in subsonic flows is supported by the measurements of reference 11. In these tone-excited jet experiments, the total acoustic energy radiation to the far field by the excitation tone and the excited instability wave was compared with the input acoustic energy from upstream of the nozzle. Over the jet Mach number range of 0.1 to 0.9 and Strouhal number range of 0.1 to 3.5 the two acoustic power levels were found to be nearly equal, so that the excited instability waves effectively produced very little additional noise.

Soon after the discovery of the large turbulence structures in free shear flows, it was proposed that vortex pairing was the dominant noise generation mechanism in subsonic jets (ref. 166). Following this proposal a sophisticated mathematical model to quantify this noise generation process was developed (ref. 162). Experimentally it was demonstrated that by acoustically exciting a subsonic jet at a moderate Reynolds number ($N_{\mathrm{Re}} = 5 \times 10^4$), vortex pairing indeed generated sound (ref. 167) . However, the measured noise spectrum contained only discrete subharmonic frequencies accompanied by a considerable reduction of the broadband noise of the jet. In light of this experiment, the issue which is of the greatest concern is whether vortex pairing is, in fact, a dominant noise generation mechanism in unforced high Reynolds number (10^6 or higher) subsonic jets. In a previous section it was pointed out that an in-depth experimental study of the large turbulence structures and the vortex pairing phenomenon in subsonic jets has been made (ref. 6). The findings reveal that as the Reynolds number of the jet increases, the length of the transition region (see fig. 2) in which vortex pairings take place decreases. These observations imply that at sufficiently high Reynolds numbers, vortex pairing might become such an infrequent event in subsonic jets that it could not be the dominant noise generation mechanism. This point of view was advocated in reference 168. Recently it was reaffirmed experimentally in reference 169.

In view of the above conflicting proposals and experimental findings, it appears that large turbulence structures and instability waves may not be the dominant direct noise sources of subsonic jets. On the other hand, our understanding of the dynamics and behavior of the large turbulence structures and instability waves is incomplete. It would, therefore, be premature to make an absolute statement. What is clear at this time is that a satisfactory self-contained subsonic jet noise theory is still very much needed.

References

1. Crow, S. C.; and Champagne, F. H.: Orderly Structure in Jet Turbulence. *J. Fluid Mech.*, vol. 48, pt. 3, Aug. 16, 1971, pp. 547–591.
2. Brown, Garry L.; and Roshko, Anatol: On Density Effects and Large Structure in Turbulent Mixing Layers. *J. Fluid Mech.*, vol. 64, pt. 4, July 1974, pp. 775–816.
3. Dimotakis, Paul E.; and Brown, Garry L.: The Mixing Layer at High Reynolds Number: Large-Structure Dynamics and Entrainment. *J. Fluid Mech.*, vol. 78, pt. 3, Dec. 7, 1976, pp. 535–560.
4. Winant, C. D.; and Browand, F. K.: Vortex Pairing: The Mechanism of Turbulent Mixing-Layer Growth at Moderate Reynolds Numbers. *J. Fluid Mech.*, vol. 63, pt. 2, Apr. 3, 1974, pp. 237–255.
5. Hernan, Miquel A.; and Jimenez, Javier: Computer Analysis of a High-Speed Film of the Plane Turbulent Mixing Layer. *J. Fluid Mech.*, vol. 119, June 1982, pp. 323–345.
6. Yule, A. J.: Large-Scale Structure in the Mixing Layer of a Round Jet. *J. Fluid Mech.*, vol. 89, pt. 3, Dec. 13, 1978, pp. 413–432.
7. Chan, Y. Y.: Spatial Waves in Turbulent Jets, Part I. *Phys. Fluids*, vol. 17, no. 1, Jan. 1974, pp. 46–53.
8. Chan, Y. Y.: Spatial Waves in Turbulent Jets, Part II. *Phys. Fluids*, vol. 17, no. 9, Sept. 1974, pp. 1667–1670.
9. Chan, Y. Y.: Spatial Waves of Higher Order Modes in an Axisymmetric Turbulent Jet. *Phys. Fluids*, vol. 19, no. 12, Dec. 1976, pp. 2042–2043.
10. Chan, Y. Y.: Wavelike Eddies in a Turbulent Jet. *AIAA J.*, vol. 15, no. 7, July 1977, pp. 992–1001.
11. Moore, C. J.: The Role of Shear-Layer Instability Waves in Jet Exhaust Noise. *J. Fluid Mech.*, vol. 80, pt. 2, Apr. 25, 1977, pp. 321–367.
12. Ahuja, K. K.; and Whiffen, M. C.: Tone Excited Jets, Part II: Flow Visualization. *J. Sound & Vib.*, vol. 102, no. 1, Sept. 8, 1985, pp. 63–69.
13. Lepicovsky, J.; Ahuja, K. K.; and Burrin, R. H.: Tone Excited Jets, Part III: Flow Measurements. *J. Sound & Vib.*, vol. 102, no. 1, Sept. 8, 1985, pp. 71–91.
14. Zaman, K. B. M. Q.; and Hussain, A. K. M. F.: Natural Large-Scale Structures in the Axisymmetric Mixing Layer. *J. Fluid Mech.*, vol. 138, Jan. 1984, pp. 325–351.
15. Laufer, John: New Trends in Experimental Turbulence Research. *Annual Review of Fluid Mechanics, Volume 7*, Milton Van Dyke, Walter G. Vincenti, and J. V. Wehausen, eds., Annual Reviews Inc., 1975, pp. 307–326.

16. Laufer, J.: Deterministic and Stochastic Aspects of Turbulence. *J. Appl. Mech.*, vol. 50, no. 413, Dec. 1983, pp. 1079–1085.
17. Roshko, Anatol: Structure of Turbulent Shear Flows: A New Look. *AIAA J.*, vol. 14, no. 10, Oct. 1976, pp. 1349–1357.
18. Cantwell, Brian J.: Organized Motion in Turbulent Flow. *Annual Review of Fluid Mechanics, Volume 13*, Milton Van Dyke, J. V. Wehausen, and John L. Lumley, eds., Annual Reviews Inc., 1981, pp. 457–515.
19. Ho, Chih-Ming; and Huerre, Patrick: Perturbed Free Shear Layers. *Annual Review of Fluid Mechanics, Volume 16*, Milton Van Dyke, J. V. Wehausen, and John L. Lumley, eds., Annual Reviews Inc., 1984, pp. 365–424.
20. Wygnanski, Israel J.; and Petersen, Robert A.: Coherent Motion in Excited Free Shear Flows. *AIAA J.*, vol. 25, no. 2, Feb. 1987, pp. 201–213.
21. Lowson, M. V.; and Ollerhead, J. B.: Visualization of Noise From Cold Supersonic Jets. *J. Acoust. Soc. America*, vol. 44, no. 2, Aug. 1968, pp. 624–630.
22. Tam, Christopher K. W.: Directional Acoustic Radiation From a Supersonic Jet Generated by Shear Layer Instability. *J. Fluid Mech.*, vol. 46, pt. 4, Apr. 27, 1971, pp. 757–768.
23. Tam, Christopher K. W.: On the Noise of a Nearly Ideally Expanded Supersonic Jet. *J. Fluid Mech.*, vol. 51, pt. 1, Jan. 11, 1972, pp. 69–95.
24. Chan, Y. Y.; and Westley, R.: Directional Acoustic Radiation Generated by Spatial Jet Instability. *C.A.S.I. Trans.*, vol. 6, no. 1, Mar. 1973, pp. 36–41.
25. Dahan, C.; Elias, G.; Maulard, J.; and Perulli, M.: Coherent Structures in the Mixing Zone of a Subsonic Hot Free Jet. *J. Sound & Vib.*, vol. 59, no. 3, Aug. 8, 1978, pp. 313–333.
26. Ozkul, Ahmet: Investigation of Acoustic Radiation From Supersonic Jets by Double-Pulse Holographic Interferometry. *AIAA J.*, vol. 17, no. 10, Oct. 1979, pp. 1068–1073.
27. Oertel, Herbert: Mach Wave Radiation of Hot Supersonic Jets Investigated by Means of the Shock Tube and New Optical Techniques. *Shock Tubes and Waves*, Assa Lipshitz and Josef Rom, eds., AFOSR-TR-80-1141, U.S. Air Force, Feb. 1980, pp. 266–275. (Available from DTIC as AD A091 010.)
28. Oertel, H.: Coherent Structures Producing Machwaves Inside and Outside of the Supersonic Jet. *Structure and Complex Turbulent Shear Flow*, R. Dumas and L. Fulachier, eds., Springer-Verlag, 1983, pp. 334–343.
29. Lepicovsky, J.; Ahuja, K. K.; Brown, W. H.; and Burrin, R. H.: *Coherent Large-Scale Structures in High Reynolds Number Supersonic Jets.* NASA CR-3952, 1985.
30. McLaughlin, Dennis K.; Morrison, Gerald L.; and Troutt, Timothy R.: Experiments on the Instability Waves in a Supersonic Jet and Their Acoustic Radiation. *J. Fluid Mech.*, vol. 69, pt. 1, May 13, 1975, pp. 73–95.
31. McLaughlin, Dennis K.; Morrison, Gerald L.; and Troutt, Timothy R.: Reynolds Number Dependence on Supersonic Jet Noise. *AIAA J.*, vol. 15, no. 4, Apr. 1977, pp. 526–532.
32. Morrison, G. L.; and McLaughlin, D. K.: The Noise Generation by Instabilities in Low Reynolds Number Supersonic Jets. *J. Sound & Vib.*, vol. 65, no. 2, July 22, 1979, pp. 177–191.
33. Troutt, T. R.; and McLaughlin, D. K.: Experiments on the Flow and Acoustic Properties of a Moderate-Reynolds-Number Supersonic Jet. *J. Fluid Mech.*, vol. 116, Mar. 1982, pp. 123–156.
34. Seiner, John M.; McLaughlin, Dennis K.; and Liu, C. H.: *Supersonic Jet Noise Generated by Large-Scale Instabilities.* NASA TP-2072, 1982.
35. Acton, E.: The Modelling of Large Eddies in a Two-Dimensional Shear Layer. *J. Fluid Mech.*, vol. 76, pt. 3, Aug. 11, 1976, pp. 561–592.
36. Delcourt, B. A. G.; and Brown, G. L.: The Evolution and Emerging Structure of a Vortex Sheet in an Inviscid and Viscous Fluid Modelled by a Point Vortex Method. *Proceedings of the Second Symposium on Turbulent Shear Flows*, Imperial College of Science and Technology (London), 1979, pp. 14.35–14.40.
37. Ashurst, W. T.: Numerical Simulation of Turbulent Mixing Layers Via Vortex Dynamics. *Turbulent Shear Flows I*, F. Durst, B. E. Launder, F. W. Schmidt, and J. H. Whitelaw, eds., Springer-Verlag, 1979, pp. 402–413.
38. Aref, Hassan; and Siggia, Eric D.: Vortex Dynamics of the Two-Dimensional Turbulent Shear Layer. *J. Fluid Mech.*, vol. 100, pt. 4, Oct. 29, 1980, pp. 705–737.
39. Hardin, Jay C.: *Analysis of Noise Produced by an Orderly Structure of Turbulent Jets.* NASA TN D-7242, 1973.
40. Fung, Y. T.; Liu, C. H.; and Gunzburger, M. D.: Simulation of the Pressure Field Near a Jet by Randomly Distributed Vortex Rings. *AIAA J.*, vol. 17, no. 6, June 1979, pp. 553–557.

41. Maestrello, L.; and Fung, Y.-T.: Quasi-Periodic Structure of a Turbulent Jet. *J. Sound & Vib.*, vol. 64, no. 1, May 8, 1979, pp. 107–122.
42. Tam, C. K. W.: Supersonic Jet Noise Generated by Large-Scale Disturbances. *J. Sound & Vib.*, vol. 38, no. 1, Jan. 8, 1975, pp. 51–79.
43. Liu, J. T. C.: Developing Large-Scale Wavelike Eddies and the Near Jet Noise Field. *J. Fluid Mech.*, vol. 62, pt. 3, Feb. 11, 1974, pp. 437–464.
44. Liu, J. T. C.; and Merkine, L.: On the Interactions Between Large-Scale Structure and Fine-Grain Turbulence on a Free Shear Flow. I. The Development of Temporal Interactions in the Mean. *Proc. Royal Soc. London*, ser. A, vol. 352, no. 1669, Dec. 29, 1976, pp. 213–247.
45. Morris, P. J.: Flow Characteristics of the Large Scale Wave-Like Structure of a Supersonic Round Jet. *J. Sound & Vib.*, vol. 53, no. 2, July 22, 1977, pp. 223–244.
46. Gatski, T. B.; and Liu, J. T. C.: On the Interactions Between Large-Scale Structure and Fine-Grained Turbulence in a Free Shear Layer. III. A Numerical Solution. *Philos. Trans. Royal Soc. London*, vol. 293, no. 1403, Jan. 4, 1980, pp. 473–509.
47. Gaster, M.; Kit, E.; and Wygnanski, I.: Large-Scale Structures in a Forced Turbulent Mixing Layer. *J. Fluid Mech.*, vol. 150, Jan. 1985, pp. 23–39.
48. Tam, C. K. W.; and Morris, P. J.: Tone Excited Jets, Part V: A Theoretical Model and Comparison With Experiment. *J. Sound & Vib.*, vol. 102, no. 1, Sept. 8, 1985, pp. 119–151.
49. Riley, James J.; and Metcalfe, Ralph W.: Direct Numerical Simulation of a Perturbed, Turbulent Mixing Layer. AIAA-80-0274, Jan. 1980.
50. Bayliss, Alvin; and Maestrello, Lucio: Simulation of Instabilities and Sound Radiation in a Jet. *AIAA J.*, vol. 19, no. 7, July 1981, pp. 835–841.
51. Maestrello, L.; and Bayliss, A.: Flowfield and Far Field Acoustic Amplification Properties of Heated and Unheated Jets. *AIAA J.*, vol. 20, no. 11, Nov. 1982, pp. 1539–1546.
52. Tam, Christopher K. W.; and Chen, K. C.: A Statistical Model of Turbulence in Two-Dimensional Mixing Layers. *J. Fluid Mech.*, vol. 92, pt. 2, May 28, 1979, pp. 303–326.
53. Plaschko, Peter: Stochastic Model Theory for Coherent Turbulent Structures in Circular Jets. *Phys. Fluids*, vol. 24, no. 2, Feb. 1981, pp. 187–193.
54. Plaschko, Peter: Axial Coherence Functions of Circular Turbulent Jets Based on an Inviscid Calculation of Damped Modes. *Phys. Fluids*, vol. 26, no. 9, Sept. 1983, pp. 2368–2372.
55. Tam, C. K. W.: Stochastic Model Theory of Broadband Shock associated Noise From Supersonic Jets. *J. Sound & Vib.*, vol. 116, no. 2, July 22, 1987, pp. 265–302.
56. Saric, William S.; and Nayfeh, Ali Hasan: Nonparallel Stability of Boundary-Layer Flows. *Phys. Fluids*, vol. 18, no. 8, Aug. 1975, pp. 945–950.
57. Crighton, D. G.; and Gaster, M.: Stability of Slowly Diverging Jet Flow. *J. Fluid Mech.*, vol. 77, pt. 2, Sept. 24, 1976, pp. 397–413.
58. Garg, V. K.; and Round, G. F.: Nonparallel Effects on the Stability of Jet Flows. *J. Appl. Mech.*, vol. 45, no. 4, Dec. 1978, pp. 717–722.
59. Plaschko, Peter: Helical Instabilities of Slowly Divergent Jets. *J. Fluid Mech.*, vol. 92, pt. 2, May 28, 1979, pp. 209–215.
60. Tam, Christopher K. W.; and Morris, Philip J.: The Radiation of Sound by the Instability Waves of a Compressible Plane Turbulent Shear Layer. *J. Fluid Mech.*, vol. 98, pt. 2, May 29, 1980, pp. 349–381.
61. Morris, Philip J.: Stability of a Two-Dimensional Jet. *AIAA J.*, vol. 19, no. 7, July 1981, pp. 857–862.
62. Tam, Christopher K. W.; and Burton, Dale E.: Sound Generated by Instability Waves of Supersonic Flows. Part 2. Axisymmetric Jets. *J. Fluid Mech.*, vol. 138, Jan. 1984, pp. 273–295.
63. Nayfeh, Ali Hasan: *Perturbation Methods.* John Wiley & Sons, Inc., c.1973.
64. Mollo-Christensen, Erik: Jet Noise and Shear Flow Instability Seen From an Experimenter's Viewpoint. *Trans. ASME, Ser. E: J. Appl. Mech.*, vol. 34, no. 1, Mar. 1967, pp. 1–7.
65. Sedel'nikov, T. Kh.: The Frequency Spectrum of the Noise of a Supersonic Jet. *Physics of Aerodynamic Noise,* A. V. Rimskiy-Korsakov, ed., NASA TT F-538, 1969, pp. 71–75.
66. Bishop, K. A.; Ffowcs Williams, J. E.; and Smith, W.: On the Noise Sources of the Unsuppressed High-Speed Jet. *J. Fluid Mech.*, vol. 50, pt. 1, Nov. 15, 1971, pp. 21–31.
67. Lighthill, M. J.: On Sound Generated Aerodynamically. I. General Theory. *Proc. Royal Soc. (London)*, ser. A, vol. 211, no. 1107, Mar. 20, 1952, pp. 564–587.
68. Lighthill, M. J.: On Sound Generated Aerodynamically. II. Turbulence as a Source of Sound. *Proc. Royal Soc. (London)*, ser. A, vol. 222, no. 1148, Feb. 23, 1954, pp. 1–32.

69. Ffowcs Williams, J. E.: The Noise From Turbulence Convected at High Speed. *Philos. Trans. Royal Soc. London*, ser. A, vol. 255, no. 1061, Apr. 18, 1963, pp. 469–503.
70. Ribner, H. S.: *Aerodynamic Sound From Fluid Dilatations—A Theory of the Sound From Jets and Other Flows.* UTIA Rep. No. 86 (AFOSR TN 3430), Inst. of Aerophysics, Univ. of Toronto, July 1962.
71. Dosanjh, Darshan S.; and Yu, James C.: Noise From Underexpanded Axisymmetric Jet Flows Using Radial Jet Flow Impingement. *Aerodynamic Noise*, Univ. of Toronto Press, c.1969, pp. 169–188.
72. Norum, Thomas D.; and Seiner, John M.: *Measurements of Mean Static Pressure and Far-Field Acoustics of Shock-Containing Supersonic Jets.* NASA TM-84521, 1982.
73. Tam, C. K. W.; Seiner, J. M.; and Yu, J. C.: Proposed Relationship Between Broadband Shock Associated Noise and Screech Tones. *J. Sound & Vib.*, vol. 110, no. 2, Oct. 22, 1986, pp. 309–321.
74. Yu, James C.: *Investigation of the Noise Fields of Supersonic Axisymmetric Jet Flow.* Ph.D. Thesis, Syracuse Univ., 1971.
75. Yu, James C.; and Dosanjh, Darshan S.: Noise Field of Supersonic Mach 1.5 Cold Model Jet. *J. Acoust. Soc. America*, vol. 51, no. 5, pt. 1, May 1972, pp. 1400–1410.
76. Tanna, H. K.: An Experimental Study of Jet Noise. *J. Sound & Vib.*, vol. 50, no. 3, Feb. 8, 1977. Part I: Turbulent Mixing Noise, pp. 405–428.
Part II: Shock Associated Noise, pp. 429–444.
77. Seiner, John M.; and Ponton, Michael K.: *Aeroacoustic Data for High Reynolds Number Supersonic Axisymmetric Jets.* NASA TM-86296, 1985.
78. Laufer, J.; Schlinker, R.; and Kaplan, R. E.: Experiments on Supersonic Jet Noise. *AIAA J.*, vol. 14, no. 4, Apr. 1976, pp. 489–497.
79. Harper-Bourne, M.; and Fisher, M. J.: The Noise From Shock Waves in Supersonic Jets. *Noise Mechanisms*, AGARD-CP-131, Mar. 1974, pp. 11-1–11-13.
80. Seiner, John M.; and Yu, James C.: Acoustic Near-Field Properties Associated With Broadband Shock Noise. *AIAA J.*, vol. 22, no. 9, Sept. 1984, pp. 1207–1215.
81. Tam, C. K. W.; and Tanna, H. K.: Shock Associated Noise of Supersonic Jets From Convergent-Divergent Nozzles. *J. Sound & Vib.*, vol. 81, no. 3, Apr. 8, 1982, pp. 337–358.
82. Westley, R.; and Woolley, J. H.: An Investigation of the Near Noise Fields of a Choked Axisymmetric Air Jet. *Aerodynamic Noise*, Univ. of Toronto Press, c.1969, pp. 147–167.
83. Yu, James C.; and Seiner, John M.: Nearfield Observations of Tones Generated From Supersonic Jet Flows. AIAA-83-0706, Apr. 1983.
84. Seiner, John M.; Manning, James C.; and Ponton, Michael K.: The Preferred Spatial Mode of Instability for a Mach 2 Jet. AIAA-86-1942, July 1986.
85. Seiner, John M.; Manning, James C.; and Ponton, Michael K.: Model and Full Scale Study of Twin Supersonic Plume Resonance. AIAA-87-0244, Jan. 1987.
86. Seiner, John M.: Advances in High Speed Jet Aeroacoustics. AIAA-84-2275, Oct. 1984.
87. Powell, A.: On the Mechanism of Choked Jet Noise. *Proc. Phys. Soc. (London)*, vol. 66, pt. 12, no. 408B, Dec. 1, 1953, pp. 1039–1056.
88. Powell, Alan: The Noise of Choked Jets. *J. Acoust. Soc. America*, vol. 25, no. 3, May 1953, pp. 385–389.
89. Powell, Alan: On the Noise Emanating From a Two-Dimensional Jet Above the Critical Pressure. *Aeronaut. Q.*, vol. 4, pt. 2, Feb. 1953, pp. 103–122.
90. Davies, M. G.; and Oldfield, D. E. S.: Tones From a Choked Axisymmetric Jet. II. The Self Excited Loop and Mode of Oscillation. *Acustica*, vol. 12, no. 4, 1962, pp. 267–277.
91. Sherman, P. M.; Glass, D. R.; and Duleep, K. G.: Jet Flow Field During Screech. *Appl. Sci. Res.*, vol. 32, no. 3, Aug. 1976, pp. 283–303.
92. Norum, T. D.: Screech Suppression in Supersonic Jets. *AIAA J.*, vol. 21, no. 2, Feb. 1983, pp. 235–240.
93. Jungowski, W. M.: Influence of Closely Located Solid Surfaces on the Sound Spectra Radiated by Gas Jets. *Mechanics of Sound Generation in Flows,* E.-A. Muller, ed., Springer-Verlag, 1979, pp. 116–122.
94. Seiner, John M.; and Norum, Thomas D.: Experiments on Shock Associated Noise of Supersonic Jets. AIAA Paper 79-1526, July 1979.
95. Westley, R.; and Woolley, J. H.: The Near Field Sound Pressures of a Choked Jet During a Screech Cycle. *Aircraft Engine Noise and Sonic Boom*, AGARD CP No. 42, May 1969, pp. 23-23-13.
96. Westley, R.; and Woolley, J. H.: Shock Cell Noise—Mechanisms, The Near Field Sound Pressures Associated With a Spinning Screech Mode. National Research Council paper presented at Conference on Current Developments in Sonic Fatigue (Univ. of Southampton, England), July 6–9, 1970.

97. Westley, R.; and Woolley, J. H.: The Near Field Sound Pressures of a Choked Jet When Oscillating in the Spinning Mode. AIAA Paper 75-479, Mar. 1975.

98. Rosfjord, Thomas J.; and Toms, Howard L.: Recent Observations Including Temperature Dependence of Axisymmetric Jet Screech. *AIAA J.*, vol. 13, no. 10, Oct. 1975, pp. 1384–1386.

99. Drevet, P.; Duponchel, J. P.; and Jacques, J. R.: Effect of Flight on the Noise From a Convergent Nozzle as Observed on the Bertin Aerotrain. AIAA Paper No. 76-557, July 1976.

100. Bryce, W. D.; and Pinker, R. A.: The Noise From Unheated Supersonic Jets in Simulated Flight. AIAA Paper 77-1327, Oct. 1977.

101. Sarohia, V.: Some Flight Simulation Experiments on Jet Noise From Supersonic Underexpanded Flows. *AIAA J.*, vol. 16, no. 7, July 1978, pp. 710–716.

102. Norum, Thomas D.; and Shearin, John G.: *Effects of Simulated Flight on the Structure and Noise of Underexpanded Jets.* NASA TP-2308, 1984.

103. Norum, Thomas D.; and Shearin, John C.: Shock Noise From Supersonic Jets in Simulated Flight to Mach 0.4. AIAA-86-1945, July 1986.

104. Lees, Lester; and Lin, Chia Chiao: *Investigation of the Stability of the Laminar Boundary Layer in a Compressible Fluid.* NACA TN 1115, 1946.

105. Lin, C. C.: *On the Stability of the Laminar Mixing Region Between Two Parallel Streams in a Gas.* NACA TN 2887, 1953.

106. Lees, Lester; and Reshotko, Eli: Stability of a Compressible Laminar Boundary Layer. *J. Fluid Mech.*, vol. 12, pt. 4, Apr. 1962, pp. 555–590.

107. Mack, Leslie M.: Linear Stability Theory and the Problem of Supersonic Boundary-Layer Transition. *AIAA J.*, vol. 13, no. 3, Mar. 1975, pp. 278–289.

108. Blumen, William: Shear Layer Instability of an Inviscid Compressible Fluid. *J. Fluid Mech.*, vol. 40, pt. 4, Mar. 9, 1970, pp. 769–781.

109. Blumen, William: Jet Flow Instability of an Inviscid Compressible Fluid. *J. Fluid Mech.*, vol. 46, pt. 4, Apr. 27, 1971, pp. 737–747.

110. Reshotko, Eli: Boundary-Layer Stability and Transition. *Annual Review of Fluid Mechanics, Volume 8*, Milton Van Dyke, Walter G. Vincenti, and J. V. Wehausen, eds., Annual Reviews, Inc., 1976, pp. 311–349.

111. Van Dyke, Milton: *Perturbation Methods in Fluid Mechanics.* Parabolic Press, 1975.

112. Whitham, G. B.: *Linear and Nonlinear Waves.* John Wiley & Sons, Inc., c.1974.

113. Cole, Julian D.: *Perturbation Methods in Applied Mathematics.* Blaisdell Publ. Co., c.1968.

114. Lighthill, M. J.: On the Energy Scattered From the Interaction of Turbulence With Sound or Shock Waves. *Proc. Cambridge Philos. Soc.*, vol. 49, pt. 3, Cambridge Univ. Press, July 1953, pp. 531–551.

115. Ribner, H. S.: *Shock-Turbulence Interaction and the Generation of Noise.* NACA Rep. 1233, 1955. (Supersedes NACA TN 3255.)

116. Howe, M. S.; and Ffowcs Williams, J. E.: On the Noise Generated by an Imperfectly Expanded Supersonic Jet. *Philos. Trans. Royal Soc. London*, vol. 289, ser. A, no. 1358, May 2, 1978, pp. 271–314.

117. Seiner, J. M.; and Norum, T. D.: Aerodynamic Aspects of Shock Containing Jet Plumes. *6th AIAA Aeroacoustics Conference,* June 1980, pp. 1–18. (Available as AIAA-80-0965.)

118. Pao, S. P.; and Seiner, J. M.: Shock-Associated Noise in Supersonic Jets. *AIAA J.*, vol. 21, no. 5, May 1983, pp. 687–693.

119. Norum, T. D.; and Seiner, J. M.: Broadband Shock Noise From Supersonic Jets. *AIAA J.*, vol. 20, no. 1, Jan. 1982, pp. 68–73.

120. Tam, C. K. W.; and Tanna, H. K.: Shock Associated Noise of Inverted-Profile Coannular Jets. *J. Sound & Vib.*, vol. 98, no. 1, 1985.
Part II: Condition for Minimum Noise, pp. 115–125.
Part III: Shock Structure and Noise Characteristics, pp. 127–145.

121. Tanna, H. K.; Brown, W. H.; and Tam, C. K. W.: Shock Associated Noise of Inverted-Profile Coannular Jets, Part I: Experiments. *J. Sound & Vib.*, vol. 98, no. 1, Jan. 8, 1985, pp. 95–113.

122. Prandtl, L.: Stationary Waves in a Gaseous Jet. *Physikalische Zeitschrift*, Jahrg. 5, Oct. 1, 1904, pp. 599–601.

123. Pack, D. C.: A Note on Prandtl's Formula for the Wave-Length of a Supersonic Gas Jet. *Q. J. Mech. & Appl. Math.*, vol. III, pt. 2, June 1950, pp. 173–181.

124. Adamson, T. C., Jr.; and Nicholls, J. A.: On the Structure of Jets From Highly Underexpanded Nozzles Into Still Air. *J. Aeronaut. Sci.*, vol. 26, no. 1, Jan. 1959, pp. 16–24.

125. Love, Eugene S.; Grigsby, Carl E.; Lee, Louise P.; and Woodling, Mildred J.: *Experimental and Theoretical Studies of Axisymmetric Free Jets.* NASA TR R-6, 1959. (Supersedes NACA RM 154L31 by Love and Grigsby; RM L55J14 by Love; RM L56G18 by Love, Woodling, and Lee; and TN 4195 by Love and Lee.)
126. Salas, Manuel D.: The Numerical Calculation of Inviscid Plume Flow Fields. AIAA Paper No. 74-523, June 1974.
127. Dash, S. M.; and Thorpe, R. D.: *A New Shock-Capturing/Shock-Fitting Computational Model for Analyzing Supersonic Inviscid Flows (The SCIPPY Code).* Rep. No. 366, Aeronautical Research Assoc. of Princeton, Inc., Nov. 1978.
128. Dash, S. M.; Wolf, D. E.; and Seiner, J. M.: Analysis of Turbulent Underexpanded Jets, Part I: Parabolized Navier-Stokes Model, SCIPVIS. *AIAA J.*, vol. 23, no. 4, Apr. 1985, pp. 505–514.
129. Seiner, J. M.; Dash, S. M.; and Wolf, D. E.: Analysis of Turbulent Underexpanded Jets, Part II: Shock Noise Features Using SCIPVIS. *AIAA J.*, vol. 23, no. 5, May 1985, pp. 669–677.
130. Tam, Christopher K. W.; Jackson, Jay A.; and Seiner, J. M.: A Multiple-Scales Model of the Shock-Cell Structure of Imperfectly Expanded Supersonic Jets. *J. Fluid Mech.*, vol. 153, Apr. 1985, pp. 123–149.
131. Fisher, M. J.; and Morfey, C. L.: Jet Noise. *Aerodynamic Noise*, AGARD-LS-80, Jan. 1977, pp. 3-1–3-23.
132. Krothapalli, A.; Hsia, Y.; Baganoff, D.; and Karamcheti, K.: The Role of Screech Tones in Mixing of an Underexpanded Rectangular Jet. *J. Sound & Vib.*, vol. 106, no. 1, Apr. 8, 1986, pp. 119–143.
133. Krothapalli, A.; Baganoff, D.; and Karamcheti, K.: On the Mixing of a Rectangular Jet. *J. Fluid Mech.*, vol. 107, June 1981, pp. 201–220.
134. Seiner, John M.; Ponton, Michael K.; and Manning, James C.: Acoustic Properties Associated With Rectangular Geometry Supersonic Nozzles. AIAA-86-1867, July 1986.
135. Tam, C. K. W.: The Shock-Cell Structures and Screech Tone Frequencies of Rectangular and Non-Axisymmetric Supersonic Jets. *J. Sound & Vib.* vol. 121, no. 1, Feb. 22, 1988, pp. 135–147.
136. Hammitt, Andrew G.: The Oscillation and Noise of an Overpressure Sonic Jet. *J. Aerosp. Sci.*, vol. 28, no. 9, Sept. 1961, pp. 673–680.
137. Morris, Philip J.; and Miller, David G.: Wavelike Structures in Elliptic Jets. AIAA-84-0399, Jan. 1984.
138. Morris, Philip J.: Instability of Elliptic Jets. AIAA-86-1868, July 1986.
139. Koshigoe, Shozo; and Tubis, Arnold: Wave Structures in Jets of Arbitrary Shape. I. Linear Inviscid Spatial Instability Analysis. *Phys. Fluids*, vol. 29, no. 12, Dec. 1986, pp. 3982–3992.
140. Koshigoe, S.; and Tubis, A.: Analysis of Compressible-Free-Shear-Layer Instabilities in Jets of Arbitrary Cross Section. AIAA-86-1919, July 1986.
141. Koshigoe, S.; Yang, V.; Culick, F. E. C.; and Tubis, A.: A New Method for Stability Analysis of a Free Jet With Arbitrary Cross Section. AIAA-86-0542, Jan. 1986.
142. Tubis, A.; and Koshigoe, S.: Instability of a Compressible Free Jet With an Arbitrary Cross Section. AIAA-86-1721, June 1986.
143. Crighton, D. G.: Instability of an Elliptic Jet. *J. Fluid Mech.*, vol. 59, pt. 4, Aug. 7, 1973, pp. 665–672.
144. Ponton, Michael K.; Manning, James C.; and Seiner, John M.: *Far-Field Acoustics of Supersonic Rectangular Nozzles With Various Throat Aspect Ratios.* NASA TM-89002, 1986.
145. Bechert, D.; and Pfizenmaier, E.: On the Amplification of Broad Band Jet Noise by a Pure Tone. *J. Sound & Vib.*, vol. 43, no. 3, Dec. 8, 1975, pp. 581–587.
146. Deneuville, P. M.; and Jacques, J. R.: Jet Noise Amplification: A Practically Important Problem. AIAA Paper 77-1368, Oct. 1977.
147. Schmidt, C.: Aerodynamic Characterization of Excited Jets. *J. Sound & Vib.*, vol. 61, no. 1, Nov. 8, 1978, pp. 148–152.
148. Jubelin, B.: New Experimental Studies on Jet Noise Amplification. AIAA-80-0961, June 1980.
149. Zaman, K. B. M. Q.: Far-Field Noise of a Subsonic Jet Under Controlled Excitation. *J. Fluid Mech.*, vol. 152, Mar. 1985, pp. 83–111.
150. Tanna, H. K.; and Ahuja, K. K.: Tone Excited Jets, Part I: Introduction. *J. Sound & Vib.*, vol. 102, no. 1, Sept. 8, 1985, pp. 57–61.
151. Ahuja, K. K.; and Blakney, D. F.: Tone Excited Jets, Part IV: Acoustic Measurements. *J. Sound & Vib.*, vol. 102, no. 1, Sept. 8, 1986, pp. 93–117.
152. Moore, C. J.; and Brierley, D. H.: Shear Layer Instability Noise Produced by Various Jet Nozzle Configurations. *Mechanics of Sound Generation in Flows*, E.-A. Müller, ed., Springer-Verlag, 1979, pp. 48–54.
153. Lu, H. Y.: Effect of Excitation on Coaxial Jet Noise. *AIAA J.*, vol. 21, no. 2, Feb. 1983, pp. 214–220.

154. Berman, C. H.: Turbulence and Noise Characteristics of Acoustically Excited Bypass Jet Flows. AIAA-81-2009, Oct. 1981.
155. Moore, C. J.: The Effect of Shear Layer Instability on Jet Exhaust Noise. *Structure and Mechanisms of Turbulence II*, H. Fiedler, ed., Springer-Verlag, 1978, pp. 254–264.
156. Tam, C. K. W.: Excitation of Instability Waves by Sound—A Physical Interpretation. *J. Sound & Vib.*, vol. 105, no. 1, Feb. 22, 1986, pp. 169–172.
157. Tam, Christopher K. W.: Excitation of Instability Waves in a Two-Dimensional Shear Layer by Sound. *J. Fluid Mech.*, vol. 89, pt. 2, Nov. 28, 1978, pp. 357–371.
158. Tam, Christopher K. W.: The Effects of Upstream Tones on the Large Scale Instability Waves and Noise of Jets. *Mechanics of Sound Generation in Flows*, E.-A. Müller, ed., Springer-Verlag, 1979, pp. 41–47.
159. Ko, D. Ru-Sue; Kubota, T.; and Lees, L.: Finite Disturbance Effect on the Stability of a Laminar Incompressible Wake Behind a Flat Plate. *J. Fluid Mech.*, vol. 40, pt. 2, Feb. 3, 1970, pp. 315–341.
160. Chan, Y. Y.: *Nonlinear Spatial Wave Development in an Axisymmetrical Turbulent Jet.* NRC No. 14756, Aeronaut. Rep. LR-585, National Aeronautical Estab., National Research Council of Canada, Apr. 1975.
161. Lau, J. C.: Mach Number and Temperature Effects on Jets. AIAA Paper 78-1152, July 1978.
162. Ffowcs Williams, J. E.; and Kempton, A. J.: The Noise From the Large-Scale Structure of a Jet. *J. Fluid Mech.*, vol. 84, pt. 4, Feb. 27, 1978, pp. 673–694.
163. Michalke, A.; and Fuchs, H. V.: On Turbulence and Noise of an Axisymmetric Shear Flow. *J. Fluid Mech.*, vol. 70, pt. 1, July 15, 1975, pp. 179–205.
164. Armstrong, Ralph R.; Michalke, Alfons; and Fuchs, Helmut V.: Coherent Structures in Jet Turbulence and Noise. *AIAA J.*, vol. 15, no. 7, July 1977, pp. 1011–1017.
165. Armstrong, R. R., Jr.: Influence of Mach Number on Coherent Structure Relevant to Jet Noise. *AIAA J.*, vol. 19, no. 6, June 1981, pp. 677–683.
166. Laufer, J.; Kaplan, R. E.; and Chu, W. T.: On the Generation of Jet Noise. *Noise Mechanisms*, AGARD-CP-131, Mar. 1974, pp. 21-1–21-6.
167. Kibens, Valdis: Discrete Noise Spectrum Generated by an Acoustically Excited Jet. *AIAA J.*, vol. 18, no. 4, Apr. 1980, pp. 434–441.
168. Hussain, A. K. M. F.: Coherent Structures—Reality and Myth. *Phys. Fluids*, vol. 26, no. 10, Oct. 1983, pp. 2816–2850.
169. Bridges, James E.; and Hussain, A. K. M. F.: Measurements Concerning Vortex Pairing as a Noise Source in Turbulent Jets. *Bull. American Phys. Soc.*, vol. 31, no. 10, Nov. 1986, p. 1684.

7 Airframe Noise

Lead author

David G. Crighton
University of Cambridge
Cambridge, England

Introduction

Airframe noise, that is, nonpropulsive noise of an aircraft in flight, became a topic of intense research interest in the 1970's following initial studies (refs. 1 and 2) in the late 1960's on the development of an ultraquiet military surveillance aircraft. Extrapolation of the noise levels measured for gliders and light aircraft to the scales relevant to heavy commercial transport aircraft indicated that then-current transports might have airframe noise levels at landing approach which were less than 10 dB below the FAR Part 36 (ref. 3) certification effective perceived noise level (EPNL). Expectations at that time were that the FAR Part 36 EPNL would be reduced by 10 dB per decade. These expectations implied a noise certification problem for new aircraft after about 1985 which could not be solved by power-plant noise control alone. Increased airframe noise was also to be expected with some of the large twin-jet transports introduced in the 1970's and 1980's, which featured much more significant interaction (even direct impingement) between the jet exhaust stream and the flaps. And further, all studies (ref. 4) in the early 1970's of short takeoff and landing (STOL) aircraft using underwing jets and blown flaps or upper-surface blowing for enhanced lift at low speeds had clearly shown that near- and far-field noise levels from such configurations would be high and would be a strong impediment to the development of all but small, low-speed STOL transports for either military or civil applications.

The extrapolated small-aircraft levels were essentially confirmed soon after in several series of full-scale tests on a variety of aircraft, both jet and propeller driven. Attempts to understand and correlate the experimental information and to predict (and ultimately control) airframe noise proceeded, naturally, along two paths. In the first, a "whole aircraft" approach was taken, in which a correlation of 1/3-octave sound pressure level (SPL) with aircraft speed, weight, wing aspect ratio, etc., was attempted for large classes of aircraft. This correlation was based on full-scale and model data with some rudimentary theoretical underpinning. Such an approach had its value, not least in providing a first estimate (based on real data for a very complex interactive noise-generating process) for whether a serious airframe noise problem was likely to exist for some proposed design or not. However, the whole aircraft approach clearly needed to be accompanied by a second "component source" approach, together with assessment of interactions between sources and between source and propagation mechanisms wherever possible. Similar approaches had already been employed in propulsive jet noise research and the limitations of each seen there (and the need to consider early the interactions in the component source approach).

Many papers on these two attacks on the airframe noise problem were presented at the AIAA Second Aeroacoustics Conference in 1975, and the third volume of the proceedings (ref. 4) of that conference is still required reading. The next 6 or 7 years saw much further work, though the publicly available data base has not been expanded to include test results for the new large twin-jet transports. Cutbacks in funding for jet noise research in the 1980's seem to have stopped airframe noise research, though there are clearly areas in commercial aircraft development where airframe noise must be expected to be important. One is in the development of still larger versions of the Boeing 747 (the airframe noise of the standard 747-100 and 747-200 differs substantially from that of the 747SP, the latter having a much simpler single-flap system instead of the triple-flap system fitted to the standard 747); a second is in the development of aircraft powered by propfans, or very large (ultrahigh bypass ratio) shrouded fans; a third is in the development of an advanced supersonic transport; and a fourth is in the development of powered-lift-assisted STOL transports.

This chapter continues now with a section giving an overview of the general results obtained in full-scale aircraft experiments, followed by a section on whole aircraft noise correlations. Then attention is paid to specific noise-generating mechanisms (see fig. 1), including flap and wing trailing edges, flap side edges, undercarriage gear sources, cavity mechanisms, and sources associated with the fuselage and wing turbulent boundary layers. Problems associated with high-lift devices and configurations are then discussed. The chapter concludes with proposals for comprehensive airframe noise prediction schemes and a discussion of possible means for alleviating airframe noise.

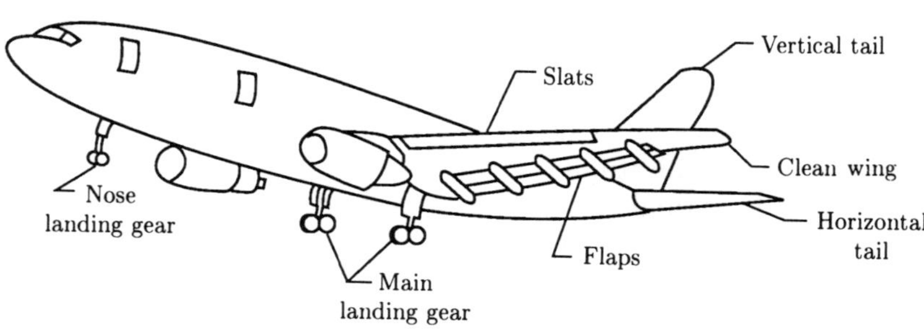

Figure 1. Sources of airframe noise. (From ref. 27.)

Overview of Experimental Results

Although the first experimental studies of airframe noise were on gliders and low-powered small reconnaissance aircraft, the need for data from large transport aircraft was very quickly made apparent in the 1971–72 NASA-sponsored Advanced Technology Transport Study Program. Design features were studied which would bring jet-engine propulsion EPNL to 20 dB below then-current FAR Part 36 regulations (ref. 3). However, correlation formulas derived from small-aircraft data indicated that airframe EPNL itself might lie 16 dB below the FAR Part 36 level at takeoff flyover, 22 dB below at sideline, but only 11 dB below at landing approach. These levels set an immediate limit on the possible community noise reduction that could be achieved by power-plant noise reduction alone. Gibson (ref. 5) very shortly after published the first airframe noise data for a (very large) transport, the C-5A Galaxy, and Healy (ref. 6) reported airframe noise levels for light aircraft and light twin-propeller transports.

Airframe noise of the C-5A is detectable over other noises only at rather low frequencies, and the C-5A has several design features (such as 12-segment flaps and 4 main landing gear assemblies each with 6 wheels) which are not typical of jet transports and which may result in untypical estimates of both intensity and spectral shape. The measured data were shown on narrow-band analysis to actually comprise a great number of essentially discrete frequencies, with a typical width of 5 Hz and each typically 5 to 10 dB above the background. A fairly prominent peak around 100 Hz was identified as noise from the "clean" wing, numerous peaks between 20 and 100 Hz were ascribed to landing gear components, and peaks below 20 Hz were ascribed to the gear cavities. The gear components were found to increase perceived noise level (PNL) by 9 to 10 dB, and 100-percent flap deflection increased PNL by a further 3 to 4 dB, for a total "clean to dirty" PNL increase of 12 to 14 dB. A similar, though slightly lower, PNL increase of 10 to 12 dB was measured at about the same time for the Boeing 747 (ref. 7). Gibson (ref. 5) obtained reasonable "prediction" of the C-5A airframe noise using a whole aircraft correlation based on glider and small powered aircraft data. The correlation, assuming a $(\text{Velocity})^4$ scaling, has even less theoretical backing than those discussed in the next section and must be considered as simply a rough first estimate of likely airframe noise levels; the idea that the C-5A (with wing span of 68 m) has a far field that can be modeled by a compact monopole is, even at frequencies of 20 Hz, not at all useful for the understanding or control of the noise.

A number of sets of airframe noise data for small piston-engine-powered transports, and also for the F-106B delta wing fighter, were published around the time of Gibson's work. Hardin et al. (ref. 8) give a compilation of data for the overall sound pressure level (OASPL) measured in these tests. More representative are the later studies by Putnam et al. (ref. 9) (Lockheed Jetstar, Convair 990, and Boeing 747) and Fethney (ref. 10) (Hawker Siddeley 125, British Aircraft Corp. 111, and British Aircraft Corp. VC10). Data taken directly below these six aircraft in the clean configuration show a $(\text{Velocity})^5$ variation of OASPL (see ref. 11, fig. 2). Fethney's work on the VC10 with various combinations of flap deflection, leading-edge slat deployment, gear deployment, and gear doors open or closed indicates an increase in OASPL of 11 dB (and in PNL of perhaps 13 dB) in the fully dirty condition, with 9 dB ascribed to either full flap or gear deployment individually, and 4 dB ascribed to having main gear doors open rather than shut (with gear deployed in both cases). Reported at about the same time were data (ref. 12) taken on a DC-10 in the clean configuration; these data showed that the flyover directivity could be fitted by the field of two correlated (lift and drag) dipoles. However, the $\sin^2(\theta/2)$ directivity of trailing-edge noise (see the subsequent section) fits the DC-10 data given in figure 10 of reference 11 just as well, where θ is the angle measured from downstream. This directivity would be associated also with the (Velocity)^5 OASPL variation that was found by Hardin for six jet-powered transports. Thus we have at least a hint that airframe noise of a clean configuration is dominated by wing and tail trailing-edge noise, with increases of 10 dB or more in the dirty configuration, associated with flaps, slats, gear, cavities, etc.

Since these early full-scale tests, most studies have examined particular components in isolation or with inclusion of interaction between two or more noise-generating components. Further full-scale data have been collected for current aircraft, in particular for the Boeing 727 (ref. 13), and both full- and model-scale data have been collected for the 747 (refs. 14 and 15). Reference 15 shows that if low-frequency peaks, whose frequencies do not change with flow velocity and probably represent cavity tones, are excluded, then the model- and full-scale airframe noise data can be scaled according to the following relations: for 1/3-octave band SPL,

$$\mathrm{SPL}_F = \mathrm{SPL}_M + 10\log\left[\lambda^{-2}\left(\frac{U_F}{U_M}\right)^5\left(\frac{r_M}{r_F}\right)^2\right] \tag{1}$$

and for band center frequencies,

$$f_F = \lambda f_M\left(\frac{U_F}{U_M}\right) \tag{2}$$

where λ is the scale factor, U_F and U_M are the flight velocities at full and model scale, and r_F and r_M are the observer distances at full and model scale at the same radiation angle. The U^5 scaling connotes the dominance of trailing-edge noise and the λ^{-2} scaling factor represents the influence of the length factors (see eq. (8)). The possibility of model to full-scale scaling simply by equations (1) and (2) is, however, unlikely to be general except in clean configurations dominated by trailing-edge noise. Its success in the dirty configuration (to within 3 dB for the 747) may simply reflect the dominance of trailing-edge noise for this particular aircraft with its long segments of triple-slotted flap; it is known, for example, that the airframe noise of the 747SP differs from that of the 747-100, the former having much simpler single-slotted flap segments.

All these full-scale studies are described in some detail in earlier reviews (refs. 8 and 11). (Later, full-scale flyover data for the McDonnell Douglas DC-9-31, in a number of configurations involving gear, flaps, and slats, also became available in refs. 16 and 17.) To simply repeat the conclusions of each test as summarized there would be pointless, nor has it turned out to be possible to interpret them all from a unified point of view. The situation is one in which there are numerous sources of tonal and broadband sound, with strong interaction between them (e.g., deployment of flaps may reduce noise from flow over the undercarriage gear, as was noted in ref. 18) and with very low acoustic power compared with jet noise at takeoff thrust. The only possible approach seems, therefore, to be based on an understanding of the separate mechanisms and their interaction. This subject is taken up in later sections. Theoretical developments are described whenever a theory exists and is likely to be relevant to the airframe noise problem. Such is the case for trailing-edge noise and for undercarriage gear noise; the basic mechanisms are understood, and in any particular configuration it should be possible to produce scaling laws and interpret experimental data. For cavity tones, however, although there is a vast body of literature on theoretical models and model-scale experiments, almost nothing seems likely to be relevant to the highly irregular flows and geometries of real aircraft, and emphasis on anything more than one or two simple ideas of "depth mode" cavity response is likely to be misleading. The results of theoretical and model-scale study should then be combined into a prediction scheme, and those predictions then checked against whole aircraft correlations and against data for similar configurations.

We end this section by noting a number of points relating to experimental study of airframe noise. Hardin (ref. 11) discusses the placement of microphones flush in a hard surface, the corrections needed to relate a glide slope path to a constant altitude flyover, the determination of aircraft position at acoustic emission time, the statistical problems arising from the aircraft flyover, and the contamination of airframe noise by residual jet noise. Full-scale difficulties are also encountered if the microphones are not in the far field; these difficulties were already seen in the C-5A flyovers of reference 5. Some of these difficulties are more severe in the use (ref. 19) of remotely piloted vehicles (RPV's) in place of real aircraft. Static model testing for airframe noise must almost certainly be conducted in an anechoic open-jet wind tunnel, with low tunnel noise and with shear layer corrections for transmission of sound from the model to the microphone outside the wind tunnel flow. These corrections are now well established (ref. 20), but are crucial even at low Mach numbers. They have been an essential ingredient of several model-scale studies of trailing-edge and gear noise.

An issue has emerged that some localized regions of a wing may be responsible for a large fraction of total airframe noise. To study these localized regions (flap side edges), source location techniques are needed and have been much developed for jet noise research in the late 1970's. Far-field correlations, far- to near-field correlations, and directional microphones have been used (refs. 21 to 23) for airframe noise work. Far- to near-field correlations require some idea of the underlying mechanisms, whereas all directional microphones, including far-field correlation arrays, merely infer an equivalent source location. These techniques can be used to best advantage at model scale, but a major problem then is to retain the correct balance between mechanisms, some of which may be highly sensitive to Reynolds number.

Whole Aircraft Correlations

In the early days of airframe noise research, a number of correlations were devised for the prediction of airframe noise from measured full-scale aircraft data, given the major aerodynamic parameters of the aircraft. These are given here essentially as a summary of those measured data and as a means of getting a first estimate of the order of magnitude for the airframe noise of an aircraft in the same class as those providing the data. The fact that the formulas often contain the velocity and directional dependence of a single compact dipole (with vertical axis) should not be taken to imply that the dominant source is a vertical dipole associated with unsteady flow over the wheels or any other particular source. The formulas are simply convenient approximate fits to data in which a number of mechanisms are probably comparably important.

Revell, Healy, and Gibson (refs. 24 and 25) analyzed data taken for the most part on older small transports (Douglas DC-3, Convair 240) but also with input from C-5A Galaxy data. They obtained

$$\text{OASPL} = 10\log\left[\left(\frac{\sin\Theta}{r}\right)^2 \frac{U^6 S}{(\text{AR})^4}\right] + 28.0\text{ dB} \tag{3}$$

where Θ is the observer angle from downstream and r the observer distance, both presumably at emission time, while U is the flight speed, S the wing area, and AR the aspect ratio; AR $= b/c$ with b the span and c the mean chord. Along with equation (3) goes a universal spectrum (fig. 5 of ref. 25) with a peak frequency $1.3U/t$ with t the mean wing thickness at the mean chord location. These relations refer to the clean (cruise) condition for which trailing-edge noise probably dominates. The prediction of OASPL which they give is in reasonable accord with estimates quoted in the next section for trailing-edge noise of clean configurations, although the functional dependence of equation (3) is quite wrong for trailing-edge noise (as discussed subsequently). For "dirty" configurations, the levels predicted by equation (3) should be raised by between 9 and 12 dB.

The background to equation (3) is described by Healy (ref. 25), who shows further that equation (3) can be applied to predict the airframe noise of a range of small to large jet transports only if the proportionality constant (28 dB) is allowed to vary significantly from one aircraft type to another—although the speed and aspect ratio exponents appear to be useful over some range of types. Equation (3) as it stands overpredicted the noise (engine noise removed) of the F-106B fighter (delta wing) by 36 dB and of a Jetstar small jet transport by 20 dB. No method—other than having a number of different proportionality constants for different types—was found to extend the whole aircraft method to cover aircraft of widely differing wing planform.

A "total aircraft correlation" similar in spirit was given by Hardin et al. (ref. 8), who regarded the airframe noise as generated by a vertical unsteady lift dipole and determined the dependence of the dipole parameters on the aircraft parameters (weight, span, aspect ratio) and on speed U through a regression analysis of data mainly for light propeller-driven transports. This scheme shares the problems of equation (3), namely reference to an inappropriate (dipole) model for the basic mechanism and a strong dependence on aspect ratio that is seen later to be erroneous. Nonetheless, these early prediction methods laid the basis for the belief that the approach airframe noise of large transport aircraft current in the early 1970's lay no lower than FAR Part 36 level minus 10 dB.

A first attempt to break the airframe noise field into constituents representing acoustic mechanisms associated with wing and flap trailing and side edges, undercarriage gear elements, wheel wells, etc., was taken in the "drag element method." The method was described in reference 24 and applied specifically to the C-5A Galaxy in reference 26 to estimate the noise spectra from the profile drag of the wing, fuselage, engine nacelles, leading-edge slats, and horizontal tail and from the induced drag of the wing/flap vortex system. From a large number of similarity scaling hypotheses, the method derives expressions for the dipole sound field associated with the various aerodynamic elements, using flight data for different configurations of the same aircraft to determine the numerous proportionality constants. The main flaw in the method is that a variety of mechanisms are forced into the same dipole straitjacket, which is a reasonable representation of only some sources and *not* a reasonable representation of sources associated with large sharp-edged surfaces. In the next sections, we therefore describe in detail the features associated with particular aerodynamic and geometrical configurations and emphasize aspects which can be clearly understood from accepted aeroacoustic theory and which are demonstrated in careful model or full-scale experiments. Mathematical developments are not given.

Hardin et al. (ref. 8) first attempted to develop such a "component analysis." That work, valuable as it was at the time, has been overtaken by improved understanding of a number of specific mechanisms (trailing-edge and gear noise) and the availability of much more model and full-scale data. A component scheme which makes use of both the additional understanding and the data is that of Fink (refs. 27 and 28), described in a subsequent section.

Trailing-Edge Noise

Theoretical work on the sound from interaction of unsteady flow with the edge of a large flat plate predates experimental proof of the relevance of it to airframe noise. The essential result, due to Ffowcs Williams and Hall (ref. 29), is that if the flat plate can be taken as semi-infinite, of negligible thickness, and lying in the planes $\theta = \pm 180°$, then the far-field intensity varies with θ as $\sin^2(\theta/2)$ and scales with a typical flow velocity U_o as U_o^5. These basic dependences are independent of the nature of the unsteady flow near the edge. The theory underlying them has been applied to the passage of a vortex ring past the edge, and the theoretical predictions have been confirmed in great detail in the papers of Kambe and his colleagues (see ref. 30, in which theory and experiment agree, with no adjustment of the predictions, to within fractions of a decibel over the full angular range).

Theoretical Half-Plane Trailing-Edge Problem

The angular variation and the velocity scaling suggest that the half-plane scattering mechanism has a non-multipole form, and this is brought out in detail in reference 30 where the 3/2-pole character is made explicit. Earlier, Powell (ref. 31) had considered the problem and obtained the U_o^5 scaling by arguing that the turbulence correlation scale should have its usual eddy size ℓ for directions normal to the edge, but a value of ℓ/M (i.e., an acoustic wavelength) for directions along the edge, where M is a flow Mach number. This incorrect argument (which could not give the results observed by Kambe for a strictly deterministic process) is another example of the many misconceptions that arose in the 1950's and 1960's about the nature of aeroacoustic sources associated with different geometrical scattering boundaries. A simple way of understanding the half-plane velocity scaling and directivity was given by Crighton and Leppington (ref. 32), who show also how the corresponding results can be quickly obtained for a wedge of arbitrary angle. The derivations in reference 32 make it clear that the non-multipole character of the field is an immediate consequence of the fact that the scattering sharp-edged surface is *noncompact* relative to the acoustic wavelength. For further discussion see references 33 to 35 for review articles.

Howe's Approach

A fairly satisfactory theoretical understanding of the trailing-edge problem has been achieved with Howe's review (ref. 36) in which numerous partially overlapping, partially conflicting theoretical approaches are reconciled and generalized to include a number of effects, including those associated with motion of the aeroacoustic sources relative to the edge and of the edge relative to the far-field observer and those associated with satisfaction, or partial satisfaction, of a Kutta condition at the edge. The essential steps in a theoretical calculation ignoring any leading-edge effects are as follows. First write down an inhomogeneous wave equation, from Lighthill's aeroacoustic theory (ref. 37), in a frame in which the plate is at rest; allow for mean flow past the plate and for the convection effects associated with the "primary" turbulence field and with any vorticity shed from the edge. Solve the wave equation by convolving the "sources" with an appropriate Green's function (essentially, after transformation, the static fluid Green's function with zero normal velocity on the plate), take the far-field limit, and transform coordinates to emission time coordinates appropriate to the flyover noise problem with static fluid at the observer location and the plate in motion. Finally, attempt to express "source" quantities in terms of measurable quantities, such as the pressure spectrum on the plate near the edge (and also determine, by imposing a Kutta condition if that is thought appropriate, the strength of any vorticity shed into the wake, and express the wake vorticity sources in terms of measurable quantities). In the first analysis of this kind (ref. 29), Ffowcs Williams and Hall ignored all convection effects and vortex shedding and obtained, by convolving a dimensionally correct model of the Lighthill quadrupole source with the static fluid half-plane Green's function, the result

$$\langle p^2 \rangle \approx \rho_o^2 u_o^2 V^2 M_v \left(\frac{L\delta}{R^2}\right) \sin\alpha \sin^2(\theta/2) \cos^3\beta \tag{4}$$

for the mean-square far-field pressure generated by all eddies within a correlation scale δ of the edge and covering a length L of the edge; $\mathbf{V}$ is a typical mean velocity (ignored except in defining the quadrupole strengths), ρ_o is ambient density, u_o is root-mean-square (rms) turbulence velocity, M_v is turbulence convection Mach number, and the angles are as depicted in figure 2, being measured, together with the distance R, at *reception* time. (See ref. 36, eq. (7), for the explanation for the predicted $\cos^3\beta$ dependence on trailing-edge sweep angle β.) Howe (ref. 36) shows how the dominant features of equation (4), that is, the (Velocity)^5 scaling and $\sin^2(\theta/2)$ directivity, are unchanged except in numerical magnitude by the imposition of a Kutta condition. They are changed only by certain Doppler factors, discussed below, by convection effects. Equation (4) underlies all recent attempts to correlate experimental data on trailing-edge noise. The aim of a more refined theory is only (1) to expose explicitly convection effects, which can be large even at landing approach conditions, (2) to estimate the uncertainty in predicted far-field level which goes with uncertainty over the edge condition, and (3) to suggest near-field quantities that might be used to get the best collapse of data for correlation and prediction purposes.

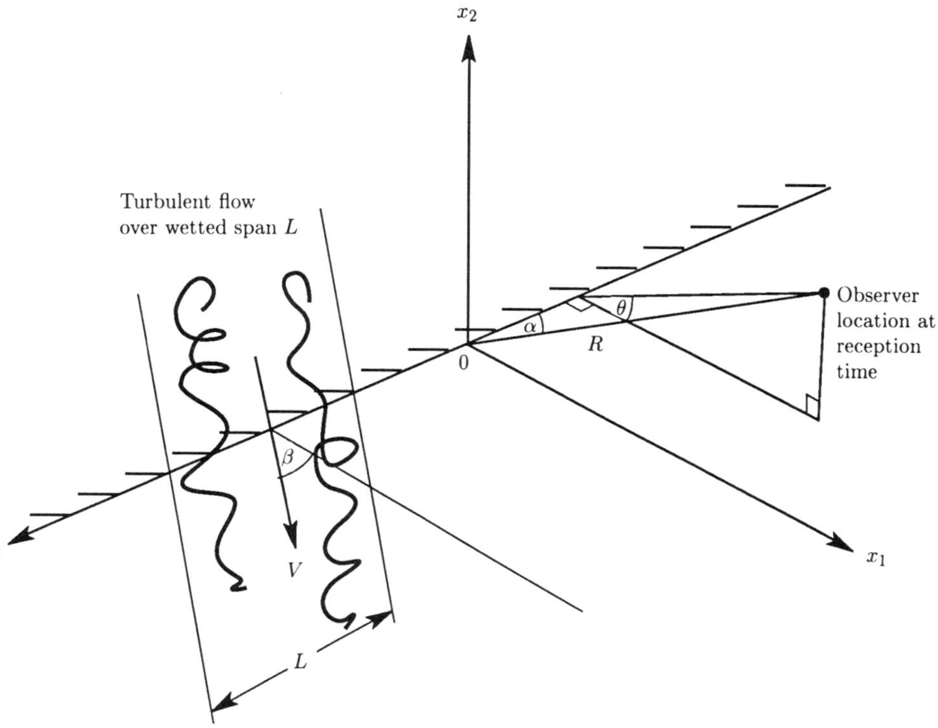

Figure 2. Turbulent flow over a wetted span L of a trailing edge. The observer has coordinates (R, θ, α) at reception time.

Howe (ref. 36) has given such a unified theory. In the absence of any vorticity shedding into the wake and for an observer fixed relative to the plate, he shows that

$$\langle p_I^2 \rangle = c_\chi \frac{\rho_o^2 u_o^2 V^2 M_v}{2\pi} \left(\frac{L\ell_3}{R^2} \right) \frac{\sin\alpha \sin^2(\theta/2)\cos^3\beta}{(1+M_{oR})^2(1-M_{vR})^2(1-M_{v1}\sin\alpha)} \tag{5}$$

in which c_χ is a numerical dimensionless longitudinal integral scale of a vorticity distribution associated with the incident turbulence, $\mathbf{V}$ is now the convection velocity, and ℓ_3 is the vorticity correlation scale parallel to the edge. He notes that if $\beta << 1$, then the principal turbulent source contribution to this estimate is what he refers to as the "principal edge noise dipole," the component normal to the half-plane of the acoustic dipole $\mathbf{Q} = \boldsymbol{\Omega} \times \mathbf{V}$, where $\boldsymbol{\Omega}$ is the flow vorticity vector. This component involves the incident vorticity parallel to the edge and the vorticity convection velocity normal to the edge. However, the scattered field induced by the interaction of this forcing dipole with the edge of the half-plane is *not dipole*, as equation (5) makes clear.

The three Doppler factors in equation (5) involve Mach numbers defined as follows: M_{oR} is the component in the observer direction of the mean flow Mach number (relative to the plate); M_{vR} denotes the component in the observer direction of the turbulence convection Mach number; and M_{v1} is the component of the turbulence convection Mach number normal to the edge in the downstream direction. Equation (5) reduces immediately to the basic Ffowcs Williams and Hall result (eq. (4)) when all convection effects are ignored. Howe's theory also supplies the power spectral density of $\langle p_I^2 \rangle$ in the form

$$S_I(\omega) \approx \langle p_I^2 \rangle \frac{1}{c_\chi} \left(\frac{\ell_1}{V_1} \right) \chi \left(\frac{\omega \ell_1}{V_1}, 0 \right) \tag{6}$$

where χ is a vorticity spectral density in wave number (wave numbers normal and parallel to the plate edge) integrated over the boundary layer thickness, ω is the circular frequency, ℓ_1 is the vorticity correlation scale normal to the edge, and V_1 is the convection velocity component normal to the edge. Equation (6) produces equation (5) with the identification

$$c_\chi = \int_0^\infty \chi(s, 0)\, ds \tag{7}$$

For flyover noise prediction we express equation (5) in terms of *emission* coordinates (r, Θ, Φ) with $\Theta = 0°$ on the downstream continuation of the plate. Then in the flyover plane $\Phi = 0°$ (and with consistent neglect of all M^2 terms throughout), we have

$$\langle p_I^2 \rangle \approx c_\chi \frac{\rho_o^2 u_o^2 V^2 M_v}{2\pi} \left(\frac{L \ell_3}{r^2} \right) \frac{(1 - M_o + M_{v1}) \sin^2(\Theta/2) \cos^3 \beta}{(1 + M_o \cos \Theta)^3 \left[1 + (M_o - M_{v1}) \cos \Theta \right]^2} \tag{8}$$

Now, if vortex shedding is allowed and its strength determined by a Kutta condition (of finiteness of all fluctuating velocity components at the plate edge), the corresponding result is

$$\langle p_K^2 \rangle = \langle p_I^2 \rangle (1 - \sigma)^2 \frac{(1 + M_o \cos \Theta)^2}{[1 + (M_o - M_{w1}) \cos \Theta]^2} \tag{9}$$

in which the wake vorticity has convection velocity W_1 normal to the edge, M_{w1} is the corresponding Mach number, and $\sigma = W/V$ is the ratio of convection velocity magnitudes for the wake and incident vorticity. Thus at $\Theta = 90°$, where convection effects are absent, imposition of the Kutta condition reduces the mean-square pressure by $(1 - \sigma)^2$. The interaction between turbulence and the edge produces *no sound if the convection velocities are equal.* Howe estimates that $M_{v1} \approx 0.7 M_o$ for sources in the upper part of a turbulent boundary layer and $M_{w1} \approx 0.5 M_o$ for the wake, and he shows that then for $M_o = 0.3$, equation (9) represents a decrease in SPL of 9.8 dB at $\Theta = 0°$, 10.9 dB at $\Theta = 90°$, 12.6 dB at $\Theta = 180°$.

The main difference between $\langle p_I^2 \rangle$ and $\langle p_K^2 \rangle$ is one of magnitude, typically 11 dB at $M_o = 0.3$, and this should be considered in correlation schemes. The status of the Kutta condition for unsteady flow is now fairly well understood (ref. 38) for laminar flow near the edge of a flat plate in the unseparated regime. The Kutta condition applies provided, essentially, that the Strouhal number is not greater than $\mathrm{Re}^{1/4}$, (Re is Reynolds number) and that the amplitude of the forcing is appropriately small (see ref. 38 for details). However there is no corresponding theory, or even a widely accepted set of data, for turbulent flow near the edge of a wing, even in the clean configuration, let alone with multiple flaps at very high angle of attack (though attention must be called to ref. 39 on a theoretical model for turbulent flow near a flat plate trailing edge). We might hope to overcome this ignorance as to the extent to which the Kutta condition is satisfied by reworking the theory in terms of a fluctuating pressure field on the plate, near the edge, as the specified source function. Howe (ref. 36) obtains the appropriate expressions, with and without the Kutta condition, but shows that precisely the same indeterminacy arises. Faced with this—and the likelihood that the edge condition may change from one configuration to another, so that no universal results could be obtained for the magnitude of $\langle p^2 \rangle$—the only course of action is to use the functional forms of equation (8) and to determine c_χ empirically for a configuration close to the actual one. The no-Kutta-condition case provides an upper bound. Model-scale experiments may not be appropriate, since the nature of the edge flow is highly sensitive to Reynolds number.

The possibility of expressing the far field in terms of surface pressure measurements near the edge is an attractive one, but can be achieved only if the Kutta condition applies (otherwise the pressure is infinite at the edge) and provided that the eddy convection velocity $\mathbf{V}$ is constant through the boundary layer (or provided that the dominant incident pressure sources are located in a region of effectively constant $\mathbf{V}$). Then Howe (ref. 36), extending the formulation of Chase (ref. 40), obtains, for the far-field spectrum $S_K(\omega)$ with Kutta condition imposed, the relation

$$S_K(\omega) = \frac{2M_v L \sin\alpha \sin^2(\theta/2) \cos\beta}{\pi R^2 (1+M_{oR})^2 (1-M_{vR})^2 (1-M_{wR})^2 (1-M_{v1}\sin\alpha)} \times \int_{-\infty}^{+\infty} \Pi_K(\mu_1, \mu_3, \omega)\, d\mu_1 \tag{10}$$

where $\mu_3 = \omega \cos\alpha / c_o$ with c_o the ambient speed of sound. In equation (10) Π_K is the wave-number–frequency spectral density of the mean-square pressure on the plate $\langle p_{K0}^2 \rangle$, so that

$$\langle p_{K0}^2 \rangle = \int_{-\infty}^{+\infty} \Pi_K(\mu_1, \mu_3, \omega)\, d\mu_1\, d\mu_3\, d\omega$$

The integral in equation (10) over μ_1 gives a quantity $\Pi_K^*(\mu_3, \omega)$ which can be determined from correlation measurements parallel to the edge, and in time, at a fixed location close to and just upstream of the edge.

Equation (10) can be simplified further if we make the approximation of reference 41 that $\omega \cos\alpha / c_o << \ell_3^{-1}$, for then

$$\Pi_K^*(\mu_3, \omega) \approx \Pi_K^*(0, \omega) = \ell_3 \phi(\omega\delta/V) \tag{11}$$

where $\phi(\omega\delta/V)$ is the (point) frequency spectrum of the surface pressures near the edge. For the far-field spectrum we then have

$$S_K(\omega) = \frac{2}{\pi} \left(\frac{L\ell_3}{R^2} \right) \frac{M_v \sin\alpha \sin^2(\theta/2) \cos\beta\, \phi(\omega\delta/V)}{(1+M_{oR})^2 (1-M_{vR})^2 (1-M_{wR})^2 (1-M_{v1}\sin\alpha)} \tag{12}$$

which reproduces the dimensional result in equation (9) for $\langle p_K^2 \rangle$ when integrated over ω and transformed from reception to emission time coordinates (and provided that the angle β is small, as in most applications). A very significant aspect of equation (12) is its prediction that the spectral shapes $S_K(\omega)$ and $\phi(\omega\delta/V)$ are the same when $\omega \cos\alpha/c_o << \ell_3^{-1}$ and ℓ_3 does not strongly depend on frequency. This prediction is remarkably well borne out in one of the experimental findings of reference 42, reproduced here as figure 3. Note that the theoretical prediction of identity of the spectral shapes requires the Kutta condition to be satisfied.

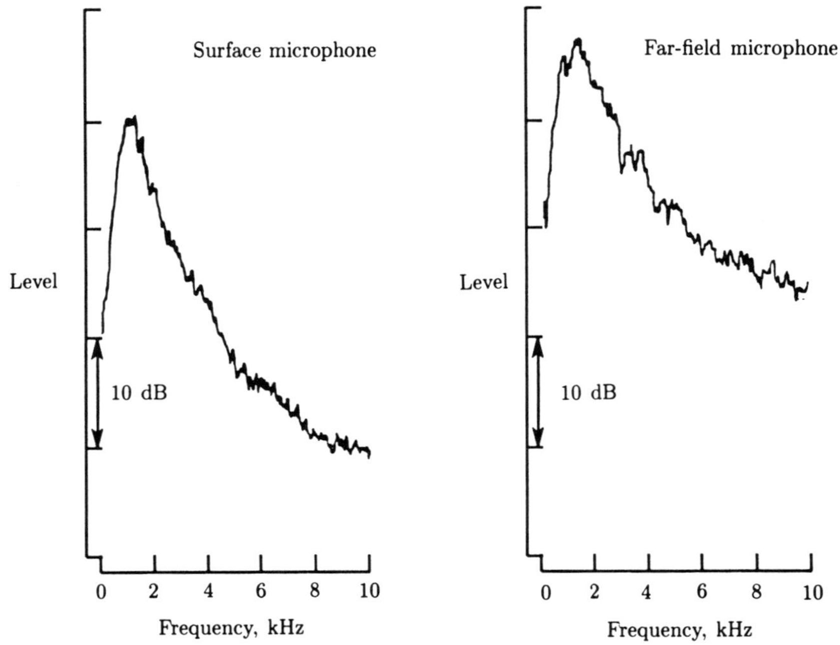

Figure 3. Comparison of spectral density functions for surface pressure fluctuations near a trailing edge and for far-field pressure fluctuations. (From ref. 42.)

Summarizing his examination and development of theoretical work on the half-plane trailing-edge problem, Howe (ref. 36) concludes that

1. All theories are in agreement on the SPL scaling at 90° to the flight path as $L\ell_3V^5(1 - M_o - M_{v1})$ with L the wetted span, ℓ_3 a correlation length parallel to the edge, V an eddy convection velocity, M_o the flight Mach number, and M_{v1} the eddy convection Mach number component normal to the edge.
2. There are large differences (≈11 dB at 90°) between the SPL's for the Kutta-condition and no-Kutta-condition cases (though only weak differences in directivity).
3. Some evidence exists (e.g., that associated with fig. 3 and eq. (12) and also a considerable number of aeroacoustic problems where acoustic forcing of an attached flow is concerned (ref. 38)) to support the Kutta condition; however

there is an urgent need for a study of the appropriate edge conditions in circumstances closer to the trailing-edge noise problem.

4. Forward flight and turbulence source motion can be represented, at least for moderate Mach numbers, by various Doppler factors given in equations (8) and (9).
5. The possibility of relating the far-field spectrum directly to measures of the surface pressure on the plate near the edge is a real one (though probably of limited applicability); the appropriate experimental input on the surface pressures is now required.
6. At high convection and wake Mach numbers, the predictions of equation (8) will remain valid provided that compressibility effects on the turbulent sources can be neglected, a neglect which needs further study, theoretical and experimental. (The argument for the correctness of the Doppler factor expressions at high Mach number does not come from ref. 36, where $O(M^2)$ terms are systematically neglected. However, for an idealized problem of so-called "diffraction radiation," all Mach number terms can be included exactly, and then the results are found to involve only precisely those Doppler factor effects which are found at low M (see ref. 43).)

Amiet's Approach

An alternative theoretical approach to trailing-edge noise has been developed by Amiet (refs. 44 to 47), who immediately takes as input the convecting surface pressure spectrum upstream of the edge. The induced loading of the airfoil is then calculated by standard gust-interaction methods, which allow incorporation of all finite Mach number effects within linear theory, but which treat the airfoil as extending to infinity upstream. In the subsequent evaluation of the radiation field by integration of the surface pressure dipole field, the integration is performed only over the region actually occupied by the finite airfoil, with the result that the acoustic field vanishes on the upstream axis (as is required of the exact solution), whereas that of equation (10) does not and indeed attains its maximum value there.

The Kutta condition is satisfied in Amiet's work, which produces expressions corresponding to equations (10) and (12). It too predicts the identity of far-field and near-field pressure spectral forms provided that $\ell_3(\omega)$ is taken to be independent of ω, as was done above. In reference 45 Amiet points out, however, that integration of the well-known data of reference 48 yields $\ell_3(\omega) \approx 2.1V/\omega$, which actually implies a faster high-frequency rolloff for the far-field spectrum than for the near-field—as indeed is seen in figure 3. Finally, mean flow Mach number effects are included exactly in Amiet's work, though since this involves an exact linearized calculation, the vorticity is always required to convect at the free-stream velocity, and differences between that velocity, the convection velocity of eddies in the boundary layer, and the convection velocity of eddies in the wake, which appear in Howe's model, are excluded. In reference 49, Amiet has returned to this issue, pointing out that the eddies in the wake can convect at a speed other than that of the free stream only if the wake itself supports a loading and that this fictitious loading may contribute to the trailing-edge acoustic field in an unintended, physically spurious way. This raises questions as to the correctness, for acoustic calculations, of models which attempt to incorporate "realistic" flow features such as differences in convection velocity from one flow region to another.

Experimental Study of Trailing-Edge Noise

Turning now to experimental study of trailing-edge noise, early work by Hayden (ref. 50) suggested that a sixth power of velocity provided the best collapse of his spectral data. However, Fink later reanalyzed these data (ref. 51) and showed (his fig. 6) that a very much better fit over the entire frequency range considered (in which the Strouhal number based on the width of a jet used to provide turbulent flow over the trailing edge of a large plate varied from 0.6 to 100) was obtained using a fifth power law. Fink (ref. 51) also conducted his own experiments, studying the surface pressures and far-field sound both from incidence fluctuations (when an airfoil was exposed to turbulent fluctuations in the incident flow) and from trailing-edge interactions (when the airfoil was faired into the nozzle wall of a large rectangular nozzle, providing a turbulent flow over the trailing edge when a grid was placed upstream in the nozzle). Although far-field data were taken at several angular stations, details were given only for $\theta = 120°$ from downstream, so that the directivity variation $\sin^2(\theta/2)$ cannot be checked. However, Fink did display 1/3-octave far-field spectra (at $\theta = 120°$) for mean flow velocities from 31.5 to 177 m/sec and for grids producing two different turbulence levels. These 1/3-octave spectra, normalized on (Turbulence level)^2 and (Velocity)^5, collapsed onto a single curve as a function of frequency normalized on mean velocity and transverse turbulence velocity correlation scale (obtained from hot-wire probes near the edge). Even better collapse is obtained if the high-frequency portions of the spectra at the highest velocity, probably dominated by background noise, are omitted. The collapsed spectra given in figure 3 of reference 51 thus provide at least a first estimate of the spectrum $\phi(\omega\delta/V)$ of equation (12). Those spectra have a decay very close to f^{-3} from $f\delta/U_o = 2$ upward, where f is frequency, δ transverse correlation scale, and U_o mean velocity. There is no theoretical basis for an f^{-3} spectrum unless the f^{-3} decay is built into some assumed model for the surface pressure spectrum ϕ in equation (12).

Before going on to prediction of trailing-edge noise from measured surface pressure data, mention should be made of references 42, 52, and 53 by Tam, Yu, and Joshi. First, in the theoretical part of reference 52 the authors emphasize the effects on the directivity of trailing-edge noise which are associated with diffraction by the leading edge. To obtain a tractable problem, they take a large finite plate in static fluid with a dipole normal to the plate and just beyond the (trailing) edge, the "half-baffled dipole" corresponding to Hayden's view (ref. 50) of the trailing-edge noise mechanism. This approach does give the correct velocity scaling and directivity for the half-plane problem if correctly handled (see ref. 36). A formal solution for the Tam and Yu problem can be written in Mathieu functions, and these determined numerically. The essential feature of the result is that the directivity oscillates increasingly rapidly with angle as the ratio of plate length to wavelength increases. A simple theory accounting for primary diffraction at the leading edge of sound generated at the trailing edge predicts the right features in a gross sense, but exaggerates the rapidity of the fluctuations, which are to some extent smoothed by multiple diffraction effects correctly accounted for in the exact solution. These results cannot be applied in detail to the full-scale problem; the interference effects depend critically on detail of the source and the diffracting body which is not properly represented in the model, and one can safely say only that in the flyover plane of principal interest for airframe noise, the infinite half-plane directivity $\sin^2(\theta/2)$ can be regarded as the *envelope* of the actual directivity functions at high frequency.

Next, Yu and Tam (ref. 42) reported flow-field observations of a wall jet issuing from a rectangular nozzle and passing over the sharp edge of the wall some 8 nozzle diameters downstream. The configuration resembles an upper-surface-blowing arrangement more than a conventional wing or flap trailing edge, but it is mentioned here because the authors suggest that the essential features somehow imply a mechanism of trailing-edge flow and noise production different from that underlying the types of theory reviewed in reference 36. The flow pattern was found to be rather coherent, dominated by a series of large, essentially two-dimensional vortices in the upper (jet) shear layer and a similar series of vortices, of opposite sign, convecting in the "wake" shear layer. In fact both shear layers are jet-type shear layers, separating the same core jet from ambient static fluid, and the dominance of large-scale vortical structures in these fully turbulent shear layers is now widely accepted. Such structures produce sound by two mechanisms. First, when such a "large eddy" structure is farther than a characteristic eddy length from the edge, it generates sound of quadrupole type at a *low* frequency of order u_o/ℓ where u_o is the rms velocity and ℓ the eddy scale; effects on the frequency and amplitude of eddy convection at a velocity $V >> u_o$ arise only in Doppler factors, leading to considerable convective amplification downstream. In contrast, the interaction between such a large eddy and the sharp edge on a large body is of precisely the kind described by trailing-edge noise theory. It has a frequency V/ℓ controlled by convection (or by the mean flow velocity U_o as the eddy is detached from the edge), an intensity scaling with velocity as V^5 rather than quadrupole V^8, and a sound field of the $\sin^2(\theta/2)$ kind peaking in the upstream direction. Trailing-edge noise theory may not yet have been specifically adapted to the upper-surface-blowing configuration, but there is certainly no contradiction between the requirements of that theory and the flow features seen by Yu and Tam.

The two mechanisms can be seen in data (figs. 12 and 13) of Joshi and Yu (ref. 53), who examined differences in the large-scale structures of the same wall-jet flow produced by profile-modifying grids upstream of the nozzle exit. They found, for all three mean jet profiles that could be achieved, a sound intensity and spectral level scaling with velocity to a power of 8.6 to 8.8 at an angle of 45° from downstream (consistent with quadrupole radiation plus some further convective amplification) and with velocity to a power of 6.1 to 6.6 at 90°. These results suggest at least the emergence of trailing-edge noise itself as the angle from downstream is increased, but a definitive conclusion could be reached only from data much farther into the forward arc, where theory predicts that quadrupole noise will reduce rapidly and trailing-edge noise will continue to increase. A further point to be made is that the phase opposition (difference of 180°) measured in references 42 and 53 at corresponding points on either side of the edge does not identify dipole radiation; the "scattered field" of trailing-edge noise theory has such phase opposition, but is not dipole if the surface concerned is large compared with the wavelength. There seems to be no reason at all for thinking in dipole terms about these experiments, and the velocity exponents of 8 and 6 at 45° and 90° suggest *only* that the quadrupole field is giving way to a lower exponent field as θ increases. The same criticism—of the irrelevance of dipole ideas in the trailing-edge noise problem—should be made of the studies of references 54 to 56 (the latter finding a V^5 law and talking sometimes in the language of Howe's trailing-edge noise theory and sometimes in dipole terms).

Prediction of Trailing-Edge Noise From Measured Surface Pressure Data

Brooks and Hodgson (ref. 56) made the first determined attempt to provide the surface pressure data called for by Howe and to relate these to far-field acoustic data. They took surface pressure data from about 0.5 to 1 hydrodynamic wavelength upstream of the edge of a symmetric airfoil, with flow on both sides and with tripped turbulent boundary layers. Cross-spectra in frequency were determined and compared very favorably with the predictions of the scattering of evanescent pressure waves. In the evanescent wave theory (refs. 36, 40, 41, and 57), a convected pressure field travels past the edge without any wetting of the surface and edge by a mean flow. The problem for a half-plane is then essentially a classical no-flow acoustic problem, and the issue of a Kutta condition at the edge simply does not arise; there is no downstream wake, and yet the pressure differential vanishes at the edge. Brooks and Hodgson (ref. 56) add that in further analysis of data taken in an earlier study (ref. 58) they had found no significant component of wake vorticity with sign opposite to that of the incident boundary layer and coherent with the incident pressure field. (They did, on the other hand, state in their earlier study that far-field noise measurements fell far below the no-Kutta-condition estimate, implying then that a Kutta condition must be in force—but that conclusion must be set aside on the basis of the later study.)

Far-field noise spectra were also taken by Brooks and Hodgson, applying well-established corrections (ref. 20) for refraction across the shear layers of the open-jet wind tunnel flow around the airfoil. The 90° OASPL was found to scale with free-stream velocity U_o to a power very close to 5 (see fig. 4), and 1/3-octave spectra, normalized on $\rho_o^2(U_o^5/c_o)(L\delta^*/R^2)$, collapsed acceptably as a function of $f_c\delta^*/U_o$, f_c being the band center frequency and δ^* the displacement thickness. Directivity of the OASPL (reproduced as fig. 5) followed very closely the predicted directivity for the no-Kutta-condition case from equation (8); significantly less good agreement would be obtained with equation (9). Also shown in figure 5 are the measured and predicted $(\sin^2\Theta/(1+M_o\cos\Theta)^6)$ directivities for the compact dipole aerodynamic sound (ref. 59) from flow over a thin circular rod replacing the airfoil; the 90° OASPL for the rod scaled very accurately with U_o^6. Note the appearance of six powers of the Doppler factor, normally associated with a quadrupole source, rather than dipole. The prediction of six powers, rather than four, was given in references 60 and 61.

Brooks and Hodgson also gave far-field 1/3-octave spectra for the (symmetric) airfoil at 5° incidence and for the airfoil at zero incidence but with a number of trailing-edge modifications. The small incidence led only to an increase of the low-frequency noise by several dB, with no significant effect on OASPL or on perceived noise level. Modifications to the edge also caused surprisingly little change, even when a flap deflected 17.5° was fitted. In all cases, cross-spectral phase data indicated that the source of noise was located at the new edge locations—but see later comments on the work reported in reference 22.

Brooks and Hodgson also made direct predictions of the far-field spectra from surface pressure measurements, essentially using the prediction of equation (10), but with allowance for a variable convection velocity through the boundary layer and with

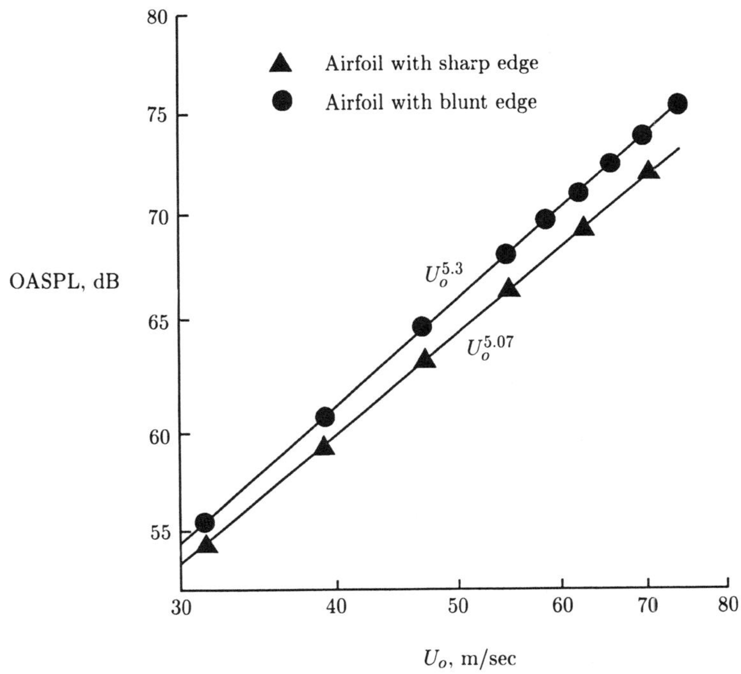

Figure 4. Overall sound pressure level of trailing-edge noise as a function of free-stream velocity U_o. $\alpha = 90°$ (flyover plane); $\Theta = 90°$. (From ref. 56.)

M_{wR} set equal to zero in accordance with their assertion that no vortex shedding coherent with the turbulent excitation was taking place. The forcing spectrum Π was obtained experimentally from correlation measurements just upstream of the edge. The approximate result (eq. (12)), with V assumed constant, was also evaluated from measurements of the point pressure frequency spectrum upstream of the edge. Figure 6, from reference 56, shows that for $\Theta = 90°$ and the two mean velocities tested, equation (10) can indeed be used to give an accurate outright prediction of the far-field spectrum at and above the peak spectral frequency (which is adequate for approach EPNL calculations) and also that the simplified model (eq. (12)) gives a sufficiently accurate result for most purposes. Overall, the conclusion of this study is that for clean laboratory conditions, surface pressure data can be used with evanescent wave theory (with no allowance for vortex shedding) to accurately predict directivities and far-field 1/3-octave spectra. In reaching this conclusion, however, Brooks and Hodgson do not reconcile their suppression of all effects associated with vortex shedding with Howe's requirement that the Kutta condition be satisfied if surface pressure measurements near the edge are to be used.

Two further pieces of work on trailing-edge noise for model configurations must now be discussed. These theoretical analyses make some allowances for the relaxation of the turbulent flow as it is convected past the edge. First, Howe (ref. 62) extends his earlier theory (ref. 36) by allowing for displacement thickness fluctuations on the upstream boundary layer. These convect past the edge and evolve into

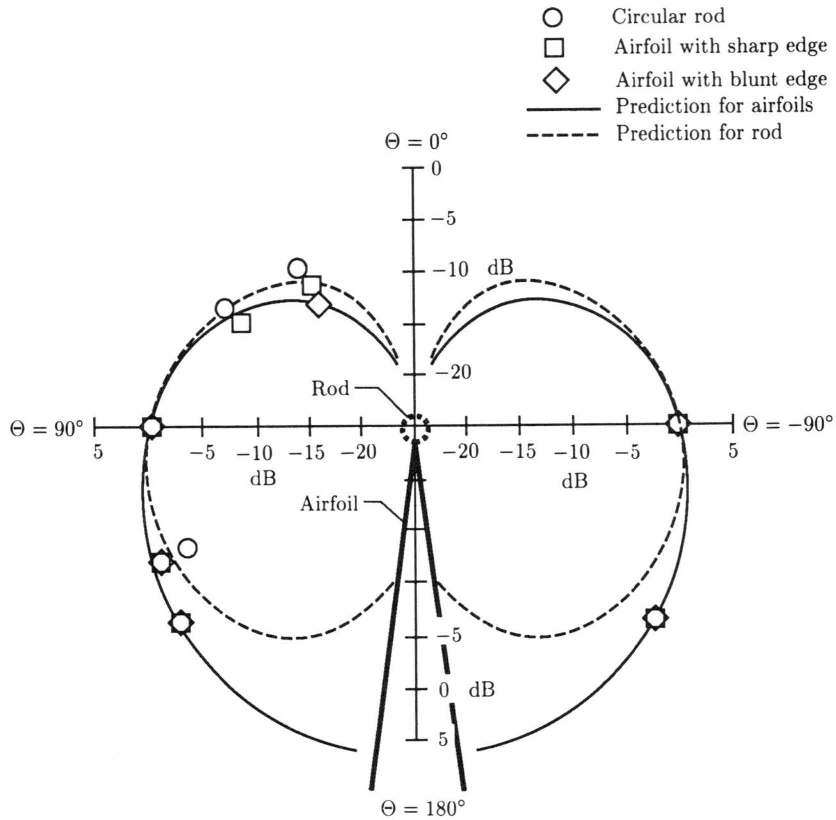

Figure 5. Directivity of trailing-edge noise OASPL in flyover plane as function of emission-time angle Θ. Data corrected for wind-tunnel shear layer refraction. (From ref. 56.)

displacement thickness fluctuations in the wake. The wake can support symmetric and antisymmetric large-scale disturbances, whose amplitudes are chosen so that no mass flux occurs out of the boundary layer plus wake and so that an unsteady Kutta condition is satisfied at the trailing edge. However, the mass fluxes associated with boundary layer and wake do not vanish individually, and they combine to produce an acoustic dipole field, with the dipole axis in the downstream direction. At Mach numbers of interest in the airframe noise problem, this dipole does not change the field below the aircraft and slightly increases the field ahead of the aircraft, with a larger increase downstream which is likely to be masked by other fields.

A quite different approach is taken by Goldstein (refs. 63 and 64), who models the leading- and trailing-edge noise problems by calculating the interaction between a gust on a nonuniform shear flow and the edges of a flat plate immersed in the flow. This generalizes the idea of convected pressure disturbances which generate an evanescent field at the plate in potential flow. It is argued that the approach deals correctly, in linearized theory, with the relaxation of the gust disturbance as it experiences the change in boundary condition at the plate edge and also with the

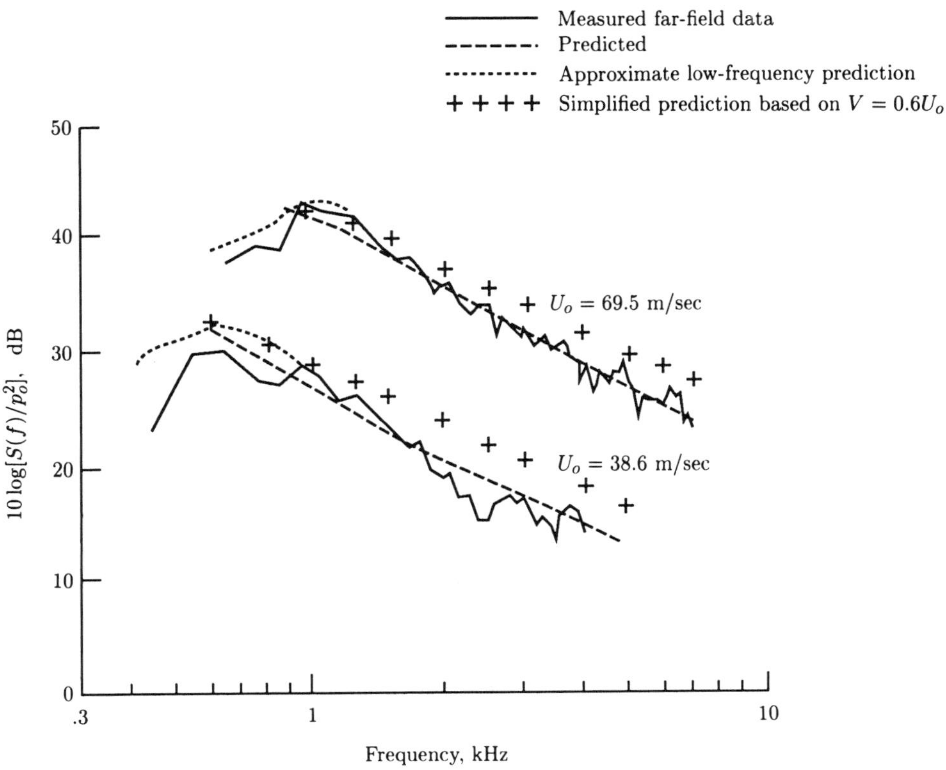

Figure 6. Measured far-field noise spectra and spectra predicted from measured surface pressure spectra. $S(f)$ is power spectral density. $\alpha = 90°$ (flyover plane); $\Theta = 90°$; $p_o = 20\ \mu Pa$. (From ref. 56.)

convective amplification and fluid shielding of the edge-generated sound passing from the edge through the flow. It is also claimed that the approach allows satisfaction of a Kutta condition at a trailing edge, in the sense of zero pressure differential across the plate edge—but in this approach there would not be the singular concentration of wake vorticity that is needed in potential flows to satisfy such a condition.

Details of the solution for the scattered field associated with some prescribed incident gust upstream depend on a Wiener-Hopf factorization, which generally depends on details of the mean flow. High- and low-frequency estimates can, however, be obtained for the distant acoustic field directivity. At low frequencies, the mean-square pressure at an observer located in static fluid and at rest relative to the plate, around which there is flow of limited extent, is

$$\langle p^2 \rangle \propto \frac{\sin^2(\theta/2)}{[1 - M(y_c)\cos\theta]^2} \tag{13}$$

where θ is the angle measured from the downstream continuation of the plate, as earlier, and $M(y_c)$ is the convection Mach number of an incident gust vorticity distribution concentrated at some particular level y_c from the plate. This is precisely the directional distribution of equation (5) or (10) if we confine attention to the flyover plane $\alpha = 90°$, ignore any effects associated with the concentrated shed wake vorticity ($M_w = 0$), and also set the quantity M_{oR} equal to zero because for comparison with Goldstein we must take the mean flow Mach number to have its value at the observer location when the observer is at rest relative to the plate.

The high-frequency solution involves a "zone of silence" centered downstream, outside which (i.e., for angles where edge noise might not in any case be masked by other noise fields):

$$\langle p^2 \rangle \propto \frac{\sin\theta}{[1 - M(y_c)\cos\theta]^2 (1 - M_o \cos\theta)^2} \left[\frac{1 - (1 + M_o)\cos\theta}{1 + (1 - M_o)\cos\theta}\right]^{1/2} \tag{14}$$

Goldstein (ref. 64) claims that the convection factors here greatly change the directivity from the basic cardioid $\sin^2(\theta/2)$ of Ffowcs Williams and Hall (ref. 29) with its increase toward $\theta = 180°$. When $M_o = M(y_c)$, which Goldstein takes, this is certainly the case, and the directivity increases rearward before finally, for small θ, being cut off by the refraction associated with the zone of silence. Only when $M_o = 0$ (where here M_o refers to the mean flow Mach number at the plate surface) does the directivity in equation (14) reduce to Howe's extension of the Ffowcs Williams and Hall result, that is, to equation (13). In reference 47 the high-frequency predictions of reference 64 for a bounded region of flow past the plate and separated by a shear layer from the observer outside are compared with Amiet's trailing-edge noise theory with the flow region extending to infinity with no shear layer. Amiet (ref. 47) shows the directivities to be identical, except for a shear refraction term arising naturally in the former configuration and absent from the latter.

The low- and high-frequency field shapes corresponding to equations (13) and (14) for a leading-edge interaction were also given in reference 64 and shown there to agree very favorably with data measured by Olsen (ref. 65), who generated leading-edge noise by inserting a large flat plate into the mixing region of a turbulent jet. The data were taken at two jet speeds high enough to make convection and shielding effects really significant and were well predicted, for six different 1/3-octave spectral levels, over almost the whole relevant angular range.

Olsen and Boldman (ref. 66) checked the Goldstein theory against experiments on a trailing edge with a wall jet upstream. At the lowest speed tested (91 m/sec) and at low frequencies, their results are broadly in agreement with equation (5) or (13), with a velocity exponent around 5; data for rear arc angles ($\theta < 90°$) also appear to be well predicted as a function of θ, but the velocity exponent is around 7, suggesting contamination of the data by jet mixing noise. At higher speeds the basic $\sin^2(\theta/2)$ directivity is in strong competition with the Doppler factors in equation (13), and as a consequence the sound pressure levels actually decrease toward $\theta = 180°$; however equation (13) still appears to predict the low-frequency field shape well. High-frequency field shapes were found to be well predicted by the high-frequency law (eq. (14)) and both the low- and the high-frequency field shapes were shown to be unchanged by quite drastic changes to the turbulence passing the edge (though the OASPL and spectra did change substantially).

One may conclude therefore that equation (13) emerges from a number of different approaches as the correct result for low frequencies, although there is considerable doubt as to the correctness of effects in equations (9) and (10), which correspond to vorticity shedding and the Kutta condition. That condition is satisfied—though in a rather different sense—by Goldstein's calculations leading to equation (13), in which the vorticity convection factors are absent; but the relationship between the edge conditions in the Howe (potential flow) and Goldstein (sheared, vortical flow) approaches is not clear. One should note also that the convection Mach number Doppler factors $(1 + M_{oR})^{-2}$ in Howe's expression (eq. (5)) are absent from Goldstein's (eq. (13)); however, the configurations which the authors have in mind differ: Howe's observer is fixed relative to the plate, being in flow of Mach number M_o; Goldstein's is in static fluid. Howe's model is appropriate for flyover noise generated by boundary layer interaction with wing and flap trailing edges; Goldstein's, for blown flaps or upper surface blowing. However, Goldstein's theory cannot be applied to flyover noise of an aircraft with wall-jet blowing of upper or lower surfaces until it is extended to include forward flight effects, represented by $(1 + M_{oR})^{-2}$ in Howe's expression (eq. (5)). The $(1 + M_{oR})^{-2}$ factors probably represent the present best estimate of flight effects, at any rate for low frequencies.

For higher frequencies the directivity seems better predicted by equation (14) than equation (13), though in the basic airframe noise problem, where all Mach numbers are small in the absence of wall-jet blowing or other interaction between a jet exhaust and the flaps, the differences are small. Forward flight effects need to be incorporated in equation (14). Howe's prediction is that these are again represented by a factor $(1+M_{oR})^{-2}$, and there is no theoretical evidence for any more significant effects than these.

Three final points must be raised. First, Dobrzynski (ref. 54) measured surface pressure fluctuations at various spanwise stations near the trailing edge of a business jet aircraft, the Messerschmitt-Bölkow-Blohm HFB 320, and of a McDonnell-Douglas DC-10-30 airliner, both at cruise and at landing approach, and found that vibration effects were very significant. At some spanwise locations there was evidence of panel vibration that increased the *broadband* surface pressure spectrum by as much as 10 dB over a large frequency range. There was also clear evidence for discrete tones corresponding to individual vibration modes. This broadband response to discrete frequency excitation is reminiscent of broadband jet turbulence and noise response to tonal forcing, even of very low amplitude (refs. 67 and 68), and Dobrzynski's work may provide the first suggestion that low-amplitude panel vibration can cause broadband increase in boundary layer turbulence near the panel. The mechanisms in the turbulence must certainly be nonlinear, but these mechanisms can possibly be excited by very low-amplitude ostensibly linear forcing. At any rate, Dobrzynski's measurements serve as a warning that the surface pressure input to a prediction like equation (12) may be drastically changed—and not necessarily uniformly along the span—by surface vibration, which is impossible to simulate in model-scale tests.

Second, Dobrzynski noted that flyover tests of a 747 in the cruise configuration and with flaps deflected 25° had revealed drastic differences in airframe noise level, but his studies showed very little difference between surface pressure data (away from the flap side edges) for the two configurations. His prediction (based on the largely empirical method of ref. 69) of trailing-edge noise thus roughly agreed at lower frequencies with 747 cruise data, but fell far short (15 dB) of 747 data for 25° flaps. Dobrzynski attributed the failure of the prediction to the possibility that the *flap side edges* actually generate the dominant noise when the flaps are deployed, and he gave some evidence based on analysis of the phase variation with spanwise location to justify that claim.

Third, Kendall and Ahtye (ref. 22) examined noise generation by a large wing-plus-flaps model with a directional microphone and also found strong evidence that trailing-edge noise itself is unimportant when flaps are deployed. With the two-part flaps differentially deployed (0° and 35° along different portions of the span), a very noisy field was found, with the most intense sources at the gap between the deflected and undeflected flap segments. If both segments were deflected 35°, the principal source was located all along the leading edge of the flap at the gap between flap leading edge and wing trailing edge. These sources became concentrated at any spanwise location occupied by a flap track fairing (bracket) if this was fitted, while for the differentially deflected flap segments, the addition of such a bracket at the spanwise gap reduced the intensity of the flap edge sources.

Summary of Trailing-Edge Noise

The following conclusions may be drawn. For configurations resembling the half-plane prototype, an extensive theory exists, capable of describing trailing-edge noise from both boundary layer and wall-jet excitation. Many features are firmly established; in particular, the low-frequency directivity and the velocity scaling and reasonable estimates of forward flight effects. The high-frequency features are less well established, and the Kutta condition issue seems quite unresolved, the balance of evidence at present being perhaps marginally against satisfaction of that condition for turbulent excitation of trailing edges. Trailing-edge noise can be predicted quite well from measured surface pressures. Such predictions seem to give reasonable estimates of trailing-edge noise for clean (cruise) conditions. At approach conditions, however, airframe noise levels are much higher, and except when surface vibration is significant, surface pressure characteristics are not greatly changed at moderate flap deflections. The increased noise seems to be radiated by highly three-dimensional flow around the flap side edges and by the slot (gap) ahead of the flap leading edge.

Flap Side-Edge Noise

The idea that a source of intense airframe noise is associated with the side edges of deployed wing flaps began in experimental studies (refs. 18, 21, 22, and 70) in the late 1970's. In reference 23, surface pressure and far-field acoustic measurements were taken on a large (15.2-m span) model wing with a part-span triple-slotted flap system, in the NASA Ames 40 × 80 foot tunnel. Surface pressure transducers were mounted along the chord of each flap at midspan and near the outer flap edge. Cross-correlations were made between the outputs of these transducers to determine the convection velocity and length scale of the surface pressure fluctuations. Cross-correlations were made also between surface pressure transducer and far-field microphone outputs to provide high discrimination against tunnel and probe noise and to some extent to discriminate between one localized correlation area of turbulent surface source and another. The cross-correlations of reference 23 indicate a far-field intensity per unit area of surface greater by 10 to 15 dB when the source is close to a flap side edge than when the source is at mid-span. Far-field intensity scaled with mean speed U_o as $U_o^{5.4}$, with a directivity biased toward the forward arc and more like the classic half-plane $\sin^2(\theta/2)$ directivity than that of a free-field dipole normal to the flaps. Surface correlations indicate that the edge turbulence responsible has a scale around half the flap chord and an intensity of about 25 percent of the mean velocity. If a flap track fairing is fitted just inboard of the edge, then the edge turbulence just outside the fairing has comparable, or even greater, intensity and somewhat reduced length scale. However, the acoustic output per unit area is reduced by typically 10 dB when the flap track is fitted. It is to be emphasized that the acoustic results are inferred from cross-correlations and that the raw data are quite insensitive to whether the flap track is fitted or not and are some 25 dB above the correlated signal from the side-edge regions. The raw data are, however, heavily contaminated by tunnel background noise, and the effect of the side-edge regions would be expected to be detectable in full-scale flight, given the known importance of the trailing-edge noise sources.

Two models have been proposed for the mechanism behind flap side-edge noise. The first is due to Hardin (ref. 71), who argues that the flow into the underside of the flap creates a higher pressure there than above the flap, and the pressure differential causes a spanwise flow around the edge—just the usual spanwise flow responsible for creation of the tip vortex, in fact. However, satisfaction of the no-slip condition on the flap undersurface leads there to the production in the boundary layers of positive vorticity in the streamwise direction on the right wing flap and negative vorticity on the left flap, as seen by an observer behind the aircraft looking forward. That vorticity migrates away from the flap edge under the action of its image, but is nonetheless swept round the edge if the large-scale spanwise flow is strong enough. Vorticity that does get swept round the edge passes very close to it, with rapid acceleration and intense sound production.

Hardin gives a two-dimensional model for this phenomenon (two dimensions in an (x, y) plane normal to the flight direction and to the flap side edge). The model provides a description of the incompressible flow due to a line vortex and a parabolic edge flow, the condition for the edge flow to sweep the vortex round the edge, and the vortex trajectory. See figure 7 for a sketch of the model configuration and of typical vortex paths. He also calculates the two-dimensional sound field by a low-frequency Green's function approach from reference 72; as expected this field has the $\sin(\theta/2)$ pressure variation found earlier by Crighton (ref. 73) for a vortex-edge interaction problem. This two-dimensional model quite well represents measured data on the structure of an airfoil side-edge flow taken in reference 74. However, while the two-dimensional model is probably adequate for the hydrodynamics, it is surely inadequate for the acoustic field, where one might actually expect the $\sin(\theta/2)$ pressure variation in the (y, z) plane containing the chord rather than the span of the flap. It is, however, not clear how to model the acoustic aspects of Hardin's theory. The flap generates a rapidly accelerating unsteady flow near the wing, which is essentially a half-plane ($y = 0, z < 0$), which would suggest from the general theory a U_o^5 acoustic intensity scaling, with directional variation $\sin^2(\theta/2)$ in the (y, z) plane; this is entirely consistent with the measurements (ref. 23).

Meecham (ref. 75) has extended Hardin's analysis, with some allowance for the finite thickness of the flap (in the y-direction of fig. 7). The right-angle geometry he assumes for the flap thickness, replacing the zero thickness model of Hardin, would be expected to lead to less violent vortex acceleration round the edge and lower noise levels. This would be consistent with experimental results on the effect of the flap fairing (ref. 23). This fairing geometry should reduce the strength of the edge flow by greatly increasing the edge thickness scale, while at the same time

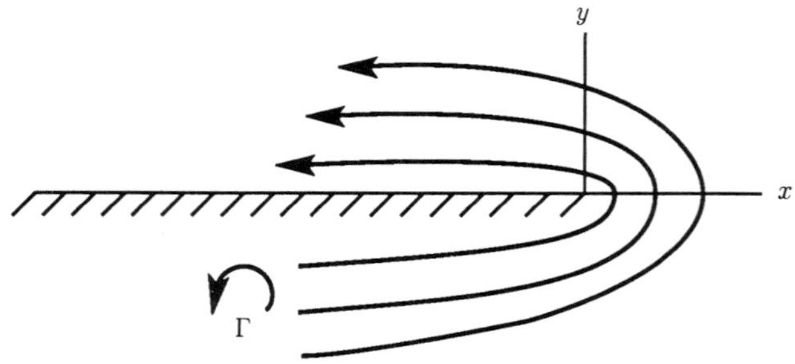

(a) Geometry of flow round side edge of a flap; lines give the streamlines of the main edge flow which are cut by a vortex Γ.

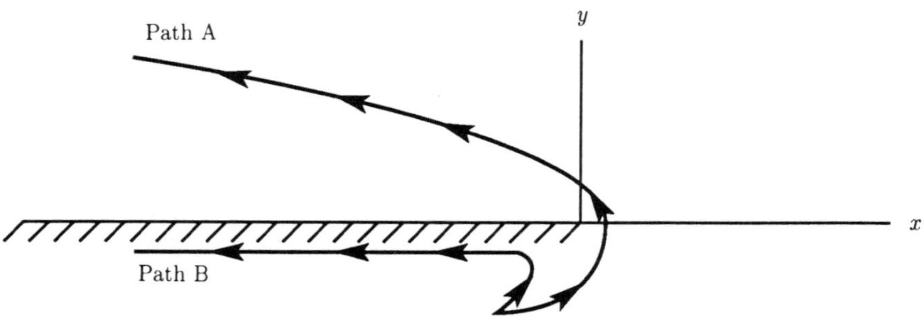

(b) Typical paths of a vortex Γ; for path A the edge flow is strong enough to sweep the vortex round the edge.

Figure 7. Typical vortex paths. (From ref. 71.)

perhaps increasing local edge turbulence intensities because of flow separation over the fairing, also consistent with reference 23.

Howe (ref. 76) offers a quite different model. Rather than explicitly referring to the details of any flow processes, he incorporates these into an equivalent surface pressure field, which might be measured or simulated numerically. He does emphasize features of the acoustic Green's function that might be associated with the flap edge. Specifically, he claims that at moderate deflections the gap between the flap edge and wing may be modeled by a finite chordwise slot in a wing with an otherwise straight trailing edge (and infinite chord as far as acoustic waves generated by the flap are concerned). The slot length is equal to the flap chord, the slot width is equal to the mean distance between the flap side edge and the side edge of the adjacent undeflected wing, and the deflection of the flap out of the wing plane is ignored. In all cases the slot width is small compared with the wavelength, but analytical expressions for the Green's function are obtained for both the compact and the noncompact slot length cases, with smooth interpolation between. Effects of uniform mean flow at subsonic Mach number M_o parallel to the flap side edge are fully accounted for.

Howe's theory predicts the far-field sound pressure spectrum generated by turbulent fluctuations near the slot in terms of quantities represented in the Green's function and characterizing acoustic mean flow and slot diffraction effects (the flap edge radiation is actually the diffraction radiation by the slot of the nonradiating turbulent boundary layer pressure field in this view) and of quantities characterizing the magnitude and spectral distribution of the turbulent fluctuations in the presence of the flap edge (slot). The latter are introduced in the theory in such a way that they can be estimated in terms of measured pressure fluctuations in a standard flat plate boundary layer and close to the flap edge in the actual wing-flap configuration (as measured in ref. 56). Howe's conclusions, from study of the asymptotic expressions and numerical evaluations, are that

1. At low Strouhal numbers based on flap chord c_F and flight speed U_o, the directivity in the flyover plane is $\sin^2(\Theta/2)/(1+M_o\cos\Theta)^3$, while at high $\omega c_F/U_o$ it is $1/(1+M_o\cos\Theta)^2$ (except close to the fore and aft directions, $\Theta = 0°$ and $180°$, where the intensity vanishes like $1/[\ln(\sin\Theta)]^2$).
2. The radiation efficiency of the side-edge sources is much greater for low values of $\omega s/U_3$ than high, where $2s$ is the slot width, and U_3 the mean flow convection velocity along the span of the flap. Thus efficient radiation (essentially corresponding to a dipole normal to the flap chord, at the trailing edge, and baffled ahead by a semi-infinite wing) occurs at low frequencies, while less efficient monopole radiation, corresponding to a mass flux through the slot, occurs at high frequencies.
3. Numerical evaluation of the predicted side-edge radiation and comparison with the predictions (see the previous section) for trailing-edge radiation suggest that in the case studied in reference 23, the edge radiation can exceed that from the whole of the flap trailing edge (17 times longer than the chord of the leading-flap segment) by more than 3 dB, consistent with the measurements reported.

The models of Hardin (ref. 71) and Howe (ref. 76) between them give a good understanding of one of the most complex aeroacoustic phenomena. The essential hydrodynamics is described by Hardin; chordwise vorticity swept round the edge by the spanwise lifting flow cuts rapidly across the streamlines of the spanwise flow and accelerates rapidly and very close to the edge. However, that idea on its own takes no account of the strong acoustic effects of the neighboring configuration, effects which Howe likens to the well-known "installation effects" in jet noise. The convected vorticity forces a mass flux through a slot formed by the edges of the undeflected wing and the deployed flap. At high frequencies there is a weak residual monopole generated by this mass flux, and except near $\Theta = 0°$ and $180°$, a convected monopole dominates (the wing causing no further diffraction effects). At low frequencies the mass flux degenerates to form a dipole source, with the dipole axis normal to the flap chord, and then the dominant directional effect is of the diffraction of the dipole field by the semi-infinite wing ahead, leading to the familiar $\sin^2(\Theta/2)$ directivity with convective modification. The source in either case is determined by the Hankel function $H_o^{(1)}(\omega s/U_3)$, which is small at high frequencies and large (though only weakly) at low frequencies, the result being that for comparable turbulence excitation the high-frequency radiation is 10 to 25 dB less intense than the low.

The slot considerations do not apply to aircraft with effectively full span flaps; neither is the slot model relevant if the flap deflection is large (although it was found in ref. 23 that the side edges of the first, least deflected, flap segment were noisiest), nor do any of these ideas apply directly when engine exhaust impinges on part-span flaps at large deflections. Much further work is therefore needed, but at least the importance of flap side-edge noise and a basic understanding of the flow and acoustic mechanisms have been established (refs. 71 and 76), with some demonstration (ref. 23) of how the noise can be reduced by changes to the edge geometry (the flap fairing).

Undercarriage Gear Noise

The components of an undercarriage are mostly bluff bodies of a wide range of shapes and aspect ratios: wheels, axles, struts, shafts. Unsteady separated flow past them (vortex shedding) and wake interactions between them give rise to broadband sound over a wide frequency range. For the largest components on a large aircraft, the typical frequency may be as low as 50 Hz and typically 200 to 300 Hz for small to medium transports; there are also many smaller details and features producing sound in the sensitive 2- to 4-kHz range. Low-frequency tones may also be produced, even in a generally turbulent flow, by the interaction of flow with the cavities forming the wheel wells (see the next section).

The general mechanism of undercarriage gear noise is well understood. At typical shedding frequencies associated with any particular component, the radiated wavelength is large compared with the component chordwise dimension d (roughly by a factor $5M_o^{-1}$ with M_o the flight Mach number, if we take the standard Strouhal number relation $fd/U_o = 0.2$ as representative). It is also fairly large compared with any other length of the component (the main wheel shaft length, for example). Under such circumstances, the field radiated is that of a compact dipole whose strength is the net unsteady force exerted on the fluid by the component concerned—the simplest application of the Curle (ref. 59) theory. A variation of far-field intensity with U_o^6 is expected.

For cylindrical shafts and struts with nearly vertical alignment, the dominant force is a side lifting force contributing to sideline noise but not to approach noise (which usually peaks when the aircraft is almost overhead). There is a smaller drag force, which contributes little to the field at points directly below the aircraft. Horizontal members, such as axles and struts not close to the vertical, experience a transverse force, which does contribute to the approach noise. So do wheel clusters upon which the separated flow from other wheels ahead impinges; here strong forces with comparable components in all directions are likely to be generated.

Significant "installation effects" may be expected for undercarriage noise. The force dipoles are generated within a wavelength (generally) of a large flat wing undersurface, equivalent to image dipoles. Dipoles in the horizontal plane will add constructively with their images, while those in the vertical will tend to cancel and degenerate to a vertical longitudinal quadrupole. For the latter an intensity variation more like U_o^8 than U_o^6 would be expected, but with greater forward arc amplification due to aircraft motion. A convective amplification factor $(1 + M_o \cos\Theta)^{-4}$ on intensity is expected for the dipoles and $(1 + M_o \cos\Theta)^{-6}$ for quadrupoles, with $\Theta = 180°$ as the flight direction and with Θ measured at emission time (though note that the quadrupole convective amplification factor is likely also to apply to the dipoles, for reasons that can be seen from the model problems analyzed in refs. 60 and 61). Such convection effects are negligible, however, around the peak approach noise point $\Theta \approx 90°$, and one must conclude that at such points installation effects due to wing proximity significantly reduce undercarriage noise below what would be expected on the basis of uninstalled dipole noise ideas. This point seems not to have been appreciated in published work; see, for example, the FAA airframe noise prediction method devised by Fink in references 27 and 28 and discussed in reference 17, where free-field dipole models are used with coefficients chosen for best fit with experimental data.

The fullest experimental study of undercarriage gear noise is due to Heller and Dobrzynski (ref. 77) who studied two-wheel and four-wheel models in an outdoor wall-jet facility and attached to the wing of a high performance glider with very low airframe noise. The models were representative of nose and main gear assemblies for all but the largest aircraft (C-5A, 747) and had main elements including wheels, shaft, diagonal strut, door, and cavity, but not typical fine scale protrusions producing high-frequency sound at full scale. Static (wall-jet) data were presented for locations in a vertical plane through the gear at right angles to the flight direction, directly below the gear, to the side, and at 45° to the sideline. Glider flight data were presented only for locations directly below the gear. A Strouhal number fD/U_o based on wheel diameter D was used, but other relevant lengths (e.g., exposed strut lengths and wheel widths) are not vastly different.

At high frequencies, $fD/U_o > 5$, the two-wheel nose gear model generated identical noise levels at the three measurement locations; essentially all the noise was generated by the side support struts (and main shaft). The peak of the spectrum directly below the gear was at $fD/U_o = 5$, but those at 45° and at the sideline occurred around $fD/U_o = 2$ and had spectral levels that were 5 and 10 dB, respectively, greater than those below. These greater levels were again dominated by the shaft and struts. As shown in figure 8, at no angle or frequency were any other components (wheels or door) significant. When speed-independent tones, related to cavity resonances, are excluded, the spectra collapse well on an (Intensity)/U_o^6 versus fD/U_o basis and the field is dominated by transverse dipoles representing side forces on nearly vertical structural members. When the nose gear was mounted on the glider wing, the predicted gear noise level (from the rig tests) exceeded the glider self-noise only for $fD/U_o > 5$ and for a limited range of lower frequencies where, however, cavity tones were also significant and prevented a direct comparison of rig and flight noise levels. For $fD/U_o > 5$ the agreement between rig and flight data was acceptable. Nose gear noise can therefore probably be adequately predicted on the basis of figure 9, from reference 77; see also the discussion in the subsequent section entitled "Development of a Comprehensive Prediction Scheme." Installation effects on nose gear noise can also be expected to be much less significant than those on noise sources associated with the main gear.

Nose gear noise is, however, generally low in level. Figure 10, from reference 77, shows that the spectral levels produced by a four-wheel main gear model are essentially identical at measuring points below and to the side of the assembly (except at high frequencies, $fD/U_o \approx 5$, where the level below is about 3 dB lower than the sideline level), and are 6 to 8 dB above levels generated by a simplified two-wheel set. These results are entirely consistent with the idea that comparable unsteady forces are generated in all directions on the assembly, with comparable dipole fields radiated below the aircraft and in the sideline direction (and presumably in the flight direction as well), but that at the higher frequencies the vertical dipoles suffer some

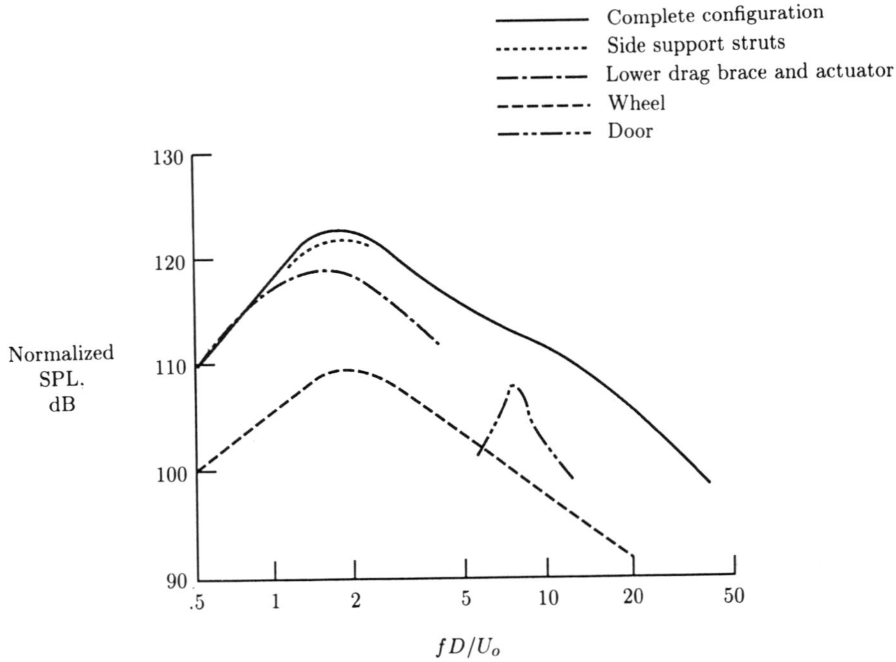

Figure 8. Normalized sideline spectra of nose gear model components. Reference flight speed, 100 m/sec. (From ref. 77.)

cancellation by their images while the sideline dipoles experience some enhancement from their images in the surface supporting the gear. In reference 77 main gear spectral levels collapse well when normalized on U_o^6, as a function of fD/U_o, except at low frequencies where cavity tones dominate. Figure 12 of reference 77 shows that scaled rig noise data agree rather well with data taken from glider tests with the main gear model attached to the glider wing; here the main gear generates a field clearly above the clean glider self-noise across the entire frequency range.

Heller and Dobrzynski (ref. 77) then used these rig and glider data to predict the gear noise of three aircraft (Jetstar, BAC 111, and HS 125) for which measurements of gear noise had been published elsewhere (obtained as level differences between gear-up and gear-down configurations). These aircraft had three two-wheel gear sets. The model data gave a rather good prediction of spectral levels for the Jetstar, and a reasonable one for the HS 125 (nose gear noise being negligible for these two cases), but a significant overprediction (by 5 dB) for the BAC 111. The rig gear did not accurately model any of these undercarriages closely. Application of the scaled four-wheel gear noise data to the larger VC10 aircraft gave prediction significantly above measurement at low frequencies and significantly below at high. However, in reference 77 the VC10 data taken from reference 10 appear to refer to the case in which the VC10 gear bay doors were closed again after the gear had been deployed, whereas the predictions in reference 77 were derived from model data in which an open wheel well was present. If the predictions are instead compared with figure 15 of reference 10 for the VC10 with open doors (5 dB higher in level than those for closed

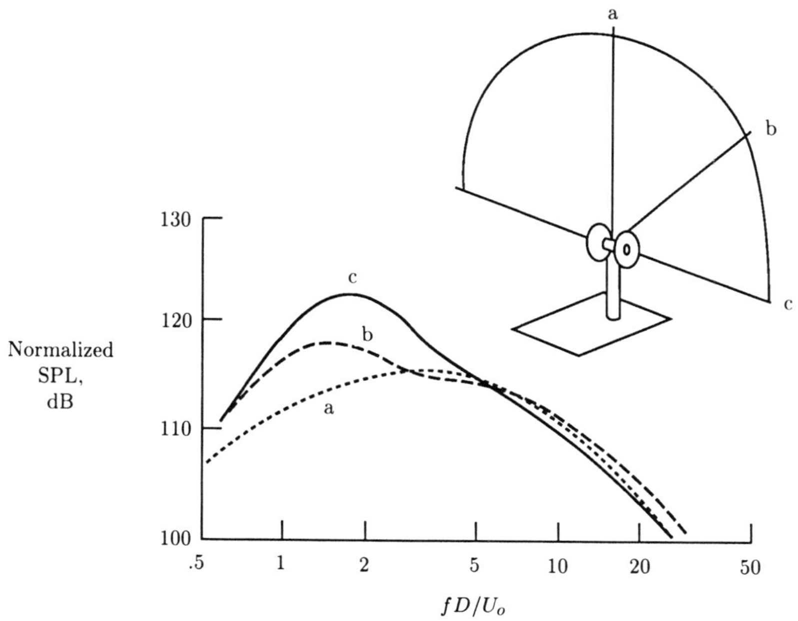

Figure 9. Normalized nose gear model spectra at three measurement points. Reference flight speed, 100 m/sec. (From ref. 77.)

doors at frequencies below about 200 Hz), one gets a much better agreement. Thus, one can conclude that high levels of *broadband* low-frequency noise are generated by main landing gear with open doors to the wheel well and that these are substantially reduced if the doors can be closed after deployment of the gear, as in the VC10. In OASPL terms (40 Hz to 1.6 kHz), Fethney (ref. 10) indicates that if the gears are deployed with closed doors, the noise of the otherwise "clean" aircraft increases by 6 dB at approach and by a further 4 dB if the doors remain open.

Much more complicated undercarriage layouts may be necessary for large aircraft, and these studies of two- and four-wheel assemblies cannot be scaled to deal with, for example, the two twelve-wheel assemblies studied in reference 78 on an advanced supersonic transport model. These assemblies retracted into shallow cavities which generated only a small amount of (high-frequency) noise themselves, when exposed to flow. The combination of cavities and deployed gear increased the clean aircraft noise levels uniformly in angle and by only 2 dB in OASPL; the increase was entirely confined to frequencies around and above the spectral peak. However, a significant interaction between gear and trailing-edge flaps was also reported in reference 78. The increase in OASPL due to gear and trailing-edge flaps being deployed together exceeded the sum of the increases due to deployment of the individual devices, presumably because of the interaction of the flap trailing (and side) edge with turbulent wake flow from the landing gear. Such an interaction was also reported in reference 18 for landing gear of two-wheel type, as for aircraft of the 727 or DC-9 standard, and a single part- or full-span flap. Here, however, a favorable (small) interaction occurred, interpreted through directional microphone readings as

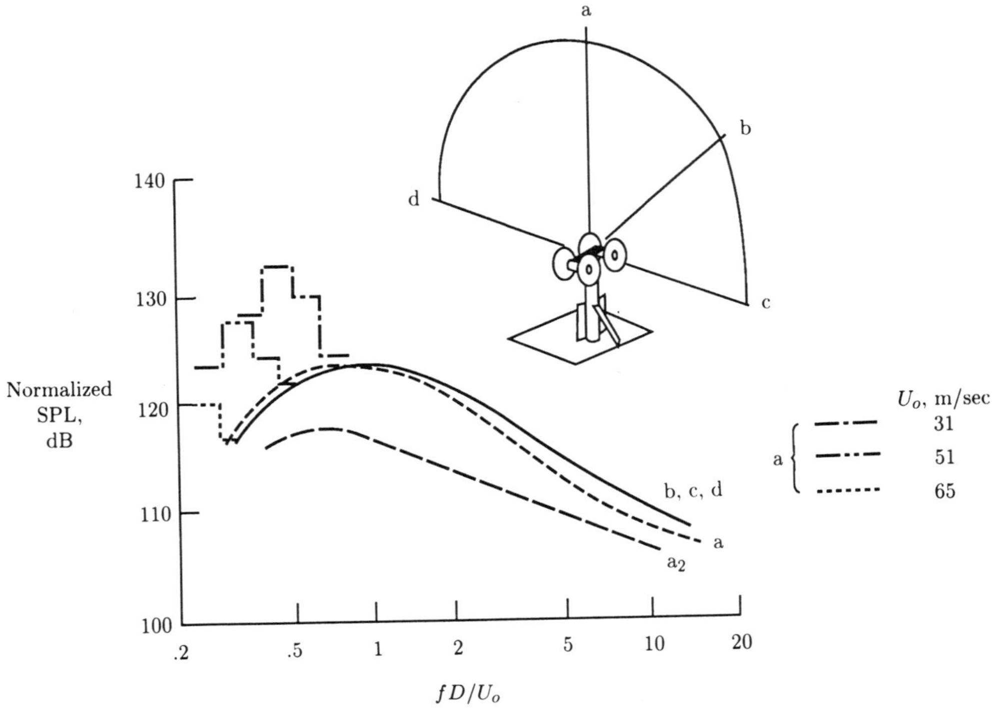

Figure 10. Normalized main gear model spectra at four measurement points. a_2 is spectrum of simplified two-wheel model at point a. Reference flight speed, 100 m/sec. (From ref. 77.)

equivalent to a somewhat increased trailing-edge noise field caused by the relatively weak wake turbulence of the two-wheel set and a more greatly decreased gear noise field caused by the reduced velocity created on the pressure surface of the wing by the deployed flap. A more substantial reduction in gear plus flap noise would result if, as suggested in reference 27, the acoustic impedance of the flaps could be tailored—for example, by porous flap edge regions, as studied in reference 79—to reduce the edge scattering by convected turbulence.

One may conclude that the results of reference 77—in particular the scaling and data of figure 9 for two-wheel gear and of figure 10 for four-wheel gear—are adequate for a prediction of the noise of similar gear at full scale, with open wheel wells and in the absence of any interaction with the flaps (and possibly with leading-edge slats and flaps). A better prediction is actually obtained from the analytical fit to figures 8 and 9 devised by Fink (ref. 27), discussed in a subsequent section. Significant benefits seem to derive from the ability to close the gear doors once the wheels have been deployed. Interactions with the flaps seem to be favorable for two-wheel main gear, but unfavorable with multiwheel sets, where the trailing-edge flap noise is significantly increased when the wheels are lowered.

Attempts to provide a more specific theoretical prediction, rather than general understanding and motivation for correlations, now seem rather misplaced. Some effort in this direction was made early by Hardin et al. in reference 8 (who give a very useful description of the vortex shedding from cylinders and other bluff bodies over a wide Reynolds number range), concentrating on aeolian tone and broadband noise production by cylinders. However, the main gear struts seem not to be important sound sources, especially in multiwheel gear, and in any case noise from them would radiate predominantly to the side and cause no significant problem for approach. Theoretical predictions are, further, of little value when the flow is known to be rather sensitive to geometrical detail, making studies of long uniform cylinders of dubious relevance.

Cavity Noise

It has been known for decades that high-speed flow over cutouts and cavities generates both tonal and broadband response with intense near-field pressures. Several types of response have been identified in a very large number of studies, theoretical and experimental. First, there are discrete frequency oscillations in a feedback cycle in which vortical disturbances generated at the leading edge of the cavity convect downstream in the shear layer, impinge on the downstream edge, and there generate acoustic waves traveling upstream, in the cavity or in the free stream, which trigger further vortex shedding on reaching the upstream cavity wall and edge. A review of such shear layer feedback cycles in a variety of different configurations is given in reference 80. Work continues on the modeling of these systems, for instance on the modeling of the vortex which may be trapped in the cavity and on determining conditions at the downstream cavity edge which determine the volume flux across the cavity mouth (see ref. 81). Second, discrete frequency oscillations occur when the cavity is driven in a "volume mode" by the unsteady pressure across its mouth. In a mode which is essentially a standing wave across the width or along the length of the cavity there is little response normal to the flow, so that such modes are hard to excite by shear layer motions, and they radiate inefficiently. In contrast, "depth modes" are both easily driven by the shear layer oscillations and efficient radiators, especially when the wavelength exceeds the length or width of the cavity. Third, there are possibly "Helmholtz resonator" modes, though these could occur only at frequencies below any of the volume-mode frequencies and only when the cavity geometry defines a reasonable mass of fluid in the neck to balance the volume stiffness.

Aircraft wheel wells are large partly open cavities with the potential to develop oscillations in any of these forms, to respond with a lower amplitude to turbulent shear-layer excitation at frequencies other than the various resonance frequencies, and to radiate efficiently at low frequencies to the far field below the aircraft on landing approach. Fortunately, two facts emerge from model and full-scale experiments which limit the possibilities drastically. First, as far as the acoustic radiation is concerned, realistic cavities respond only in their depth modes. Other discrete frequency modes, corresponding to feedback cycles, to length or width modes, and to Helmholtz resonator modes, are apparently unimportant. Second, the far-field level scales simply with the internal cavity pressures at the frequencies involved in depth modes and has an essentially isotropic (monopole) directivity. Cavity tones should therefore be reasonably predicted over the range of typical low approach speeds, given the internal cavity pressure spectrum for the cavity complete with landing gear.

The basis for this comes from references 82 and 83, dealing with internal and far-field pressure measurements from model flow-excited cavities and with similar measurements from cavities with landing gear at both model and full scale. In reference 82, rectangular cavities, with varying length L and depth D and of fixed width W, set in a large plate, were exposed to grazing boundary layer flow exhausting over the plate from a large rectangular nozzle. Flow velocities of 43 m/sec and 80 m/sec were used, and L/D ranged from 1 to 4. Microphones outside the flow were used (with shear layer refraction corrections to give free-field acoustic data), and a single microphone monitored the internal cavity pressure fluctuations.

For all but the highest L/D, the far-field spectra were dominated by a single tone around 1 kHz. Near-field spectra showed additional tones between 2 and 4 kHz, but these radiated very poorly, the reason being that they correspond to length modes that are poorly coupled to the sound field. The depth mode nature of the 1-kHz tone was demonstrated by its invariance under change of L at fixed D and under change of velocity (except for shallow cavities, $L/D = 2$ or 4). A rough estimate of the frequency of this lowest depth mode can be obtained by arguing that if the shear layer presents a low impedance to the cavity, then the lowest mode is the lowest mode of a cavity with all walls rigid and of depth $2D$. This gives a frequency of the right order, and a better estimate can be obtained if a model of the actual impedance presented by the shear layer is used; see reference 84 and figure 3 of reference 82, where the first few depth modes are predicted to have frequencies close to those apparent from the spectra.

The directivity patterns measured in reference 82 were all essentially isotropic (monopole) and dominated by the 1-kHz tone. Deviations from isotropy were generally small, and the author has found no consistent way of explaining them through the appropriate Doppler factors for the mean flow (even at a velocity of 86 m/sec) and through a small axial dipole term arising from the force on the downstream part of the cavity.

Attempts were made to alter the internal and far-field sound levels by increasing shear layer turbulence levels through the action of a roughness strip on the inside of the nozzle supplying the flow. Some reduction was achieved, the reduction diminishing at higher flow speeds and for longer cavities—consistent with the view that the relevant excitation is the spectral component of the shear layer fluctuation at the tone frequency and at zero wave number in the plane of the cavity mouth.

Low-frequency tones of this kind (frequencies in the 50- to 500-Hz range at full scale, say) are often seen in airframe noise data, but the measured far-field noise levels are usually well below those that might be inferred from scaling (as described below) "clean cavity" noise levels to the full scale. The discrepancy is largely attributable to the presence of landing gear components in, and protruding from, the full-scale wheel well, rather than to any difference in the fundamental mechanism. Indeed, in the typical real aircraft the dominance of resonant volume (depth) modes seems even more certain than in clean model experiments. However, the levels, both internal and far-field, are drastically changed by the presence of gear components, as was shown in the experiments of reference 83 at model scale, where internal tone levels decreased by more than 10 dB (30 dB in one case) when representative gear components were inserted into the cavity. The physical reason for this reduction is not clear. Certainly it is not a matter of acoustic scattering, because the waves are very long; possibly the cavity fluctuations are so intense that they lead to vortex shedding from the gear components at the expense of the sound energy (as in many cases of vortex shedding from sharp edges analyzed in ref. 85). In any event, data from an empty cavity may be used to predict frequencies, because the depth modes are insensitive to internal detail, but they cannot be used to estimate levels, which seem to be quite sensitive to the gear arrangement in the cavity.

The model-scale experiments of reference 83 involved nose and main landing gear components and cavities typical of large transport aircraft, with acoustic and cavity internal pressure measurement. Narrow-band analysis of the low-frequency spectrum revealed the dominance of two tones whose frequencies (and indeed the entire spectral shape) were unchanged over a doubling of the speed of the external flow and whose internal cavity sound pressure level increased with U_o^4. Higher frequency components of the spectrum varied in spectral shape with U_o and were believed to be generated directly by the landing gear components themselves, as in the previous section. The far-field SPL's for the low-frequency tones (at frequencies of 270 Hz and 770 Hz for 1/13 model scale) were found to differ from the cavity levels by the same amount for both tones and for all M_o up to 0.4. Thus the frequency-squared weighting associated with a monopole at the cavity mouth is almost exactly offset by some other mechanism not as yet understood. Certainly, it seems that 270 Hz and 770 Hz are the first two depth mode frequencies of a cavity in which the pressure gradient vanishes at the hard wall $z = D$ and, to a first approximation, the pressure vanishes at the mouth $z = 0$ covered by the shear layer. In this model the radian frequencies would be given by $\omega D/c_o = (n - \frac{1}{2})\pi$ with $n = 1,2, \ldots$, the pressure in the cavity by $p = p_n \cos[\omega(z - D)/c_o]$, and the velocity magnitude at the mouth by $|u(z = 0)| = |p_n|/\rho_o c_o$. For $p_1 = p_2$, the far-field levels would then be expected to differ by a factor 9, or 9.5 dB, whereas figures 5(a) and (b) of reference 83 indicate essentially identical levels. This discrepancy can be explained only by arguing that the depth mode idea may be adequate for rough prediction of frequencies, but a more detailed analysis of the shear layer motion across the mouth is needed to predict the monopole strength. For a cavity with $L/D = O(1)$, the hydrodynamic wavelength is small at low Mach numbers compared with the streamwise length L, and at higher frequencies there may be substantial cancellation over the mouth area because of oscillations of the shear layer, this offsetting the monopole frequency-squared increase. This seems a point worth further examination; at present the experimental evidence is that there is a constant difference between internal and far-field tone SPL's regardless of frequency, a result found in reference 83 for both nose and main landing gear cavities.

Experiments over a wide speed range were also conducted in reference 83 on a DC-10-30 airliner, together with ground tests in which the cavity excitation was provided by broadband engine noise. Distinct low-frequency tones (frequencies less than 200 Hz) were observed for both acoustic and shear layer excitation, the latter excitation being much more significant. Cavity tone pressure levels were, however, typically 6 dB lower for the nose landing gear and 10 dB lower for the main landing gear cavity than in scaled model tests (even when the models contained simple landing gear components). The cavity SPL's were found again to scale with U_o^4 for the low frequencies, and with U_o^6 for higher frequencies which might well be generated by the gear components themselves. Pressure spectra normalized by U_o^4, as a function of frequency normalized by the sound speed c_o and the cube root of the cavity volume, were found to be crudely similar for the three types of gear cavity used on the DC-10-30, though between 10 dB and 15 dB lower for the large main gear cavities than at model scale.

The transfer function, relating internal wheel well SPL's to those in the far field, having been determined for simple model-scale representations of nose and main gear wheel wells, and internal SPL's having been measured at full scale on the DC-10-30, the full-scale far-field noise can be estimated. The necessary far-field airframe noise data do not exist for the DC-10-30, but were reported in reference 10 for a VC10 with nose and main gear cavities similar to those of the DC-10-30, and the estimated DC-10-30 far-field wheel well noise is compared with VC10 data in figure 14 of reference 83. Only the nose gear higher frequency tone has a high enough frequency to fall in the range for which VC10 data exist, and then the nose gear tone was about 6 dB lower than VC10 data and an estimated 12 dB lower if 6 dB are subtracted from the prediction to allow for the "damping" effects of realistic wheel wells. The predicted main gear cavity frequency was very low, and the level perhaps comparable with the VC10 data if cavity damping is ignored, but 10 dB below the VC10 data if 10 dB is subtracted as a typical estimate of "damping" in the larger cavity.

The data base is not large enough for any firm conclusion to be drawn. All the data refer to conventional transport aircraft nose and main gear cavities, and the situation could differ for aircraft with different gear layouts. All indications at the moment, however, are that the tonal noise from gear cavities is not a dominant contributor to the practical airframe noise problem. The levels for realistic cavities are not very high, being limited very strongly by some cavity "damping" mechanisms whose nature is far from clear. Also, the frequencies are very low, so that tonal cavity noise cannot be significant in perceived noise terms. The only circumstances in which it might be important arise at locations upstream and downstream, early and late in the flyover, where the monopole directivity might allow cavity noise to dominate other more highly directional fields. There is evidence, however, from the data reported in reference 10 for the VC10, that considerable broadband noise may be generated by an open cavity at frequencies above the first few cavity resonances and that this may significantly contribute to perceived noise levels at approach. From this one example, there appear to be clear advantages in gear arrangements which allow the cavity doors to be closed again after deployment of the gear.

Turbulent Boundary Layer and Panel Vibration Noise

The theoretical and experimental determination of the far-field sound from turbulent boundary layer flows over a homogeneous surface remains very difficult. Such sound is of quadrupole type, the Curle surface pressure dipole vanishing identically for both rigid and flexible uniform homogeneous plane surfaces (refs. 86 and 87). Howe (ref. 88) extended those arguments to show further that the surface dipole associated with viscous shear stresses should also be regarded as affecting only the propagation of quadrupole sound and not the generation of a more intense (dipole) field. Apart from these facts and the consequent prediction of U_o^8 variation for the intensity, almost no quantitative theoretical estimates are available. In underwater applications the U_o^8 variation gives such low SPL's as to rule out the direct quadrupole field, but the position for aeronautical applications is not so clear, although most fields of airframe noise have a much lower velocity exponent (4 to 6) and any higher exponent is usually from the obvious contamination of the data by residual jet noise.

To estimate crudely the possible SPL's associated with the boundary layers, square and average the far-field solution

$$(\rho - \rho_o)(\mathbf{x}, t) = \frac{1}{4\pi c_o^4 x} \frac{\partial^2}{\partial t^2} \int T_{ij} \left(\mathbf{y}, t - \frac{x}{c_o} \right) \, d\mathbf{y} \tag{15}$$

to the Lighthill (ref. 37) inhomogeneous wave equation, in which the quadrupole density T_{ij} will be assumed to be distributed statistically in a homogeneous fashion over a disk-shaped volume formed by a large plane surface of area S and with thickness δ. The double integral then gives a factor δ^3 from the correlation volume and a factor $S\delta$ from the physical volume occupied by the eddy quadrupoles, δ being the turbulence length scale, of order of the boundary layer thickness. The quadrupole density is estimated as $T_{ij} \approx \rho_o u_o^2$ where u_o is the rms turbulence velocity, though there are arguments in favor of the larger estimate $\rho_o u_o U_o$ where U_o is the free-stream velocity. Time derivatives should certainly be estimated as $\partial/\partial t \approx u_o/\delta$ and (for turbulence over a homogeneous surface) not as U_o/δ. Then one arrives at the estimate

$$\langle(\rho - \rho_o)^2\rangle_Q \propto \pi^2 \rho_o^2 (S/x^2)(u_o/c_o)^8 \tag{16}$$

or, at most,

$$\langle(\rho - \rho_o)^2\rangle_Q \propto \pi^2 \rho_o^2 (S/x^2)(u_o/c_o)^6 (U_o/c_o)^2 \tag{17}$$

which is to be compared with the estimate

$$\langle(\rho - \rho_o)^2\rangle_{TE} \propto (1/2\pi) \rho_o^2 (L\delta/x^2)(u_o/c_o)^2 (U_o/c_o)^3 \tag{18}$$

for trailing-edge noise when the same turbulence passes over an edge of span L. The ratio of equation (17) to equation (18) is overestimated if we take $u_o/U_o = 0.1$, $U_o/c_o = 0.3, L = 50$ m, and $\delta = 0.1$ m; even then equation (17) exceeds equation (18) only if S exceeds $10L^2$ (or 10^3L^2 if eq. (16) is used for the quadrupole radiation). Thus it seems safe in the airframe noise context to ignore the pure quadrupole radiation from the boundary layers, in comparison with that from edges and other inhomogeneities, even for very large aircraft.

A probably more significant source of airframe noise is associated with panel vibration driven by turbulent pressure fluctuations. This was noted early on in the study of reference 5 of airframe noise of the C-5A Galaxy, where narrow-band far-field spectra and wheel well door vibration spectra both contained large amplitude peaks at two frequencies in the 20- to 40-Hz range. Significant vibration of flap sections on full-scale aircraft, including a DC-10-30, was also reported in reference 54. The possibility that this might be a significant noise source was raised earlier in reference 11 and has been mentioned in several other reports. Hardin (ref. 11) suggests that panel vibration may be the source of a high-frequency peak observed in the noise spectrum of the 747, CV-990, and Jetstar aircraft, noting that the peak frequency (around 1500 Hz) is considerably higher than expected for trailing-edge noise mechanisms and rather insensitive to flight speed. A similar high-frequency peak was observed in experiments reported in reference 10 for a number of aircraft (HS 125, BAC 111, and VC10) and certainly does not seem to be associated with flap deflection and gear deployment, which both tended to produce increases at rather lower frequencies. The peak is not narrow band, and in fact in figure 11, from reference 54, the increase of surface pressure believed to be caused by panel vibration is substantial and broadband, though centered on a particular frequency.

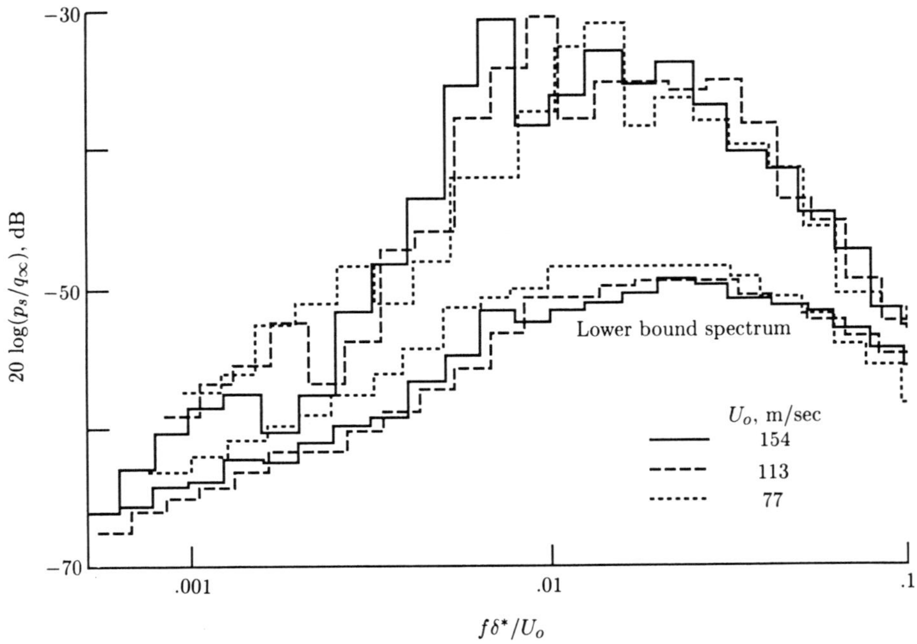

Figure 11. Measured surface pressure spectra p_s at two spanwise stations near the trailing edge of an HFB 320 flap in cruise configuration. q_∞ is free-stream dynamic pressure. (From ref. 54.)

Dowell (ref. 89) has considered vibration as a source of airframe noise and given a simple estimate of the far-field radiation from the lowest order mode of a rectangular panel under turbulent excitation. For a panel of area 3 ft^2, he finds a lowest eigenfrequency of 37 Hz and an SPL of 97 dB at a range of 300 ft. However, the estimate is very sensitive to the modeling of the surface pressure field (which we have noted may itself be significantly changed by the vibration), especially if the dominant radiation is from a high order panel mode, as seems necessary to explain the high-frequency peak. The relevant spatial scales are then likely to be much smaller than the typical panel dimension, leading to substantial cancellation of the driving pressure and a decrease of the monopole source strength below that estimated by Dowell for the lowest mode. On the other hand, typical panel dimensions are likely to be no more than comparable with the acoustic wavelength even at 1 kHz, so that the frequency-squared weighting for the monopole should still operate and offset the partial cancellation of the driving pressure.

A much more detailed theory of panel vibration and radiation in response to turbulent boundary layer pressures has now been given by Howe (see ref. 90 and references to previous work given there). His theory is specifically for when acoustic waves are short compared with the panel dimension. It identifies two principal mechanisms for the conversion of hydrodynamic energy into sound. In one, low-wave-number components of the turbulence excite flexural waves on the panel which generate sound by scattering at the edges; in the other, the convective peak of the turbulence near-field wave-number spectrum scatters directly (or secondarily, via flexural waves) into sound at the edges. The relative importance of the two types of "wave-number conversion" depends on the fluid loading and on the mechanical conditions at the edges. However, the configuration analyzed (a single flexible panel set between semi-infinite plates of the same properties as the panel) is not close to configurations of interest in the airframe noise context. Also the emphasis in reference 90 on fluid loading aspects is appropriate to steel in water, but not to aluminum in air. Nevertheless, this study shows how quantitative estimates can be obtained for the far-field sound with modeling of the surface pressures developed from the work of reference 91. The approach might be adapted, largely with neglect of fluid loading, for the aircraft panel problem. However, the principal sites of vibration are likely to be associated with regions in which the surface pressure modeling is probably inaccurate—for example, the doors of undercarriage wheel wells, the flaps at high deflection with separated suction surface flow, impingement by jet exhaust, and highly unsteady three-dimensional edge flows. Theoretical work seems of little help in those circumstances; even the prediction of the frequencies of the typical high-order panel modes that might be excited is unlikely to be possible. However, it seems worthwhile to stress vibration as a probably significant source of airframe noise, particularly with large flap deflections and with gear down; to note its essentially monopole directivity for baffled panels, its dipole directivity for free panels such as gear doors, and its cardioid directivity for vibrating panels baffled to one side only, as at the trailing edge of a flap; and to note the *broadband* increase that can apparently occur in surface pressure spectra when a panel vibrates predominantly in one or two discrete modes. We might also note the general effect of *surface roughness* in breaking the Powell cancellation mechanism for a smooth plane wall and leaving a residual surface dipole with U_o^6 far-field scaling. Howe (refs. 92 and 93) has worked out a theory for the generation of this dipole field by turbulent boundary layer flow over a plane surface roughened by a random distribution of hemispherical bumps. The theory gives the expected U_o^6 scaling and a spectral shape in remarkably close agreement with measured spectra for the sound from the interior of a sand-roughened pipe carrying turbulent flow. There is no immediate application to the airframe noise problem, though Howe asserts that "it is likely that a substantial fraction of the airframe noise of a wide-bodied jet transport plane flying in the 'clean' configuration (landing gear and high lift devices retracted) is associated with surface roughness dipole sources." The more significant approach noise problem is, however, certainly dominated by more specific sources than general small-scale surface roughness.

Noise of Powered-Lift Configurations

A great deal of work, mainly experimental, has been carried out, principally in the 1970's, on the noise from powered-lift devices of two kinds: the upper-surface-blown flap (USB) and the externally blown flap (EBF).[1] (Much less work has been done on the internally blown flap, in which the flap is closely integrated with the propulsion nozzle; see ref. 94.) For the EBF the interest is purely in the added lift, with applications to STOL aircraft. For the USB the interest is in both the powered-lift benefits and the noise shielding achieved in an engine-over-the-wing configuration. In all cases, however, there is a noise penalty, generally at low frequencies, associated with fluctuating lift on the flap segments, with increased trailing-edge and perhaps side-edge noise, and perhaps with a monopole type of radiation driven by unsteady mass flow through slots between the flap segments.

Much of the work was done in the early 1970's and is referenced in the eight papers on STOL noise published in reference 4. The situation is more complicated than in standard airframe noise because large flap deflections are involved, mean flow Mach numbers near the flaps are much higher than the flight approach Mach number of a conventional aircraft (typically 0.9), and the flow over the flaps may be a highly nonuniform wall jet issuing from a rectangular or D-shaped nozzle. In limiting cases the naively expected picture is found. At frequencies for which the flap chord is smaller than a wavelength, the flap-generated field is essentially that of a dipole normal to the flap chord with 90° intensity scaling as (Velocity over flap)^6 and spectrum shape a function of Strouhal number based on that velocity and flap chord. Convective effects (associated with the *aircraft* flight Mach number) might be expected to be represented by the dipole amplification factor $(1 + M_o \cos\Theta)^{-4}$ with Θ measured at emission from the downstream axis and M_o being the flight Mach number. The dipole directivity is expected to be as $\sin^2\Theta'$ with Θ' measured from the flap chord, provided that the diffracting and scattering effects of the dipole field by the wing ahead are small. These predictions are reasonably in accord with data presented in reference 4 for USB and EBF configurations at appropriate frequencies. They represent a considerable oversimplification in that the scattering effects of the nearby wing are completely ignored. They may be accurate only if the wing chord is also smaller than the acoustic wavelength or if, as in many experiments, the main wing is absent altogether. Further critical remarks along these lines are given in the next section. More substantial convective amplification effects may also be possible. For example, Crighton (ref. 60) and Howe (ref. 61) have shown that when the near field of a source is scattered by a compact body, the scattered field has dipole directivity combined with the quadrupole amplification factor $(1 + M_o \cos\Theta)^{-6}$ on mean-square pressure.

There are only limited data for the case of the acoustic wavelength being short compared with chord, but those that exist are roughly in accord with the expectations of the earlier section on trailing-edge noise. For example, in the studies of a rectangular jet exhaust over a USB flap (ref. 95), the directivities (see fig. 12) show a systematic change from a pattern close to nonconvected dipole (the observer is at rest relative to the flap) to one described by equation (5) (namely, $\sin^2(\theta/2)/(1 - M_{v1}\cos\theta)^2$, with $M_o = 0$) in which the basic half-plane cardioid is strongly modified by eddy convection downstream relative to the flap edge. Such a

[1] See also the chapter entitled "Propulsive Lift Noise" by Martin R. Fink.

pattern still has the upstream angle $\theta = 180°$ as a stationary value, but this gives a local minimum when $M_{v1} > 1/3$ and the peak of the directivity is then at angles on either side of the flap given by $\cos^2(\theta/2) = (3M_{v1} - 1)/2M_{v1}$. Reddy (ref. 95) observes a good collapse of spectral data at the higher frequencies using a $(\text{Velocity})^5$ intensity scaling and a Strouhal number based on a flow scale at the flap edge—all consistent with the idea that the dipole scalings and directivity progressively give way to those of trailing-edge noise theory as the wavelengths become short. These ideas have been confirmed experimentally in very fine detail in reference 30, where the disturbance generating the dipole or trailing-edge noise was a vortex ring passing close by the surface without impingement. This also shows that it is not helpful to distinguish, as has sometimes been done (ref. 96), between "scrubbing noise," when the flow impinges directly on the flap, and that generated by the interaction of the near field of eddies with the surface or edge. The scaling laws are the same for both cases, though the magnitude of the local forces may be larger in one case than in the other.

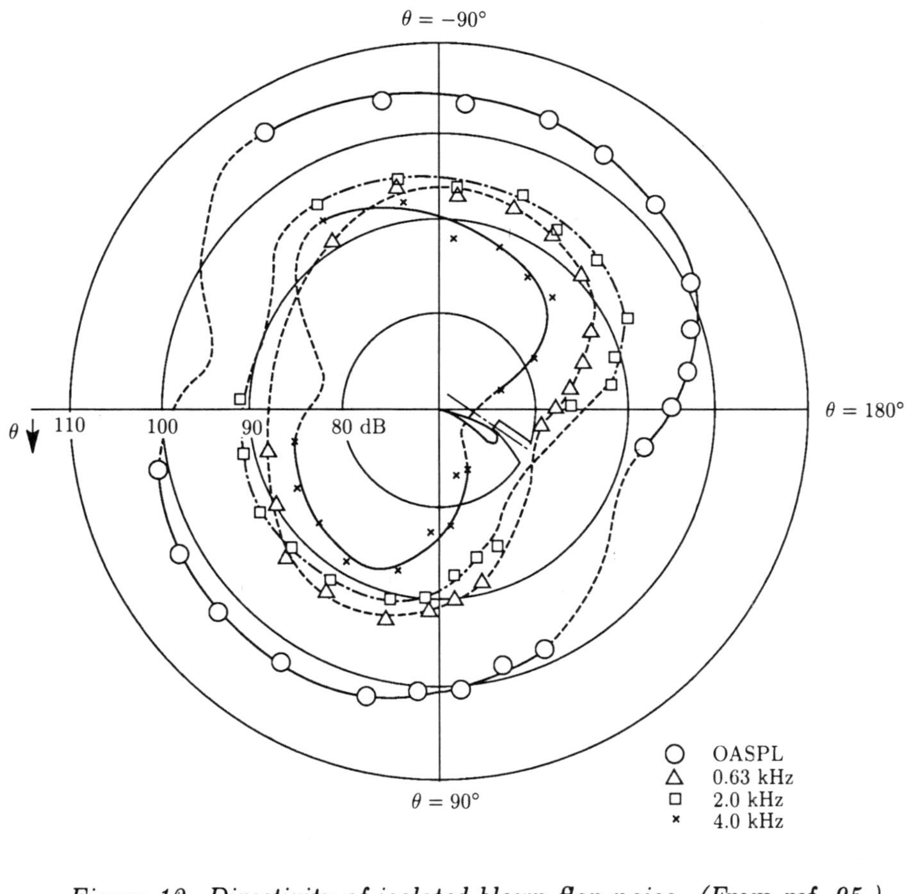

Figure 12. Directivity of isolated blown-flap noise. (From ref. 95.)

An important feature revealed by several studies of USB configurations is the presence of very strong tones, apparently a fundamental and first harmonic, apparently generated by the kind of feedback loop much studied in low-speed laminar shear layers; see reference 80 for a review and the work of reference 97 on wall jets modeling the USB arrangement. In reference 94 these tones were observed for a wall jet with exit Mach number of 0.9 and ratio δ of wall extension length to nozzle height of 6.85, when shadowgraphs also clearly revealed large-scale vortex concentrations in the primary and secondary shear layers. For δ close to 10, similar tones were observed. There is still (despite the denial of the authors) clear evidence of enhanced coherent vortex activity in both shear layers. At $M_o = 0.9$ this configuration gave the largest OASPL. Evidence of similar tones is to be found also in reference 51, and such intense tones very likely create substantially increased *broadband* turbulence and noise, as has been found to occur in the jet noise (refs. 67 and 68) and panel vibration (ref. 53) problems. Tone-like spectral features have also been noted in the EBF configuration (ref. 98). A possible source was thought to be the flow through the slot between the first and second segments of the blown flap; certainly the observed frequency is consistent with that expected, given the flow speed and slot width. The slot configuration would also, however, be expected to produce enhanced broadband sound associated with the unsteady mass flow through the slot when a turbulent eddy passes either through the slot or stays entirely to one side of it. This broadband field has not been examined experimentally in any systematic way, and no attempt appears to have been made to separate it out in EBF studies. One theoretical model exists (ref. 99) which shows that above-the-wing turbulence not passing through the slot generates a monopole field in the flyover plane, with four inverse powers of aircraft Doppler factor enhancing the field ahead of the aircraft and a typical fourth to fifth power of velocity scaling law. Essentially similar qualitative results hold for the more important cases of below-wing turbulence and turbulence convected through the slot, though the detailed results are more complicated. The noise field toward the sideline has vertical dipole characteristics. In most conditions these slot fields are likely to be at least as powerful as the force dipole or trailing-edge sources associated with the flap segments and pose a noise hazard that should be avoided—though more systematic experimental study is really needed here.

It is appropriate to add here a few remarks on leading-edge noise. When turbulence interacts with the leading edge of an isolated airfoil, a strong field is radiated, indeed much stronger than for a trailing-edge interaction. There, the edge singularity of attached flow can be relieved by vortex shedding into the wake, a process which greatly reduces the change in conditions experienced by an eddy passing over the edge. No such mechanism is available at a leading edge which, for a thin blade, experiences an inverse-square-root pressure loading. If the airfoil is isolated and if the high-frequency limit is relevant, the field produced has the familiar cardioid basic directivity of equation (4), but now peaks on the downstream axis (and is modified by convection effects). If the airfoil is a flap segment near a large wing surface, the essential noise mechanism probably involves the unsteady mass and momentum flux through the narrow gap (the slot mentioned above), and the isolated airfoil leading-edge noise ideas are irrelevant. Configurations may be found, however, in which isolated leading edges of plates may be subjected to incident turbulence from upstream. For these a theoretical treatment exists (for waves short compared with the plate dimensions), in which Amiet (ref. 44) solves a gust interaction problem for the leading edge and determines the unsteady loading and from that the radiation, just as for trailing-edge noise (except that no Kutta condition is imposed). Some model needs to be assumed for the wave-number spectrum of the rigidly convected incident turbulence fluctuations (e.g., the Von Kármán spectrum), but then once the scales are fixed, the theory predicts absolute noise level and directivity which compare very favorably with experiment in reference 100.

This approach is likely to be quite adequate for an estimate of the contribution of isolated leading-edge noise to airframe noise for appropriate components such as landing gear struts and perhaps gear cavity doors—but it is again emphasized that the predictions are not relevant to a flap segment near a wing.

Development of a Comprehensive Prediction Scheme

Several airframe noise prediction schemes have been put forward in the literature, including whole aircraft correlations (refs. 24 and 25), the "drag element" method (ref. 26), and two schemes (refs. 8 and 27) based on prediction of various component fields. Of these, that due to Fink (ref. 27) is, in the opinion of the present author, the one representing most nearly the state of the art in its combination of a wide range of full-scale and model data with what appear at the moment to be the theoretical ideas most widely accepted as correct. Accordingly, this section presents the main points of Fink's proposals, with comments on other more recent evidence that lends or removes support from them.

The scheme starts with clean airframe noise, with all gear and high-lift devices retracted. This is assumed to be entirely associated with turbulent boundary layer flow over the trailing edges of the wings and horizontal tail (for noise in the flyover plane). Here, for aircraft with nonswept trailing edges, we have from the section on trailing-edge noise, leaving aside all convection effects and vorticity shedding,

$$\langle p^2 \rangle \propto \left(\frac{u_o}{U_o}\right)^2 \left(\frac{\delta b}{h^2}\right) U_o^5 \sin^2\Theta \sin^2(\Theta/2) \tag{19}$$

for a flyover ($\alpha = 90°$) at altitude h; here u_o/U_o is the turbulence intensity, b the wing span (or tail span), and δ the boundary layer thickness at the trailing edge. Equation (19) predicts peak noise at $\Theta = 109°$, corresponding to peak emission before the aircraft is overhead; in contrast, Fink notes that any vertical dipole modeling gives a peak at $\Theta = 90°$ and reception of peak noise when the aircraft is well past the overhead position, features that have never been reported. The simplest prediction from equation (19) assumes δ to be proportional to chord and independent of U_o (actual exponents close to 0.8 and -0.2, respectively), and states that peak OASPL $-10\log(S_w/h^2) - 50\log U_o$ should be a constant for all aircraft, with S_w the wing area, and figure 13 (from ref. 27) shows that this is a good first approximation for 16 aircraft ranging from high-performance gliders through small propeller and jet transports to the largest transports flying (747, C-5A). Aspect ratio does not feature in equation (19) and the collapse in figure 13 includes a more than 10:1 variation in aspect ratio, indicating that the $(\mathrm{AR})^4$ variation postulated in earlier work (see eq. (3)) is probably incorrect and misleading. An improved prediction is obtained by taking $\delta = 0.37(S_w/b)(U_o S_w/b\nu)^{-0.2}$ with ν the viscosity; this gives (cf. eq. (6) of ref. 27 and the related discussion)

$$\text{OASPL} = 50 \log U_o + 10 \log(\delta b/h^2) + 7.5 \text{ dB} \tag{20}$$

for jet aircraft (U_o measured in knots).

The spectrum shape recommended by Fink is an empirical fit:

$$\text{SPL}_{1/3} - \text{OASPL} = 10 \log \left\{ 0.613 \left(\frac{f}{f_m} \right)^4 \left[\left(\frac{f}{f_m} \right)^{3/2} + \frac{1}{2} \right]^{-4} \right\} - 0.03 \left| \frac{f}{f_m} - 1 \right|^{3/2} \tag{21}$$

to data taken on an upper-surface-blown-flap rig. The peak frequency f_m should be taken as $f_m = (0.1)U_o/\delta$, and Fink (ref. 27) argues that these equations should be adequate for civil transports with wings of modest taper ratio, without any need for modification for different values of lift coefficient. Spectra calculated from equation (21) agree well with those (ref. 8) derived from the NASA Aircraft Noise Prediction Program (ANOPP) (which uses the drag element method of ref. 26) and with measured spectra.

Directivity predicted by equation (19) in the flyover plane was shown by Fink (figs. 5 and 6 of ref. 27) to fit clean DC-10 airframe noise better than the dipole modeling used in reference 101 to represent that noise. No convection effects were recommended for inclusion in Fink's comparison, and given the uncertainty discussed with regard to trailing-edge noise, it is not clear what convection effects should be included (although the usual convection Doppler shift on frequencies should apply whatever amplitude changes take place).

An additive correction is proposed to take account of about 8 dB difference in OASPL between "very clean" aircraft (high performance gliders and jet aircraft with simple trailing-edge flaps and few flap track fairings) and aircraft in the clean configuration, but with numerous flap fairings and complicated multisegment flaps; see reference 27 for details.

Fink takes the azimuthal directivity of trailing-edge noise to be

$$\langle p^2 \rangle \propto \sin^2 \alpha \tag{22}$$

with $\alpha = 90°$ as the flyover plane, and he (fig. 7 of ref. 27) uses this to explain a phenomenon noted in several full-scale tests, in which a moderately high-frequency speed-independent peak was observed, dominating the OASPL. Fink shows that the OASPL and high-frequency spectral levels do not change with azimuthal angle α, but that the low-frequency levels vary according to equation (22); the inference is that low-frequency levels are set by trailing-edge noise, with scaling according to equation (22), while the higher frequency levels are associated with noise from the idling jet engines, with no dependence on α. Attractive as this explanation is, however (and remarkably well supported by fig. 7 of ref. 27), equation (22) is incorrect for trailing-edge noise; the (3/2)-pole nature of that source gives a variation of p with $\sin(\theta/2)$ in the flyover plane, and with $(\sin \alpha)^{1/2}$ in the azimuthal angle, both non-multipole. The $(\sin \alpha)^{1/2}$ factor can be regarded as confirmed to a high degree of accuracy by the experiments of Kambe (ref. 30) on sound from vortex rings

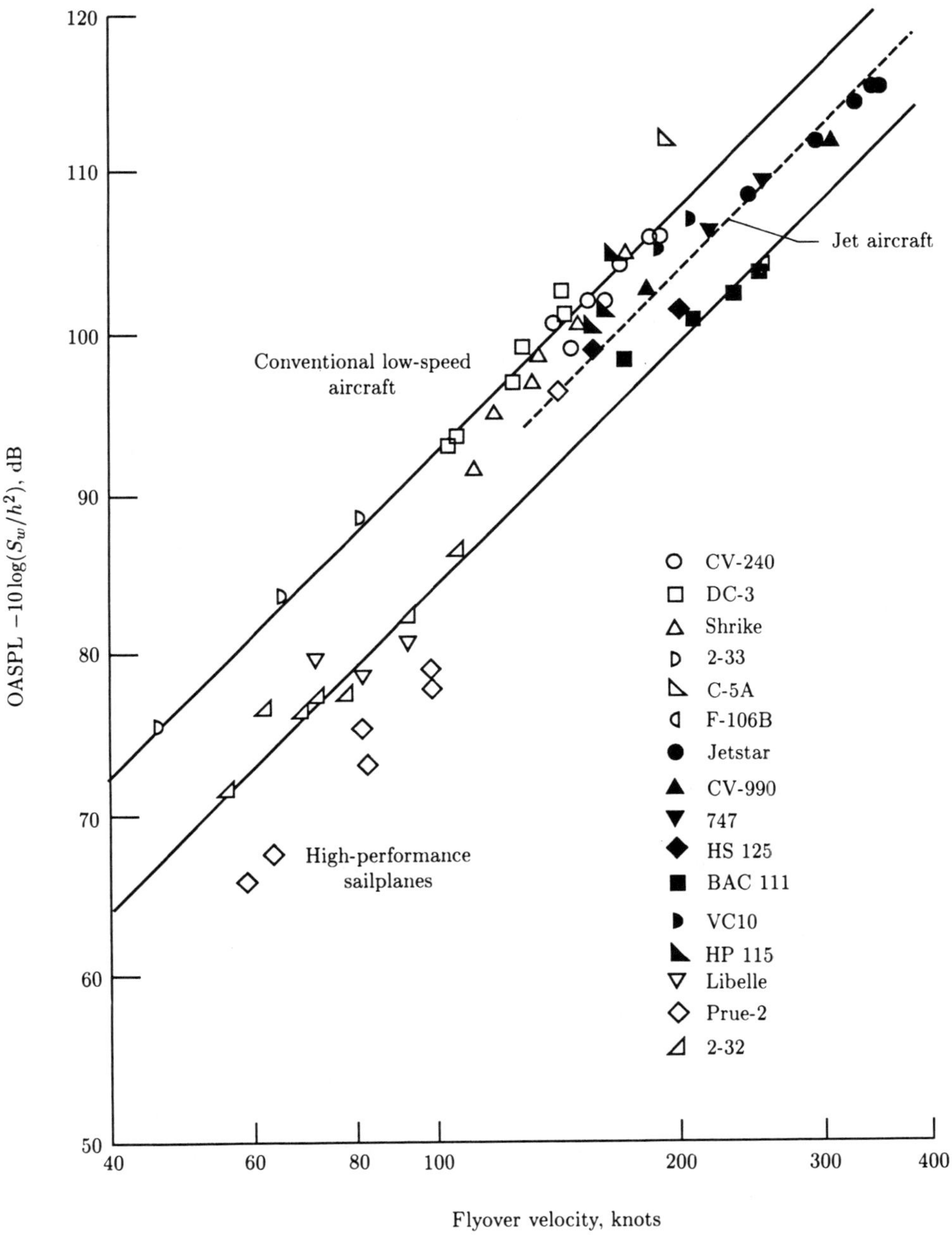

Figure 13. Airframe noise (maximum OASPL for clean configuration at flyover altitude h) normalized for wing area S_w as suggested by trailing-edge noise theory. Solid lines correspond to the U_o^5 variation predicted by trailing-edge noise theory. (From ref. 27.)

near trailing edges, and it is independent of the detailed unsteady flow mechanism considered. Thus the correct azimuthal variation can now be taken as

$$\langle p^2 \rangle \propto \sin\alpha \tag{23}$$

and although this has the right tendency to be consistent with Fink's interpretation, it gives only one-half the decibel reduction with angle that goes with equation (22) and which was plotted from experiment by Fink. Thus the azimuthal directivity of low-frequency trailing-edge noise is a topic needing further experimental study. (The convection factor $(1 - M_{v1}\sin\alpha)^{-1}$ in eq. (5) reduces the discrepancy between eq. (23) and measurement, but not by much at typical low M_{v1}.)

Now, to the trailing-edge noise predictions for the clean wing and horizontal and vertical tails (note that the higher frequency noise from the horizontal tail significantly contributes to perceived noise levels in the flyover plane), Fink's noise component method adds predictions for the noise of nose and main landing gear and of wing flaps and leading-edge slats. With the exception of the leading-edge slats, for which very few data are available, *no interactions* are included between these fields. The basis for this comes from reference 10 where flights of a VC10 with deployment, one at a time, of slats, flaps, and gear could be used to determine single-component additions to the cruise configuration noise levels (engines idling) and could then be simply added to obtain an approach noise spectral prediction in very close agreement with measurement for the fully dirty configuration (see fig. 14). The assumption of weak interaction is a natural first approximation, but other examples can be quoted (refs. 18 and 78) which show significant interactions, both favorable and unfavorable, between the gear and leading- and trailing-edge devices. We should also note that the wheel well doors of the VC10 are closed again after the gear are deployed, which both removes cavity noise and reduces gear-flap interaction noise.

The peak noise from gear components is predicted in reference 27 from analytical fits to curves in reference 77 for two- and four-wheel gear sets (see figs. 8 and 9). It is assumed that the nose gear contributes a drag force, giving a field independent of azimuthal angle, and a side force dipole dominating the sideline field, as discussed in the section on gear noise. For the four-wheel main gear there is a similar split, but now the field vertically below the aircraft is stronger, relative to the sideline field, presumably because of strong vertical forces arising from the interaction between the first and second wheel rows. The analytical fits to the spectra measured on models in reference 77 reproduce the measured data well in the flyover plane and in the sideline direction, but with a slightly larger error (3 dB) at an intermediate angle. Measured spectra (gear down and flaps and slats retracted) are available for the peak approach noise of the VC10 (ref. 10) and 747 (see table 6 of ref. 27). These spectra show oscillatory features, but are quite well predicted by Fink's method based on model data, up to frequencies as high as 5 kHz full scale. The noise of the 747 is underpredicted, however, by as much as 5 dB over the entire range from 1 to 4 kHz, but this may be due to residual engine noise (which would not affect the good prediction of the OASPL for the clean 747). A much larger underprediction, by more than a further 5 dB, results from the use of the method devised by Heller and Dobrzynski on the basis of their own data (ref. 77). As noted previously, that method also seriously underpredicts VC10 gear noise at high frequencies, where the Fink method is much better and is accordingly to be preferred for PNL calculations.

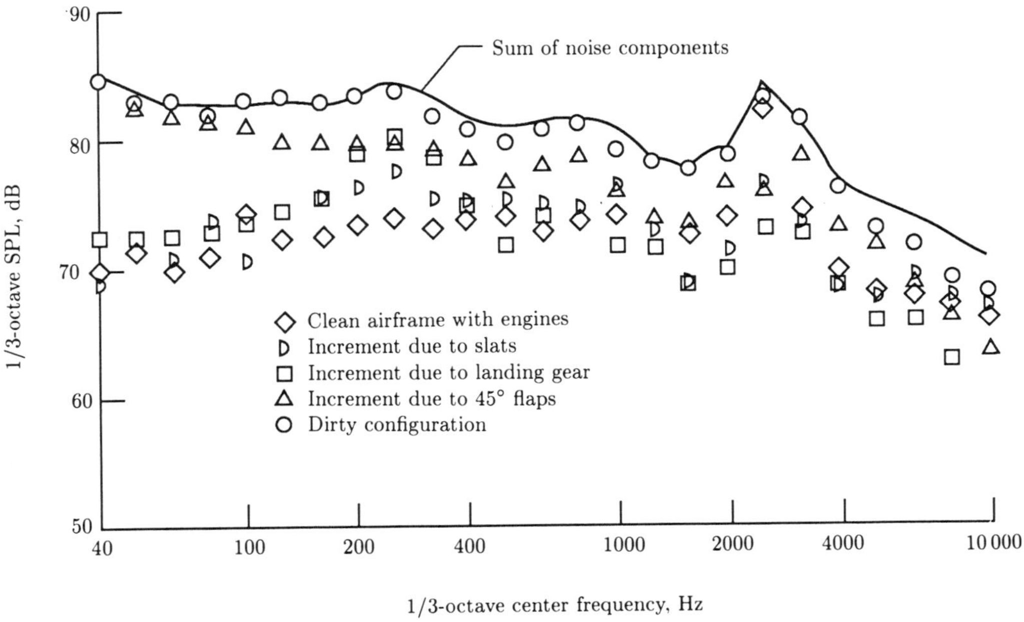

Figure 14. Comparison of sum of measured noise components with measured total noise for VC10 at approach (ref. 10). (From ref. 27.)

Trailing-edge flap noise is modeled in reference 27 as a single lifting dipole field, the dipole being generated by wing turbulence forces on the flap. An equivalent flap chord c_F is defined for the whole (doubly or triply slotted) flap system, and it is asserted that spectral collapse should occur on a Strouhal number basis with c_F as the length scale and that spectral amplitudes should scale with $(\text{Velocity})^6$. It is also argued, and shown rather convincingly from VC10 data at three different flap deflections δ_F, that flap SPL's should scale with $\sin^2 \delta_F$, and a small correction $\cos^2(\delta_F/2)$ is also allowed for rotation of the dipole directivity pattern at the peak noise condition. Figure 15, from reference 27, shows the results of these scalings, mainly for large jet transports, and shows also the difference at high frequencies between aircraft with double- and triple-slotted flap systems. The prediction method of reference 27 represents figure 15 by the straight lines shown and gives them in analytical form.

These results, based as they are on much full-scale data, are probably adequate for prediction in generally similar circumstances. However, they are quite unsatisfactory at a fundamental level, and therefore possibly misleading if applied to rather different configurations. The flap segment is not an isolated airfoil generating compact dipole noise in response to incidence fluctuations associated with the oncoming turbulent flow. The ability of a given unsteady pressure distribution over the airfoil to generate sound is strongly influenced by the geometry of nearby bodies, and the presence near the flap of a wing of large chord compared with the wavelength must affect both the level and directivity of the flap-generated sound. Indeed, one might argue

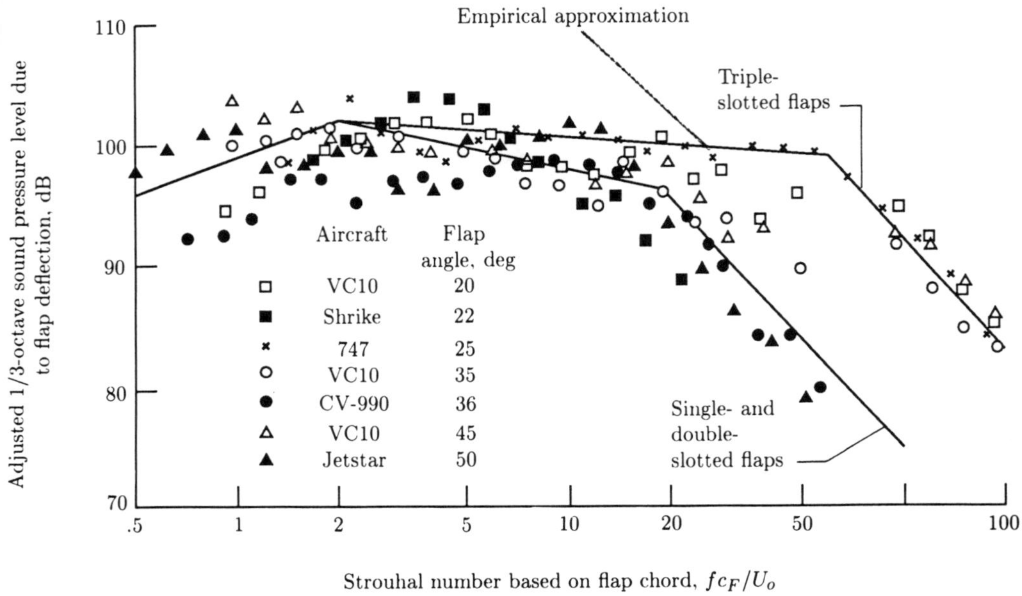

Figure 15. Measured normalized spectra of trailing-edge flap noise. (From ref. 27.)

that for moderate flap deflections the idea that the flap is separate from the wing is irrelevant and that noise is generated by the flap in just the normal trailing-edge way, the deflection of the flap merely leading to different turbulence velocity and length scales near the edge from those pertaining to the clean configuration. And if the flap deflection is large, one might argue that the unsteady loading of the wing itself can be neglected and that the dominant source is associated with force dipoles on the flap. Then the problem is again that of finding the acoustic field of a dipole source near a half-plane, for which again the scaling of $\langle p^2 \rangle$ with velocity involves the fifth power and the directivity is again the half-plane $\sin^2(\theta/2)$ (refs. 32 and 102). The directivity is *fixed* and does not rotate with the dipole attached to the airfoil; but the amplitude in any fixed direction *does* change as the strength and direction of the dipole change with change in the flap deflection. Some more detailed study of these variations would be worthwhile, but would be unlikely to change the conclusion that the flap noise mechanism is misrepresented by dipole scalings and that the proper scalings are those of half-plane trailing-edge noise. Since most of the emphasis in reference 27 is on peak noise levels, the directivity problem as yet is not properly explored, but the enhanced forward directivity of half-plane noise could be significant, as could the (Velocity)^5 rather than (Velocity)^6 law in the case of EBF or USB configurations.

These conclusions are reinforced by the study of reference 103 on the sound generated by the convection of vorticity past the trailing edge of a half-plane, downstream of which is a thin flat plate in the same plane as the half-plane and separated from it by a slot which is narrow on the acoustic wavelength scale. The flat plate, modeling a flap segment, may be compact or noncompact relative to the acoustic wavelength. The most important conclusions are that in all cases the directivity of the scattered field is similar to that due to turbulence at the edge of a large rigid plate, and that this casts serious doubt on correlation procedures which use ideas related to the gust response and acoustic field of a single isolated airfoil.

Data for leading-edge devices—slats or flaps—are scarce, and the only theoretical work is that of reference 99 on the convection of vortices through a slot, which was not intended to model the leading-edge slat device. There are several possible mechanisms: dipole or trailing-edge noise from the slat itself, with strong influence from the large wing nearby; noise from flow past exposed actuators and tracks in the slat gap, again heavily influenced by the local geometry over an acoustic wavelength; monopole radiation from unsteady volume flow through the slat gap; and modifications to the wing mean and unsteady flow. Fethney (ref. 10) gives data for VC10 flyovers with and without leading-edge slat deployment. Typically deployment increases OASPL by 5 dB, the increase being roughly constant up to 1 kHz where engine and clean airframe noise then dominate. Boeing 747 data have been taken (ref. 15) for a large model in a wind tunnel. In the model the leading-edge devices could be operated as flaps with no gap between flap and wing or as slats with a gap. Slats were noisier, typically by 3 dB. A much larger excess of slat noise over leading-edge flap noise was reported, at model scale, in reference 18, in wind-tunnel studies using a directional microphone. These studies indicated that the distribution of slat noise sources was uniform along the slat span, but concentrated at the trailing edge. Scaled model slat noise was 5 dB higher than the measured VC10 levels, so the mechanisms here may be sensitive to Reynolds number. No information is available on the directivity of slat-generated noise. In the absence of any more data, Fink (ref. 27) used an ad hoc modification of his expression for clean wing noise to fit the VC10 slat data; the modification implies an interaction between slats and wing, though there is no postulated mechanism for this. Figure 16 of reference 27 contains the available full-scale data; no theory is available to suggest appropriate SPL, spectral scalings, or directivities.

This Federal Aviation Administration (FAA) noise component scheme of reference 27 was then compared with flyover measurements of the noise from a range of aircraft types (glider, light twin-prop, business jet, large jet transport, and fighter) and with the NASA ANOPP whole aircraft predictions of reference 8 and the NASA ANOPP drag element method predictions of reference 24. A small selection of the results is given below in figures 16 to 20. Considerable variations can be seen between the different predictions; for example, the whole aircraft (ref. 8) and noise component (ref. 27) methods differ by almost 20 dB in OASPL for the 747 in the clean configuration. The noise component method underestimates the OASPL by 8 dB, the spectral levels above 1 kHz by more than 20 dB, and the PNL by 15 dB. More relevant, however, are the predictions for the dirty configuration, where the noise component method reproduces measured OASPL and PNL to within 1 dB. In general the noise component method is reasonably accurate for the OASPL, and more importantly for the PNL, in the dirty configurations of all aircraft studied except the F-106B delta wing fighter, for which the noise component and drag element methods underpredicted OASPL by nearly 10 dB and the whole aircraft method overpredicted OASPL by more than 20 dB (see fig. 20). Since data from the F-106B were entirely excluded from the regression analysis that led to the scheme of reference 8, one can only conclude that delta wing designs are not adequately covered by anything but a large ad hoc modification of any of the existing prediction schemes. Note, however, that all three schemes gave a directivity in good agreement with measurement for the F-106B (and conforming to the half-plane directivity law), so that it appears that the scale and intensity of trailing-edge fluctuations on delta wings are not correctly modeled by those on conventional wings of moderate taper ratio.

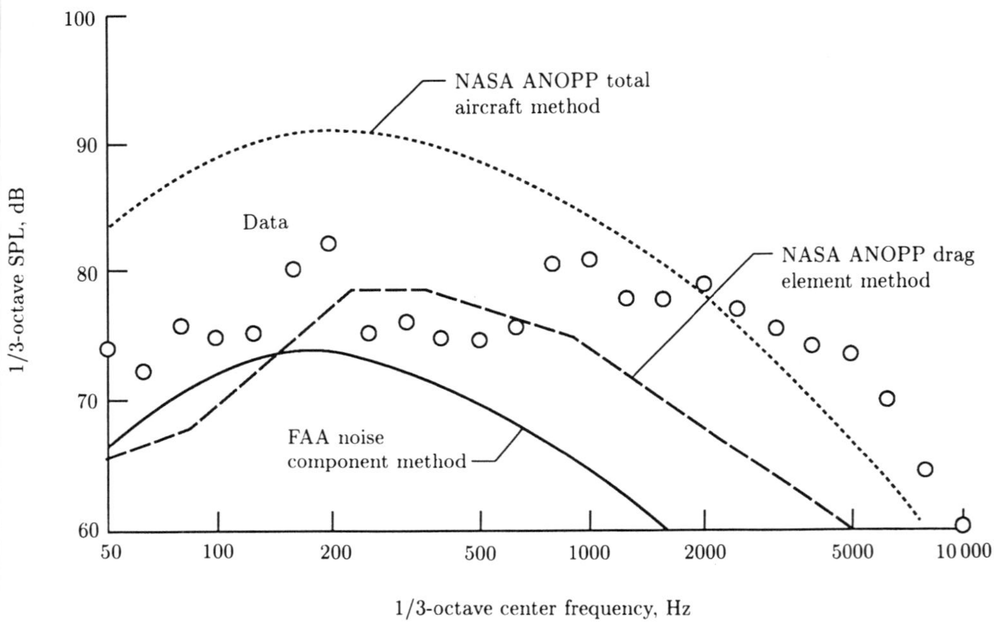

Figure 16. Comparison of measured and predicted flyover noise spectra for Boeing 747 in clean configuration at 233 knots. (From ref. 27.)

Conclusions

We have reviewed current understanding of airframe noise as represented by experiment at model and full scale, by theoretical modeling, and by empirical correlation methods. The principal component sources are associated with the trailing edges of wing and tail, deflected trailing-edge flaps, flap side edges, leading-edge flaps or slats, undercarriage gear elements, gear wheel wells, fuselage and wing boundary layers, and panel vibration, together with many minor protrusions like radio antennas and air-conditioning intakes which may contribute significantly to perceived noise. There are also possibilities for interactions between the various mechanisms. With current engine technology, the principal airframe noise mechanisms dominate only at low frequencies, typically less than 1 kHz and often much lower, but further reduction of turbomachinery noise in particular may make airframe noise the principal element of approach noise at frequencies in the most sensitive range.

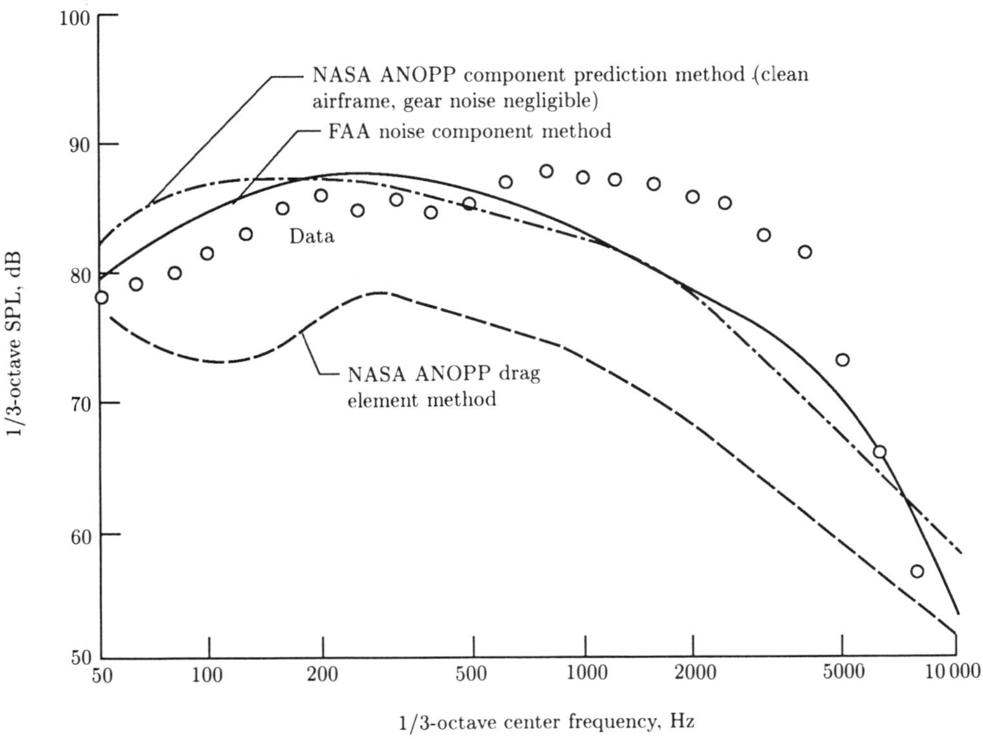

Figure 17. Comparison of measured and predicted flyover noise spectra for Boeing 747 with landing gear extended, at 222 knots. (From ref. 27.)

In some areas there is a good basic understanding—for example, in the case of gear noise, though here the only use of that understanding is likely to be in the adoption of appropriate scalings for correlation schemes. The work described here from references 27 and 77 represents a successful example of such correlation; more sophisticated theoretical study is not needed and, indeed, likely not to be helpful. For trailing-edge noise there is a substantial body of theory, and theory and experiment are in agreement on overall features, but apparently in conflict over several points: one is the matter of the Kutta condition, which controls the level of trailing-edge noise and modifies its directivity; another is the issue of whether the prediction of reference 36 from an aeroacoustic analogy can deal properly with high-frequency noise from trailing edges, or whether a modification (to deal with the aircraft flyover situation) of the theory of reference 64 is needed. Flap side edges appear to be important sources, and much further work, theoretical and experimental, is needed here; none of the complete prediction schemes yet incorporates side-edge acoustic sources. Significant effects have also been reported for leading-edge devices, but there is a great need for more data here, and for theoretical modeling, as there is also for the question of noise radiation from the vibration of panels in undercarriage gear doors and on deflected flap segments. For undercarriage gear cavities, the basic response mode at low frequencies is understood, but damping mechanisms that are

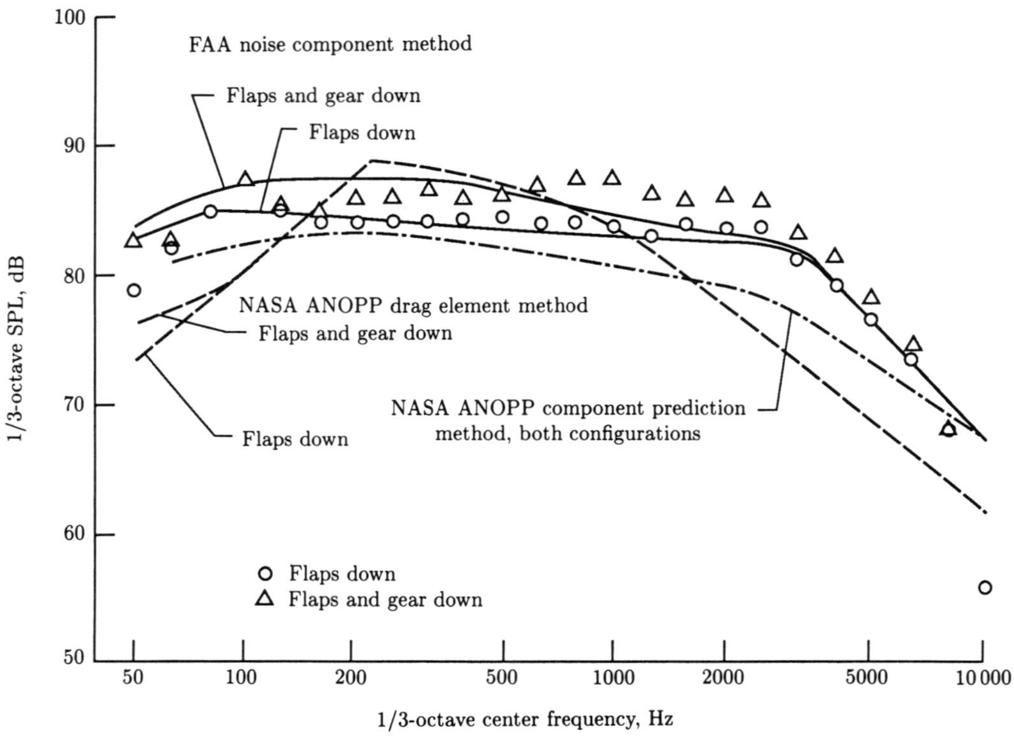

Figure 18. Comparison of measured and predicted flyover noise spectra for Boeing 747 with trailing-edge flaps and gear extended, at 204 knots. (From ref. 27.)

not understood seem to limit noise of realistic cavities to levels well below those observed in idealized experiments, and probably below other airframe noise fields. Significant benefits are achieved, though, if the wheel well door can be closed after deployment of the landing gear (ref. 10). For USB or EBF configurations, even the basic mechanisms are not understood, and the combination of large flap deflection, multiple flaps and slots, and high Mach number flow makes it difficult to suggest appropriate scalings, directivity, and spectral features.

We must stress that the generally available data base is extremely limited, in terms of the aircraft included, configurations and speeds flown, and directional characteristics measured. Not surprisingly there are many diametrically contradictory findings reported, and it has not been possible here to present a consistent view of much of the work. These contradictions may be due in part to the fact that many of the airframe noise mechanisms are Reynolds number dependent (to a greater extent than in turbomachinery and jet noise), in part to the fact that the noise levels under discussion are very low indeed, and in part to the difficulties of experimental technique for flyovers of large aircraft at low thrust.

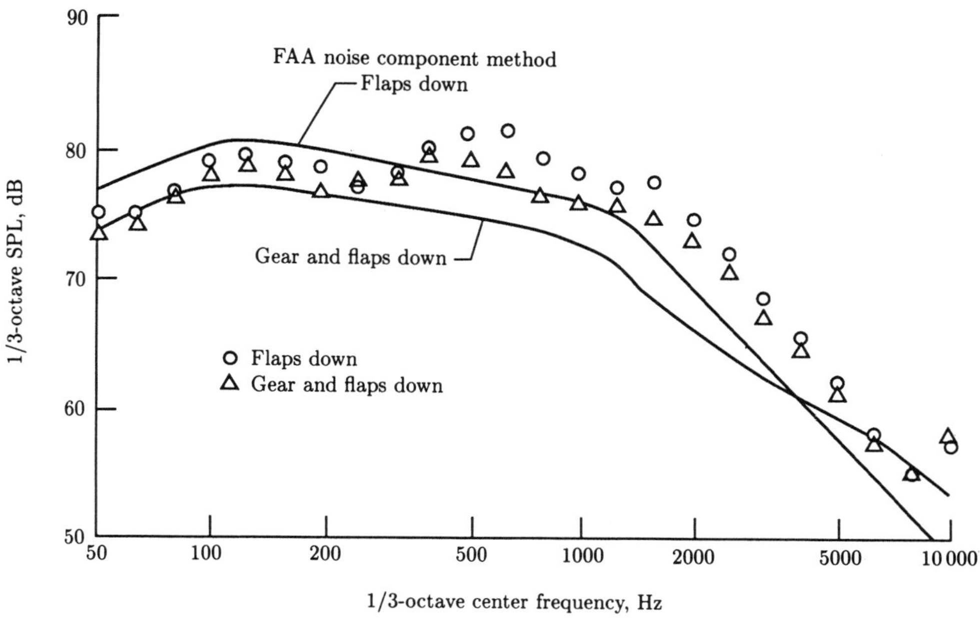

Figure 19. Comparison of measured and predicted flyover noise spectra for Convair CV-990, with trailing-edge flaps extended (36°) at 189 knots and flaps and gear extended at 160 knots. (From ref. 27.)

Turning now to the possibilities for airframe noise reduction, these lie entirely at the moment in the modification of elements contributing to the noise of the dirty configuration. The noise from these elements can be somewhat reduced by major change to the basic airframe to achieve lower approach speed. Doubling wing area could reduce speed by a factor 0.7 and reduce gear noise by 8 dB (ref. 27), but such modification has major operating disadvantages that would probably be unacceptable. Modifications to individual elements should therefore be sought at constant typical approach speeds.

For trailing-edge and flap noise, numerous proposals have been made (see references in ref. 27) and tested at small scale. These involve either a change to the surface impedance (over the whole surface or near leading or trailing edges) to minimize the discontinuity felt by a convected eddy passing over an edge, or a change to the turbulence properties near an edge by the action of a screen or grid. Porous trailing edges were found to have no effect on the noise from an isolated airfoil exposed to grid turbulence, whereas porous leading edges reduced noise by more than 6 dB at frequencies significant to PNL at full scale. However the recommendation (ref. 27) for adoption of porous leading edges on flap segments must be treated with great caution. The noise of a chordwise compact airfoil is dipole and determined by the gust-response lift of the airfoil, which in turn is determined mainly by the leading-edge peak pressures, so that some significant reduction would be expected from any impedance change at the leading edge. By contrast, the trailing-edge loading is zero

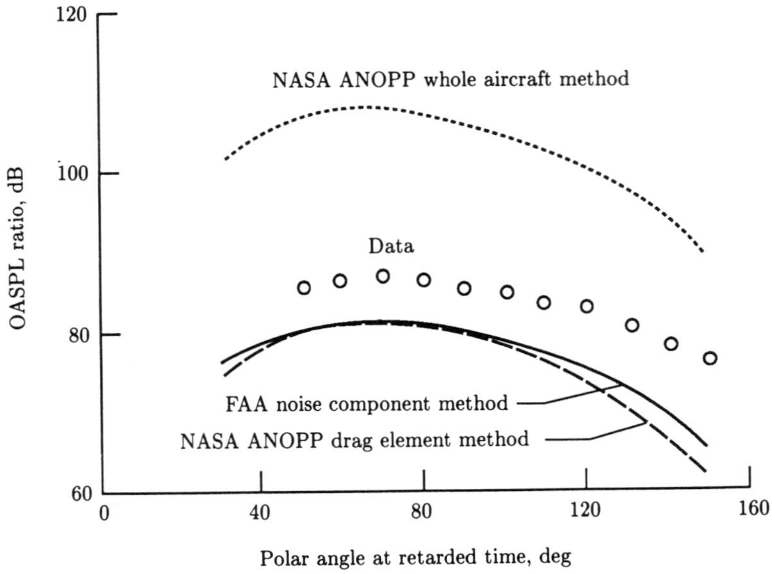

Figure 20. Comparison of measured and predicted OASPL directivity for Convair F-106B. (From ref. 27.)

if the Kutta condition is applied, and any impedance change at the trailing edge is likely to be less important. These arguments do not, however, apply directly to the flap noise problem, where the presence of the wing ahead controls the extent to which a given force distribution on the flap can generate noise. The porous flap studies need to be performed not on an isolated airfoil, but on an airfoil coupled to a wing. In view of the importance of flap side-edge sources, attention should also be given to changes of surface impedance near those side edges. Such studies have in fact been conducted in connection with USB and EBF tests, and substantial benefits have been found (ref. 104) from the use of porous trailing-edge regions—benefits typically of 5 dB over a wide frequency range and of as much as 10 dB over narrow ranges.

Serrated leading and trailing edges have been studied (see ref. 27) as a means of obtaining a more gradual impedance change at the edge. So far no noise reductions have been reported with such edges, though small noise increases have been.

Trailing-edge noise is reduced if the trailing edge is swept relative to the mean flow (eq. (5)). Significant sweep of the whole wing is obviously not feasible in many applications, but one might hope that the same effect might be achieved by a sawtooth trailing edge, with alternating portions with large forward and backward sweep. Limited tests have been carried out on models with a sawtooth trailing edge (ref. 105), and noise reductions of 3 to 6 dB were obtained for sweep angles of 60°; therefore this appears to be a promising idea for further development, perhaps also when combined with the idea of a porous trailing-edge region mentioned above. Because the two concepts exploit different properties of trailing-edge noise, the benefits should be additive. Both have been tested at larger scale, and the sawtooth trailing edge has been tested at full scale in flight (ref. 106), where it reduced EPNL by 2 dB.

Flap noise may be controlled somewhat by minimizing the number of side edges and by preventing direct jet impingement on the side edges. The effects of flap track fairings may also be important, and narrow-chord vanes forming part of a triple-slotted flap system may generate high-frequency noise contributing to PNL. Interaction between the gear and flaps may be beneficial if the flap deployment reduces velocity around the gear, but unfavorable if the gear wake impinges on the flap trailing and side edges.

Ideas for controlling cavity noise have been examined in reference 107, among them one for preventing cavity shear layer interaction with the downstream cavity lip by fluid injection into the cavity from the base. This may be helpful for long cavities, but it seems to go against the idea (ref. 81) that an interacting flow might lead to the generation of vorticity from the downstream lip and that this would tend to reduce the volume flow across the cavity mouth and hence reduce the radiated sound. But in any event, typical aircraft cavities seem to operate in a resonant depth mode, producing only low-frequency noise heavily damped by cavity fittings and gear and not significant for PNL. The landing gear elements themselves are, however, strong noise sources, and large clusters of wheels are likely to produce intense noise below and to the side of the aircraft. Vortex shedding from cylindrical members can be greatly reduced by the presence of splitter plates. These were recommended in reference 27, but appear feasible only when mounted vertically behind vertical struts, where they reduce sideline, but not approach, noise. Control of broadband aerodynamic interactions between gear components producing noise below the aircraft seems the most difficult problem in airframe noise. Much might, however, be accomplished, as one of the reviewers has pointed out, by relatively simple efforts to generally "clean up" the flap and gear components, the actuators and doors, and the leading and side flap edges, from the aerodynamic point of view.

References

1. Smith, D. L.; Paxson, R. P.; Talmadge, R. D.; and Hotz, E. R.: *Measurements of the Radiated Noise From Sailplanes.* AFFDL-TM-70-3-FDDA, U.S. Air Force, July 1970. (Available from DTIC as AD 709 689.)
2. Kroeger, Richard A.; Grushka, Heinz D.; and Helvey, Tibor C.: *Low Speed Aerodynamics for Ultra-Quiet Flight.* AFFDL-TR-71-75, U.S. Air Force, Mar. 1972. (Available from DTIC as AD 893 426.)
3. *Noise Standards: Aircraft Type and Airworthiness Certification.* FAR Pt. 36, Federal Aviation Adm., June 1974. (Consolidated Reprint Aug. 12, 1985.)
4. Schwartz, Ira R., ed.: *Aeroacoustics: STOL Noise; Airframe and Airfoil Noise.* American Inst. of Aeronautics and Astronautics, c.1976.
5. Gibson, John S.: *Non-Engine Aerodynamic Noise Investigation of a Large Aircraft.* NASA CR-2378, 1974.
6. Healy, Gerald J.: *Measurement and Analysis of Aircraft Far-Field Aerodynamic Noise.* NASA CR-2377, 1974.
7. Blumenthal, V. L.; Streckenbach, J. M.; and Tate, R. B.: Aircraft Environmental Problems. AIAA Paper 73-5, Jan. 1973.
8. Hardin, Jay C.; Fratello, David J.; Hayden, Richard E.; Kadman, Yoran; and Africk, Steven: *Prediction of Airframe Noise.* NASA TN D-7821, 1975.
9. Putnam, Terrill W.; Lasagna, Paul L.; and White, Kenneth C.: Measurements and Analysis of Aircraft Airframe Noise. *Aeroacoustics: STOL Noise; Airframe and Airfoil Noise,* Ira R. Schwartz, ed., American Inst. of Aeronautics and Astronautics, c.1976, pp. 363–378.
10. Fethney, P.: An Experimental Study of Airframe Self-Noise. *Aeroacoustics: STOL Noise; Airframe and Airfoil Noise,* Ira R. Schwartz, ed., American Inst. of Aeronautics and Astronautics, c.1976, pp. 379–403.

11. Hardin, Jay C.: *Airframe Self-Noise—Four Years of Research.* NASA TM X-73908, 1976.
12. Pendley, Robert E.: Recent Advances in the Technology of Aircraft Noise Control. AIAA Paper 75-317, Feb. 1975.
13. Hayden, R. E.; Kadman, Y.; Bliss, D. B.; and Africk, S. A.: Diagnostic Calculations of Airframe-Radiated Noise. *Aeroacoustics: STOL Noise; Airframe and Airfoil Noise,* Ira R. Schwartz, ed., American Inst. of Aeronautics and Astronautics, c.1976, pp. 179–201.
14. Shearin, J. G.; and Block, P. J.: Airframe Noise Measurements on a Transport Model in a Quiet Flow Facility. AIAA Paper 75-509, Mar. 1975.
15. Shearin, John G.; Fratello, David J.; Bohn, A. J.; and Burggraf, W. D.: Model and Full-Scale Large Transport Airframe Noise. AIAA Paper 76-550, July 1976.
16. Bauer, A. B.; and Munson, A. G.: *Airframe Noise of the DC-9-31.* NASA CR-3027, 1978. (Available as AIAA Paper 77-1272.)
17. Block, P. J. W.: Assessment of Airframe Noise. *J. Aircr.,* vol. 16, no. 12, Dec. 1979, pp. 834–841.
18. Fink, Martin R.; and Schlinker, Robert H.: Airframe Noise Component Interaction Studies. *J. Aircr.,* vol. 17, no. 2, Feb. 1980, pp. 99–105. (Available as AIAA Paper 79-0668.)
19. Fratello, David J.; and Shearin, John G.: A Preliminary Investigation of Remotely Piloted Vehicles for Airframe Noise Research. AIAA Paper 75-512, Mar. 1975.
20. Amiet, Roy K.: Correction of Open-Jet Wind-Tunnel Measurements for Shear Layer Refraction. *Aeroacoustics: Acoustic Wave Propagation; Aircraft Noise Prediction; Aeroacoustic Instrumentation,* Ira R. Schwartz, ed., American Inst. of Aeronautics and Astronautics, c.1976, pp. 259–280.
21. Ahtye, Warren F.; Miller, Wendell R.; and Meecham, William C.: Wing and Flap Noise Measured by Near- and Far-Field Cross-Correlation Techniques. AIAA Paper 79-0667, Mar. 1979.
22. Kendall, James M.; and Ahtye, Warren F.: Noise Generation by a Lifting Wing/Flap Combination at Reynolds Numbers to 2.8×10^6. AIAA-80-0035, Jan. 1980.
23. Miller, W. R.: Flap Noise Characteristics Measured by Pressure Cross Correlation Techniques. Ph.D. Thesis, Univ. of California, 1980.
24. Revell, James D.; Healy, Gerald J.; and Gibson, John S.: Method for the Prediction of Airframe Aerodynamic Noise. *Aeroacoustics: Acoustic Wave Propagation; Aircraft Noise Prediction; Aeroacoustic Instrumentation,* Ira R. Schwartz, ed., American Inst. of Aeronautics and Astronautics, c.1976, pp. 139–154.
25. Healy, Gerald J.: Aircraft Far-Field Aerodynamic Noise: Its Measurement and Prediction. *Aeroacoustics: STOL Noise; Airframe and Airfoil Noise,* Ira R. Schwartz, ed., American Inst. of Aeronautics and Astronautics, c.1976, pp. 203–219.
26. Revell, James D.: Induced Drag Effect on Airframe Noise. *Aeroacoustics: STOL Noise; Airframe and Airfoil Noise,* Ira R. Schwartz, ed., American Inst. of Aeronautics and Astronautics, c.1976, pp. 221–235.
27. Fink, Martin R.: *Airframe Noise Prediction Method.* FAA-RD-77-29, Mar. 1977. (Available from DTIC as AD A039 664.)
28. Fink, Martin R.: Noise Component Method for Airframe Noise. *J. Aircr.,* vol. 16, no. 10, Oct. 1979, pp. 659–665.
29. Ffowcs Williams, J. E.; and Hall, L. H.: Aerodynamic Sound Generation by Turbulent Flow in the Vicinity of a Scattering Half Plane. *J. Fluid Mech.,* vol. 40, pt. 4, Mar. 9, 1970, pp. 657–670.
30. Kambe, T.: Acoustic Emissions by Vortex Motions. *J. Fluid Mech.,* vol. 173, Dec. 1986, pp. 643–666.
31. Powell, Alan: On the Aerodynamic Noise of a Rigid Flat Plate Moving at Zero Incidence. *J. Acoust. Soc. America,* vol. 31, no. 12, Dec. 1959, pp. 1649–1653.
32. Crighton, D. G.; and Leppington, F. G.: On the Scattering of Aerodynamic Noise. *J. Fluid Mech.,* vol. 46, pt. 3, Apr. 13, 1971, pp. 577–597.
33. Crighton, D. G.: Basic Principles of Aerodynamic Noise Generation. *Prog. Aerosp. Sci.,* vol. 16, no. 1, Mar. 1975, pp. 31–96.
34. Ffowcs Williams, J. E.: Hydrodynamic Noise. *Annual Review of Fluid Mechanics, Volume 1,* William R. Sears and Milton Van Dyke, eds., Annual Reviews, Inc., 1969, pp. 197–222.
35. Ffowcs Williams, J. E.: Aeroacoustics. *Annual Review of Fluid Mechanics, Volume 9,* Milton Van Dyke, J. V. Wehausen, and John L. Lumley, eds., Annual Reviews Inc., 1977, pp. 447–468.
36. Howe, M. S.: A Review of the Theory of Trailing Edge Noise. *J. Sound & Vibration,* vol. 61, no. 3, Dec. 8, 1978, pp. 437–465.
37. Lighthill, M. J.: On Sound Generated Aerodynamically. I. General Theory. *Proc. Royal Soc. London,* ser. A, vol. 211, no. 1107, Mar. 20, 1952, pp. 564–587.

38. Crighton, David G.: The Kutta Condition in Unsteady Flow. *Annual Review of Fluid Mechanics, Volume 17,* Milton Van Dyke, J. V. Wehausen, and John L. Lumley, eds., Annual Reviews Inc., 1985, pp. 411–445.
39. Bogucz, E. A.; and Walker, J. D. A.: The Turbulent Near Wake at a Sharp Trailing Edge. *J. Fluid Mech.,* vol. 196, Nov. 1988, pp. 555–584.
40. Chase, David M.: Sound Radiated by Turbulent Flow Off a Rigid Half-Plane as Obtained From a Wavevector Spectrum of Hydrodynamic Pressure. *J. Acoust. Soc. America,* vol. 52, no. 3, pt. 2, Sept. 1972, pp. 1011–1023.
41. Chandiramani, K. L.: Diffraction of Evanescent Waves, With Applications to Aerodynamically Scattered Sound and Radiation From Unbaffled Plates. *J. Acoust. Soc. America,* vol. 55, no. 1, Jan. 1974, pp. 19–29.
42. Yu, J. C.; and Tam, C. K. W.: Experimental Investigation of the Trailing Edge Noise Mechanism. *AIAA J.,* vol. 16, no. 9, Sept. 1978, pp. 1046–1052.
43. Howe, M. S.: The Effect of Forward Flight on the Diffraction Radiation of a High Speed Jet. *J. Sound & Vibration,* vol. 50, no. 2, Jan. 22, 1977, pp. 183–193.
44. Amiet, R. K.: Acoustic Radiation From an Airfoil in a Turbulent Stream. *J. Sound & Vibration,* vol. 41, no. 4, Aug. 1975, pp. 407–420.
45. Amiet, R. K.: Noise Due to Turbulent Flow Past a Trailing Edge. *J. Sound & Vibration,* vol. 47, no. 3, 1976, pp. 387–393.
46. Amiet, Roy K.: High Frequency Thin-Airfoil Theory for Subsonic Flow. *AIAA J.,* vol. 14, no. 8, Aug. 1976, pp. 1076–1082.
47. Amiet, R. K.: A Note on Edge Noise Theories. *J. Sound & Vibration,* vol. 78, no. 4, Oct. 22, 1981, pp. 485–488.
48. Corcos, G. M.: The Structure of the Turbulent Pressure Field in Boundary-Layer Flows. *J. Fluid Mech.,* vol. 18, pt. 3, Mar. 1964, pp. 353–378.
49. Amiet, R. K.: Gust Response for Flat-Plate Airfoils and the Kutta Condition. To appear *AIAA J.,* July 1990.
50. Hayden, Richard E.: Sound Generation by Turbulent Wall Jet Flow Over a Trailing Edge. M.S. Thesis, Purdue Univ., Aug. 1969.
51. Fink, Martin R.: Experimental Evaluation of Theories for Trailing Edge and Incidence Fluctuation Noise. *AIAA J.,* vol. 13, no. 11, Nov. 1975, pp. 1472–1477.
52. Tam, Christopher K. W.; and Yu, J. C.: Trailing Edge Noise. *Aeroacoustics: STOL Noise; Airframe and Airfoil Noise,* Ira R. Schwartz, ed., American Inst. of Aeronautics and Astronautics, c.1976, pp. 259–280.
53. Joshi, M. C.; and Yu, J. C.: Reduction of Wall Jet Trailing Edge Noise by Mean Flow Modification. *J. Aircr.,* vol. 17, no. 9, Sept. 1980, pp. 633–640.
54. Dobrzynski, Werner M.: Trailing-Edge Airframe Noise Source Studies on Aircraft Wings. *J. Aircr.,* vol. 18, no. 5, May 1981, pp. 397–402.
55. Heller, Hanno H.; and Dobrzynski, Werner M.: Unsteady Surface Pressure Characteristics on Aircraft Components and Far-Field Radiated Airframe Noise. *J. Aircr.,* vol. 15, no. 12, Dec. 1978, pp. 809–815.
56. Brooks, T. F.; and Hodgson, T. H.: Trailing Edge Noise Prediction From Measured Surface Pressures. *J. Sound & Vibration,* vol. 78, no. 1, Sept. 8, 1981, pp. 69–117.
57. Chase, David M.: Noise Radiated From an Edge in Turbulent Flow. *AIAA J.,* vol. 13, no. 8, Aug. 1975, pp. 1041–1047.
58. Brooks, T. F.; and Hodgson, T. H.: Investigation of Trailing-Edge Noise. *Mechanics of Sound Generation in Flows,* E.-A. Müller, ed., Springer-Verlag, 1979, pp. 76–84.
59. Curle, N.: The Influence of Solid Boundaries Upon Aerodynamic Sound. *Proc. Royal Soc. (London),* ser. A, vol. 231, no. 1187, Sept. 20, 1955, pp. 505–514.
60. Crighton, D. G.: Scattering and Diffraction of Sound by Moving Bodies. *J. Fluid Mech.,* vol. 72, pt. 2, Nov. 25, 1975, pp. 209–227.
61. Howe, M. S.: The Generation of Sound by Aerodynamic Sources in an Inhomogeneous Steady Flow. *J. Fluid Mech.,* vol. 67, pt. 3, Feb. 11, 1975, pp. 597–610.
62. Howe, M. S.: The Displacement-Thickness Theory of Trailing Edge Noise. *J. Sound & Vibration,* vol. 75, no. 2, Mar. 22, 1981, pp. 239–250.
63. Goldstein, M. E.: Characteristics of the Unsteady Motion on Transversely Sheared Mean Flows. *J. Fluid Mech.,* vol. 84, pt. 2, Jan. 30, 1978, pp. 305–329.
64. Goldstein, M. E.: Scattering and Distortion of the Unsteady Motion on Transversely Sheared Flows. *J. Fluid Mech.,* vol. 91, pt. 4, Apr. 27, 1979, pp. 601–632.

65. Olsen, W. A.: Noise Generated by Impingement of Turbulent Flow on Airfoils of Varied Chord, Cylinders, and Other Flow Obstructions. AIAA Paper 76-504, July 1976.
66. Olsen, W.; and Boldman, D.: Trailing Edge Noise Data With Comparison to Theory. AIAA Paper 79-1524, July 1979.
67. Bechert, D.; and Pfizenmaier, E.: On the Amplification of Broad Band Jet Noise by a Pure Tone Excitation. *J. Sound & Vibration,* vol. 43, no. 3, Dec. 8, 1975, pp. 581–587.
68. Moore, C. J.: The Role of Shear-Layer Instability Waves in Jet Exhaust Noise. *J. Fluid Mech.,* vol. 80, pt. 2, Apr. 25, 1977, pp. 321–367.
69. Hayden, R. E.; Fox, H. L.; and Chanaud, R. C.: *Some Factors Influencing Radiation of Sound From Flow Interaction With Edges of Finite Surfaces.* NASA CR-145073, 1976.
70. Kendall, James M.: Measurements of Noise Produced by Flow Past Lifting Surfaces. AIAA Paper 78-239, Jan. 1978.
71. Hardin, Jay C.: Noise Radiation From the Side Edges of Flaps. *AIAA J.,* vol. 18, no. 5, May 1980, pp. 549–552.
72. Howe, M. S.: Contributions to the Theory of Aerodynamic Sound, With Application to Excess Jet Noise and the Theory of the Flute. *J. Fluid Mech.,* vol. 71, pt. 4, Oct. 14, 1975, pp. 625–673.
73. Crighton, D. G.: Radiation From Vortex Filament Motion Near a Half Plane. *J. Fluid Mech.,* vol. 51, pt. 2, Jan. 25, 1972, pp. 357–362.
74. Francis, Michael S.; and Kennedy, Donald A.: Formation of a Trailing Vortex. *J. Aircr.,* vol. 16, no. 3, Mar. 1979, pp. 148–154.
75. Meecham, W. C.: Aerosound From Corner Flow and Flap Flow. *AIAA J.,* vol. 21, no. 2, Feb. 1983, pp. 228–234.
76. Howe, M. S.: On the Generation of Side-Edge Flap Noise. *J. Sound & Vibration,* vol. 80, no. 4, Feb. 22, 1982, pp. 555–573.
77. Heller, Hanno H.; and Dobrzynski, Werner M.: Sound Radiation From Aircraft Wheel-Well/Landing-Gear Configurations. *J. Aircr.,* vol. 14, no. 8, Aug. 1977, pp. 768–774.
78. Preisser, John S.: Airframe Noise Measurements on a Supersonic Transport Small-Scale Model. *J. Aircr.,* vol. 17, no. 11, Nov. 1980, pp. 795–801.
79. Bohn, A. J.: Edge Noise Attenuation by Porous-Edge Extensions. AIAA Paper 76-80, Jan. 1976.
80. Rockwell, D.; and Naudascher, E.: Self-Sustained Oscillations of Impinging Free Shear Layers. *Annual Review of Fluid Mechanics, Volume 11,* Milton Van Dyke, J. V. Wehausen, and John L. Lumley, eds., Annual Reviews Inc., 1979, pp. 67–94.
81. Howe, M. S.: On the Theory of Unsteady Shearing Flow Over a Slot. *Philos. Trans. Royal Soc. London,* ser. A, vol. 303, no. 1475, Nov. 3, 1981, pp. 151–180.
82. Block, Patricia J. W.; and Heller, Hanno: *Measurements of Farfield Sound Generation From a Flow-Excited Cavity.* NASA TM X-3292, 1975.
83. Dobrzynski, Werner M.; and Heller, Hanno H.: Are Wheel-Well Related Aeroacoustic Sources of Any Significance in Airframe Noise? *Zeitschrift Flugwiss. & Weltraumforschung,* Bd. 2, Heft 1, 1978, pp. 35–42.
84. Plumblee, H. E.; Gibson, J. S.; and Lassiter, L. W.: *A Theoretical and Experimental Investigation of the Acoustic Response of Cavities in an Aerodynamic Flow.* WADD-TR-61-75, U.S. Air Force, Mar. 1962.
85. Howe, M. S.: The Dissipation of Sound at an Edge. *J. Sound & Vibration,* vol. 70, no. 3, June 8, 1980, pp. 407–411.
86. Powell, Alan: Aerodynamic Noise and the Plane Boundary. *J. Acoust. Soc. America,* vol. 32, no. 8, Aug. 1960, pp. 982–990.
87. Ffowcs Williams, John E.: Sound Radiation From Turbulent Boundary Layers Formed on Compliant Surfaces. *J. Fluid Mech.,* vol. 22, pt. 2, June 1965, pp. 347–358.
88. Howe, M. S.: The Role of Surface Shear Stress Fluctuations in the Generation of Boundary Layer Noise. *J. Sound & Vibration,* vol. 65, no. 2, July 22, 1979, pp. 159–164.
89. Dowell, E. H.: Radiation From Panels as a Source of Airframe Noise. *AIAA J.,* vol. 13, no. 11, Nov. 1975, pp. 1529–1530.
90. Howe, M. S.: Diffraction Radiation Produced by Turbulent Boundary Layer Excitation of a Panel. *J. Sound & Vibration,* vol. 121, no. 1, Feb. 22, 1988, pp. 47–65.
91. Chase, D. M.: Modeling the Wavevector-Frequency Spectrum of Turbulent Boundary Layer Wall Pressure. *J. Sound & Vibration,* vol. 70, no. 1, May 8, 1980, pp. 29–67.
92. Howe, M. S.: On the Generation of Sound by Turbulent Boundary Layer Flow Over a Rough Wall. *Proc. Royal Soc. London,* ser. A, vol. 395, no. 1809, Oct. 8, 1984, pp. 247–263.

93. Howe, M. S.: The Influence of Viscous Surface Stress on the Production of Sound by Turbulent Boundary Layer Flow Over a Rough Wall. *J. Sound & Vibration,* vol. 104, no. 1, Jan. 8, 1986, pp. 29–39.
94. Patterson, Grant T.; Joshi, M. C.; and Maus, James R.: Experimental Investigation of the Aeroacoustic Characteristics of Model Slot Nozzles With Straight Flaps. *Aeroacoustics: STOL Noise; Airframe and Airfoil Noise,* Ira R. Schwartz, ed., American Inst. of Aeronautics and Astronautics, c.1976, pp. 41–57.
95. Reddy, N. N.: Propulsive-Lift Noise of an Upper-Surface-Blown Flap Configuration. *Aeroacoustics: STOL Noise; Airframe and Airfoil Noise,* Ira R. Schwartz, ed., American Inst. of Aeronautics and Astronautics, c.1976, pp. 27–39.
96. Fink, Martin R.: Scrubbing Noise of Externally Blown Flaps. *Aeroacoustics: STOL Noise; Airframe and Airfoil Noise,* Ira R. Schwartz, ed., American Inst. of Aeronautics and Astronautics, c.1976, pp. 3–25.
97. Horne, C.; and Karamcheti, K.: Some Flow Features of Wall Jet Tones. AIAA Paper 79-0606, Mar. 1979.
98. Goodykoontz, J.; Von Glahn, U.; and Dorsch, R.: Forward Velocity Effects on Under-the-Wing Externally Blown Flap Noise. *Aeroacoustics: STOL Noise; Airframe and Airfoil Noise,* Ira R. Schwartz, ed., American Inst. of Aeronautics and Astronautics, c.1976, pp. 147–174.
99. Howe, M. S.: *The Aerodynamic Noise of a Slot in an Aerofoil.* TR 77129, Royal Aircraft Establ., Aug. 1977.
100. Paterson, Robert W.; and Amiet, Roy K.: Acoustic Radiation and Surface Pressure Characteristics of an Airfoil Due to Incident Turbulence. AIAA Paper 76-571, July 1976.
101. Munson, A. C.: A Modeling Approach to Nonpropulsive Noise. AIAA Paper 76-525, July 1976.
102. Meecham, W. C.; Bui, T. D.; and Miller, W. R.: Diffraction of Dipole Sound by the Edge of a Rigid Baffle. *J. Acoust. Soc. America,* vol. 70, no. 5, Nov. 1981, pp. 1531–1533.
103. Howe, M. S.: Aerodynamic Sound Generated by a Slotted Trailing Edge. *Proc. Royal Soc. London,* ser. A, vol. 373, no. 1753, Nov. 25, 1980, pp. 235–252.
104. Hayden, Richard E.: USB Noise Reduction by Nozzle and Flap Modifications. *Powered-Lift Aerodynamics and Acoustics,* NASA SP-406, 1976, pp. 283–305.
105. Filler, L.: Swept Edge To Reduce the Noise Generated by Turbulent Flow Over the Edge. *J. Acoust. Soc. of America,* vol. 59, no. 3, Mar. 1976, pp. 697–699.
106. Bohn, A. J.; and Shovlin, M. D.: Upper Surface Blowing Noise of the NASA Ames Quiet Short-Haul Research Aircraft. *J. Aircr.,* vol. 18, no. 10, Oct. 1981, pp. 826–832.
107. Sarohia, V.; and Massier, P. F.: Control of Cavity Noise. *J. Aircr.,* vol. 14, no. 9, Sept. 1977, pp. 833–837.

8 Propulsive Lift Noise

Lead author

Martin R. Fink
Norden Systems
Norwalk, Connecticut

Introduction

Propulsive lift noise is the increase in noise that occurs when airframe surfaces are placed in the propulsive system's exhaust to increase their lift force. Increased local flow velocities and turbulence levels, due to the propulsive system exhaust gases passing along the airframe lifting surfaces, cause an increase in maximum lift coefficient. The airplane's flight speed required for takeoff and landing can then be significantly reduced, allowing operation from shorter runways than those of conventional commercial airports. Unfortunately, interaction of high-velocity turbulent exhaust flow with the airframe's solid surfaces generates additional noise radiation.

Aeroacoustic processes that cause propulsive lift noise also are present in airframe noise and propulsive system installation noise. However, airframe noise varies with the flight velocity, while propulsive lift noise depends weakly on the flight velocity and strongly on the exhaust or slipstream velocity. Turbulence levels and scale lengths that determine airframe noise often are those for attached turbulent boundary layers rather than the much larger values within free shear layers in the propulsive exhaust. For both of these reasons, propulsive lift noise is much stronger than airframe noise.

Short-takeoff-and-landing (STOL) propulsive lift concepts were examined (ref. 1) during the early and mid 1970's. They were intended to accommodate the predicted increase in airline passenger-miles without overcrowding the existing airports. These new transport aircraft were expected to operate from a multitude of new small airports (STOLports) that would each occupy a small land area. STOL could be commercially acceptable only if takeoff, approach, and sideline noise of these new aircraft was acceptably low and fuel costs for these inherently less fuel-efficient aircraft remained small. Instead, fuel costs increased greatly, and commercial operation of such aircraft would have been economically impractical.

Research studies of propulsive lift noise led to development of improved methods for predicting noise radiation from surfaces in turbulent flows. These analyses were applied in the late 1970's to prediction and understanding of airframe noise. Knowledge gained about causes of propulsive lift noise has also been useful in reducing propulsive system installation noise.

A 1984 study of potential technology opportunities that might be possible in the future (ref. 2) had listed, as opportunities under the subject of aero/propulsion interaction, "Doubling of the usable maximum lift through propulsive lift concepts that employ low engine bleed rates." This would be applied to commercial subsonic transports having greater community acceptability due to reduced noise during takeoff and approach. The subject of propulsive lift noise will reappear for such aircraft. As is discussed later, one useful starting place would be to perform experiments using imaging microphone techniques to identify physical locations from which this noise is radiated at takeoff and approach flight speeds. These experimental techniques had not existed when most propulsive lift noise data were obtained.

Description of Propulsive Lift Vehicles

STOL performance can be achieved by use of either small airplanes having low wing loading or conventionally sized airplanes using the propulsive system exhaust gases to generate increased lift at low flight speeds. The first approach leads to high operating costs and poor ride qualities. In the early 1970's, the second approach had the problem of then-unknown noise levels. At least four types of propulsive lift STOL concepts (fig. 1) were developed and test flown, each with its own advantages and difficulties. Physical and aeroacoustic properties of each type are summarized below.

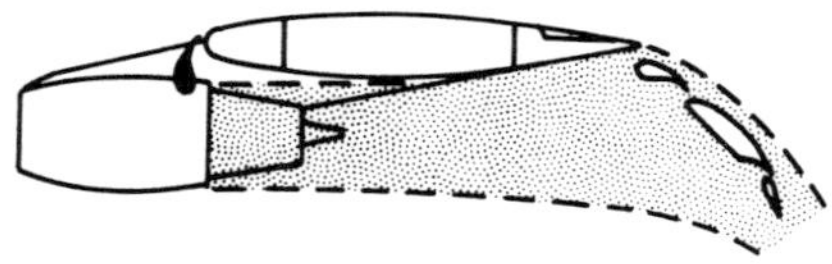

(a) Engine under the wing.

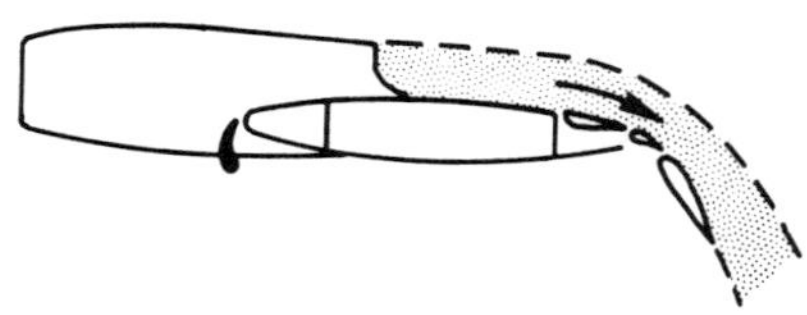

(b) Upper surface blowing (Coanda) flow.

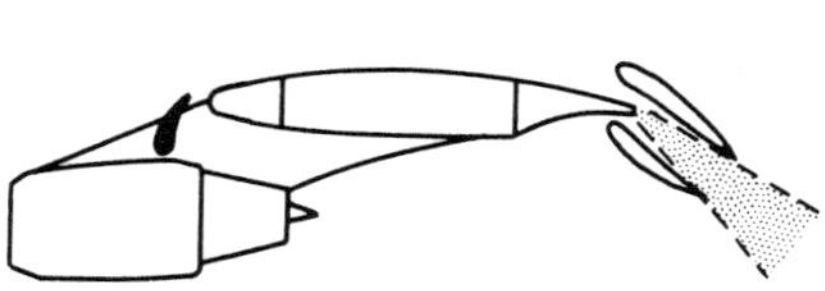

(c) Augmentor wing.

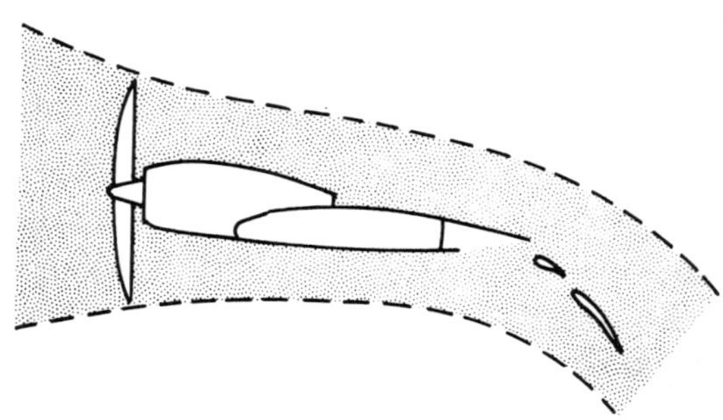

(d) Wing in propeller slipstream.

Figure 1. Types of propulsive lift STOL configurations. (Based on ref. 32.)

Under The Wing (UTW) Turbofan

The first type of turbofan STOL configuration to receive detailed attention was the UTW concept (fig. 1(a)). Its engines are mounted under the wing and ahead of the torsional axis as for a conventional turbofan, but they are not mounted as far beneath the wing. The wing trailing edge flaps extend across the exhaust region instead of having cutouts at the exhaust jet locations. Relative positions are such that the upper edges of the spreading exhaust jets attach to the wing lower surface a short distance ahead of the first flap slot.

When the trailing edge flaps are deflected, part of the exhaust gas passes through the slots. This high-velocity high-turbulence flow attaches to the flap upper surfaces despite the presence of strong local adverse pressure gradients that otherwise would separate the local external airflow in forward flight. Turbulent mixing of the exhaust flow and the external flow causes the external flow to follow the deflected flap panels. Increased lift force and reduced drag are achieved on the deflected flaps. An upward component of thrust force also is generated, because the flap lower surfaces turn most of the impinging exhaust gases to nearly the flap deflection angle.

Aeroacoustic data were first obtained for this type of configuration at small scale and zero forward speed (refs. 3 and 4). A wing and trailing edge flap being scrubbed by the exhaust jet was found to increase the noise radiation beneath the wing. This noise, referred to as "scrubbing noise," clearly would increase takeoff and approach flyover noise of a UTW airplane. UTW airplanes such as the McDonnell-Douglas YC-15 and C-17 resemble conventional turbofan transports designed for short field operation and have little additional drag penalty in cruise flight. However, increased noise is likely to be radiated at takeoff and approach flight speeds from the large number of slots and edges immersed in high-velocity turbulent exhaust flow at low flight speeds.

Upper Surface Blowing (USB) Turbofan

One reaction to the high measured noise levels beneath the UTW aircraft was development of the upper surface blowing (USB) configuration, originally called over the wing (OTW) (refs. 5, 6, and 7). This configuration is sketched in figure 1(b). It has shields over the trailing edge flap slots that can be retracted to allow slot airflow in case of engine failure. The engines are mounted above and ahead of the wing. This combination of engine location and unslotted aft wing causes propulsive system aft-radiated noise and some of the propulsive lift noise to be acoustically shielded from locations beneath the aircraft's flight path.

The engine exhaust can be spread over a larger portion of the wing span by use of exhaust nozzles with rectangular or D-shaped cross sections. Exhaust gas flow tends to remain attached to the wing and deflected flap upper surface because of the Coanda effect. Mixing with the external air above the flap then causes that air to be deflected. USB airplanes include the Boeing YC-14 and Quiet Short-Haul Research Aircraft (QSRA), the National Aerospace Laboratory (Japan) "Asuka" Quiet STOL Research Aircraft, and the Antonov An-72 STOL transport.

Aerodynamic disadvantages include increased viscous drag in cruise flight because more of the wing upper surface is scrubbed by the exhaust jet during cruise and thicker, heavier wing upper surface skin is required to resist structural fatigue caused by adjacent engine exhaust flow. Variable geometry devices may be needed on the nozzle and wing upper surface to maintain attached flow over the full range of exhaust pressure ratios. Aeroacoustic advantages include reduced downward-radiated noise because aft-radiated engine tone noise and much of the jet mixing region is shielded by the wing upper surface.

Augmentor Wing (AW)

The augmentor wing, sketched in figure 1(c), is a concept (ref. 8) in which part of the engine exhaust gas is spread spanwise into a large number of narrow ejector nozzles within an ejector formed by upper and lower segments of the trailing edge flap. One De Havilland of Canada DHC-5 Buffalo aircraft was extensively modified to demonstrate this concept. It avoided flap upper surface flow separation by using the ejector as a suction source to rotate and ingest such flow. The deflected high-momentum exhaust of the ejector acted as a jet flap to induce lift on the wing and achieve an upward component of thrust.

Noise was expected to be reduced by shielding the ejector jet mixer within acoustically absorbing upper and lower flap panels. The dominant ejector exhaust noise was intended to have high frequency because of the small primary nozzle jet height, so that much of the outward-radiated noise would be attenuated by the atmosphere. Disadvantages include weight and complexity of the trailing edge flap panels and the high-pressure high-temperature spanwise ducts and mixer nozzles.

The augmentor wing is one example of the general category of STOL aircraft that use a trailing edge jet flap to achieve increased lift at constant airspeed and incidence. Many such concepts require such large momentum coefficients that the exhaust jet is supersonic and causes additional noise.

Wing in Propulsive Slipstream

The earliest and most conventional type of propulsive lift STOL airplane is sketched in figure 1(d). It uses the propeller slipstream to generate increased relative airspeed between the wing and its upstream airflow. Such airplanes include small single-engine airplanes with large-diameter propellers, such as the Helio Courier, and larger four-engine airplanes, such as the Lockheed C-130 and De Havilland of Canada DHC-7. The four-engine airplanes, of course, can generate larger increases in lift coefficient because a larger fraction of the wing span is immersed in the propeller slipstream.

Propeller slipstreams contain the blade viscous wakes and potential-flow wakes. Rotation of the slipstream causes these regions of altered velocity to sweep past the wing and induce lift fluctuations on the wing. Acoustically noncompact dipole noise radiation is generated, and such noise can be predicted (ref. 9) by available analytical methods. Tone noise is generated at frequencies related to the shaft revolution rate and its harmonics, as with a turbomachine rotor followed by a stator.

Disadvantages include slightly increased propeller noise and slightly reduced propeller efficiency both caused by the wing's induced azimuthally nonuniform flow field at the propeller. For conventional propeller-driven aircraft, the increased propeller noise has been found to dominate any increase in noise radiated from the wing. Data are not yet available to determine whether this also is true for aircraft powered by high-disk-loading prop-fans.

Slipstreams can also be obtained as the exhaust of high-bypass-ratio turbofan engines. This variant had been tested at model scale (ref. 10) as an engine in front of the wing (EFW) configuration. EFW noise can be predicted using the same methods as for wing broadband noise in propeller slipstreams, except that the turbulence is that of the jet mixing region at the wing leading edge axial location.

Experimental Observations of Propulsive Lift Noise

Under The Wing Configurations

Overall sound pressure level (OASPL) directivity in the flyover plane for noise radiation from a small UTW model is shown in figure 2 for nozzle pressure ratios of 1.25 and 1.7. These data from reference 3 are for zero forward speed and 10° to 20° flap deflection, as would be used in takeoff. This noise is nearly omnidirectional in much of the flyover plane. Maximum levels occur almost vertically above and below the wing. Minimums occur in the aft direction of the deflected jet and in the forward upper quadrant opposite of that direction. By examining the variation of directivity shape and amplitude with exhaust velocity, it can be realized that OASPL near the deflected jet is dominated by jet exhaust noise.

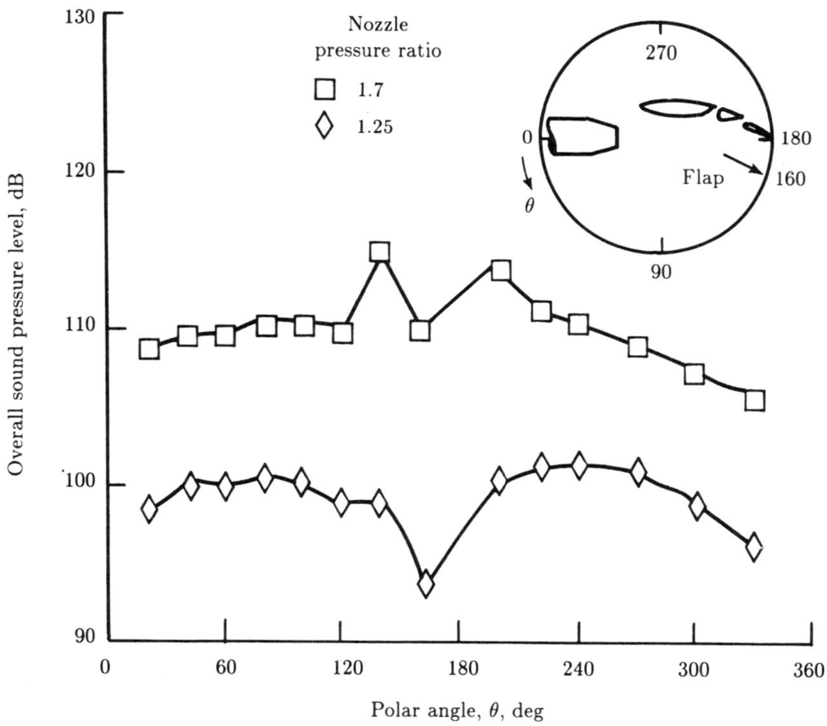

Figure 2. OASPL directivity in flyover plane for a small UTW model at takeoff flap deflection (20°) and zero flight speed.

Amplitude of this noise near the deflected jet is larger than that of an isolated jet at the same exhaust velocity. Noise beneath the wing would be expected to exceed that for an isolated jet because of noise reflection from the wing lower surface. This effect should have decreased, not increased, the noise measured above and behind the wing. The combination of the exhaust jet and an adjacent wing had caused additional noise radiation both above and below the wing.

OASPL directivity of an isolated exhaust jet, rotated to the measured exhaust deflection angle, can be empirically increased several decibels to match measured OASPL amplitudes near the deflected exhaust. Subtracting this empirical quadrupole noise contribution from the measured OASPL directivity gives a shape that resembles a compact lift dipole in the two aft quadrants. However, it has considerable amplitude in the forward direction where a lift dipole would be predicted to approach zero amplitude.

OASPL directivity in the flyover plane at approach flap deflection (30° to 60° for the aft flap panel) is shown in figure 3 for the same model. These directivity shapes have greater angular variation than those for takeoff flap deflection. Noise radiation perpendicular to the highly deflected flap surfaces and quadrupole noise from the deflected exhaust jet are more easily recognized to have rotated with increased flap deflection.

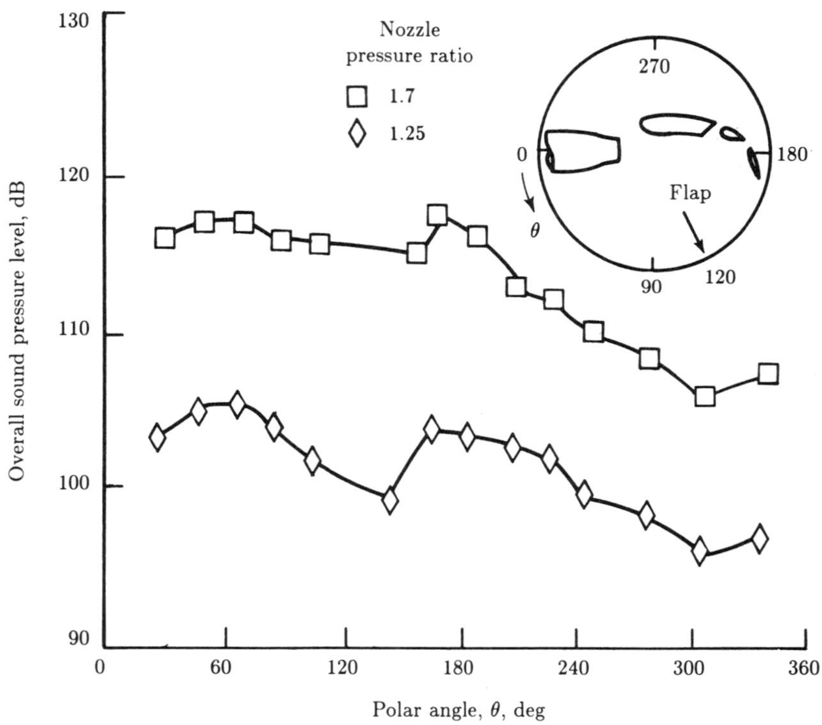

Figure 3. OASPL directivity in flyover plane for a small UTW model at approach flap deflection (60°) and zero flight speed.

OASPL's at sideline directions have larger variation with measurement angle. This is most noticeable (not shown) at takeoff flap deflection and sideline angles of 60° to 70° from the flyover plane, typical of a sideline position beyond the end of the runway. Sideline noise is greatly decreased as the polar angle increases in the aft direction from the 90° (wing tip) to the 120° region. The directly radiated noise then changes from that which originates beneath the wing (a mixture of surface-radiated noise and volume-radiated quadrupole noise) to noise originating above the wing and having a more rapid decay with sideline angle.

Full-scale nozzle exit diameters would be relatively large and exhaust velocities relatively low. Frequencies that strongly influence the full-scale perceived noise levels correspond to Strouhal numbers greater than 10. Relatively large models had to be tested so that frequencies corresponding to those Strouhal numbers were not too large. Otherwise, corrections for atmospheric attenuation would have introduced excessive error.

Upper Surface Blowing Configurations

Unlike UTW configurations, whose directivity in the two aft quadrants is almost symmetrical about the deflected exhaust jet, this directivity for USB configurations is unsymmetrical. OASPL directivities in the flyover plane are shown in figure 4 for a USB model with a D-shaped exhaust nozzle and jet flow deflector (ref. 5) at takeoff and approach flap deflections and zero flight speed. (The cross section of a D-shaped nozzle has a straight lower edge next to the wing and curved upper and side edges, resembling the letter D rotated 90° counterclockwise.) Measured OASPL clearly is larger above the wing than below.

OASPL directivity shape beneath a USB configuration has less angular variation than that beneath a UTW configuration. Levels are nearly constant in the lower quadrants until the deflected exhaust jet is approached. Then the levels and shape closely resemble those for a UTW configuration with the same aft flap deflection. Directivity in the sideline plane is similar except that for the USB configuration, OASPL abruptly increases at large sideline angles in the aft quadrant. For those directions, the far field position is not shielded from the forward part of the exhaust region by the wing. (Instead, it has a direct line of sight to that strongly noise-radiating exhaust region.)

Augmentor Wing

Acoustic data for small-scale and large-scale augmentor wing models are given in reference 8 for zero flight speed. The large-scale model was tested both with hard-wall surfaces and with acoustic absorbing panels. Scaled results for the small model generally agreed with those for the large model. Only data for the large-scale model are discussed here.

OASPL directivity in the flyover plane at takeoff flap deflection is shown in figure 5 for nozzle pressure ratios of 1.6 (subsonic) and 2.6 (supersonic). The solid symbols are data obtained with hard-wall surfaces. Open symbols are for acoustic panels on the inner surfaces of the augmentor intake door, shroud, and flap.

Acoustic treatment reduced the noise levels up to 3 dB within about 60° of the deflected exhaust. Noise radiation ahead of this angular region was nearly omnidirectional in the limited region measured. Below the wing, it was about 3 dB higher than above the wing. Pressure amplitudes near the overhead direction varied approximately with nozzle jet velocity to the fifth to sixth power, as would be expected for noise radiation from large surfaces in turbulent flow with a small turbulence scale length.

OASPL directivity patterns in the flyover plane at approach flap deflection are shown in figure 6 for nozzle pressure ratios of 1.6 (subsonic) and 2.0 (slightly

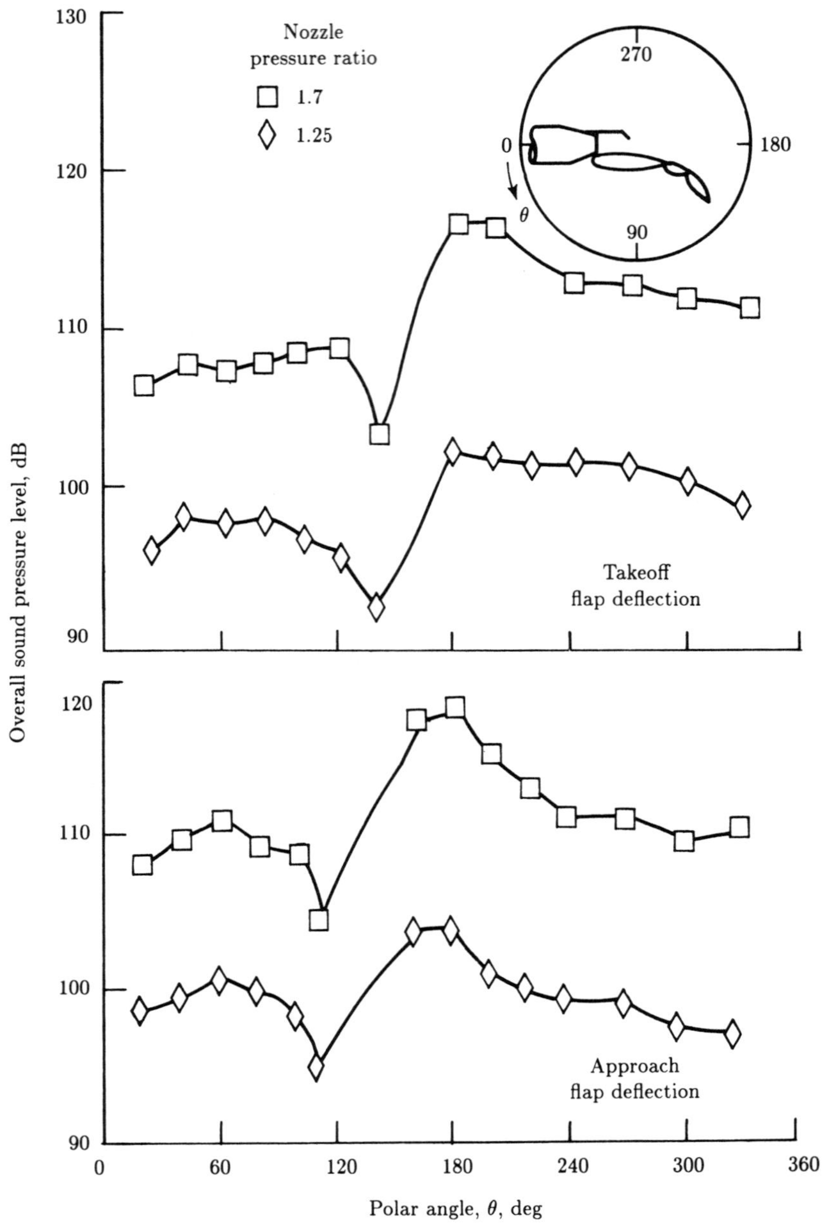

Figure 4. OASPL directivity in flyover plane for a small USB model with a D-shaped exhaust nozzle and jet deflector, at takeoff and approach flap deflections and zero flight speed.

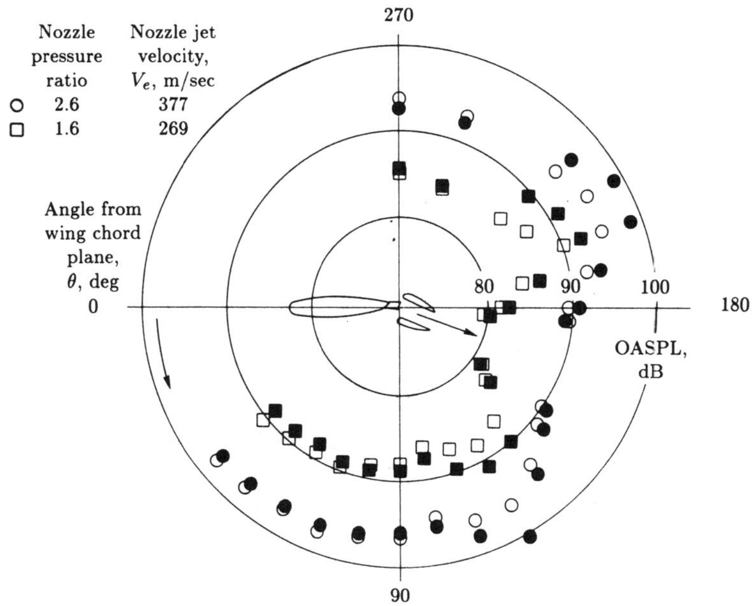

Figure 5. OASPL directivity for augmentor wing at takeoff flap deflection (20°) and zero forward speed. Open symbols denote acoustic treatment; solid symbols denote hard walls. (From ref. 8.)

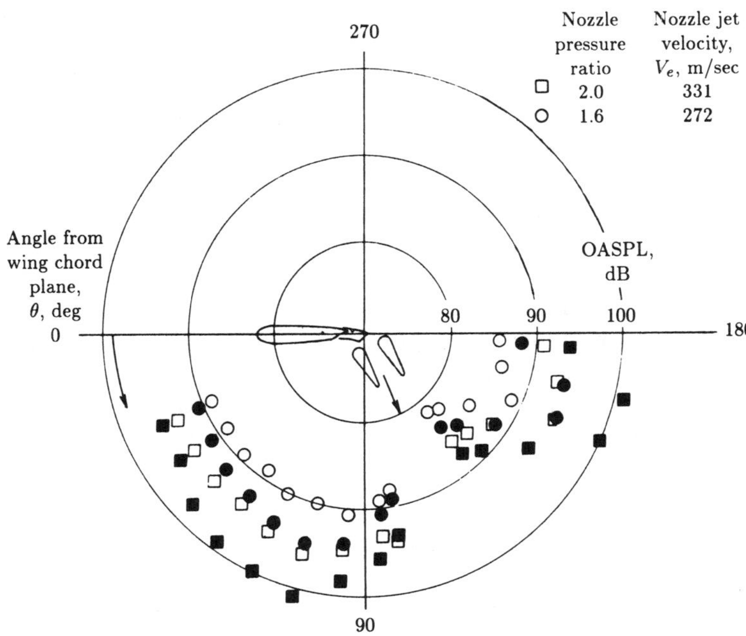

Figure 6. OASPL directivity for augmentor wing at approach flap deflection (65°) and zero forward speed. Open symbols denote acoustic treatment; solid symbols denote hard walls. (From ref. 8.)

supersonic). Here, peak flyover noise levels occurred at about 40° from the deflected exhaust jet direction. Noise amplitudes varied with exhaust velocity to the 8th to 10th power, and were reduced 4 to 5 dB by acoustic treatment. Noise levels at these directions of practical interest for approach flight were dominated by augmentor jet noise. The angular region for which the acoustic absorbing panels caused large noise reduction was increased as nozzle pressure ratio increased above that for sonic exhaust velocity. Acoustic lining apparently suppressed or absorbed some of the exhaust jet noise caused by shock waves.

Acoustic spectra for this large-scale model, corrected for atmospheric attenuation, are shown in figure 7 for 30.5 m distance in the flyover plane at 70° (at takeoff flap deflection) and 75° (at approach flap deflection) from forward, below, and and slightly ahead of the wing. These are typical of the spectra that would dominate perceived noise levels beneath the aircraft flight path.

Other types of full-scale powered lift configurations radiate their highest sound pressure level (SPL) at low frequencies. Unlike them, the augmentor wing has strong noise radiation at high frequencies and at the highly objectionable mid frequencies. Use of small nozzle lobes for augmentor mixing causes small turbulence length scales, with inherent generation of mid- and high-frequency quadrupole noise in addition to mid- and low-frequency noise radiation induced from adjacent surfaces. Noise at higher frequencies is more strongly attenuated by the atmosphere and more easily absorbed by fuselage wall treatment. However, keeping within the annoyance-rated noise limits at STOLport boundaries may be more of a problem for the augmentor wing than for the UTW or USB configuration.

Wing in Propulsive Slipstream

Beneath an aircraft with wings in the propulsive slipstream, flyover noise in the plane of symmetry slightly exceeds that for the isolated propellers and engines. This added noise includes (1) installation noise caused by upstream interference of the wing's aerodynamic flow field on propeller loading variations around the disk and (2) aft-radiated propeller noise reflected from the wing and flap lower surfaces. It also would contain propulsive lift noise, as discussed in the previous section entitled "Description of Propulsive Lift Vehicles." This additional noise is so small that data are not available for validating prediction methods.

A high-turbulence approximation to this type of installation was tested at model scale as an engine in front of the wing configuration (ref. 10). The propulsive system slipstream for those tests was produced by a jet exhaust nozzle rather than a propeller. Noise radiation was similar to that of a UTW configuration with the same exhaust nozzle and trailing edge flaps.

Noise Mechanisms and Underlying Physical Concepts

Propulsive lift noise differs from conventional airframe, propulsive system, and installation noise because of its strong noise radiation from airframe surfaces. This noise is caused by high-intensity, large-scale-length turbulence generated in the propulsive exhaust shear layer mixing region. The turbulence is convected past the wing and flaps at moderate to high subsonic relative velocities, in flow patterns

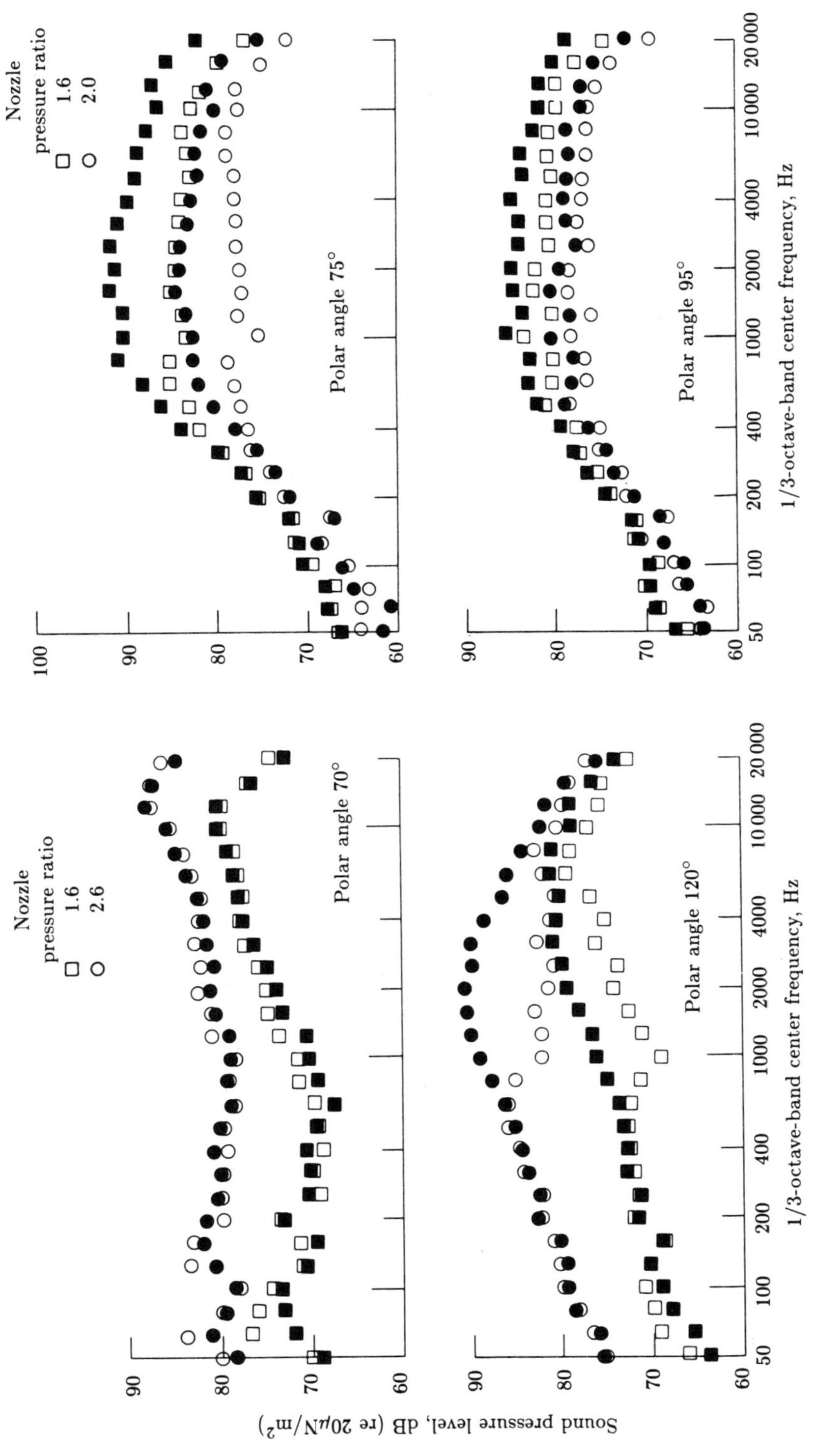

(a) *Takeoff flap deflection.*

(b) *Approach flap deflection.*

Figure 7. 1/3-octave-band sound pressure levels for augmentor wing. Open symbols denote acoustic treatment; solid symbols denote hard walls. (From ref. 8.)

that can be altered by the aircraft's low subsonic flight speed during takeoff and approach.

Much of the work on propulsive lift noise had ended before analytical solutions were developed for predicting noise directivity and spectrum shapes of acoustically noncompact surfaces at moderate subsonic speeds. Analytical solutions such as those of references 9 and 11 were not yet available. Semiempirical models for simple noise sources at low subsonic speeds therefore had to be used when inferring noise generation processes. The following discussion follows the historical approach, which combined the analytical models for several simple acoustically compact noise sources in low subsonic flow. This combination provided qualitative explanations of the actual complex situation.

Overall mean square acoustic pressure directly beneath both UTW and USB configurations, at low and moderate subsonic exhaust velocities and zero flight speed, was found to vary approximately with exhaust velocity to the sixth power. This velocity dependence is expected for dipole noise radiation. However, an acoustically compact lift dipole in low subsonic flow would generate intense lobes of noise above and below the radiating surface. This expected highly directive pattern did not match the observed weakly directional shapes.

OASPL directivity, amplitudes, and spectra for UTW models with retracted flaps at directions near the jet exhaust were easily understood. They were what would be expected for the isolated jet, slightly increased in amplitude under the wing by reflection from the wing lower surface and reduced above it by wing shielding. The resulting estimate of the jet exhaust noise contribution to OASPL at each direction angle was subtracted from measured OASPL to obtain an approximate measurement of surface-radiated noise. Directivity of this noise resembled the sum of a classical lift dipole and another noise source that radiated primarily in the forward direction.

This forward-radiated noise was further examined by subtracting from the measured OASPL directivities the adjusted jet noise and also a lift dipole noise with amplitude assumed to vary with exhaust velocity to the sixth power. The amplitude of the remaining noise was nearly constant in the forward upper and lower quadrants and varied approximately with exhaust velocity to the fifth power. This directivity shape and velocity dependence had been predicted (refs. 12 and 13) for noise caused by turbulence convected past the trailing edge of a semi-infinite plate.

Decomposition of OASPL directivity for this simplest powered lift configuration therefore led to its noise being analytically modeled as a sum of three simple noise components. These were jet exhaust noise, lift fluctuation noise, and trailing edge noise. Physical locations and directivity patterns of these noise mechanisms are sketched in figure 8.

Lift Fluctuation Noise

For exhaust velocities of practical interest, noise radiation directly below UTW aircraft during takeoff and approach would be dominated by surface-radiated noise. This noise, called "scrubbing noise" by NASA, was conceptually modeled as a pure lift dipole (upper sketch in fig. 8) appropriate for very low subsonic flow.

Turbulence scale length and acoustic noncompactness were known to affect the spectrum shape of lift fluctuation noise. This would cause short-chord trailing edge flap panels in the presence of exhaust jet shear layer turbulence scale lengths to

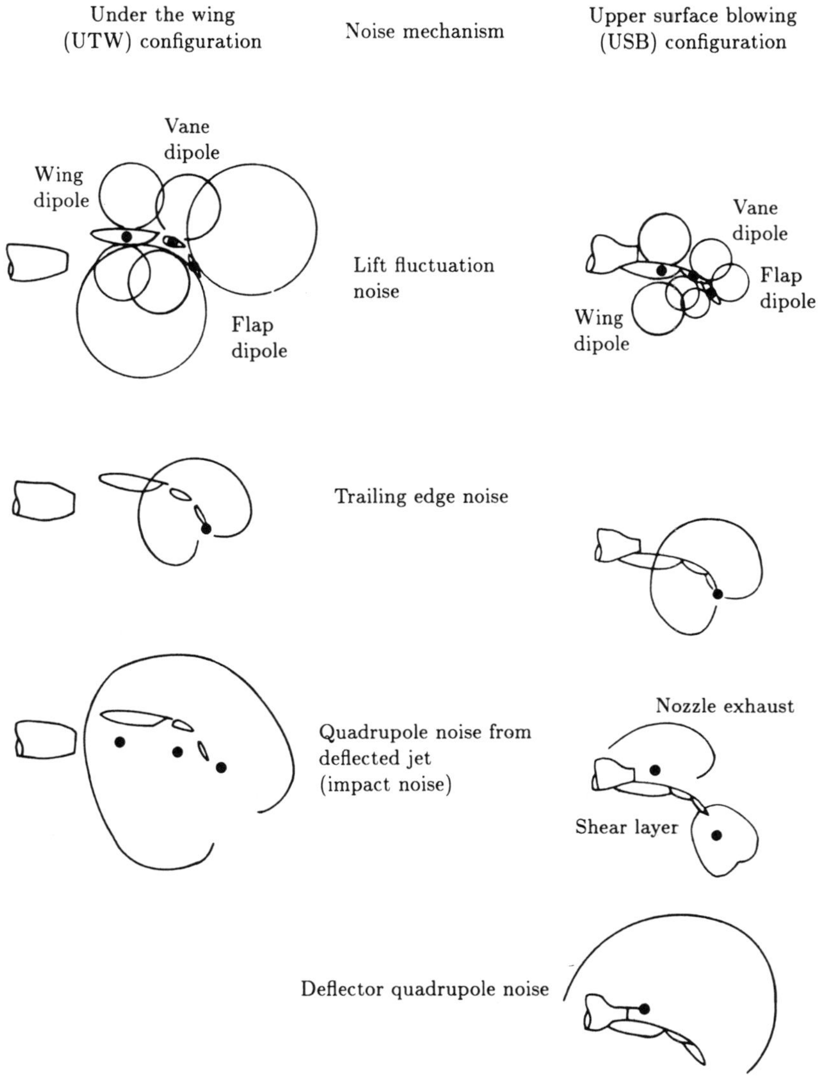

Figure 8. Sketch of directivity patterns for different assumed mechanisms of externally blown flap noise. (From ref. 18.)

radiate turbulence-induced noise more efficiently than the larger chord main wing. Also, directivity of noise radiated from the flap panels would be rotated as the flaps were deflected. The observed noise level increases in the lower forward quadrant at large flap deflections (typically, 30° vane and 60° aft flap) used during approach could then be explained by forward tilt of each segment's lift dipole.

Experimental studies to evaluate the assumed lift dipole mechanism for scrubbing noise used cross correlation of surface pressure transducers and far field microphones. (Imaging microphone techniques for locating noise source positions had not yet been developed.) Typical results, such as those of reference 14, showed that surface pressure fluctuations had opposite phases on the two sides. This result would be predicted for all kinds of surface-radiated noise.

Delay times between surface and far field measurements provided an unexpected result. Turbulence convected along the scrubbed side of the airfoil did not induce noise until it approached the trailing edge. The resulting sound waves then traveled forward to transducers on both surfaces of the airfoil and to the far field microphones. The surface-radiated noise was neither that of a pure compact lift dipole (upper part of fig. 8) nor the pure cardioid (second part of fig. 8) predicted for trailing edge noise from a semi-infinite plate. Its observed directivity is now known to arise from diffraction of sound waves by the airfoil and partial phase cancellation in the near field.

From experiments such as these, it was realized that available rigorous analytical aeroacoustic solutions failed to describe the observed noise-generating process. This result did not help solve the practical engineering problem of developing useful semiempirical methods to predict propulsive lift noise. Such prediction methods continued to be developed and evaluated. Simplified noise mechanisms often gave useful physical insight. They provided guidance for changing an aeroacoustic test model's geometry to reduce its noise while retaining good aerodynamic performance.

Cross correlations were not available for surface pressure measurements on opposite sides of trailing edge flap panels immersed in the spreading shear layer of the exhaust jet on a UTW configuration. Nozzle exit location usually is chosen so that the high-turbulence, high-velocity mixing region passes through the flap slots when the flap is deflected. This flow energizes the deflected flap's upper surface boundary layer in forward flight. The boundary layer then can more easily withstand the strong aerodynamic adverse pressure gradient at large lift coefficients and low flight velocities, while remaining attached to the flap upper surface.

Directional microphone measurements, conducted as part of later studies of airframe noise (ref. 15), subsequently identified the noise source locations for conventional slotted trailing edge flaps at small and moderate deflection. There, the wing lower surface turbulent boundary layer is ingested by the flap slot and convected past the flap. Noise source was shown to be strongest at or near each flap segment's leading edge, as with noise radiation from isolated airfoils in turbulent flow. It is likely that deflected trailing edge flap panels, immersed in the exhaust jet mixing region, would also produce conventional lift fluctuation noise. The level of this noise would vary as the deflected aft flap panel moved within the mixing region to regions of different mean velocity and turbulence level.

Another airframe noise source, produced by turbulence within the flap side edge vortex as it rolls up and is convected past the flap trailing edge, is important for highly deflected conventional trailing edge flaps. This noise generation process causes a nonlinear increase in noise amplitude with increased flap deflection. Such noise may also be produced near the side edges of the jet exhaust mixing region for UTW and USB externally blown (EBF) configurations. As discussed later under "Flight Effects," this process is a major source of turbofan engine installation noise.

Directional microphone and other imaging microphone techniques were developed after funding for research in propulsive lift noise had been greatly decreased. These experimental techniques have not been applied to propulsive lift models at typical takeoff and approach flight Mach numbers. Such tests should be performed. They would identify the important noise source locations for UTW and USB configurations, as a guide to developing future methods for noise prediction and reduction.

Trailing Edge Noise

Trailing edge noise was analytically modeled in propulsive lift noise prediction methods as that for a semi-infinite flat plate rotated to the aft flap panel deflection angle at a very low subsonic Mach number. This noise mechanism has been analytically investigated in many studies such as those of references 12 and 13. Its distinctive features include a dependence on velocity to the fifth rather than the sixth power, a cardioid directivity shape (second part of fig. 8) which is strongest in the forward direction and decreases to zero amplitude in the aft direction, and a 1/3-octave spectrum shape that decays approximately inversely with frequency cubed at high frequencies.

Subsequent studies of trailing edge noise have shown analytically (ref. 11) and experimentally (ref. 16) that directivity shape of trailing edge noise strongly depends on the ratio of acoustic wavelength to flat plate length. It also is changed by an increase of convection Mach number from near zero to moderate subsonic values. The analytical solution given in reference 11 would be a recommended starting place for development of future powered lift noise prediction methods.

Quadrupole Noise

Jet exhaust noise directivity at zero trailing edge flap deflection is changed from that of an isolated jet, because some noise radiated toward the wing is reflected by that surface. More important for EBF noise, the amplitude of quadrupole noise is increased when the jet is deflected by a solid surface. This noise is called impact noise in reference 18.

This increased jet mixing noise is different from trailing edge noise or other surface-radiated noise. It was investigated experimentally (ref. 17) in tests with an exhaust jet directed at a very large flat surface such that wall-jet velocity was small at the solid edges. Empirically, the increase in mean square acoustic pressure at flow deflections typical of trailing edge flaps was a factor of about 6 times sine squared of the deflection angle (ref. 18).

Slotted and slotless UTW configurations at zero simulated flight speed have been observed to produce approximately equal peak amplitudes of quadrupole noise at small and moderate angles above and below the deflected jet. This was thought to be quadrupole noise produced by turbulence generated in the high-deflection region under the last flap segment. Increased turbulence levels would spread to the shear layer above the deflected exhaust jet downstream of the flap trailing edge. Upward noise radiation from that aft region would not be shielded by the flap surfaces. This noise would vary with local mean flow velocity at the trailing edge rather than depend explicitly on exhaust velocity. It would be more intense for UTW than for USB configurations at the same exhaust velocity.

UTW quadrupole noise appears to be that for the jet exhaust, increased in amplitude by flap deflection and rotated to the exhaust deflection angle (third portion of fig. 8). Normalized spectrum shapes at each angle from the deflected jet centerline are approximately those for an isolated jet at the same relative direction.

Some USB installations used a jet deflector at the aft upper part of the exhaust nozzle. The device was rotated downward to ensure that the exhaust jet would attach to the wing upper surface ahead of the flap. It was expected that the jet would also remain attached along the deflected flap upper surface in flight. Quadrupole noise radiation measured above the wing was increased by this device. The increased noise (bottom sketch in fig. 8) generally was shielded from the ground by the wing upper surface. Amplitude of the noise increase over the wing was adequately predicted by the empirical equation for increased quadrupole noise of a UTW jet deflected through the same angle.

Still another source of increased noise radiation attributed to the jet exhaust noise was described in reference 19. At some test conditions, a feedback interaction can exist between a deflected flap surface and the UTW exhaust jet airflow. Jet exhaust noise in several 1/3-octave bands was increased by this feedback process, which can also occur (ref. 20) if a small obstruction is placed several diameters downstream of an exhaust nozzle.

Flight Effects

As noted in reference 21, there are two major independent types of flight effects on EBF noise. One of them occurs because exhaust velocity relative to the external air is decreased although exhaust velocity relative to the wing surface is not changed. This decrease reduces the turbulence level in the portion of the shear layer that is farther from the wing surface and increases its convection velocity relative to the wing surface. OASPL amplitude is decreased and spectrum peak frequency may be increased.

The other major flight effect is dynamic amplification. This is the change in apparent acoustic directivity pattern, and Doppler shift of frequency, between measurements by an observer moving with the noise source and those measured by a stationary observer.

Additional flight effects can be caused by changes in the exhaust jet flow pattern and location due to external airflow. Those changes would alter the jet mixing region's velocity, turbulence level, and physical location relative to airframe surfaces. Also, the propulsive lift exhaust flow may energize some of the airframe noise generation mechanisms.

Noise prediction methods that use a noise component approach can use the established equation for dynamic amplification of each component. Empirical methods must assume one dominant noise generation process and use its equation. The required equations are given in reference 21. For example, OASPL of a compact lift dipole moving at aircraft flight velocity V_a relative to an observer at polar angle θ (measured from the forward direction to the observer position, with 90° directly beneath the aircraft in level flight in the flyover plane) would be changed by adding

$$\Delta\text{OASPL} = -40 \log \left(1 - \frac{V_a}{c} \cos\theta\right) \tag{1}$$

where c is the atmospheric speed of sound. Mean square acoustic pressure would be Doppler shifted from frequency f to

$$f_v = \frac{f}{1 - (V_a/c)\cos\theta} \tag{2}$$

where f_v is the frequency measured by a stationary observer for noise radiated at frequency f from a source moving past that observer at velocity V_a.

Forward flight effects on some of the noise-generating processes were discussed in reference 22. The root-mean-square turbulence velocity in a turbulent shear layer is proportional to the velocity difference across the layer. Convection velocity of this turbulence past flap surfaces of UTW configurations would be determined by the exhaust velocity, with little effect of flight speed. OASPL amplitude of a UTW aircraft with exhaust velocity V_e, flying at velocity V_a, would then be expected to vary as

$$\Delta\text{OASPL} = 20\log\left(1 - \frac{V_a}{V_e}\right) \tag{3}$$

with no shift in frequency other than the convective amplification Doppler shift. As is shown in a later section entitled "Comparisons," this prediction has been validated using limited wind tunnel acoustic data.

Spectrum amplitudes beneath USB configurations behave differently in simulated flight. They were found to decrease by about twice the amount given by equation (3) at Strouhal numbers less than about 0.2. For Strouhal numbers greater than 0.5, they seemed to be independent of flight velocity. Peak amplitudes of the 1/3-octave spectra were decreased and were shifted to higher frequency. An alternate interpretation was that all amplitudes were decreased by the increment given by equation (3), but were shifted upward in frequency by the factor $(1 + V_a/V_e)$. Spectrum slopes at large and small Strouhal numbers were such that moving the measured zero flight speed spectra down in amplitude and up in frequency would produce the observed effects.

Analytical studies, model tests at zero and low simulated flight speeds, and full-scale flyover tests (refs. 23 and 24) of transport aircraft have shown additional flight effects on jet-airframe interaction noise. The tested configuration represented a high-bypass-ratio turbofan engine mounted close under the aircraft wing and a trailing edge flap with a cutout downstream of the spreading jet exhaust. Installation noise was represented as a sum of wing lift fluctuation noise, trailing edge noise from the large-chord wing and the short-chord flap, and reflection of jet mixing noise from the wing lower surface. These noise mechanisms were selected by analogy with propulsive lift noise.

There were substantial differences between local aerodynamic flow fields and major noise sources measured in static tests and in forward flight. At zero flight speed, the major noise source was shown to be convection of shear layer turbulence past the wing trailing edge downstream of the exhaust nozzle. In simulated forward flight, the dominant airframe noise was induced by the vortex that was shed from the flap side edge cutout and convected past the flap trailing edge. This noise apparently was strengthened by the presence of the turbofan engine exhaust jet shear layer. It was stated to be the dominant noise source in full-scale flight. The same velocity scaling law would not necessarily apply for both of these noise processes.

Noise Reduction

Powered lift noise can be reduced by either of two approaches. Local properties of the flow field which affect noise, such as mean velocity, turbulence scale length, and turbulence level, can be changed. Surface properties also can be changed to decrease or absorb the induced acoustic pressures. These changes must be achieved without worsening the wing high-lift system's aerodynamic performance at takeoff and approach flight speeds. Unfortunately, aeroacoustic model tests sometimes were conducted only at zero flight speed and aerodynamic effects could not be measured.

Tests of several UTW noise reduction concepts at zero flight speed were described in reference 25. The exhaust nozzle was moved aft, so that the spreading exhaust passed through the second flap slot (for a double slotted flap) but not the first slot. This change in geometry reduced the low-frequency noise but increased the more annoying mid- and high-frequency noise. Noise radiation from the wing and the flap vane probably was reduced, but the aft flap was subjected to higher local mean velocity. Moving the exhaust nozzle so that the jet no longer passed through the first flap slot would be expected to reduce the wing maximum lift coefficient at approach flight speeds and flap deflections.

A ramp screen ahead of the aft flap trailing edge was tested in an attempt to reduce the local turbulence intensity and scale length. It caused little or no noise reduction and decreased the lift and thrust at zero flight speed.

Also tested were plugs that filled the slots of the deflected flaps in the vicinity of the exhaust jet. These plugs reduced OASPL by up to 10 dB, with most of the reduction occurring at low Strouhal numbers. Exhaust jet turning angle at zero flight speed was decreased. More important, the presence of such plugs would ensure that the wing upper surface airflow would separate from the deflected flap surface in flight. Trailing edge flaps of conventional aircraft have slots to delay upper surface flow separation. The UTW jet exhaust is directed through the flap slots to delay upper surface flow separation at higher lift coefficients. These noise-reducing plugs would reduce lift and increase drag in flight, eliminating the aerodynamic benefits of propulsive lift trailing edge flaps.

Another flow modification, described in reference 26, changed the exit velocity profile of a simulated USB slot exhaust nozzle. This is not related to the use of a mixer nozzle on turbofan engines to provide a nearly uniform exhaust velocity profile rather than a high-velocity core flow surrounded by a lower velocity fan exhaust. Instead, a nominally uniform velocity profile was replaced by ones with constant lower velocity near the wing surface and constant higher velocity in the upper half of the slot exhaust and vice versa. Both variations decreased the turbulence length scale. Noise was reduced about 6 dB near peak level by use of only ± 10 percent differences in mean velocity.

Porous flap leading edge regions and thin perforated forward surfaces with bulk acoustic absorbing backing reduced noise by about 3 dB when tested on EBF models (ref. 27) and conventional wing models (ref. 28) at typical takeoff and approach flight speeds. These surfaces probably decreased the fluctuating pressures induced by incident turbulence. Porous trailing edges on a USB model (ref. 6) achieved up to an 8-dB peak reduction at a Strouhal number of 0.2 based on porous edge extension length, with about a 3-dB decrease at larger Strouhal numbers.

Sawtooth trailing edge shapes (ref. 6) reduced noise levels by 2 to 3 dB on a full-scale USB aircraft at typical flight speeds. The sawtooth size was much larger than the expected turbulence length scale. Model tests showed that noise reduction was affected by lateral position of the spreading jet boundary relative to the tooth corners. A blowing slot at the trailing edge of a full-scale USB wing (ref. 7) reduced noise levels by about 2 dB in flight by modifying the turbulence flow field at the edge.

The augmentor wing model of reference 8 had acoustically lined inner surfaces of its inlet door, shroud, and flap. These panels absorbed shock wave noise at supersonic nozzle exhaust velocities and noise generated by the augmentation mixing process. Maximum noise level reductions were about 5 dB for supersonic and 3 dB for subsonic nozzle jet velocities.

Prediction Methods

Easily used semiempirical methods were needed for predicting UTW and USB noise. These predictions would be combined with predictions of propulsive system and airframe noise to obtain estimates of total aircraft noise radiation. In the absence of exact solutions for propulsive lift noise, it was necessary to use simple crude approximations.

One approach, used at NASA (ref. 21) and also at Lockheed-Georgia (refs. 29 and 30), developed empirical normalized directivity shapes that depended only on whether the flaps were at takeoff or approach deflection. A quick summary of major equations for the Aircraft Noise Prediction Program (ANOPP) method of reference 21 is given below. This simple, easily used prediction method agrees relatively well with much of the UTW and USB data. Comparisons with data that were not part of its original data base are given in the subsection "Model Data Versus Predictions" of the following section entitled "Comparisons."

OASPL directly beneath the wing was scaled with jet velocity to an empirical exponent, the ratio of nozzle diameter squared to far field distance squared, and an empirical function of flap deflection angle. Effective exhaust velocities were defined for use with coaxial or mixed exhaust jets. Different normalized spectrum shapes were used for the flyover plane and for the wing tip sideline direction.

In this prediction method, measured OASPL directivity shapes including noise from the exhaust jet were approximated by empirical directivity curves. OASPL for UTW configurations was taken to be independent of polar angle in the flyover plane for angles 30° to 130° from the forward direction. Levels were taken to be 2 dB lower at a (forward) polar angle of 0° than at 90° and 6 dB lower near the centerline of the deflected jet. OASPL amplitude at the 90° flyover direction and zero flight speed is given in the method of reference 21 by the following equation:

$$\text{OASPL} = 83.6 + 0.14\delta_f + 10\log\left[(A/A_o)(R_o/R)^2\right] + 67\log(V_e/V_o) \qquad (4)$$

Here, δ_f is the flap deflection angle, A the exhaust nozzle exit area, R the actual far field distance, V_e the effective exhaust velocity, A_o the nozzle exit reference area (0.093 m^2), R_o the reference far field distance (30.5 m), and V_o the reference exhaust velocity (152.5 m/sec).

UTW turbofan engine installations generally have used separate coannular nozzles for the core and fan exhaust flows. Effective exhaust velocity was defined by

$$V_e = \left(\frac{A_c V_c^6 + A_f V_f^6}{A_c + A_f} \right)^{1/6} \tag{5}$$

where A_c and A_f are the core and fan exit areas and V_c and V_f are ideal fully expanded velocities at the core and fan nozzle exits. For a fully mixed exhaust, as with an internal mixer nozzle, effective exhaust velocity is taken as the mass-averaged velocity

$$V_e = \frac{\text{BPR } V_f + V_c}{\text{BPR} + 1} \tag{6}$$

where BPR is the mass-flow bypass ratio. Such installations are more likely to be used for USB rather than for UTW configurations.

Empirical correlation curves for normalized 1/3-octave spectra beneath UTW configurations at zero flight speed are plotted in figure 9, also taken from reference 21. These spectra, and all other spectra shown in this chapter, are based on data that have been corrected to remove the effects of atmospheric attenuation and ground reflection. For Strouhal numbers fD/V_e larger than 1, predicted amplitudes increase about 1 dB per 20° increase in flap deflection angle. Peak levels occurred at Strouhal numbers of about 0.3 to 0.4 and decayed at about 10 dB per decade at large Strouhal numbers.

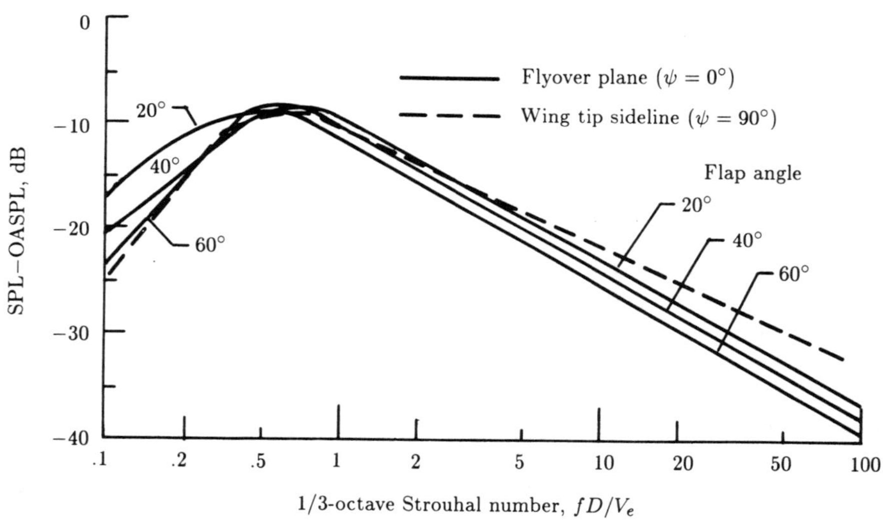

Figure 9. Normalized 1/3-octave-band spectra for UTW configurations at all polar angles, used with the ANOPP noise prediction method (ref. 21).

OASPL amplitude directly beneath USB configurations was predicted in the method of reference 21 by the following equation:

$$\text{OASPL} = 85.1 + 0.01\delta_f + 10 \log\left[(A/A_o)(R_o/R)^2\right] + 60 \log(V_e/V_o) \tag{7}$$

The velocity exponent is smaller (60 in eq. (7) rather than 67 in eq. (4)) for USB than for UTW configurations. USB configurations generally can produce the same flyover noise as UTW configurations while operating at larger exhaust velocities.

Normalized 1/3-octave spectra for predicting USB noise are shown in figure 10, taken from reference 21. These spectra decay at the same rate as UTW spectra for Strouhal numbers up to about 10 and at a higher rate at larger Strouhal numbers. As with UTW spectra, these curves show a small increase with increased flap deflection at Strouhal numbers larger than about 5.

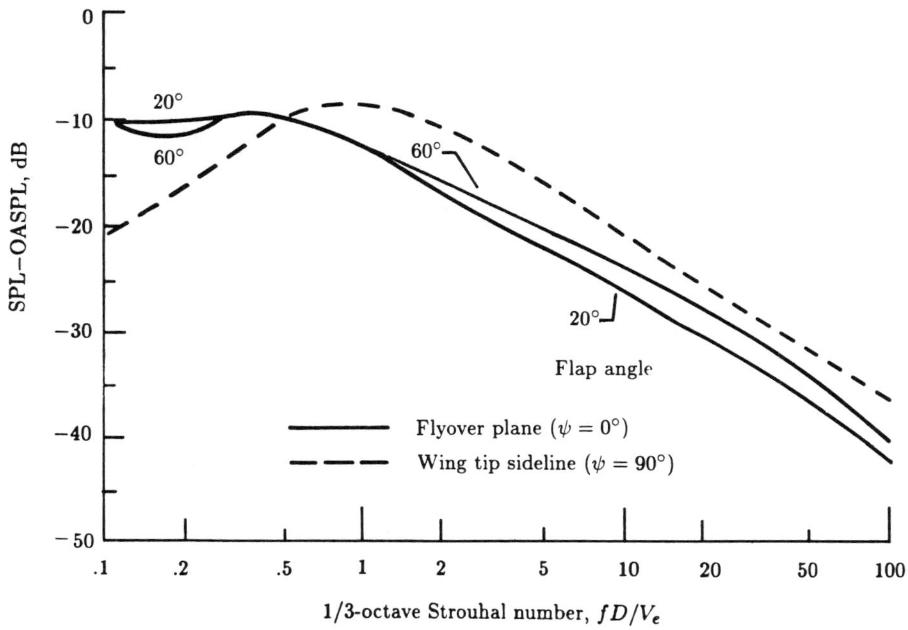

Figure 10. Normalized 1/3-octave-band spectra for USB configurations at all polar angles, used with the ANOPP noise prediction method (ref. 21).

This type of prediction method was useful for scaling data over a limited range of exhaust velocities at constant ratio of wing and flap chord to nozzle diameter. It cannot be used as a design tool for determining the effects of flap segment size, position, and deflection on noise radiation. Semiempirical methods therefore were developed (e.g., refs. 18 and 31) that tried to represent several hypothesized noise mechanisms in terms of details of the local flow field.

It was concluded in references 18 and 32 that the noise component method of reference 18 gave best agreement with data for a variety of configurations and test conditions selected by NASA. Those data were not part of the data base from which either method had been developed. Neither method has been adequately validated at takeoff and approach airspeeds because large-scale flyover data were not (and are not yet) available.

The UTRC (United Technologies Research Center) method of reference 18 approximated most of the UTW surface-radiated noise with lift dipoles for the wing and for each separate flap panel. Each dipole was rotated as its flap panel deflected, changing the directivity. Local flow velocities at each flap panel were calculated by an approximate empirical method as a function of axial and radial location relative to the exhaust nozzle. These were used for predicting that panel's noise. Estimates of local velocity were necessary for predicting noise levels of test model configurations that had unusually long wing and flap chords. The normalized 1/3-octave spectrum for this lift fluctuation noise was an empirical equation matched to data from which the quadrupole noise portion had been analytically removed.

Trailing edge noise was modeled in reference 18 by use of an empirical turbulence level and the calculated local exhaust flow velocity at the trailing edge. It was assumed to radiate only from the trailing edge of the last flap segment. Because this calculated noise beneath the wing was weaker than the calculated lift fluctuation noise, its effect on predicted flyover noise was small. Trailing edge noise was included only to fill out the predicted OASPL directivity curve in the forward upper and lower quadrants to match the data. Its normalized spectrum was chosen as an empirical equation that matched the analytically predicted asymptotic decay rates for trailing edge noise at low and high frequencies.

Quadrupole noise for UTW configurations was computed in that noise component method by starting with the noise of an isolated exhaust jet rotated to the measured or estimated static-thrust jet deflection angle. This noise was increased in amplitude by the amount described in the previous section on quadrupole noise to account for deflection by the UTW wing and flap lower surface. The normalized spectrum was an analytical curve fit to the spectra of isolated exhaust jets at directions near peak amplitude.

The same two approaches (empirical directivity and spectra, and semiempirical combinations of analytical noise component directivity shape with separate semi-empirical normalized spectra) were also applied to USB noise prediction. For the noise component method of reference 18, turbulence convection velocity at the shear layer above the attached wall jet was predicted to decay with increasing distance along the surface. This caused a decrease in predicted USB noise radiation from the highly deflected aft section of the flap, relative to that predicted for UTW configurations. A lift dipole noise approximation then predicted moderate noise source strength for the undeflected wing panel and reduced rather than increased source strength for the aft flap segments.

Trailing edge noise and quadrupole noise beneath the deflected exhaust jet were calculated using the attached wall jet's predicted peak velocity at the flap trailing edge. Calculated quadrupole noise radiation above the wing included the increased noise radiation caused by the USB nozzle jet deflector.

Several other powered lift noise prediction methods were evaluated in reference 32. They are not cited herein because either they poorly predicted the data or their range of applicability was too limited.

OASPL directivity shapes and velocity dependence for the augmentor wing are consistent with lift dipole noise radiation from the shroud and flap along with jet exhaust noise from the deflected augmented jet. Turbulence levels and scale lengths in this ejector region differ from those of a conventional exhaust jet. If these quantities were known, it is likely that noise prediction methods described above for other types of externally blown flaps could be extended to augmentor wings.

The reader should remember that propulsive lift noise is only one of the general categories of noise radiated by a STOL airplane. Engine noise other than the exhaust noise, and predicted airframe noise from surfaces other than portions of the trailing edge flap immersed in the exhaust jet, should be added to the estimated propulsive lift noise. Noise radiation from one of these other sources may dominate at some flyover angles.

Comparisons

Model Data Versus Full-Scale Data

Model tests of propulsive lift noise often were conducted at small scale. Generally they used exhaust nozzle diameters of 5.08 cm, far field distances of 3.28 m, unheated exhaust flow, and full-scale exhaust velocities. Far field microphones were mounted on supports above a hard reflecting ground surface, and the wing spanwise direction was perpendicular to that surface. Some moderate-scale models with exhaust diameters of 33 cm also were tested at zero flight speed, using unheated exhaust flow and a similar installation.

Nominal half-scale models were tested in the NASA Ames Research Center 40×80 ft wind tunnel at zero airspeed and in simulated forward flight. These models (e.g., ref. 33) were powered by Pratt & Whitney JT15D turbofan engines, which have a bypass ratio of 3, maximum thrust of 8900 N, and fan nozzle exit diameter of about 0.64 m.

Tests of complete half-scale models in large wind tunnels eliminate many of the scaling and reflection-plane problems. However, other problems are introduced by background noise from the wind tunnel fan drive system and from airflow past the microphones and their support struts. Noise reflection from unlined wind tunnel walls often constrains the measurement positions to less than far field distances. Model construction costs and facility operational costs are high. Within these limitations, large-scale wind tunnel tests have the advantages that they can provide propulsive lift acoustic and aerodynamic data free from uncertainties about exhaust flow simulation, model geometric shape, and atmospheric attenuation at high frequencies.

Full-scale configurations generally were tested at zero flight speed with a General Electric TF-34 turbofan engine. It was mounted in a nacelle which contained extensive acoustic treatment (ref. 34) to suppress engine fan and core noise. This engine has a bypass ratio of 6.5 and thrust of 42 000 N. Typical nozzle exhaust diameter was 127 cm for a UTW unmixed coannular nozzle and 95 cm for a USB internal mixer nozzle.

Spectrum measurements at low and moderate frequencies contained both directly radiated noise and noise reflected from the ground plane. The two signals could reinforce or partly cancel, depending on their phase difference at the microphone. Resulting measured 1/3-octave spectra were locally wavy. For many small-scale EBF models, this waviness occurred near the Strouhal number for maximum 1/3-octave SPL. Tests at different scales were likely to use different ratios of far field distance and height above the ground plane to nozzle diameter. Measured OASPL, and details of the spectrum shape, could then differ for tests at different scales. After this problem was recognized, it was avoided in small-scale model tests by use (ref. 4) of an acoustically absorbing ground plane.

Also, spectrum measurements at high frequencies can be greatly affected by atmospheric attenuation. Apparent changes in measured small-model spectrum slope between center frequencies of 10 and 40 kHz were likely to be caused by errors in the attenuation correction. However, these model-scale frequencies correspond to only 0.5 to 2 kHz at full scale. Extrapolation of small-model spectra to 1/3-octave bands that strongly affected perceived noise levels at full scale sometimes gave misleading results.

Model- and full-scale acoustic data sometimes differed because aerodynamic details of the flow field were not reproduced. This was noticeable in the augmentor wing data of reference 8 for a small and a large model. Either the flow through the ejector nozzles varied greatly with test Reynolds number or the small nozzles did not match the shape of the large nozzles. Measured velocity profiles within the augmentor at takeoff flap deflection significantly differed at small and large scales. Spectra for the small model, scaled to the large-model size and test conditions, underpredicted the 1/3-octave SPL by about 2 dB at low frequencies and overpredicted by up to 10 dB at center frequencies greater than 2 kHz. At approach flap deflection, velocity profiles for the two models were similar and the spectra generally matched within 2 dB.

Data obtained by NASA Lewis Research Center with different diameter exhaust nozzles and unheated flow were used for validating the acoustic scaling laws. These validations formed the basis for empirical prediction methods such as that of reference 21. Differences between full-scale hot-exhaust data and moderate-size unheated-exhaust data were used in developing equations for effective exhaust velocity. These were presented in reference 21 and in earlier versions of that prediction method and are given as equations (5) and (6) herein.

Differences between USB noise data for two sizes of models with unheated-exhaust jets were discussed in reference 5. Dimensions differed by a factor of 6.5. Normalized spectra for the larger model decayed less rapidly with increased exhaust velocity at Strouhal numbers from about 2 to 20. The ratio of USB upper shear layer turbulence scale length to nozzle diameter may vary with test Reynolds number. This change would affect the rapid decay of surface-radiated noise at large Strouhal numbers until the spectra become dominated by jet mixing noise. OASPL directivities measured under the wing, between 60° and 140° from the inlet direction, were found to scale very well.

Model Data Versus Predictions

Comparisons of extensive acoustic data from selected model tests and predictions from several prediction methods were given in references 18 and 32. These comparisons have the drawback that they include the data base from which each method was developed.

A comparison was given in reference 4 with data for a small UTW model that differed in configuration and flap positions from previous models. The test apparatus, test techniques, and data reduction were chosen by NASA to minimize effects of ground reflection, atmospheric attenuation, and background noise. Predictions were computed before the data were made available for comparison. All tests were conducted at zero forward speed. Figures shown here were taken from reference 18, which contains corrections of some computation errors not recognized when reference 4 was prepared, and from reference 32.

Measured OASPL directivity patterns at takeoff flap setting are given in figure 11 for exhaust velocities V_e of 122 and 226 m/sec. These triple slotted flaps extended farther into the jet exhaust at takeoff deflection than flaps of most other UTW configurations.

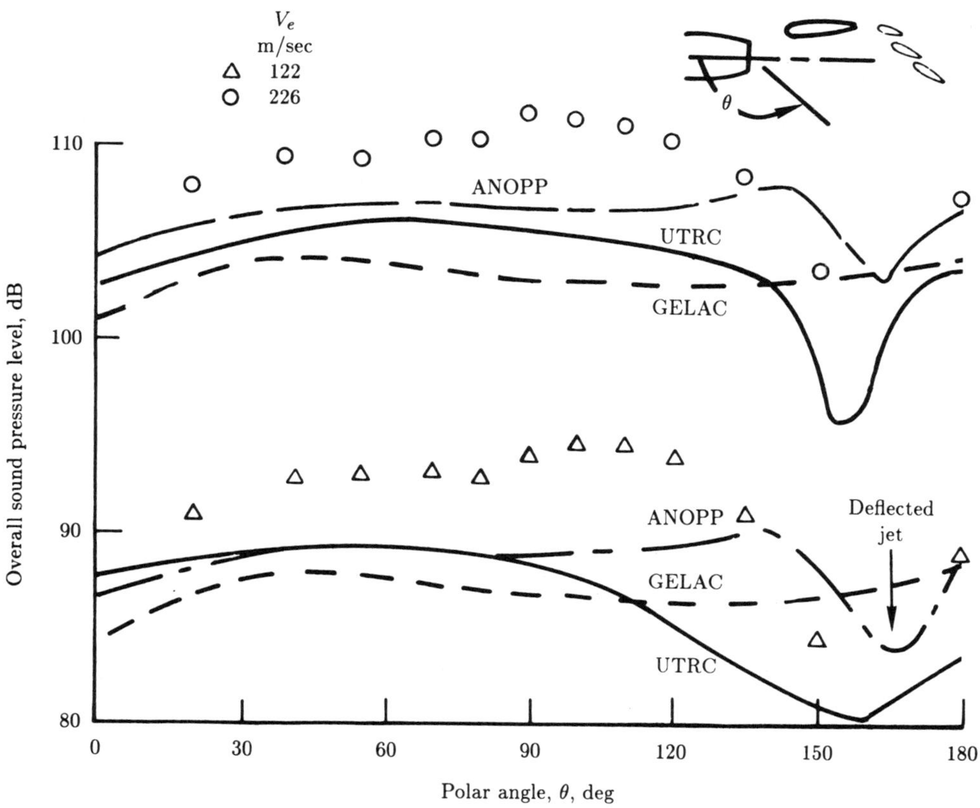

Figure 11. Predicted and measurd OASPL directivity in flyover plane for three-flap UTW model at takeoff flap deflection. (From ref. 18.)

Noise radiation patterns computed by the NASA ANOPP method of reference 21 matched the general shape of the data but were 3 to 8 dB low in level. This method has no adjustment for relative position of the wing and flap at constant deflection angle. The GELAC (Lockheed Aircraft Co., Georgia) method of reference 30, which does include such adjustment, was 5 to 10 dB low. It predicted that maximum OASPL would occur considerably forward of the measured angular location. Neither method predicted the size of the OASPL decrease near the flap exhaust. The UTRC noise component method of reference 18 was about 3 to 5 dB low and best predicted the measured directivity shape.

A similar comparison is given in figure 12 for approach flap deflection. Here, relative geometry of the flap and jet exhaust was similar to most other UTW configurations. Predictions by the ANOPP and GELAC methods bracketed the data and generally were within 2 dB in level. The UTRC method was about 2 dB low at the higher exhaust velocity and 4 dB low at the lower one.

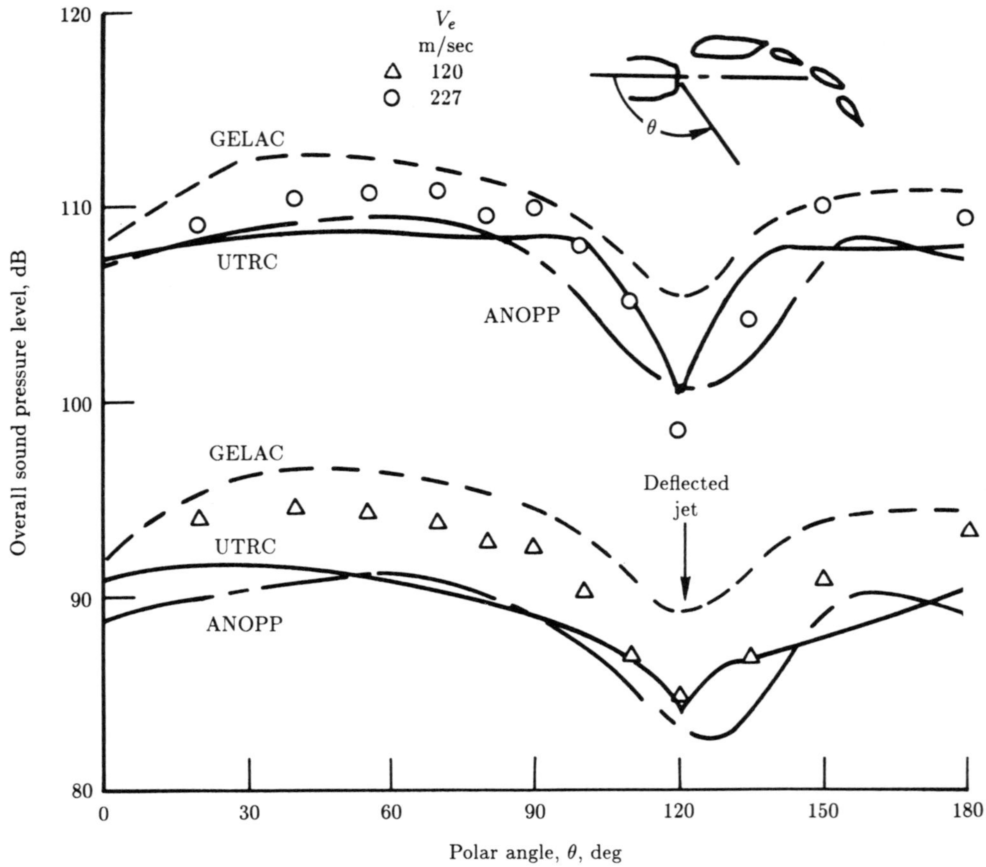

Figure 12. Predicted and measured OASPL directivity in flyover plane for three-flap UTW model at approach flap deflection. (From ref. 18.)

All three methods generally gave about the same predicted normalized 1/3-octave spectrum shape for this UTW model and test conditions. For this reason, comparisons of predicted and measured normalized spectra are not shown here.

Effects of sideline azimuth angle on OASPL directivity were predicted more accurately (not shown here) by the UTRC method than by the other two methods. That method predicts a decay with cosine squared of the angle from the plane of the surface-radiating noise sources, to a noise floor set by axisymmetric quadrupole noise of the deflected jet. It gave the correct maximum decrease of OASPL (10 to 15 dB at polar angles of 90° to 120°) and general shape of the decrease, but it missed some details of the directivity. The ANOPP and GELAC methods predicted about half the measured reductions of sideline noise.

The baseline data of reference 3 were used in developing each of these three prediction methods. Those acoustic data, and data for larger scale models which have the same proportions, generally were matched within 2 dB in OASPL and normalized spectrum level by all three UTW propulsive lift noise prediction methods.

The improved test techniques described in reference 4 were used for testing some USB configurations at small scale. Those configurations were intended to be tested later at nominal full scale and zero flight speed. They included a QCSEE (Quiet Clean STOL Experimental Engine) USB installation having a slot nozzle with a nominal 2:1 aspect ratio, tested with an equivalent nozzle diameter of 14 cm.

The same wing also was tested with a circular nozzle equipped with an external vane deflector that would be retracted in cruise flight. External air then could pass between the exhaust jet and the wing upper surface during cruise, reducing the wing's viscous drag relative to that for conventional USB configurations. These comparisons between data and predictions also were taken from references 18 and 32.

Directivity shapes in the flyover plane for the QCSEE USB model at takeoff flap deflection with exhaust velocities V_e of 152 and 220 m/sec are shown in figure 13. The UTRC method of reference 18 generally matched the OASPL data within 2 dB. The ANOPP method (ref. 21) also matched the shape, but it was about 5 dB low. The GELAC method of reference 29 was about 5 dB low near a polar angle of 90°, and its directivity shape gave poor agreement aft of that direction. The comparison was similar (not shown) at approach flap deflection, with predictions by the ANOPP and UTRC methods bracketing the data.

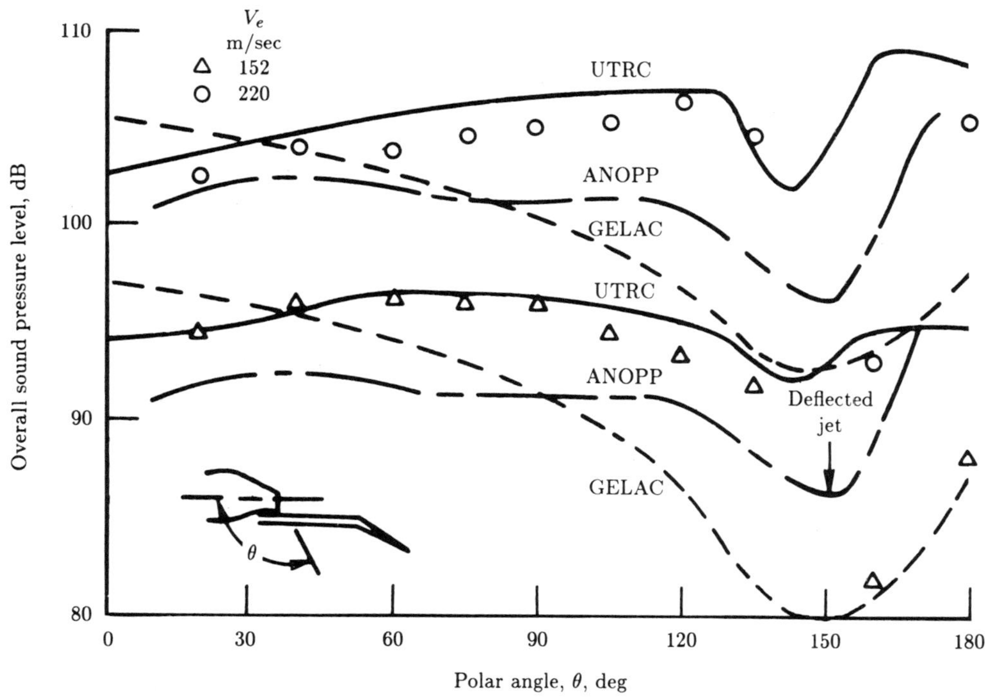

Figure 13. Predicted and measured OASPL directivity in flyover plane for QCSEE USB model at takeoff flap deflection. (From ref. 18.)

OASPL directivity data in the flyover plane for the USB model with a circular exhaust nozzle and vane deflector at exhaust velocities V_e of 145 and 253 m/sec and takeoff flap deflection are compared with predictions in figure 14. For this configuration, the ANOPP method generally was in best agreement with the shape and levels of data. The GELAC method matched the general amplitudes in the aft quadrant but greatly overpredicted them in the forward quadrant, poorly matching the directivity shape. The UTRC method was about 8 dB high in amplitude but matched the general shape. Similar comparisons were obtained at approach flap deflection.

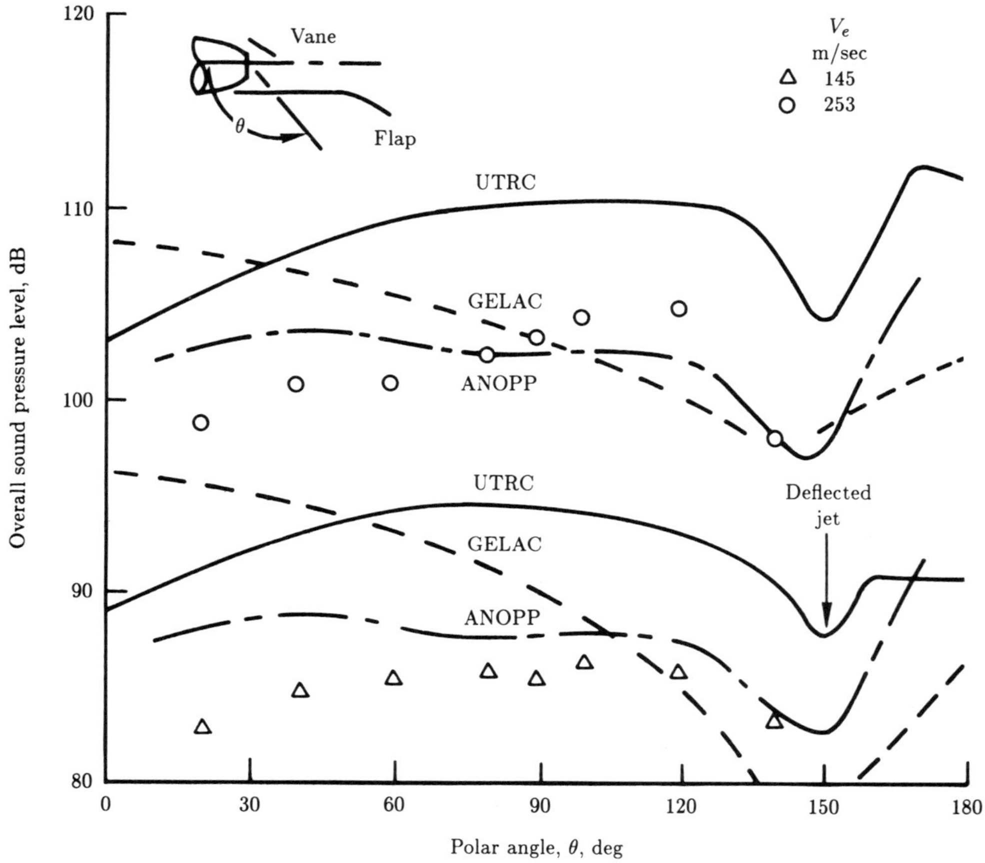

Figure 14. Predicted and measured OASPL directivity in flyover plane for USB model with circular nozzle and vane deflector at takeoff flap deflection. (From ref. 18.)

All three methods had been developed using early baseline USB data, and they closely matched those data. Clearly, none of the three prediction methods proved to be highly accurate for both of these less conventional USB configurations.

Normalized 1/3-octave spectra in the flyover plane for two exhaust velocities and two polar angles at takeoff and approach flap deflections are plotted in figure 15. Both the ANOPP and the UTRC method generally matched the data for Strouhal numbers larger than 1. The GELAC method predicted a larger variation of normalized spectrum shape at moderate and large Strouhal numbers than was measured.

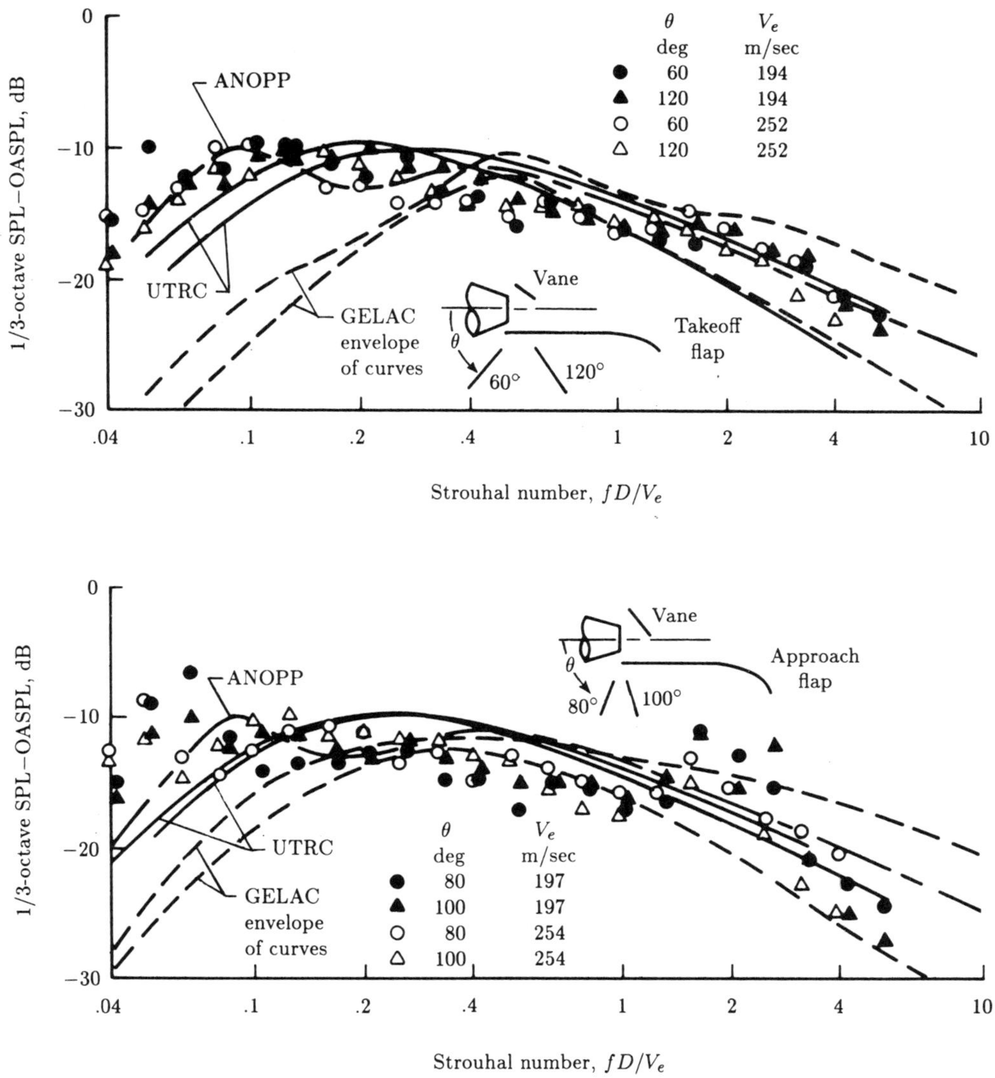

Figure 15. Predicted and measured normalized 1/3-octave-band spectra in flyover plane for USB model with circular nozzle and vane deflector. (From ref. 18.)

An evolved version of the GELAC methods of references 29 and 30 was evaluated in reference 32. Predictions by that method, and by the ANOPP and UTRC methods, are compared in that report with test cases designated by NASA. The UTRC method best matched the UTW data; few comparisons with USB data were shown. It was concluded in that study that the UTRC method gave best results for the full range of propulsive lift noise installations.

An alternate conclusion, based on results given in references 18, 30, and 32, is that the GELAC method of reference 30 should be used for most USB configurations in the flyover plane. The UTRC method of reference 18 should be used for the effects of USB sideline angle and for UTW and EFW configurations in which the exhaust jet passed under or in front of the wing and flaps. For rapid prediction of UTW and USB noise levels and spectra at a flyover position near a polar angle of 90° and at exhaust pressure ratios of 1.4 to 1.8, the ANOPP method of reference 21 generally is almost as accurate as the other methods and is much easier to use.

Flight Effects

A limited comparison of predicted and measured simulated flight effects on propulsive lift noise was given in reference 22. Spectra were measured (ref. 33) in the NASA Ames Research Center 40 × 80 ft wind tunnel under a UTW model powered by four JT15D small turbofan engines. Data were obtained at the low exhaust velocity of 115 m/sec to ensure that propulsive lift noise could be observed above engine noise. Tests were conducted at airspeeds of 0 and 31 m/sec, giving ratios of forward velocity to exhaust velocity of 0 and 0.27.

Measured spectra are plotted in figure 16 along with the curve obtained by decreasing the zero flight speed spectrum by the calculated OASPL increment. Measured dynamic pressure in the jet exhaust upstream of the flap was about 7 percent larger than for zero tunnel airspeed, which is predicted to cause about 0.9 dB higher noise level. Decreased turbulence intensity, caused by reduced velocity difference across the shear layer, is predicted to reduce this noise by 20 log $(1-V_a/V_e)$ or about 2.7 dB. The sum of these two calculated increments of OASPL agrees with the observed 2-dB decrease in amplitude at frequencies below those dominated by engine fan noise.

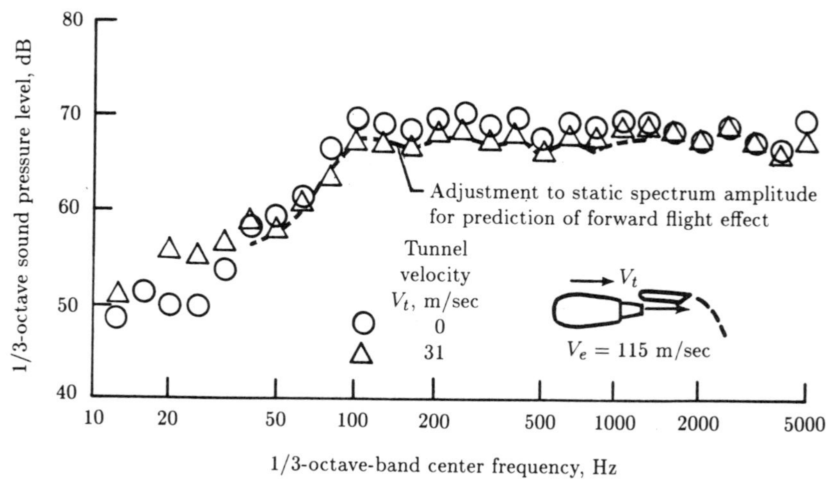

Figure 16. Predicted and measured effect of wind tunnel velocity on 1/3-octave-band spectrum for nominal half-scale UTW model at approach flap deflection.

Effects of forward speed on noise radiation also were measured in the same wind tunnel with a USB model having two JT15D engines with aspect-ratio-5 slot nozzle exit ducts. The model was tested at forward speeds of 0 and 40 m/sec at an exhaust velocity of 241 m/sec, giving velocity ratios of 0 and 0.17. Spectra measured at these test conditions are plotted in figure 17. The predicted 1.6-dB decrease in level and about 0.7 of one 1/3-octave-band increase in center frequency caused the adjusted static test spectrum to match the data for simulated forward flight.

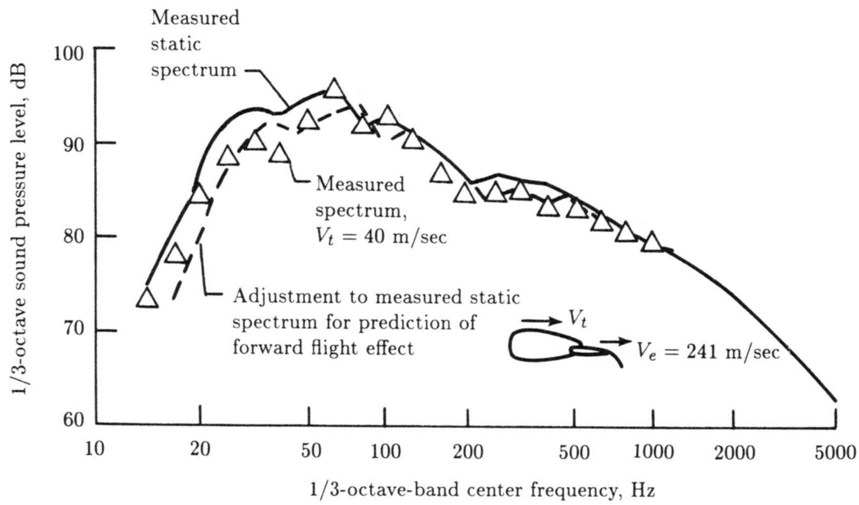

Figure 17. Predicted and measured effect of wind tunnel velocity on 1/3-octave-band spectrum for nominal half-scale USB model at takeoff flap deflection.

Spectra measured for a smaller USB model with exhaust velocity of 250 m/sec, at wind tunnel velocities of 0 and 62 m/sec, are shown in figure 18. This velocity ratio of 0.25 was predicted to cause a 2.5-dB decrease in level and one 1/3-octave increase of frequency, relative to that for zero tunnel speed. Again, good agreement was obtained. The observed effects of forward speed for USB noise were a decrease of level at low frequencies and negligible change at higher frequencies that affect annoyance-weighted noise levels.

Full-Scale Implementations

Only a relatively small number of full-scale propulsive lift aircraft have been built and flown. These were listed in the section of this chapter entitled "Description of Propulsive Lift Vehicles." Most of them are military transports which must be able to operate from short runways near a combat zone. Low external noise radiation is not an important consideration for such aircraft. Noise data during takeoff, climb, approach, and landing have not been made available for comparisons with predictions.

Limited data exist for the two civilian research aircraft in this group (refs. 6 and 7). These aircraft were completed at a time when fuel costs and financial interest rates had greatly increased. The economic penalty caused by purchase and use of STOL aircraft with oversized wings and engines, which increase both initial cost and operating cost relative to conventional aircraft, was larger than the economic value of greatly reduced noise.

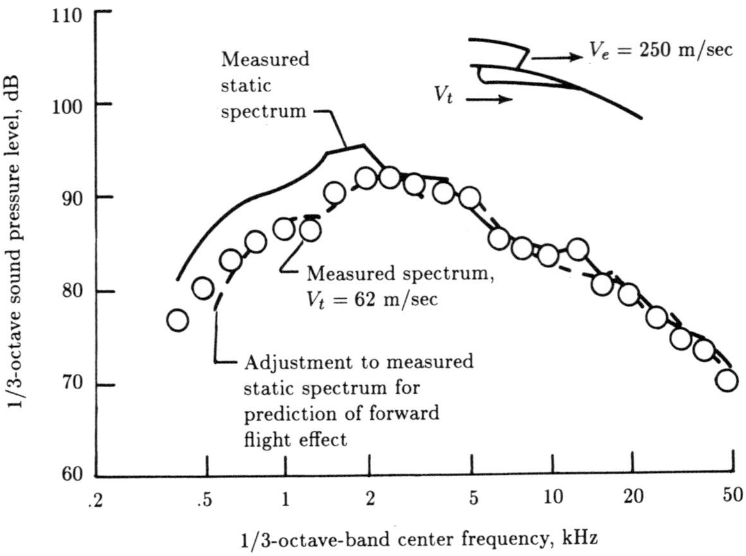

Figure 18. Predicted and measured effect of wind tunnel velocity on 1/3-octave-band spectrum for a USB configuration at takeoff flap deflection.

Overcrowding of commercial airport runways has been temporarily overcome by development of larger conventional aircraft powered by efficient high-bypass-ratio turbofan engines. These aircraft can operate from existing moderate-length runways rather than only from an airport's longest runway. The combination of decreased engine fuel consumption per unit thrust and decreased airframe weight per passenger gives increased efficiency at cruise. Noise levels of these aircraft satisfy the current Federal Aviation Administration (FAA) and International Civil Aviation Organization (ICAO) regulations.

Quiet propulsive lift aircraft have not yet been implemented for commercial use. The situation may change when population has increased such that airports become overcrowded and noise constraints become more important.

References

1. Conlon, John A.; and Bowles, Jeffrey V.: Powered Lift and Mechanical Flap Concepts for Civil Short-Haul Aircraft. *J. Aircr.*, vol. 15, no. 3, Mar. 1978, pp. 168–174.
2. Rosen, Cecil C., III; Burger, Robert J.; and Sigalla, Armand: Aeronautical Technology 2000: A Projection of Advanced Vehicle Concepts. AIAA-84-2501, Oct.–Nov. 1984.
3. Olsen, William A.; Dorsch, Robert G.; and Miles, Jeffrey H.: *Noise Produced by a Small-Scale, Externally Blown Flap.* NASA TN D-6636, 1972.
4. Fink, M. R.; and Olsen, W. A.: Comparison of Predictions and Under-the-Wing EBF Noise Data. AIAA-76-501, July 1976.
5. Reshotko, Meyer; Goodykoontz, Jack H.; and Dorsch, Robert G.: Engine-Over-the-Wing Noise Research. *J. Aircr.*, vol. 11, no. 4, Apr. 1974, pp. 195–196.
6. Bohn, A. J.; and Shovlin, M. D.: Upper Surface Blowing Noise of the NASA Ames Quiet Short-Haul Research Aircraft. *J. Aircr.*, vol. 18, no. 10, Oct. 1981, pp. 826–832.

7. Maita, Masataka; and Torisaki, Tadao: Acoustic Characteristics of the External Upper Surface Blowing Propulsive-Lift Configuration. *J. Aircr.*, vol. 18, no. 8, Aug. 1981, pp. 695–701.
8. Goodykoontz, Jack H.: *Acoustic Tests of Augmentor Wing Model.* NASA TM X-3519, 1977.
9. Amiet, R. K.: Acoustic Radiation From an Airfoil in a Turbulent Stream. *J. Sound & Vib.*, vol. 41, no. 4, Aug. 1975, pp. 407–420.
10. Karchmer, Allen M.; and Friedman, Robert: *Noise Tests on an Externally Blown Flap With the Engine in Front of the Wing.* NASA TM X-2942, 1973.
11. Goldstein, M. E.: Scattering and Distortion of the Unsteady Motion on Transversely Sheared Flows. *J. Fluid Mech.*, vol. 91, pt. 4, Apr. 27, 1979, pp. 601–632.
12. Powell, Alan: On the Aerodynamic Noise of a Rigid Flat Plate Moving at Zero Incidence. *J. Acoust. Soc. America*, vol. 31, no. 12, Dec. 1959, pp. 1649–1653.
13. Ffowcs Williams, J. E.; and Hall, L. H.: Aerodynamic Sound Generation by Turbulent Flow in the Vicinity of a Scattering Half Plane. *J. Fluid Mech.*., vol. 40, pt. 4, Mar. 1970, pp. 657–670.
14. Fink, M. R.: *Additional Studies of Externally Blown Flap Noise.* NASA CR-135096, 1976.
15. Fink, Martin R.; and Schlinker, Robert H.: *Airframe Noise Component Interaction Studies.* NASA CR-3110, 1979.
16. Olsen, W.; and Boldman, D.: Trailing Edge Noise Data With Comparison to Theory. AIAA-79-1524, July 1979.
17. Olsen, William A.; Miles, Jeffrey H.; and Dorsch, Robert G.: *Noise Generated by Impingement of a Jet Upon a Large Flat Board.* NASA TN D-7075, 1972.
18. Fink, Martin R.: *A Method for Calculating Externally Blown Flap Noise.* NASA CR-2954, 1978.
19. Brown, W. H.; and Ahuja, K. K.: Jet and Wing/Flap Interaction Noise. AIAA-84-2362, Oct. 1984.
20. Neuwerth, Günther (Leo Kanner Assoc., transl.): *Acoustic Feedback Phenomena of the Supersonic and Hypersonic Free Jet Impinging on a Foreign Body.* NASA TT F-15719, 1974.
21. Dorsch, Robert G.; Clark, Bruce J.; and Reshotko, Meyer: *Interim Prediction Method for Externally Blown Flap Noise.* NASA TM X-71768, [1975].
22. Fink, Martin R.: Forward Flight Effects on Externally Blown Flap Noise. *J. Aircr.*, vol. 15, no. 9, Sept. 1978, pp. 545–546.
23. SenGupta, G.: Analysis of Jet-Airframe Interaction Noise. AIAA-83-0783, Apr. 1983.
24. Miller, W. R.: Flight Effects for Jet-Airframe Interaction Noise. AIAA-83-0784, Apr. 1983.
25. McKinzie, D. J., Jr.: EBF Noise Suppression and Aerodynamic Penalties. AIAA-78-240, Jan. 1978.
26. Joshi, M. C.; and Yu, J. C.: Reduction of Wall Jet Trailing Edge Noise by Mean Flow Modification. *J. Aircr.*, vol. 17, no. 9, Sept. 1980, pp. 633–640.
27. Pennock, A. P.; Swift, G.; and Marbert, J. A.: *Static and Wind Tunnel Model Tests for the Development of Externally Blown Flap Noise Reduction Techniques.* NASA CR-134675, 1975.
28. Fink, M. R.; and Bailey, D. A.: *Airframe Noise Reduction Studies and Clean-Airframe Noise Investigation.* NASA CR-159311, 1980.
29. Reddy, N. N.; Blakney, D. F.; Tibbets, J. G.; and Gibson, J. S.: *V/STOL Aircraft Noise Prediction (Jet Propulsors).* FAA-RD-75-125, June 1975. (Available from DTIC as AD A028 765 6.)
30. Brown, W. H.; Searle, N.; Blakney, D. F.; Pennock, A. P.; and Gibson, J. S.: *Noise Characteristics of Upper Surface Blown Configurations; Experimental Program and Results.* NASA CR-145143, 1977.
31. Guinn, Wiley A.; Blakney, Dennis F.; and Gibson, John S.: *V/STOL Noise Prediction and Reduction.* FAA-RD-73-145, Aug. 1973.
32. Reddy, N. N.: *Blown Flap Noise Prediction.* NASA CR-158978, 1978.
33. Falarski, Michael D.; Aiken, Thomas N.; Aoyagi, Kiyoshi; and Koenig, David G.: Comparison of the Acoustic Characteristics of Large-Scale Models of Several Propulsive-Lift Concepts. *J. Aircr.*, vol. 12, no. 7, July 1975, pp. 600–604.
34. Jones, W. L.; Heidelberg, L. J.; and Goldman, R. G.: Highly Noise-Suppressed Bypass 6 Engine for STOL Application. *Aeroacoustics: Fan, STOL, and Boundary Layer Noise; Sonic Boom; Aeroacoustic Instrumentation*, Henry T. Nagamatsu, ed., American Inst. Aeronautics and Astronautics, c.1975, pp. 141–156.

9 Combustion and Core Noise

Lead author

J. Robert Mahan
Virginia Polytechnic Institute and State University
Blacksburg, Virginia

Contributing author

Allen Karchmer
NASA Lewis Research Center
Cleveland, Ohio

Introduction

Two types of aircraft power plant are considered in this chapter: the gas turbine and the reciprocating engine. The gas turbine engine uses a jet of air or a propeller, or a combination of the two, to develop thrust. The jet may consist entirely of gas that has passed through the core of the engine, or it may be a mixture of this gas with bypass air ducted around the engine core. When a high bypass ratio is used, the engine is called a fanjet, or turbofan, because much of the thrust is produced by a fan in the bypass duct. The engine types considered in this chapter are illustrated in figure 1 (see ref. 1). They are (a) the reciprocating engine, (b) the turbojet engine, (c) the turboprop engine, and (d) the turbofan engine.

This chapter deals mostly with combustion noise in gas turbine engines, although reciprocating-engine combustion noise is treated briefly at the end of the chapter. An exhaustive review of the general combustion noise literature is contained in reference 2 and a more recent comprehensive interpretive review of the gas turbine engine combustion and core noise literature is included in reference 3 on the aerothermodynamics of aircraft engine components.

Combustion noise in gas turbines is classified according to source mechanism as either direct or indirect. Direct combustion noise is produced by the combustion process itself, while indirect combustion noise occurs when the hot products of combustion pass through the turbine and exhaust nozzle. The combination of direct and indirect combustion noise in a gas turbine engine makes up an important part of what is generally called core noise. Depending on the authority cited, core noise

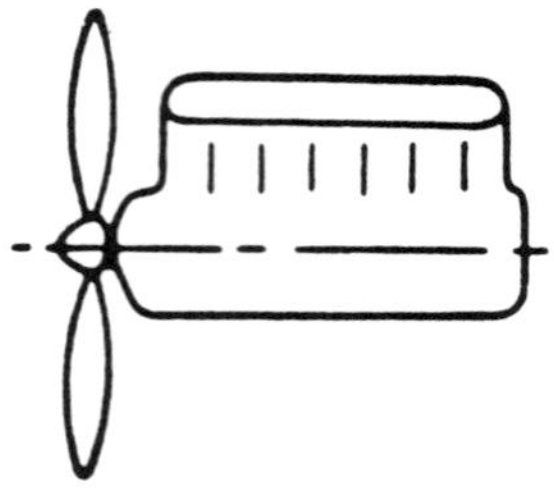

(a) Reciprocating engine.

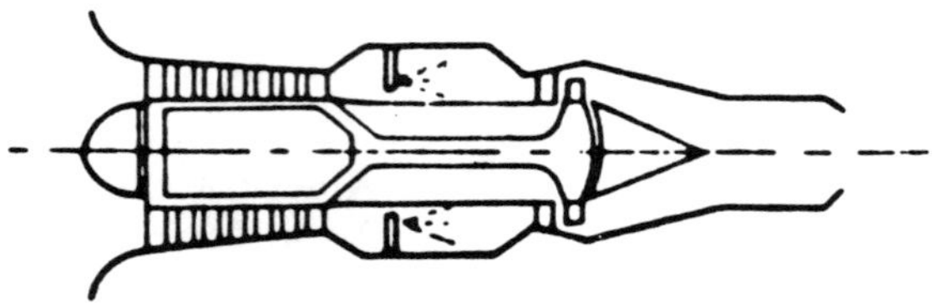

(b) Turbojet engine.

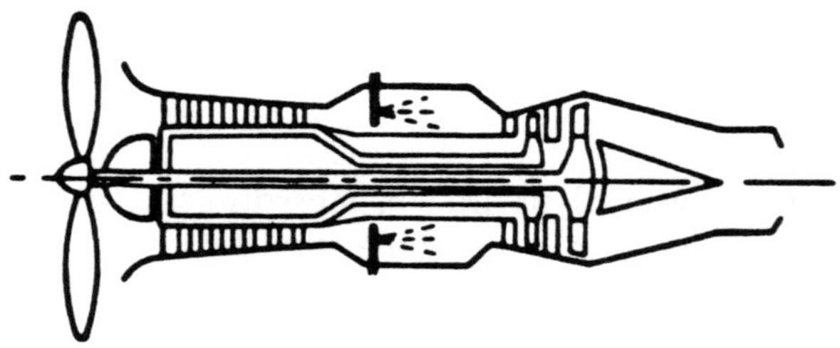

(c) Turboprop engine.

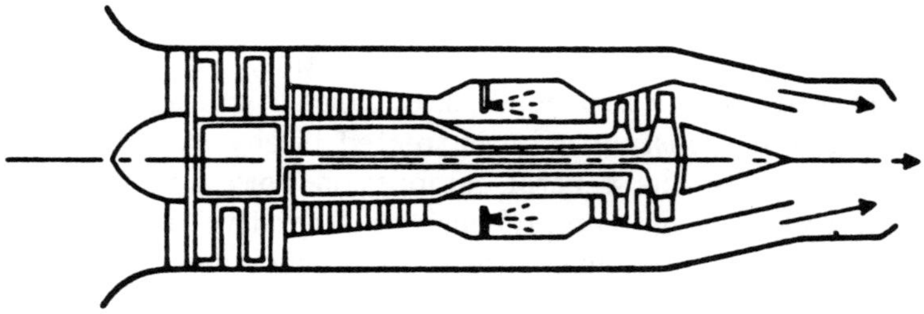

(d) Turbofan engine.

Figure 1. Types of aircraft power plant considered. (From ref. 1.)

may or may not also include compressor noise as well as components of turbine and flow noise not associated with the combustion process. In some quarters, core noise is defined as all noise exclusive of jet noise emitted into the rear arc of a gas turbine engine. In any case, compressor noise is not considered in this chapter, and only those components of turbine and flow noise are considered which would not be present in the absence of the combustion process. In view of these restrictions, it might be argued that the chapter should be entitled simply "Combustion Noise." However, "Core Noise" has been retained in the title because in many, if not most, cases core noise is dominated by, and therefore synonymous with, combustion noise, especially if compressor noise is excluded as a separate category, as is often done.

The importance of core noise is illustrated in figure 2, which shows low-frequency acoustic power radiated from a gas turbine engine as a function of effective exhaust jet velocity (ref. 4). The solid curve with triangular symbols represents the overall acoustic power radiated to the far field, while the short-long dashed line represents jet noise, which is known to increase as the eighth power of jet velocity. The two curves merge as jet velocity increases, indicating that the overall radiated power is dominated by jet noise at high jet velocities. However, at relatively low jet velocities, such as would occur at engine idle, during taxiing, and at approach and cruise conditions, the overall noise level exceeds that predicted by jet noise theory. This "excess noise" is generally attributed to core noise. In figure 2, the solid curve with the square symbols is core noise measured at the engine nozzle exit.

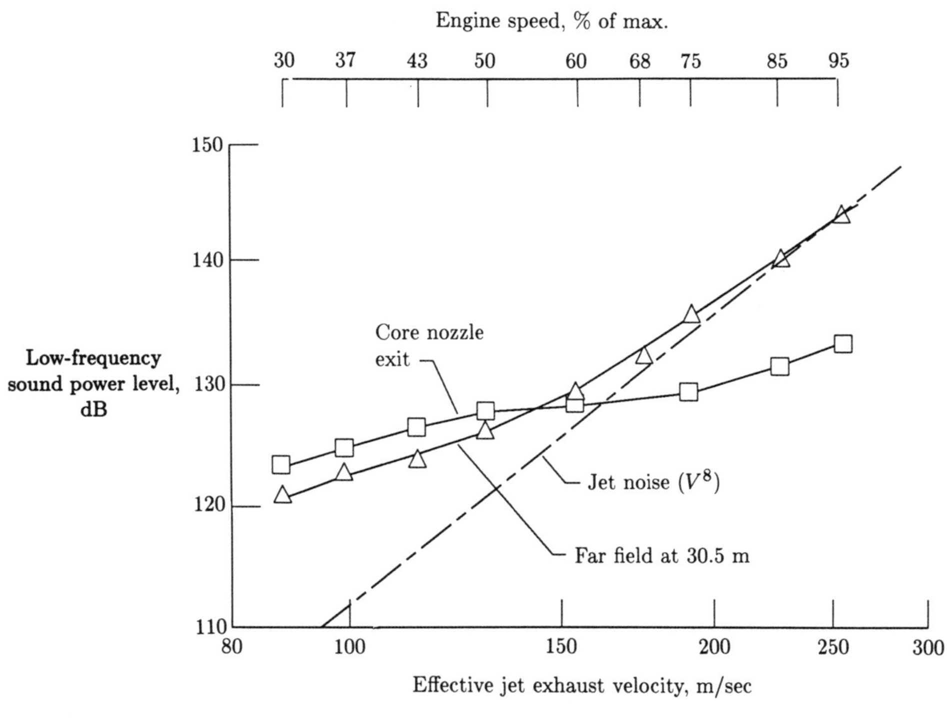

Figure 2. Low-frequency sound power in the far field and at the core nozzle exit of a turbofan engine as a function of effective jet velocity. (From ref. 4.)

Configuration Variables

The physical geometry of the hot gas path through a gas turbine engine, as well as the pressure, temperature, and combustion heat release distributions along this path, is an important controlling factor for both direct and indirect combustion noise. The combustor, the fuel and air induction system, the interface between the combustor and the turbine, and the turbine itself together form an acoustic circuit. The response of this circuit to combustion noise source activity depends on the acoustic responses of the individual components and the manner in which they are interconnected.

A cutaway view of a typical turbofan engine is shown in figure 3 (ref. 5). The present discussion of combustion and core noise is limited to activities which take place in the segment of the gas path between the compressor diffuser and exhaust nozzle exit planes. These boundaries have been chosen to exclude compressor, fan, and jet noise, which are beyond the scope of this chapter. In any complete analysis of combustion and core noise it is necessary to specify the acoustic conditions at these boundaries.

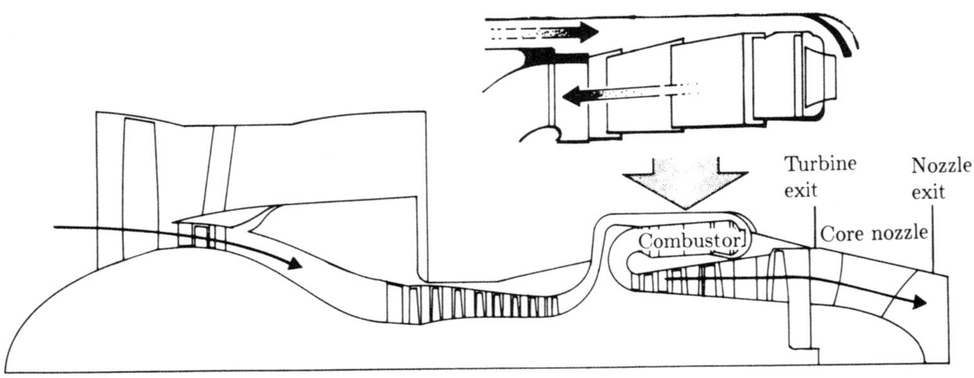

Figure 3. Cutaway view of turbofan engine showing the combustor, turbine exit, and core nozzle. (From ref. 5.)

Many variations on the more or less generic combustion system of figure 3 are possible. The diffuser might be long and narrow, with a relatively low exit velocity, or it might be a short “dump” diffuser, with a relatively high velocity jet within the receiver. The combustor itself might consist of an array of individual chambers, called cans, or combustion might occur in a continuous annular region which wraps around the engine. Variations include “canular” combustors, which, as the name implies, are hybrids of the can-type and annular combustor ideas. The distribution of combustion, dilution, and cooling air along the length of the combustor varies from one design to the next. Finally, the geometry of the transition duct that connects the combustor to the first stage turbine nozzle is often quite complex and is highly variable with engine design.

Operational Variables

The combustion process in a practical gas turbine combustor is turbulent. There are several reasons why this is desirable. First, because turbulence encourages mixing of the fuel and air with each other and with the hot products of combustion, a turbulent flame is more compact. This permits the engine itself to be smaller and lighter. Next, the turbulent wake downstream of a swirler or behind a flame holder anchors the flame in a well-defined location. Also, the enhanced mixing of the hot products of combustion with dilution air introduced downstream of the primary combustion zone leads to a shorter secondary zone and a more uniform temperature field entering the turbine. Finally, turbulence-enhanced mixing ensures more complete combustion, thereby improving efficiency and reducing some harmful emissions. Unfortunately, turbulent flames are inherently noisy.

Combustion noise is characterized by its overall radiated power and by its spectral and directional distribution. The key parameter for determining the overall radiated sound power is the thermoacoustic efficiency, defined as the fraction of the combustion heat release which is radiated away as acoustic energy. The sound spectrum of an open turbulent flame is rich in information about turbulence scales, burner geometry, convective velocities, flame speeds, and so forth. This has led to the development of a school of combustion diagnostics which exploits details of the spectral shape out to frequencies of several thousand hertz. However, the radiation of acoustic power by a turbulent flame is dominated by a limited portion of the spectrum which rises slowly to a single blunt peak somewhere between 300 and 600 Hz and then decreases more or less monotonically. Figure 4 (ref. 6) shows typical open-flame sound pressure spectra corresponding to the turbulent premixed and diffusion-flame burning of hydrocarbon fuels. Departures from the spectral shape of figure 4 for combustion noise radiated from an engine are mostly due to the resonant modes of the acoustic circuit defined by the combustion system. Then the controlling factors for direct combustion noise are (1) those which determine the thermoacoustic efficiency of the combustion process and (2) those which determine the acoustic response of the combustion system.

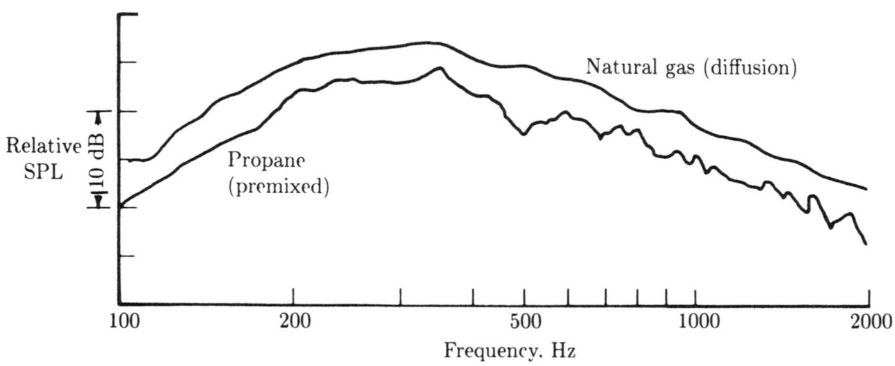

Figure 4. Typical sound pressure level spectra in the far field of open, turbulent, premixed and diffusion flames with gaseous hydrocarbon fuels. (From ref. 6. Copyright AIAA. Reprinted with permission.)

The thermoacoustic efficiency of an open turbulent flame generally changes when the same flame is placed within an enclosure. There are two reasons for this change. First, the enclosure modifies the steady flow through the flame, thereby producing fundamental changes in the combustion process. It is easy to imagine, for example, how an enclosure might dictate the way the fuel and air are mixed and then brought together with the hot products of combustion. Also, as the enclosure becomes smaller relative to the original volume of the open flame, the average steady flow velocity, and with it the turbulence intensity, must increase.

The second reason is more subtle. In certain situations, coupling may exist between the acoustic response of the combustion system and the heat release process. Such coupling can occur in any number of ways. For example, experiments (ref. 7) have demonstrated that acoustic energy introduced to a gas turbine combustor at the proper frequency can enhance mixing downstream of the flame zone so that the turbine inlet temperature spread is significantly reduced. Other experiments have been conducted with laboratory burners in which periodic acoustic signals have been introduced either upstream of the flame zone in a mixing chamber (ref. 8) or directly in the flame zone itself (ref. 9). The former establishes a clear relationship between acoustically induced mixing and enhancement of thermoacoustic efficiency, while the latter may be interpreted as evidence of another kind of enhancement, discussed in the next paragraph. In view of the relationship between acoustically induced mixing and enhanced thermoacoustic efficiency, it is not difficult to believe that the acoustic pressure field produced by the turbulent combustion process itself can enhance the mixing process and thereby influence the thermoacoustic efficiency of the combustion process. But the presence in the flame zone of an acoustic particle velocity antinode (pressure node) at the critical frequency required to make this idea work is determined by the combustion system geometry.

One classical coupling mechanism of pedagogical, if not practical, interest is that first described by Lord Rayleigh (ref. 10). If the combustion heat release process is periodic, or at least has a periodic component, the resulting acoustic pressure waves emanating from the flame are periodic with the same frequency. Because a turbulent flame produces a broad, nearly random noise spectrum, significant acoustic energy is present at virtually all frequencies below about 1000 Hz. The presence of an enclosure can cause pressure waves created in the flame zone to be returned to the flame zone with a time delay that depends on the length and average sound speed in the combustor. Energy is added to the pressure wave at any frequency for which the instantaneous peak in acoustic pressure in the flame zone coincides with the instantaneous peak in heat release. When this critical situation holds, the wave grows in amplitude with each cycle until a limit cycle is reached where losses and nonlinearities arrest further growth. Similarly, energy is removed from a pressure wave at any frequency for which the wave is 180° out of phase with the periodic heat release in the flame zone.

In principle the pressure variation associated with unsteady combustion can become so strong that it actually modulates the flow of fuel and/or air into the combustor. When this happens, there is nearly always a frequency for which the total phase angle between the pressure oscillation in the flame zone and the corresponding heat release oscillation is an integer multiple of 360°. This automatically satisfies Rayleigh's criterion and leads to a type of combustion instability which produces very large pressure amplitudes, as explained in reference 11. While the authors know of no incidence of this so-called feed instability in an aircraft gas turbine engine, the first author has observed it in a gas-fueled industrial gas turbine (where the fuel nozzle pressure drop is significantly lower than in liquid-fueled gas turbines). In this case the resulting periodic pressure fluctuation was sufficient to drive the combustion hardware to fatigue failure. The problem was fixed by increasing the fuel nozzle pressure drop.

The operational variable having the greatest influence on combustion and core noise in a gas turbine engine is engine power. Both the mass flow through the combustion system and the combustor temperature level increase with power level. If the thermoacoustic efficiency remained constant, the radiated acoustic power would be proportional to the power developed by the engine. However, the tendency is for the thermoacoustic efficiency to increase with the mass flow rate through the combustor.

The spectral distribution of the radiated acoustic power can also depend on engine power level through the combustor temperature, although this is usually a minor effect. As the temperature in the combustor increases with power level, so does the speed of sound (approximately with the square root of temperature). As the speed of sound increases for a given combustor geometry, the resonant modes shift to higher frequencies. In principle this could even result in either a slight increase or a slight decrease in thermoacoustic efficiency, depending on how the change in sound speed affects the timing between the pressure waves and the heat release distribution in the combustion zone. In practice, this effect is probably unimportant, however.

Characteristics of Combustion and Core Noise

Any discussion of the characteristics of combustion and core noise in the engine environment must begin at the noise source, that is, in the combustor. As already stated, most practical gas turbine combustion systems consist either of an array of individual combustion chambers, called cans, or of a continuous annular chamber that wraps around the engine. Although variations of these two ideas are possible, for example, the canular combustor, familiarity with the combustion noise characteristics of these two main combustor types is sufficient for basic understanding of gas turbine combustion and core noise.

The sound pressure spectra for the open turbulent flames of figure 4 are very similar to the sound spectrum produced by a turbulent jet. This is not surprising in view of contemporary direct combustion noise theory, in which the source mechanism is attributed to the turbulent mixing of fuel and air with the hot products of combustion. In fact, the shape of the curve and the frequency at which it peaks for a given fuel are surprisingly insensitive to size of the burner, the power level, and flame temperature, even though the overall sound pressure level is sensitive to these factors.

As shown in figure 4, significant combustion noise is limited to frequencies on the order of a few hundred hertz. Consequently, the wavelength of the pressure disturbances in the combustor associated with combustion noise is generally large compared with the transverse dimensions of the engine. In this case the combustion noise propagates into the surroundings as a plane wave. Two-dimensional modes are present within annular combustors for which the circumference is comparable to the length. However, some of these modes often are cut off in the smaller diameter turbine and exhaust nozzle and thus are unable to propagate efficiently into the surroundings as sound.

The combustion system geometry and temperature distribution, the acoustic conditions at either end, and the source activity combine to define a series of resonant frequencies not unlike those observed in an organ pipe. Also, it has already been explained that a pressure disturbance is selectively amplified if it occurs at a frequency which satisfies Rayleigh's criterion. Therefore, a typical core noise spectrum measured in the far field of an engine has the general form of the open flame spectrum of figure 4, but with a superimposed series of relatively sharp peaks corresponding to the resonant frequencies and, though rarely, to one or more frequencies satisfying Rayleigh's criterion.

Typical 1/3-octave band pressure spectra obtained within a can-type combustor and in the far field of its exhaust are shown in figure 5 (ref. 12). These spectra are for a single can exhausting through a relatively short nozzle rather than through a turbine. The peaks in the spectra, which cause them to deviate from the general trend of figure 4 for an open flame, are due to combustor resonance.

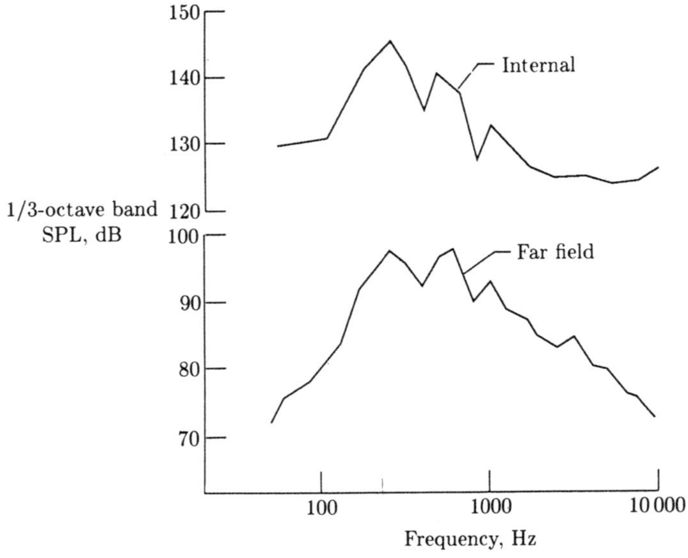

Figure 5. One-third-octave band sound pressure levels measured in a can-type combustor and in its far field. (From ref. 12. Copyright AIAA. Reprinted with permission.)

Higher order modes involving the axial and circumferential degrees of freedom are available to annular combustors because of their two-dimensional geometry. Typical 1/3-octave band pressure spectra obtained in the combustor, at the turbine exit, at the core nozzle exit, and in the far field of a gas turbine engine with an annular combustor are shown in figure 6 (ref. 4). The activity above about 1500 Hz is attributable to the rotating machinery. It is clear from comparison of the spectra below 1500 Hz that not all the modes in the combustor are able to propagate to the far field as sound. In particular, some higher order modes that are present in the combustor are cut off in the turbine and tail pipe and are thus unable to propagate into the surroundings.

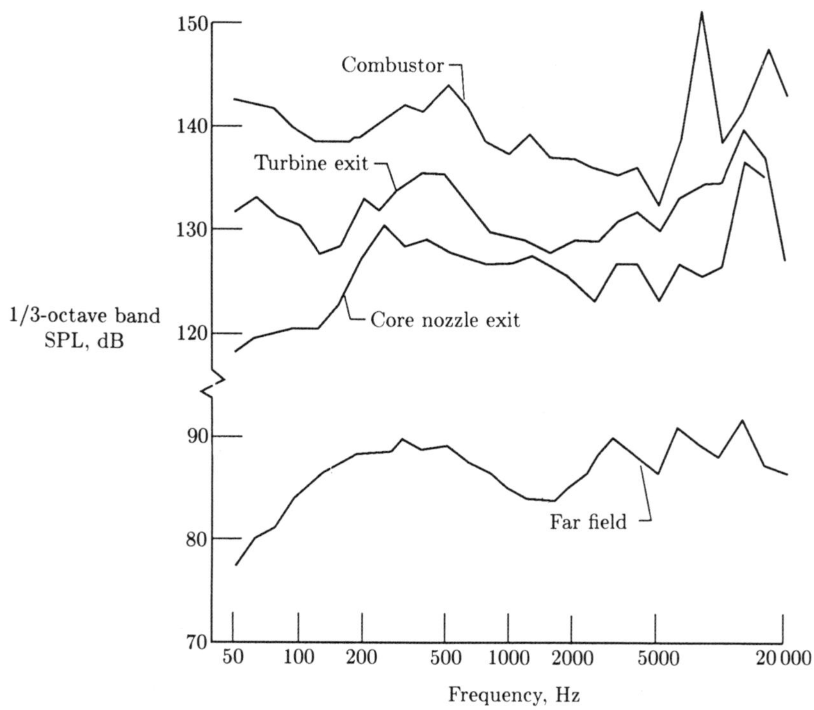

Figure 6. One-third-octave band sound pressure levels measured in the combustor, at the turbine exit, at the core nozzle exit, and in the far field of a turbofan engine having an annular combustor. (From ref. 4.)

Correlations of overall sound power for both laboratory-scale burners and engines are remarkably similar. In general, radiated sound power varies somewhat as the mass flow rate to a power between two and three and as the temperature rise across the burner to an exponent of about two. The thermoacoustic efficiency, defined in the previous section, then varies somewhat as the square of mass flow and linearly with the temperature rise. Typical values of the thermoacoustic efficiency are found to range from 10^{-6} to 10^{-5}.

The directivity of core noise is determined by the exhaust conditions rather than by the source activity, although theoretically there should also be a slight frequency effect. In general, a spherical pattern is observed at low exhaust velocities and frequencies, but as the velocity and frequency are increased, a nonspherical directivity pattern begins to emerge. As a practical matter, the deviation from a spherical directivity pattern associated with increasing frequency is not important in gas turbine core noise because of the relatively low frequencies involved. Figure 7 shows directivity patterns for a range of engine speeds for a gas turbine engine (ref. 13). As the exhaust velocity increases with engine speed, the peak in the directivity pattern shifts in the flow direction and its magnitude increases. The combustion noise peak typically occurs at an angle of about 120°, measured with respect to the engine inlet, and corresponds to a deviation of about 10 dB from the minimum value.

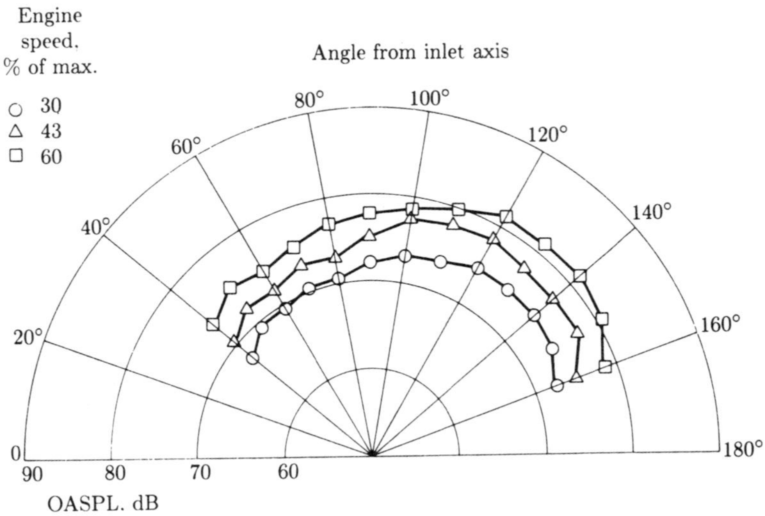

Figure 7. Combustion noise overall sound pressure level (OASPL) directivity patterns. (From ref. 13.)

Sources of Combustion Noise

It has already been noted that combustion noise can be categorized as being either direct or indirect depending on its source characteristics. While direct combustion noise is related to the creation of local hot spots in the combustor, indirect combustion noise is related to the subsequent flow of these hot spots through the steep pressure gradients encountered in the turbine and exhaust nozzle.

Direct combustion noise results when a volume of mixture expands at constant pressure as it is rapidly heated by combustion. This local expansion causes the cooler surrounding gas to be pushed back; that is, the expanding gas does work on its surroundings. This work in turn produces waves in the surrounding gas which propagate into the far field as sound. Such sources are called acoustic monopoles. The strength of an acoustic monopole produced in this way depends on the net work done as it expands and on the rate of doing this work. The amount of work done depends on the thermal energy deposited in the volume element by the combustion process and on the efficiency with which this energy is converted to work.

In a practical aircraft gas turbine combustor, the fuel and air are introduced separately. The liquid fuel must be atomized and vaporized to produce volatile gases, which must then mix with the combustion air before they can be burned. The physical processes by which the atomization, evaporation, and mixing occur are necessarily turbulent; otherwise the combustor would be prohibitively long.

A reasonable model (ref. 14) for turbulent combustion is shown in figure 8. The model encompasses two possible scenarios which represent two extremes, with reality lying somewhere in between. In one scenario, a turbulent eddy consisting of pure vaporized fuel enters a region of hot combustion air; in the other, a turbulent eddy consisting of a mixture of vaporized fuel and combustion air enters a region of hot products of combustion. As the eddy penetrates the hot region, it entrains the surrounding hot gases, creating a sort of pinwheel, or "flamelet," consisting of alternating layers of either (1) volatile mixture and hot products of combustion or (2) fuel and hot combustion air, as the case may be. The rapid radial expansion of the hot gases generated by the burning of this flamelet produces the monopole source behavior described above.

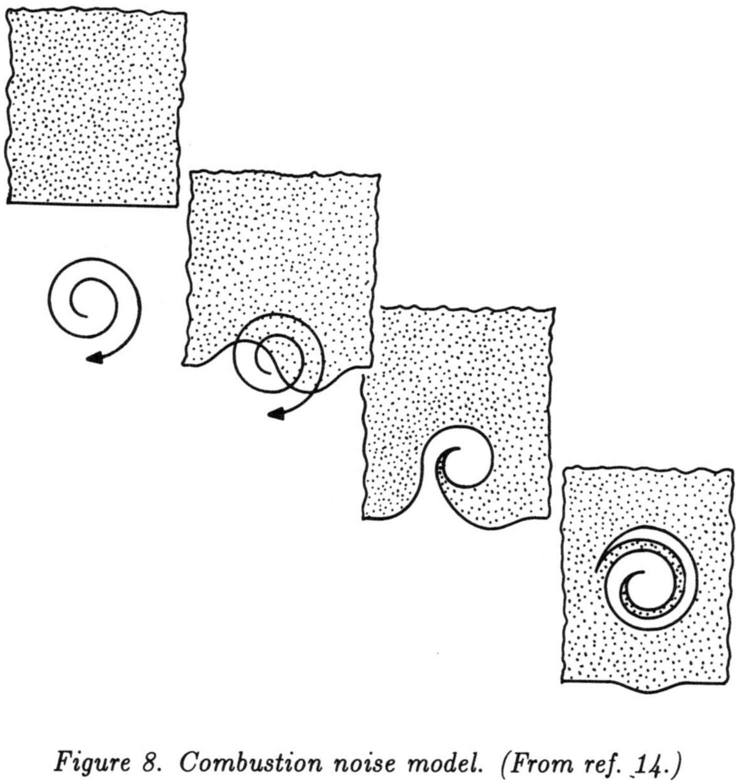

Figure 8. Combustion noise model. (From ref. 14.)

In the first scenario, consumption of the eddy is paced by the local diffusion rate at the air-fuel interface, while in the second scenario, the combustion rate depends only on the local laminar flame speed. The peak frequency of the direct combustion noise produced by the transient combustion of many such eddies, randomly distributed in time and space throughout the flame zone, might be expected to vary somewhat as the inverse of the time it takes to burn one of these typical eddies. However, when the statistical nature of turbulence is taken into consideration (ref. 15), the relationship between the peak in the combustion noise spectrum and the average survival time of a turbulent eddy of fuel-air mixture may not be that simple. In any case, two factors cause the sound produced to be dispersed in a spectrum about the peak frequency. First, as already noted, the eddies have a size distribution that depends on the statistical distribution of turbulent mixing lengths. This would lead to a distribution of eddy survival times for a fixed burning rate. Second, for a given eddy size there is a distribution of burning rates between the two extremes described in the above model. This, too, leads to a distribution of survival times and thus to a further dispersion of frequencies about the peak. In fact, it is the uncertainty about which mechanism dominates the burning rate, flame speed or diffusion, which most contributes to differences in combustion noise theories.

Fortunately, the total combustion heat release in an actual gas turbine combustor is distributed among a very large number of small uncorrelated flamelets. As a simple model, consider the quasi-periodic pressure field produced by the continuing life cycle of the flamelets, which form, burn, and then reform, more or less regularly, in a given locality of the combustion zone. The Fourier series representation of this pressure field consists of a mean component and an infinite series of periodic components whose mean values are zero and whose amplitudes decrease monotonically with frequency. The locality is then acoustically equivalent to a compact source consisting of many point sources of mass (acoustic monopoles), each oscillating at a different fixed frequency. There are many such uncorrelated localities in a turbulent flame, each producing more or less the same pressure spectrum, but with phases unrelated to those of other localities. The net effect is that destructive interference occurs within the source region, so that the overall source mechanism is rather inefficient. In fact, as already mentioned, it has been observed that the overall radiated acoustic power associated with gas turbine combustion noise is on the order of only a few parts per million of the total thermal power.

The contribution of indirect combustion noise to gas turbine core noise was first reported in reference 16 in 1972, and the term "indirect combustion noise" was coined in reference 6 in 1973. An excellent development of the theory is given by Marble (ref. 17), who distinguishes between kinematically unsteady flows, in which velocity fluctuations produce pressure fluctuations, and thermodynamically unsteady flows, in which temperature fluctuations, through the associated density fluctuations, lead to pressure fluctuations. Marble demonstrates that thermodynamic unsteadiness can be equally as effective as kinematic unsteadiness in its ability to produce unsteady loading on an obstacle in the flow field. Briefly, indirect combustion noise, or entropy noise as it is sometimes called, results when relatively large-scale temperature nonuniformities generated by turbulent combustion are convected through pressure gradients in the turbine. This produces an entropy fluctuation in conformity with the first and second laws of thermodynamics. Because the density of an ideal gas depends on any two independent thermodynamic properties, say entropy and pressure, a density fluctuation occurs whenever an entropy nonuniformity is convected through the pressure drop associated with a stage of the turbine. Just as in the case of the direct combustion noise mechanism, this density fluctuation produces waves that propagate through the surrounding gas.

The relative importance of the direct and indirect components of gas turbine combustion noise is yet to be definitively established. Both source mechanisms probably contribute significantly to engine core noise, with their relative dominance depending on such things as engine power setting and combustor and turbine design considerations. The dominant frequencies associated with both combustion noise source mechanisms are determined by the rate of production and the size distribution of hot spots within the combustor. The frequency spectrum of indirect combustion noise should not be sensitive to the convective velocity through the turbine because as a hot spot is accelerated through a turbine stage, it elongates so that the ratio of its velocity to its length, and therefore its characteristic frequency, remains essentially constant.

The similarity in the sound spectra of direct and indirect combustion noise makes it difficult to separate the two source mechanisms simply by studying the core noise sound pressure spectrum. However, in one laboratory experiment involving a single gas turbine combustor can operated at reduced pressure, Muthukrishnan et al. (ref. 18) were able to separate near-field combustion noise into its direct and indirect components over the dominant part of its spectrum. They accomplished this by correlating the near-field sound pressure signal with a second signal, either from a pressure probe in the combustion can or from a high-frequency-response thermocouple in the exit plane of the can. In the study, a high correlation between the signals from the near-field microphone and combustion-can pressure probe is interpreted to mean that the near field is dominated by direct combustion noise, while a relatively high correlation between signals from the near-field microphone and the exit-plane thermocouple is interpreted to mean that the near field is dominated by indirect combustion noise. Subject to this interpretation, it was found that as the pressure drop increased across a nozzle or orifice plate downstream of the combustion can, dominance of the near-field sound pressure shifted from direct to indirect combustion noise. From this the authors concluded that direct combustion noise dominates the near field for small pressure drops, while indirect combustion noise dominates for large pressure drops. Also, they interpreted the low coherence observed between the signals from the combustion-can pressure transducer and the exit-plane thermocouple above 200 Hz to indicate that direct combustion noise and indirect combustion noise are statistically independent at higher frequencies. On the other hand, observed high correlation between these two signals below 100 Hz seemed to indicate that the two combustion noise mechanisms are inseparable at low frequencies.

Muthukrishnan et al. concluded that indirect combustion noise should dominate in an actual gas turbine engine because of the relatively large turbine pressure drop. However, the validity of this conclusion must be tempered somewhat by two facts. First, the temperature fluctuations observed at the combustion-can exit plane were up to six times those expected in an actual gas turbine combustor. This would presumably produce higher indirect combustion noise levels in the near field of the experimental burner than in the near field of an actual engine. Second, the laboratory burner was operated at a total pressure on the order of 10 percent of that typical of an actual gas turbine engine. This could conceivably lead to lower levels of direct combustion noise than would be produced by an engine.

Reference 19 presents a theoretical development which predicts that indirect combustion noise power should vary as the square of the pressure drop across each turbine stage. If true, this means that the indirect combustion noise can be reduced for a given turbine pressure drop by increasing the number of stages. Finally, reference 20 shows good agreement between the 1/3-octave band combustion noise spectrum measured in the rear arc of an actual gas turbine engine and the spectrum predicted on the basis of an indirect combustion noise theory. The theory underpredicts the measurement only at the low- and high-frequency extremes of the spectrum, with the measured spectrum being well predicted between 100 and 500 Hz.

Introduction to Combustion Noise Theory

Several competing combustion noise theories have emerged in recent years based more or less on the fundamental source mechanisms described above. Many of these theories provide reasonable estimates of observable trends in combustion and core noise behavior. Indeed, the differences between the most successful of these theories are frequently only superficial; the underlying physics and mathematics are often essentially the same.

Perhaps the most pedagogically sound and rigorously complete direct combustion noise theories are those that have been inspired by Lighthill's (ref. 21) aeroacoustics theory. While Lighthill's theory explains jet noise in terms of acoustic quadrupoles produced by turbulent mixing in a shear layer, many of the more promising direct combustion noise theories, while attributing combustion noise to equivalent acoustic monopole activity, draw heavily on Lighthill's formalism.

Unfortunately, a quantitative prediction of combustion noise from first principles is not yet possible because the equations describing turbulent flow, which are central to Lighthill's analogy, cannot yet be solved. At best, combustion noise theories inspired by Lighthill's theory can be used to predict noise trends only when simplifying assumptions are made about the turbulence structure and its relationship to the unsteady heat release. Even in some of these developments, no attempt is made to solve the equations. Instead, the principles of dimensional analysis are relied on to imply the dependence of thermoacoustic efficiency and peak frequency on the combustor design and operating conditions.

Not all theoretical developments which successfully predict observed combustion noise trends have been inspired by Lighthill's theory. An alternative approach involves postulating a physical model for the dependence of the unsteady volumetric combustion heat release distribution on the local flow and thermodynamic variables. This is done on the basis of either physical arguments, similar to those presented in the previous section, or experimental results. This combustion heat release term is introduced into the appropriate energy equation which, together with the continuity and momentum equations and an equation of state, describes the resulting unsteady flow in the combustor.

Two continuations are possible after the unsteady volumetric heat release term has been defined and the governing equations have been established. In the first, the equations are linearized (by assuming small perturbations of the flow variables), cast in the form of a wave equation, and then solved, usually numerically. Alternatively, the equations are sometimes simplified by rejecting certain terms on the basis of order-of-magnitude arguments. In this case it is often possible to obtain a closed-form analytical expression for the acoustic pressure in terms of the source term and other physical variables. The radiated power is then estimated consistent with the assumed form of the volumetric heat source term.

Regardless of the theoretical approach used to predict combustion noise, the results obtained ultimately depend on the assumed form of the unsteady volumetric heat release term. Also, most theoretical developments attempt to describe the sound power radiated from an open flame rather than from the interior of a combustor. In these cases modifications must be made to account for the acoustic response of the combustion system and the transmission loss through the turbine and exhaust nozzle.

In spite of the differences among the various theoretical approaches, several of them predict experimentally observed trends with acceptable accuracy.

The three principal results sought from combustion noise theory are the radiated sound power (or the thermoacoustic efficiency), the peak frequency of the radiated sound, and the directivity. All three are usually presented in the form of a regression formula in which exponents of the parameters determine their influence on a given result. Several parameters are common to nearly all theoretical developments: the total mass flow rate through the burner, the burner length and cross-sectional area, the air-fuel ratio, and some measure of the fuel reactivity. Alternatively, some theories present their results in terms of burner pressure drop or burner temperature rise, which themselves are relatable to the above parameters.

In fact, most combustion noise theories reported in the literature have developed alongside of, and been strongly influenced by, contemporary experimental studies. In some cases, experimentally determined coefficients have been used in the "theoretical" development. At the very least, the availability of experimental results has aided the theoretician in selecting the best theory from among several attractive candidates. For this reason, very nearly all of them correctly predict at least some aspect of the observed trends. In view of the synergistic relationship which exists between theory and experiment, the results of both are presented together in the next section.

Combustion Noise Theory and Comparison With Experiment

The earliest known combustion noise theory is that of S. L. Bragg (ref. 22). Bragg's theory is based on the direct combustion noise source model, described earlier, in which the flame zone is assumed to consist of a region of uncorrelated flamelets, created by turbulent mixing, which produce monopole-type sound upon burning. The theory appeals to purely physical reasoning to deduce that the sound power radiated from a turbulent flame should vary as the fuel reactivity and as the square of the mixture flow velocity. A thermoacoustic efficiency of about 10^{-6} is predicted with a peak frequency around 500 Hz for a typical hydrocarbon fuel.

Thomas and Williams (ref. 23) measured the sound power radiated from burning soap bubbles filled with combustible mixtures. The radiated sound power can be calculated exactly for this simple geometry, and their measurements are in excellent agreement with theory. The measured and predicted thermoacoustic efficiencies for centrally ignited bubbles were shown to vary with flame speed over a range of about two orders of magnitude centered about a value of 10^{-5}. Both theory and experiment also indicate that the efficiency decreases by about an order of magnitude when the bubble is ignited at or near the outer surface, a situation more nearly like what must occur in an actual turbulent flame. Also, it has already been stated that a source region composed of individual uncorrelated monopoles would be less efficient than a single monopole of the same combined strength because of destructive interference. For these reasons the thermoacoustic efficiency of 10^{-6} predicted by Bragg's simple theory is consistent with the rigorously correct results of Thomas and Williams.

In an extension of Thomas and Williams' work, Hurle et al. (ref. 24) postulate, on the basis of simple monopole source theory, that the sound pressure radiated from an open turbulent premixed flame should vary as the time rate of change of light emission by certain free radicals in the reaction zone. A key element in the development is their demonstration that the intensity of emission by these free radicals increases directly as the flow rate of combustible mixture for both laminar and turbulent flames. They interpret their experimental confirmation of this idea for an ethylene-air flame as supporting both the monopole source nature of combustion noise and the flamelet, or wrinkled flame, model of turbulent flames. This result is significant because it establishes a direct relationship between the radiated sound pressure and the combustion heat release fluctuation.

Strahle's direct combustion noise theory is prominent among the most elaborate of those that have been influenced by Lighthill's theory and developed along the lines indicated in the preceding section. Instead of invoking the energy and entropy principles from thermodynamics to rigorously account for density fluctuations in the turbulent reaction zone, Strahle (ref. 25) proposes a version of the flamelet model which permits the introduction of two time scales, one due to convection and one due to diffusion. This leads to an expression for the thermoacoustic efficiency estimate having two adjustable exponents whose values depend on the relative dominance of the two rate processes. Strahle demonstrates that the experimental trends from the literature for open premixed flames can be predicted by this theory if appropriate values of the two adjustable exponents are chosen. The expression reduces, to within a constant multiplicative factor, to Bragg's of result if all Bragg's assumptions are invoked. In a more mature version of his theory, Strahle (ref. 26) gives an expression for the acoustic component of the density ρ' at a far-field point r outside the region undergoing turbulent combustion:

$$\rho' = \frac{1}{4\pi c_o^2 r}\frac{\partial^2}{\partial t^2}\int_V \rho_T(r_o, t - \frac{r}{c_o})\, dV(r_o) \tag{1}$$

where c_o is the sound speed outside the flame zone, r_o is a point within the flame zone, t is time, and V is the volume of the flame zone. This result assumes that the acoustic component of the density fluctuation ρ' in the combustion zone is small compared with the density fluctuation ρ_T due to turbulent combustion.

The most practical version of Strahle's theory, reported in reference 27, predicts that the sound power in watts radiated from a can-type combustor which is perfectly impedance matched to the surroundings should be given by

$$P_{\text{eq}} = a_1 p^{a_2} V_{\text{ref}}^{a_3} T_i^{a_4} F^{a_5} N_f^{a_6} A_e^{a_7} (A_e^{1/2}/\ell)^{a_8} \tag{2}$$

where p is the combustor mean pressure in psia (kPa), V_{ref} is the mean flow velocity in ft/sec (m/sec), T_i is the combustor inlet temperature in °R (K), F is the fuel-air ratio, N_f is the number of fuel nozzles, A_e is the cross-sectional area at the combustor exit in in^2 (cm^2), and ℓ is the combustor length in in. (cm). The values of the exponents are given in table 1. In reference 27, combustor rig data from a wide variety of sources were used to develop a multiple regression formula having the same general form as the above relation, and the corresponding exponents are also given in table 1. If quantities in the regression relation are expressed in SI Units instead of British Engineering Units, the factor a_1 should be 0.047.

From comparison of the exponents in table 1, it is clear that except for the exponents of chamber pressure and mean velocity, there is excellent agreement between theory and experiment. In reference 27 the higher experimentally determined exponents on these two quantities were attributed to jet noise present in the experiments. A close relative of this combustion noise theory is the basis for a successful engine core noise prediction method described in the next section.

Table 2 gives the regression exponents on combustor mass flow rate, inlet temperature, and temperature rise for the radiated power results obtained by several experimentalists (refs. 12, 28, 29, and 30). The tendency seems to be for theory to favor an exponent of about 2 for the mass flow (or velocity), while experiment seems to favor an exponent closer to 3. It could very well be, as suggested in reference 27, that experimental results tend to be contaminated by jet noise, which is known to vary as velocity to the eighth power. It is interesting to note that the original form of the theory of reference 27 also predicts a dependence on combustor temperature rise with an exponent of about 2, although in the final form this dependence is suppressed, evidently by lumping it with the inlet temperature. The dependence of combustion noise on temperature or temperature rise remains unclear.

Table 1. Comparison of Theory With Regression Analysis Results

[From ref. 27]

	a_1	a_2	a_3	a_4	a_5	a_6	a_7	a_8
Experiment	0.91	1.9	3.4	−2.5	1.3	−0.78	1.0	1.0
Theory	*	1	2	−2 to −3	2	0 to −1	1	1

*No theoretical value.

Table 2. Regression Exponents Obtained by Several Investigators

[From ref. 12]

		Exponents on—		
Reference	Burner type	Air mass flow	Temperature rise	Inlet temperature
Shivashankara and Crouch (ref. 12)	Can	3.4	2.4	0.8
Kazin and Emmerling (ref. 28)	Annular	3.0	2.0	
Ho and Tedrick, modified (ref. 29)		1.0	2.0	
Strahle and Shivashankara (ref. 30)	Can	2.3 to 2.7	0 to 1.5	

Available Prediction Methods

Most available gas turbine engine combustion and core noise prediction methods have been derived from engine data, although some have been influenced by near-scale combustion rig data and by the theoretical developments described above. In general, the accuracy of a noise prediction method decreases as it becomes more universal: while it is relatively easy to develop an accurate noise prediction method for parametric variations within a given engine design, it is significantly more difficult to develop a method of comparable accuracy which is valid for a range of engine designs. Even so, universal prediction methods exist which provide 3- to 5-dB accuracy in overall sound power level while requiring knowledge of remarkably few design and operating parameters.

According to the core noise prediction method used at General Electric Co. (ref. 31), the overall sound power level (OAPWL) is given by

$$\mathrm{OAPWL} = 10 \log \left(\frac{\dot{m} c_o^2}{P_{\mathrm{ref}}} \right) + 10 \log \left[\left(\frac{T_{\mathrm{out}} - T_{\mathrm{in}}}{T_{\mathrm{in}}} \right)^2 \left(\frac{p_{t_{\mathrm{in}}}}{p_o} \right)^2 \left(\frac{\Delta T_{\mathrm{des}}}{T_o} \right)^{-4} \right] - 60.5\,\mathrm{dB} \tag{3}$$

where $\dot{m}$ is the combustor mass flow rate, T_{in} is the combustor inlet temperature, T_{out} is the combustor outlet temperature, ΔT_{des} is the design point temperature drop across the turbine, $p_{t_{\mathrm{in}}}$ is the combustor inlet total pressure, and P_{ref} is the reference power, 10^{-12} W. The subscript o refers to standard sea level conditions. Core noise data for a range of turbojet, turboshaft, and turbofan engines are compared with this function in figure 9, and the corresponding "universal" spectrum shape and directivity pattern are shown in figures 10 and 11, respectively.

Although all the engines represented in figures 9 to 11 have "traditional" combustion systems (can-type or annular), the prediction method has also been used to correlate combustion noise from an engine with a radially staged annular combustion system. In this case, however, the method had to be modified to account for the differences between the traditional and new combustion system designs. Specifically, during operation of the pilot stage only, an effective pilot stage exit temperature was used rather than the mixed-mean combustor exit temperature.

Because the General Electric prediction method is based entirely on engine data, it necessarily contains several empirical constants. Its chief advantage is that it involves relatively few parameters while achieving remarkable universality. A theoretically based prediction method has been developed at Pratt and Whitney which relies on only two empirical constants, one associated with the source activity and the other associated with transmission loss through the turbine. Because it involves more parameters than the General Electric relation, including several operating and geometrical variables, it is potentially even more universally applicable. The fact that it successfully predicts both the peak frequency and the overall sound power level for a wide range of engine designs and operating conditions tends to verify the theoretical notions upon which it is based, thereby further enhancing its value.

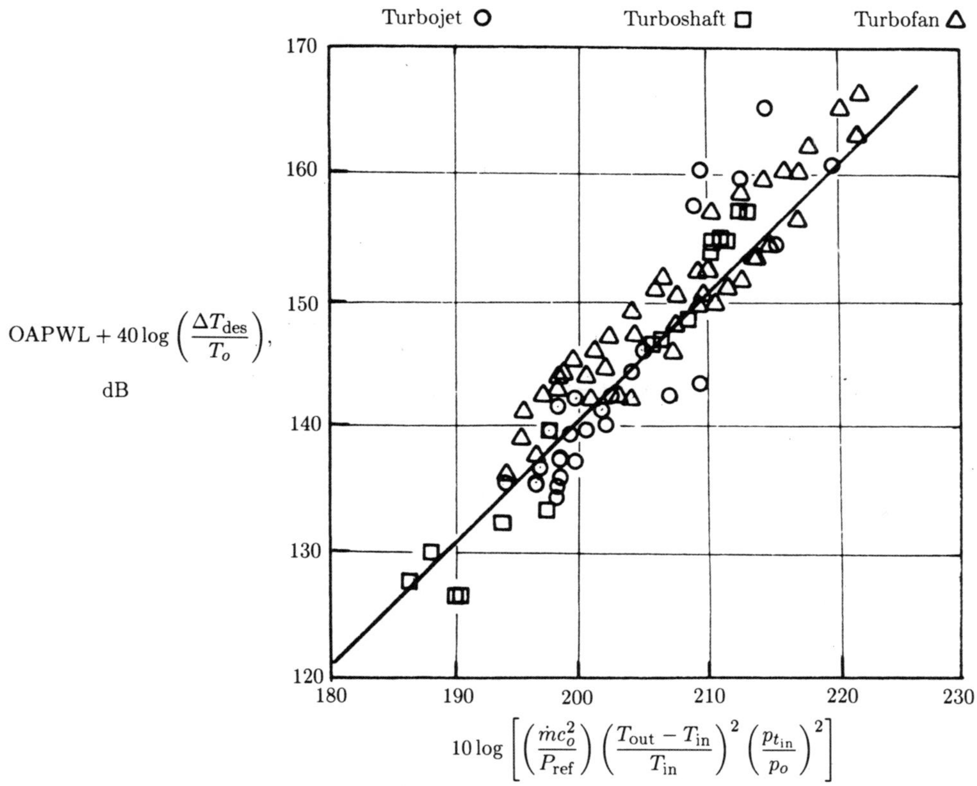

Figure 9. Comparison of the core noise overall sound power level correlation represented by equation (3) with experimental data. (From ref. 31. Copyright AIAA. Reprinted with permission.)

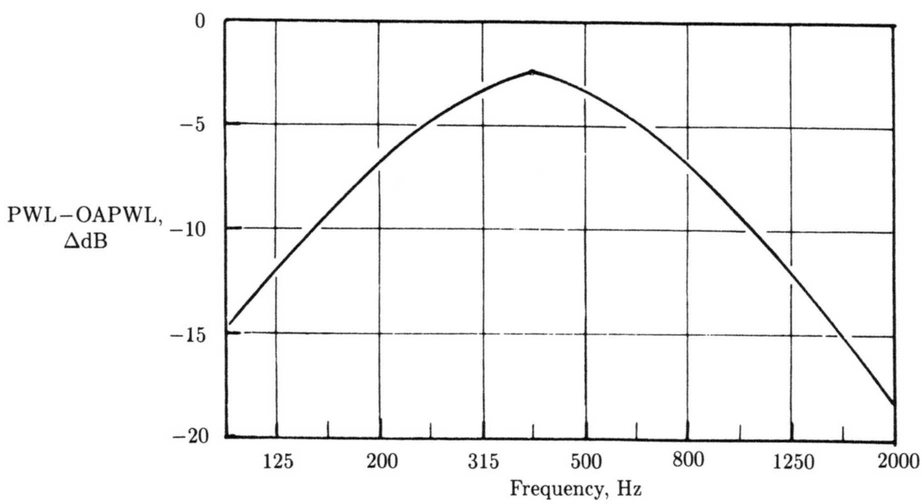

Figure 10. Universal gas turbine combustion noise spectral shape. (From ref. 31. Copyright AIAA. Reprinted with permission.)

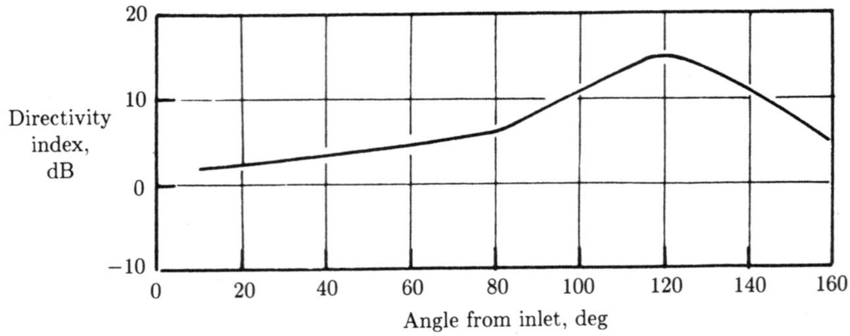

Figure 11. Universal gas turbine combustion noise directivity pattern. (From ref. 31. Copyright AIAA. Reprinted with permission.)

The Pratt and Whitney relation (ref. 32) is

$$\mathrm{OAPWL} = 10 \log \left[\frac{1}{N_f} A^2 p_{t_{\mathrm{in}}}^2 \left(\frac{\dot{m}\sqrt{T_{t_{\mathrm{in}}}}}{p_{t_{\mathrm{in}}} A} \right)^4 \left(1 + \frac{H_f F_{\mathrm{st}}}{c_p T_{t_{\mathrm{in}}}} \right)^2 F^2 \right] + K_3 - \mathrm{TL} \qquad (\text{ref. } 10^{-12}\ \mathrm{W}) \tag{4}$$

where the transmission loss is

$$\mathrm{TL} = 10 \log \left[\frac{(1+\varsigma)^2}{4\varsigma(L/\pi D)} \right]$$

where N_f is the number of fuel nozzles, A is the combustor cross-sectional area, $p_{t_{\mathrm{in}}}$ is the total pressure at the combustor inlet, $\dot{m}$ is the combustor air mass flow rate, $T_{t_{\mathrm{in}}}$ is the total temperature at the combustor inlet, H_f is the heating value of the fuel, F_{st} is the stoichiometric fuel-air ratio, c_p is the specific heat at constant pressure in the flame zone, F is the combustor fuel-air ratio, ς is the ratio of characteristic impedances across the turbine, L is the circumferential extent of a correlated source, and D is the outer diameter of the turbine at its interface with the combustor. All these quantities are in British Engineering Units. The theoretical development leading to this result draws heavily on Strahle's theory described earlier (refs. 25 and 26).

In all cases the ratio $L/\pi D$ has been found to be 0.20 and the constant K_3, which in theory should be a function of fuel type and combustor wall and exit impedances, has been found to be 132. Therefore, the only two quantities whose dependence on design and operating conditions is not directly determined from theory seem to be constant within the broad family of engines used to establish the correlation.

Measured overall sound power levels for a range of engines and operating conditions are plotted against the prediction relation in figure 12. The measured values are scattered about the prediction with a standard deviation of 1.7 dB.

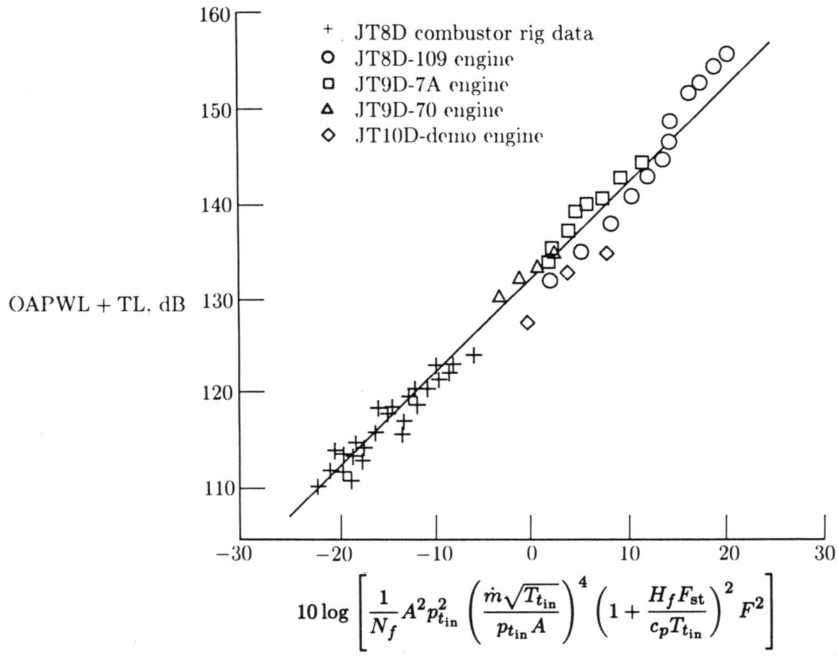

Figure 12. Comparison of core noise overall sound power level correlation represented by equation (4) with experimental data. (From ref. 32. Copyright AIAA. Reprinted with permission.)

The Pratt and Whitney method predicts the peak frequency to be

$$f_c = K_f \frac{RH_f}{c_p}\left(\frac{\dot{m}_f}{p_{t_{in}}}\right)_{ref}\frac{1}{A\ell} \tag{5}$$

where R is the gas constant for air, $\dot{m}_f$ is the fuel mass flow rate, ℓ is the combustor length, and K_f is an empirical constant. The ratio $\left(\dot{m}_f/p_{t_{in}}\right)_{ref}$ is evaluated at the design point of the burner corresponding to near takeoff conditions.

Figure 13 shows the measured peak combustion noise frequency plotted against the prediction relation with K_f set equal to unity. The data fall along one of two lines depending on the type of combustor: can or annular. However, if K_f is chosen to be 8 for can-type combustors and 3 for annular combustors, all the data fall along the same line.

Diagnostic Techniques

Combustion and core noise studies require measurement of the total sound power radiated from a gas turbine engine, as well as its spectral distribution and directivity. The total radiated sound power must also be separated into its various components according to source (i.e., combustion noise, jet noise, etc.). Further, the source activity must be isolated and characterized. This requires measurement of the

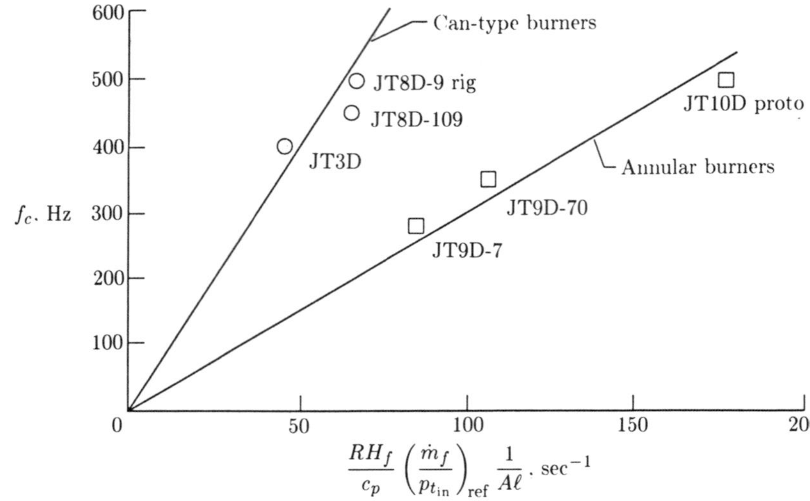

Figure 13. Comparison of core noise peak frequency correlation represented by equation (5) with experimental data. (From ref. 32. Copyright AIAA. Reprinted with permission.)

dynamic pressure in the engine. (It is common practice in the context of gas turbine combustion noise studies to refer to the unsteady component of pressure in an engine as the "dynamic" pressure. Because this usage is widespread and because there is little chance for confusion of the unsteady component of pressure with the true dynamic head, this term is used in this chapter.) Finally, various engine operating characteristics must be measured so that they can be correlated with combustion and core noise. Because these latter measurements are routine in engine performance studies and thus are not unique to combustion noise studies, they are not discussed here. The material in this section is divided into three parts: measurement techniques, data interpretation, and example applications.

Measurement Techniques

Standard microphones are used to measure the sound power in the far field. The combustion rig or engine should exhaust into an anechoic chamber for these measurements. If this is not possible, the rig or engine should be mounted in a very large room, or even outside, to minimize the effects of reflections. Then the microphones are usually mounted flush with the hard floor or ground and the measurements are then halved to account for reflections. If these precautions are not taken, reflections must be treated analytically, a tedious procedure that inevitably compromises the confidence of the results obtained.

The dynamic pressure in a source region is made up of an acoustic component and a nonacoustic component. The acoustic component is governed by a wave equation and thus propagates at the local sound speed, while the nonacoustic component is a local pressure disturbance that does not propagate. In order to identify source activity that leads to propagating sound, the dynamic pressure within the engine must be measured and separated into its acoustic and nonacoustic components.

Measurement of the dynamic pressure within the combustor of a gas turbine engine requires a probe which can provide reliable data at high temperatures. Its calibration must be thermally stable, or at least must be easily correctable for temperature variations, and it must have a small temperature sensitivity. Two general approaches have been used to meet these requirements. In the first, a specially designed high-temperature transducer is introduced directly into the combustor, while in the second, a transducer intended for use at room temperature is somehow thermally isolated from the hot combustion gases.

Transducers capable of stable and reliable operation at typical gas turbine combustion temperatures usually must be custom-built. As a consequence, they are prohibitively expensive and often difficult to use. The chief problem is that the mechanical and electrical properties of most common transducer materials change rapidly with temperature as combustion temperatures are approached. Special materials must be used just to ensure the mechanical and electrical survivability of the transducer. The transducer must be calibrated over the anticipated temperature range, and its temperature must be monitored during use so that the calibration can be subsequently applied. The extremely limited availability of high-temperature dynamic pressure calibration sources added to the high initial cost and the necessity for post-processing of data make the routine use of this type of transducer unattractive.

The most widely used alternative to the dynamic pressure measurement techniques described above is an acoustic waveguide to transmit the dynamic pressure signal to an externally mounted "room temperature" transducer. Such a system is shown in figure 14. It consists of a pressure-transmitting tube, or probe, with a pressure transducer mounted in a side branch. The probe is as short as possible while still providing adequate thermal isolation of the transducer from the hot products of combustion. The probe is continued beyond the transducer by a long coiled tube

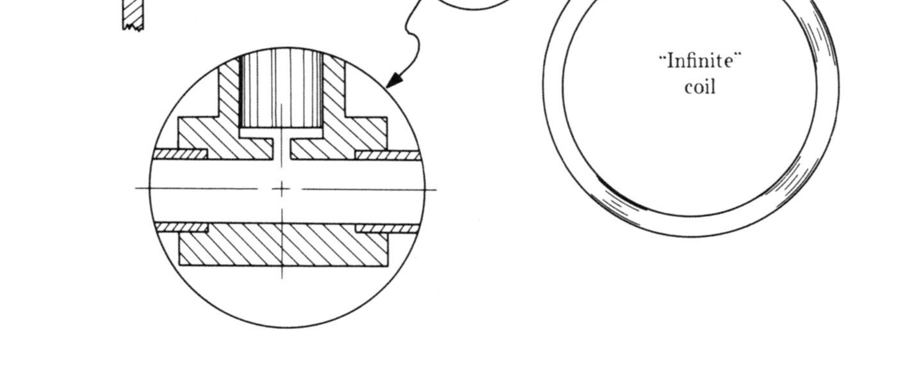

Figure 14. Waveguide pressure probe for use in hot gas path.

of the same inside diameter as the probe. The purpose of this coil is to prevent the signal at the transducer from being distorted by reflections which would occur if the tube was terminated at or near the transducer. By the time that pressure waves entering the coil reach the sealed end, are reflected, and return to the transducer, they have been attenuated to the point that their contribution to the signal measured by the transducer is negligible. The end of the "infinite" coil is either sealed or connected to a source of high-pressure inert gas such as nitrogen. The inert gas helps cool the transducer while keeping the transmitting tube clear of contaminants such as soot and unburned liquid fuel.

An example of the distortion which can occur if the coil is too short is illustrated in figure 15. The two pressure autospectra shown were obtained simultaneously in a can-type combustor using probes like the one in figure 14. The probes were identical except for coil length. The series of peaks (at 11.25-Hz intervals) clearly visible in the spectrum corresponding to the 50-ft (15-m) long coil are due to half-wave resonance of the system, probe plus coil. In contrast, the half-wave resonances (at 3-Hz intervals) are barely discernible in the spectrum corresponding to the 150-ft (46-m) long coil.

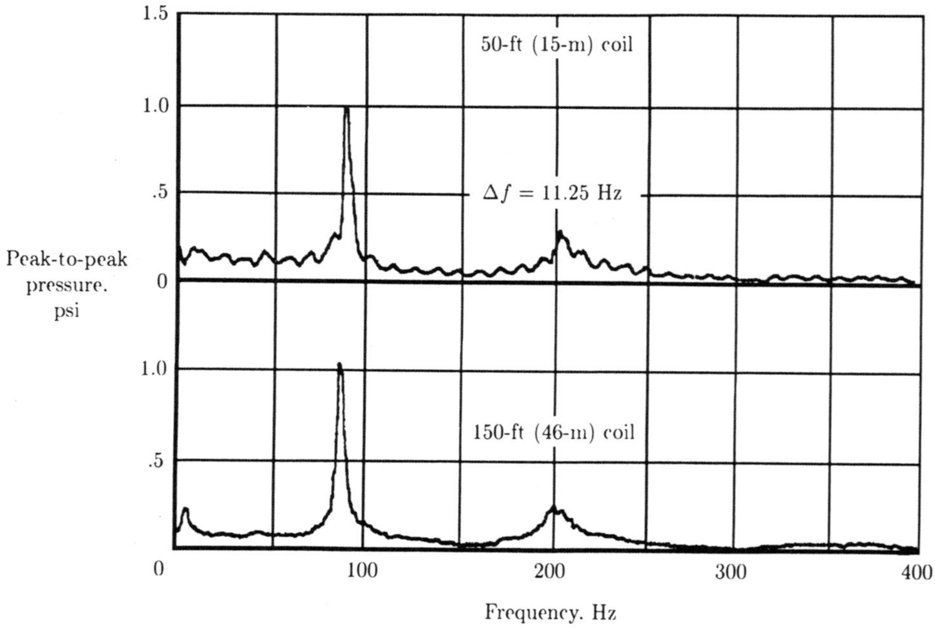

Figure 15. Comparison of pressure autospectra obtained simultaneously using two waveguide probes with different "infinite" coil lengths. 1 psi = 6.89 kPa.

A second type of distortion occurs in probe systems as a result of the distance from the open end of the probe to the transducer. Figure 16 shows the magnitude and phase of the pressure transfer function for a typical probe system. The undulation evident in the magnitude and visible in the phase is due to the alternating alignment of pressure nodes and antinodes with the transducer as the frequency is increased.

At certain frequencies, there is a pressure node at the transducer and a pressure antinode at the open end of the probe. When this occurs, the pressure at the open end has a diminished influence at the transducer, resulting in a reduction in the magnitude of the transfer function. This produces a measurement error that depends on frequency and the length of the probe. However, the form of the error is completely predictable from elementary duct acoustic theory, and its magnitude is usually negligible in typical gas turbine combustion noise applications. For example, in the case illustrated in figure 16, which is representative of typical applications, the maximum error due to this phenomenon is about 2 dB.

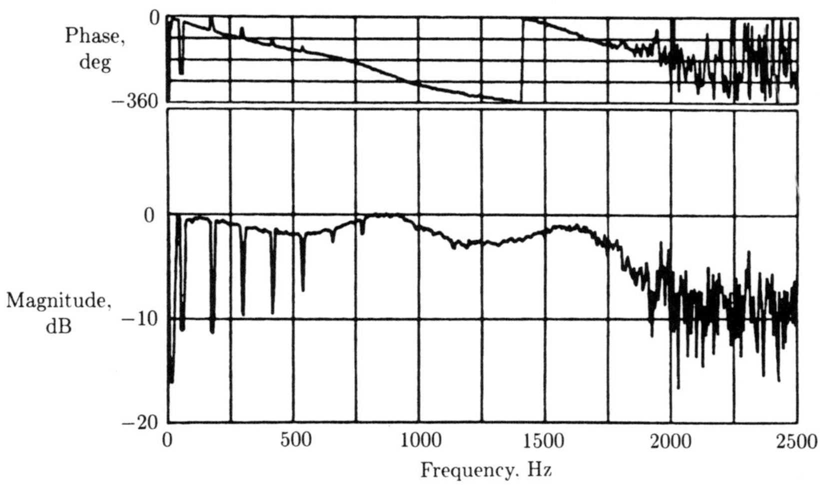

Figure 16. Magnitude and phase of transfer function of a typical 8-in. (20-cm) waveguide probe with 60-ft (18-m) coil.

A third and fourth type of measurement error associated with this type of probe system are also apparent in figure 16. The most obvious of these is the phase error due to the finite time required for a pressure wave to travel from the tip of the probe to the transducer. Though large, this error is completely correctable if the gas temperature distribution in the probe is known or can be reasonably well estimated. The estimate can be fairly rough since the sound speed in the probe varies as the square root of absolute temperature. The final type of error is that due to attenuation of the wave as it travels from the probe tip to the transducer. This error can be made negligible by using a sufficiently large probe inside diameter (1/4 in. (0.6 cm) is typical) and a sufficiently short probe (8 to 18 in. (20 to 40 cm) is typical). Note that this error increases with frequency, as the number of wavelengths traveled in the tube increases. In the typical case illustrated in figure 16, the attenuation error at 1000 Hz, which is near the upper limit of combustion noise, is on the order of 1 dB.

In summary, all the errors inherent in the probe-type dynamic pressure measurement system are either negligible or correctable. Such systems have the advantage that they can be calibrated at room temperature and then used at combustion temperatures. Also, the transducers themselves are relatively inexpensive and easy to use.

Two-wire thermocouple probes, of the type developed in reference 33, have been used to measure the dynamic component of temperature in the hot gas path of gas turbines and combustion rigs. Such measurements are required, for example, for the study of entropy noise. A typical two-wire probe is shown in figure 17 (ref. 34). It consists of two small thermocouple beads made of different size wires so that their time constants are different. The junctions are sufficiently close together to ensure that they are exposed to the same fluctuating temperature field. This permits the time constants of the two thermocouples to be determined experimentally by exploiting the fact that any difference between their response to a fluctuating temperature must be due to the difference in time constants. Once the time constants are known, the frequency-domain signal from either of the thermocouples can be corrected using the relation,

$$T_m = \frac{T_g}{1 + i\omega\tau} \tag{6}$$

where T_m is the measured temperature at frequency ω, T_g is the actual gas temperature at that frequency, τ is the time constant, and i is the imaginary operator.

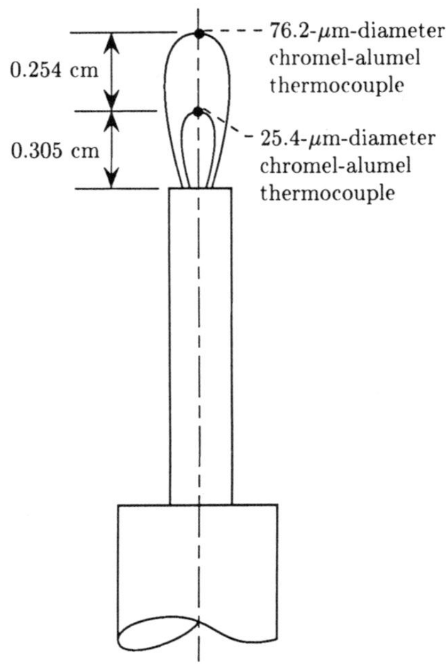

Figure 17. Two-wire thermocouple probe. (From ref. 34.)

Data Interpretation

The usual starting point in combustion noise data interpretation is to convert the time-domain pressure signal into a frequency-domain signal. This is most often done using a fast Fourier transform (FFT) analyzer, which performs a quasi-real-time Fourier analysis of the time-domain signal. Two assumptions are made in the analysis. The first is that the time-domain signal is stationary (i.e., that its power spectrum is independent of when the signal is sampled) and periodic with a period equal to the sampling interval. The second assumption requires that the sampling interval be at least as long as the period of the lowest frequency component expected to be present in the signal. Most FFT analyzers incorporate antialias filtering to reject frequency components which are undersampled at the specified sampling rate.

The principal result of FFT analysis of a pressure signal is the pressure autospectrum, which shows the distribution of pressure with frequency. Two examples of pressure autospectra are given in figure 15. However, most FFT analyzers also include the possibility for extensive statistical intercomparison of two or more signals. Of the statistical functions typically available on modern FFT analyzers, the most useful for combustion noise studies are the cross-correlation, the cross-spectrum, the transfer function, and the coherence.

The cross-correlation gives a direct measure of the time delay between incidences of an event common to two signals. It is therefore useful for measuring the time required for a pressure signal to propagate from one transducer to another. A typical application would be to determine whether a disturbance is acoustic or nonacoustic. If the signal travels between the two transducers at the mean sound speed, it is acoustic, but if it travels at the mean convective velocity, it is nonacoustic.

The cross-spectrum between two signals is the Fourier transform of their cross-correlation. As such, it contains the same information as the cross-correlation but in a more convenient form. The cross-spectrum has a real part and an imaginary part or, alternatively, it can be expressed as a magnitude and a phase. The magnitude at a given frequency represents the degree to which two signals have common harmonic content at that frequency, and the phase is the true phase angle between the two signals at that frequency. The cross-spectrum is very useful for studying wave propagation and for identifying the presence of standing waves.

The utility of the cross-correlation is greatly enhanced when it is used with the coherence. The coherence is another measure of the commonality of two signals and is usually expressed as a number between zero and unity. The higher the coherence at a given frequency, the greater the probability either that one signal is causing the other or that the two signals are caused by the same agent. At any frequency for which the coherence is low, either the signals are relatively independent of each other, or the signal-to-noise ratio on one or both channels is low. A third possibility leading to low coherence is a nonlinear relation between the two signals. In combustion noise work, a coherence of about 0.1 may be considered high, depending on the length of the data record available.

In a typical application, one channel is a dynamic pressure transducer in the combustor and the other a microphone in the near or far field of the exhaust. Then if, for example, at a given frequency the coherence between the two signals is high while the cross-spectral density is low and the phase angle is consistent with acoustical propagation, it can be deduced that an acoustic component present in the combustor does not propagate well into the far field. The reasonable explanation would be that the transfer function of the turbine is low at that frequency. This can be checked directly and independently using the transfer function option usually available on an FFT analyzer. The transfer function is the complex ratio of the Fourier components at each frequency. The interested reader is referred to any of a number of excellent books which treat FFT analysis in detail, for example, reference 35.

Example Applications

It has already been mentioned that two-signal correlations were used in reference 18 to separate direct and indirect combustion noise. In an important application of waveguide-type dynamic pressure probes and FFT-based statistical treatment of the data (ref. 36), Karchmer obtained cross-spectra between pairs of probes displaced an angle θ around the circumference of a full-scale annular combustor. The data, of which figure 18 is typical, were used to verify that the combustion noise source region is a homogeneous collection of random, uncorrelated monopole sources. The reconstructed characteristics are least-square fits to the data of an acoustic wave model based on these assumptions. Their excellent agreement with the measured characteristics is then taken as direct verification of the model. The model, once verified and properly scaled by least-squares fitting to the cross-spectral density magnitude and phase, can then be used to predict the individual independent modes which together make up the pressure autospectrum in the combustor. Figure 19 shows the measured pressure autospectrum and the contributions of the individual modes, as predicted from the model. The important result here, other

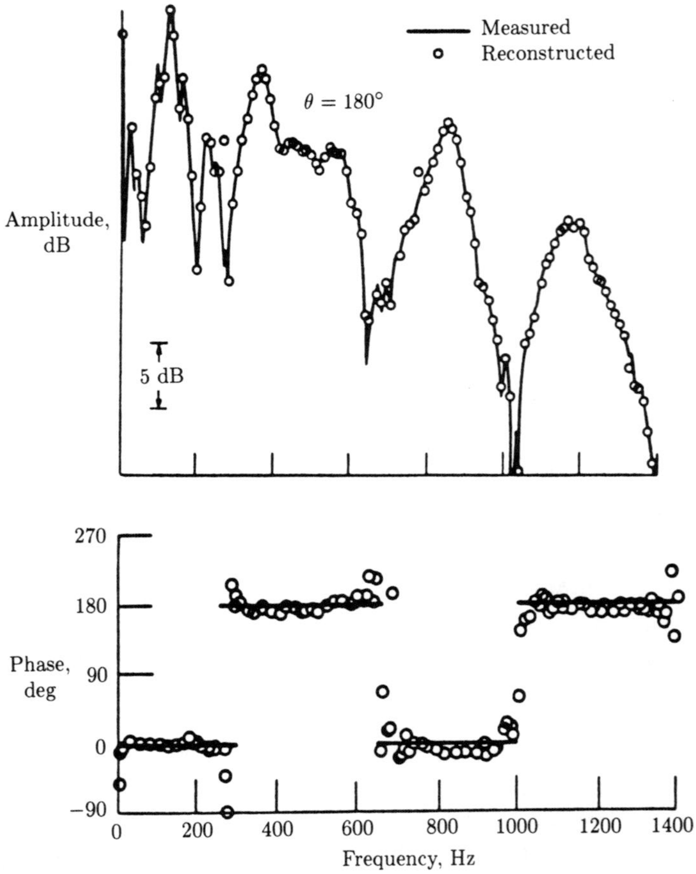

Figure 18. Measured and reconstructed cross-spectra between two waveguide probes in an annular gas turbine combustor. (From ref. 36.)

than confirmation of the combustion noise source model, is that individual features of the combustion noise spectrum can be uniquely related to a single specific mode.

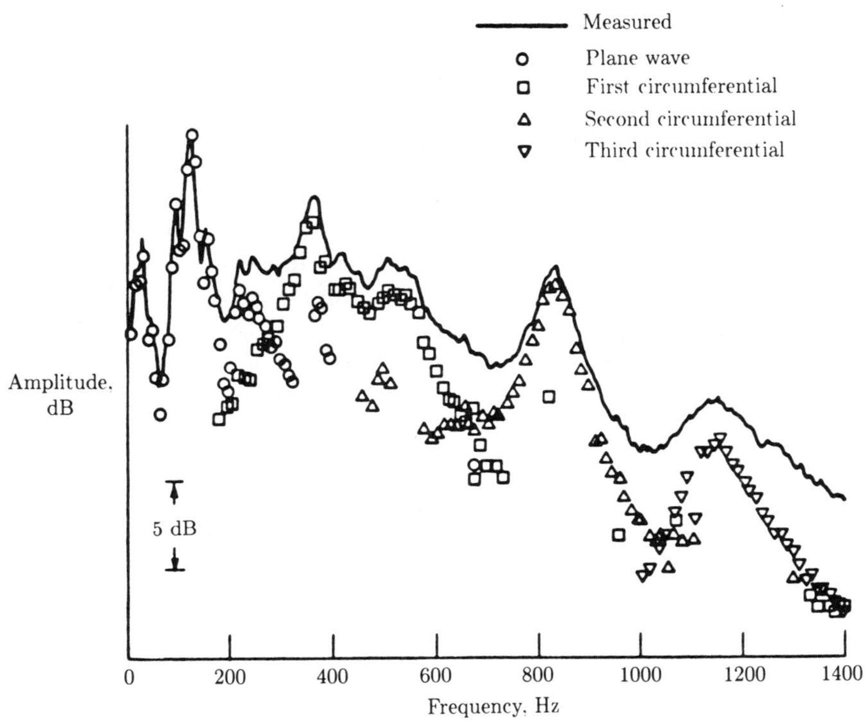

Figure 19. Measured pressure autospectrum and predicted contributions of individual modes in an annular gas turbine combustor. (From ref. 36.)

In another example application, Krejsa (ref. 5) has developed a three-signal coherence technique for separating core noise from the other gas turbine noise sources at a particular far-field location. Three simultaneous measurements are made, two in the engine core and one at the far-field location of interest. The two probes in the engine core are separated sufficiently in the streamwise direction to ensure that any local pseudo-noise present is uncorrelated between the two probes. (Pseudo-noise results when a local component of unsteady flow, usually associated with turbulence, stagnates on the active element of a pressure transducer. In this case the resulting "pressure" fluctuation would not even occur if the pressure transducer was not present.) The autospectrum of the core noise at the field point is then computed as

$$|{}_3p_{F_c}(\omega)|^2 = \frac{|G_{p_{\mathrm{exit}}p_F}(\omega)||G_{p_c p_F}(\omega)|}{|G_{p_c p_{\mathrm{exit}}}(\omega)|} \tag{7}$$

where $|G_{p_{\mathrm{exit}}p_F}(\omega)|$ is the magnitude of the cross-spectral density between the pressure fluctuations at the core exit and at the field point, $|G_{p_c p_F}(\omega)|$ is the magnitude of the cross-spectral density between the pressure fluctuations in the combustor and at the field point, and $|G_{p_c p_{\mathrm{exit}}}(\omega)|$ is the cross-spectral density between the pressure fluctuations in the combustor and at the core exit.

The results obtained using the technique may be compromised if an acoustic mode present in the combustor does not propagate to the far field. This would be the case, for example, if a higher order mode was present in an annular combustor but cut off in the turbine and tail pipe. The technique assumes that this does not occur, or at least that it occurs only to a negligible degree. The results actually obtained tend to justify this assumption, in spite of the evidence already cited (see fig. 6) that some high order modes present in the combustor do not reach the far field. Evidently the error associated with this effect in practice is small.

An application of the three-signal coherence technique is illustrated in figure 20. Shown is the variation with engine speed of the total overall sound pressure level measured in the far field, 120° from the engine inlet, and the component of this total due to core noise alone, as determined using the three-signal coherence technique. Also shown is the predicted overall sound pressure due to jet noise (ref. 37). The "excess noise" at low engine speeds, which earlier investigators had always attributed to combustion noise, is clearly shown to be a mixture of core noise and fan noise in this case. More importantly, the contribution of core noise to the overall sound pressure level at engine speeds where it is effectively masked by jet noise is clearly recovered here.

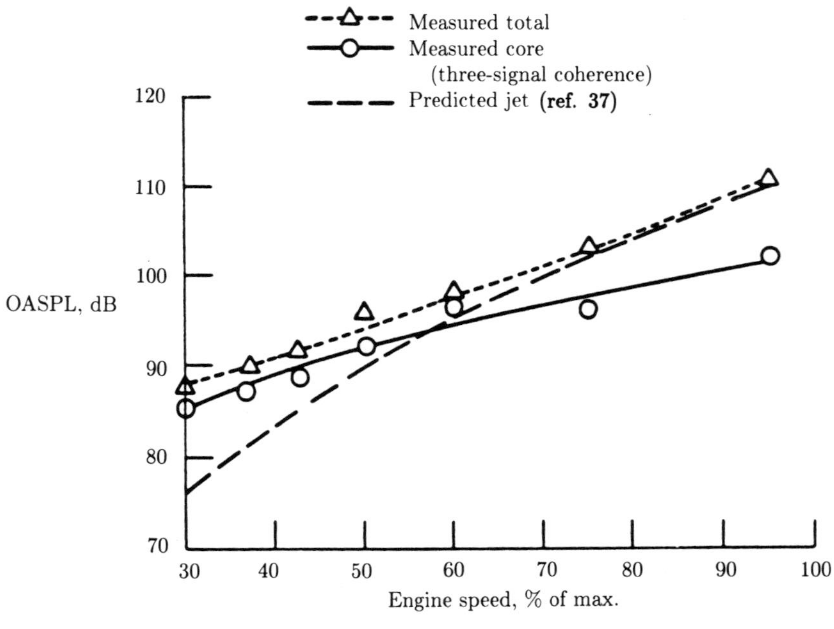

Figure 20. Engine core noise as a function of engine speed measured directly using the three-signal coherence technique. (From ref. 5.)

An important joint application of the waveguide pressure probe and the two-wire thermocouple probe is described in reference 34. A pressure probe and a temperature probe were located near each other in each of two planes of a can-type gas turbine combustor. The upstream measurement plane was in the combustion zone and the downstream plane was at the combustor exit. The respective time delays between the pressure signals and the temperature signals at the upstream and downstream stations were obtained by cross-correlating pressure with pressure and temperature with temperature. This resulted in two different downstream propagation velocities, a sonic velocity associated with the pressure signal and a lower velocity associated with the bulk downstream convection of temperature nonhomogeneities. Next, by cross-correlating the temperature and pressure signals at the downstream station it was determined that the temperature signal was coherent with the pressure signal, but with a phase lag consistent with the difference in the two velocities found previously. Figure 21 shows a typical cross-spectrum between pressure and temperature at the combustor exit.

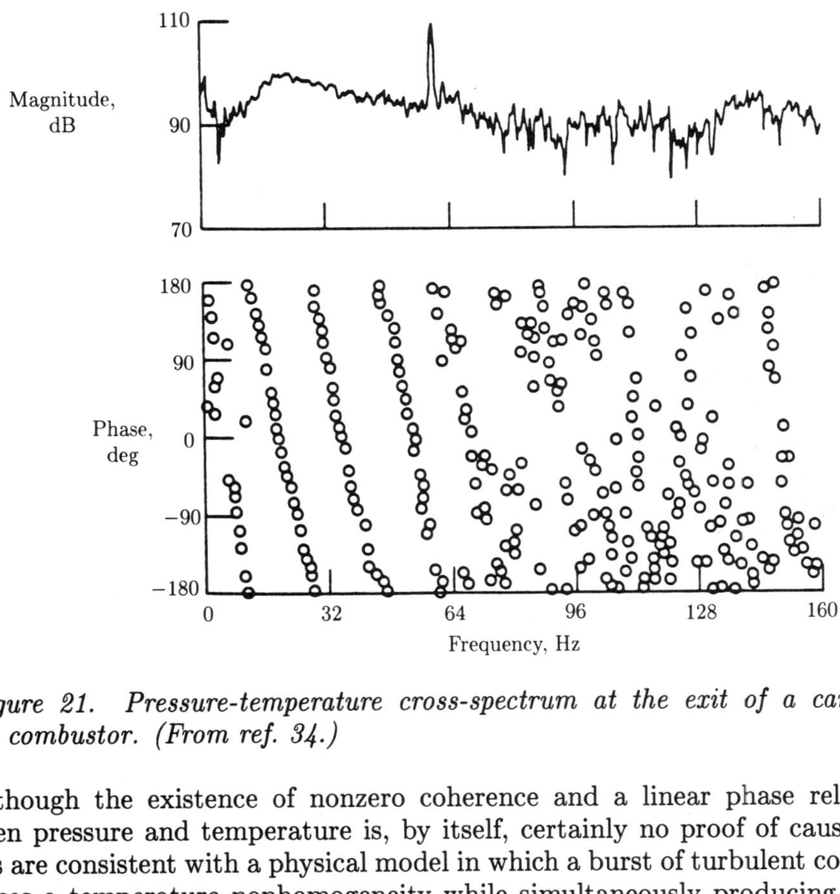

Figure 21. Pressure-temperature cross-spectrum at the exit of a can-type combustor. (From ref. 34.)

Although the existence of nonzero coherence and a linear phase relationship between pressure and temperature is, by itself, certainly no proof of causality, the results are consistent with a physical model in which a burst of turbulent combustion produces a temperature nonhomogeneity while simultaneously producing noise, as discussed previously in this chapter. The noise then propagates downstream at the speed of sound and the turbulent eddy convects downstream at a lesser velocity but with sufficient identity remaining when it reaches the downstream location that it still at least partially correlates, with the appropriate time delay, with the pressure signal measured there. This experiment is significant because it supports the relation, assumed in most contemporary theories, between combustion noise and the creation of temperature nonhomogeneities by turbulent combustion.

Reciprocating-Engine Combustion Noise

Combustion noise in reciprocating engines is effectively masked by propeller noise. There are two reasons why this is true, the first being that exhaust mufflers are quite effective, and the second being that propellers are very noisy. Also, both sources increase in strength at about the same rate with engine speed.

Figure 22 shows the sound pressure autospectra measured at three different locations on the fuselage of a reciprocating-engine, propeller-powered airplane (ref. 38). Spectrum (a) was measured in front of the propeller plane, spectrum (b) in the propeller plane, and spectrum (c) opposite the exhaust. Both propeller and exhaust harmonics are clearly visible in all three spectra, with the propeller harmonics being dominant in spectra (a) and (b) and the exhaust harmonics in spectrum (c). The frequencies at which both types of harmonics occur are easily predictable, because both are simply related to engine speed and to either the number of cylinders, or the number of propeller blades. The point of figure 22 is that, except in the vicinity of the exhaust itself, propeller noise dominates combustion noise.

The combustion noise source mechanism in a reciprocating engine is distinctly different from that in a gas turbine engine. In fact, almost no direct combustion noise radiates from a reciprocating engine. During the actual combustion process, the pressure in the cylinder increases at constant volume. Although there are deflagration waves which propagate through the mixture as it burns, the mass and stiffness of the piston and cylinder walls confine the associated acoustic energy to the combustion volume, where it is eventually absorbed. The power stroke begins only after combustion is essentially complete, and the sound pressure associated with the piston being rapidly displaced is absorbed in the crankcase, which is a massive, stiff-walled, sealed enclosure. This absorption is aided by the fact that while some pistons are moving into the crankcase, others are moving out.

The noise associated with the combustion activity is indirect, occurring when the products of combustion are forced from each cylinder during the exhaust stroke. The exhaust manifold consists of an elongated chamber which communicates with the individual cylinders through exhaust valve ports in its walls. These valve ports then act as monopole sources powered by the periodic bursts of hot exhaust products from the cylinders. The number and distribution of these compact sources depend on the number of cylinders and their physical arrangement. The cylinders, exhaust manifold, muffler, and exhaust pipe together make up an acoustic circuit whose response to this source activity depends on the acoustic behavior of the individual components and the manner in which they are interconnected.

Combustion noise reduction in reciprocating engines then becomes a matter of muffler design. While a detailed treatment of muffler design is beyond the scope of this chapter, a few of the basic principles are worth mentioning. The fundamental idea is to trap the dominant acoustic waves in a chamber where they can then be dissipated by the resistive component of impedance. The resistance generally takes the form of an array of small holes in a plate positioned in the silencer at a location or locations where the normal component of the acoustic particle velocity is high. The wave trap is usually an elongated chamber whose dimensions and end impedances create a resonant volume at the fundamental source frequency and its harmonics.

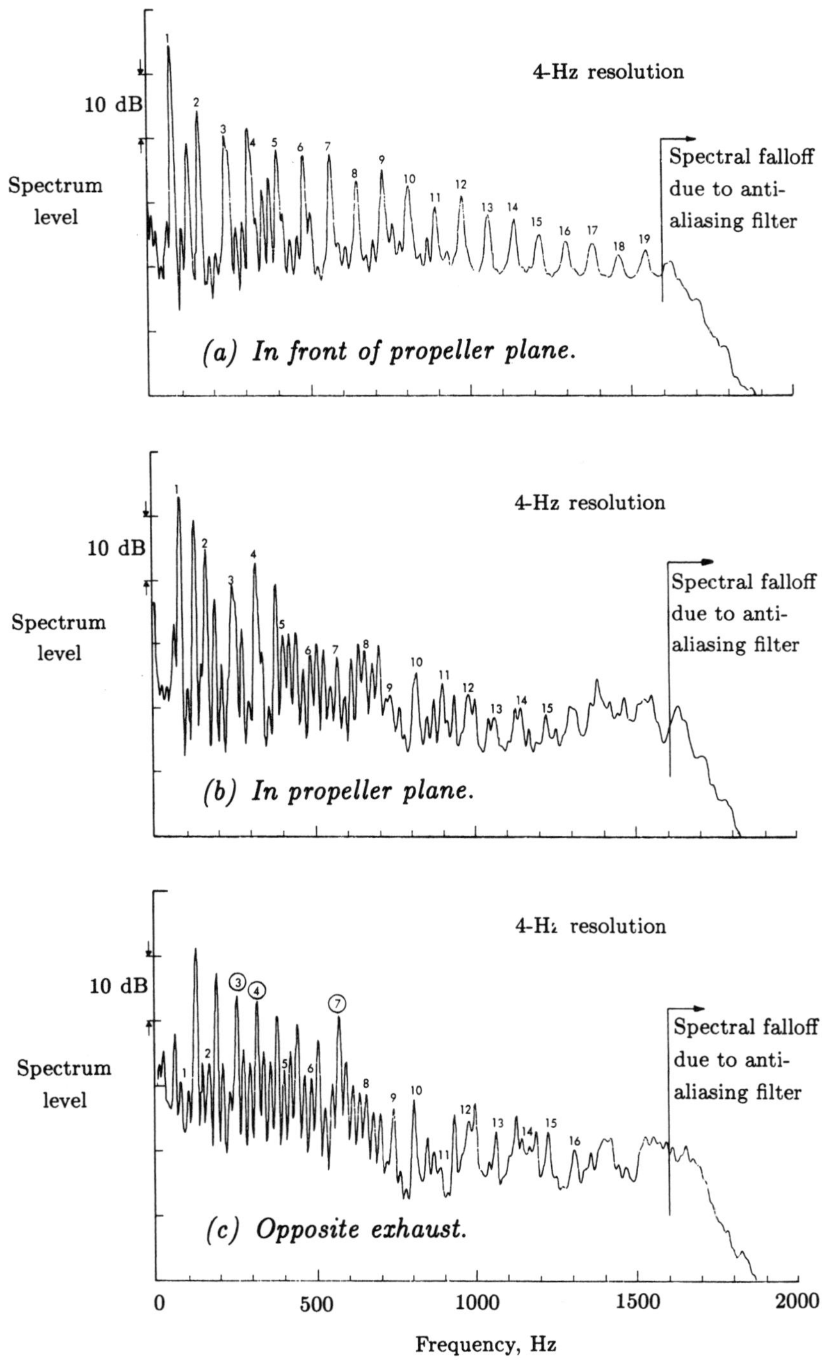

Figure 22. Sound pressure spectra showing propeller and exhaust noise on the fuselage of a reciprocating-engine powered airplane. Circled numbers indicate data heavily contaminated by exhaust tones. (From ref. 38.)

The chamber may be "folded" to pack the correct acoustic length into the available physical space. The walls are generally multilayered, with the inner layer or layers perforated and the outer layer solid.

References

1. Cohen, Henry; Rogers, Gordon Frederick Crichton; and Saravanamuttoo, H. I. H.: *Gas Turbine Theory,* Second ed. (Second Impression With Corrections). Longman (London), 1974.
2. Strahle, W. C.: Combustion Noise. *Prog. Energy & Combust. Sci.,* vol. 4, no. 3, 1978, pp. 157–176.
3. Cumpsty, Nicholas A.: Engine Noise. *Aerothermodynamics of Aircraft Engine Components,* Gordon C. Oates, ed., American Inst. of Aeronautics & Astronautics, Inc., c.1985, pp. 475–548.
4. Reshotko, Meyer; and Karchmer, Allen: *Core Noise Measurements From a Small, General Aviation Turbofan Engine.* NASA TM-81610, 1980.
5. Krejsa, Eugene A.: *New Technique for the Direct Measurement of Core Noise From Aircraft Engines.* NASA TM-82634, 1981.
6. Strahle, Warren C.: A Review of Combustion Generated Noise. AIAA Paper No. 73-1023, Oct. 1973.
7. Vermeulen, P. J.; and Odgers, J.: *Acoustic Control of the Exit Plane Thermodynamic State of a Combustor.* 79-GT-180, American Soc. of Mechanical Engineers, Mar. 1979.
8. Riley, John; Goldschmidt, V. W.; Leonard, R. G.; and Baade, P. K.: Noise Generation of Gas Flames Due to Feedback Excited Oscillations. Purdue Univ. and Carrier Corp. paper presented at the Third Conference on Natural Gas Research and Technology (Dallas, Texas), Mar. 1974.
9. Valk, M.: Acoustic Power Measurements of Oscillating Flames. *Combust. & Flame,* vol. 41, no. 3, June 1981, pp. 251–260.
10. Rayleigh, (Lord): *The Theory of Sound,* First American ed., Volumes I and II. Dover Publ., 1945. (Original edition published in 1896.)
11. Baade, Peter K.: Design Criteria and Models for Preventing Combustion Oscillations. Paper presented at the ASHRAE Symposium on Combustion-Driven Oscillations (Atlanta, Georgia), Jan.–Feb. 1978.
12. Shivashankara, Belur N.; and Crouch, Robert W.: Noise Characteristics of a Can-Type Combustor. *J. Aircr.,* vol. 14, no. 8, Aug. 1977, pp. 751–756.
13. Karchmer, A. M.; Reshotko, M.; and Montegani, F. J.: Measurement of Far Field Combustion Noise From a Turbofan Engine Using Coherence Functions. AIAA Paper 77-1277, Oct. 1977.
14. Mahan, J. R.: *A Critical Review of Noise Production Models for Turbulent, Gas-Fueled Burners.* NASA CR-3803, 1984.
15. Mahan, J. Robert: A Turbulent Combustion Noise Model. *High Technology for Noise Control—NOISE-CON 87 Proceedings,* Jiri Tichy and Sabih Hayek, eds., Noise Control Foundation, c.1987, pp. 201–206.
16. Candel, Sebastien M.: *Analytical Studies of Some Acoustic Problems of Jet Engines.* DOT-TST-76-104, U.S. Dep. of Commerce, May 1976. (Available from NTIS as PB 264 918.) (Originally published as Ph.D. Thesis, California Inst. of Technology, May 1972.)
17. Marble, F. E.: Response of a Nozzle to an Entropy Disturbance. *Unsteady Aerodynamics, Volume 2,* Univ. of Arizona, 1975, pp. 699–717.
18. Muthukrishnan, M.; Strahle, W. C.; and Neale, D. H.: Separation of Hydrodynamic, Entropy, and Combustion Noise in a Gas Turbine Combustor. *AIAA J.,* vol. 16, no. 4, Apr. 1978, pp. 320–327.
19. Pickett, G. F.: Core Engine Noise Due to Temperature Fluctuations Convecting Through Turbine Blade Rows. AIAA Paper 75-528, Mar. 1975.
20. Cumpsty, N. A.; and Marble, F. E.: Core Noise From Gas Turbine Exhausts. *J. Sound & Vib.,* vol. 54, no. 2, Sept. 22, 1977, pp. 297–309.
21. Lighthill, M. J.: On Sound Generated Aerodynamically. I. General Theory. *Proc. Royal Soc. London,* ser. A, vol. 211, no. 1107, Mar. 20, 1952, pp. 564–587.
22. Bragg, S. L.: Combustion Noise. *J. Inst. Fuel,* Jan. 1963, pp. 12–16.
23. Thomas, A.; and Williams, G. T.: Flame Noise: Sound Emission From Spark-Ignited Bubbles of Combustible Gas. *Proc. Royal Soc. London,* ser. A, vol. 294, no. 1439, Oct. 18, 1966, pp. 449–466.
24. Hurle, I. R.; Price, R. B.; Sugden, T. M.; and Thomas, A.: Sound Emission From Open Turbulent Premixed Flames. *Proc. Royal Soc. London,* ser. A, vol. 303, Mar. 19, 1968, pp. 409–427.
25. Strahle, Warren C.: On Combustion Generated Noise. *J. Fluid Mech.,* vol. 49, pt. 2, Sept. 29, 1971, pp. 399–414.

26. Strahle, W. C.: Some Results in Combustion Generated Noise. *J. Sound & Vib.*, vol. 23, no. 1, July 8, 1972, pp. 113–125.
27. Strahle, Warren C.; and Muthukrishnan, M.: Correlation of Combustor Rig Sound Power Data and Theoretical Basis of Results. *AIAA J.*, vol. 18, no. 3, Mar. 1980, pp. 269–274.
28. Kazin, S. B.; and Emmerling, J. J.: *Low Frequency Core Engine Noise.* 74-WA/Aero-2, American Soc. of Mechanical Engineers, Nov. 1974.
29. Ho, P. Y.; and Tedrick, R. N.: Combustion Noise Prediction Techniques for Small Gas Turbine Engines. *INTER-NOISE 72 Proceedings,* Malcolm J. Crocker, ed., Inst. of Noise Control Engineering, c.1972, pp. 507–512.
30. Strahle, W. C.; and Shivashankara, B. N.: *Combustion Generated Noise in Gas Turbine Combustors.* NASA CR-134843, 1974.
31. Ho, P. Y.; and Doyle, V. L.: Combustion Noise Prediction Update. AIAA Paper 79-0588, Mar. 1979.
32. Mathews, D. C.; and Rekos, N. F., Jr.: Prediction and Measurement of Direct Combustion Noise in Turbopropulsion Systems. *J. Aircr.*, vol. 14, no. 9, Sept. 1977, pp. 850–859.
33. Strahle, Warren C.; and Muthukrishnan, M.: Thermocouple Time Constant Measurement by Cross Power Spectra. *AIAA J.*, vol. 14, no. 11, Nov. 1976, pp. 1642–1643.
34. Miles, J. H.; Wasserbauer, C. A.; and Krejsa, E. A.: *Cross Spectra Between Temperature and Pressure in a Constant Area Duct Downstream of a Combustor.* NASA TM-83351, 1983.
35. Bendat, Julius S.; and Piersol, Allan G.: *Random Data: Analysis and Measurement Procedures.* Wiley-Interscience, c.1971.
36. Karchmer, Allen M.: *Acoustic Modal Analysis of a Full-Scale Annular Combustor.* NASA TM-83334, 1983.
37. Stone, James R.: *Interim Prediction Method for Jet Noise.* NASA TM X-71618, 1974.
38. Piersol, A. G.; Wilby, E. G.; and Wilby, J. F.: *Evaluation of Aero Commander Propeller Acoustic Data: Static Operations.* NASA CR-158919, 1978.

10 Sonic Boom

Lead authors

Domenic J. Maglieri
Eagle Engineering, Inc.
Hampton, Virginia

Kenneth J. Plotkin
Wyle Laboratories
Arlington, Virginia

Introduction

The introduction of the century-series fighter aircraft in the 1950's, with their ability to fly at supersonic speeds in level flight, brought into prominence the sonic boom phenomenon. This phenomenon, which is now well understood from a physical standpoint, was heretofore quite infrequent and was usually associated with aircraft which had to dive in order to attain slightly supersonic speeds. Concerted efforts in the 1960's, in conjunction with increased operations of high-performance military aircraft, the proposed (and later canceled) U.S. supersonic transport (SST), and the eventual entry of the British-French Concorde into commercial service, have provided significant insight into the generation, propagation, and prediction of sonic booms and their effects on people, animals, and structures (refs. 1 to 7). Even so, sonic booms continue to be a community acceptance problem for aircraft operations at supersonic speeds. In fact, commercial supersonic flight over land in the United States is prohibited (ref. 8). The Concorde confines its supersonic operations to overwater routes only.

Sonic boom studies continue to play a role in the formation of environmental impact statements regarding the establishment of military operational training areas and the Space Shuttle program. Recent research in long-range hypersonic vehicles, such as the "Orient Express," recognizes that the sonic boom will loom large as a serious threat to complete success.

This chapter is intended to provide a status of the knowledge of sonic booms, with emphasis on their generation, propagation, and prediction. For completeness, however, material relating to the potential for sonic boom alleviation and the response to sonic booms is also included. The material is presented in the following five sections: *Nature of Sonic Booms, Review and Status of Theory, Measurements and Predictions, Sonic Boom Minimization, and Responses to Sonic Booms.*

Nature of Sonic Booms

This section begins with a description of the shock flow fields surrounding bodies moving at supersonic speeds and the manner in which sonic booms are observed. A description of the sonic boom carpets, both primary and secondary, is given for a typical aircraft operation. The role of the atmosphere in establishing and influencing the primary and secondary booms is discussed.

Shock Flow Field

Any body which moves through the air at speeds exceeding the local speed of sound has associated with it a system of shock waves, as shown in figure 1. A simple body of revolution (i.e., a projectile) generally has two waves, one attached to the front called the bow wave and the other emanating from the rear called the tail wave (fig. 1(a)). More complicated configurations, such as the small aircraft model in figure 1(b), produce whole systems of shock waves. At very large distances from the body, the wave system tends to distort and steepen, ultimately coalescing into a bow and a tail wave as in the case of the simple projectile.

Figure 1(c) shows a schematic diagram representing the far-field wave patterns typical of the projectile and wind-tunnel data. At the bow wave a compression occurs in which the local pressure p rises to a value Δp above atmospheric pressure. Then a slow expansion occurs until some value below atmospheric pressure is reached, after which there is a sudden recompression at the tail wave. Generally, the bow and tail shocks are of similar strengths and the pressure decreases linearly between the two. This nominal sonic boom signature is called an N-wave. It moves with the aircraft and is associated with continuous supersonic flight, not just with "breaking the sound barrier." One speaks of a sonic boom "carpet," whose width depends on flight and atmospheric conditions, swept out under the full length of a supersonic flight. Receivers within the carpet detect the sonic boom—that is, the N-wave—once as the aircraft passes.

If these waves were sweeping by an observer on the ground, the ear's aural response would be as shown schematically in the sketch at the bottom of figure 1(c). Since the ear detects changes in pressure only above a certain frequency, it would respond to the steep part of the wave and not to the portion which is changing slowly. If the time interval Δt between those two rapid compressions is small, as for a bullet, the ear would not be able to distinguish between them and they would seem as one explosive sound. If the time interval is on the order of 0.10 sec or greater, as is the case for an aircraft at high altitude, the ear would probably detect two booms.

Some of the characteristics of the pressure signatures within the flow field surrounding the XB-70 aircraft are shown in figure 2. These in-flight measurements (ref. 9) were obtained by probing the flow field above and below the XB-70 with an instrumented aircraft. The XB-70 was flying at $M = 1.5$ at 37 000 ft above ground level, and in-flight surveys were made at 2000 ft above and at 2000 and 5000 ft below the aircraft. Also shown is the corresponding signature measured at ground level.

The measured signatures are shaded to highlight the individual pressure peaks. These pressure peaks are associated with details of the aircraft geometry (wings, inlets, canopy, empennage, and so on). It is shown that more complex signatures are measured close to the aircraft and that the individual shock waves from the aircraft tend to coalesce as distance from the aircraft increases, although in this case an ideal N-wave has not yet evolved. It is also shown that the shock wave signature above the aircraft differs markedly (in shape and amplitude) from that below the aircraft at a comparable distance. This signature difference results from the difference in the detailed geometry of the aircraft and the manner in which the volume and lift components interact.

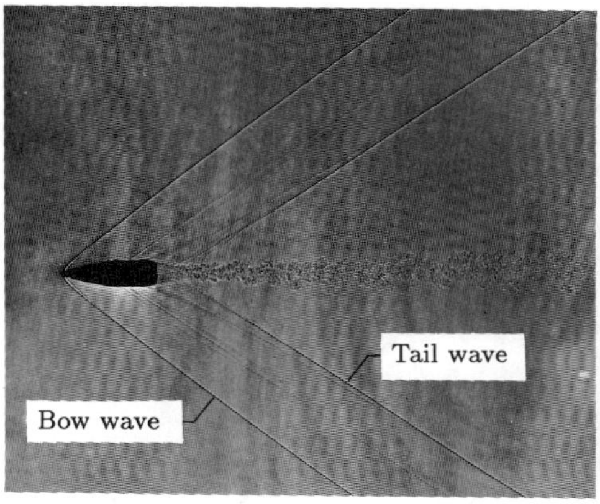

(a) Flow field for projectile.

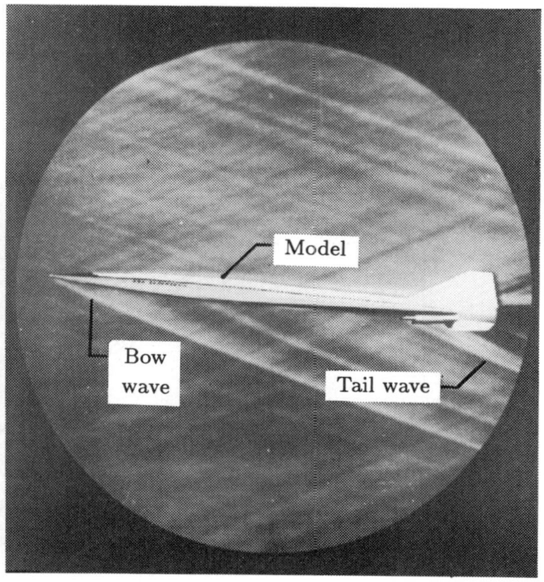

(b) Flow field for aircraft.

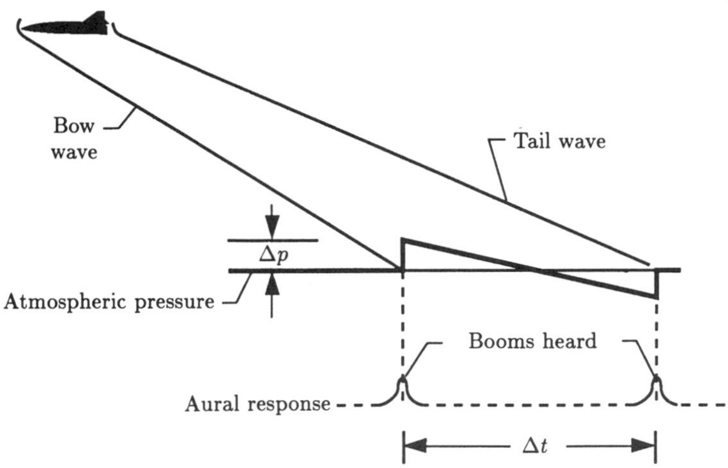

(c) Far-field wave patterns.

Figure 1. Shock flow fields.

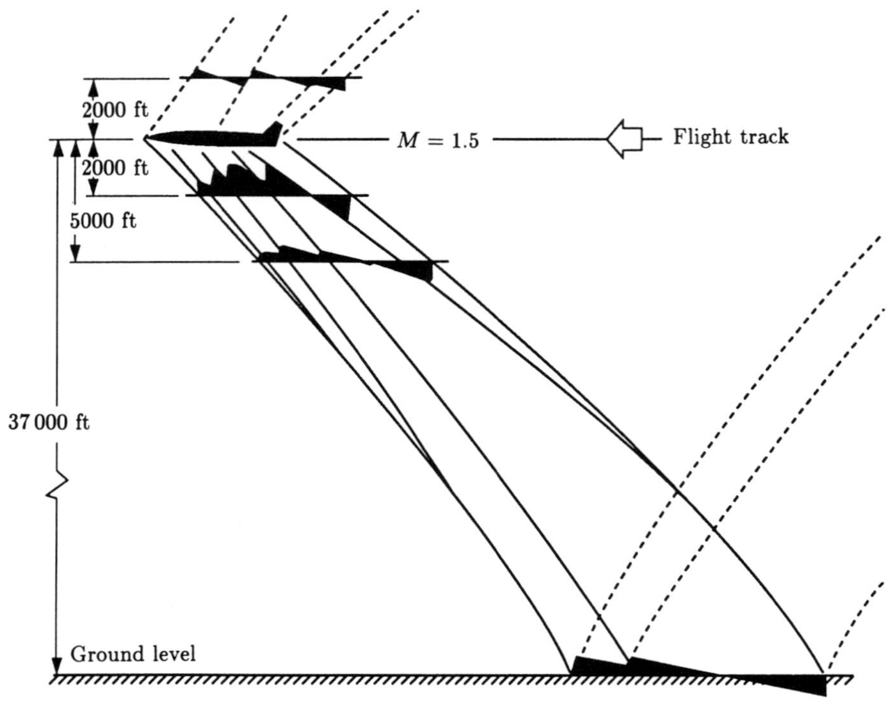

Figure 2. Measured signatures above and below XB-70 aircraft.

Description of Sonic Boom Carpets

Figure 3 shows schematically the nature of the sonic boom carpets for a flight such as that of the Concorde, during which the aircraft flies a large portion of the distance supersonically and without maneuvers. Two ground exposure patterns in which booms are observed are shown. The primary boom carpet contains the normally observed sonic boom overpressures and results from wave propagation through only that part of the atmosphere below the aircraft. Secondary boom carpets may exist which involve the portion of the atmosphere above the aircraft as well as that below the aircraft. Between the primary and secondary carpets exists a region in which no sonic booms are observed. The secondary boom carpets are more remote from the ground track and the overpressure levels are much less intense than in the primary carpet.

The waveform characteristics of the boom signatures can vary widely at the different observation points, as indicated in figure 3. In the region of the primary boom carpet, on or near the ground track, N-wave signatures are typically observed. For typical high-altitude cruise conditions, these are usually of the order of 1 to 3 lb/ft^2 in intensity and from 0.10 to 0.30 sec in duration. At the fringes of the primary boom carpet, near the lateral cutoff, the signatures degenerate into weak sound waves and they lose their N-wave characteristics. In the region of the secondary boom carpet, the disturbances tend to be very weak in intensity (of the order of 0.02 to 0.20 lb/ft^2) but persist over longer periods of time (refs. 10 to 15).

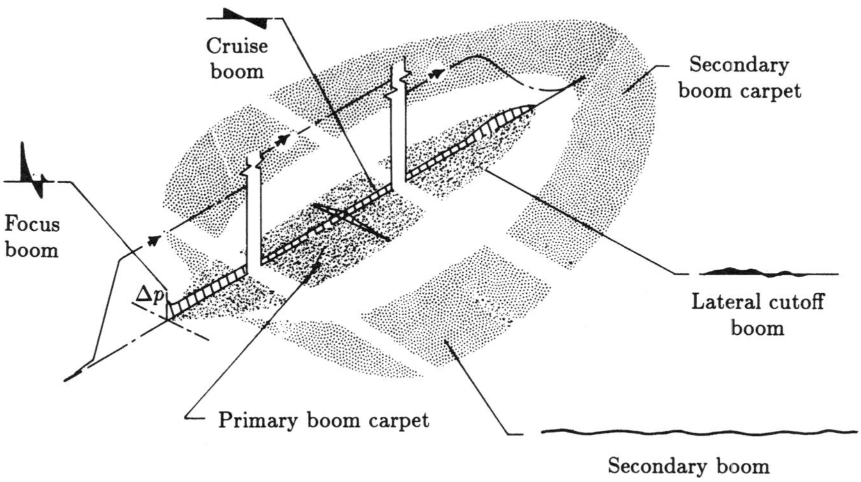

Figure 3. Nature and extent of sonic boom ground exposure carpets and waveforms associated with supersonic aircraft operations.

It should be noted that the higher overpressure N-wave sonic booms have caused community acceptance problems. On the other hand, the lateral cutoff booms and the secondary booms, which do not have an N-wave character and are lower in intensity, tend to be more of a curiosity and are not apt to be the source of serious community response problems. Near the lateral cutoff, primary booms usually resemble low rumbles or rolling thunder. Secondary booms, however, are generally not audible (0.1 to 1.0 Hz), but can cause building vibrations which are readily observed.

Another type of pressure signature, that of a focus boom (shown in the lower left of fig. 3), can be observed when any aircraft accelerates from subsonic to supersonic speeds. These “acceleration” focus booms are followed by regions on the ground in which multiple booms are observed. The focus booms enhance the booms generated in steady, level flight operations.

Sonic boom footprints from military operations, particularly air combat maneuvers, can be quite complex. They have the same essential features as shown in figure 3, but can have a very short cruise component (because of the brief nature of supersonic combat maneuvers) and can be distorted by turns.

Role of the Atmosphere

The manner in which the atmosphere above and below the aircraft is involved in developing the primary and secondary boom carpets is shown in more detail in the ray diagram of figure 4. On the right-hand side of figure 4 are examples of temperature and wind profiles for a normal atmosphere. Of note is that there is a portion of the higher atmosphere in which the temperature increases as altitude increases, and the associated wave propagation speed thus increases compared with that in the lower portions of the atmosphere. Similarly, the wind may participate in such a way as to further increase the wave propagation speed in certain directions.

On the left-hand side of figure 4 is a ray diagram for an aircraft at an altitude of 60 000 ft, traveling toward the viewer. The downward-propagating rays, shown by the solid lines, impact the ground to form the primary carpet, as indicated in

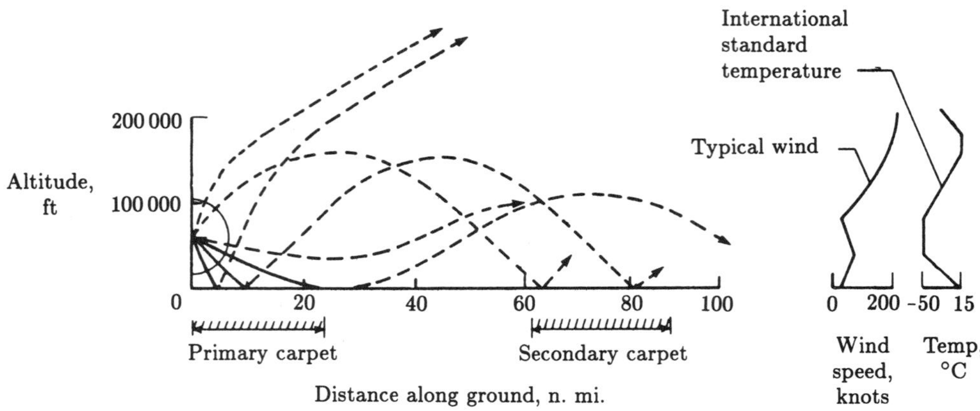

Figure 4. Propagation paths of sonic boom disturbances from an aircraft and associated ground exposure carpets.

the figure. At some point—about 28 n.mi. in the example shown—the rays refract away from the ground and thus define the lateral extent of the primary carpet. Also indicated is a secondary carpet, at about 65 to 85 n.mi. from the flight track, in which the dashed-line rays impact. These dashed-line rays arrive in two different ways: they either travel directly to the secondary carpet as a result of bending in the upper atmosphere, or they may first impinge in the primary carpet, reflect upward from the surface, and then bend downward after traveling through a portion of the upper atmosphere. The representation of the secondary carpet in this illustration is probably oversimplified, because there is reason to believe that it could consist of several well-defined impact areas (refs. 12 and 13). Variations in atmospheric wind and temperature profiles, however, could cause these impact areas to lose their identities. Some of the steep-angle rays above the aircraft may travel in such a way that they are dissipated without ever approaching the ground.

The atmosphere, particularly the first few thousand feet of the Earth's boundary layer, plays another very significant role relative to the sonic boom signature waveforms. Figure 5 presents examples of sonic boom waveforms that were measured in the primary carpet for three different types of aircraft. The tracings of measured waveforms for the F-104 aircraft are for a time duration of about 0.10 sec. The waveforms vary from the nominal N-wave shape previously described, varying from a sharply peaked to a gently rounded shape. Similar tracings are shown for the B-58 and XB-70 aircraft. The B-58 signatures are roughly 0.20 sec in duration and the XB-70 signatures are approximately 0.30 sec in duration. The main differences between waves for a given aircraft occur at the time of the rapid compressions. The largest overpressures are generally associated with the sharply peaked waves. Such differences in the sonic boom waveform result primarily from the turbulence and the thermal activities in the lower layer of the atmosphere (ref. 9).

Review and Status of Theory

In this section the theory is developed, beginning with the acoustic source and including atmospheric effects and nonlinear steepening. Sonic boom computations are sufficiently complex to necessitate computerization. A discussion of a number of

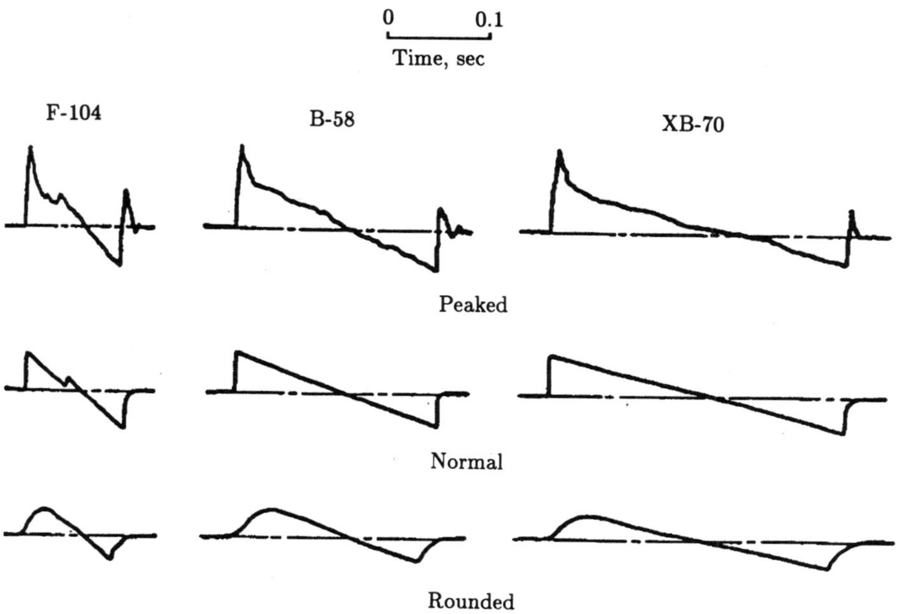

Figure 5. Variation of measured sonic boom waveforms at ground level for small, medium, and large aircraft in steady, level flight.

these programs is presented. Maneuvering flight and the potential for sonic boom focusing is addressed, along with the different types of focus conditions that may be encountered. Finally, a discussion of the applicability of the theory to hypersonic speeds is presented.

Sonic Boom Theory

A slender axisymmetric body in uniform supersonic flow, as shown in figure 1(a), generates a cylindrical acoustic wave field with overpressures $\Delta p = p - p_0$ given by

$$\Delta p(x - \beta r, r) = p_0 \frac{\gamma M^2 F(x - \beta r)}{(2\beta r)^{1/2}} \tag{1}$$

where

p	pressure
p_0	undisturbed ambient pressure
x	axial coordinate (body fixed)
r	radius
γ	ratio of specific heats
M	Mach number
β	Prandtl-Glauert factor, $\sqrt{M^2 - 1}$

and

$$F(x) = \frac{1}{2\pi} \int_0^x \frac{A''(\xi)}{(x - \xi)^{1/2}} \, d\xi \tag{2}$$

where A is the cross-sectional area of the body as measured by cutting planes aligned with the Mach angle and ξ is a dummy variable of integration. The quantity $F(x)$ was introduced by Whitham (ref. 16) and is generally referred to as the Whitham F-function. The F-function has an implicit dependence on Mach number because of A being defined on Mach tangent cutting planes, so that, in principle, Mach number dependence is not limited to the explicit factors in equation (2). In practice, the Mach number dependence of equation (2) is relatively weak, so an F-function computed at one Mach number can be considered to be "the" F-function over a reasonable range of conditions.

Equations (1) and (2) are derived from linearized supersonic flow and area-rule theory for axisymmetric bodies. They can be shown to be valid for nonaxisymmetric vehicles if the actual area $A(x)$ is replaced with an equivalent area which is a function of azimuthal angle about an axis through the body in the flight direction. The equivalent area consists of two components: the actual area as cut by a plane tangent to the Mach cone at the azimuth plus an effective area directly proportional to the axial distribution of lift in that direction. This formulation follows from the linearized supersonic flow and area-rule results of Hayes (refs. 17 and 18) and Lomax (ref. 19) and was applied by Walkden (ref. 20) to Whitham's basic sonic boom analysis (ref. 21, discussed below), leading to the analysis of sonic booms in terms of volume and lift components. When generalized to asymmetric bodies, the locally asymmetric version of equation (1) is valid at distances which are large compared with body dimensions (i.e., $r \gg x - \beta r$). A very good presentation of the equivalent area formulation is given in reference 22, which also contains a more detailed presentation of sonic boom theory than the current synopsis.

The complete role of the aircraft configuration in sonic boom generation is embodied in the F-function. Analysis of minimization concepts generally centers on calculating F-functions for various configurations. At hypersonic speeds, for which linearized flow theory is not accurate, the problem is that of obtaining the F-function by means other than equation (2); equation (1) is always valid beyond some radius r at which $\Delta p/p_0$ is sufficiently small. These two topics are discussed in detail subsequently. For now, it suffices to note that the aircraft source is defined by the F-function.

Pressure signatures at large distances do not retain the fixed shape of equations (1) and (2); explosions and supersonic artillery projectiles were long known to generate far-field shock wave signatures. Landau (ref. 23) showed that weak nonlinear effects (second order in overpressure) cause the far-field signature of a projectile to have a dual-shock N-wave shape. The mechanism is that air in the positive-pressure pulse has an elevated temperature and a forward velocity, so that local propagation speed is faster than ambient sound speed and the wave steepens, eventually forming a shock. Landau obtained the result that shock strength in the axisymmetric case follows an $r^{-3/4}$ law rather than the $r^{-1/2}$ law of equation (1). DuMond et al. (ref. 24) performed a series of measurements on small-caliber projectiles, clearly demonstrating the N-wave and the $r^{-3/4}$ law.

The theory supporting this mechanism was set forth in a consistent manner by Whitham (refs. 16 and 21) who showed that second-order nonlinear steepening could be viewed as a uniform first-order solution: the linear solution (eq. (1)) provides the correct amplitude to the first order, but the location ($x-\beta r$, representing propagation at the ambient sound speed) is correct only to the zeroth order. Correcting propagation speed to the first order (based on the linear solution) provides the required second-order solution.

In the form used by Whitham (ref. 21), the acoustic overpressure may be written as

$$\frac{p - p_0}{p_0} = \frac{F(\tau)}{\sqrt{S}} \tag{3}$$

where

τ	$t - (s/c_0)$
t	time
s	distance along a ray
S	ray-tube area
c_0	undisturbed ambient sound speed

where a ray-fixed coordinate system has been adopted. Figure 6 shows the relation between a wave-fixed viewpoint, as shown in figures 1 and 2, and a ray-fixed viewpoint, as sketched in figure 4. The wave front exists at a given time, whereas the rays represent the path that the boom will take after being generated at some time. The ray-tube area term $1/\sqrt{S}$ is a generalization of the cylindrical wave quantity $\gamma M^2/(2\beta r)^{1/2}$ in equation (1). For plane waves S is a constant, and for spherical waves it is proportional to r^2 or s^2. In a general nonuniform atmosphere, an acoustic impedance factor is present and S is the geometrical acoustic ray-tube area (to be discussed subsequently).

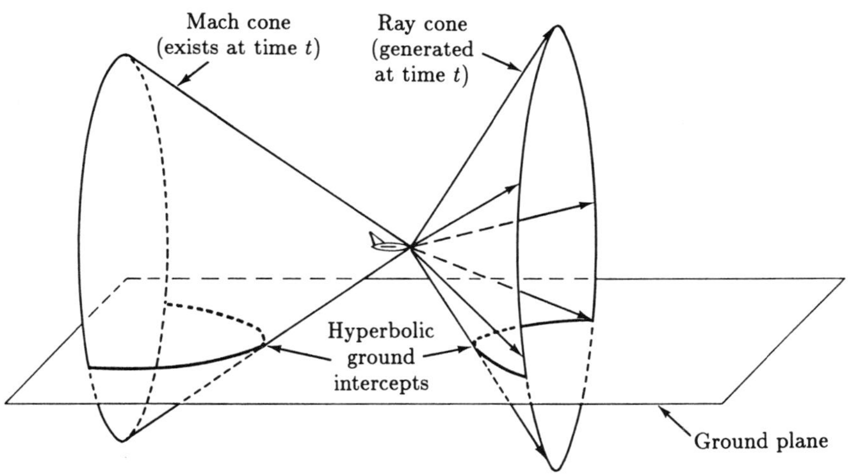

Figure 6. Mach and ray cones in supersonic flight.

Whitham's rule calls for replacing c_0 in τ by $c + u$, the perturbed sound speed c plus the velocity perturbation u. The normalized perturbations $(c - c_0)/c_0$ and u/c_0 are both proportional to $(p - p_0)/p_0$. For an isentropic acoustic wave, the propagation speed is

$$c + u = c_0 \left(1 + \frac{\gamma + 1}{2\gamma} \frac{p - p_0}{p_0}\right) \tag{4}$$

The parameter τ represents a point on the acoustic wave, and t is its arrival time at location s. The arrival time may be obtained by integrating the reciprocal of equation (4). To the first order in $\Delta p/p_0$, this arrival time is

$$t = \tau + \frac{s}{c_0} - \frac{\gamma + 1}{2\gamma c_0} F(\tau) \int_0^s \frac{ds}{\sqrt{S}} \tag{5}$$

Equation (5) has been written in terms of t, rather than τ, to present an explicit relationship.

The physical interpretation of equations (3) and (5) is illustrated in figure 7. A signature near the aircraft (fig. 7(a)) undergoes an amplitude change because of the ray-tube area factor (fig. 7(b)) and undergoes a steepening distortion (fig. 7(c)) as given by equation (5). One point in the signature is highlighted in the figure and traced through this process. Note that the advance of each signature point (the last term of eq. (5)) is proportional to its F-function value and a quadrature which is independent of F. The quadrature term is part of the ray geometry solution and, in various normalized forms, has been denoted as the age parameter (ref. 25) or the advance factor (ref. 26).

Parts of the aged signature constructed in figure 7 are triple valued. This is physically impossible. At some earlier time the aging process would have caused the signature slope to be vertical, at which point there would be a discontinuous pressure jump. Propagation of this jump must be handled as a shock wave rather than as an isentropic wave. Linearizing the Rankine-Hugoniot relations gives the following speed u_s for a weak shock of strength Δp:

$$u_s = c_0 \left(1 + \frac{\gamma + 1}{4\gamma} \frac{\Delta p}{p_0}\right) \tag{6}$$

This is slower than the isentropic wave speed behind it, so that the original signature is absorbed into the shock. The shock is sketched in figure 7(c). In general, the linearized shock speed is equal to the average of the isentropic wave speeds ahead of and behind the shock. This leads to the "area balancing" rule for fitting shocks: construct the steepened isentropic signature, then eliminate triple-valued areas by fitting shocks such that total area is conserved. In figure 7(c), the shaded areas ahead of and behind the shock are equal.

Figure 7(c) is similar to sketches by Landau (ref. 23) and Whitham (ref. 21) showing the evolution of an N-wave signature. Key quantities for an N-wave are the shock overpressure and the total duration. Concentrating on the forward, positive-overpressure portion of the N-wave, they matched equations (3), (5), and (6) to obtain a closed-form solution.

Whitham's final result for the far-field bow shock overpressure is

$$\Delta p_{\text{shock}} \approx \frac{p_0}{S} \left[2 \int_0^{\tau_0} F(\tau)\, d\tau\right]^{1/2} \left(\frac{\gamma + 1}{2\gamma c_0} \int_0^s \frac{ds}{\sqrt{S}}\right)^{1/2} \tag{7}$$

where τ_0 is the value of τ corresponding to the end of the positive phase of the F-function. For a uniform atmosphere, where $S \propto r \propto s$, equation (7) reduces to an $r^{-3/4}$ law. A similar result for the duration of an N-wave follows an $r^{1/4}$ law.

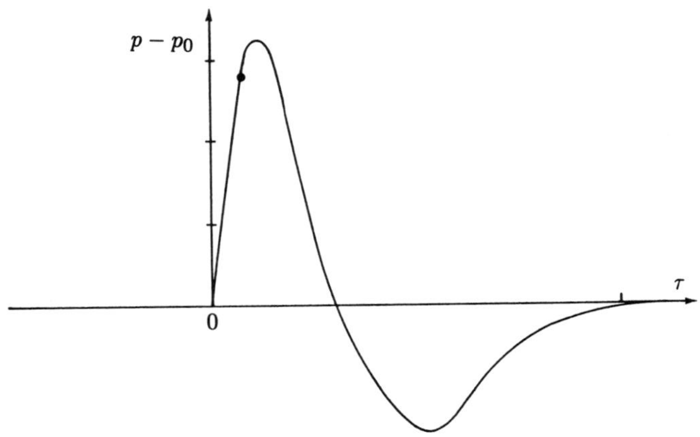

(a) Signature near aircraft (F-function shape).

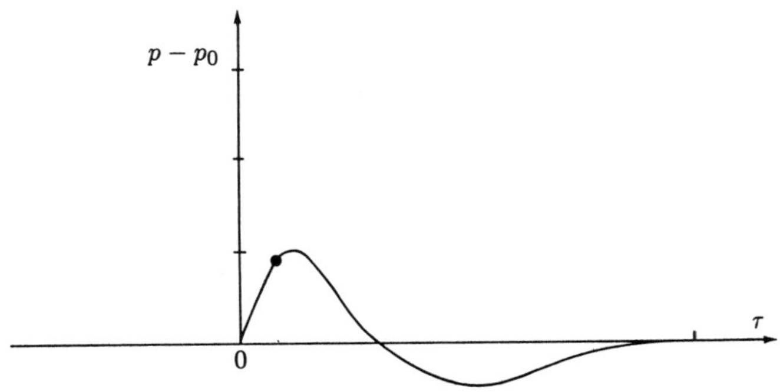

(b) Far-field acoustic amplitude change (F-function shape).

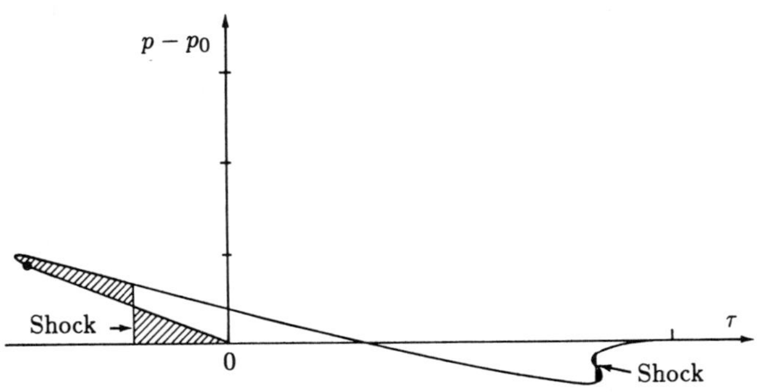

(c) Far-field steepened (aged) signature.

Figure 7. Evolution and steepening of sonic boom signature.

Equation (7) is very simple; it contains terms related to the ray-tube area dependence of the acoustic overpressure, with the aircraft geometry embodied in a simple integral of the F-function. This suggests that a far-field sonic boom is not particularly sensitive to fine details of the aircraft, and flight test results indeed show that N-waves for various conventionally shaped aircraft of similar size and weight are virtually the same. In this N-wave regime, the effect of size is that boom overpressures decrease as a function of the aircraft length to the one-fourth power. Lift-induced boom varies as the square root of aircraft weight, and the boom is relatively insensitive to Mach number.

Early calculations of sonic booms exploited this behavior. One expression used for volume-induced sonic boom was (ref. 27)

$$\Delta p = K_r K_s \sqrt{P_v P_g} (M^2 - 1)^{1/8} \frac{D}{l^{1/4}} \, r^{-3/4} \tag{8}$$

where

K_r	ground reflection coefficient (usually 2)
K_s	aircraft shape constant, typically 0.4 to 0.8
D	equivalent aircraft diameter
l	aircraft length
P_v, P_g	ambient pressure at the vehicle and the ground

Based on the Walkden theory (ref. 20), similar formulas were developed for lift-induced sonic boom.

The $\sqrt{P_v P_g}$ factor in equation (8) is a partial adjustment for the fact that the atmosphere is not uniform. A complete adjustment for the atmosphere utilizes the theory of geometrical acoustics. This theory accounts for curvature of shock waves and rays (as in figs. 2 and 4 and compared with the straight lines of fig. 6) and the variation in sound speed and air density. A full derivation of geometrical acoustics was presented by Blokhintzev (ref. 28). Two other noteworthy derivations are those in references 29 and 30. Geometrical acoustics applies for waves which are short compared with atmospheric gradients. Ray shapes depend on sound speed and wind gradients and are computed by methods directly analogous to those of geometrical optics. Figure 8 shows typical ray curvatures for a sonic boom under standard atmospheric conditions. Figure 8(a) shows rays under the flight track. At a given time there are rays directed at various azimuthal angles ϕ, as shown in figure 8(b). A ray-tube area, as sketched in figure 8(a), is computed to account for the effect of curvature on amplitude. The effect of the ray calculation and the variation in air density and sound speed is that the quantity S in the acoustic solution (eq. (3)) is replaced with a quantity B given by

$$B = \frac{\rho_v c_v}{\rho_0 c_0} \, S \tag{9}$$

where S is the ray-tube area, $\rho_0 c_0$ is the local acoustic impedance of air, and $\rho_v c_v$ is the impedance at the vehicle. The quantity B is slightly more complex than this if there are winds.

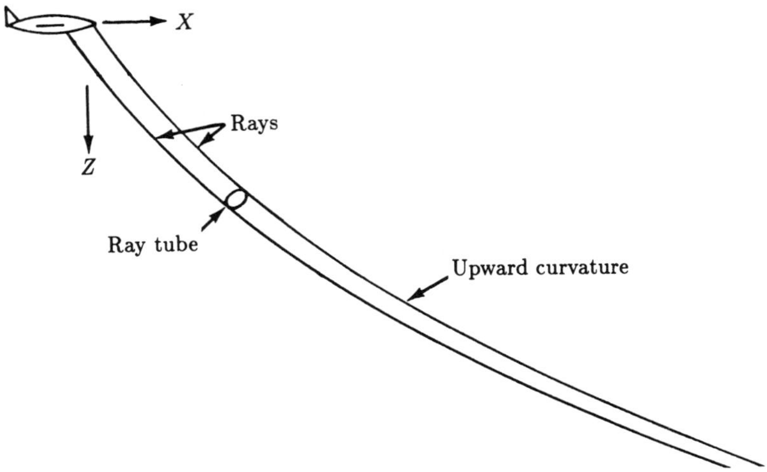

(a) Rays under flight track.

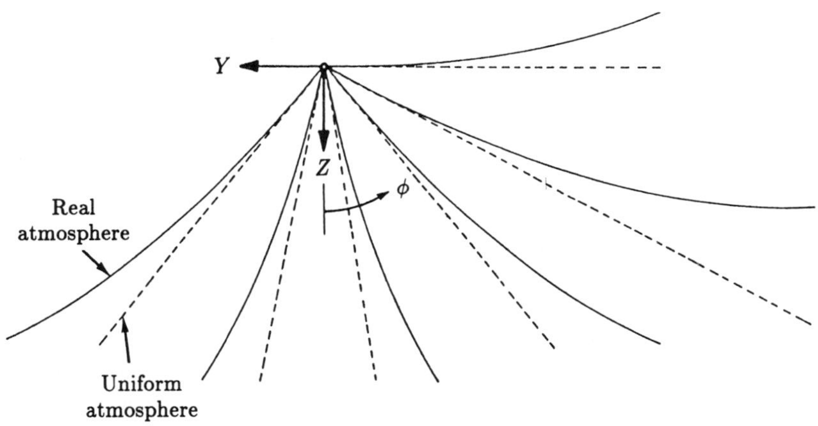

(b) Rays across flight track.

Figure 8. Curvature of sonic boom rays in atmosphere.

The ray calculation depends only on flight parameters and the atmosphere. Figure 9 shows a typical definition of four rays which outline a rectangular ray tube. Each ray lies on a ray cone. The effect of maneuvers is automatically included by use of the local flight velocity and the Mach angle at each time point. Once the ray calculation is completed, the rest of the boom calculation proceeds exactly as outlined earlier, except for the use of B instead of S throughout.

A final step in boom calculation is that, for a receiver on the ground, the perceived boom is enhanced by reflection from the ground. This reflection generally is a factor of 2. It can be less for soft ground, and it can be higher if there are multiple reflectors such as the corner between the ground and a wall.

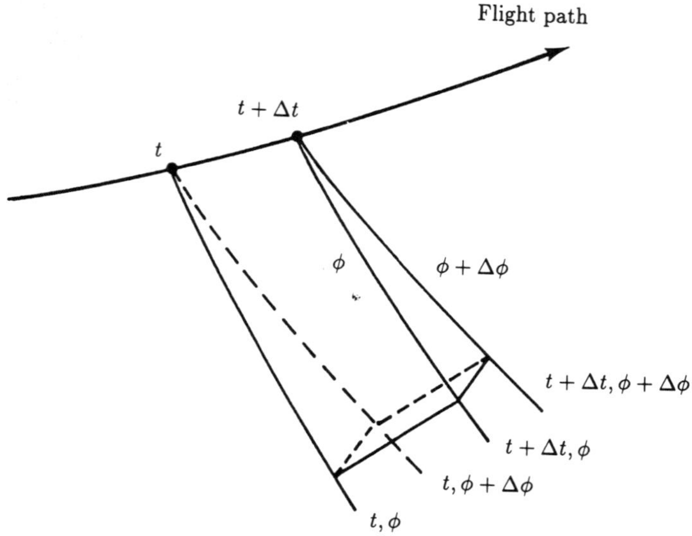

Figure 9. Ray tube outlined by four corner rays Δt and $\Delta\phi$ apart.

Computation of Sonic Booms

The theory outlined above is presented as a collection of components. The acoustic source signature is given by equations (2) and (3), with atmospheric effects included via equation (9). Nonlinear steepening is calculated by equation (5) and applied as shown in figure 7. Except for equation (8), which is very simplified, no formulas are presented by which the reader can compute sonic boom. The process is sufficiently complex that a computerized implementation is generally required. Figure 10 shows the computational flow of such a program. A number of computer programs have been written (refs. 25 and 31 to 34, for example). They all perform the same basic calculations, but each has particular capabilities and features added for specific applications. Reference 35 contains a review of the various program capabilities. All these programs were originally developed for mainframe computers. However, because of the current interest in sonic booms, it is expected that personal computer versions will be available soon.

A very useful calculation procedure for steady-flight booms is the simplified model developed by Carlson (ref. 36). He noted that the computerized geometrical acoustics calculations could be performed once for a range of flight parameters and implemented as an extension of formulas such as equation (8). His formulas for an N-wave are

$$\Delta p_{\max} = K_p K_r \sqrt{P_v P_g} (M^2 - 1)^{1/8} h_e^{-3/4} l^{3/4} K_s \tag{10a}$$

$$\Delta t = K_t \frac{3.42}{c_v} \frac{M}{(M^2 - 1)^{3/8}} h_e^{1/4} l^{3/4} K_s \tag{10b}$$

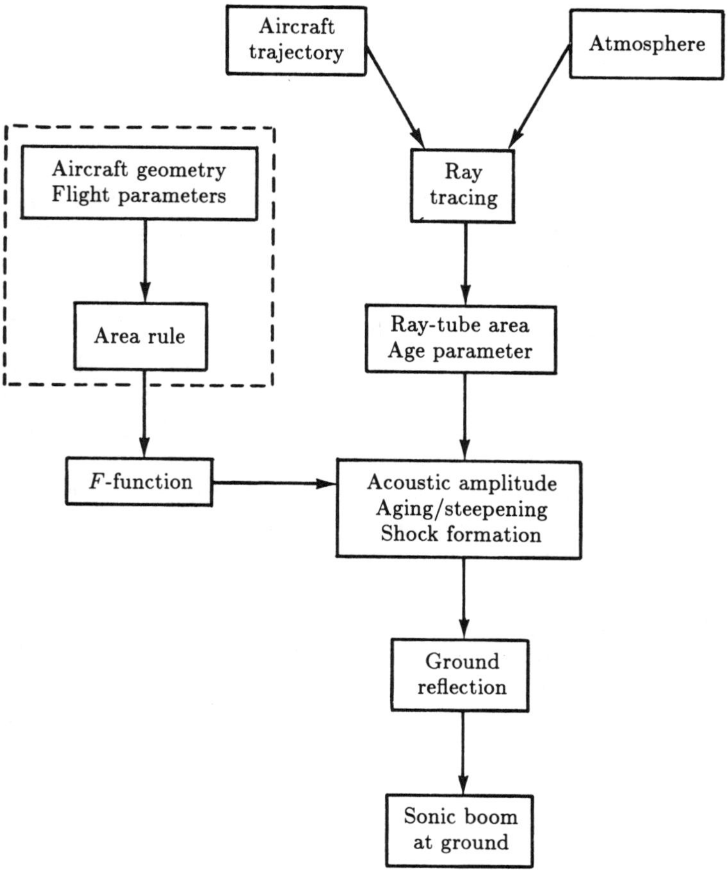

Figure 10. Logical flow of sonic boom calculation.

where

$\Delta p_{\max}$	shock strength
Δt	N-wave duration
h_e	effective altitude
l	aircraft length
K_p	pressure amplification factor
K_r	ground reflection factor (nominally 2.0)
K_s	aircraft shape factor
K_t	signature duration factor

Charts of K_p and K_t are presented in reference 36 for various flight altitudes and Mach numbers. A procedure is also presented for computing K_s based on aircraft type. The K_s procedure can be used to estimate an N-wave F-function for input to a full sonic boom model with which maneuver effects can be calculated. For steady, level flight in the standard atmosphere under conditions where N-wave sonic booms occur, Carlson's method is generally within 5 percent of computer calculations.

Maneuvers and Focusing

Under certain conditions, converging ray patterns can exist which produce focused "superbooms." Studies of cylindrical implosions and intuitive concepts of lens-like focusing give expectations of very high amplitudes. In practice, those sonic boom foci which do occur tend to be low-order types with moderate amplifications typically no more than two to five times the carpet boom shock strength.

Figure 11 illustrates a focus condition for acceleration. As the Mach number increases, the Mach angle decreases and rays converge to a focus at some distance from the aircraft. Only infinitesimally separated rays cross at a given point; the focus tends to move farther from the aircraft as M increases. The focus is thus smeared out over a line generally referred to as a caustic. There are three orders of focus to consider: a simple focus corresponding to a smooth caustic (as shown in fig. 11), a superfocus corresponding to a cusp between two smooth caustics, and a perfect lens-like focus (ref. 32). When a sonic boom focus occurs, it is predominately or completely a simple focus. Superfoci can occur for transient maneuvers such as turn entry and mark the initial point of the associated simple focus. Perfect foci do not occur for any credible supersonic maneuver.

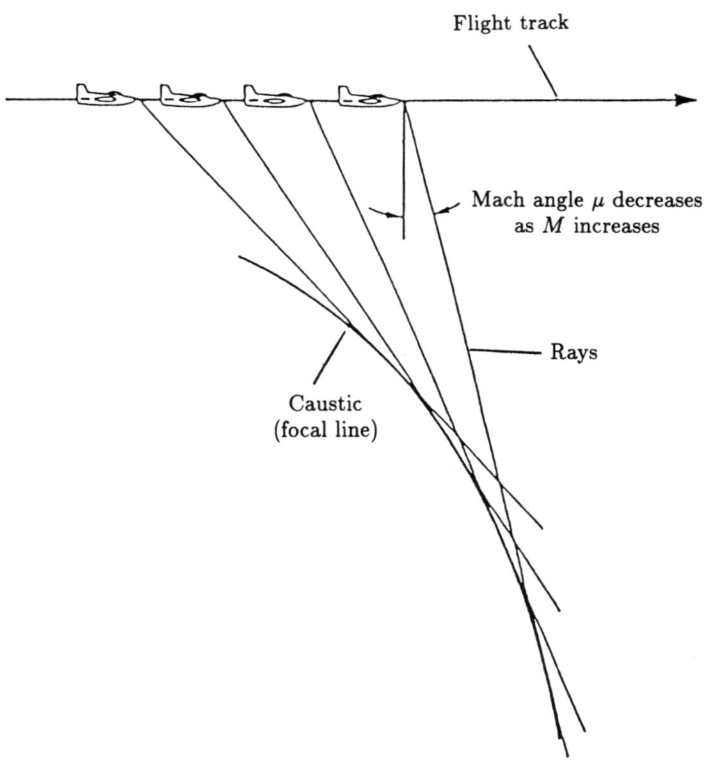

Figure 11. Sonic boom focusing due to acceleration.

In three dimensions, a caustic is a two-dimensional surface. A focal zone at the ground is a line representing the ground-caustic intersection. In three dimensions, a superfocus cusp is a line, with the superfocal zone at the ground being a point.

Figure 12 illustrates two other basic focusing maneuvers, that at sonic cutoff and that in the plane of a steady, level turn. The ray and caustic topology of these two cases and of the acceleration case of figure 11 are similar. Close to the caustic, the wave behavior depends on the relative geometry of the rays and caustics, and the three cases are mathematically interchangeable. Because the caustic represents a boundary to the wave field, the focus amplitude is limited by diffraction effects. This is a solved problem for the linear acoustic case. The linear solution is singular. The equations describing nonlinear behavior at a caustic were written by Guiraud (ref. 37), who derived a similitude and corresponding scaling law. Guiraud's scaling law leads to the following simple form for the maximum shock pressure at a simple focus:

$$\frac{p_{\max}}{p_{\text{ref}}} = C\left[\frac{y_{\text{ref}}}{(\gamma+1)p_{\text{ref}}R}\right]^{1/5} \tag{11}$$

where p_{ref} is the incoming N-wave boom pressure at a normal distance y_{ref} from the caustic, R is the relative curvature between the rays and the caustic, and C is a constant. If a focused signature is available for one smooth caustic case, Guiraud's similitude can be used to adapt it to any other simple focus. The similitude also defines the size of the focal zone, within which standard boom theory is invalid. Focal zones are very narrow, with amplification significantly above the boom carpet typically within a region less than 300 ft from the focal line. Although standard boom theory can detect a focus (by virtue of ray-tube area vanishing), calculation of the focus requires that the caustic be traced and its curvature determined.

Numerical focus solutions for a single shock wave have been obtained by Gill and Seebass (ref. 38) and Gill (ref. 39) and more recently in reference 40. The Gill-Seebass solution and the scaling law have been incorporated into one of the sonic boom computer programs described previously (refs. 32 and 35). Figure 13 shows a typical focus solution for an incoming N-wave. The shocks are amplified more than the rest of the signature (typical of diffraction, which tends to wash out low frequencies more than high), so that focused signatures typically have U-shaped waves. Focus factors, based on shock amplifications, range from 2 to 5, both from calculations (ref. 41) and from flight tests (refs. 42 and 43). Calculations from this theory and flight test data are in good agreement (ref. 44).

Cusped caustic focal zones, sometimes termed "superfoci" or "super-superbooms," have been observed in flight tests (ref. 43), with shock focus factors of almost 10 at a point. Such a superfocus is limited to a region a few hundred feet in size. Theory has been formulated for cusped superfoci (refs. 45 and 46), but no results comparable to those of references 38–40 are yet available.

Hypersonic Speeds

Most parts of sonic boom theory work well at all Mach numbers, but calculation of the F-function from slender-body theory (i.e., eq. (2)) fails at high Mach numbers (above about Mach 3 for slender transport-type aircraft) or for blunt bodies. At hypersonic speeds, some other theory is required. Three approaches have been

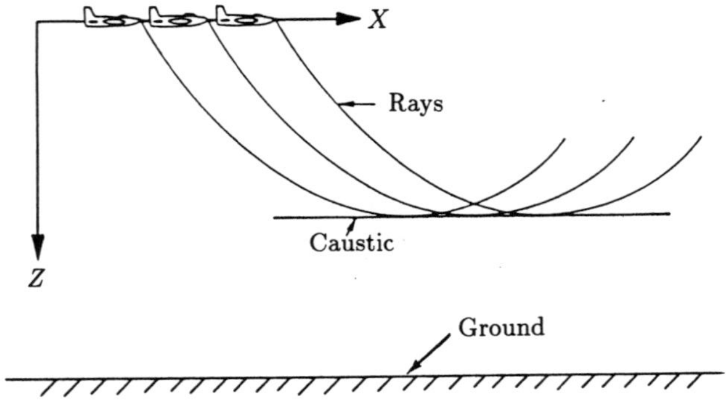

(a) Sonic cutoff.

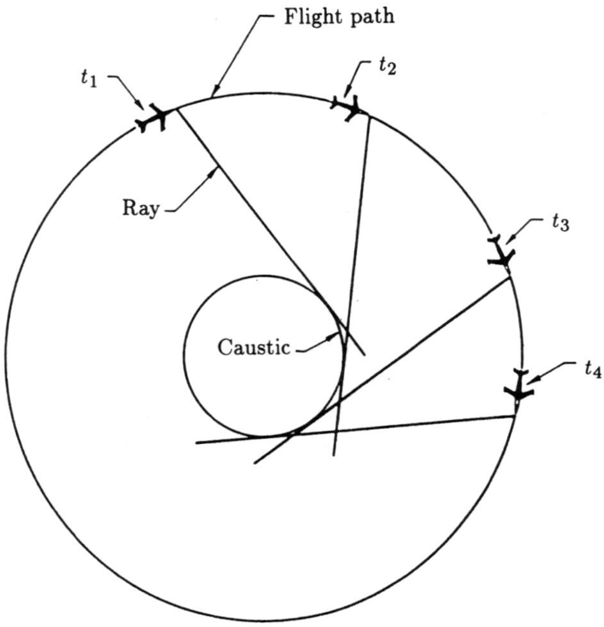

(b) Steady turn.

Figure 12. Focusing due to sonic cutoff and in-plane steady turn.

utilized: hypersonic finite-difference calculations for the near-field flow, wind-tunnel measurements to obtain purely empirical F-functions, and theoretical analyses based on tractable simplified conditions.

Wind-tunnel measurements of launch and reentry vehicles (refs. 47 and 48) have clearly demonstrated the ability to measure F-functions at hypersonic speeds and for blunt bodies. Subsequent use of these F-functions as inputs to boom calculations has been very successful. An associated task of this type of study was the calculation of selected points via a finite-difference computer code. In references 48 and 49, a 1970's vintage code was utilized, with very good agreement with wind-tunnel data.

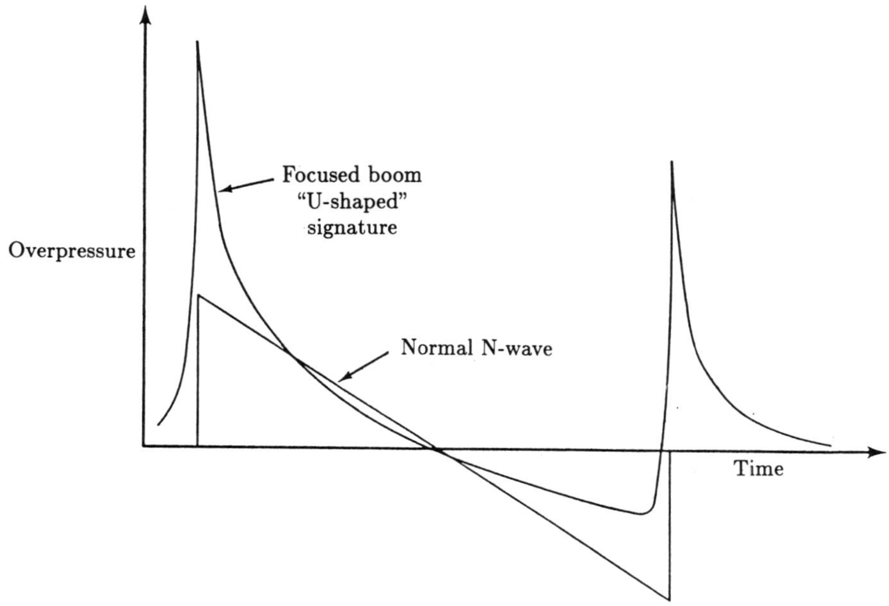

Figure 13. Focused and unfocused boom signatures.

With continued improvements in computational fluid dynamics and ever-lowering computer costs, it should be practical to use Euler codes to compute complete F-functions for hypersonic speeds. No such application has yet been made, but it is an expected development for current hypersonic projects.

One theory for hypersonic booms is available, that developed in reference 50. It is based on a concept by Seebass (ref. 51) that even slender hypersonic vehicles effectively have blunt noses (both physically, because of heating considerations, and aerodynamically, because of the entropy layer), and the resultant drag dominates the sonic boom. The model of such a vehicle is a spherical nose on a very slender, infinite afterbody, very much like the physical model used for the hypersonic-boom wind-tunnel study of reference 49. The far-field wave pattern of such a body can be computed by means of a blast wave analogy. Reference 50 contains a careful analysis of this configuration, identifying the significant terms in the hypersonic flow equations, writing the appropriate similitude-scaling laws, and matching near-field flow (where entropy layer effects are important) with the far field (where entropy layer effects can be argued to be negligible). Quantitative results for the far field were presented, with constants incorporated from a numerical solution to the equivalent blast wave. The analogy is valid for the positive-pressure phase of the far-field N-wave Δp (the positive impulse I) and also provides an estimate of the location of the trailing shock. The final far-field sonic boom results were combined with geometrical acoustics atmospheric corrections for an isothermal model atmosphere. Figure 14 shows the final prediction; this figure is based on reference 52, which contains a synopsis of reference 50. The only vehicle parameter is drag, as might be expected from the nature of the theory. In reference 50, agreement with reentry data for the blunt-body spacecraft is reasonably good.

The reference 50 theory predicts hypersonic transport (HST) sonic boom levels which are apparently lower than those established for supersonic transports. Shown for comparison in figure 14 are Δp and I for a nominal 400 000-lb SST at $M = 2.7$. The SST boom is significantly greater than the HST boom. This type of comparison has led to speculation that hypersonic transports may have a sonic boom advantage. However, there are two points to consider. First, the theory is effectively a volume-only model and does not account for vehicle lift. Almost 40 percent of the example SST boom is due to lift. An extension of the theory to account for lift-induced boom, analogous to Walkden's theory at supersonic speeds, would be very useful but has not yet been attempted. Second, the drag-dominated theory implicitly assumes a short body. This assumption results in durations considerably shorter than those calculated for the SST, with a correspondingly lower impulse.

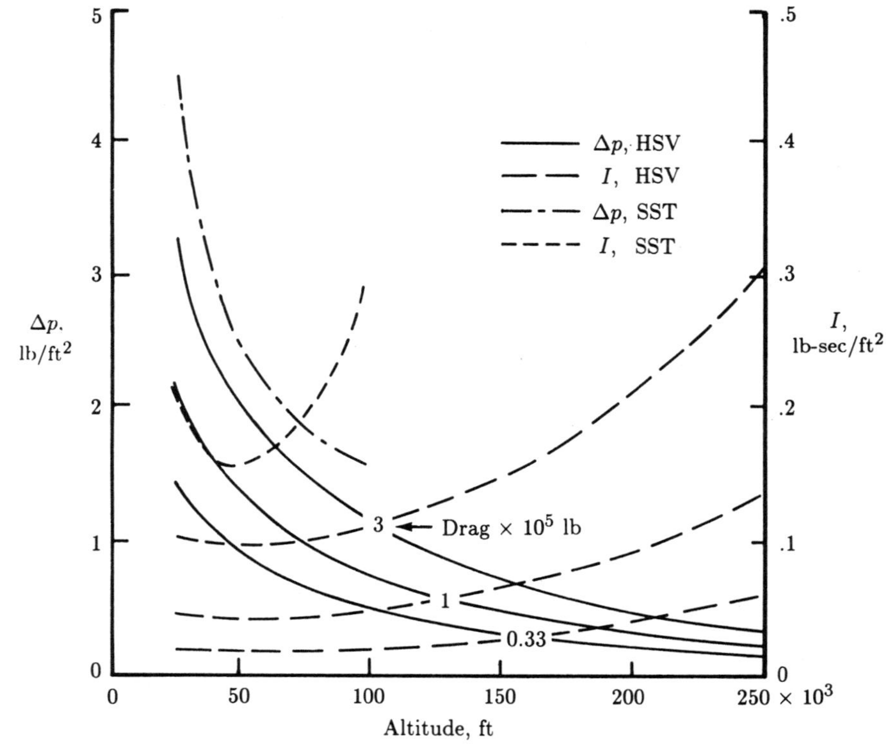

Figure 14. Overpressure and impulse as function of altitude for drag-dominated hypersonic vehicle. Conventional SST shown for comparison. (Based on ref. 52.)

Measurements and Predictions

This section deals with the primary boom carpets for both steady, level flights and for aircraft in maneuvers. For steady, level flight, both on-track and lateral measurements and comparisons with predictions are presented. Variability in the sonic boom measurements as a result of the atmosphere is presented and changes in waveform and probability distributions of measured-to-predicted boom levels are shown. Focus booms associated with various types of operations, their ground patterns, and the pressure buildups are described. Finally, recent experience with secondary boom carpets is discussed, including the signature characteristics and amplitudes of the booms.

Primary Boom Carpets for Steady, Level Flight

On-Track Measurements

A considerable number of studies have been conducted which were aimed at defining the peak amplitudes (overpressures) of the signatures for primary boom carpets for a wide range of vehicles and flight conditions. A summary of these results is shown in figure 15. Predicted and measured on-track sonic boom overpressures are plotted as a function of altitude for several aircraft of various sizes and weights (including Concorde) along with measured data for the launch and reentry phases of the Apollo 15, 16, and 17 spacecraft (ref. 53), and the Space Shuttle ascent and reentry flights (ref. 54). Measured and predicted values of overpressure correlate well for the aircraft cases. The sonic boom levels in general increase with increasing aircraft size and decrease with increasing altitude. The theory is valid for direct booms of conventional aircraft.

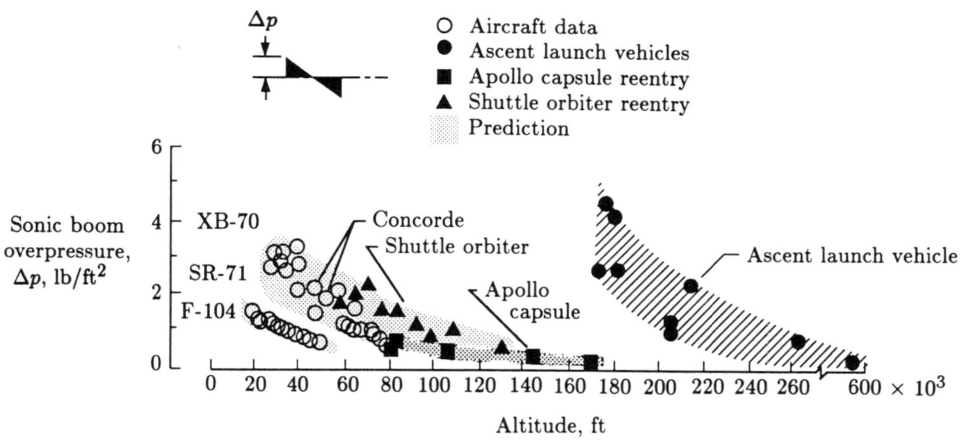

Figure 15. Measured and predicted on-track sonic boom overpressures in primary carpet area for several aircraft and spacecraft.

For measurements made during the reentry of spacecraft, the measured data are consistent with data obtained for aircraft in that they appear to be comparable in magnitude to extrapolated levels for fighter or medium bomber aircraft and they display a similar decrease with increased altitude.

In general, the measured overpressures for the launch and ascent portion of spacecraft flights indicate the same trend of decreasing pressure with increasing altitude. However, the magnitudes of the overpressure values are much greater than those of the reentry case. Since the launch vehicle is considerably larger than the reentry vehicle, higher boom levels can be expected. However, the largest portion of the increased overpressure from launch vehicles results from the "effective body" produced by the rocket exhaust plume. Note that disturbances from the ascent phase, with engines thrusting, were measured at ground level for the vehicle operating at altitudes up to about 600 000 ft (ref. 55). Simplified methods for prediction of spacecraft sonic booms are discussed in references 36 and 48.

Lateral Spread Measurements

Considerable attention has also been given to defining the lateral extent of the primary boom carpet for steady flights of aircraft at various Mach numbers and altitudes. The calculated and measured primary carpet data for 13 flights of the XB-70 at $M \approx 2.0$ and an altitude of 60 000 ft are shown in figure 16. These data are also typical of other aircraft and operating conditions. At the top of the figure an approaching supersonic aircraft is schematically shown, along with the downward-propagating rays. The extent of the primary carpet is the point at which the ray refracts away from the ground (the cutoff distance). This lateral cutoff point is independent of aircraft type and is only a function of the aircraft altitude, the Mach number, and the characteristics of the atmosphere below the aircraft.

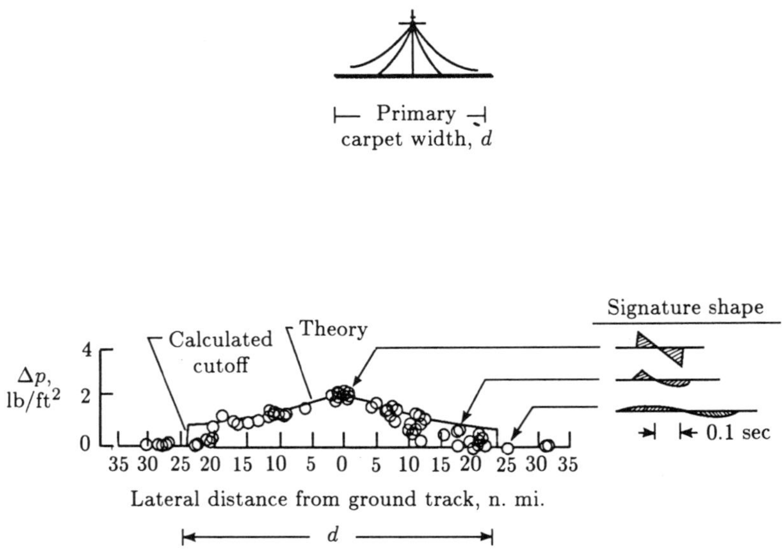

Figure 16. Sonic boom overpressures for XB-70 aircraft at an altitude of 60 000 ft as function of lateral distance. $M \approx 2.0$.

Comparisons of the calculated and measured lateral extent of the sonic boom patterns as a function of aircraft altitude and Mach number for steady flight in a standard atmosphere are given in figure 17. The data points represent averages of a number of measurements involving various aircraft. The widths of the sonic boom carpets on the ground increase with increasing altitude and Mach number. For example, at an altitude of 20 000 ft and $M = 1.5$, the total width of the pattern is 20 n.mi. At 60 000 ft and $M = 2.0$, the pattern width is about 60 n.mi. However, as is illustrated by the two sketches at the top of the figure, supersonic flights at low altitudes result in narrow carpets having higher overpressures, whereas at higher

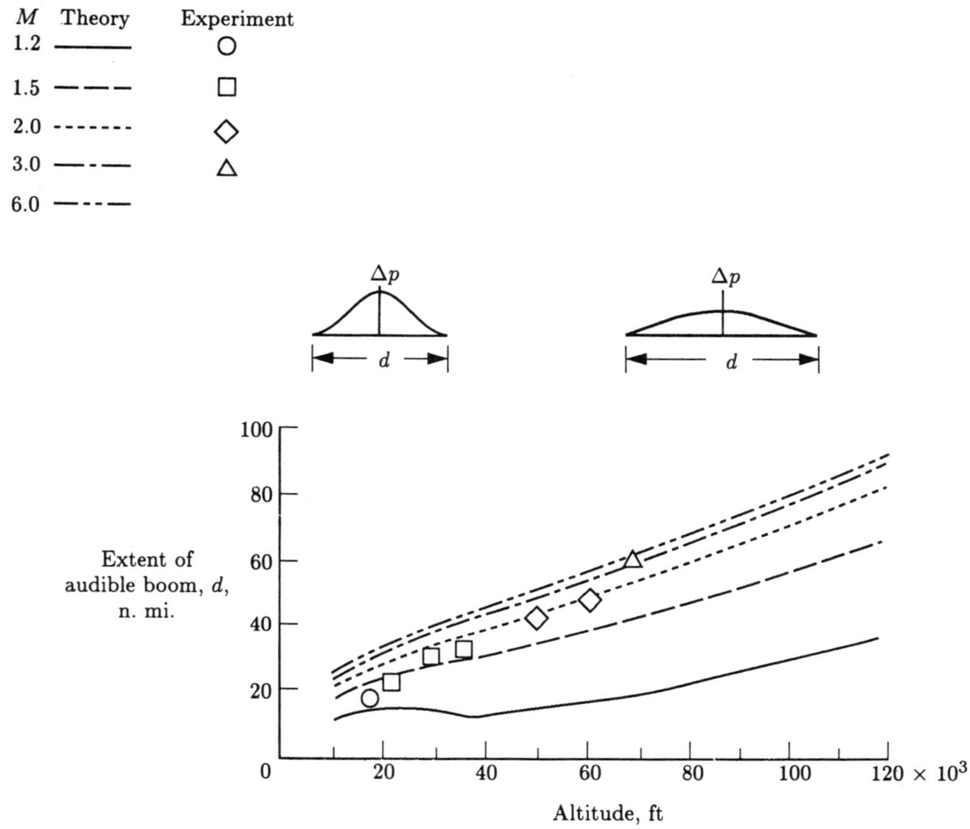

Figure 17. Width of area of audible sonic boom pattern on ground.

altitudes the carpet widths are much broader but with lower ground overpressures. Good agreement exists between measured and calculated values. The hypersonic aircraft will operate at altitudes and Mach numbers beyond the current experience. However, there is no reason to believe that theory would not provide reasonable estimates of the carpet width for this anticipated flight regime.

Variability Due to Atmosphere

The boom signatures associated with the on-track and lateral measurements were measured under fairly stable atmospheric conditions. It has been noted previously that atmospheric variations, particularly those in the first few thousand feet above the Earth surface, can be very influential in bringing about distortions of the sonic boom signature (see fig. 5), changing it from the normally expected N-wave to a "peaked" or "rounded-type" signature (ref. 56). Higher overpressures result when the signatures are peaked, whereas lower pressures are associated with rounded signatures. This peaking and rounding of the boom signatures is statistical in nature and occurs as a function of either time or distance.

A summary of the variations of the on-track overpressures resulting from the atmosphere for steady, level flight is given in figure 18. This statistical analysis comprises most of the planned sonic boom experiments that have been conducted in the United States. Data are included for a wide range of aircraft, a Mach number range of 1.2 to 3.0, and an altitude range of about 10 000 to 80 000 ft. A total of 12 406 data samples have resulted from 1625 supersonic flights.

Aircraft	No. flights	Data samples
B-70	43	655
SR-71	22	704
F-101	549	549
F-104	789	5063
B-58	222	5435
	1625	12 406

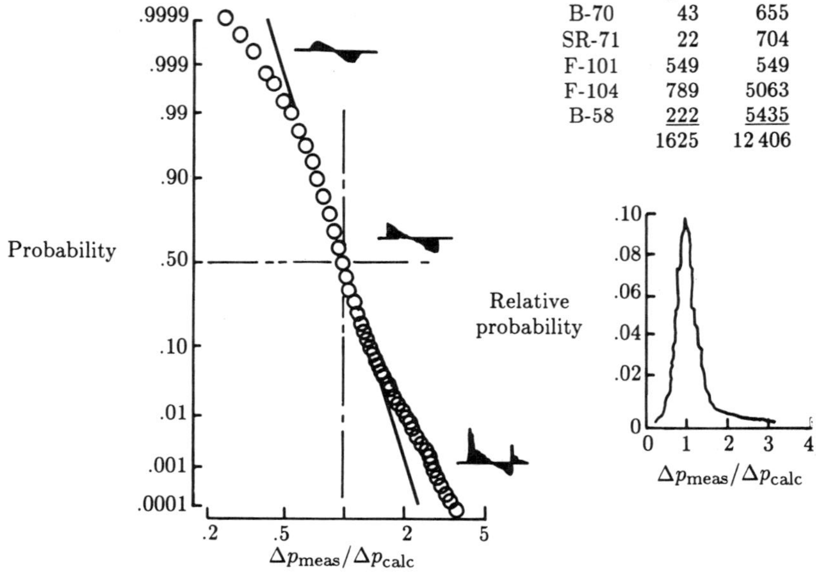

Figure 18. Statistical variation of sonic boom waveforms and overpressures resulting from atmosphere for steady, level supersonic flight.

Plotted on figure 18 is a relative cumulative frequency distribution and histogram for on-track measurements showing the probability of equaling or exceeding the ratio of the measured overpressure to the calculated or nominal overpressure for steady flight in standard atmosphere. For this type of presentation, all the data would fall in a straight line if the logarithm of the data fit a normal distribution. Rounded signatures of the waveform sketched in the figure are usually associated with overpressure ratios less than 1. Nominal or N-wave signatures are observed on the average, and peaked signatures of higher overpressures are observed usually at ratios greater than 1. The data of figure 18 indicate that variation in the sonic boom signatures as a result of the effects of the atmosphere can be expected during routine operations.

Primary Boom Carpets for Maneuvers

Any rapid deviation of a vehicle from steady, level flight conditions can produce considerable modifications in the location, number, and intensity of the ground shock wave patterns. This maneuvering phenomenon is illustrated in figure 19, which shows the shock wave ground-intersection patterns for two flight conditions of an aircraft (ref. 57). For simplicity, only the bow shock wave is shown.

At the left in figure 19 the lateral spread pattern on the ground for an aircraft in steady, level flight is shown. The ray paths on the ground, represented by the angled lines, are generally parallel to each other, and the shock wave ground-intersection patterns, as represented by the heavy lines, are essentially hyperbolic in shape. The pattern to the right is for an aircraft experiencing a lateral acceleration. The ray paths are no longer parallel; in fact, in some regions they tend to converge and in others to diverge. Likewise, the shock wave ground-intersection pattern is no longer hyperbolic and contains some irregularities, including a shock fold in which multiple booms would be observed and a cusp formation in which the pressures are higher than for steady flight conditions. Such pressure buildups correspond to focused superbooms discussed previously.

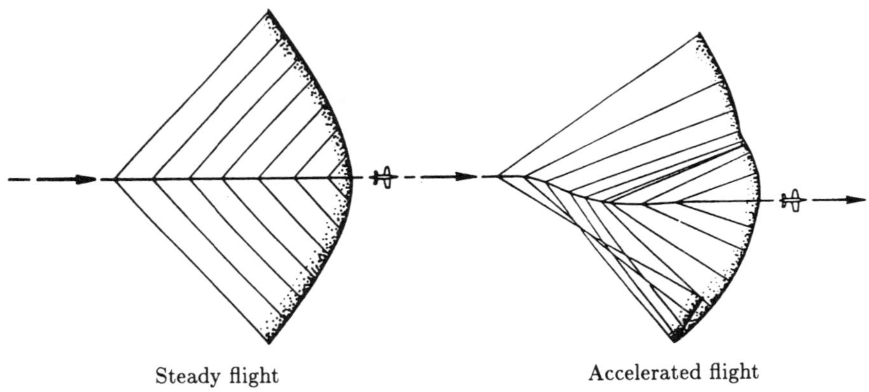

Figure 19. Shock wave ground-intersection patterns for aircraft in steady and accelerated flight at constant altitude. (From ref. 57.)

Sonic boom enhancement can result from a variety of aircraft maneuvers. Figure 20 illustrates three types of maneuvers which could result in pressure buildups at ground level: a longitudinal acceleration, a 90° turn, and a pushover maneuver. In each maneuver, pressure buildups occur in the localized regions suggested by the shaded areas in the sketches. It is very important to remember that although the aircraft and shock waves are moving, these localized areas on the ground in which pressure buildups occur are fixed and do not move with the aircraft. The localized regions, incidentally, are on the order of 1000 ft or less in width. The pressure buildups in these focus areas are a function of the type of maneuver and the acceleration involved and are noted in the *Review and Status of Theory* section to be 2 to 5 times the boom carpet values. As noted previously, pressure buildups will always result for the longitudinal maneuver when the aircraft accelerates from subsonic to supersonic speeds. The pressure buildup areas associated with turns and pushover maneuvers can be minimized or avoided by reducing acceleration (or decelerating) or by simply avoiding the maneuver.

In scheduled commercial operations, longitudinal acceleration from subsonic to supersonic speeds is the only maneuver of significance from a ground exposure point of view. Experience has demonstrated that the focus boom region associated with this acceleration can be placed to within about 2 miles of the designated area.

It is important to note that any randomness of the atmosphere, which brings about waveform distortions discussed in connection with figure 18, may decrease the focus factor value and, for certain situations, may eliminate the focus altogether.

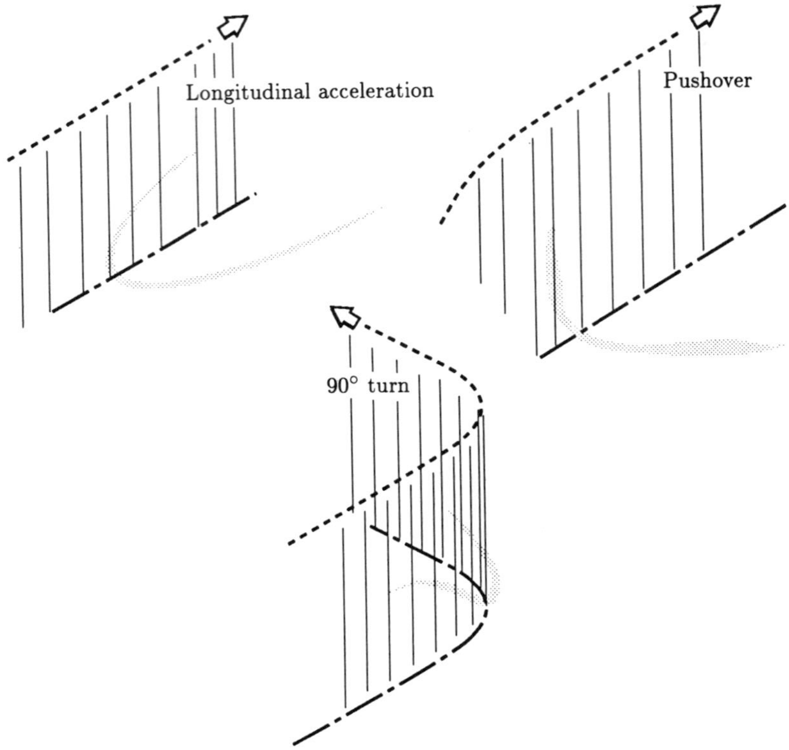

Figure 20. Areas on ground exposed to superbooms resulting from three maneuvers.

Secondary Boom Carpets

Secondary sonic booms, or so-called "over-the-top" boom disturbances (refs. 10 to 15), are quite distinct from primary booms, not only in the manner in which they are propagated from the aircraft to the ground but also in the way their signatures are shaped. The characteristics of these secondary sonic booms are illustrated in figure 21. An overall pressure time history (ref. 15) from the Concorde for a secondary boom is shown in figure 21(a). Note the signal is complex in that a number of disturbances are observed at this particular measurement location over a period of 1.5 minutes with a maximum peak-to-peak pressure of about 0.2 psf. Three sections of the overall pressure signature at A, B, and C are presented with expanded time scales in figures 21(b) to 21(d) to provide an indication of the frequency of these signals. Note that the fundamental frequency is about 1.5 to 2.0 Hz. For secondary boom signatures, the pressure changes very slowly and is in the subaudible frequency range. This, combined with the very low amplitudes, makes it difficult for the ear to sense this sound. These secondary booms are heard, however, and descriptions vary from the rumbling of far-off thunder to startling. Indoors, of course, such a pressure signature can be more noticeable since it vibrates the structure and causes rattling and motions.

Secondary booms have existed since the beginning of supersonic flight capabilities. Each of the major sonic boom flight programs sent secondary booms propagating through the upper atmosphere. These secondary booms went essentially unnoticed

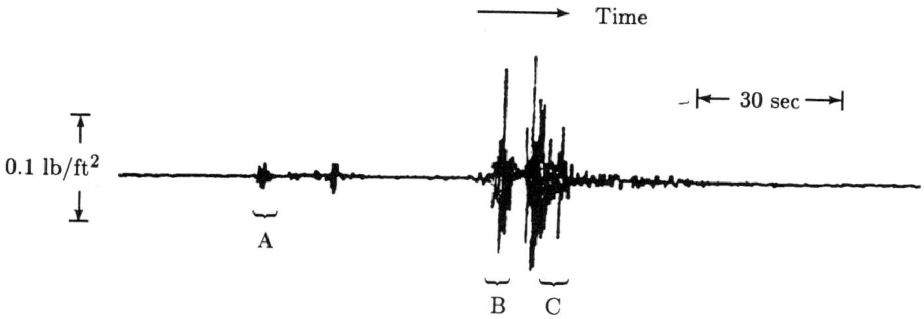

(a) Overall pressure time history.

(b) Expanded pressure time history at A.

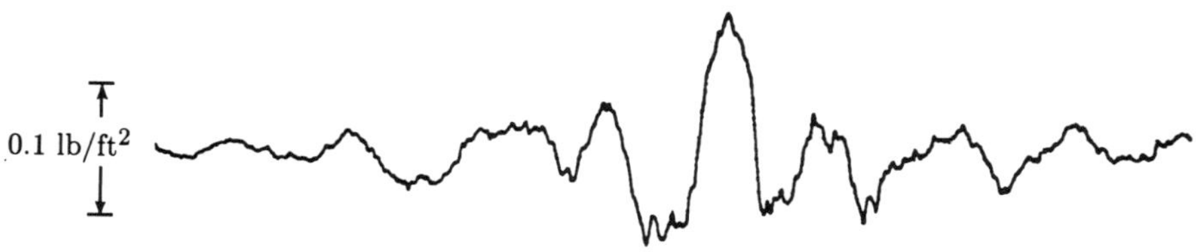

(c) Expanded pressure time history at B.

(d) Expanded pressure time history at C.

Figure 21. Characteristics of secondary sonic booms.

until the 1977 to 1978 time period, when the Naval Research Laboratory investigated the east coast acoustic disturbances (ref. 58), which were initially reported to be strange and mysterious in origin. The Concorde had entered into commercial service in mid-1976, of course, and scientific use was being made of its secondary boom, which was a consistent and known source, for determining the characteristics of the upper atmosphere (refs. 12 and 13). Concorde secondary booms thus became more evident and complaints to this effect were received. However, in every case of Concorde-generated secondary sonic booms, rerouting of the flight tracks and changes in operating conditions, depending upon atmospheric and seasonal variations, eliminated the problem.

The secondary sonic boom carpet and the disturbances experienced within it are not as well defined as for the primary sonic boom, and only fragmentary observations and measurements are available. These disturbances are known to involve both the upper and lower levels of the atmosphere during propagation, to have very low overpressure values, and to have a very low frequency content. Propagation distances greater than 100 miles are common and relatively large ground areas are exposed, but the significance from a community response standpoint is not well-defined.

Sonic Boom Minimization

This section addresses sonic boom minimization through aircraft design and aircraft operation. Minimization through design involves modification of the aircraft size, weight, and shape in order to reduce the overpressure or to alter the waveform. Minimization through aircraft operation relates to flying the aircraft at a Mach number and altitude combination so that a boom does not reach the ground. The atmosphere plays a significant role in this process. Mention is made of sonic booms from aircraft at transonic Mach numbers ($0.89 \leq M \leq 1.00$) and relatively low altitudes, and the associated waveform and boom levels are discussed.

Design

Sonic boom effects are minimized through increased distance between the aircraft and the ground. Minimizing sonic booms through aircraft design modifications has also been investigated and lower bounds have been established (refs. 52 and 59 to 62). Some of the approaches that have been considered are illustrated in figure 22. Sonic boom minimization can be achieved through a reduction in the overpressure or an increase in the signature rise time, each of these parameters being significant with regard to human response (refs. 63 and 64) and to structural response (ref. 65). Altering the overpressure and the rise time also results in changes in impulse. As illustrated in the lower sketches of the figure, reduced overpressures can be obtained by reducing the size of the aircraft (that is, lower aircraft weight and volume) or by proper shaping of the aircraft geometry to provide a modified (i.e., flattop) signature. The minimum impulse signature is generally that of an N-wave. The minimum overpressure is that of the flattop and flat-bottom N-wave. These two approaches have been given consideration in the past, and reductions in bow wave overpressures of about 30 percent to perhaps as much as 40 percent appear to be obtainable.

Other minimization techniques involving increasing rise times have also been investigated (refs. 66 and 67). If the rise time of the signature could be increased to the point where a sine wave would result instead of an N-wave, the sine wave pressure signature should not be audible to an outdoor observer. However, this avoidance of the shock wave altogether would result in an increased impulse, and such a signature would still vibrate buildings such that people indoors would react. To obtain even small increases in rise times, the aircraft length would have to be increased by at least a factor of 3 over the greatest length now being considered (to about 1000 ft). This in itself is an impractical approach. Another means would be to alter the airstream so that the same beneficial effects associated with the increased length are obtained. This would be accomplished by the addition of heat or other forms of energy. Studies of the airstream alteration or the "phantom body" concept suggest that large amounts of heat or energy (at least the equivalent of the output from the onboard propulsion systems) are required to obtain increased-rise-time signatures. This approach therefore also appears to be impractical.

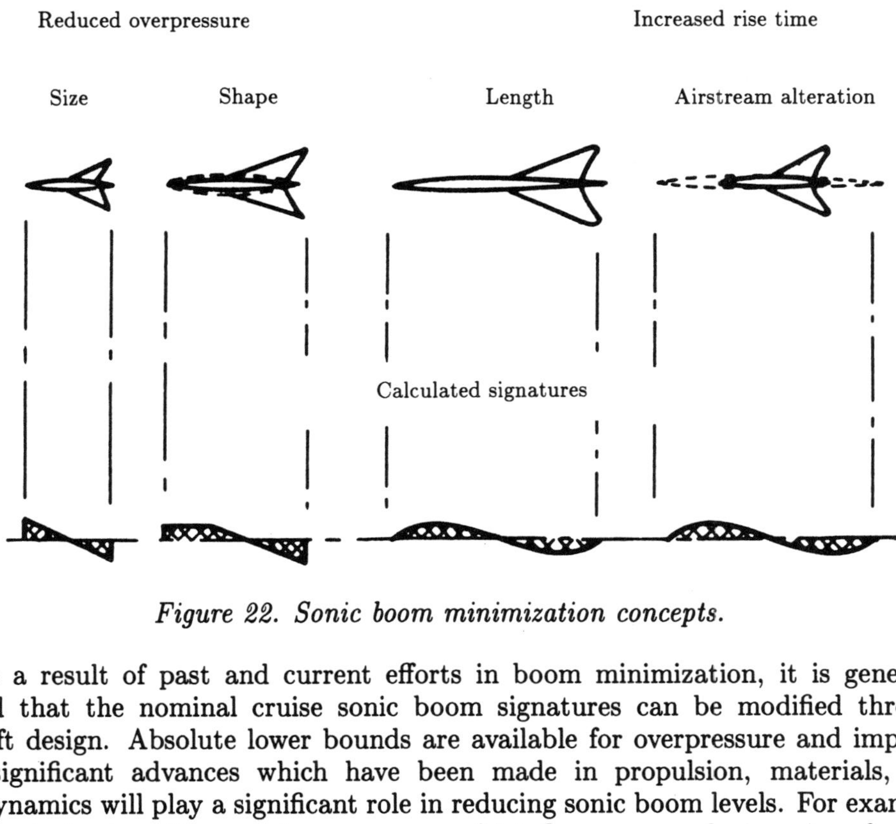

Figure 22. Sonic boom minimization concepts.

As a result of past and current efforts in boom minimization, it is generally agreed that the nominal cruise sonic boom signatures can be modified through aircraft design. Absolute lower bounds are available for overpressure and impulse. The significant advances which have been made in propulsion, materials, and aerodynamics will play a significant role in reducing sonic boom levels. For example, supersonic laminar flow exerts a very powerful influence on reducing aircraft gross weight and increasing cruise altitude, both of which lower the sonic boom level. In-house NASA studies (ref. 68) suggest the feasibility of a long and light SST having a sonic boom overpressure level of less than 1.0 lb/ft^2 (about half the overpressure estimated for the canceled U.S. SST) during cruise flight for domestic ranges. The analytic tools for defining the required aircraft characteristics are available and have been validated with wind-tunnel experiments on small wing-body configurations representing transport aircraft. Means for evaluating the trade-offs for boom minimization in terms of actual aircraft design are available. The most desirable signature shape, from the point of view of minimum response of an outdoor and indoor observer and of structural response, has not yet been established.

"Boomless" Flight Operations

In addition to sonic boom minimization, sonic boom avoidance can also be realized through operation of the aircraft at low supersonic Mach numbers such that the shock waves extend down toward, but do not intersect, the ground because of atmospheric refraction or cutoff, as suggested by the sketch in the upper left of figure 23. The range of Mach numbers and altitudes over which operations at cutoff Mach number M_{co} can be performed is shown in the figure for steady, level flight in a standard atmosphere with no wind. Flights at Mach numbers to the left of the hatched curve will result in no booms reaching the ground, whereas flights at Mach numbers to the right of the curve will result in booms reaching the ground. The highest speed at which the aircraft could operate in a standard atmosphere without producing booms at the ground is about $M = 1.15$. In the real atmosphere, variations in the speed of sound do exist because of temperature and winds. Climb or descent angles would also permit an increase or decrease in M_{co}, respectively. The practical range of M_{co} varies from 1.0 to about 1.3 for steady, level flight for a fairly wide range of atmospheric conditions.

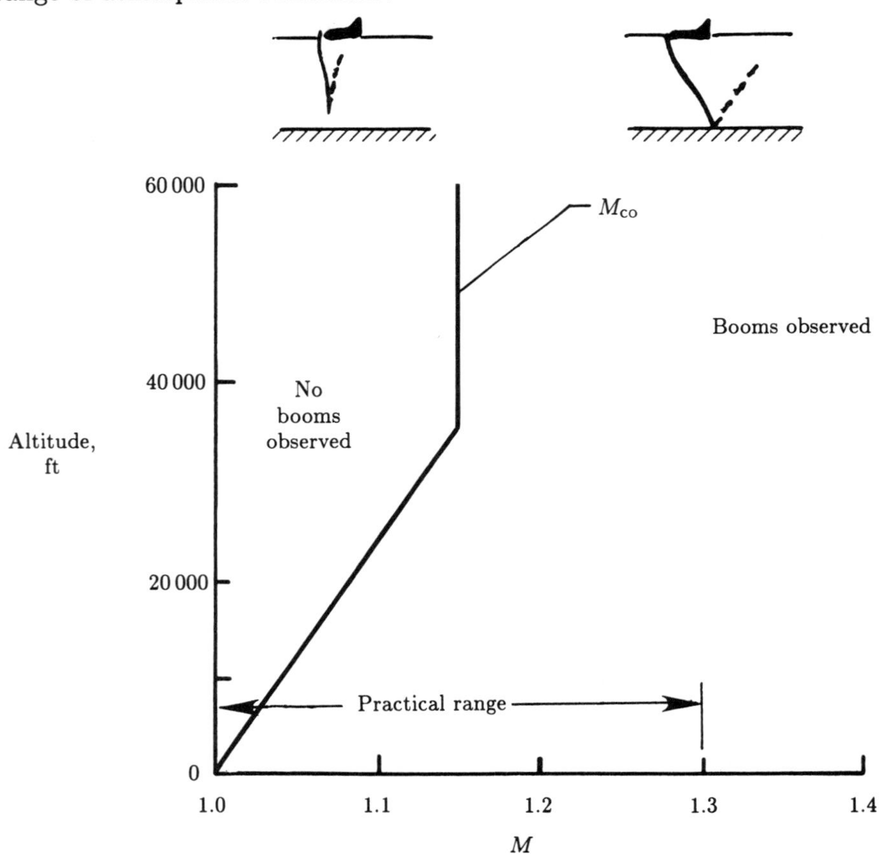

Figure 23. Combinations of Mach number and altitude for boomless flight.

Two points should be made about boomless flight operations. Boomless flight is independent of aircraft configuration and is a function of the aircraft operating conditions and the atmosphere. Aircraft configuration is, of course, important from the standpoint of efficiency of operation at these low Mach numbers; for example, flying a high Mach cruise design at the off-design M_{co} is less desirable than flying an aircraft designed to operate at $M = 1.15$. In either case, boomless flight operations, unlike the stationary focus from an aircraft maneuver, always result in a continuous caustic or focal line where the overpressures can be higher than those of the steady-flight boom from the aircraft flying at the same altitude but at the higher cruise Mach number. Therefore, in order to assure the shocks, and thus the boom, for flight at M_{co} will terminate at some safe height above ground level, a margin of safety in the form of reduced Mach number is required.

Low-Altitude Transonic Flight

It can generally be stated that as long as the aircraft speed over the ground is less than the speed of sound at the ground, then boomless flights at low supersonic Mach numbers can be achieved. It may be further stated that as long as the aircraft speed is less than $M = 1.00$, no sonic booms should be experienced at ground level. This is true for aircraft flying at altitudes of 100 body lengths or greater. Experience, confirmed by measurements (ref. 42), indicates that booms can be observed at ground level from aircraft in steady, level flight at Mach numbers from about 0.95 to 0.99 at altitudes of about 300 to about 2000 ft above ground level. This phenomenon is similar to that observed for airfoils in wind tunnels (as shown in fig. 24(a)) at high transonic Mach numbers, for which localized shocks occur at the maximum thickness (where the flow accelerates to $M = 1.00$ or greater). These shocks extend for some distance from the airfoil before dissipating into acoustic disturbances. This extended shock is also shown in the unusual photograph of an aircraft in flight at an indicated Mach number of 0.89 and an altitude of about 300 ft (fig. 24(b) from ref. 69). The explanation for why the shock waves are visible is given in reference 69 as follows. "Aircraft is flying in a cloud of water vapor condensed by a shock wave created when the local Mach number reaches or exceeds 1.0 at a point on the fuselage aft of the cockpit, where the shock attaches." The basic mechanism involved in the flight picture is the lower pressure behind the shock front causing the moisture in the air to condense.

The aircraft, like the airfoil, has a maximum thickness (equivalent area distribution) such that the local flow can equal or exceed $M = 1.0$ at some given free-stream transonic Mach number. These localized shocks have been observed to extend outward and downward as much as 30 airplane body lengths. The intensity of the booms is substantial because of the very low altitudes, and the signature, shown in figure 25, is considerably different in nature from the normally observed N-wave-type signature associated with a fully developed supersonic flow field.

The detailed analysis of low-altitude transonic flight test data (ref. 42) has indicated that existing meteorological conditions influence the vertical extent of attached shock waves produced at nearly sonic flight. Aircraft Mach number also has a direct influence on the vertical extent of the attached shock waves. The extension of these attached shock waves to lower altitudes may explain several "accidental" sonic booms produced by low-altitude, marginally subsonic aircraft (although Machmeter and altimeter errors may also be responsible).

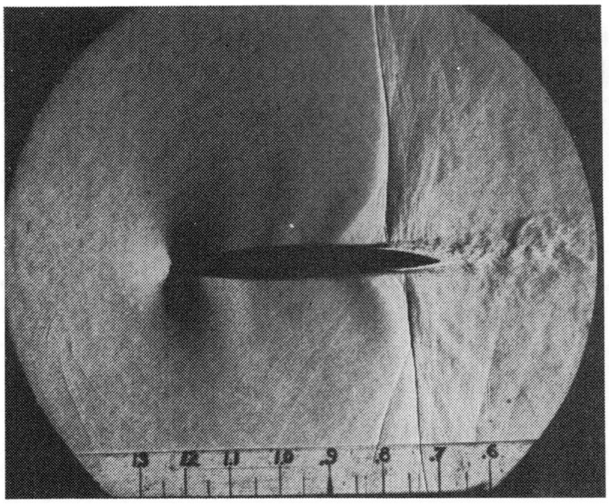

(a) NACA 16-212 airfoil in wind tunnel. M = 0.90.

(b) Aircraft in transonic flight. M = 0.89. (From ref. 69. Copyright Paul A. Ludwig.)

Figure 24. Transonic flow fields.

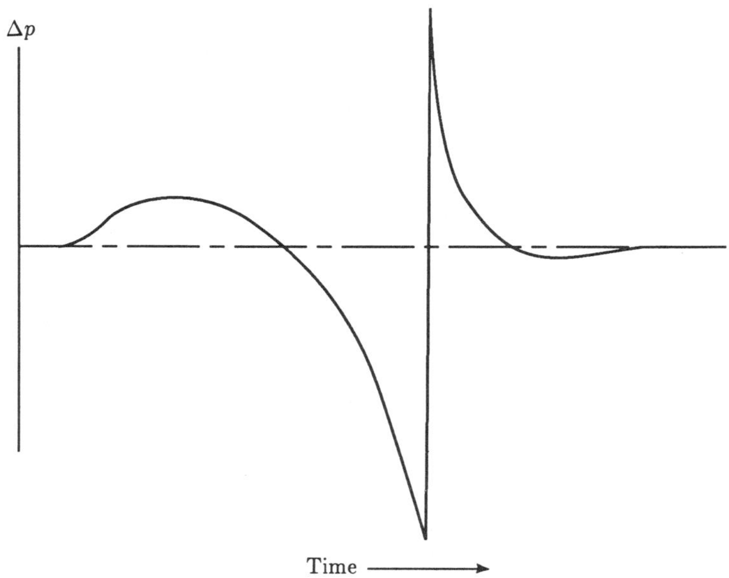

Figure 25. Character of transonic boom signature.

Responses to Sonic Booms

This section begins with a description of the factors involved in boom exposure, including the air path, ground path, and building vibration. Outdoor and indoor stimuli include audible, vibratory, and visual cues. A discussion of damage complaints, relative to primary and secondary structural members, is given as a function of the range of boom exposure levels. People responses to booms include startle, annoyance, effects on sleep, and long-term effects on health. Other responses to sonic booms cover animals, birds, and fish, and also include seismic effects, avalanches, landslides, and other subsonic aircraft.

Factors Involved in Boom Exposures

There is considerable concern about the manner in which people and structures respond to sonic booms and how such responses will affect community acceptance of overland operations. The nature of the response problem is illustrated in figure 26 (from ref. 70). The sketch at the top of the figure suggests two different exposure situations for people. In one case, the person is outdoors and is impinged on directly by the waves. In the other case, the observer is inside a building and the waves impinge first on the building. The building then acts as a filter which determines the nature of the exposure stimuli reaching the inside observer. The ingredients of this indoor exposure situation are included in the chain diagram at the bottom of the figure. The sonic-boom-induced excitation which causes the building to vibrate may arrive either through the air or through the ground. It is generally conceded that the air path is the more significant one in most cases and is thus designated the primary path in figure 26. The ground path is considered secondary and is designated by a dashed line in figure 26. Building vibrations can be observed directly by the subject. A person may also observe vibration-induced noise or, in the extreme case, associated superficial damage of the structure.

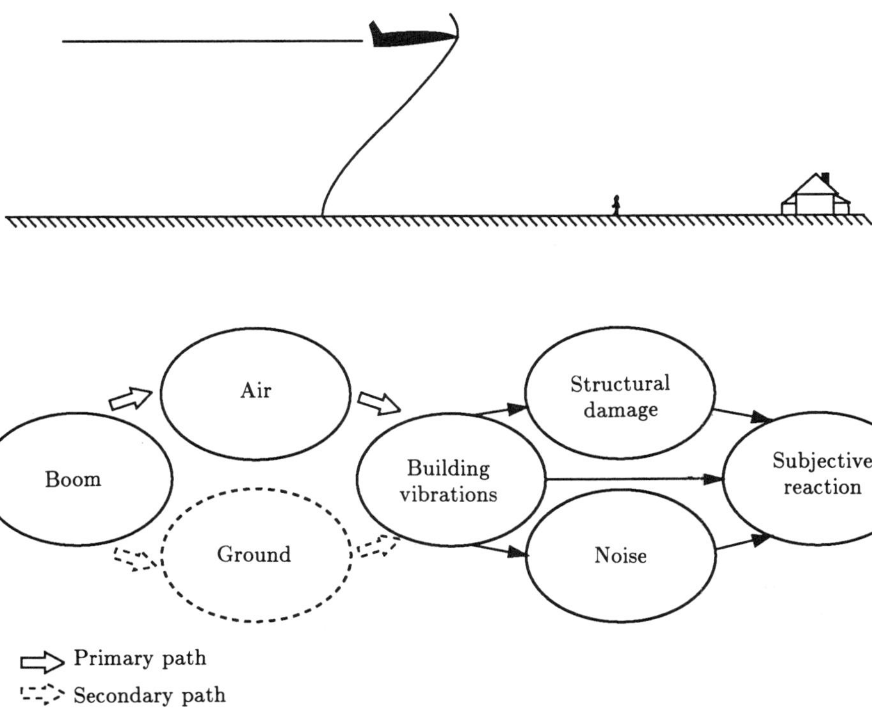

Figure 26. Factors involved in boom exposures. (From ref. 70.)

Outdoor and Indoor Stimuli

A person inside a building is exposed to a rather complex series of stimuli, including auditory, visual, and vibratory inputs. The nature of the auditory and vibratory inputs is illustrated in figure 27 (from ref. 70). The top trace is a sample outdoor pressure exposure as measured for one particular case. This wave is of the N-wave type, but it differs from an N-wave in some of its details, as do many of the waves measured in the field. The three bottom traces represent corresponding indoor exposure stimuli. The topmost of these traces represents the pressure variation inside the building owing to vibratory motions of the building and the cavity resonances. Although this is a pressure disturbance, it generally occurs in a frequency range that is not audible to humans. The audible portion of this signal, as measured with a separate microphone system, has the characteristic shape of the middle trace and is an order of magnitude lower in amplitude. It is believed that this audible portion of the pressure signal is associated with the rattling of the building structure and furnishings because of the primary mode responses in the building. Finally, the bottom trace represents the vibration of the floor that would be sensed by a person either directly or through the furniture. A person indoors therefore can be influenced because of an auditory, vibratory, or visual cue. At the present time, the indoor exposure situation is not understood well enough to permit the relative importance of each of these stimuli to be determined, although it is believed that in certain situations each one is significant.

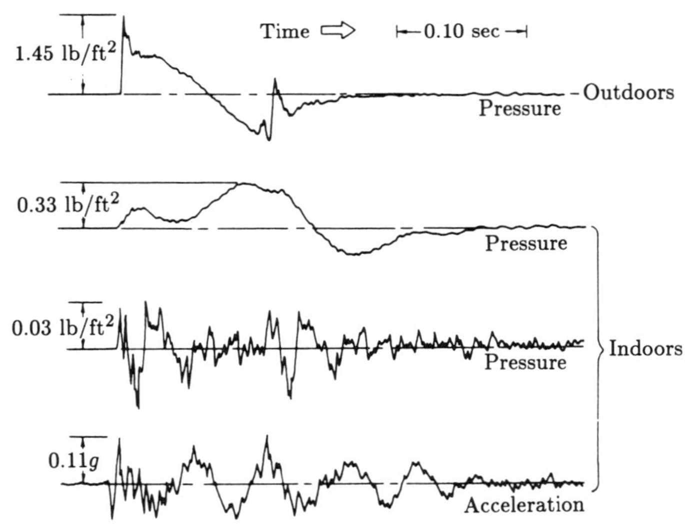

Figure 27. Outdoor and indoor sonic boom stimuli. (From ref. 70.)

Damage Complaints and Range of Boom Exposures

Experience has shown that supersonic flights over communities have resulted in complaints regarding damage because of sonic booms. The nature of the reported damage is fairly well represented by the data in figure 28, which shows the frequencies with which certain types of damage are mentioned (ref. 6). Plaster cracks, the type of damage reported most frequently, are mentioned in 43 percent of the complaints. Other reported damage includes cracks in window glass, walls, and tile. Structures reportedly damaged by sonic booms are mostly brittle surfaces and are secondary structural components.

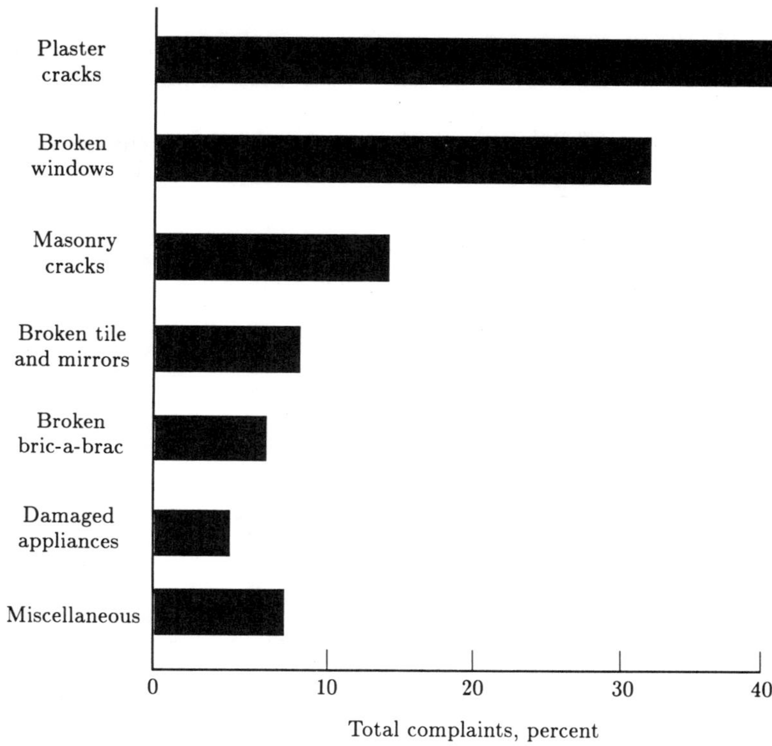

Figure 28. Sonic boom damage complaints. (From ref. 6.)

The relation between sonic booms and damage has the same complexity as the relation between sonic booms and indoor responses: a rigorous relationship depends on the frequency content of the boom and the frequency response of the structure. A practical, simple measure of the boom (for correlation with damage) would be the energy content in frequencies around the fundamental response frequency of structures, since this is where the greatest response occurs (refs. 65 and 71). Previously in this chapter the argument was presented that aural response to sonic booms can be quantified by the peak overpressure, since that was associated through audio frequency components of booms. Similarly, it can be argued that structural response involves low frequencies, so that the impulse of the boom is an appropriate quantity. For this reason, boom impulses, as well as peak overpressures, are often reported. The boom impulse tends to represent the lowest frequency components, in the range of several hertz. The fundamental frequency response of buildings is typically 10 to 30 Hz (ref. 72), however, so impulse alone may not be an adequate measure.

The vast majority of experience with sonic booms has been with N-waves 200 to 300 msec in duration. Usually only the overpressure has been reported or correlated with damage. One would, however, expect the relation between impulse, spectra, and overpressure to be fairly consistent for booms of such similar shapes. Care must be taken when these data are applied to significantly different types of boom signatures, but these correlations of boom damage with overpressure should certainly be self-consistent and well worth examining.

As expected, the reported damage varies depending upon the intensity of the boom. This is illustrated in figure 29, in which sonic-boom-induced incidents per flight per million people are shown for various overpressure ranges (ref. 6). The ranges of boom levels up to about 3.0 lb/ft^2 are fairly representative of the majority of booms associated with controlled supersonic flight operations. It may be significant that no damage incidents occurred for boom exposures below about 0.8 lb/ft^2, although a smaller number of data samples were available in this range.

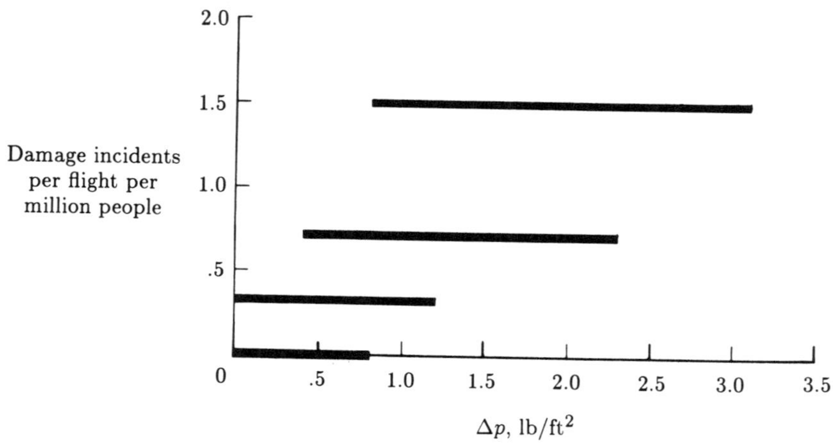

Figure 29. Sonic-boom-induced incidents for various overpressure ranges.

The nature of the sonic-boom-induced damage problem can be illustrated with the summary plot of figure 30 (from ref. 6). The number of damage incidents for a given type of structure increases as the overpressure increases, and this is particularly evident at the higher overpressure values. Also shown in the figure is a schematic illustration of the amplitude distribution of the overpressures. Even though the nominal, or predicted, overpressure for a given aircraft at specific flight conditions has a value which is generally lower than that at which building damage might be expected, there is a distribution of pressure amplitudes such that a small percentage of the total amplitude values occurs in the relatively high overpressure range. These high values, which occur only occasionally because of either atmospheric effects or focus booms due to maneuvers, may be sufficient to trigger incipient damage in existing structures. Two points can be made from this figure. It is obvious that a lower nominal value is desirable because of the reduced probability of building damage. However, though the nominal overpressure is established at a relatively low value, no assurance can be given that the triggering of damage can be completely avoided.

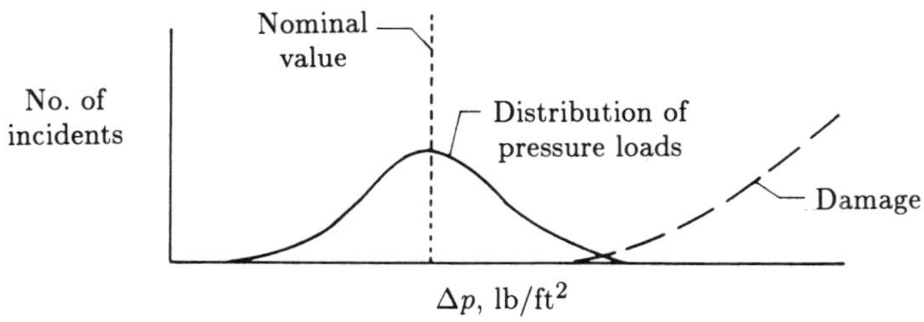

Figure 30. Nature of sonic-boom-induced damage problem. (From ref. 6.)

People Responses

It has been shown that people are annoyed by sonic booms because of concerns that the booms may damage their property. This suggests that the annoyance of booms might be diminished if the public could be convinced that boom levels from military or commercial supersonic operations are well below the damage threshold. However, sonic boom effects on people are difficult to pinpoint because of the subjectivity of the people's responses and because of the diversity of variables affecting their behavior. Responses depend on previous exposure, age, geographic location, time of day, socioeconomic status, and many other variables.

Research and experimentation have turned to several findings about sonic boom phenomena related to humans. These findings indicate that booms do not adversely affect human hearing, vision, or circulation (ref. 6). The human psychological response is more complex, involving attitudes and habituation to booms and their sources.

These findings have also turned up a number of other points. Long-term effects on health of repeated daily booms have not been investigated. Possible long-term effects on sleep of repeated night booms are unknown. Although existing evidence suggests booms of 1.0 lb/ft^2 or greater are unacceptable to a significant portion of the population, a level of acceptability of sonic booms has not been determined. Values of sonic boom overpressure of 0.5 to 1.0 lb/ft^2 have been suggested, but with no scientific support.

Finally, the possibility exists that human responses to booms measured one to two decades ago may differ from those recorded in the next decade (ref. 6).

Other Responses

To date, it has been difficult to make detailed interpretations of the effects of sonic booms on different animal species. However, research on chickens, chicken and pheasant eggs, pregnant cows, racehorses, sheep, wild birds, and mink suggests that boom effects on domestic and wild animals are negligible (ref. 6). Like humans, animals are startled by loud noises, but this diminished during testing. In any case, our dependence on animals for food (livestock), companionship (pets), relaxation (horses), and aesthetic value (wildlife) strongly suggests they receive more attention and research regarding the effects of sonic boom exposures.

The effects of sonic booms on aquatic life may not have been investigated to any great extent. This is a result of the initial findings that the attenuation of the sonic boom in water suggests that these effects should be small (ref. 6). In particular, boom overpressures dissipated to about a tenth of their initial value at a depth of about 60 ft and so appear to pose no threat to aquatic life, including the capacity of fish eggs to hatch. This experience is associated with aircraft traveling at Mach numbers of 3.0 or less. At hypersonic Mach numbers, the aircraft speed can equal or exceed the speed of sound in water (about 4.5 times that in air), thereby greatly increasing the potential penetration of the sonic booms.

Sonic booms produced by aircraft moving at supersonic speeds apply moving loads to the Earth's surface. Although the ground motion recorded was about 100 times the largest natural, steady seismic noise background, it was still less than 1 percent of the accepted seismic damage threshold for residential structures (ref. 6). Experiments have shown that sonic booms probably cannot trigger earthquakes, but they might precipitate incipient avalanches or landslides in exceptional areas which are already stressed to within a few percent of instability. Research efforts on the effects of booms in areas prone to avalanches and landslides have been recommended. Furthermore, the differences between triggering snow and earth avalanches need to be better understood. Once again, the experience thus far is associated with aircraft speeds of Mach number 3.0 or less. The situation may be different at hypersonic speeds.

Questions have been raised concerning the effects of sonic booms on other subsonic aircraft, transport and general aviation types, both in flight and on the ground. Controlled tests (ref. 73) have shown that the sonic-boom-induced accelerations, which were structural rather than rigid-body motions, were small relative to those induced by such commonly encountered phenomena as runway roughness and moderate air turbulence. The general conclusions were that sonic booms constitute no serious concern for the safety of all types of subsonic aircraft in flight.

Summary

During flight at supersonic speeds, shock waves are formed which propagate outward in all directions; some may extend to the ground and cause objectionable noise. For vehicles operating at high altitudes, the shock wave patterns coalesce into a bow shock at the front of the vehicle and a tail shock at the rear. The passage of these shock waves past an observer results in rapid changes in atmospheric pressure in the form of an N-wave signature and is interpreted by the ear as two explosive-type sounds, commonly referred to as sonic booms. In a typical supersonic mission the shock waves, which are moving with the aircraft, generate sonic boom "carpets" on the ground whose width depends on flight and atmospheric conditions. These carpets are made up of primary and secondary booms. The primary boom carpet contains the normally observed sonic boom overpressures and results from wave propagation through only that part of the atmosphere below the aircraft. Secondary boom carpets may exist which involve the portion of the atmosphere above the aircraft as well as that below the aircraft. Between the primary and secondary carpets exists a region in which no booms are observed. The secondary boom carpets are more remote from the ground track and the overpressure levels are much less intense than in the primary carpet.

Sonic boom theory, in general, is well established. The evolution of the sonic boom signature from its pattern near the aircraft to the pressure signature received on the ground can be accurately predicted in terms of overpressure level, number and location of shocks, and duration. The complete role of the aircraft configuration in sonic boom generation is embodied in the F-function. Analyses of minimization concepts generally center on calculating F-functions for various configurations. For typical slender supersonic vehicles, the F-function may be computed directly from vehicle geometry via linearized supersonic flow theory. At hypersonic speeds, for which linearized flow theory is not accurate, the problem is that of obtaining the F-function by other means, such as wind tunnel tests or computational fluid dynamics (CFD) codes.

Utilization of the theory of geometric acoustics allows for the inclusion of the real atmosphere and nonlinear steepening on the shock wave system as it propagates to ground. It also allows for the calculation of the number and location of multiple booms resulting from maneuvering flight and the location of "superboom" focal zones. Focus boom signatures can be computed for simple "smooth caustic" foci, but similar results are not yet available for rarer, higher order "superfoci."

A number of sonic boom computer programs have been written. They all perform the same basic calculations, but each has particular capabilities and features added for specific applications. All these programs were originally developed for mainframe computers, but it is expected that personal computer versions will be available.

The primary boom carpet and the disturbances that are experienced within it have been intensely researched. A considerable experimental data base has been accumulated for a wide range of vehicles, Mach numbers, and altitudes. Agreement between measurements and predictions is quite good for both on-track and lateral locations for steady, level flight conditions. Sonic boom overpressures are noted to increase with increasing vehicle size and to decrease with increasing attitude. The lateral extent of the primary boom carpet increases with increasing altitude.

Atmospheric variations, especially those in the first few thousand feet above the Earth's surface, can be very influential in bringing about distortions of the sonic boom signature, changing it from the normally expected N-wave to a "peaked" or "rounded-type" signature. Higher overpressures result when the signatures are peaked, whereas lower overpressures are associated with rounded signatures. This peaking and rounding of the signature is statistical in nature and occurs as a function of either time or distance. As such, variations in the sonic boom signature can be expected during routine vehicle operations.

Rapid deviations of a vehicle from steady, level flight can produce considerable modifications in the location, number, and intensity of the ground shock wave patterns. Thus, multiple booms and focused booms may be observed. The pressure buildups in these focus areas are a function of the type maneuver and the accelerations involved and can be 2 to 5 times the nominal levels of steady flight. It is important to note that these focused booms are very localized and do not move with the vehicle. They can be placed to within a few miles of the designated location. Pressure buildup areas associated with turns, pull-up, or pushover maneuvers can be minimized by reducing accelerations. Atmospheric randomness will also decrease or, in some cases, even eliminate focusing altogether.

Secondary sonic booms are quite distinct from primary booms not only in the manner in which they are propagated from the aircraft to the ground but also in the way their signatures are shaped. For secondary boom signatures, the pressure changes very slowly and is in the subaudible frequency range. This lack of audibility combined with the very low amplitudes makes secondary booms difficult to sense outdoors, but they can be noticeable indoors. The secondary sonic boom carpet and the disturbances experienced within it are not as well-defined as those for the primary boom and only fragmentary observations and measurements are available.

Sonic booms may be minimized through aircraft design and operation. Minimization through design involves modification to the aircraft size, weight, and shape in order to reduce the overpressure or alter the signature waveform. The analytical tools for defining the required aircraft characteristics and means for evaluating trade-offs for boom minimization in terms of aircraft design are available. The most desirable signature shape, from the point of view of minimum response of an outdoor and indoor observer and of structural response, has not yet been established.

Low supersonic "boomless" flight operations are feasible and provide a means for domestic overland flight. The atmosphere plays a significant role in these types of operations, and considerable care must be exercised to assure that the shock waves associated with boomless flights do not extend to the ground. The practicality of such operations is very questionable.

Booms can be observed at ground level from aircraft in steady, level flight at high transonic Mach numbers and relatively low altitudes. The intensity of the boom is substantial because of the very low altitudes, and the signatures are considerably different in nature from the normally observed N-wave associated with fully developed supersonic flow.

The effects of sonic booms (particularly the responses they invoke) are not completely known, even through a considerable data base has been accumulated over the years. Many factors are involved in boom exposure, including the air path, ground path, and building vibrations. Outdoor and indoor stimuli include audible, vibratory, and visual cues. Human response to booms include startle, annoyance, effects on sleep, and long-term effects on health.

Structures reportedly damaged by sonic booms are mostly brittle surfaces and are secondary structural components. The number of damage incidents increases with increasing boom intensity. It may be significant that no damage incidents are reported for boom exposures less than 0.8 lb/ft^2, although the data sample is small.

Sonic booms do not adversely affect human hearing, vision, or circulation. The human psychological response is more complex, involving attitudes and habituation to booms and their sources. Long-term effects on health of repeated daily booms and effects on sleep of repeated night booms are not known. Although boom levels of 1.0 lb/ft^2 or greater are apparently unacceptable to a significant portion of the population, a level of acceptability has not been determined.

To date, it has been difficult to make detailed interpretations of the effects of sonic booms on different animal species. Research suggests that boom effects on domestic and wild animals are negligible; however, it is strongly suggested they receive more attention and research.

Since sonic booms attenuate rapidly in water, they appear to pose no threat to aquatic life, nor do they affect the capacity of fish eggs to hatch. At hypersonic Mach numbers, the aircraft speed can equal or exceed the speed of sound in water, thereby greatly increasing the potential penetration of the sonic boom.

Ground motions associated with sonic booms are less than 1 percent of the accepted seismic damage threshold for residential structures. Sonic booms probably cannot trigger earthquakes, but they might precipitate incipient avalanches or landslides in exceptional areas which are already stressed to within a few percent of instability. The situation may be different at hypersonic speeds.

Sonic booms constitute no serious concern for the safety of all types of subsonic aircraft in flight. The boom-induced accelerations are small relative to those induced by runway roughness and moderate air turbulence.

References

1. Proceedings of the Sonic Boom Symposium. *J. Acoust. Soc. America,* vol. 39, no. 5, pt. 2, May 1966, pp. S1–S80.
2. Carlson, Harry W.: Experimental and Analytic Research on Sonic Boom Generation at NASA. *Sonic Boom Research,* A. R. Seebass, ed., NASA SP-147, 1967, pp. 9–23.
3. Garrick, I. Edward: Atmospheric Effects on the Sonic Boom. *Second Conference on Sonic Boom Research,* Ira R. Schwartz, ed., NASA SP-180, 1968, pp. 3–17.
4. Morris, Odell: Experimental Studies of Sonic Boom Phenomena at High Supersonic Mach Numbers. *Third Conference on Sonic Boom Research,* Ira R. Schwartz, ed., NASA SP-255, 1970, pp. 193–203.
5. Maglieri, Domenic J.; Carlson, Harry L.; and McLeod, Norman J.: Status of Studies on Sonic Boom. *NASA Aircraft Safety and Operating Problems, Volume I,* NASA SP-270, 1971, pp. 439–456.
6. Ribner, H. S.; and Hubbard, H. H., eds.: *Proceedings of the Second Sonic Boom Symposium.* American Inst. of Physics, 1972.
7. Hubbard, Harvey H.; Maglieri, Domenic J.; and Stephens, David G.: *Sonic-Boom Research—Selected Bibliography With Annotation.* NASA TM-87685, 1986.
8. *Civil Aircraft Sonic Boom.* Aeronautics and Space, Code of Federal Regulations, Title 14, Part 91, Federal Aviation Adm., Jan. 1, 1987, pp. 169–170.
9. National Sonic Boom Evaluation Off.: *Sonic Boom Experiments at Edwards Air Force Base.* NSBEO-1-67 (Contract AF 49(638)-1758), Stanford Research Inst., July 28, 1967.
10. Gardner, John H.; and Rogers, Peter H.: *Thermospheric Propagation of Sonic Booms From the Concorde Supersonic Transport.* NRL Memo. Rep. 3904, U.S. Navy, Feb. 14, 1979. (Available from DTIC as AD A067 201.)
11. George A. R.; and Kim, Y. N.: High-Altitude Long-Range Sonic Boom Propagation. *J. Aircr.,* vol. 16, no. 9, Sept. 1979, pp. 637–639.
12. Balachandran, Nambath K.; Donn, William L.; and Rind, David H.: Concorde Sonic Booms as an Atmospheric Probe. *Science,* vol. 197, no. 4298, July 1, 1977, pp. 47–49.
13. Donn, William L.: Exploring the Atmosphere With Sonic Booms. *American Sci.,* vol. 66, no. 6, Nov.–Dec. 1978, pp. 724–733.
14. Liszka, Ludwik: Long-Distance Focusing of Concorde Sonic Boom. *J. Acoust. Soc. America,* vol. 64, no. 2, Aug. 1978, pp. 631–635.
15. Rickley, Edward J.; and Pierce, Allan D.: *Detection and Assessment of Secondary Sonic Booms in New England.* FAA-AEE-80-22, May 1980.
16. Whitham, G. B.: The Flow Pattern of a Supersonic Projectile. *Commun. Pure & Appl. Math.,* vol. V, no. 3, Aug. 1952, pp. 301–348.
17. Hayes, Wallace D.: *Linearized Supersonic Flow.* Ph.D. Thesis, California Inst. of Technology, 1947.
18. Hayes, Wallace D.: Brief Review of the Basic Theory. *Sonic Boom Research,* A. R. Seebass, ed., NASA SP-147, 1967, pp. 3–7.
19. Lomax, Harvard: *The Wave Drag of Arbitrary Configurations in Linearized Flow as Determined by Areas and Forces in Oblique Planes.* NACA RM A55A18, 1955.
20. Walkden, F.: The Shock Pattern of a Wing-Body Combination, Far From the Flight Path. *Aeronaut. Q.,* vol. IX, pt. 2, May 1958, pp. 164–194.
21. Whitham, G. B.: On the Propagation of Weak Shock Waves. *J. Fluid Mech.,* vol. 1, pt. 3, Sept. 1956, pp. 290–318.
22. Seebass, R.: Sonic Boom Theory. *J. Aircr.,* vol. 6, no. 3, May–June 1969, pp. 177–184.
23. Landau, L.: On Shock Waves at Large Distances From the Place of Their Origin. *J. Phys. USSR,* vol. 9, no. 6, 1945, pp. 496–500.

24. DuMond, Jesse W. M.; Cohen, E. Richard; Panofsky, W. K. H.; and Deeds, Edward: A Determination of the Wave Forms and Laws of Propagation and Dissipation of Ballistic Shock Waves. *J. Acoust. Soc. America,* vol. 18, no. 1, July 1946, pp. 97–118.
25. Hayes, Wallace D.; Haefeli, Rudolph C.; and Kulsrud, H. E.: *Sonic Boom Propagation in a Stratified Atmosphere, With Computer Program.* NASA CR-1299, 1969.
26. George, A. R.; and Plotkin, Kenneth J.: Sonic Boom Waveforms and Amplitudes in a Real Atmosphere. *AIAA J.,* vol. 7, no. 10, Oct. 1969, pp. 1978–1981.
27. Maglieri, Domenic J.; and Carlson, Harry W.: *The Shock-Wave Noise Problem of Supersonic Aircraft in Steady Flight.* NASA MEMO 3-4-59L, 1959.
28. Blokhintzev, D. I.: *Acoustics of a Nonhomogeneous Moving Medium.* NACA TM 1399, 1956.
29. Officer, C. B.: *Introduction to the Theory of Sound Transmission With Application to the Ocean.* McGraw-Hill Book Co., Inc., c.1958.
30. Keller, J. B.: Geometrical Acoustics. I. The Theory of Weak Shock Waves. *J. Appl. Phys.,* vol. 25, no. 8, Aug. 1954, pp. 938–947.
31. Thomas, Charles L.: *Extrapolation of Sonic Boom Pressure Signatures by the Waveform Parameter Method.* NASA TN D-6832, 1972.
32. Plotkin, K. J.; and Cantril, J. M.: Prediction of Sonic Boom at a Focus. AIAA Paper 76-2, Jan. 1976.
33. Haber, Jerold M.; and Drake, James F.: *Sonic Boom Risk Management: SABER & SABERPLOT User's Manual.* Contract No. F04703-79-C-0004, Task No. 10/79-16, J. H. Wiggins Co., July 15, 1980.
34. Taylor, Albion D.: *The Traps Sonic Boom Program.* NOAA Tech. Memo. ERL ARL-87, U.S. Dep. of Commerce, July 1980.
35. Plotkin, Kenneth J.: *Sonic Boom Focus Conditions Due to Tactical Air Operations.* WR 84-8, Wyle Lab., Feb. 1984.
36. Carlson, Harry W.: *Simplified Sonic-Boom Prediction.* NASA TP-1122, 1978.
37. Guiraud, Jean-Pierre: Acoustique Geometrique, Bruit Balistique des Avions Supersoniques et Focalisation. *J. Mech.,* vol. 4, no. 2, June 1965, pp. 215–267.
38. Gill, Peter M.; and Seebass, A. Richard: Nonlinear Acoustic Behavior at a Caustic: An Approximate Analytical Solution. *Aeroacoustics: Fan, STOL, and Boundary Layer Noise; Sonic Boom; Aeroacoustic Instrumentation,* Henry T. Nagamatsu, ed., American Inst. of Aeronautics and Astronautics, c.1975, pp. 353–386.
39. Gill, Peter Maxwell: *Nonlinear Acoustic Behavior at a Caustic.* Ph.D. Thesis, Cornell Univ., 1974.
40. Fung, K.-Y.: Shock Wave Formation at a Caustic. *SIAM J. Appl. Math.,* vol. 39, no. 2, Oct. 1980, pp. 355–371.
41. Plotkin, Kenneth J.: *Focus Boom Footprints for Various Air Force Supersonic Operations.* WR 85-22, Wyle Lab., Oct. 1985.
42. Haglund, George T.; and Kane, Edward J.: *Analysis of Sonic Boom Measurements Near Shock Wave Extremities for Flight Near Mach 1.0 and for Airplane Accelerations.* NASA CR-2417, 1974.
43. Wanner, Jean-Claude L.; Vallee, Jacques; Vivier, Claude; and Thery, Claude: Theoretical and Experimental Studies of the Focus of Sonic Booms. *J. Acoust. Soc. America,* vol. 52, no. 1, pt. 1, July 1972, pp. 13–32.
44. Plotkin, Kenneth J.: *Evaluation of a Sonic Boom Focal Zone Prediction Model.* WR 84-43, Wyle Lab., Feb. 1985.
45. Cramer, M. S.; and Seebass, A. R.: Focusing of Weak Shock Waves at an Arête. *J. Fluid Mech.,* vol. 88, pt. 2, Sept. 27, 1978, pp. 209–222.
46. Cramer, M. S.: Focusing of a Weak Three-Dimensional Shock Wave. *AIAA J.,* vol. 19, no. 10, Oct. 1981, pp. 1363–1365.
47. Hicks, Raymond M.; Mendoza, Joel P.; and Thomas, Charles L.: *Pressure Signatures for the Apollo Command Module and the Saturn V Launch Vehicle With a Discussion of Strong Shock Extrapolation Procedures.* NASA TM X-62,117, 1972.
48. Ashby, George C., Jr.: *Near-Field Sonic-Boom Pressure Signatures for the Space Shuttle Launch and Orbiter Vehicles at Mach 6.* NASA TP-1405, 1979.
49. Ashby, George C., Jr.: *A Study of the Sonic-Boom Characteristics of a Blunt Body at a Mach Number of 6.* NASA TP-1787, 1980.
50. Tiegerman, B.: *Sonic Booms of Drag Dominated Hypersonic Vehicles.* Ph.D. Thesis, Cornell Univ., 1975.
51. Seebass, A. Richard: *Hypersonic Boom.* Tech. Commun. 030, Flight Sciences Lab., Boeing Scientific Research Lab., June 1970.

52. Seebass, A. R.; and George, A. R.: Design and Operation of Aircraft To Minimize Their Sonic Boom. *J. Aircr.*, vol. 11, no. 9, Sept. 1974, pp. 509–517.
53. Henderson, Herbert R.; and Hilton, David A.: *Sonic-Boom Measurements in the Focus Region During the Ascent of Apollo 17.* NASA TN D-7806, 1974.
54. Garcia, Frank, Jr.; Jones, Jess H.; and Henderson, Herbert R.: A Comparison of Measured and Theoretical Predictions for STS Ascent and Entry Sonic Booms. *Shuttle Performance: Lessons Learned, Part 2,* James P. Arrington and Jim J. Jones, compilers, NASA CP-2283, Pt. 2, 1983, pp. 1277–1301.
55. Cotten, Donald; and Donn, William L.: Sound From Apollo Rockets in Space. *Science*, vol. 171, no. 3971, Feb. 12, 1971, pp. 565–567.
56. Plotkin, Kenneth J.; and George, A. R.: Propagation of Weak Shock Waves Through Turbulence. *J. Fluid Mech.*, vol. 54, pt. 3, Aug. 8, 1972, pp. 449–467.
57. Maglieri, Domenic J.; and Lansing, Donald L.: *Sonic Booms From Aircraft in Maneuvers.* NASA TN D-2370, 1964.
58. *NRL Investigations of East Coast Acoustics Events 2 December 1977—15 February 1978.* Dep. of Navy, Mar. 10, 1978.
59. Seebass, R.: Minimum Sonic Boom Shock Strengths and Overpressures. *Nature,* vol. 221, no. 5181, Feb. 15, 1969, pp. 651–653.
60. Jones, L. B.: Lower Bounds for Sonic Bangs in the Far Field. *Aeronaut. Q.*, vol. XVIII, pt. 1, Feb. 1967, pp. 1–21.
61. Jones, L. B.: Lower Bounds for the Pressure Jump of the Bow Shock of a Supersonic Transport. *Aeronaut. Q.*, vol. XXI, pt. 1, Feb. 1970, pp. 1–17.
62. Mack, R. J.; and Darden, C. M.: Some Effects of Applying Sonic Boom Minimization to Supersonic Cruise Aircraft Design. *J. Aircr.*, vol. 17, no. 3, Mar. 1980, pp. 182–186.
63. Niedzwiecki, A.; and Ribner, H. S.: Subjective Loudness of N-Wave Sonic Booms. *J. Acoust. Soc. America,* vol. 64, no. 6, Dec. 1978, pp. 1617–1621.
64. Niedzwiecki, A.; and Ribner, H. S.: Subjective Loudness of "Minimized" Sonic Boom Waveforms. *J. Acoust. Soc. America,* vol. 64, no. 6, Dec. 1978, pp. 1622–1626.
65. Cheng, D. H.; and Benveniste, J. E.: Transient Response of Structural Elements to Traveling Pressure Waves of Arbitrary Shape. *Int. J. Mech. Sci.*, vol. 8, no. 10, Oct. 1966, pp. 607–618.
66. McLean, F. Edward; Carlson, Harry W.; and Hunton, Lynn W.: *Sonic-Boom Characteristics of Proposed Supersonic and Hypersonic Airplanes.* NASA TN D-3587, 1966.
67. Miller, David S.; and Carlson, Harry W.: *A Study of the Application of Heat or Force Fields to the Sonic-Boom-Minimization Problem.* NASA TN D-5582, 1969.
68. Driver, Cornelius; and Maglieri, Domenic J.: The Impact of Emerging Technologies on an Advanced Supersonic Transport. *ICAS Proceedings, 1986—15th Congress of the International Council of the Aeronautical Sciences,* P. Santini and R. Staufenbiel, eds., International Council of Aeronautical Sciences, c.1986, pp. 213–220. (Available as ICAS-86-2.2.3.)
69. Vapor Condensation Makes F-14 Shockwaves Visible. *Aviation Week & Space Technol.*, vol. 107, no. 21, Nov. 21, 1977, pp. 46–47.
70. Hubbard, Harvey H.; and Mayes, William H.: Sonic Boom Effects on People and Structures. *Sonic Boom Research,* A. R. Seebass, ed., NASA SP-147, 1967, pp. 65–76.
71. ARDE Assoc.: *Response of Structures to Aircraft Generated Shock Waves.* WADC Tech. Rep. 58-169, U.S. Air Force, Apr. 1959. (Available from DTIC as AD 229 463L.)
72. Sutherland, Louis C.: *Low Frequency Response of Structures.* WR 82-18, Wyle Lab., May 1982.
73. Maglieri, Domenic J.; and Morris, Garland J.: *Measurements of the Response of Two Light Airplanes to Sonic Booms.* NASA TN D-1941, 1963.

Glossary of Terms

Acoustic compactness—Exists when the acoustic wavelength is much greater than the source dimensions.

Acoustic (signature) waveform—The time history of the acoustic pressure.

Aeolian tones—See **Strouhal number**.

Aerodynamic interaction noise—Arises from the coupling between two blade rows due to wakes from the upstream blade row convecting into the downstream blade row and the potential flow field of the downstream row acting on the upstream row.

Aerodynamic noise—Noise generated by the flow of unsteady air. Noise generated by flowing liquids is referred to as *hydrodynamic noise.*

Airfoil trailing-edge noise—Noise generated when the turbulent boundary layer developing over the surface of an airfoil is convected into the wake of the airfoil past its trailing edge.

Airframe interaction noise—In certain aircraft and engine layouts, there are interactions between the jet exhausts and the airframe. These affect the entrainment of air into the jets as a result of the proximity of the airframe. As a result of these interactions, the jet noise radiated to the far field is increased. This increase is referred to as *airframe interaction noise.*

Airframe noise—The nonpropulsive noise of an aircraft in flight.

Anechoic environment—Nonechoing environment provided in test cells and open jet wind tunnels by wall treatments that are highly absorptive (nonreflecting) for the acoustic wavelengths involved.

Augmentor wing (AW) vehicles—Short-takeoff-and-landing configurations that make use of turbofan engines for which part of the secondary exhaust flow is ducted through ejector nozzles at the wing trailing edge to enhance overall lift.

Blade slap noise—Term applied to the impulsive noise generated by a rotating blade over a portion of the rotor disk because of blade-vortex interactions or localized compressibility effects.

Blade-vortex interaction (BVI) noise—Generated when helicopter rotor blades interact with tip vortices shed during previous blade passages. The strength of each interaction is governed by the local strength of the tip vortex, the core size of the tip vortex, the local interaction angle of the blade and the vortex line, and the vertical separation between the vortex and the blade.

Blade-wake interaction (BWI) noise—Generated when helicopter rotor blades interact with aerodynamic wakes shed during previous blade passages. This noise is most evident during nearly level flight and is drastically reduced in climbing flight.

Boomless flight—Accomplished when an aircraft is operated at supersonic speeds in such a manner that the associated shock wave pattern does not extend to the ground. The criterion for boomless flight is that the ground speed of the aircraft does not exceed the speed of sound at the ground.

Broadband noise—Produced by sources that are random both in time and in location, such as the interactions of rotating blades with turbulence, and distributed continuously in frequency.

Boundary-layer trip—A small protuberance or area of roughness that is placed on an aerodynamic surface to promote transition of the boundary-layer flow from laminar to turbulent.

Buzz-saw noise—See **Multiple pure tones**.

Caustic—In geometrical acoustics, an envelope of rays. Because of the shrinkage of ray tube areas to zero at the caustic, such regions have infinite or high acoustic intensity and are considered mathematical singularities.

Cavity noise—Noise (discrete frequency and broadband) generated by high-speed airflow over cavities such as wheel wells.

Combustion thermoacoustic efficiency—The fraction of the combustion heat release that is radiated away as acoustic energy.

Compact (point) acoustic source—Source for which the differences in emission times (for one observer time) of points on the source are negligible compared with the periods of fluctuations of the source. Note that for a moving source, compactness is not only a property of the source itself, but may also depend on the observer position.

Continuous spectrum—Spectrum for which components are distributed continuously over a range of frequencies.

Counterrotation (contrarotation)—Involves the addition of a second rotating blade row behind the first. The second blade row rotates in the opposite direction to that of the first. It imparts its own swirl, but in the opposite direction to that of the first, thus tending to cancel it. The result is increased efficiency (thrust) compared with a single blade row.

Cusp—A point of discontinuity of the caustic where two segments of the caustic are tangent to each other.

Delocalization—The process by which shock waves on a rotating blade surface become connected to the acoustic radiation field.

Dipole (doublet) source—Represented by two equal monopole sources that are a small distance apart and are pulsating 180° out of phase with each other. The associated radiation pattern is highly directional and has a "figure-8" shape. Dipoles are used to model sources that impart forces, such as lift and drag, to the medium.

Direct combustion noise—Results when a volume of mixture expands at constant pressure as it is rapidly heated by combustion.

Doppler effect—The changing of observed frequency of a sound whose source is moving with respect to a stationary observer. The frequency of the sound is observed to increase or decrease, respectively, depending on whether the sound source is moving toward or receding from the observer.

Doppler factor—Expressed as $1/(1\text{-}M)$, where M is the Mach number of the component of the flow velocity in the direction of radiation. In radiation equations, it appears raised to various powers, depending on source type, to account for the changes in amplitude of acoustic pressure from moving sources. In aeroacoustics, the Mach number is frequently defined as the flow speed divided by the ambient speed of sound, whereas in aerodynamics, Mach number is defined in terms of the local speed of sound.

Eddy convection speed—The average speed at which the larger scale eddies in a turbulent flow are transported by the flow. When different convection speeds are ascribed to different eddy scales, then the correct descriptor needs to be added, such as small eddy or large eddy convection speed. Convection speeds are typically 0.5 to 0.7 times the free-stream value.

Energy spectrum—In the analysis of turbulence and noise, relates to the distribution with respect to frequency of the fraction of the total energy, on average, that can be allocated to a given frequency.

Engine core noise—The summation of noise produced by the combustion process itself and the noise produced as the hot products of combustion pass through the turbine and exhaust nozzle. (In some writings, core noise is defined to also include turbomachinery components).

Entrainment—The property of the mixing regions of viscous and turbulent flows such that, with increasing distance downstream, fluid external to the mixing region is continuously being engulfed into the mixing region, with the result that its mass flux also increases.

Excess noise—Usually refers to all extraneous sources of noise, both external and internal, existing upstream of the nozzle exit that modify the characteristics of jet noise from those associated solely with the generation of noise from the jet mixing regions.

Externally blown flap (EBF) vehicles—See **Under-the-wing (UTW) vehicles**.

Flap side-edge noise—Noise generated by airflow over the edges of deployed aerodynamic flaps; it is a component of airframe noise.

Flow-acoustic interaction—Occurs when the sound field generated by an unsteady fluid motion crosses a nonuniform field of flow. Normally for sound fields up to moderate intensity the properties of the flow field are unaffected by the interaction, whereas the directivity of the sound field can be greatly distorted.

Frequency-domain methods—Eliminate time from the wave equation by means of Fourier transformation.

Harmonic noise—The periodic signal component, that is, its time signature can be represented by a pulse that repeats at a constant rate.

Helicopter—Vehicle that utilizes only rotor(s) to create lift during hovering and forward flight.

High-speed impulsive (HSI) noise—Arises from shock wave formations associated with localized supersonic flows on helicopter blades. They are generated during only a portion of the revolution of each advancing blade.

Impulsive noise—A well-recognized characteristic of helicopters, it is periodic in nature, has an intensive low-frequency content, and consists mainly of discrete frequency components. Possible sources are blade-vortex interactions and localized compressible flows on the blades.

Indirect combustion noise—Results when relatively large-scale temperature nonuniformities generated by turbulent combustion are convected through pressure gradients in the turbine and exhaust nozzle.

Inflow turbulence noise—Noise generated when rotor blades interact with inflow turbulence from the atmosphere, from wakes of upstream fixtures, and from disturbances induced by the flow into the engine inlet.

Instability waves—Spatially or temporally amplifying waves generated by naturally occurring perturbations of the fluid or by artificial excitations in an inherently unstable flow. In jets or shear layers, the observed instability waves are found to be consistent with spatially amplifying instability wave theory.

Intermittent periodic noise—Arises from the encounter by a helicopter rotor of an atmospheric disturbance having a duration of several blade passages. The spectrum shows discrete frequency components that exist for a short time, disappear, and then recur at a later time. In frequency domain analysis, this phenomenon appears as a narrow-band random noise.

Jet mixing region—That portion of the jet exhaust flow downstream of the nozzle exit in which mixing occurs with the ambient air.

Karman vortex shedding—Arrangement of eddies, in the wake of a bluff body, which at some Reynolds numbers has a definite geometric pattern and an associated tone. (See **Strouhal number**.)

Kutta condition—As applied in airfoil potential flow theory, specifies that normally the rear stagnation point must be at the trailing edge. The exception is the airfoil having a cusped trailing edge, where the velocity at the trailing edge is finite and the velocities on the upper and lower surfaces at the trailing edge are the same. This uniquely specifies the circulation around the airfoil and thus its lift. In all cases, a more general definition that applies in subsonic flows, both inviscid and viscous, is the condition that no normal load exists at the trailing edge, and hence the upper and lower pressures at the trailing edge are the same.

Laminar flow—Viscous flow; normally applies to the nonturbulent flow in the boundary layer that exists around a body placed in a moving fluid at small to moderate Reynolds numbers.

Landing gear noise—Noise generated by vortex shedding and wake interactions of the wheels, axles, struts, and shafts subjected to unsteady separated flows.

Lighthill's acoustic analogy—The assumption that the unsteady fluid flow in the jet mixing region may be replaced by a volume distribution of equivalent acoustic sources assumed to be embedded in a uniform medium at rest.

Line spectra—Spectra that contain only discrete frequency components.

Loading noise—Results from the forces on a rotating airfoil.

Mach number—For an object in flight it is the dimensionless ratio $\frac{V}{c}$, where V is the speed of the object and c is the ambient speed of sound. The Mach number in a fluid flow, correspondingly, is the local speed divided by the local speed of sound. However (see **Doppler effect**), in aeroacoustics it is often more convenient to define an acoustic Mach number in the noise-generating source region of a moving fluid flow, which is given by the local speed divided by the ambient (constant) speed of sound external to the flow and which is therefore the speed of sound in the radiation field.

Mach wave radiation—Referred to as the acoustic radiation generated by supersonically traveling disturbances. For supersonic jets the dominant part of Mach wave radiation is believed to be associated with the instability waves and large turbulence structures of the jet flow.

Monopole (simple) source—Represented by a pulsating sphere whose dimension is small compared with the wavelength of sound emitted. If it is assumed all parts of the sphere are moving in phase, the radiation is essentially the same in all directions. Monopoles are used to model volume sources, such as fuel ignition and rotor blade thickness.

Multiple pure tones—Spectral components of ducted rotor noise occurring at multiples of shaft rotation frequency. They are referred to also as "buzz-saw" noise, are present only at supersonic rotor tip speeds, and are associated with the unsymmetrical rotating shock wave patterns locked to the rotor blades.

Noncompact acoustic sources—A source is noncompact if it cannot be treated as a compact (point) source.

N-wave shock patterns—Observed in the distance flow fields of objects traveling at supersonic speeds in the atmosphere. They consist of bow waves and tail waves and an expansion region with a linear pressure decrease between them. At the bow wave a compression occurs in which the local pressure rises to a value $\Delta\rho$ above ambient pressure. Then a slow expansion occurs until some value ($\approx \Delta\rho$) below ambient pressure is reached and there is a sudden recompression at the tail wave.

Potential flow region—That portion of the jet exhaust flow, downstream of the exhaust nozzle, in which the mean flow velocity is approximately equal to that at the nozzle exit. For a circular nozzle exit, the potential flow region is conical in shape.

Primary sonic boom carpet—The ground area exposed to sonic booms that propagate only through that portion of the atmosphere below the aircraft. This carpet area includes the ground track and extends laterally in both directions from it.

Propulsive lift (scrubbing) noise—The increase in noise that occurs when airframe surfaces are placed in the exhaust of the propulsion system to enhance the overall lift.

Quadrupole source—Noise produced by fluid flow in a volume in which no sources of mass or momentum are present. They can be modeled by the superposition of four equal monopole sources that oppose each other in pairs. Quadrupole noise is the main component of aerodynamic noise arising from turbulent flow in the absence of solid boundaries. It is also the noise generated by acoustic sources distributed in the volume surrounding a rotating blade to represent nonlinear effects that occur when the flow over the blade is transonic.

Retarded source position—The position of a source when it produced the sound that is being heard at the present time.

Retarded time—The time at which a sound wave traveling at the local speed of sound would have to leave the source to arrive at the observation point at the present time.

Reynolds number—A dimensionless quantity $\frac{V\ell}{\nu}$, where V and ℓ are the characteristic velocity and length scales, respectively, and ν is the kinematic viscosity of the air.

Rotorcraft—Vehicles that utilize unducted rotors to create enough lift to achieve hovering flight out of ground effect. (See **Helicopter**, **Tilt-rotor aircraft**, and **Tilt-wing aircraft**.)

Screech noise—Discrete frequency shock associated noise that arises because of feedback loops that are established between shock cells in the mixing region of a supersonic jet and the nozzle lip.

Secondary sonic boom carpets—The ground areas exposed to sonic booms that propagate through that portion of the atmosphere above the aircraft. These areas are remote from the ground track and are separated from the primary sonic boom carpet, on each side, by regions in which no booms are heard.

Self-preserving flow—Exists when the properties of a developing flow, in particular the average properties of a convected turbulent flow, at various stations along the flow are similar when referenced to a characteristic length and velocity scale.

Shock-associated noise—Generally referred to as the broadband noise radiation arising from the passage of coherent turbulence structures through the quasi-periodic shock cells in the mixing layer of a supersonic jet.

Shock waves—Discontinuities in a supersonic flow through which the fluid undergoes a finite decrease in velocity accompanied by marked increases in pressure, density, temperature, and entropy while satisfying the conservation laws of mass, continuity, and momentum.

Short-takeoff-and-landing (STOL) vehicles—Vehicles designed with integrated lift and propulsion systems such that lift augmentation is provided by beneficial interaction of the exhausts with the wing flow field.

Sonic booms—Boom-like sounds that arise from the passage of shock waves from objects traveling in the atmosphere at speeds that exceed the local speed of sound.

Sonic boom carpet—An area on the ground within which sonic booms may be observed for a particular flight. Its length corresponds roughly to the length of the supersonic portion of the flight and the width increases with increases in altitude and Mach number. An observer anywhere within the carpet will experience the passage of an N-wave disturbance one time during the flight.

Stopped-rotor vehicle—Vehicle for which the lift during transition flight is provided mainly by a rotor that is stored in the stopped position during cruise flight. Lift during cruise is then provided by a fixed wing.

Strouhal number—A dimensionless quantity that characterizes the frequency of steady vortex shedding from blunt bodies. Strouhal number is $\frac{fV}{d} \approx 0.2$, where f is the shedding frequency, V is the free-stream velocity, and d is the thickness. For some combinations of the above parameters, the shedding frequencies have a tonal quality and are referred to as *aeolian tones.*

Superbooms—Intense sonic booms associated with converging ray patterns or caustics that result from accelerated flight. Associated focusing may result in sonic boom pressure buildups of 2 to 5 times nominal values in cruising flight at constant speed and Mach number.

Synchrophasing—Involves the phasing of the rotation position of the propeller blades on opposite sides of the fuselage so that the impinging sound has a certain phase relationship that promotes noise cancellation within the cabin.

Thickness noise—Arises from the periodic displacement of the air by the volume of a passing blade. Acoustic source strength is related to the normal velocity at the blade surface and becomes important at high speeds. Thickness noise can be represented by a monopole source distribution.

Tilt-rotor aircraft—Vehicle for which the lift for takeoff and landing is supplied mostly by the rotor and the lift for forward flight is supplied mostly by the wing. Wing-tip power plants rotate from the vertical to the horizontal and back during transition flight.

Tilt-wing aircraft—Vehicle for which the wing and power plants (rotors) rotate from vertical orientation for takeoff and landing to horizontal for cruise.

Time-domain methods—Methods used to solve the wave equation directly in terms of the space-time variables.

Turbomachinery noise—Comprises the tonal and broadband noise components generated by the fan, compressor, and turbine sections of jet engines.

Turbulence ingestion noise—Noise generated when rotor blades interact with turbulence arising from wakes of an upstream blade row, blade boundary layers, and the ambient atmosphere.

Turbulent-boundary-layer noise—Noise generated by turbulent-boundary-layer flows on a homogeneous surface.

Turbulent flows—Flows characterized by the chaotic nature of the small-scale motions of the flow, the onset of which occurs at a critical value of Reynolds number.

Turbulent-mixing noise—The noise generated aerodynamically outside of the nozzle of a jet engine in the region where the high-velocity jet exhaust flow mixes with the ambient air.

Under-the-wing (UTW) vehicles or externally blown flap (EBF) vehicles—Short-takeoff-and-landing UTW or EBF configurations make use of turbofan engines that exhaust under the wing and interact with the wing flow field to enhance the overall lift.

Upper-surface-blowing (USB) vehicles—Short-takeoff-and-landing configurations that make use of turbofan engines that exhaust over the wing and interact with the wing flow field to enhance the overall lift.

Visual source position—The position of a source at the time the sound is heard, as opposed to the retarded source position.

Vortex (eddy)—A group of fluid particles having a common circular motion about an axis. Turbulent motion may be considered as a superposition of eddies of different sizes and vorticities all in irregular motion.

Vorticity—The state of a fluid in vortical motion.

Wing-in-slipstream vehicles—Conventional short-takeoff-and-landing configurations in which the propeller slipstream is used to generate increased relative speed between the wing and its upstream airflow and thus increase the overall lift.

Zone of silence—With reference to jet noise radiation, a zone of silence is observed in the downstream direction in the far field on the jet longitudinal axis, mainly due to refraction effects.

Index